Tables

Tables

Acknowledgments

In the preparation of this book and in the conduct of the research upon which it is based, we have been helped by many individuals and groups. In particular we would like to thank: the USAF Institute of Aeromedicine for their financial support (USAF grant number AFOSR-83–0148), the Civil Aviation Authority and the British Airline Pilots' Association not just for their financial support but also for their advice and co-operation. We wish to express our gratitude to those pilots and their wives who contributed directly to the project.

Finally we would like to thank Lesley Dutton, Sally Jarratt and Christine Long, all of whom have been involved at some stage or other in preparing this text.

Chapter 1
An overview of stress in the workplace

Hans Selye (1946) was one of the first to try to explain the process of stress-related illness with his 'General Adaptation Syndrome' theory. In it he described three stages an individual encounters in stressful situations:

(1) The *alarm reaction* in which an initial shock phase of lowered resistance is followed by countershock during which the individual's defence mechanisms become active;

(2) *resistance*, the stage of maximum adaptation and, hopefully, successful return to equilibrium for the individual. If, however, the stressor continues or the defence does not work, he will move on to the third stage;

(3) *exhaustion*, when adaptive mechanisms collapse.

Since Selye first postulated this process of environmental stressor and bodily reaction, a great deal of research work has been undertaken in the field of occupational stress. From the growing research literature, it is felt that the available data can be organised into the model shown in Figure 1.1.

Most research indicates that depending on the particular job and organisation, one or some combination of the sources of stress in this model, together with certain personality traits, may be predictive of a variety of stress manifestations, such as coronary heart disease, mental ill-health, job dissatisfaction, marital disharmony, excessive alcoholic intake or other drug taking, etc. Just to set the scene, it would seem appropriate to discuss some aspects of this model. The six major sources of occupational stress will therefore be discussed: factors intrinsic to the job; role in the organisation; career development; relationships at work; organisational structure and climate; and home/work interface (Cooper, 1983). This should provide an interesting basis from which to start and a framework around which we can put our data on pilots into perspective.

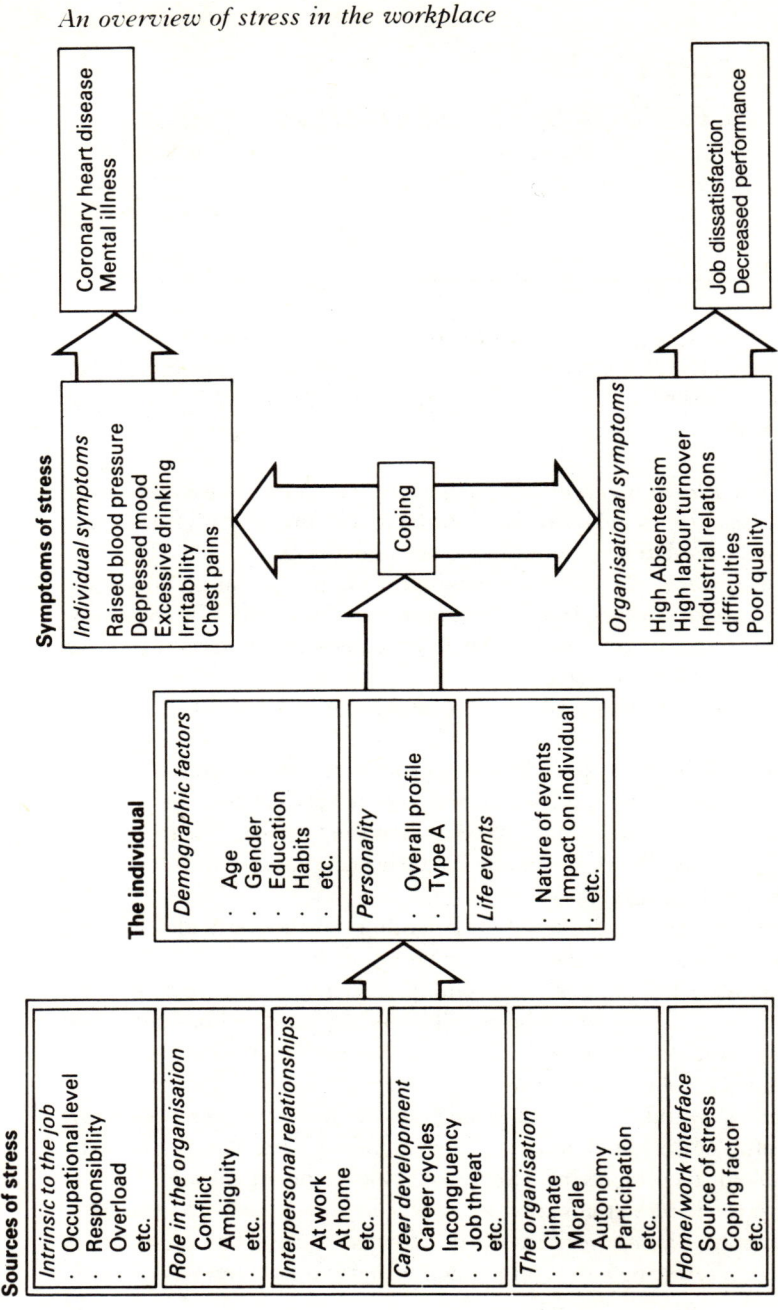

Source: Based on C.L. Cooper in Ivan Robertson and Cary L. Cooper (eds), *Human Behaviour in Organisations,* Plymouth, Macdonald & Evans, 1983.

Figure 1.1 A basic model of occupational stress

Factors intrinsic to the job

By 'intrinsic' factors we refer to those aspects which are integral parts of the job. In other words, fundamental characteristics of the job that affect other aspects of the job and the individual's working life. Across a variety of occupations such sources of stress intrinsic to the job include: (1) poor physical working conditions; (2) shift work; (3) work overload or underload; (4) physical danger; (5) person-environment fit (P-E) and job satisfaction (Cooper, 1981).

Poor physical working conditions

Poor physical working conditions can enhance stress at work. In regard to nuclear power plant operators, for example, Otway and Misenta (1980) believe that the design of the control room itself is an important variable in terms of worker stress. They propose that control room designs need to be updated, requiring more sophisticated ergonomic designs. Furthermore, Otway and Misenta provide various illustrations of the effects of such issues. An important stress factor in the Three Mile Island Accident, for example, was the distraction caused by excessive emergency alarms. Poor physical working environments are a particular problem for blue collar workers (Cooper and Smith, 1985). In a study carried out by Kelly and Cooper (1981) on the stressors associated with casting in a steel manufacturing plant, poor physical working conditions were found to be a major stressor. Many of the stressors were concentrated, as one might expect, in the physical aspects of noise, fumes and, to a lesser extent, heat, plus the social and psychological consequences of isolation and interpersonal tension. A further possible source of stress was seen to reside in the lack of job satisfaction, particularly arising from the stressors above, and partially endemic to the nature of casting of liquid steel, in a continuous process lasting some seventy minutes. For 75 per cent of this time cycle, the casters were exposed to (and by the nature of their task unable to move away from) very high levels of noise (up to 110 dB for much of the time), and also to periodic and unpleasant air pollution caused by the activities of other workers and machines in their proximity. These conditions necessitate the wearing of ear protection, in the form of ear muffs or cotton wool swabs, which effectively, in plant conditions, isolate the wearer from those around him.

Shift work

Numerous occupational studies have found that shift work is a common occupational stressor, as well as affecting neurophysiological rhythms, such as blood temperature, metabolic rate, blood sugar levels, mental efficiency and work motivation, which may ultimately result in stress related disease (Selye, 1976). A particular occupational study by Cobb and Rose (1973) on air traffic controllers found four times the prevalence of hypertension, and also more mild diabetes and peptic ulcers, among the experimental subjects than in their control group of second class airmen. Although these authors also identified other job stressors as being instrumental in the causation of these stress-related maladies, shift work was isolated as a major problem area.

Nevertheless, although there are stressors associated with shift work, one needs to take note of Selye's (1976) conclusion on the issue. He suggests that most investigations agree that shift work becomes physically less stressful as individuals can (and often do) habituate to the condition. Even so, being 'excluded from society' is a common complaint among shift workers.

Job overload

This is probably one of the most 'obvious' occupational stressors. It is commonplace for us to use the expression, 'I'm under a hell of a lot of pressure at the moment at work.' By this, we are referring to job overload. This however is only part of the picture. French and Caplan (1972) see work overload as being either quantitative (i.e. having too much to do) or qualitative (i.e. being too difficult), and certain behavioural malfunctions have been associated with job overload (Cooper and Marshall, 1976). For example, in another study on air traffic controllers, Crump, Cooper and Maxwell (1981) found that one of the primary short-term but uncontrollable stressors was 'being overloaded'. They devised a unique and complicated method of measuring job stress (Figure 1.2). This was called the Repertory Grid technique, which allowed them to assess the sources of stress among air traffic controllers in terms of a number of paired constructs (e.g. controllable/uncontrollable and long-term/short-term stress).

In investigation of stress among British police officers, Cooper, Davidson and Robinson (1982) found that work overload was a major stressor among the lower ranks, particularly police sergeants. In

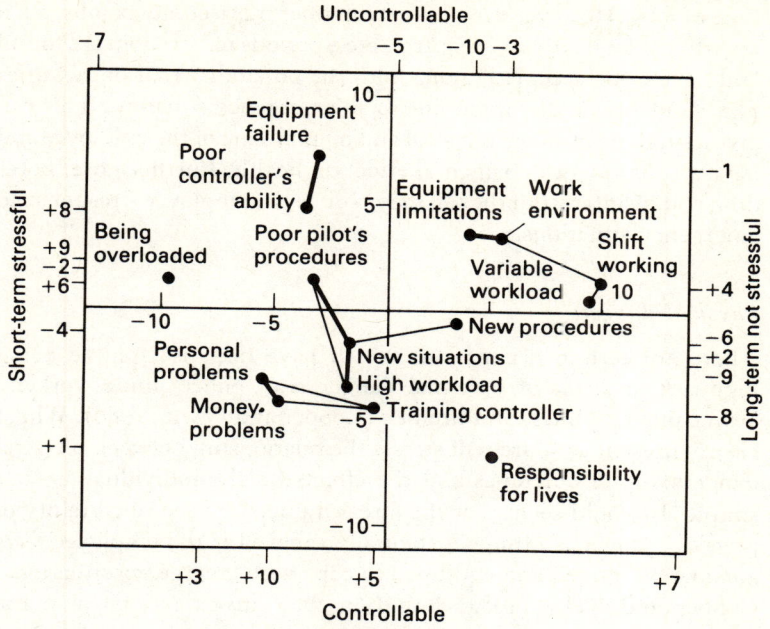

Source: Crump, J., Cooper, C.L. and Maxwell, V.,
Journal of Occupational Behaviour 2(4), 1981.

Figure 1.2 Job stress and the air traffic controller

particular, sergeants who scored high on the depression scale of the
Middlesex Hospital Questionnaire (a well-used measure of mental ill-
health) tended to be older operational officers who believed they were
overloaded and who perceived a number of bureaucratic and outside
obstacles to effective police functioning. They complained about the
long hours and heavy workload, as well as the increased paperwork,
lack of resources and the failure of the courts to prosecute offenders.

Job underload

Whilst having too much work, or work that is too difficult, is bad,
having too little also has negative effects for us as individuals. Job
underload associated with repetitive, routine, boring and understi-
mulating work has been associated with ill health (Cox, 1980). We
would normally expect to find these qualities in the machine-minding

5

type of jobs. However even in certain other more complex jobs, such as airline pilots, air traffic control, etc., periods of boredom are found and have to be accepted, along with the possibility that one's duties may suddenly be disrupted due to an emergency situation. This can give a sudden jolt to the physical and mental state of the employee and have a subsequent detrimental effect on health. Furthermore, boredom and disinterest in the job may reduce the employee's response to emergency situations.

Physical danger

There are certain occupations which have been isolated as being high-risk in terms of potential danger, e.g. police, mine workers, airline pilots, soldiers and firemen (Cooper and Payne, 1980). Whilst clearly present as sources of stress, the relationship between physical dangers, stress outcomes and the effects on the individual are not simple. It would seem that the stress induced by the uncertainty of physical danger is often substantially relieved if the employee *feels* adequately trained and equipped to cope with emergency situations. Cooper and Kelly (1984) found in their investigations of crane operators, for example, that one of the major explanations of accidents involving tower cranes was in the poor selection and training of the operators, together with feelings of isolation from other workers.

The person-job fit

There will of course be a whole host of different variables associated with the job itself that could specifically affect any single individual doing any specific job. Clearly, it is not possible to review here the factors that are idiosyncratic to particular jobs other than those recurrent ones we have just mentioned. One way, however, of summarising such idiosyncratic factors is to look at the relationship between the individual and the job itself.

This measure of job satisfaction and related variables which deserves mention is the measure known as person-environment fit (P-E), (Caplan *et al.*, 1975). According to McMichael (1978), P-E fit can be defined as an interaction between an individual's psychosocial characteristics and objective environmental work conditions. Consequently, a score of P-E fit can be attained by subtracting the amount/degree of a particular job factor (e.g. workload) preferred by a person from the actual amount in that same person's job environment. The overall hypothesis is that stress can occur and result in such problems as anxiety, depression, job dissatisfaction and physiological maladies if there is a P-E misfit.

Role in the organisation

An individual who occupies any job is effectively playing a 'part in a play', i.e. there is a series of expectations that other people have about the nature of that job and the sort of behaviour that is expected from whoever occupies it. Not surprisingly this process has been dubbed as one's job 'role'. A person's role at work has been isolated as a main source of occupational stress, involving role ambiguity (i.e. lack of clarity about one's job) and role conflict (i.e. conflicting job demands), as well as responsibility for people and conflicts stemming from organisational boundaries (Cooper and Marshall, 1976). Authors such as French and Caplan (1972), Beehr, Walsh and Taber (1976) and Shirom *et al.* (1973) have indicated that these organisational stressors stemming from role ambiguity and conflict can result in such stress-related illnesses as coronary heart disease (CHD). Furthermore, Cooper and Marshall conclude that less physical occupations, such as managerial, clerical and professional jobs, are more prone to occupational stress related to role conflict.

After reviewing the relevant literature, Kasl (1973) concluded that correlations between role conflict and ambiguity and components of job satisfaction tend to be strong (on the other hand, correlations with mental health measures tend to be weak). However, personality differences are important determinants in how an individual reacts to role conflict, and greater job-related tension is produced in introverts than in extroverts. French and Caplan also hold that 'flexible' people show greater job-related tension under conditions of conflict than do 'rigid' individuals.

An inherent quality of job role is responsibility. The degree of responsibility for people (in particular) and their safety also appears to be a potentially significant occupational stressor. There is variation across different occupations in the dominance of this as a stressor, even in jobs that are almost exclusively 'people'-orientated. Kroes (1976), for example, sees responsibility for people as a potential stressor for police, although not to the extent it is for air traffic controllers. This has recently been verified by a study of occupational stress in air traffic controllers which isolated responsibility for

7

Table 1.1   *The impact of personality and job stressors on raised diastolic blood pressure among dentists*

Personality and job stressor	Multiple R^2	R^2	R^2 Change
Age	.36	.13	.13
Dentist as inflictor of pain	.40	.16	.04
Coping with difficult patients	.45	.20	.04
Administrative duties	.49	.24	.04
Too little work	.52	.27	.03
16PF factor Q11 'anxiety'	.54	.29	.01
Sustaining and building a practice	.55	.31	.02
Job interfering with personal life	.57	.32	.02

$F = 6.88$ $p < .01$

Source: Cooper, C.L., Mallinger, M. and Kahn, R., *Journal of Occupational Psychology*, 51 (3), 1978.

people's safety and lives as a major long-term occupational stressor (Crump, Cooper and Smith, 1980).

Such role-related issues extend into the professions. For example, the problems that role conflicts can generate were amply demonstrated by Cooper, Mallinger and Kahn (1978) in their investigation into dentists (Table 1.1). It was found that the variables which predicted abnormally high diastolic blood pressure among dentists were factors related to the role of the dentist, that is, that he considers himself to be 'an inflictor of pain' rather than 'healer'; has to carry out non-clinical tasks such as administrative duties, and sustaining and building a practice; and his role also interfered with his personal life, primarily in terms of time commitments. It is interesting to note that part of this image of being 'professional' must inevitably involve coping, especially in public, with the stress that such conflicts create.

Career development

The next group of environmental stressors is related to career development, which Cooper has found to be a fundamental stressor at work. This refers to the impact of over-promotion, under-promotion, status incongruence, lack of job security and thwarted ambition. Status congruency, for example, or the degree to which there is job advancement (including pay grade advancement), was found by Erickson, Pugh and Gunderson (1972), in their large sample of

Navy employees, to be positively related to military effectiveness and negatively related to the incidence of psychiatric disorders. However, in terms of pay, Otway and Misenta (1980) postulate that large increases in workers' pay would not necessarily mean simultaneous increases in job satisfaction and might result in personnel remaining in jobs which no longer give them satisfaction. When longer-term career issues are involved, there are some things that money cannot buy.

Career development blockages are most notable among women managers, as a study by Cooper and Davidson (1982) revealed. In this investigation, the authors collected data from over 700 female managers and 250 male managers at all levels of the organisational hierarchy and from among several hundred companies. It was found that women suffered significantly more than men on a range of organisational stressors, but the most damaging to their health and job satisfaction were the ones associated with career development and allied stressors (e.g. sex discrimination in promotion, inadequate training, male colleagues treated more favourably, not enough delegation to women).

The increase in women who work is just one illustration of the ways in which our industrialised society is changing. Given this and other factors such as the impact of new technology, clearly career-related stress might well be expected to increase significantly over the next two decades.

Relationships at work

We all have some idea about how well or badly we 'get on' with others we work with. These interpersonal relationships, which include social support from one's colleagues, boss and subordinates, have also been related to job stress (Payne, 1980). According to French and Caplan (1972), poor relationships with other members of an organisation may be precipitated by role ambiguity in the organisation, which in turn may produce psychological strain in the form of low job satisfaction. Moreover, Caplan *et al.* (1975) found that strong social support from peers relieved job strain and also served to condition the effects of job stress on cortisone, blood pressure, glucose and the number of cigarettes smoked. It is interesting to note that among air traffic controllers, greater help and social support (as assessed by the repertory grid) were provided by friends and colleagues than by those in supervisory positions at work. Generally speaking, however, it is

9

one's relationship with one's superiors that seems to be a recurrent critical factor in most jobs.

However, even those at the top can experience interpersonal sources of stress, as Cooper and Melhuish (1980) discovered in their study of 196 very senior male executives. It was found that the combination of male executives' predispositions (e.g. outgoing, tough-minded, etc.) with their relationships at work were central to their increased risk of high blood pressure. They were particularly vulnerable to the stresses of poor relationships with subordinates and colleagues, lack of personal support at home and work, and to the conflicts between their own values and those of the organisation. A possible link might be in the degree of discretion one has in the job. Indeed, lack of autonomy shown by one's boss or employer is a major stressor at work. Cooper and Roden (1985) found that a major source of pressure on income tax officers is their feeling of 'lack of influence' or autonomy' in planning and implementing their work and responsibility.

Organisational structure and climate

The organisation provides a backcloth in front of which individuals must work. In doing so, its characteristics will pervade almost all aspects of daily working life. Organisations are usually created to be efficient in their design and the way they function. But things can go wrong, especially in terms of those organisational features that are perceived such as structure and climate. These include such factors as office politics, lack of effective consultation, lack of participation in the decision-making processes and restrictions on behaviour. Margolis, Kroes and Quinn (1974) and French and Caplan (1972) found that greater participation led to higher productivity, improved performance, lower staff turnover and lower levels of physical and mental illness (including such stress-related behaviours as escapist drinking and heavy smoking). However attempts to combat these organisational sources of stress are not always successful and are often difficult to implement in practice. Additionally, of course, organisational variables by definition are wide in scope, allowing much variation within any single organisation or company.

Home/work pressures

Another danger of the current economic situation is the effect that work pressures (such as fear of job loss, blocked ambition, work overload and so on) have on the families of employees. At the very best of times young managers, for example, face the inevitable conflict between organisational and family demands during the early build-up to their careers, as the British Institute of Management survey entitled *The Management Threshold* (1974) illustrates. But during a crisis of the sort we are currently experiencing, the problems escalate in ever-increasing proportions as managers strive to cope with some of their basic economic and security needs. As Pahl and Pahl (1981) suggest in their book on *Managers and Their Wives*, most male managers under normal circumstances find home a refuge from the competitive and demanding environment of work, a place where they can get support and comfort. However, when there is a career crisis (or stress from job insecurity as many employees are now facing), the tensions the individuals bring with them into the family affect the wife and home environment in a way that may not meet their 'sanctuary' expectations. It may be very difficult, for example, for the wife to provide the kind of supportive domestic scene her husband requires at a time when she is beginning to feel insecure, when she is worried about the family's economic, educational and social future. Many individuals like to think that they keep their domestic and professional lives separate. But the evidence clearly tells us that the two are inextricably entwined – particularly in terms of stress.

Dual career stress

Most would agree that it is difficult for a housebound wife to support her breadwinning husband and at the same time cope with all family demands. Increasingly, however, women are seeking full-time careers as well. According to the US Department of Labor, the 'typical American family' with a working husband, a homemaker wife and two children now makes up only 7 per cent of the nation's families. In fact, in 1975, 45 per cent of all married women were working, as were 37 per cent of women with children under 6; in 1960 the comparable figures were 31 per cent and 19 per cent respectively. Although simple economics might force a wife into work, there is a price to pay. It is claimed by many psychologists and sociologists that dual-career family development is the primary culprit of the very large increase in the divorce rate over the last ten years in the United States and countries in Western Europe (Cooper and Davidson, 1982).

The problems this creates for the male worker are enormous. It affects almost all aspects of his life at work. For example, many professional men (e.g. managers, pilots, etc.) are expected, as part of their job, to be mobile, that is, to be readily available for job transfers, both within and between countries. A crucial element of his promotional prospects depends wholly on availability and willingness to accept promotional moves. In the 1980s and 1990s, as women themselves begin to pursue full-time careers as opposed to 'part-time jobs', the prospects of professional men being available for rapid deployment will decrease substantially. In the past, these men have, with few exceptions, accepted promotional moves almost without family discussion. Future such decisions will create major obstacles for both breadwinners in the family. We are already seeing this happen throughout Europe and the United States, and this is particularly exacerbated by the fact that corporations have not adapted to this changing social phenomenon. Few facilities are available in organisations to help either of the dual-career members of the family unit.

Conclusion

As this chapter has suggested throughout, the sources of job stress cannot be understood by positing single causative agents like 'long hours' or 'a lousy boss' or 'an unpleasant work environment'. It is usually more complex, involving a number of factors interacting at the same time.

Some jobs, however, are acknowledged as potentially very stressful by virtue of the intrinsic nature of their tasks, technology and hours of work. Wilby (1985) recently asked a number of distinguished stress researchers to assess 'on a ten point scale' a number of different occupations, drawing on their experience and knowledge of stress research (Table 1.2).

It was found that the independent ratings of stress experts were remarkably similar, both in their absolute assessments and in their relative position. It was also interesting to note that although a great deal of research has been done on many of the 'top-ranked' stressful jobs, little has been done on the psychological life of airline pilots (available research will be reviewed in chapter 2).

Stress on airline pilots has become particularly topical recently, with a number of crashes and near misses that have been associated with pilot error, possibly as a result of stress. Two recent examples of this were the 1982 DC8 crash in Tokyo Bay, in which twenty-four passengers were killed; and the 1983 747 crash in which 181 people were killed in Madrid. In the Japanese incident the pilot was found to have a history of psychosomatic disorders, which manifested itself in the pilot pulling his control column forward and plunging into the bay. In the second incident in Spain, the pilot misread his altimeter

Table 1.2 *Stress experts' ratings of various occupations*

Financial areas		Health	
Accountancy	4.3	Chiropodist	4.0
Banking	3.7	Dentistry	7.3 * * *
Building societies	3.3	Dietetics	3.4
Insurance	3.8	Environmental health	4.6
Actuary	3.3	Doctor	6.8 * *
Stockbroker	5.5 *	Nursing midwifery	6.5 * *
		Occupational therapy	3.7
Average	4.0	Optician	4.0
		Osteopath	4.3
Commerce/management		Pharmacist	4.5
Advertising	7.3 * * *	Vets	4.5
Management	5.8 *	Physiotherapy	4.2
Marketing/export	5.8 *	Radiographer	4.0
Market research	4.3	Remedial gymnast	3.5
Personnel	6.0 * *	Speech therapy	4.0
Public relations	5.8 *		
Purchasing and supply	4.5	Average	4.6
Sales and retailing	5.7 *		
Secretary	4.7 *	Environment	
Company secretary	5.3 *	Farming	4.8
Work study/O and M	3.6	Forestry	3.7
		Horticulture	3.8
Average	5.3	Nature conservancy	3.2
Arts and communications		Average	3.9
Art and design	4.2		
Broadcasting	6.8 * *	Public administration	
Journalism	7.5 * * *	Civil Service	4.4
Museums	2.8	Diplomatic Service	4.8
Photographer	4.6	Local government officer	4.3
Publishing	5.0 *	Town and country	
Musician	6.3 * *	planning	4.0
Actor	7.2 * * *	Sports/recreation	
Film production	6.5 * *	admin	3.5
Professional sport	5.8 *		
(footballer, etc)		Average	4.2
Librarian	2.0		
Average	5.3		

Uniformed professions			Personal service industries		
Armed forces	4.7		Catering/hotels, etc.	5.3	*
Pilot (civil aviation)	7.5	* * *	Travel	4.8	
Merchant Navy	4.8		Hairdresser	4.3	
Fireman	6.3	* *	Beauty therapy	3.5	
Police	7.7	* * *			
Prison officer	7.5	* * *	Average	4.5	
Ambulance service	6.3	* *			
			Public service industries		
Average	6.4		Post and		
			telecommunications	4.0	
Caring professions			Gas	4.0	
Nursery nurse	3.3		Electricity	4.6	
Social worker	6.0	* *	Water	4.0	
Teacher	6.2	* *	Public transport	5.4	*
Youth and community					
work	4.2		Average	4.5	
Church	3.5				
Psychologist	5.2	*			
			Industrial production		
Average	4.7		Ceramic technology	4.0	
			Food technology	4.0	
Professional services			Printing	5.6	*
Architecture	4.0		Plastics and rubber	4.5	
Barrister	5.7	*	Textiles/clothing		
Solicitor	4.3		technology	4.5	
Surveyor	3.7		Timber/furniture		
Estate agent	4.3		technology	4.3	
			Leather/footwear		
Average	4.4		technology	3.8	
			Mining	8.3	* * *
Technical specialities			Construction/building	7.5	* * *
Biologist	3.0		Brewing	4.0	
Chemist	3.7				
Computer	3.8		Average	5.1	
Engineer	4.3				
Geologist	3.7				
Lab technician	3.8				
Metallurgist	3.8				
Operational research	3.8				
Packaging	3.8				
Patent work	4.2				
Physicist	3.4				
Biochemist	3.6				
Statistician	4.0				
Linguist	3.6				
Astronomer	3.4				
Average	3.7				

Average scores

6.4	Uniformed professions
5.3	Commerce/management
5.3	Arts/communications
5.1	Industrial production
4.7	Caring professions
4.6	Health
4.5	Personal services
4.5	Public services
4.4	Professional services
4.2	Public administration
4.0	Financial areas
3.9	Environment
3.7	Technical specialities

     * * * = jobs which are very high risk
       * * = jobs which are high risk
         * = jobs which are above average risk

and ran into the side of a mountain, minutes after he was talking about a domestic crisis at home with the co-pilot in the cockpit.

Whilst 'pilot stress' has a long history, research studies in the past have focused primarily upon the physical experience of flying the aircraft. In chapter 2 we shall see that it is not possible to draw up a 'stressor profile' for airline pilots that embraces the sort of topics highlighted in previous research described throughout this chapter. If we refer back to figure 1.1 as our basic model of stress, it is not possible to accurately complete the boxes – especially those concerning possible sources of stress.

It was for these reasons that we felt a thorough investigation into the link between the pilot's job and domestic life and his performance, stress/satisfaction and health was needed. This book reports on a large-scale study carried out by the authors on commercial airline pilots and their families.

Chapter 2
The third factor: a review of research on emotional stress and pilot behaviour*

Most individuals feel that they know something about 'pilot stress'. It is a topic we tend to take for granted. Generally, we expect pilots to be able to cope with such stresses. We rationalise this by 'glamorising' the job and point to various aspects of pilot lifestyle in an attempt to justify a job fraught with stress.

Due to this intrinsic quality of the job, it is not surprising that pilots should be subject to much investigation. Aerospace psychologists have identified three types of human stress: physical stress and two types of emotional stress, cognitive and affective (Haward, 1977). Historically, the study of physical stress, as defined by Flack (1918), comes first. This deals with the effects of extremes of heat and cold, vibration, oxygen deficiency, etc. that aircrew can experience. As Haward points out, this aspect of stress has been the subject of half a century of research in aeromedical centres around the world, and the human response to physical stress has, to a large extent, been well mapped out. However, the effects of some common stressors such as sleep deprivation and 'jet lag' are still imperfectly understood and, Haward considers, underrated. Evidence about this type of physical stress is emerging all the time. Particularly important is the research on circadian rhythms, flying and work scheduling (Folkard and Monk, 1985). But as Yanowitch (1977) points out, even the early First World War studies into pilot behaviour and accidents hint at the involvement of a third, undefined factor besides mechanical or biomedical failures. These early studies, he argues, identify an area that has been almost neglected for too long. In terms of accident causation, he points out that statistics show that mechanical failures account for approximately 10 per cent of accidents and biomedical factors for some 5 to 10 per cent, leaving some 80 per cent of accident causation factors that can only be generally categorised as pilot behaviour.

* This chapter is based on Nimick, B., Cooper, C.L. and Sloan, S.J., 'Emotional stress and pilots: a review', *Inst. J. Aviat., Safe.*, 493–7, 1984.

Goerres (1977) accounts for this 80 per cent in terms of 'psycho-physical workload', the effects of the grand total workload on the human organism, human behaviour and subjective feeling. It appears that psychophysical workload induced by an activity will not only depend on duration and intensity of stressing stimuli, but also on factors associated with the individual stressed subject himself such as physical features, functioning of the sensory organs, present state of health, job-related knowledge, abilities, skills, need for achievement, experience and emotional stress. Simonov, Frollov and Ivanov (1980) take up this point and comment that among the factors that have a marked influence on the state and work efficiency of aerospace operators (pilots, cosmonauts, ground crew personnel), emotional stress plays an important role.

Clearly it is difficult to put a figure on the 'importance' of such stress, but one would certainly expect it to be of major significance. It is easier, however, to be more specific about the other characteristics of this type of pilot stress. Emotional stress, as an issue in flying, has found therefore general acceptance. This can be subdivided into two general concepts: cognitive and affective stress. Cognitive stress can be applied to the 'intellectual', that is, the non-emotional and imper-sonal functions of the aircrew. Cognitive stress can be defined objec-tively as the nature of the task presented to the operator, excessive cockpit workload being one of its most frequent forms. Parry and Fokkeman (1958) explain that when Bleriot crossed the Channel in 1909, he had no instruments and performed only four actions apart from operating the flying controls. Half a century later, a DC4 pilot flying over the same route required 195 actions together with sixty-six by the co-pilot, a sixty-six-times increase in workload. A second form of cognitive stress is inadequacy of information on which some critical decision has to be reached. Landing manoeuvres provide this sort of scenario, for example. Landing in bad visibility on a small airfield with few or malfunctioning ground aids would illustrate this phe-nomenon (Haward, 1977).

The study of pilot stress transcends political and cultural bound-aries even into East Bloc countries. Although there is some interesting Soviet literature, for example, on emotional stress, it has not con-centrated greatly on affective stress. By contrast, in Western research, it appears to be taking an ever-increasing prominence. Haward (1977) regards it as being predominantly subjective in nature, its effects only being meaningfully interpreted by reference to the life history and the personality of the person under stress and its intensity measurable by

17

the strain produced. It is usually hidden from others, is often insidious in its onset and is unpredictable in both intensity and duration. Unlike cognitive stress, which gradually produces increasing impairment of efficiency, affective stress, within seconds of its onset, can bring about a complete breakdown in rational behaviour, as in the form of paralysis known as 'freezing at the controls'. Carrol (1969) correlates this form of paralysis with shock showing, for example, that affective stress accounts for many passengers found dead in their seats, but with no impact injuries, after an aircrash involving fire. A survey by Aitken (1969) stated that 71 per cent of all the service pilots questioned in the study had suffered a significant degree of affective stress and suggests that in commercial pilots the incidence may be even higher. On balance most would agree that it is a neglected area of primary importance (Yanowitch, 1977). While recognising this, Alkov and Borowsky (1980), in appreciating that its effect can only be predicted by reference to the life history and personality of the individual, feel that attempts to elicit such information may be ethically precluded. They do, however, agree that some method must be found to procure such information, supporting Haward (1977) in his belief that affective stress is the most dangerous kind. Shuckburgh (1975) supports this, identifying affective stress as being both additive and cumulative. So, a long-distance pilot with sleep deprivation and time zone changes, together with an unsettled domestic life, will have a lower tolerance to affective forms of stress. The Civil Aviation Authority (1975) does appreciate this problem and emphasises the need to avoid cumulative stress. However, the same emphasis is not given some seven years later in their new guide to the avoidance of excessive fatigue in aircrews (1982).

Clearly the distinction between cognitive and affective stress is, in practice, a difficult one to pin down. Methodologically it is a highly complex task to design a study to isolate these sorts of processes. The research has tended to be conducted within specific contexts. Indeed, research into the effect of emotional stress on aircrew has been undertaken to explain the relationship between stress and accidents. On balance the research subdivides into two approaches: those dealing with the relationship between the pilot personality and stress, and those dealing with adverse life events and stress.

Personality predispositions and stress

Christy (1975) quotes Sullivan's definition of personality: 'the relatively enduring pattern of recurrent interpersonal situations which characterise a human life'. He regards personality as a dynamic process involving, amongst other factors, the mastery of methods and techniques for coping in problem situations. To Christy, the factors of personality among the aircrew that he is interested in are (1) those which motivate a person to enter into flying, and (2) the factors which produce tension. While Green (1977) generally agrees with the definition used by Christy, he approaches the question of personality by trying to elicit those personality differences that lead to differences in behaviour when different individuals are subject to the same situations. What is common to both approaches is the conviction that stress, personality and accidents are, in some way, linked.

The equation is a complex one. Indeed, there are several approaches to the identification of the pilot personality. Christy (1975) sees pilots as requiring a need for mastery, prestige, control, aggression, expression and competition. He believes the problem is that many of these traits may be assets when learning to fly but if excessive can be contributory, in the longer term to the conflict and tensions that produce unsafe flying practices. This is due to pilots' tendencies towards (1) overorganisation, (2) overconscientiousness, (3) perfectionism, and (4) inability to relax. Green (1977), whilst accepting this argument, sees the main problem as one of personality measurement. He points out that there is a certain amount of self-selection in pilots and that this has the effect of decreasing the validity of empirical measures. In addition, he finds it difficult to determine which aspects of an individual's behaviour may be attributed to his personality and which to his experience. Despite these doubts, he comes to the same conclusions as Alkov and Borowsky (1980) about the 'pilot personality' profile. Reflecting Christy's findings, Alkov cites the pilot personality type as being heterosexual, and orientated to demonstrating strength and to competency tasks. In short, the type of individual, whether male or female, who is attracted to aviation is the stereotype pilot of myth, folklore and motion picture. However, for the psychologist, the problem is one of demonstrating the existence of a relationship between this pilot personality type, stress and accidents.

To Ursano (1980), Selye's General Adaptation Theory (referred to in chapter 1) helps to provide part of the required link. Extending the theory to Canon's (1929) conceptualisation of reaction to threat situations – 'fight or flight' – he argues that when neither of these two alternatives is available then Selye's alarm, adaptation and exhaustion process will come into operation. Stress, he argues, represents a mentally or emotionally disruptive influence, a change in the homeostatic balance, and adaptation, therefore, becomes the adjustment of an organism to its environment. Stress, in this context, is therefore considered to be any event which requires the individual to undergo change, and too much stress in too short a time is associated with the onset of illness. Like Christy (1975), he believes that the overall psychological strength highlighted in the pilot personality is modified by recognising that strength in certain areas may lead to weaknesses in others. He comments, and is supported in this by Alkov and Borowsky (1980), that the pilot tends to avoid and deny his internal emotional life. He perceives his inner feelings as external to himself and tries to cope with this by changing the situation around him but lacks insight and introspection. Additionally, people who fail to cope with stress tend to internalise their feelings when under pressure. In such a situation the pilot may possess few coping strategies for dealing with it. The results of this inability are externalised in depressive and self-destructive behaviour. Such a profile fits that of the accident-involved pilot described by Reinhardt (1966) as 'the action orientated individual who acts out his frustrations rather than verbalizing them and whose error prone behaviour is manifest in all areas of his life; personal, social and professional'.

This sort of relationship between personality and stress coping is accepted by Haward (1977), who believes that there is a clear relationship between the degree of stress experienced and impairment of flying skill. Although this link is disputed to some extent by Yanowitch (1977), who argues for a more multi-causal approach, there is an agreement that personality and stress coping are important components of pilot behaviour in accident situations. Haward (1977) goes on to identify two different subsets of response to affective stress. The Type I pilot is characterised by a marked increase in activity of the sympathetic nervous system and shows the greatest impairment of flying efficiency when his emotions are at first aroused but his efficiency gradually improves as his feelings abate. The Type II pilot, however, is characterised by minimal autonomic nervous activity but excessive mental rumination. He may show a lower level of flying impairment than the Type I pilot, but this will tend to persist for longer.

What is apparent is that many psychologists feel that there is a link between inadequate stress coping, the pilot personality and accidents. Personality inadequacies in coping with stress do seem to result in some form of flying impairment. But this is only half the story. Although this inadequate coping mechanism may respond to cognitive and some types of physical stress, it is at its most dangerous in conjunction with affective stress, the source of which lies in the pilot's personal and emotional life. Whilst individuals' lives may provide a rich source of such stress, there will be much variation. In general, most of us can identify particular events or circumstances that we found stressful or traumatic. These particular stressors would seem a most likely source of affective stress even to pilots. There is, therefore, a need to consider, as McCarron (1982) insists, a requirement to examine personality in conjunction with these 'life changes'.

Life events and stress

In a paper on life changes, Alkov and Borowsky (1980) hypothesised that ongoing life changes described by a US Navy psychiatrist as affecting the health of Navy men might also impair their performance, leading to aircraft accidents. Studies were carried out over a period of several years and demonstrated that life changes in the two years prior to deployment are correlated with events later during the deployment, on particular health changes. The hypothesis is that these life events create internal stress, which demands coping strategies and behaviours. When the system is overloaded or the coping system is inadequate, then it may break down and poor health, both mental and physical, results. Green (1977) finds that this is intuitively reasonable; however, while agreeing that life events such as domestic stress are not desirable for flight safety, he is not convinced that a statistical relationship between domestic stress and accidents has been readily established. In many ways this is to be expected since, as we have established, it is difficult in practice to pin down much of the psychological processes going on. This of course does not mean that there is no relationship – merely that it is not amenable to simple measurement. Green does, however, acknowledge the existence of a link between life changes and illness.

Some studies have been successful in establishing a link. Aitken (1969), for example, tried to establish such a relationship between domestic situations and the accident rates of a number of RAF squadrons. He showed, indeed, that more pilots from squadrons with higher accident rates had worries, particularly about flying and bereavement and about their wives and love lives, than had pilots from squadrons with lower accident rates. Shuckburgh (1975) analysed commercial aircraft accidents attributable to human error on the flight deck. He shows that, on average, around 45 per cent of commercial flying accidents were due to pilot error. Having established this figure, he offers several explanations as to why such errors should occur, such as (a) fatigue, (b) environmental factors such as vibration, temperature, etc., (c) pilot workload including physical stressors such as irregular sleep, long hours of night flying, and (d) the pilot's unsettled domestic life – the relationship between a man's private and professional life and the effect that worry or unhappiness in one has on the other. It would seem that, from an aircraft accident point of view, it is the home-to-work direction that is the more important. But, what might the sources of such stress be?

Ursano (1980) defines as a cause of stress any event which requires the individual to undergo change. He believes that life events such as the death of a spouse, divorce, marital separation, pregnancy, marriage or vacations represent environmental events necessitating a change in the individual's usual adaptive relationship with his environment and that it is this necessity to change that presents significant stress to the individual. Illness is certainly one of the results of such changes and he advocates the use of procedures designed to elicit such life change factors, so that their effects in terms of accidents might be calculated. While only hinting at a direct casual relationship with accidents, by the use of three case histories Ursano (1980) demonstrates a direct relationship between life events and psychosomatic illness.

It was in an attempt to determine whether this direct causal relationship between accidents and life events existed or not that Alkov and Borowsky (1980) undertook a study of aircrew in the US Navy. A questionnaire approach was developed to determine the effects of both personality and life changes on pilot behaviour. The ultimate goal was to see if such behaviour could be predicted by examination of the questionnaire results. The questionnaire contained fifty items; twenty-five of these were concerned with the pilot's personality, such as judgment, maturity, leadership abilities and professionalism, and the remaining twenty-five were extracted from a well-used measure of life events (Holmes and Rahe, 1967). The questionnaire was given to two groups of pilots, one consisting of pilots who had been involved in accidents and the other of those with accident-free records. The results demonstrated two main predictors of error-prone behaviour: first, involvement in a recent major decision regarding the future, and second, the likelihood of having suffered from difficulties with interpersonal relationships. Whilst being a step in the right direction, Alkov and Borowsky concluded that further research was required to determine more precisely the relationship between accident proneness, life change and personality.

Indeed, it was Alkov, Borowsky and Gaynor (1982) who initiated a follow-up study. Working on a hypothesis that pilots who contribute to aircraft mishaps are more likely to be identified as having troubles with interpersonal relationships and, by implication, shown to show symptoms of inadequate stress coping, a twenty-two-item questionnaire was designed and submitted to accident investigators. All of the pilots in the study had been involved in some sort of aircraft accident or incident. Alkov knew, however, a priori, which pilots had been 'at fault' in these incidents or accidents. He was able therefore to identify two subgroups which, in turn, allowed the possibility of meaningful comparisons between the two. Generally, it was found that pilots who were 'at fault' were likely to have marital problems (Haward, 1977); they were likely to have difficulty with interpersonal relations, (Cooper, 1981); to have recently become engaged or married (Flack, 1918); and finally, to have made a recent major decision regarding their future. As a result, the researchers believed their hypothesis to be upheld in that they had identified at least part of the personality/life changes/pilot behaviour equation.

Others were not so certain. McCarron and Haakonson (1982) also started from the premise that a majority of aircraft accidents have a significant human factor element and that, therefore, a tool is required to assess and predict accident proneness among pilots, which proportionally increases their personal stress levels. As such, they believed that the study of such life events provided a reasonable point from which to start designing a screening process. Based upon Canadian Air Force personnel, Rahe's Recent Life Change Questionnaire (RLCQ) (Rahe, 1978) was given to three populations, two of them designed to be controls. The seventy-six life changes on the RLCQ were classified into five main variables: (1) health, (2) work, (3) family, (4) social, and (5) financial. After analysis of the results, McCarron and Haakonson concluded that, although there were correlations between the onset of illness and stressful life events, the relationship with aircraft accidents was not so clear. However, they did feel that there were indicators to support the premise that an excessive amount of life change does contribute to personal levels of stress and that this, in turn, may increase accident liability in the pilot. In fact, it was found that some pilots were operating at a life crisis level that would normally predict health changes in 50 per cent of the general population and that, therefore, pilots should be classified as a high-risk group when it comes to the amount of life changes that they are subject to on a routine basis. McCarron and Haakonson concluded, therefore, that the effect of excessive life changes is at the moment only a contributory factor to personal stress and illness and merits further attention.

Some concluding remarks

Throughout this chapter we have reviewed the material that falls under the umbrella title of 'pilot stress'. Whilst this is interesting material, the content of the material reviewed does not conform to the lay person's image of 'pilot stress'. What we have done is to present that material which is most akin to the stress literature presented in chapter 1 and applied it to pilots. However this 'third factor' seems rather elusive. It is reasonably acceptable to make comments like '45 per cent of accidents are due to human error'; however it is a much more difficult task to actually pin down the causal factors that contribute to such statistics. Part of the problem is the fact that much of the area of study seems so subjective. The very basis of 'affective' stress is the emotional experience of the pilot. This will inevitably entail the conscious perception of the situation by the pilot and the success or failure of his attempts to cope. But even this may not go far enough since the effects or outcomes of stresses (even though the individual may well be able to identify the source) may well be unconscious and beyond the pilot's conscious perception and control.

Of the different types of stress identified, it would seem that it is this affective form of emotional stress that is the more volatile and subject to the least control by the pilot. Whilst much of the work has been in the context of accidents and incidents, it seems that it is easier to pin down the outcomes of such stress, in terms of ill-health. As indicated, this is probably for methodological reasons. In other words, owing to the complexity of accidents and incidents it is difficult to partial out all those other things that might have an effect. Certainly it seems rather cavalier to try to put a percentage figure on to such factors since these, by definition, are merely broad sweeping indicators. It would seem that, at best, the background literature has attempted to tap recurrent themes such as personality and life events.

The former, personality characteristics, are well documented. Whilst it is, of course, a gross overgeneralisation to state that pilots tend to conform to the stereotype, certainly there do seem to be a series of traits that are recurrent. With the identification by some researchers of two pilot types, the overall message would seem to be that there is some link between personality and flying impairment. Such a link seems to be provided in the nature of how the pilot copes with stress.

In terms of the latter, life events or changes, the relationship with accidents seems more elusive. However such precipitating factors might well be important. They do seem to form a logical and coherent theme. Additionally their impact on pilot ill-health seems fairly well established.

In conclusion there are several characteristics that make the literature outlined briefly in this chapter rather distinctive. The first is the fact that the pilot studies have attempted not just to examine the sources of stress in pilots, but have also attempted to isolate the processes or mechanisms by which these may work. This has also been done elsewhere for other occupations though the data outlined in chapter 1 concentrated primarily on sources. A second issue is the fact that these sources of stress in pilots do not seem to be expressed in the same terms as those outlined in chapter 1. As previously indicated, it does not appear to be possible to draw a psychosocial stress profile for the job of commercial pilot, as it is for many other occupations.

Chapter 3
Investigating pilot stress: our study

The pool of literature that we have reviewed in the last chapter is relatively small. This reflects the fact that although there is a large body of research on pilots' behaviour *per se*, most of the work does not adopt a psychosocial orientation. The data that has been collected has been collected within the context of aircraft accidents or incidents with a view to attributing causality. However, we do not know much about the 'day-to-day' effects of stress. In investigating the more volatile affective stress, research has taken two routes: pilot personality and domestic stress (in the form of life events). Whilst the data on personality permits us to erect a fairly satisfactory profile, the data on life events has been of only limited utility.

Outcome variables (i.e. the effects of stress) seem rather ill-defined. We are not really in a strong position to suggest which domestic or occupational stresses affect pilots' mental and physical ill-health or performance. The literature, having concentrated on short-term effects of stress (such as the effects on attention, vigilance, etc.), has done so at the expense of the broader, longer-term outcomes, especially ill-health.

Finally, the deficiency in our knowledge about pilots is clearly demonstrated by examination of Figure 1.1 (page 2). Examination of the pilot data illustrates that we cannot easily fill in the cells of Figure 1.1. From a psychosocial point of view, we do not really know much about occupational stresses in the job itself, career variables, interpersonal issues, and so on. We do not have a picture of the pilots' domestic scene nor do we know how his working life and home life are interrelated. Finally, we do not know how pilots cope with the stresses they encounter. It was with these and many other unanswered questions in mind that we decided to undertake the present study with an overall aim of exploring and investigating the issues we have just mentioned. The goal, therefore, is to see if we can uncover the contents of Figure 1.1 as applied to pilots and to highlight the salient variables and their complex interrelationships. In particular, we are

26

interested in two further aims. The first is, can we not only identify sources of stress for pilots but also identify themes or trends in these sources? The second is, can we identify the effects of these stressors on pilots?

The research project

Overview of what we did and how we did it

Basically, the research we have conducted may be divided into three parts. The first part was a series of preliminary interviews that were conducted to provide a starting point and a springboard from which we could delve into the subject in greater depth. This more intensive examination was the second part of the study and took the form of a large-scale survey involving around 450 pilots. The third part was rather unusual and was in reality performed concurrently with the second. It concerned the examination of data from pilots' wives and took the form of an interview plus survey.

In this chapter, we intend to give an account of how we conducted the pilot interviews, the pilot survey and the wives study. In particular we intend not only to 'home in' on the background reasons why these were conducted, but also on the methods we used and how we analysed the data collected.

Study 1: preliminary background interviews with pilots

Aims and objectives

There were three overall aims for conducting a series of background investigative interviews. The first was to set the scene and provide us with an opportunity to familiarise ourselves with this unique occupational group. This was an important goal. The reader will recall no doubt that although we already knew a lot about pilots in terms of their physical or ergonomic environment, and something about their personality, we knew very little about their psychosocial life (e.g. domestic stress, interpersonal and work relationships, career prospects, etc.). The second aim, therefore, was to focus our attention on these most important areas for further investigation as suggested by the pilots themselves. This latter aspect is crucial, for it is on the psychosocial factors identified by the pilots that we really want to focus. Our third objective was an extension of the second. Not only

27

was it desirable for our attention to be focused on specific areas of concern, but it was also desirable that pilots identify particular issues within each area that could be more rigorously tested in subsequent research.

Our sample

We obtained our preliminary sample of interviewees through several means. The main source was through the British Airline Pilots Association (BALPA), which is the pilots' professional association based in London. Additional samples of pilots were available through contacts at Manchester Airport. In both sources, pilots were simply chosen at random from available lists. So as to avoid a biased sample, before interviewing began we ensured from information already available that as wide a cross-section of characteristics as possible were represented, such as long haul/short haul, rank, seniority, etc. These pilots were approached by letter. All interviewees were volunteers.

There were fifty-four pilots in our sample of preliminary interviews. A cross-check confirmed that a fairly wide cross-section of pilot characteristics was represented. There were no major obvious sources of bias. All were commercial airline pilots flying for British companies and were a mixture of long haul, short haul, captains, co-pilots, seniors and juniors in status and of varying ages.

The questions we asked

Since item generation was our third and especially important objective, we paid particular attention to asking open-ended questions on relevant topics.

Questions structure for preliminary interview of pilots

(1) Which are the most important domestic stressors? Why? What sort of effects or consequences do they have for you?

(2) What are the key elements of domestic stress? What sort of things make you tense/worried/anxious in your life? How do you know when you are experiencing stress?

(3) Which are the most important job stressors? Why? What sort of effects or consequences do they have for you?

(4) Which are the most important job stressors which affect home life? What is the nature of the relationship between work and home?

(5) Which are the most important home stressors which affect work life? What is the nature of the relationship between home and work?

(6) Have you experienced any life events? What was the last one? Can you describe the events and your feelings before, during and after the event?

(7) How do you cope with stresses you experience and what are the consequences of failure to cope?

(8) Can you design a self-report measure of performance for use in a postal survey?

(9) Anything else relevant?

The general line of questioning was designed to tap as many areas as possible that had been indicated by deficiencies in previous research outlined earlier in this chapter. This involved questions not just about stressors that were solely occupational or domestic, but also those that transcend the home-to-work and work-to-home relationship. You will note that we examined life events and included a purely investigative series of questions about how pilots try to cope with the stresses they experience. Finally, we had previously identified a methodological problem in that there was no self-report measure of pilot performance which could be utilised as one of a number of criteria measures of stress outcomes. We therefore took this opportunity to ask the pilots themselves to assist us in the design of such a measure. This will be discussed further below. In order to induce as much reliability into the process as possible, we adhered fairly closely to the above question schedule. Naturally, however, we asked pilots to develop their own ideas and themes where relevant.

The analysis

We were also fairly rigorous in our analysis of interview data. Data were divided into nine groups relating to each series of questions asked; the individual items were recorded along with their frequency of occurrence throughout the fifty-four interviews. The items were then grouped together into clusters of items that seemed to relate to the same issue. Again to induce some element of reliability, several individuals not connected with the project regrouped the items for us. There was close agreement between the clusters identified by them and by us. The end result, therefore, was that the descriptive interview data were transformed into an array of items rated by their relative frequency of occurrence. It was these items that formed the basis of the questionnaire employed in the main survey. We will discuss this in detail later though a copy of the questionnaire is included in the Appendix. We feel that the primary objectives of the preliminary work were achieved. The interviews gave us the opportunity to explore the area of research. They also shed light on an area of pilot research which had previously been relatively neglected. The data we collected was exceedingly rich. In fact, chapter 5 is a fairly detailed account of our preliminary findings. However it was the next stage (see chapter 6) that really forms a major thrust of the research.

Study 2: the major survey of British commercial pilots

Aims and objectives

After conducting the first study, we had amassed a large amount of qualitative data. This sort of data is very appealing, since it is expressed in everyday terms in the words of the pilots themselves. The great drawback in trying to draw conclusions from such data is the fact that its superficial appearance and appeal can overshadow its underlying (and more important) reliability. As has just been described, we made some attempt to make the situation more rigorous than it would otherwise be in a simple examination of interview quotations. What is really required, however, is a more scientifically organised study that examines statistically the responses of a large and robust sample of pilots. It is only by the use of a more complex methodological approach and the application of appropriate analytical procedures that our two main goals of identifying specific pilot stressors and highlighting their potential effects may be achieved. To conclude, because only a limited amount of relevant previous research had tried to adopt a psychosocial approach to pilot stress, particular attention was paid to the investigative quality of the research. This is especially true of the relationships between the pilots' home life and work, an area which seemed to be devoid of rigorous in-depth examination.

The sample and how we achieved it

Around 50 per cent of British commercial pilots are members of BALPA. With a potential total population of 5,000 to 6,000 pilots, it was clear that this would be an expedient means through which to collect a sample. We obtained a potential sample of 1,000 pilots by a two-stage randomisation process. First, approximately every third name was selected from the BALPA list of members, providing 1,500 names. We then selected from these 1,000 at random. We then sent explanatory literature to these 1,000 pilots together with a questionnaire. Completed questionnaires were received from 523 pilots – a response rate of 52.3 per cent. Of these, twenty were unusable and sixty-one arrived too late to be analysed, hence the sample included here for analysis was of 442 pilots.

Although relatively straightforward, conducting a survey in this way is not particularly efficient in the sense that there is a lot of wastage. Traditionally, a response rate of more than 25 per cent is considered good. So it is clear, therefore, that the response rate in our study was excellent. This is a fairly good indication of the interest generated in the area of research. We mentioned earlier that we had obtained the sample on a random basis. This was to combat any underlying trends which might have biased our sample. Of course, the very fact that we only used BALPA members may in itself have induced a degree of bias. It could be argued that members are subject to different influences to non-members or are particularly sensitive to particular issues. We could, of course, have contacted airline companies directly and tried to obtain a sample of non-members in that way. However, this is time-consuming and brings with it other difficulties (such as confidentiality, for example). In any case, the main point to consider is this: given the psychological nature of the study we propose here, there were no reasons to suppose that the members and non-members would be significantly different in terms of the subject matter being examined. It seems reasonable to conclude therefore that the method of obtaining respondents employed by us did not create a biased sample.

How we measured stress, its causes and effects

The questionnaire we sent to pilots was a battery of inventories, designed to create a complete and efficient investigative package. As may be seen in Table 3.1, the contents closely reflect our main areas of interest. Generally, the first parts are devoted to setting the scene and pinpointing sources of stress, whilst the latter parts are geared more towards measuring stress outcomes. Since Table 3.1 contains the structure and the Appendix contains a copy of the questions themselves, there is little point in discussing these in much depth. We

Table 3.1   *The structure of the questionnaire used in the main pilot survey*

Section	Title	Number of parts	Number of parts
1	Biographical data	7	56
	(i) the pilot and his family		
	(ii) pilot interests and hobbies		
	(iii) exercise and fitness		
	(iv) smoking		
	(v) eating habits		
	(vi) drinking		
	(vii) work history		
2	Problems, occurrences, issues that are entirely domestically orientated	1	29
3	Problems, occurrences, issues that are entirely occupationally orientated	1	30
4	Relationship between work and home	1	16
5	Relationship between home and work	4	
	(i) home factors that may affect work		12
	(ii) nature of effects of home factors on work		12
	(iii) factors generally affecting pilot performance		
	(iv) relative similarities and differences in sources of home and work stress		16
6	Life events	3	
	(i) life events experienced by the pilot		6
	(ii) life events generally		8
	* (iii) life events (based upon Alkov, Borowsky and Gaynor, 1982)		22
7	Coping	1	33
8	* Job satisfaction (Warr, Cook and Wall, 1979)	1	16
9	*Mental ill-health (Crown and Crisp, 1979)	1	48
10	Performance	1	15

All instruments were designed by the authors especially for the present study. Instruments marked thus* were designed by previous researchers.

A copy of the questionnaire is included in the Appendix.

would, however, like to say something about the questionnaires that had been designed elsewhere and employed by us, and also about our measures of self-reported performance.

Life events. The aerospace medical research we discussed earlier identified life events as a potentially fruitful area for research. The major studies in this area had, however, taken a fairly sophisticated questionnaire-type approach to the study of the impact of life events on pilots. We know, however, from research elsewhere that other authors advocate a more simplistic approach. It will be noted from Table 3.1 that we, in fact, employed both methods. In addition to asking simple questions about the nature and impact of life events, we employed the questionnaire used by Alkov and his colleagues (1982). However, we modified the twenty-two-item inventory in three ways. First, instead of asking respondents to use the items to describe a specific pilot, we asked pilots to use the items to draw a portrait of a typical accident-prone pilot based on their own experience and knowledge of other pilots. Second, we expanded the answering scale from a simple 'yes/no' scale to a five-point Likert-type rating scale. Third, we asked pilots to use all items and not just a selection of items. The above changes are in recognition of the fact that we could not employ Alkov's particular experimental design in the context of the present study. Additionally, we felt that by expanding the scale and asking for responses based on all items, our data would be all the more interesting and robust.

Job satisfaction. The relationship between the effects of occupational stress on job satisfaction are not as straightforward as might at first be thought. The psychological literature contains both positive and negative relationships and also records no relationships in some instances. On balance, however, most would accept that as stress increases perceived job satisfaction decreases. Since job satisfaction is itself of major concern as a psychological state we decided that it would be a suitable outcome variable to examine. The measure we chose to employ was British, relatively recent in its design and has been tested for validity and reliability (Warr, Cook and Wall, 1979). It also contains seven subscales. These are defined below:

(i) *Intrinsic job satisfaction:* the contribution to job satisfaction that is derived by the person in actually doing the job such as taking responsibility, demonstrating skills and abilities, and so on.

(ii) *Extrinsic job satisfaction:* the contribution to job satisfaction that is derived by the person from the context in which the job is carried out such as fellow workers and rate of pay.

(iii) *Job itself intrinsic satisfaction:* intrinsic satisfaction that is derived from specific features of the actual job.

(iv) *Working conditions extrinsic satisfaction:* issues such as hours of work, job security.

(v) *Employee relations satisfaction:* the experience of working as a team and the quality of interpersonal relationships.

(vi) *Total job satisfaction:* a combination of all fifteen items.

(vii) *Overall job satisfaction:* the general level of job satisfaction measured by a single item taking everything generally into account.

We know from the massive medical and psychological research literature that the experience of stress will have significant implications for the well-being of the individual. The physical effects of stress can be identified from short-term biochemical and physiological responses through to longer-term physical ill-health. Although the implication of stress can be recorded in all sorts of illnesses, by far the greatest concern is with coronary heart disease (CHD). Many studies have highlighted a clear causal link between stress and CHD (Cooper, 1983).

Mental health. The measurement of physical effects of stress presented us with a twofold problem. The first is the relatively straightforward one of how to actually measure physical ill-health within the aims and confines of a study of this sort. In fact, it was for this reason that we decided not to include a measure of physical ill-health. However, it was the second conceptual issue that proved to be the more worrying. This was as follows. All of the pilots in the sample had valid medical certificates and hence were deemed 'fit' to fly. We presume also that they were fairly 'healthy' in the normal sense. It is, of course, possible for health to fluctuate and still meet any minimum standards that are required in medical checks. Whilst not being able to measure these in a physical way, it is possible to measure them in terms of mental ill-health.

Mental ill-health or 'neuroticism' is the other major stress outcome recorded in the literature. The instrument we selected is a well-used British screening device of mental well-being (Crown and Crisp, 1979). It yields six subscales plus a total mental health index, each of which may be treated as an outcome scale in its own right. These were

(i) Anxiety: state of apprehension in which the source is not usually as specific as it is in fear. Vague but persistent feeling of tension, worry or fear in the absence of any realistic cause or source.

(ii) Phobia: pathological fear of an object or situation. The individual may realise the fear is irrational but be unable to control it.

(iii) Obsessionality: persistent idea or behaviour that the individual realises is irrational but feels compelled to dwell on and is closely related to compulsion.

(iv) Psychosomatic disorders: physical symptoms of disorders that have no specific physical cause that can be attributed to them. Their origins therefore are assumed to be psychological.

(v) Depression: disproportionate feelings of sadness and apathy.

(vi) Hysteria: a generalised term used to describe emotional excitability, excessive anxiety and psychologically caused sensory or motor disturbances.

(vii) Overall total mental health: summation of the above six scales.

Pilot performance. Whilst both job satisfaction and mental ill-health are crucial aspects of the pilots' psychological states, we are also interested in whether or not an issue identified by pilots will have an effect on their perceived performance. For practical reasons we could not administer an objective test of performance. Similarly we did not have access to individual performance data, such as simulator checks or route checks. Such sensitive material is of course confidential and is not normally available. We decided therefore to design our own measure of performance, which would yield data that could be treated as a stress outcome, in the same way as mental ill-health and job satisfaction.

The initial preliminary interviews contained questions about this issue of self-reported performance. We asked pilots how they might go about the task of designing such a measure. The result, as illustrated in the Appendix, was an array of fifteen items. The resultant questionnaire was not only geared towards assessing performance, but was expressed in terms that the pilots in the survey would find meaningful. Finally, in order to try to build into the measure some degree of rigour, the scores for each respondent were not just simply totalled for the fifteen items. We used a sophisticated statistical technique for developing weighting coefficients. These were created on the basis of the perceived contributions each of the fifteen items made to the concept of measuring performance. The total, therefore, was the summation of responses to each of the fifteen weighted items. This total was treated in the same way as the other two outcome variables, i.e. it was later included in analyses to see which stressors were associated with it and which predicted it.

Study 3: pilots' wives

As has been previously indicated, one of the main themes of the study was to try to build up an overall picture of the pilot, his work, his home life, and the relationship between the latter two. The role played by the pilot's wife will be very important. Not only will the pilot's wife play a crucial role in determining his quality of life, but also she too will be affected by whatever stresses the pilot is experiencing. Previous research seemed to have left this potentially rich source of data completely unexamined. An additional reason for adopting this rather unique approach is that the pilots themselves during interviews clearly indicated the important role played by their wives.

To conclude, the aim of this third study was to gain insights into the quality of pilots' lives as perceived by their wives. This study was conducted concurrently with the survey study of pilots previously mentioned. The format of the study was in two parts, interviews followed by a large-scale survey in which responses from 250 wives were analysed. Measures were taken of domestic stresses experienced by wives; factors from work that affect the pilot at home; the effects of work factors on the pilot; and life satisfaction. Details of these measures and overall conduct of the study are given in chapter 8.

Data analysis

It ought to be apparent by now that we have tried to adopt a fairly strict approach to the collection of data from pilots and their wives. This underlying theme of methodological rigour was applied in the analysis of data collected. Reliance upon a questionnaire-type approach makes some attempt to induce standardisation and consistency into responses. These are two of the major aspects one would expect to find in a 'scientific' study. Another requirement, that of validity, is also met since the questionnaires are based directly upon the responses of pilots elicited in the preliminary background interviews.

The analyses were performed by computer, since we were dealing with high numbers of subjects involving a high number of items. Even the most simple of statistical tasks can, under such conditions, require a massive amount of processing. Generally speaking the analyses followed three stages: univariate analyses, i.e. the analysis of single items such as averages, percentages and so on; bivariate analyses, i.e. the analysis of pairs of items such as comparisons of averages, breakdown of groups, distributions, etc.; and finally multivariate analysis, i.e. the examination simultaneously of large numbers of interrelationships between items.

This latter category is quite crucial to the outcome of the study so it might be appropriate to raise one or two further points. There are two techniques in particular that crop up fairly regularly in the later chapters of this book. They are called Factor Analysis and Multiple Regression. In the first technique, the method examines a large series of items and uncovers the existence of statistical *trends* in the data. It also points out which items are significant and to which trends they are related. The second technique is different in that it takes a specific variable (usually an outcome variable such as mental ill-health, for example) and analyses it with reference to other sets of variables. The aim is to see which items *predict* the outcome variable. These themes of *trends* and *predictors* are important and we will refer to them later in the text where appropriate.

How we have presented results

Whilst we adopted a fairly strict approach to the collection and analysis of data, we have attempted to present the findings in a way that avoids lists of complicated statistics. We present in the text that follows not just the numerical results, but also data from the interviews. This is useful, since we can use the statistics to provide a structure and illustrate the various points further by reference to the interview data.

This descriptive account of the studies we performed concludes the overall introduction to the subject and our studies. Effectively, the next chapter (chapter 4) marks the beginning of a 'results' section. Whilst covering broad descriptive information, the following two chapters are not just to set the scene. They are to introduce and to develop important and relevant issues. Chapter 4 for example contains a demographical account of pilots, their leisure interests, work history and lifestyle habits. The next chapter (chapter 5) contains our preliminary thoughts and conclusions about sources of stress. Reliance is mainly upon interview-type data. However, the issues raised about the domestic and occupational sources of psychosocial stress are those that are further developed and elaborated in chapter 6. The emphasis here is upon identifying trends and discovering the predictive value of the items examined in the manner just described. Subsequent major chapters concern life events, coping and the data collected from pilots' wives. Throughout, the data is presented and discussed by reference to both interview and statistical data. Finally the aim of the last section, chapter 10, is to draw together underlying ideas and themes and to draw together those issues raised in preceding chapters. Additionally we try to tease out what seem to be the most reasonable conclusions on the basis of the data collected in the present study and in the light of the deficiencies in previous research already mentioned at the beginning of this chapter.

Chapter 4
Pilots: who are they?

An appropriate way to commence would seem to be to focus our attention on 'individual' behaviour. We said earlier that we have to account for underlying characteristics in whatever sample of subjects we are examining before any meaningful conclusions can be drawn. Since the size of the sample of pilots in our study is just under 450, we are in a fairly good position to highlight the important and possibly confounding trends that characterise the sample. The overall goal, therefore, is to start to build up our portrait of the profession and job of a pilot by identifying and summarising pilots' basic biographical and demographical features.

Based upon experience, we have divided these characteristics into four categories. The first concerns 'pilots and their families'. This was to gain some insight into basic biographical facts such as age and marital status. It was also to identify other features such as whether the pilot's wife worked or not, or how many children he had. The second category, 'interests and hobbies', was to examine the sort of activities pilots engage in when not working. Often, the behaviours of the individual outside work will have as great an effect on their 'well-being' as whatever they experience in their job. Before we can make judgments about psychological and particularly health outcomes it is necessary to identify outside influences on the results. Typically we would be interested in four such features: smoking, drinking, eating and exercise. These 'lifestyle habits' form the third part of our examination of biographical characteristics. Finally, it is also necessary to pin down work-related trends. The fourth section, 'work experience', summarised these work characteristics, such as rank, seniority and flying experience. Such features are especially necessary in the examination of outcome effects. You will recall that these concern job satisfaction, mental ill-health and performance. To complete our portrait we have included this data towards the end of this chapter.

To conclude, because we have an extensive database to work from, we are in a unique position of being able to do two things. First, we can identify and discuss biographical data that is not normally recorded in this form or in such detail. Second, compatible with our original intentions, we can further our understanding of the relative contributions that are made by basic descriptive features of pilots.

Pilots and their families

As one might expect, all 442 pilots in the study were male. Although the trend is changing, the number of female pilots who fly for commercial airlines is still only a handful (Table 4.1). Indeed, it is unlikely to change significantly since various attempts in former years to attract them into the industry have failed. For whatever other reasons, they do not seem to be attracted to the job. Similarly, the majority of pilots fell into the 31–40- or 41–50-year-old ranges. From an analytical point of view, this was advantageous since this means that the 83 per cent who fell into these categories also were in the age ranges which might be considered as being the main 'active period' of the pilots' working life. This is because the training is long and, with experience perhaps hard to come by, the 21–30-year-old group are still very much in the early part of their careers. There is also the fact that by far the largest employer represented was British Airways. Since they stopped recruiting pilots in the mid 1970s, the youngest pilot who flies for them must be in his very early thirties. Hence those in the 21–30 category must fly for the other independent airlines who were represented in the sample. Similarly, pilots usually retire at around 55, so those at this end of the spectrum are very much in the later part of their careers.

The vast majority of pilots were married. Certainly, there were generally only low percentages of pilots who were either single, divorced or cohabiting. The impression of broken marriages throughout the aviation industry may be misguided. It must be remembered, of course, that the 'married' category may well include those who have been divorced and then re-married.

We will discover later that there is only a very limited scope for a pilot's wife to have a career of her own. This is reflected in the clear split that exists in the data. The distribution of those pilots whose wives worked was fairly equal to those whose wives did not work, though on balance the latter formed the larger group. Of those wives who work, commitment is equally divided between those who work full-time and those who work part-time. The data on the other

Table 4.1   *Biographical data: descriptive statistics*

	Number	Percentage
Sex		
Male	442	100%
Female	0	0%
Age		
21–30	30	6.8%
31–40	242	54.8%
41–50	123	27.8%
51–55	35	7.9%
55+	12	2.7%
Marital status		
Married	386	87.3%
Single	16	3.6%
Divorced	14	3.2%
Widowed	2	0.5%
Separated	7	1.6%
Cohabitating	17	3.8%
If married now (either formally or in common law), does your wife work?		
Yes	179	40.5%
No	220	49.8%
If yes, does she work		
Occasionally?	34	17.4%
Part-time?	72	41.3%
Full-time?	74	41.3%

	Average	Standard deviation
Number of children	2.30	1.91
Ages of children		
Number of children over 18 years old	0.82	0.25
Number of children under 18 years old	1.96	0.18
Number of dependants	2.48	1.39

Note: In certain instances percentages reported are of the total population of pilots. In other instances, percentages reported are expressed as a function of subgroup depending upon applicability of questions.

biographical features such as number of children, their ages or number of pilots' dependants revealed no surprising trends.

Pilots' interests and hobbies

The examination of 'out of work' leisure-time activities did, however,

Table 4.2 *Pilots' leisure activities*

	Number	Percentage
Do you find time to relax and 'wind down'?		
Always	219	49.5%
Sometimes	180	40.7%
Only when possible	33	7.5%
Not usually	10	2.2%
Do you have an interest or hobby?		
Yes	418	94.6%
No	24	5.4%
If yes, is it related to work?		
Yes	23	5.2%
No	397	89.8%
In general, do you mix socially with other aviation colleagues outside work?		
Yes	186	42.1%
No	256	57.9%

produce some interesting and surprising findings. Many of us express our perceptions of stress and work and pressure as a function of being able to 'relax and wind down'. So we asked pilots to complete this sort of scale. As Table 4.2 indicates, nearly 50 per cent said that they 'always' found time to relax and wind down. We shall confirm from other sources that this is probably a function of the fact that being at work and being not at work is a relationship that is more characteristically different for pilots than it is for other occupational groups. This is reinforced by the fact that a relatively high percentage (40.7 per cent) said they 'sometimes' found time to relax and wind down. Only small percentages fell into the lower categories. It would appear, therefore, that 'rest' plays a particularly important role in the life of the pilot.

Overwhelmingly high percentages (94.6 per cent) had a hobby which was generally not related to work. This can be construed as not only again emphasising the importance of resting and leisure time in the life of pilots, but also as revealing the valued separation of work- and non-work-related activities. In addition, pilots do not seem to mix socially with colleagues outside work. In fact, the apparently high percentage of pilots who said 'yes' (42.1 per cent) could well be a function of pilots who include the post-flight rituals of relaxation with fellow aircrew when away on a trip. This is relevant, particularly in the case of long haul pilots.

Lifestyle habits

It has been noted that an overall theme in this book is to examine the nature of stress and ill-health in commercial pilots. Background research tells us that there is a growing awareness, in the examination of ill-health, of the role played by underlying habits of the individual. In fact, some authors believe that it is these sorts of issues that have a greater influence on ill-health than any other factor (Fletcher and Payne, 1980). In other words, basic human activities such as drinking, smoking, exercising and eating habits will increase the likelihood or promote the chances of subsequent ill-health, as a direct function of the degree to which the individual indulges in them.

The examination of 'pilot stress' has a long history. On that basis, therefore, we should be in a position to state that we know a lot about pilots and the way they behave. This idea is further supported by another factor that renders the profession of pilot distinctively different from other professions. This is of course the fact that its members are subject to annual scrutiny as a matter of routine, i.e. medical checks. There is a massive database that has been built up over the years. Indeed, as was said earlier, much of the previous 'pilot stress' material is expressed in physiological or biochemical terms, which feeds readily into the medical model. However, we don't know quite so much about those relatively simple features that probably would not be included in a medical check.

One final issue to mention is why we chose these particular behaviours to examine. There were, as suggested, exploratory intentions which were to see if there were any surprises in the behaviours and lifestyles of this group – a group about which we *supposedly* know a lot. Quite apart from these relatively elementary motives there was another more serious intention. The data on the aetiology of coronary heart disease clearly indicates that the most frequently reported sources are genetic/hereditory factors; smoking; diet; lack of exercise; and stress. The latter is a theme being examined concurrently throughout the study. However, the other causes are clearly of interest in themselves. It should be noted that genetic factors were not accounted for since it was not possible to reliably measure them in a survey of this type.

The general aim, therefore, of this part of the study was to try to 'home in' on those factors that might well be underlying determinants of ill-health outcomes. Finally, it should be noted that no measure was taken of physical ill-health since, again, reliable measures were not available which could be integrated reliably into the survey. Since all pilots were currently flying, at the time of the study, they must have all passed their medical checks. Hence the assumption is that they are 'healthy' (at least in a physical sense).

Exercise and fitness

As may be seen from Table 4.3, the questions involved the examination of weight maintenance, frequency and extent of exercise and the active use of leisure time.

A readily visible consequence of not exercising (and consequently of reduced fitness) is one's weight. We can readily see that the majority of pilots do regulate their food intake and exercise to maintain some sort of prescribed or 'ideal' weight – in fact just under 70 per cent did so. Only 11.1 per cent said they never tried to avoid either over- or underweight. Superficial impressions of this distribution of analyses are clearly encouraging. However, one must be careful in the interpretation of such statistics. It would be appealing to conclude that pilots are fairly health-conscious and one of course assumes that to a certain degree they are. An alternative explanation is that maintenance of a desired weight is simply a characteristic sought in medical checks. In other words, one gets the impression that this finding could be the result of adhering to a medically prescribed

Table 4.3 *Pilots' exercise- and fitness-related behaviours*

	Almost always	Sometimes	Almost never
I maintain a desired weight avoiding overweight or underweight	308 69.7%	85 19.2%	49 11.1%
I do vigorous exercises for 15–30 minutes at least three times a week	84 19.0%	117 26.5%	241 54.5%
I do exercises that enhance my muscle tone for 15–30 minutes three times a week	67 15.2%	109 24.7%	266 60.2%
I use part of my leisure time participating in individual, family or team activities that increase my level of fitness	117 26.5%	183 41.4%	142 32.1%
I do some type of gentle stretching exercises at least three times a week to improve flexibility	89 20.1%	126 28.5%	227 51.4%

criterion, rather than a positive awareness of, and voluntary indulgence in, health-related habits. Unfortunately, examination of further answers might well tend to support this view.

Since the increase in interest in exercise and fitness in Britain over the last two years (and marginally longer elsewhere like the United States), there has been a flood of advice purporting to present programmes of daily and weekly activity sufficient to promote 'good' health. When reduced to their basics, most such programmes would agree that a 'rule of thumb' to adhere to is that exercise for around fifteen to thirty minutes repeated on average three times a week is required to make a *significant* contribution to good health and relieve the effects of stress. On the assumption that pilots are health-conscious one would, therefore, have expected them to indulge in this fairly elementary exercise schedule, even if only to a limited degree. Yet it was found that just over a quarter of the sample said they sometimes did rigorous exercise of this duration and of this frequency. More surprisingly, however, over half (54.5 per cent) of pilots said they never did exercise of this type and frequency.

Two further aspects of exercise and fitness concerned stretching exercises to promote flexibility and other exercises to enhance muscle tone. Generally, only 15 to 20 per cent said they did these regularly. In fact 50 to 60 per cent said they never performed exercises of this kind. One could well hypothesise that pilots may have a propensity towards muscular skeletal disorders, due to the basic body posture and changes in this which result from the performance of certain physical tasks. Given this background, one would have thought that pilots would have recognised the benefits of exercise in preventing such disorders, i.e. that 'prevention is better than the cure'.

A healthier picture was portrayed by the degree to which pilots participated in group activities that promoted fitness. There was, however, a spread of scores, with the number of pilots who never participated in these (32.1 per cent) exceeding those who always did these (26.5 per cent). One can see that whilst these results may be described as 'better', this is really only a relative description. In absolute terms the pilots portray themselves as really only being mediocre. This is true when one takes account of a further methodological issue. This is the general tendency for pilots to include relatively 'normal' family activities in their positive responses, as opposed to particular activities that are specifically geared towards promoting fitness. The implication of this is that the percentage of 'always' is inflated. This ultimately renders the groups as rather 'mediocre'.

To summarise, therefore, the overall portrait with respect to exercise and fitness is poor. This was particularly true of exercise associated with enhancing muscle tone, improving flexibility and regular rigorous exercise. In each of these, the percentage of pilots who 'never did them' (over 50 per cent) was consistently larger than the percentage of those who said they 'always did them'. This also applies to indulgence in fitness-enhancing team and family activities.

These questions yielded some interesting and unexpected insights. One would have anticipated that the pilots were fairly 'fit', in that exercise would play an integral part in their lifestyle. This does not appear to be the case. The pilots do not appear to be an especially active group. They do not appear to engage in preventative activities to avoid ill-health. Unfortunately, the dominant trend in the data forces us to conclude that they do not appear or wish to attain a level of fitness that is any better than maintaining a desired body weight.

Smoking

The unhealthy links associated with smoking have been clearly demonstrated. Such data can be classified on the basis of occupation or level in the socio-economic hierarchy. This indicates that those who smoke tend to be lower down the socio-economic hierarchy. This means that one would expect professional groups such as pilots to indulge less in the habit. This was, indeed, found to be the case. Only a minority (just under 22 per cent) of pilots were found to smoke (Table 4.4).

Although there were variations in the extent of smoking, the pilots were found to have been smoking for around twenty years on average. Hence they may be regarded as fairly 'mature' smokers.

Most questions were about cigarette smoking which carries the highest risk. Although there was a broad distribution of pilots who smoke, but who avoided smoking cigarettes, reliably we can only conclude that around 36 per cent avoided cigarettes. Furthermore, there is a tendency for pilots to smoke low-tar cigarettes, pipes or cigars. Further data indicated that of those who do smoke cigarettes, the degree of indulgence is moderate, i.e. most smoke less than half a pack daily. Generally, therefore, this information portrays the pilots as a group who are aware of the health risks associated with smoking.

Our attention must be focused, however, on this specific minority who smoke since they are the group most at risk. The cynic might

Table 4.4 *Pilots' smoking habits*

		Number	Percentage
Do you smoke now?	Yes	96	21.7%
	No	346	78.3%
If yes, how long have you smoked for?	x = 20.38		sd = 7.57 years

	Almost always	*Sometimes*	*Almost never*
I avoid smoking cigarettes	35	29	32
	36.4%	27.3%	36.3%
I smoke only low tar cigarettes or	51	21	24
I smoke only a pipe or cigars	53.0%	21.8%	25%

	Yes		*No*	
	Number	Percentage	Number	Percentage
I smoke less than half a pack daily	60	63.6%	35	36.4%
I smoke more than a pack daily	21	21.8%	74	78.2%
I have recently reduced my smoking	27	28.1%	69	71.7%
I have plans to quit smoking	47	48.8%	49	51.1%
I am currently attempting to quit	13	13.5%	83	86.3%
If you do not smoke now, have you ever smoked?	123	35.5%	83	64.5%
If yes, how long ago did you stop?		x = 10.22		sd = 11.46 years

x = Average
sd = Standard deviation

observe that all smokers have plans to quit! However, the pilots were evenly split on whether they had or had not plans to stop smoking (in fact, on balance, marginally more pilots said that they did not have plans than those who did). More extreme trends were apparent in the fact that large majorities had neither recently reduced their smoking nor were presently trying to stop. Finally, one should note that just over one-third of those who did not smoke were themselves once smokers. On average they stopped around ten years ago, though again a wide spread in scores was observed.

Pilots: who are they?

It is pertinent to introduce a methodological caveat. As individuals responding to questionnaires of this type, we tend to unconsciously adjust our answers in response to different types of questions. Amongst the sort of questions that are subject to this unconscious manipulation are those about topics which the individual regards as being emotive or questions which make the individual feel 'guilty'. Traditionally individuals tend to downgrade their estimates (we shall see later that these comments also apply to 'emotive' questions about drinking habits).

Ask any smoker to give their first initial estimate of how much they smoke per day, then count the number of stubs they produce. The discrepancy is nearly always in the expected direction. The first initial answer of the smoker will more often than not be less than the actual number of stubs. Such self-report data is also subject to variation in the opposite direction. In other words, smokers tend to upgrade their estimations of their intentions to stop or reduce smoking. Whilst these effects are very subtle, the overall implication is that one must regard the present findings as 'optimistic', portraying a picture that is marginally better than reality. The overall conclusion is that pilots are as aware as one might expect about the risks associated with smoking – the moderate indulgence by those who did smoke reflects this. However, surprisingly high percentages of smokers did not have plans to stop, or were not currently trying to reduce or stop smoking. This was doubly depressing in the light of the methodological issues mentioned above.

Eating

The role played by one's diet has been well documented and popularised recently in the media. Most discussion contrasts the positive role played by fibre with the negative effects of a variety of foods, and is expressed in terms of physical ill-health – most being related to coronary and circulatory disorders.

The data on eating habits (Table 4.5) portrayed the sample of pilots as a group who are aware of the role that diet can play in the prevention of ill-health and the maintenance of good health.

Eating habits are not the sort of behaviours one would expect to change quickly. However, as a nation Britain is notoriously lax in adopting any such trends. The nature of our culture dictates that our traditional habits persist in the face of opposing logic. However, as previously indicated, it is those who are higher up the socio-economic

Table 4.5   *Pilots' eating habits*

	Frequently	Sometimes	Almost never
Do you eat a variety of foods each day such as fruits, whole grain breads, cereals, lean meats, fish and poultry, etc?	337 76.2%	97 21.9%	8 1.8%
Do you eat foods high in fat, saturated fat and cholesterol (e.g. fatty meats, eggs, butter, cream, organ meats such as liver)?	98 22.2%	293 66.3%	51 11.5%
Do you eat salty foods, add salt at the table or use a lot of salt in cooking?	104 23.5%	162 36.7%	176 39.5%
Do you eat a large amount of sugar (especially sugary snacks, desserts and soft drinks)?	29 6.6%	154 34.8%	259 58.6%
Do you eat a high fibre diet including lots of whole grain bread, cereals, fresh fruit and vegetables?	281 63.6%	141 31.9%	20 4.5%

ladder who are most likely to respond to such information. The picture of eating habits in the sample of pilots reflects this. All of the pilots apart from a tiny minority either 'frequently' or at least 'sometimes' tried to eat a variety of foods. Similarly, a high percentage (63.6 per cent) regularly tried to consume a high-fibre diet. No doubt these high percentages reflect previously mentioned British trends (amongst some circles anyhow!).

Unfortunately, examination of remaining answers illustrates no overall clearly positive picture. There were some surprises which provide further and rather disturbing insights. This is illustrated by examination of the data on foods that are high in cholesterol and foods high in sugar. In accordance with previous comments, large majorities ate diets relatively free from these substances. However, what was surprising was the percentage who frequently ate foods high in cholesterol. This was just over 22 per cent. (A similarly high percentage (23.8 per cent) report frequently eating foods high in salt.) The overall message is, therefore, that whilst portraying an overall good picture, there are some features in the data that could give cause for concern.

This is further illustrated in the reported consumption of sugary foods. A cursory examination reveals that only 6.6 per cent reported that they 'frequently' eat such foods. One must remember, however that examination of previous data tells us that pilots are weight-conscious. This of course is a good result. However, what are the background reasons for it? These will, in part, be due to an awareness of the negative implications of consuming a high amount of sugar. Although the reported habits are good, the underlying reasons for them may not necessarily be due to a heightened awareness of health risks. In the case of sugar intake, the low consumption may well be simply related to the desire to avoiding weight gain, for example.

One final point worth making is that no record was made of the effects of occupational factors on eating habits. There is little doubt that such issues as time zone changes or working a variable shift will have major disruptive effects on eating habits. It was impossible, however, to incorporate the collection of such data in an intelligible way into the study.

Drinking habits

Examinations of pilot incapacitation statistics indicate that alcoholism is not an especially dominant cause of pilot licence loss. The group really of concern, however, is not these extreme cases, but rather the pilots who don't make the record books. In other words, those who might be at risk due to their heavy drinking but who manage to handle it without drawing attention to themselves or allowing it to have a noticeable effect on their behaviour (either in work or out of it). An observed increase in alcohol consumption (like increased smoking) is traditionally regarded as a *response* to stress. As is the case with respect to smoking, for example, individuals are indulging in behaviours which in themselves have negative health implications. A background knowledge of pilots' drinking habits would therefore seem a doubly suitable topic for examination.

It is clear that all but a tiny minority of pilots (1.4 per cent) consume alcohol (Table 4.6).

This is a disturbing finding, and indicates that pilots are a 'high risk' group to alcoholism. As was mentioned with respect to smoking, one must again regard these figures as being conservative estimates, since individuals tend to downgrade their reported assessments of drinking habits. Support for this deduction is provided by the answers to more detailed and relatively more emotive questions.

There was a marked reticence by pilots to use the 'often' or even 'yes' answers to questions about feeling the need to cut down and so

Table 4.6   *Pilots' drinking habits*

Do you drink alcohol?	Yes	436	98.7%
	No	6	1.4%
If yes, how many days per week do you drink?		x = 2.12	sd = 1.96
On those days on which you drink, on average how many drinks do you have?		x = 4.89	

	Often	*Yes*	*No*
Do you drink more than two drinks	85	145	206
per day?	19.2%	32.8%	46.6%
Do you use alcohol as a way of dealing with stressful situations	6	52	378
or life problems?	1.4%	11.8%	85.5%
Have you ever felt the need to	6	107	323
cut down on drinking?	1.4%	24.2%	73.1%
Have you ever had guilty feelings	0	62	374
about drinking?	0%	14.0%	84.6%
Has anyone ever told you they	0	56	380
think you drink too much?	0%	12.7%	86.0%

on. The majority of pilots' answers were 'no' throughout. To specific questions about feeling guilty about drinking too much, using alcohol to deal with stressful situations or even being told by others that they drink too much, only around 12 per cent replied 'yes'. However when the questions were expressed in terms of the need to reduce their alcohol intake, the percentage of pilots who replied 'often' or 'sometimes' rose to a surprising 26 per cent.

Most of us would advocate a sensible approach to alcohol consumption, which can act as an excellent aid to unwinding and set the scene for breaking down social conventions at work (such as the post-flight get-togethers). There are, however, still over 50 per cent of the pilots who admit to drinking at least two drinks every day. This is particularly worrying in the light of the fact that these are probably conservative estimates. We know too that, on balance, pilots do not mix much with colleagues outside work. Hence, most of their drinking might well be 'at home'. Estimates of consumption of alcohol at home are traditionally unreliable, since the size of drinks vary (i.e. 'home' measures are usually larger than 'pub' measures).

Pilots' work experience and history

By far the biggest employer was British Airways (58.4 per cent), which was expected. Also apparent from Table 4.7, however, was the

Table 4.7   *Pilots' employers*

	Number	Percentage
Present employer		
British Airways	258	58.4%
British Caledonian	51	11.5%
Britannia	45	10.2%
Dan Air	16	8.8%
Logan Air	16	
British Midland Airways	15	
Air UK	11	11.1%
CAA	1	
British Air Ferries	1	
Manx Airlines	1	

Table 4.8   *Types of aircraft flown by the pilots*

	Number	
Trident	101	
B737	84	
BAC 1-11	71	
B747	57	
B707	21	
DC10	20	
L1011 (Tristar)	19	
F27	13	
HS748	9	
Shorts SD3–30/360	9	
B727	8	
DC9	7	
Concorde	6	
EMB110/DHC6	6	(pilots flew both types of aircraft)
B757	3	
Viscount	3	
BAE146	1	
BN2	1	
PA31	1	
Islander	1	
Herald	1	
	N = 442	

Table 4.9   *Pilots' work experience*

	Average	Standard deviation
Number of years with present employer	13.06	6.28
Number of years' experience on aircraft type	6.06	4.18
Number of hours' experience on aircraft type	3,454.24	3,111.38
Total flying hours' experience	7,912.95	4,518.31
Average flying hours' experience per month	45.72	29.28
Number of landings performed in aircraft type	1,247.16	911.78
Total number of landings performed	3,321.67	3,030.11
Average number of landings performed per month	19.32	16.90
Average length of sector flown (hours)	2.88	2.38

fact that the three largest independent operators were also represented, each by approximately 10 per cent of the sample. Seven other
smaller independents were also included. On average, length of time
with present employers was around thirteen years. Pilots of twenty-
one different aircraft types were represented (Table 4.8). Additionally, the average number of years' experience on present aircraft type
was around six years.

The interpretation of this sort of data is relatively straightforward;
however, one should exercise caution in the examination of data of a
more detailed kind. Consider Table 4.9, for example. Highlighted are
the average number of hours' experience, landings, length of sector
flown, and so on. It will be noted, however, that range of scores, as
measured by the standard deviation, is massive. Given a sample of
this size, this is to be expected. The presence of a wide range of scores
is not 'bad', but one must remember its existence when making
comments that involve these data.

A breakdown of pilots in other ways (Table 4.10) revealed the ratio
of short haul to long haul pilots as being greater than two to one.

Only 7 per cent of the sample flew domestic routes only. Again,
these proportions were, to a certain extent, to be expected since they
partly reflect the composition of BALPA membership, through
which the sample was obtained in the first place. The sample was
rather 'top heavy' in that just over half were of the rank of captain and
just under 40 per cent were senior first officers. It must be noted,
however, that these figures will reflect the fact that some airlines do
not use a three-rank system (i.e. there are merely captains and first

Table 4.10 *Pilots' status*

	Number	Percentage
Long haul	116	26.2%
Short haul	295	66.7%
Domestic only	31	7.0%
Captain	227	51.4%
Senior first officer	176	39.8%
First officer	39	8.8%
Other functions?		
Yes	61	13.8%
No	381	86.2%
Seniority		
Very high	28	6.3%
High	123	27.8%
Middle	181	41.0%
Low	82	18.6%
Very low	28	6.3%

officers). A relatively small percentage said that they had functions other than pilot. These were generally base or route training positions, or supervisory posts. Finally, the distribution of pilots in terms of seniority reflected a wide spread of levels though, on balance, there were higher proportions of those in the higher seniority categories.

Outcomes of stress

The portrait of pilots would be incomplete if we did not attempt to include information which was directly applicable to stress and its effects. It has been previously stated that the examination of physical effects of stress was not performed and that major attentions were focused upon psychological and behavioural outcomes. It seems appropriate, therefore, to try to answer three basic questions: (1) how do pilots feel about their jobs; (2) how do pilots 'feel' generally; and finally (3) how do factors associated with their jobs, their home lives and general well-being affect their levels of performance at work? These three questions embrace job satisfaction, mental ill-health and job performance. It should be noted that these are important themes which will be mentioned again and which will reoccur throughout this text. This is particularly the case when later we try to relate or attribute specific causes to specific outcomes. Also, these three specific outcomes measured here have all been recorded in the past as being adversely affected by psychosocial stress.

Job satisfaction

Superficially one might expect airline pilots to be fairly satisfied with their jobs. Its constituent components contain those features which most of us find desirable in work, i.e. use of skills and abilities, discretion, recurrent feedback, and so on, plus the more 'obvious' features such as the status, the salary, and flying to different places around Europe and the rest of the world. The only features that the public might identify as negative are the inherent responsibility associated with the job and also the phenomenon of 'pilot stress', a subject which, as we have already mentioned, the general public mainly assume to be part of the job (although having no real conception of the make-up of this phenomenon).

The pilots were asked to complete a recently designed British inventory that measured 'job satisfaction'. It must be remembered that this is more than simply how 'satisfied one is with one's job'. When we refer to 'job satisfaction' we refer to a comprehensive psychological state that has some sort of emotional or experiential quality.

The greatest source of satisfaction for pilots was found to be that associated with working conditions. In this context, the term 'extrinsic' means that which is derived from the context in which the job is carried out (Table 4.11). We shall go on to see that this was a rather unusual result for, indeed, issues associated with physical working conditions, fellow pilots and supervisors were all issues that pilots

Table 4.11   *Job satisfaction: descriptive statistics*

	Average	Standard deviation
Total job satisfaction	65.97	11.83
Intrinsic job satisfaction	29.71	6.75
Extrinsic job satisfaction	36.27	6.31
Job itself intrinsic satisfaction	18.50	4.18
Working conditions extrinsic satisfaction	24.62	3.69
Employee relations satisfaction	22.86	6.14
Overall job satisfaction	5.22	1.17

later identified as being sources of stress! Probably the best way to interpret this is by saying that the scale was rated as highest on the basis of its relative position to other subscales.

If one looks at the other end of the distribution, however, the 'least satisfactory' scale was compatible with later findings. Sources of least satisfaction were issues such as recognition for good work, levels of pay, promotion chances and industrial relations between management and pilots.

In some ways, from these data, it is difficult to conclude whether pilots were or were not experiencing job satisfaction. Indeed, if one examines the scores as expressed as a percentage of available points (Table 4.12), one might conclude that the pilots were fairly satisfied but not excessively so.

Whilst it is interesting to make comments about pilots in general, further examination of the data revealed that different groups of pilots expressed different feelings. Those variables which were found to be most influential in the examination of job satisfaction were those that defined status such as rank, seniority, and so on. In general, job satisfaction was found to be positively related to status.

Furthermore, if one examines job-related factors (i.e. by relating the job satisfaction scores with their results elsewhere), occupational items were found to be negatively related (i.e. sources of dissatisfac-

Table 4.12   *Job satisfaction: comparisons of subscales*

	Total marks available	Pilot's score	Percentage
Total score (15 items)	105	65.97	62.8%
Intrinsic job satisfaction (7 items)	49	29.71	60.6%
Extrinsic job satisfaction (8 items)	56	36.27	64.7%
Job itself intrinsic job satisfaction (4 items)	28	18.50	66.0%
Working conditions extrinsic job satisfaction (5 items)	35	24.62	70.3%
Employee relations job satisfaction (6 items)	42	22.86	54.4%
Overall job satisfaction (1 item)	7	5.22	74.5%

Note: To facilitate comparisons between subscales, pilots' scores were expressed as a percentage of total marks available for each particular subscale.

tion). Greatest dissatisfaction was expressed with general rather than specific aspects of work, particularly managerial issues, seniority systems and career-related factors.

On balance, the groups as a whole would appear fairly satisfied. However, since the examination of subscales revealed the groups to only score on the upper side of 'average' scores, this does not seem really to be consistent with the idea that flying is a vocation and those who fly feel self-fulfilled about their work.

Comparing different elements of the scores tells us which facets of job satisfaction are of value compared with others. However, such an analysis does not tell us the level of job satisfaction experienced by the pilots in absolute terms. To come to some final decision on the pilots' perceived job satisfaction, it is necessary to compare them with other groups. In other words, we need to erect some yardstick or reference point for comparison. These are called 'norms'. Unfortunately, normative values of the scales could not be located which were satisfactorily compatible. So it was impossible to compare like with like (i.e. to compare the pilots in this study with other pilots in other studies). It was decided, therefore, to use the group for which the measuring device was originally designed (Warr, Cook and Wall, 1979) as a comparison. This group were males, doing blue-collar jobs in British manufacturing industry. It may be seen from Table 4.13

Table 4.13 *Job satisfaction: comparisons of pilots' scores with expected norms*

	Pilots' average scores	Norm average scores	Decile	Range
Total job satisfaction	65.97	70.53	3	63.6
			4	68.0
Intrinsic job satisfaction	29.71	32.61	3	28.1
			4	31.4
Extrinsic job satisfaction	36.27	37.99	4	36.2
			5	38.2
Job itself intrinsic satisfaction	18.50	20.32	3	17.7
			4	19.2
Working conditions extrinsic satisfaction	24.62	25.89	4	24.6
			5	25.8
Employee relations satisfaction	22.86	24.40	3	20.0
			4	23.0
Overall job satisfaction	5.22	5.33	5	5.1
			6	5.4

that in all instances the pilot mean scores fell below those of the norm group. In fact, in four of the comparisons, pilots' average scores fell into the 3–4 decile; in two comparisons pilots' average scores fell into the mid-range 4–5 decile, and in only one comparison did pilots' average scores fall into the upper 5–6 decile.

The overall conclusion is that the pilots' job did not provide them with particularly high levels of satisfaction. Of course, the result must be construed in the light of the nature of the comparison groups. Although the normative group of blue-collar males in manufacturing industry is clearly inappropriate, this was hardly a result one would have expected to find. The indications are that pilots in the sample are portrayed as not having particularly high levels of job satisfaction. Certainly they do not conform to the anticipated 'satisfaction' stereotype.

Mental ill-health

Another underlying theme of this investigation is ill-health. There are, however, several issues which should be mentioned before proceeding with an examination of the pilots' mental ill-health profiles. First, it must be noted that simplistic examination of descriptions data (such as averages) is of only limited utility. This is because the measuring device is one of 'psychiatric' as opposed to 'psychological' factors. The problem associated with this is that it is difficult to establish exactly what is meant by 'normal'. We stated earlier that the measuring device had been used extensively in many other studies (Crown and Crisp, 1979). However, comparison between results of these studies with those of this study of pilots tends to be precluded by factors that underlie variations in mental health data, e.g. gender, age, socio-economic status, and so on.

A second issue is that one must treat reported data as conservative estimates. With respect to smoking and drinking we noted that respondents in surveys tend to modify their answers to certain kinds of questions. In terms of mental ill-health respondents also downrate their answers. This is because some subjects may spot fairly quickly what the questionnaire is trying to measure and manipulate their replies accordingly. In addition, one might expect a population such as pilots to be cautious in their responses. This may be due at one level to their personalities. Alternatively one might also expect caution because of perceived potential threats to their livelihoods (e.g. suspension or removal of licence on grounds of mental ill-health).

Table 4.14 *Mental ill-health: descriptive statistics*

	Average	Standard deviation
Anxiety	2.41	2.23
Phobia	2.05	1.86
Obsessionality	4.91	2.95
Psychosomatic disorders	2.31	1.91
Depression	2.55	2.24
Hysteria	3.98	2.90
Overall neuroticism	18.21	9.07

Although the study was confidential, the perceived threat is always there.

Given our previous comments, it will be fairly clear that Table 4.14 by itself is not particularly illuminating. More interesting results were produced by inspection of *distributions* of scores. Simple tabulation of distributions enables us to discover how many pilots were above or below certain cut-off points. Using different cut-off points (Table 4.15) it was found that just under 28 per cent had mental ill-health scores which were greater than one would expect to find in a male industrial population. These were certainly surprising results and, in fact, further surprises were in store as different cut-off points were examined. When the cut-off points were made more rigorous, surprisingly high percentages of pilots were found to be *above* these levels.

Table 4.15 *Mental ill-health: distributions of average overall scores*

	Cut-off point score	Number of pilots above cut-off point
'Normals' (Crown and Crisp, 1966)	27.80	62
'Psychiatric outpatients' (Crown and Crisp, 1966)	40.8c	9
'Males invalidity study' (Crisp *et al.*, 1978)	22.47	123
'Normal/abnormal'[1]	28.60	54

[1] This was calculated by taking the normal score and adding one standard deviation, taking the abnormal score and subtracting one standard deviation and then calculating the mid-point between the two.

For example, a normal/abnormal cut-off point was developed. This was the point of ill-health scores which divided those who were 'normal' (but at the upper end of the average distribution) and those who were clinically 'ill'. It was found that just over 12 per cent of pilots had ill-health scores above this point! In fact, nine pilots had mental ill-health scores greater than one would expect to find in psychiatric outpatients! Whilst nine pilots form only 2.2 per cent of the total sample, it must be remembered that all of these pilots are supposedly fit and healthy. They had all passed medical checks and had valid commercial licences to fly aircraft at the time of investigation!

To conclude, the majority of pilots scored well within the normally expected range. However, it is fairly clear that there are minorities that can be readily identifiable as scoring in the abnormal, unhealthy category. Subsequent analysis shows that, generally, issues such as the degree of compatability of home and work life, practical issues, problem identification and solving, and attaining self-set levels of flying performance were found to be the predictors of these scores. The overall theme seemed to be achievement and success. As expected, those pilots who had difficulty in relaxing and 'winding down' reported higher ill-health scores as did those who did not reduce their stress by the social support systems provided by mixing with friends and colleagues. Age, too, was a significantly related underlying characteristic. The fluctuations of mental ill-health with age conformed to previously recorded and expected psychiatric conventions.

Performance

The final element in drawing up our portrait of the pilots is to examine their levels of performance. Pilots are subject to rigorous tests of their knowledge and flying ability. Hence all pilots in the study were performing within standards accepted by the Civil Aviation Authority.

The measure we diagnosed for use in this study was not used, therefore, to identify and compare 'good' with 'bad', but 'optimal' with 'suboptimal' performance. We are ultimately interested in which issues predict the levels of subsequent performance in pilots. These will be discussed later (chapter 6). Unfortunately the complex construction of the measure does not lend itself to meaningful simple analyses. For the present, we must be content with the conclusion that the pilots appeared to attain satisfactory levels of self-perceived performance.

Conclusions

We stated at the beginning of this chapter that our aim was to establish some fundamental facts about pilots and the type of people they are. We have tapped many different sources of material and drawn some conclusions. However some final comments seem appropriate.

Examination of biographical features reveals four surprises. The age of the sample portrays the pilots are being in the main categories of the working population (though as noted this may be due to other aspects of the sample). Other features such as marriage, children, etc. reveal the pilots to be fairly ordinary though we shall see later how home life plays a crucial role in pilot coping, for example (chapter 9). Rest and relaxation seem especially important also.

We made some forceful comments however in our examination of lifestyle habits. We made some negative remarks about exercise, for example. We do not mean to be too disparaging. It must be remembered that compared to other occupations, pilots are good! However by definition 'good' in this case is a relative measure and the picture is not quite so positive when one remembers that other occupations are dreadful! The absolute standards of 'good' are represented in the positive responses to the questions asked. It was on the relative absence of these that our attention was focused. Similarly, in terms of smoking we focused upon relatively 'mature' smokers. These are probably pilots who have tried to stop and failed. Eating habits were also examined though we did not try to assess the impact of the operational effects of flying which will have a major impact on the pilots' food intake. Again, we made some rather negative comments about the reasons why pilots consume the foods they do. This was especially true for example not just in terms of the minorities who consumed unhealthy foods but also those who did try to 'eat well'. We offer these as only potential reasons and on balance one must give the benefit of doubt to the pilots. Finally the monitoring of drinking habits seems to work well and the very heavy drinkers are accounted for. However the next level of 'high' consumers still exists within the sample of pilots as it does in other jobs, the main difference being of course that the consequences of failing to cope with heavy drinking are very different. One final set of biographical data was that describing employment and work experience. The main point to make here was that whilst such data seem easy to interpret, on balance, they reflect some complexity in the form of a wide variation in data recorded.

Three 'outcome' variables were discussed in the form of job

satisfaction, mental ill-health and performance. The first outcome, job satisfaction, was an important issue in that it was possible to isolate specific aspects of the job as sources of satisfaction or not. The meaning of those issues relating to the 'job itself' are worth mentioning. If one defines 'job itself' in its strictest sense in terms of actually flying an aircraft, then it is probably true to say that nearly all pilots would identify this as a massive source of satisfaction. In terms of the questionnaire, 'job itself' is defined in a wider sense in terms of factors associated with the job itself that seem to affect it most. One result that was surprising however was the relatively low levels of satisfaction exhibited generally by the pilots.

Similar care must be taken in the interpretation of mental health data, particularly the comparisons between pilots and other occupations. Fluctuations in mental health occur in the same way that they do for physical health. The implications of this is that pilots who had bad scores now may not show such scores when examined at a later date. Of course it must be remembered that the opposite can occur! Generally speaking however we are concerned with those characteristics that are more deeply entrenched. The final comment must be that, given all the monitoring of pilots and the exacting parameters to which they must fly, it was surprising to find some of the results, though, of course, one must exercise caution in not sensationalising these. Later in this text we shall make some comments about pilot performance. For the present we shall conclude that since all of the pilots in the sample were flying, they must, by definition, be performing satisfactorily. We would like to think that this chapter has expanded the overall background picture of commercial pilots. Having established this base, in the next chapter we turn our attention to our preliminary findings and the sorts of things pilots say about their jobs.

Chapter 5
What pilots say about their job and home

Domestic stressors

Our attention was focused initially on those sources of stress which originated in the pilots' homes. In many ways these issues will have idiosyncratic qualities, because everyone's home life is obviously different. However, it ought to be possible to identify a profile of domestic stressors that typically characterise the life of a pilot.

We are not talking about 'life events'. We shall see later that those are large traumatic issues (such as divorce, moving house, etc.) which one can identify as 'milestones' or turning points in our lives. The factors we refer to here are more specific and integral parts of the pilot's ordinary domestic scene. An additional characteristic one might expect is the degree of width or range of issues that potentially may be identified. Again, it is *recurrent* issues that typify pilots that we are primarily interested in. In an initial interview, lasting several hours, pilots were asked to identify those issues, concerns and pressures in their home life that may adversely affect them in their job. The four sets of stressors that we include here are only *illustrations* of the type of material that we will examine in greater detail in later chapters. By way of introduction we shall now focus our attention on the pilots' domestic scene.

1 Disagreements, arguments, differences of opinion
2 Quality of marital relationship with partner
3 Degree to which household is 'geared to flying'
4 Family health
5 Nature of the home social environment
6 Lack of money
7 Dependability
8 Potential for extra marital relationships
9 Build-up of tasks, duties and things to do
10 Issues associated with children (health, education, etc.)

11 Domestic situations that aren't clear-cut
12 Worries on behalf of others
13 Conflicts of interests and resulting compromises
14 The 'good' use of time at home and how it is spent
15 Inability of spouse to fulfil her own ambitions
16 Absence of calm, stability and dependability in home life
17 Constant, ongoing irritations
18 Disappointment when others fail to meet expectations
19 Degree to which your personal goals and aims in life have been achieved
20 Success or failure of one's efforts to achieve
21 Inability to identify problems (and hence solution)
22 New and unfamiliar experiences
23 Others not obeying or things that go wrong
24 Enforced or adapted roles at home
25 Responsibilities of home activities (e.g. PTA, Councillor, etc.)
26 Spouse's lack of understanding about the job
27 Not having someone to talk to about your work
28 Interpersonal relationships
29 The degree to which home life is the way you want it

We found twenty-nine domestic issues that pilots suggested may be sources of stress for them. Whilst these are a miscellaneous array of domestic stressors, it is possible to identify simple trends in them. We have included an array of comments which illustrate the rich context in which these stressors were identified.

The first theme concerns that relatively 'normal' array of issues such as *arguments and disagreements* which inevitably arise in any household.

'Having a family has added a source of stress. I can't necessarily get to sleep when I want to and they tend to wake up early while I might not want to.'

Children [are a source of stress], whether they are good or bad, whether my wife is under pressure by them.'

'Sometimes the wife will want to go out or have friends around and I'm working or have been called out. It can cause a bit of friction.'

There is nothing especially unusual about this category of domestic stressors. In fact they are the sort of things that many of us would identify with. Many pilots felt these issues to be important, but were not sure of their impact, as one senior pilot indicated:

'I don't think that many of these problems actually stop a chap who is a good pilot being able to perform but . . . what he's thinking about in the evening when he's mulling it over, I just don't know.'

We will return later to the sort of impact stressors of various origins can have on pilots and, indeed, how they cope with them. However, based on such comments, items included in the questionnaire were fairly straightforward such as 'conflicts of interests and resulting compromises', 'disagreements, arguments, differences of opinion', or 'constant ongoing irritations'.

A second theme concerns the *quality of interpersonal relationships* that the pilot experiences. It would appear also that the quality of marital relationships is an especially important source of domestic stress.

'I'd like to think that it doesn't affect me. But, it depends upon the degree. If one was in a situation with a marriage breaking up or you have problem kids I think perhaps, yes it does; in fact, I've seen it.'

This perhaps is a somewhat extreme example, however, it does serve to illustrate that marital relationships can be influential. Indeed, we shall see just how much reliance pilots place on such issues, in terms of coping. Additional problem areas in this group of issues were 'having someone to talk to about your work', 'others not obeying when things go wrong', 'disappointment when others fail to meet expectations', or 'spouse's lack of understanding about the job'.

A third theme that could well be extracted may be identified as embracing *the way the pilot organises his home life*. The disruption of 'normal' domestic routine is illustrated well by the experiences of a long haul pilot who was working away from home for a period of time:

'They are common everyday problems, but you either can't cope with them or are less able to cope with them if you are spending substantial parts of every month away. It's all a question of having one's organisation disrupted. Home stressors tend to be less of a problem than job stressors in some ways, but when you are away, they increase in magnitude and effect. It's a break in routine.'

Not only is there the obvious build-up of domestic jobs that have to be completed upon return, but there is also a general disruption of home life. Additional problems in this theme of items were 'degree to which household is geared to flying', 'build-up of tasks, duties or things to do', or 'the "good" use of time at home and how it is spent'.

A fourth theme that is apparent is that which concerns the *qualities* of the pilot's domestic scene. The following are fairly typical examples of such issues.

'My daughter has had post-natal depression and recently separated from her husband. She was silly – only had the child because her friends were having them – it was fashionable. The main thing is that the normal routine and balance in the household has been completely disrupted.'

'My wife is a housewife, we have two children. I live eighty miles from the airport – a fairly small house, large garden and so on. She has recently become pregnant again which is forcing us to move house. We moved six or seven years ago to our present house which required a lot of work. We have spent the last five years working on the house and now we will have to move. The home atmosphere has completely changed. It has become a tie.'

Items derived from such comments were 'domestic situations which aren't clear-cut', 'absence of calm, stability and dependability in home life', 'responsibilities of home life activities' and 'enforced or adopted roles at home'.

It should be emphasised that the preceding themes illustrate the sort of issues which pilots identified as being potentially stressful associated with their domestic lives. They are merely to set the scene and give us a flavour of the basis from which we developed further our study of psychosocial stress in pilots. For this reason, it would appear appropriate to apply a similar strategy in the examination of job stressors.

Occupational stressors

As was indicated in chapter 1, the sources of psychosocial stress in other jobs are well documented. The reader will no doubt recall that the following, for example, have been identified as important and recurrent sources:

(i) Responsibility – for things (such as machinery, money, etc.), but especially for people.
(ii) Level – how high or low the individual is in the organisational hierarchy.
(iii) Overload – usually of two types: quantitative (i.e. having too much work) or qualitative (i.e. work that is too difficult).
(iv) Role – the problems that ensue in 'playing the part' associated with the job one occupies in the organisation, such as conflict and ambiguity.

There are, of course, other issues such as those associated with one's career (reaching a career ceiling or lack of career opportunities) or with the organisation one works for (psychological climate, etc.). The interviews generated about thirty occupational stress issues for pilots.

1 Career opportunities and lack of potential advancement
2 Seniority systems
3 Impending major career change or threat (e.g. redundancy, redeployment, etc.)
4 Not enough hours actually spent flying
5 Sharing of work evenly
6 Scheduling
7 Patterns of flying (i.e. relative times you are asked to fly)
8 Interpersonal problems with aircrew
9 Interpersonal problems with cabin staff
10 Style of management
11 Lack of management support
12 The whole experience (before and during) of medical checks
13 The whole experience (before and during) of checks on your flying ability
14 Misuse of time (e.g. low amount of preparation time, delays)
15 The aggregate or cumulative effects of minor tasks when flying
16 Fulfilling role expectations (either as a captain or first officer)
17 Changes in experience of flying (e.g. conversion course)

18 Impact of a lack of flying (practice effects)
19 Future career uncertainty
20 Tiredness and fatigue
21 Morale and organisational climate
22 Conditions of employment
23 Factors not under your direct control
24 Periods in flight of high workload
25 Inherent responsibility of your job
26 Making important decisions
27 Ambiguous factors or difficulties in problem identification
28 Attaining your own personal levels of performance
29 Situations that are ongoing
30 Effects of being overfamiliar (with routes, type, routines, etc.)

To illustrate further the sort of data collected and identified as being potential sources of stress originating at work, an initial analysis revealed roughly four themes. The first theme that emerges concerns *career* variables.

> 'I have contemplated changing fleets – going back to long trips. I fly Concorde – the job is very routine, no variety – stale. Flying the plane itself is interesting but going to the same place on the same routes at more or less regular times in the month. I'm not really interested in short haul, but it's the only way I'll get a command: I'm not sure which direction my career is heading.'

Comments such as these led us to derive items such as 'career opportunities and lack of potential advancement', 'impending career change or threat', or 'future career uncertainty'.

The general thrust of the second theme seemed to concern the way in which the *work itself was organised* – particularly in terms of how the period of being 'on duty' does not comply to a 'normal' routine. The fact that the job of pilot does not comply to a 'Monday to Friday, nine to five' pattern is an inherent part of the job of a commercial pilot. However, the incompatibility is more fundamental even than this. The way that work is scheduled is potentially a fundamental source of stress. Naturally, there are legal limits to which airlines must adhere; however, even these can be extremely taxing.

> 'I wonder how many holidaymakers, coming back from, say, Palma or Corfu . . . back into a dirty, windy, low-cloud, poor-

visibility and stormy situation . . . realise that the guy up front could well have been on duty for fourteen hours – legally.'

'If you are going to carry on working like that extensively, then it can do nothing but induce stress, if not in the short term, then in the long term.'

Of course, not everyone identifies timetabling as being the key issue but, on balance, most agree that it is an important issue.

'I like my flying, I like the routes. I like the odd day off here and there but, yes I must be honest, I prefer my time off at home. I like to choose what I'm going to do. I prefer to go away, do the work and then relax at home.'

These feelings were translated into items such as 'sharing of work evenly', or 'scheduling or patterns of flying'.

Third, factors associated with airline companies appear to cluster together. Pilots identified *organisational issues* in the following ways, for example:

'My main stressors occur on the ground; getting the damn thing into the air . . . once I'm in the air and totally in control of what's going on, I find that there is very little stress.'

'. . . management at all levels – their interference stops me doing what I consider a proper job.'

Items such as 'lack of management support', 'morale and organisational climate' or 'conditions of employment' illustrate this theme further.

Finally, as one might expect, issues intrinsic to the job of being a pilot emerged as being important. Generally, however, since pilots must fly to such exacting standards, the actual process of flying the aircraft does not lend itself to much comment. For example:

'I don't really have that much problem flying the aircraft. It's all so exacting. Perhaps, though, the number of errors one makes and their extent could be an indication that it's not that easy. Night-flying, approach and landings are sometimes tricky. Also if I'm tired – otherwise it's routine.'

though, of course, pilots may in fact welcome a challenge . . . .

'Occasionally, you come upon some quite exciting weather.'

'I feel the adrenalin wind up most when I have to find my way through thunderstorms or severe turbulence.'

Items based on such comments were, for example, 'interpersonal problems with aircrew', 'the aggregate or cumulative effects of minor tasks when flying', 'periods in flight of high workload', or 'situations that are ongoing'.

Again, it should be emphasised that the above are merely extracts of the issues identified during the interviews with the pilots. It should be recalled also that these items were included in the questionnaire where one loses the richness of the context in which they were quoted. However, such loss is inevitable. Certainly, by this stage, it should be fairly clear that the 'pilot stress' data collected in the study is very different from the material presented earlier (chapter 2.)

The issues discussed so far have been divided into two parts, i.e. those stresses originating from home, and those stressors originating from work. In addition, the background questions used to elicit these items specifically requested the identification of issues which had an impact either at home or at work. It is apparent, however, that the quality of the relationship between home and work will in itself be of critical significance as a source of psychosocial stress. In other words, there will be a range of issues that originate at home which will have an impact on the pilot at work. Similarly, there will be an array of work stressors that will have carryover effects into the pilot's home life. Such stressors form the basis for a third level of analysis.

The home/work relationship

We have divided the third and final part of this chapter into three parts. In the first two we present some of the preliminary findings about home-to-work and work-to-home stressors in much the same way as we did in the previous two sections above. The third part of this section, however, is slightly different, because we attempt to answer two further questions. The first is, what are the perceived differences and similarities in the nature of home and work stressors? The second is, what is the nature of the effects that domestic stressors have on the pilot at work? Reasons for concentrating on these two questions and their answers will be discussed later in this section.

Our first comments are directed towards those stressors that originate at home, but which may have 'carryover' effects into the pilot's work life and could be a source of stress. For this reason, we would expect the range of issues identified by pilots to be relatively large.

Pilots were asked, 'Can you identify any issues or situations that occur at home that have an effect on you when you are at work?' When we refer to 'at work' we include the total period of being on duty and not only the period of actual flying. Most pilots would agree that the job of pilot consists not just of the period spent in the cockpit, but also the periods of the flight preparations and post-flight checks. (Although on balance the latter are of lesser importance, they are still an integral part of the job.)

Consider the following comments:

'I work at a slow pace. My wife works at double pace and she expects me to work at her pace. If I do something, I like to get it right first time and to make sure it's a good job. It's like having to make compromises at home that makes me feel irritable and dissatisfied with home life. This irritation can bug me sometimes when I think things over when I'm at work.'

These are fairly typical of the reactions our questioning evoked. The data acquired was reduced to twelve issues.

1　Efficiency of pre-flight preparation time (i.e. period at home *not* on duty)
2　Issues or situations that are ongoing or left unresolved
3　Overall satisfaction with home life
4　How satisfied one is on how things have been left
5　Lack of stability
6　Division of loyalties
7　Indirect results of home life activities
8　Marital problems
9　Spouse's attitude to flying
10　Length of time spent at home
11　Serious events that occur
12　Particular arrangements that have been disrupted

They appear to break down into three themes as follows:

What pilots say about their job and home

(i) overall satisfaction with home life, i.e. issues that are relatively long-term such as 'how satisfied one is with how things have been left', 'spouse's attitude to flying', or 'marital problems'.

(ii) topics involving an element of dilemma or forced choice by the pilot such as 'issues or situations that are ongoing or left unresolved', 'division of loyalties', or 'particular arrangements that have been disrupted'.

(iii) specific qualities or features of the pilot's home such as 'lack of stability', or 'efficiency of pre-flight preparation time'.

Whilst the number of issues identified was lower than was at first anticipated, one must remember that these twelve items represent common elements expressed by the group of fifty or so pilots, all of whom, by definition, will have home lives each of which is entirely different to the next.

The range of items identified by pilots as having an effect in the opposite direction was marginally greater, in that there were sixteen issues that pilots identified as 'bridging the gap' from work to home.

1 Tiredness and fatigue
2 How time of work determines when to sleep
3 Returning home and time of arrival
4 Scheduling and rosters
5 Patterns of flying
6 Unpredictability of when you are asked to fly
7 Social problems associated with rosters
8 Seasonal fluctuations in workload (and hence effects)
9 Changes in family life due to job (e.g. promotions, base change, etc.)
10 Anxiety of courses and checks
11 Preparation necessary for courses and checks
12 Factors out of your control (i.e. delays, cancellations)
13 How long a single period of flying lasts (trip)
14 How many sectors you are asked to fly
15 Effects of minor day-to-day things
16 Carryover effects of personality clashes or interpersonal issues

This was surprising. On an a priori basis one would have speculated that only a small array of issues would have been relevant. This is because (superficially) one of the features that makes the job of pilot fairly unique among professional jobs is the fact that the major part of the job (i.e. flying) can only be performed when the pilot is at work (i.e. actually seated in the cockpit). It is not the sort of job where the individual brings work home with him. The 'at work/not at work' boundaries are, in theory, fairly clearly delineated. Since the range of pilots represented in the study was wide, one would expect a spread of relatively varied and miscellaneous occupational stressors that have an effect at home. The agreement, however, on the sixteen items included for further investigation was surprisingly high. In fact, the following are fairly typical reactions:

'Yes. It would be an odd statement I think for someone to say that the job didn't affect home life in some way. Checks tend to restrict time for domestic things. More common is sleeping patterns that are different from everyone else. Socially mixing is difficult – trying to arrange things in advance. Also – coming back from a trip – I affect the household greatly – I'm usually very tired and in a bad mood!'

Once again, it seems reasonable that we divide the pilots' interview comments into various common themes. There seem to be three:

(i) Issues involving the way that flying is organised, such as 'schedules and rostas', or 'unpredictability of when you are asked to fly'.
(ii) Issues that are also associated with flying but which refer to specific rather than general qualities, such as 'factors out of your control (i.e. delays, cancellations)', or 'how long a single period of flying lasts'.
(iii) Issues that consist of flying-related problems, such as 'tiredness and fatigue', or 'how time at work determines when to sleep'.

It will be apparent that the range of issues identified is wide, from 'anxiety of courses and checks' to 'carryover effects of personality clashes or interpersonal issues' and from 'returning home and time of arrival' to 'how many sectors you are asked to fly'.

So far we have briefly introduced the material that pertains to psychosocial stressors that are wholly domestic in origin, those that are occupationally or work-orientated and those that formed the home-to-work and work-to-home relationship. We intend to develop these issues further in the next chapter; however, it is appropriate to pursue at this point this latter category of stressors (i.e. the home/

work relationship), since this forms one of the basic underlying themes of the study.

There are two basic questions which will illustrate the point further. First, what is the nature of the effect that home-orientated stressors have on pilots, when they are at work? The answer to this question is particularly interesting for several reasons. The home-to-work directional stress is ultimately of great concern, at least from an aircraft safety point of view. Also, previous research (albeit small in quantity) has tended to concentrate upon the relationship in the other direction (i.e. work-to-home directional stress). The second question is, what are the basic attitudinal differences or similarities in how pilots perceive home- and work-orientated stressors?

Table 5.1 *Nature of effects of home factors on work: descriptive analysis*

	I can usually tell when I'm experiencing stress because at work I react this way . . .				
	Always	Usually	Some-times	Seldom	Never
Tendency to worry	4 0.9%	33 7.5%	139 31.4%	212 48.0%	54 12.2% Number Percentage
Experience of tiredness due to disrupted sleep	13 2.9%	59 13.3%	184 41.6%	144 32.6%	42 9.5%
Increased alcohol consumption when not flying	— —	13 2.9%	38 8.6%	157 35.5%	234 52.9%
Recurrence of item in thoughts during periods of low workload	11 2.5%	47 10.6%	156 35.3%	181 41.0%	47 10.6%
Make errors without knowing why	1 0.2%	11 2.5%	115 26.0%	241 54.5%	74 16.7%
Make errors of omission	3 0.7%	9 2.0%	155 35.1%	238 53.8%	37 8.4%
Decreased quality of pre-flight preparation	2 0.5%	15 3.4%	108 24.4%	230 52.0%	87 19.7%
Tendency to talk about the issue at work	2 0.5%	28 6.3%	79 17.9%	175 39.6%	158 35.7%
Slows one down	2 0.5%	18 4.1%	107 24.2%	200 45.2%	115 26.0%
Tendency not to listen as intently	2 0.5%	38 8.6%	161 36.4%	161 36.4%	80 18.1%
One's mind becomes detached from tasks in hand	1 0.2%	31 7.0%	125 28.3%	178 40.3%	107 24.3%
Decreased concentration	8 1.8%	29 6.6%	170 38.5%	192 43.4%	43 9.7%

Effects of domestic stressors at work

The background interviewing produced at least twelve main effects. These were presented to the main study group as part of the question-naire and the results from this are presented in Table 5.1.

These are summarised in Table 5.2 which highlights which effects occur most and least frequently. Most important is the 'tiredness', which is a function of disrupted sleeping patterns, as one senior captain suggests:

'When I am on an unusual sleep pattern, my co-ordination isn't as good and I know I have to be more careful in the air!'

This, however, is a behavioural effect. In other words, the pilot experiences tiredness (which is, of course, a physical state) as a result of behavioural stresses (i.e. not being able to sleep continuously at regular times).

Many pilots are, however, not as specific as this. Many just simply identify that home stresses can have a general impact on the pilot. One first officer proclaims:

'I bloody well do think family pressures can affect the way people

Table 5.2   *Nature of effects of home factors on work: most and least important issues*

	Number	*Percentage*
Most important ('Always' and 'Usually')		
Experience of tiredness due to disrupted sleep	72	16.2%
Recurrence of item in thoughts during periods of low workload	58	13.1%
Tendency not to listen as intently	40	9.1%
Tendency to worry	37	8.4%
Decreased concentration	37	8.4%
Least important ('Seldom' and 'Never')		
Increased alcohol consumption when not flying	391	88.4%
Tendency to talk about the issue at work	333	75.3%
Decreased quality of pre-flight preparation	317	71.7%
Make errors without knowing why	315	71.2%
Slows one down	315	71.2%

perform. I've seen its influence. I think it would be silly to dismiss the impact of home stresses, because what one man can cope with will gnaw away in the mind of another.'

Clearly, however, Table 5.2 indicates that the mental or cognitive element is fundamental. On balance, this theme of mental consequences is probably the one that, as psychologists, we would want to pursue further in any future work. It would appear that these cognitive effects are comprised of three elements. The first is the recurrence of the item in the pilot's thoughts. This tends to apply only when the pilot does not have much to do, usually during the period of 'cruise' (i.e. the period of ordinary flying between take-off and landing). In other words, his thoughts turn to home in the way that it does for most of us who are not pilots. The potential extent, however, of this fairly ordinary effect is apparent when the two remaining components of this cognitive system are examined. These are 'decreased concentration' and the 'tendency not to listen' as intently as he should.

Finally, it should be noted that pilots do appear to be conscious of the effects such stressors will have on them, in terms of their mental functioning:

'There was an incident once [involving family health] . . . I was very concerned. I was flown back – I didn't fly. I would have considered that I was not fit to fly at that time . . . . I would not have flown if I'd been rostered.'

Additionally, the item concerning 'one's mind becoming detached from the tasks in hand' was highly rated by pilots, as illustrated by a senior captain:

'I wouldn't necessarily say that it [domestic pressure] makes you dangerous in terms of lack of concentration at the time – it may do but it's not noticeable. Where it is probably most noticeable is that something like that can make you bad-tempered and if you get bad-tempered you get a bad atmosphere on the flight deck, which can be very serious.'

Interesting insights can also be made by examining the opposite end of the scale in Table 5.2, i.e. those effects that were least important. Very few pilots identified 'an increase in their consumption of alcohol' during non-duty hours. Smoking and drinking are traditional stress outcome effects. Generally, apart from isolated specific instances, it is the slow and gradual build-up in consumption that is of interest, i.e. the 'heavy smoker' or the 'heavy drinker'. Whilst only a minority were shown to fall into such categories, it was concluded in chapter 3 that even if some pilots did tend to drink more heavily during stressful periods, there was a tendency for them to be reticent in admitting it.

The effects of home stressors did not appear to influence the quality of the pilot's pre-flight preparation. To some extent this is surprising, since one would have expected this to suffer, being the first in the series of tasks the pilot must perform. Three other effects which were relatively unimportant were 'the tendency to talk about the issue with colleagues and friends at work', 'to make errors without really knowing why', or for the pilot 'to feel that his reactions are slowed down'.

By way of conclusion, we wish to reiterate three interesting aspects of these data that should be noted. The first is that pilots were asked to assess the frequency of occurrence of the effects listed. From this we inferred their importance. Pilots were not asked to assess each of the items on their magnitude of effect or impact on the pilot. Second, inspection of Table 5.2 reveals that, on balance, pilots tended to rate items using the lower ends of the answering scales. This might well be explained by a number of factors: the pilots may not in reality be able to identify any *specific* effects; the items identified by the interview sample may not have been sufficiently comprehensive to allow pilots in the survey study to express their own reactions; or, finally, pilots simply do not express their reactions to domestic stressors in this way.

The third overall aspect that is interesting from a psychological point of view is the apparent distinction between the highly rated and mainly cognitive effects, and the generally lowly rated behavioural effects. By cognitive effects we refer to complex mental processing and 'thinking'. The low rating of behavioural effects is to be expected to some extent since being less subtle, less volatile, and fairly 'ordinary', they may well be dismissed by the pilot as being unimportant. On the other hand, the more complex cognitive effects are comparable with those previously identified, introspective aspects of the pilot personality.

What pilots say about their job and home

Attitudinal similarities and differences between work and home stresses

The overall objective here was to discover how pilots broadly discriminated between those stresses which originated at work and those whose origins lay in the pilots' domestic life. Rather than discuss individual issues, we were really interested in adopting a holistic approach. Therefore, we statistically extracted the underlying trends in the data. The solution is presented in Table 5.3.

This was an interesting result. Background statistics (which are not reported here) indicated that the first trend extracted was very

Table 5.3 *Trends in pilot attitudes towards the differences and similarities between domestic and occupational stresses*

	Loadings
Trend 1 (51.10%)	
Home factors have greater effects on me	0.77
Home factors are more far-ranging in their implications for me	0.76
One can become more emotional about domestic things than job factors	0.56
Job stresses occur in short bursts	0.36
Although problems may arise from work, it is always possible to leave and get away from them, which one cannot do with problems that occur in home life	0.27
Trend 2 (17.60%)	
Home stresses tend to occur at a less intense level	0.72
Job stresses occur in short bursts	0.35
Job stresses occur in an acute and intense level	0.30
Trend 3 (13.40%)	
Chains of command which do not exist at home help to deal with job pressures	0.5
Stress at work is under my control, unlike things at home	0.47
Stress (from either home or work) can ultimately only be overcome if I have a job	0.46
Problems at work tend to be 'cut and dried' requiring 'yes' or 'no' answers	0.40
Home stresses tend to be ongoing	0.33

powerful, and, in fact, completely dominates the attitude pilots have about stressors from different origins.

We called the first trend 'Magnitude and type of effects'. Stressors which originate at home appear to have a greater effect on the pilot than do occupational stressors. Not only do these appear to consist of a wider range of issues, but they tend to invoke a more 'emotional' response in the pilot. This former characteristic is to be expected, since everyone's home background is different. The latter, too, was also an anticipated result, since the sort of issues one might expect to face at work might not be as positively 'charged' as domestic issues. Although this is very much a common sense finding, such confirmations are gratifying.

Part of this greater domestic reaction might well be explained by two issues. The first is that job stresses occur in 'bursts', i.e. there are sporadic episodes or limited periods that the pilot can clearly identify as 'stressful'. We may infer, therefore, that home stresses occur over a longer span of time. Second, there are limits too in the job of pilot from another perspective. By the nature of the work, even if there are bursts of stressful experiences, it is always possible to leave such problems behind. A similar avoidance is not available within the home context. The overall conclusion that one can draw from this trend (which dominated the finding) is that pilots make broad discriminations between stresses that originate at home and those that originate at work. The basis of the difference appears to be the greater degree or range of implications that domestic stressors can have for the pilot, as opposed to work stresses. One might add that from a theoretical point of view, despite this database on 'pilot stress' from a psychosocial perspective, it appears to be domestic issues that are more important.

We referred to two other trends that were apparent, but which failed to meet up to our rigorous selection criteria. Their contents were, however, quite meaningful, so they both might be worth some further comment. Table 5.3 indicates again that job stresses occurring in short bursts are important. There is a suggestion that *level of effect* may well be important. From the first trend we inferred that the range and implications of domestic stresses were more important than those associated with work. The general conclusion was that effects were greater. Indeed, we can add that the extraction of this trend confirms intensity to be a key issue. The trend clearly contrasts the more intense job stresses with the less intense domestic stresses. We must remember, however, that it is the latter that seem to be the more damaging and not the former!

The third trend appears to concern coping with the two sets of stressors. The key element appears to be the 'degree of control' that the pilot can exercise over the situation. At work, chains of command operate (either by assuming control or by others assuming control) that help to combat stressors. Clearly such structures do not exist at home. Additionally, the nature of the issues faced might well be important. The stresses occurring at home tend to be 'ongoing', in that they are not readily identifiable. Also problems that arise at home tend to be more 'messy', because they do not lend themselves to easy or specific solutions.

Finally, an interesting perspective was forthcoming about the role of *work* in coping. The work ethic is well recognised as part of our Western industrialised culture. Indeed, its importance to us psychologically is apparent in the negative experience of unemployment. Such feelings are more extreme in pilots, since the skills of a pilot are not readily transferable to other jobs. When this is combined with the fact that there is already a glut of qualified pilots, it is not surprising to find that their ability to cope with both domestic and occupational stresses is a function of whether they are in employment or not.

Conclusions

The objectives of this chapter were twofold. The first was to illustrate the nature of domestic and occupational stressors that pilots identify during informal interviews. These stressors will be discussed in greater depth in the next chapter. However, the second aim was to introduce the interview material in its raw state. Pilots do seem to be able to identify three sources of psychosocial stress: those that are domestic in origin, those that are occupationally orientated and those that transcend the home/work, work/home interface. As one might expect, many of the stressors identified as originating from the pilots' home life were very ordinary. They were the sort of issues that we all could identify with, such as 'arguments/disagreements' or 'relations with other people'. The fact that these are 'ordinary' does not make them any less important. Indeed, on balance, the fact that they can be related to by most of us makes the results all the more interesting because we will be able to see how the very complex job of airline pilot is affected by such issues. Other apparent issues were identified in the interview data; for example, the way the pilot's home life is *organised* seems to be instrumental in the generation or otherwise of domestic stress. The pilot seems to value certain qualities in his home life, such as tranquillity or refuge from work.

Our preliminary findings on occupational sources of stress portray a very different picture to that which one normally associates with the term 'pilot stress'. Apart from stresses relating to the job itself, which are to be expected, three other issues predominated. The first concerns 'career-related' issues, the second concerns 'scheduling' and issues surrounding the 'organisation of flying', and the third concerns managerial/organisational structure issues.

There was a third category of preliminary findings which were more complex than the above two. This third set were those that examined the 'relationship between home and work'. This third category was itself divided into three parts. The first concerns those stressors that originate in the pilot's home life, but which have carryover effects into the pilot's working life. Potentially almost anything could transcend the two; however, overall satisfaction, specific qualities and issues forcing the pilot to choose between work and home, seem to best summarise the findings. A similar triad concerning 'rosters', 'specific flying issues' and 'general issues related to the job of pilot' summarise the relationship in the work-to-home direction.

The final point addressed two questions concerning, first, the impact of home stressors on pilots at work, and second, the underlying difference between home and work stresses. The answer to the former question revealed the importance of cognitive effects. The answer to the latter revealed that the differences lie in the fact that domestic stressors were regarded as having more emotion, were longer-lasting and had wider-ranging consequences than job stressors.

This completes our examination of preliminary data. Many of the points raised here will be discussed in greater depth in the next chapter. Hopefully, however, this chapter has set the scene for what is to follow. The following chapter takes the issues raised here one step further by statistically extracting trends in the items and by calculating the complex inter-relationships between the items themselves and also their predictive relationships with the various major outcome issues examined in the study.

Chapter 6
The sources of psychological stress among pilots

This chapter is quite important because it contains two sets of results that may be described as forming the major thrust of our research. In previous chapters we have set the scene by describing the background rationales of the studies, how the studies were performed and a profile of the sample of pilots who were studied. The last chapter contained material from the interview part of the investigation. The emphasis was upon presenting data that illustrated further the nature of stresses experienced by pilots. The main disadvantage of discussing this sort of data is that, whilst it is superficially very appealing and convincing, it does not comply with normal accepted scientific criteria. In other words, it is only after the items are tested across a wider population and are subject to more rigorous analysis involving more complex statistical procedures that we would attempt to draw firm conclusions. The last chapter was primarily descriptive in its orientation. With application of more powerful statistical techniques, we can extend the analysis to include an inferential quality also.

It will be recalled that the major points from chapter 5 were as follows:

(i) Pilots are able to identify sources of psychosocial stress that originate at home and at work, and which transcend the home-to-work and work-to-home relationship.
(ii) Whilst it is possible to draw up a distinctive profile of stressors that potentially might be of importance, the conclusions drawn so far are tentative.
(iii) Pilots indicate that the relationship between sources of stress and their effects are complex.
(iv) Pilots indicate that some items may be of greater importance than others in their contribution to particular outcome variables (such as job satisfaction, mental ill-health, etc.)

The logic behind the two-part format of this chapter is relatively straightforward. We shall consider each in turn. The first part refers to the examination of underlying trends in the data. It will be recalled that at the very beginning of our studies we had interviewed a preliminary sample of pilots. We then summarised their responses by drawing up a collection of items, which we presented to elicit responses in the main sample. Attention was focused upon the way in which the items were grouped together by us for presentation. In the last chapter, we merely took these groupings, which were a function of the background questioning, and selected apparent trends. Whilst this was satisfactory to provide a vehicle for introducing the material, there is no guarantee that the pilots themselves see the items to be grouped in this way. Additionally, of course, such a basis for grouping items together is very unreliable in that the sheer number of variables renders it impossible to uncover the precise (or even estimated) relationship of one variable with each of the others at any one moment in time. In other words, any one issue may be related to another issue, which will itself be related to another issue, and so on. The appropriate techniques to do this are available and were employed. The resultant trends are reported in the first part of the chapter.

The second part of this chapter concerns the relationship between variables in a very different sense. Within this second context, we are concerned with the relationships between the stressors (either in the form of individual issues or trends) and the outcome measures taken. We are interested in the degree of predictive association between the two. For complex theoretical reasons we cannot say that one thing *causes* another. But we can certainly infer causality if a strong predictive relationship exists between two sets of variables. As previously indicated, the three outcome variables of primary interest are mental ill-health, job satisfaction and self-reported pilot performance.

Underlying themes in pilot psychosocial stress

It will be noted from the previous chapter that the four primary groups of stressors were divided as follows:

29 domestic sources
30 occupational sources
16 work-to-home sources
12 home-to-work sources

In total there is a list of eighty-seven potential sources of stress identified by pilots

Using the responses from all 442 pilots, we *simultaneously* analysed

Table 6.1   *Complex statistical characteristics of underlying stressor trends*

Trend	Eigenvalue	Percentage of variance	Cumulative percentage
1	18.90	39.4%	39.4%
2	4.54	9.5%	48.9%
3	3.96	8.2%	57.1%
4	3.22	6.7%	63.8%
5	2.85	5.9%	69.7%
6	1.82	3.8%	73.5%
7	1.59	3.3%	76.8%
8	1.32	2.8%	79.6%
9	1.25	2.6%	82.2%
10	1.13	2.4%	84.6%
11	1.04	2.2%	86.8%

the inter-relationships between all eighty-seven items. Normally accepted statistical criteria were employed to select which trends were significant and which items were significantly related to these trends. Those readers with an interest in statistics will note from Table 6.1 that we relaxed the criteri˙ for the selection of trends on the basis of explained variance, because those trends made psychological and common sense.

It will be noted that there were, in fact, *eleven* apparent trends that were statistically extracted from the pilots' data – in contrast to the simple four-way division indicated in the last chapter. The actual contents of each of the trends are reported in Table 6.2.

It should be recalled from previous explanations that all of the items included in Table 6.2 are significant in the sense that they all contribute to the meaning of the trend. It should also be noted, however, that the higher the 'loading' the greater the importance of a particular item. Just to recap, here is a list of the eleven trends extracted.

1 Control
2 Scheduling and rostering
3 Anxiety of courses and checks
4 Home-to-work interface
5 Career and achievement
6 Insufficient flying
7 Responsibility and decision making

84

8 Interpersonal problems
9 Management and organisational issues
10 Domestic status
11 Fatigue and flying patterns

By way of introduction, it might be worth briefly highlighting the most salient features of each trend, before discussing the results in greater depth.

Trend 1: control

Since this was the first trend extracted we may infer its relative importance compared to the other ten trends. There seems to be an overall pilot preference for stability and balance. The trend reflects a concern with a lack of order or continuity, with a disruption of domestic routine and with events or processes over which the pilot has no immediate form of control.

Trend 2: scheduling and rostering

Unpredictability of flying, social problems and relative times of flying were the three main issues identified. The unpredictability feature of schedules and rosters might well be explained by the high proportion of British Airways pilots in the sample. We will discuss this in greater detail later; however, it is sufficient here to state that the scheduling system operated by British Airways is called 'Bidline'. Whilst being the system operated by most of the major US air carriers, it is distinctively different from that employed by other UK-based operators. The second issue may well be a function of the first and the third issue may be regarded as including adjustments for local time and time zone changes, night flying and so on.

Trend 3: anxiety of courses and checks

Constant scrutiny and an ongoing programme of change and updating of knowledge are two distinctive features of the job of pilot. Contents of the trend confirm that it is the whole experience of these that is important, though, of course, 'fear of failure' is the underlying common denominator. In addition to relatively normal participation, there seems to be a personal standard that each pilot holds and can identify, and which must be attained to secure success.

85

Table 6.2   *Underlying trends in pilot stress*

	Loadings
Trend 1 (39.40%) Control	
Others not obeying or things that go wrong	0.78
New and unfamiliar experiences	0.44
Disappointments when others fail to meet expectations	0.29
Disagreements, arguments, differences of opinion	0.28
Enforced or adapted roles at home	0.26
Inability to identify problems (and hence solutions)	0.26
Trend 2 (9.50%) Scheduling and rostering	
Unpredictability of when you are asked to fly	0.52
Social problems associated with rosters	0.52
Scheduling and rosters	0.50
Scheduling	0.44
Patterns of flying (relative times you are asked to fly)	0.34
Trend 3 (8.20%) Anxiety of courses and checks	
Anxiety of courses and checks	0.96
The whole experience (before and during) of checks on your flying ability	0.81
Preparation necessary for courses and checks	0.79
Changes in your experience of flying (e.g. conversion course)	0.54
The whole experience (before and during) of medical checks	0.31
Attaining your own personal levels of performance	0.29
Trend 4 (6.70%) Home/work interface	
Overall satisfaction with home life	−0.76
Lack of stability	−0.74
How satisfied one is on how things have been left	−0.69
Spouse's attitude towards flying	−0.63
Marital problems	−0.62
Indirect results of home life activities	−0.62
Division of loyalties	−0.60
Lengths of time spent at home	−0.56
Serious events that occur	−0.54
Issues or ongoing situations left unresolved	−0.47
Particular arrangements that have been disrupted	−0.40
Efficiency of pre-flight preparation time (at home)	−0.36
Trend 5 (5.90%) Career and achievement	
Career opportunities and lack of potential advancement	0.78
Future career uncertainty	0.68
Degree to which your personal goals and aims in life have been achieved	0.64
Seniority systems	0.57
Success or failure of one's efforts to achieve	0.41
Impending major career change or threat (redundancy, etc.)	0.40

	Loadings
Trend 6 (3.80%) Insufficient flying	
Not enough hours actually spent flying	0.77
Sharing of work evenly	0.63
Impact of lack of flying (practice effects)	0.62
Trend 7 (3.30%) Responsibility and decision making	
Making important decisions	−0.80
Inherent responsibility in your job	−0.77
Periods in flight of high workload	−0.52
Ambiguous factors or difficulties in problem identification	−0.49
Fulfilling role expectations	−0.32
Trend 8 (2.80%) Interpersonal problems	
Interpersonal problems with cabin staff	0.58
Interpersonal problems with aircrew	0.39
Trend 9 (2.60) Management and organisational issues	
Style of management	0.90
Lack of management support	0.86
Morale and organisational climate	0.63
Conditions of employment	0.57
Factors not under your direct control	0.37
Scheduling	0.34
Trend 10 (2.40%) Domestic status	
Family health	0.58
Issues associated with children (health, education)	0.47
Trend 11 (2.20%) Fatigue and flying patterns	
*Tiredness and fatigue	0.76
How time to work determines when to sleep	0.75
Returning home and time of arrival	0.67
*Tiredness and fatigue	0.62
Patterns of flying	0.48
How long a single period of flying lasts	0.37

* From Section 4 and Section 3 respectively; see Appendix.

Trend 4: home-to-work interface

This was a surprising and, in many ways, gratifying result. There were twelve items that concerned the home-to-work relationship that were analysed as part of the eighty-seven items presented. All twelve items emerged here as one single trend. This confirms that they all appear to measure the same underlying issue (i.e. home stressors that have an impact on the pilot at work). All of the items being present dictated that a broad label was appropriate. However, the order of the items indicates a concentration on overall satisfaction and stability in home life.

Trend 5: career and achievement

Concern with our careers is closely tied up with our sense of achievement of personal goals and ambitions in life. The trend indicated that stress is due to 'blocked career pathways', 'structures that do not offer a realistic mode of progression' and 'malaise' of thwarted career goals. By implication, reduced potential for advancement is a function of seniority systems used by airlines. However, the effects of organisational change and redundancy policies are also perceived threats.

Trend 6: insufficient flying

As a result of recent commercial and consumer trends, the aviation industry has been generally in the 'doldrums'. The knock-on effect has simply meant that too many pilots have been chasing too little work. Although within some airlines work is seasonally determined, few would disagree with the above statement. The apparent stress is a function of the frustrating experience of flying only infrequently. Additionally, a simple 'lack of practice' brings its own anxieties.

Trend 7: responsibility and decision making

To the layperson this is one of the 'expected' pilot stressors. However, as Table 6.2 indicates, these were negatively loaded. This means that this sort of stress is positive and is 'welcomed'. Consistent with the type of person who thrives on a challenge, periods of high workload, resolving ambiguities and fulfilled role demands are also identified by pilots as a satisfying experience.

Trend 8: interpersonal problems

Only two items explicitly referred to how pilots establish working relationships with others. The two items referred to the most immediate groups of reference – cabin staff and other aircrew. Interestingly the former sources of stress were rated more highly than the latter.

Trend 9: managerial and organisational issues

Since a number of different airlines were examined, it is only possible to generalise about the two most highly rated issues: 'style of management' and 'lack of management support'. Issues such as 'conditions of employment' and 'scheduling', with their resultant decrease in the quality of morale and psychological climate, are all also implicated.

Trend 10: domestic status

Consisting of only two items, the trend was difficult to label. The items involving children's education, family health and so on were construed as simply referring to the general degree of disruption experienced at home.

Trend 11: fatigue and flying patterns

The trend clearly embraces 'tiredness' and 'fatigue'. The most important issues associated with this are the matching of sleeping patterns with the pattern of preparing for, and recovering from, a period of flying. Also important is how long a particular trip actually lasts.

The above findings confirm that the pilots do not perceive the range of eighty-seven items as four trends, but as eleven individual trends. Of the original four-way formation, only the twelve items concerning the sources of domestic stress that affect the pilot at work remained intact. Otherwise, items from the remaining three original groupings permeated their way through the final multi-trend solution. We gave the trends generally work-related labels, as befitting their dominant contents; however, domestically related items were in all but a small number of trends. It will be recalled that the original four-way division was as follows:

Our original conception of items

Section   I Domestic stressors
            II Occupational stressors
           III Occupational stressors that have effects at home
           IV Domestic stressors that have effects at work

The sources of psychological stress among pilots

As the result of our more complex analysis the division of items now looks like this:

Pilots' conception of items

Trend	Composition
I	Domestic stressors (I)
2	Work-to-home stressors (III)
3	Work-to-home stressors (III)
4	Home-to-work stressors (IV)
5	Occupational (II) with high domestic content (I)
6	Occupational stressors (II)
7	Occupational stressors (II)
8	Occupational stressors (II)
9	Occupational stressors (II)
10	Domestic stressors (I)
11	Work-to-home stressors (III)

(The figures I to IV refer to the dominant contents of the trends using the original classification.)

There were only two trends that had an entirely domestic content; these were the first concerning 'control' and the tenth concerning 'domestic status'. The results from this first factor, 'control', are important. Not only did it dominate the findings from a statistical point of view, but it was confirmation of a personality characteristic well recorded by previous research. It would seem that pilots do indeed prefer to exercise some sort of control over their domestic environments. However, the same applies to the way they regard work. Some see the job itself as controllable, as this pilot illustrates:

'The job is different from office work where problems are ongoing. When I go to work there's a start and a finish to it and that's it. You don't think about the job after a day's work. Even if we have a bad day – go to the bar, talk it out and that's it over. It's that final.'

Whilst identifying control as an important issue, several pilots adopt a fairly philosophical approach and realise they cannot possibly control everything.

'You cannot control everything and therefore I see no point in
jumping up and down. As long as you realise that flying is a
stressful job . . . it doesn't matter which parts of it . . . the bits
you can't control, etc., you just have to accept that.'

We have said previously that fairly 'ordinary' issues such as arguments and disagreements were important and, indeed, these crop up
here in the context of 'control'. However, the need seems to be more
deeply rooted than just personal preference. The frustration of others
not 'obeying', 'things that go wrong', 'the anxiety of new or unfamiliar
domestic situations', 'the disappointment when others fail to meet
expectations', or just simply 'not being able to put one's finger on the
problem' are all functions of abortive attempts to control the domestic
scene. It should be added that when we use the term 'control', we
do not mean it in just its simplistic practical sense. We also refer
to psychological 'control', i.e. the extent to which the individual
perceives that he is able to significantly manipulate events. So,
this *need for control* is a dominant feature of the pilot's stressor
profile.

The other trend that was exclusively domestic in content concerned
'the status of home life', with particular reference to family health and
issues associated with children such as education and, again, health.
The issues involved are difficult to identify – for some it is just the
general status of home life.

'What affects me? I find it difficult to relate to any specific
incident to home life when performing in the aeroplane. When
flying I try to devote 100 per cent of my attention to what's
happening – only in cruise do you think about things at home.
There have been times when I've had an argument or left under a
cloud and you are looking forward to the prospect of ten days plus
before it can be resolved. I find it's more a general thing.'

Being the penultimate trend extracted, it is interesting but of less
significance.

There was a group of four trends that were composed of stressors
that were entirely occupational; these were 'insufficient flying', 'responsibility and decision making', 'interpersonal problems' and 'managerial and organisational issues'. The fact that pilots complain of not
flying enough is an interesting one. We said earlier that too little work
was spread around too many pilots. This led to anxiety and frustration

simply due to a lack of practice. The following comments illustrate the concerns of the pilots:

> 'Major things in your own personal performance. In long haul you don't fly a lot therefore the only way to maintain standards is to criticise yourself a lot. Every six months you are checked – three or four bad sectors and it's the end of your career – not like other jobs. This winter I'll do just the legal minimum in flying which I don't find sufficient to keep me in practice and I don't feel as confident as I would if I were in practice. If you haven't flown for a while you aren't completely confident that your performance will be adequate.'

> 'There is too much reliance upon machinery in flying – I find it especially in Tristar. I tend to fly without the automatics so that I'm practised. A lot of pilots, especially the new intake, have been taught to fly with automatics. I think we should go back to basic airmanship and basic flying – that requires practice.'

> 'The lack of flying is bad. I feel, to a certain extent, out of practice. Also you can well find yourself flying with other pilots who haven't flown recently – it's especially important in stressful situations such as making a landing approach in bad weather – you need to be fairly sharp.'

However, it would appear that even overpractice can be an issue.

> 'I give myself lower limits than others – keep a watch on myself throughout the flight. It's important because you can become complacent – familiarity breeding contempt and all that. I would like to see regular changes of aircraft so you become a flyer of aircraft and not just of a certain type.'

In other words, not only is insufficient flying an issue that may have potentially negative effects, but its opposite also is important. The pilot can be so familiar with his aircraft type, his routes or indeed both that this lack of variety and overfamiliarity can have negative effects also.

Three underlying trends exacerbate the situation, especially the problem of insufficient flying. The first is the overall decrease in passenger traffic as a result of the general economic recession. The second is that the carrying capacity of aircraft has increased, hence when an aircraft does fly, it carries more people thus reducing the number of flights. The third is the advance in technology, and the change in aircraft that formerly required three in the cockpit now only requiring two. Additional consequences for airlines are an increased number of pilots who fly either on a stand-by basis or on a recency basis – the latter being the minimum number of hours to maintain a valid flying licence. Quite simply, therefore, the work that does exist does not seem to be equally or equitably distributed.

The inherent responsibilities in the job are probably stressors that most readily come to mind in most of us when we consider 'pilot stress', and they indeed formed a trend. In particular, periods of high workload such as take-off and landing or circumstances that are not clear-cut are especially important tests of the pilot's decision-making capacities, as these two captains illustrate:

'Making decisions can be stressful, especially a series of decisions in special conditions such as technical failure. It's the responsibility of it all. After the event you have to justify what you've done. I enjoy being a pilot. There is some rivalry between groups of pilots in which they think that they are the best and most professional in terms of taking responsibility or making decisions. Perhaps it's some kind of need.'

'I'm very new at command – it was a major change and *superb* fun. A quantum jump mentally and physically. You can put some of your own things into practice. Planning ahead and delegation are vital – anyone who doesn't might not fail but they will certainly stumble. It's just great.'

It should be added, therefore, that all such demands are seen in a positive light as a rewarding challenge.

The responsibilities the pilot must face are part of the role he has to play. It will be recalled from chapter 1 that a person's job role is simply the series of behaviours he is expected to display (not unlike the 'role' an actor adopts in a play). These role demands are traditionally regarded by psychologists as people-orientated issues, i.e. the 'roles' are, in fact, a series of expectations held by other people about the occupant of a particular job – in this case, the job of pilot. These interpersonal demands too, however, were welcomed by the pilots. This is in stark contrast however to the third occupational trend, 'interpersonal problems', which highlighted a *different* kind of stress that results from working with other people. The sources of this type of interpersonal demand seem to be cabin staff, as this pilot suggests:

'Yes, sometimes the cabin staff are annoying. I sometimes get aggravated when stewards who are too big for their boots – some of the senior stewards like to regard themselves as "cabin captains". I have found too that they often come up front and ask questions like "how long to go?" I guess that's reasonable enough, but the fact is they all come up! It's as though they don't speak to each other! It's part of the "management" job of the captain to try to keep everyone sweet – that involves cabin staff.'

Cabin staff form only one half of the picture; other aircrew, too, make an important contribution.

'You can get really wound up by the people that you fly with. If you fly with a "funny" guy, that can wind you up from the word go. It is rare, but it does happen. The minute you get on [the aircraft], there is an atmosphere.'

'Getting on with the rest of the flight deck is very important. When you are working with someone who is ratty, it is amazing how it can turn down the whole atmosphere on the flight deck. For example, the flight last night. The captain was ratty. So, the engineer and I just carried out our duties. Everything was technically correct but a good relaxed working atmosphere wasn't there. The engineer and I talked about this at the end of the trip last night. It's the captain who determines the flight-deck atmosphere.'

This issue of the job being technically correct but a good atmosphere being absent is a good illustration of the fact that although pilots must work to technically precise parameters, intangible interpersonal issues still have an important part to play in the cockpit.

Previous research indicates that it is not just the job that provides sources of occupational stress, but also the way in which the job fits into other jobs (and of course vice versa). It is the organisation that is the source of the stress. This is often inextricably entwined with the nature of management. Organisational problems and their consequences which are frequently highlighted are 'managerial style' and

'perceived psychological climate and morale'. These are both identified by pilots as being important. There seems, in some quarters, to
be a great lack of morale over the last few years – because the job as
they know it is constantly changing, 'the future is an unknown
quantity' is a feeling that is quite widespread especially in junior
pilots. This was a finding that is contrary to our impression that the
'job of pilot' is simply to climb into the cockpit and fly the plane. Our
comments are limited by the fact that pilots from different airlines
were examined. One assumes, therefore, that these problems were
experienced in different airline companies though probably in different ways. However, since the companies all work within the same
industry, changes in one organisation may well have knock-on effects
on others, forcing them to follow suit to remain competitive, as this
pilot suggests:

'Many attitudes depend upon the particular organisation you work
for. We can identify decisions that companies make that we don't
agree with. But to be fair, many airlines are under a lot of
commercial pressure at the moment which I guess greatly
influences their decision making.'

And finally, a fifth trend concerns 'career path and development'.
Although this is, by definition, an occupationally related issue,
significant contributions are made to the meaning of the trend by
domestic issues. We have noted elsewhere the important effects of the
malaise the airlines have found themselves in. Another consequence
of this is that opportunities for acceptable advancement of a pilot's
career are greatly diminished. Pilots identify also that the seniority
system used by some airlines does not lend itself to a swift or clear
career progression. For example:

'Given the amount of time I've been in the company, I'm lucky to
have been given a command. Given the company career structure
things will probably stagnate now. Because of the merger (even
though it was some time ago), many will be coming into the 747
fleet above me. Therefore I shall not increase in seniority for a
while. There is a joint seniority list – something that's never
happened before. I've been trying for years to get into training –
that's not going to be realised because of this seniority issue.'

What is perhaps even more important is the fact that we know that one's career progression is positively related to good psychological health.

'. . . people are in it because it's their way of life. They are in it because they want to be in it; they don't want to be in anything else and would probably be happy to do it, even if they were not getting paid for it – well, almost! It's not really a job, it's part of one's life.'

Like most professionals, therefore, pilots have a tremendous need to achieve. The recorded stress in this instance is due to a sense of frustrated motivation to achieve.

The above discussion has concentrated on trends that were based upon items of either domestic or work origin. Additionally these items were not construed to have any carryover effects. It is on such stressors that we now focus our attention, i.e. the home-to-work, work-to-home stressors.

There were three trends that seemed to summarise the main carryover features of psychosocial stress from work into the pilot's home life; these were 'scheduling and rostering', 'anxiety of courses and checks' and 'tiredness and fatigue'.

Of major importance (second trend of the eleven extracted) is the way in which the pilot's *duty hours* are organised. We indicated earlier that the system of scheduling operated by many airlines in the UK is a fairly simple one, i.e. it is worked out in advance who flies what, where and when; these schedules are published and are more or less fixed and implemented as published. The disadvantage of this system is that the pilot has no control over it other than the changes he can informally negotiate. The big advantage, however, is that it is simple to operate, clear-cut and everyone knows where they stand well in advance. Many of the larger American airlines and British Airways operate a 'bidding' system. In this system of rostering, a pilot chooses (or 'bids') when and where to fly (within limits, of course, of his fleet, experience and so on). Depending on such factors the pilot may have to make a low or high number of bids before obtaining a desired flight. The theory is, therefore, that the pilot has some sort of say over the schedule he flies. In reality, many pilots must make many bids before ever achieving success – the pilot does not in fact have as much say in determining his timetable as the theory of the system might indicate. A major factor also is the times the pilots are asked to work, as one first officer reflects:

'Night working is OK if you get into a routine or pattern. But it is chopped and changed so often, there is nothing whatsoever consistent about it. It's impossible to plan anything, especially during summer months. There are, of course, added difficulties such as delays and so on.'

Additionally, the previously mentioned factors such as stand-by and recency flying exacerbate the situation. The carryover effect is the unpredictability inherent in the system. Consequently, social problems also arise. In a different sense the *relative* times of when the pilot is scheduled to fly also has an impact on the pilot's home life.

The anxiety associated with *tests and examinations* of skill and ability is something most of us have experienced at some stage in our lives. Regular scrutiny is part of the ordinary working life of the commercial pilot. So, too, is attending courses, whether it be for an update on technical knowledge or a complete change in flying (such as from short haul to long haul or from first officer to captain). Checks may be technical checks of knowledge and flying ability or medical checks. All are part of the way the commercial flying system is regulated and governed. The stress is, of course, due to the fear of failure.

'Something major was obtaining the job with X. I was a flying school instructor – it was a major change in lifestyle for me – new job, new house, etc. At the end of the training, you are under a massive pressure to pass especially first time. I took a long time to recover from the change – it no doubt affected my performance especially in the early stages. The same is true of doing checks and so on.'

This fear does not just affect the pilot in terms of completing courses and checks, but the processes that precede such checks (such as preparation) are identified as stressful periods. Perhaps the key issue is not just fear of failure, but also the knowledge that the sheer pressure involved in checks makes the pilot not perform to the level he knows he can achieve, as a senior captain and a first officer suggest:

'The simulator is stressful – and I think I know why. It's psychological. I'm a completely different animal in simulator – in flight instead of fight. I don't think that I have ever been seen properly in a simulator environment.'

'. . . so much depends upon it. You have to pass and there is someone looking over your shoulder. You don't really enjoy it in the same way that you normally enjoy flying. What really gets up my nose is when I know that I'm being watched!'

An integral part of a test of proficiency or fitness is some satisfactory level of attainment. Such criteria are erected, regulated and administered by national aviation authorities. Thus in this way, these can be regarded as being 'external' criteria (i.e. external to the pilot). Implicated also are the pilot's own standards of performance. Such 'self-set' standards not only act as a yardstick or level to be attained, they also act as a powerful motivator.

The final main theme that the pilots identified as affecting them at home was *tiredness and fatigue*. One experienced pilot expressed concern about this rather well:

'I think flying long haul over a period of time must damage your health if repeated over the years. I certainly don't think it does your health any good when you are over 50 – it's the odd hours. It can be tiredness or fatigue. I think they are both different. Tiredness is mental, fatigue is both mental and physical. I tend to experience tiredness. It's due to disrupted sleep and flight patterns and so on. Pre-flight depends upon departure time. I find it difficult to sleep – I just lie around and eat well. It's tempting just to fit things in and do things because you are up and about. Also, of course, you've got flying at the back of your mind. Morning departure is best – just get up and go. Post-flight – late return is no problem – just go straight to bed. But if you arrive early morning, there's too much time to catch up on – therefore you don't want to go to bed.'

Other pilots identified timetabling:

'Major effects on performance are a function of tiredness – determined by night flying perhaps, on a one-off basis. I don't care what the authorities say – I can't switch routines like that. It's better though if you work in a series.'

The schedules and patterns of flying are again implicated in that they dictate: (a) how long a simple period of flying lasts, (b) the time that the pilot can rest and recover, and (c), on a more practical note, when the pilot can actually return home and his time of arrival relative to the rest of the world.

We complete our brief discussion of the trends by highlighting some aspects of the fourth trend, 'home-to-work interface'. You will recall that there were twelve home stresses that pilots identified as having an effect on them at work. All twelve items emerged as a single factor in our statistical analysis. In some ways, the trend resembles the first about control of the home environment. However, since these items cluster together to form a separate factor, it is fairly clear that pilots identify the distinguishing characteristic as the 'from-home-to-work' effect. One pilot highlights the importance of these stressors:

'Home life does affect performance at work. I think it depends upon where one derives greatest satisfaction. A lot of people are unhappy at home but still carry on – perhaps drinking too much and not doing the best they could. I think if a guy does endure, say, a bad marriage after several years it's bound to creep through. It doesn't even have to be anything as dramatic as this, but it does tend to be ongoing things I think. I feel that my performance is being affected by things that occur at home at the moment. Not just moving house or the unresolved situation but little things also. It's getting to the stage of mental weariness – I feel a bit depressed about things. It's the fact that I'm not in control of what's happening and having to put up with a situation for a long time.'

However, it is the order of the items that really tells us what is of most concern to the pilots. The upper part of the factor confirms that it is the overall state that is important – how satisfied the pilot is generally with the way things are left when he goes to fly. We will find out later that the smooth functioning of home life is very much a function of the spouses' ability and skill. Hence, the pilots' wives' attitudes are important. Consequently any marital problems that arise are likely to concern the pilot, long after they occur, just in the same way as they do for most of us.

At the opposite lower end of the trend, the lower four items can be explained by other reasons; for example, pre-flight preparation is often simply a matter of routine; although disrupted plans are a source of annoyance identified elsewhere, there is little the pilot can do about them; unresolved issues are deliberately confronted and in the event of something serious occurring, the pilot may well opt not to fly. One final point is that although we divide the trend into upper and lower parts on the basis of importance, *all* the items seem to be of concern to the pilots.

Predictors of mental ill-health, job satisfaction and performance

The earlier part of this chapter examined the underlying trends of stressors identified by pilots. To analyse results we used complex statistical methods that told us how items were clustered together. A major strength in the method was that a high number of inter-relationships were examined at any one time. In this second part of the chapter we are again concerned with clustering, though in this case we use techniques to examine a specific issue (such as mental ill-health, job satisfaction or performance) and see which combination of other issues (such as stressors) *predicts* most efficiently the particular outcome being examined. In turn, by examining different profiles of stressors that predict the different outcomes, we can, by applying some logic and our skill as psychologists, arrive at some conclusions about which stressors have which effects. Since there were three major outcomes examined in the study, an appropriate mode of presentation would simply be to discuss each in turn.

In these predictive analyses we included not only the eleven trends identified here but also some 'life event' trends (chapter 7) and some 'coping trends' (chapter 9). Additionally, the following biographical variables were entered (chapter 4):

(i) age
(ii) marital status
(iii) partner work?
(iv) number of children
(v) relax and wind down
(vi) total flying hours' experience
(vii) long haul/short haul/domestic only
(viii) rank
(ix) seniority.

The significant predictive trends that refer to coping and life events are included in the presentation of results here, although these are discussed in greater detail in the appropriate chapters.

Mental ill-health

The particular measure of mental ill-health we employed yields seven scores (six subscales and the total). Each of these was treated as an outcome variable in its own right, as illustrated in Table 6.3.

This was an efficient result in several ways. Only one of the scales (phobia) did not have a significant solution. All of the predictors that were identified as significant (in a statistical sense) were also intelligible. The results for anxiety and overall mental ill-health were especially clear though all were statistically quite sound.

The trend concerning fatigue and flying patterns was confirmed as the best predictor of *anxiety*. This confirms that such issues are fundamental in determining mental ill-health manifest in the most readily identifiable form of anxiety. Courses and checks on skill and ability, and medical checks, are all confirmed as being sources of anxiety. From the interview material which identified 'fear of failure' as an underlying issue, it was not surprising to find the anxiety experienced by pilots was fuelled by concerns about their career prospects and sense of achievement.

Pilots indicate that anxiety can be offset by opportunities to 'relax and wind down' – a theme we will be returning to later. One would have expected social support from others to be a negative predictor (i.e. reducer of anxiety). The results are such that the 'support given by others' to the pilot does, in fact, act to help reduce anxiety (see chapter 9 for a full explanation of this). It would appear, however, that it is the most senior pilots who suffer most in terms of anxiety. This was quite surprising and tells us that achieving a higher level of seniority as a pilot is not necessarily by itself the answer to the anxiety-related problems. Indeed, we suspect that it is not seniority by itself that is important, but other factors that are associated with seniority such as changing career prospects or changes in home life.

We define *obsessionality* as meticulousness, adherence to routine, punctuality and dislike of sudden change. In exhibiting these behaviours, an individual is effectively responding to and manipulating events around him. Consequently, the overall important issue of control was not a surprising predictor of obsessionality. An equally expected predictor was stability of relationships and home life. This is consistent with the fact that the pilots previously identified the ideal concept of home life as a series of stable routines, a psychological

The sources of psychological stress among pilots

Table 6.3   Predictors of mental ill-health

	Multiple R	R Square	Overall F	P
Anxiety				
Trend 11: Fatigue and flying patterns	0.31	0.10		
Relax and wind down	0.39	0.15		
Trend 3: Anxiety of courses and checks	0.41	0.17		
*–Trend 17: Social support	0.43	0.18		
Seniority	0.44	0.19		
Trend 5: Career and achievement	0.46	0.21	4.42	<0.001
Obsessionality				
Trend 2: Scheduling and rostering	0.24	0.06		
–Trend 7: Responsibility and decision making	0.26	0.08		
Age	0.33	0.11		
Trend 6: Insufficient flying	0.39	0.13		
Relax and wind down	0.37	0.11		
*Trend 15: Stability of relationship and home life	0.39	0.15	3.61	<0.001
Psychosomatic disorders				
Trend 11: Fatigue and flying patterns	0.27	0.07		
Age	0.37	0.14		
Relax and wind down	0.41	0.17		
Trend 8: Interpersonal problems	0.42	0.18	4.08	<0.001
Depression				
Trend 11: Fatigue and flying patterns	0.26	0.07		
Age	0.36	0.13		
–Trend 4: Home/work interface	0.38	0.15		
Relax and wind down	0.40	0.16	4.36	<0.000
Hysteria				
–Trend 7: Responsibility and decision making	0.21	0.04		

Table 6.3—*Contd.*

	Multiple R	R Square	Overall F	P
Trend 8: Interpersonal				
problems	0.25	0.06		
–Age	0.27	0.07		
Total flying hours'				
experience	0.29	0.09	2.30	<0.001
Overall mental ill-health				
Trend 11: Fatigue and				
flying patterns	0.36	0.13		
Trend 7: Responsibility				
and decision making	0.40	0.16		
Relax and wind down	0.43	0.19		
Age	0.45	0.20		
*–Trend 17: Social				
support	0.47	0.22	4.74	<0.000

* See chapter 9
– Denotes a negative predictor

platform or collection of props upon which the pilot can rely, on an ongoing basis.

Disruptive aspects of pilot experiences help to quell obsessional behaviour. In particular, scheduling and rostering, insufficient flying and opportunities to relax and wind down all detract from this outcome, in the sense that they all detract from the concept of regularity and routine. Indications are, however, that a major way of reducing the unhealthy mental processes of compulsive behaviour is provided by the pilots exerting authority, making decisions and assuming responsibility. These emphasise the qualities of power and discretion, both of which are instrumental in an individual reducing obsessionality. We know that males tend to become more obsessional as they become older, so age being a significant predictor in this case is consistent with previously recorded trends in mental ill-health.

A similar situation is true in the case of *psychosomatic disorders*, with older pilots reporting greater incidence of these symptoms. Somatic disorders by definition are *physical* manifestations of *mental* ill-health. It was not surprising to find that fatigue and the effects of working a varying array of flying patterns were significant predictors, since these too are physically experienced by the pilot. Once again, it is the pilots who do not have as many opportunities to relax and 'wind down' who report these symptoms occurring most often. Although both of these predictors are significant here, we know that they are also significant predictors of other scales. This is true of fatigue and flying-pattern tiredness. Hence it can be added that although all are physically orientated, such effects are not seen solely in such terms. Confirming the fact that physical stress-related symptoms can have psychological origins, it was found that interpersonal problems might trigger off such responses in pilots.

Comments regarding the predictive power of age, fatigue and flying patterns and relaxation raised above also apply to *depression*. However, by far the most important aspect was the fact that the large twelve-item trend concerning the home-to-work stressors (trend number 4 above) was a significant negative predictor of depression. This might seem like a strange result, but can be explained in the following way. The deduction in this instance is *not* that the negative predictive power of the trend indicates a reduction in depressive reaction. If we refer back to Table 6.2, you will note that the relationships of all of the items to the trend ('loadings') were negative. For complex statistical reasons, this is why we ended up with a negative predictive statistic here. The conclusion is, therefore, that the effects of domestic stressors on the pilot at work *are* confirmed as sources of mental ill-health, particularly in terms of depression.

Hysteria is a complicated and difficult facet of mental ill-health. Within the context of this study, the term is used to refer to those symptoms of mental ill-health which are emotional, a separation of thoughts and actions, and also a feeling of experiencing a highly emotional state. Such individuals are portrayed as displaying susceptibility to suggestions and a general (emotional) overdependence on other people. One way of combatting this overdependence is to exercise control over them, that is, to welcome responsibility and a decision-making role. The result confirmed that pilots may be able to reduce these reactions by such mechanisms.

Another characteristic of hysteria is that those exhibiting such symptoms tend to be unsteady and unreliable in their personal relationships. Consequently, this was confirmed by the significance of the trend concerning interpersonal problems. This and the other comments concerning hysteria should be interpreted in the light of the fact that such symptoms characteristically decrease with age. This too was confirmed by our results.

It might be apparent at this stage that there is a number of recurrent predictors of the mental ill-health subscales. Since *overall mental ill-health* is a composite of the subscales, it was consequently predicted by these recurrent issues. To conclude, those most at risk are likely to be those pilots who are older, who are experiencing tiredness and fatigue as a result of different patterns of flying, who do not have enough opportunities to relax and wind down, who do not perceive enough opportunities to assume responsibility to make decisions, and finally, who do not receive enough social support from others.

Job satisfaction

The measure of job satisfaction we employed in the study yields a range of seven subscales. After applying the predictive analyses, it may be seen from Table 6.4 that there was a fairly clear picture of those stressors that significantly determined levels of job satisfaction.

To illustrate this we will discuss 'intrinsic' and 'extrinsic' satisfaction only. We define 'intrinsic' job satisfaction as the sort of gratification that accrues from accomplishment, from the expression of abilities, from exercising decisions and so on.

The frustration of thwarted career aspirations and unfulfilled ambitions are confirmed as reducing the degree of intrinsic satisfaction pilots derive from the work itself. However it would appear that not all pilots indicate this. Those of higher ranks report a higher level of intrinsic job satisfaction. Clearly those two predictors make logical sense, since those of higher rank have clearly fulfilled some of their career goals. We previously indicated that such issues were a function of managerial and organisational variables. The results here, as well, indicate that intrinsic job satisfaction is an inverse function of managerial style, perceived psychological climate and morale. This is an interesting result, since it has often been stated that pilots derive intrinsic satisfaction by actually flying. This, no doubt, is true. However, we can now add that the *context* in which this occurs will greatly limit the intrinsic satisfaction derived.

Probably one of the most interesting aspects to be revealed about 'intrinsic job satisfaction' is the fact that it was positively predicted by the anxiety of courses and checks. Previous data (Table 6.2 and Table 6.3) confirmed that this contributed significantly to negative mental health. However, the result here indicates that attending courses and completing medical and technical checks (one assumes successfully) are tremendous sources of satisfaction to the pilot. No doubt, too, feelings of achievement are also experienced.

Table 6.4 *Predictors of job satisfaction*

	Multiple R	R Square	Overall F	P
Total job satisfaction				
–Trend 5: Career and achievement	0.57	0.33		
–Trend 9: Management and organisational issues	0.63	0.39		
Trend 7: Responsibility and decision making	0.64	0.41		
–Trend 10: Domestic status	0.65	0.42	13.78	<0.001
Intrinsic job satisfaction				
–Trend 5: Career and achievement	0.59	0.35		
–Trend 9: Management and organisational issues	0.60	0.36		
Rank	0.62	0.39		
*–Trend 17: Social support	0.64	0.41		
Trend 3: Anxiety of courses and checks	0.65	0.42	12.66	<0.000
Extrinsic job satisfaction				
–Trend 9: Management and organisational issues	0.53	0.28		
–Trend 5: Career and achievement	0.56	0.32		
Trend 7: Responsibility and decision making	0.57	0.33	9.31	<0.000
Job itself intrinsic satisfaction				
–Trend 5: Career and achievement	0.50	0.25		
*–Trend 17: Social support	0.52	0.27		
Rank	0.54	0.30		
–Trend 2: Schedules and rostering	0.56	0.32		
Trend 3: Anxiety of courses and checks	0.58	0.34	9.62	<0.000

Table 6.4—*Contd.*

	Multiple R	R Square	Overall F	P
Working conditions extrinsic satisfaction				
–Trend 9: Management and organisational issues	0.40	0.16		
–Trend 11: Fatigue and flying patterns	0.43	0.19		
–Trend 5: Career and achievement	0.45	0.20		
–Trend 10: Domestic status	0.46	0.21	5.08	<0.000
Employee relations satisfaction				
–Trend 5: Career and achievement	0.57	0.33		
–Trend 9: Management and organisational issues	0.64	0.41		
–Trend 7: Responsibility of decision making	0.66	0.44		
–Trend 10: Domestic status	0.67	0.45	14.18	<0.000
Overall job satisfaction				
–Trend 5: Career and achievement	0.43	0.19		
Trend 3: Anxiety of courses and checks	0.46	0.22		
–Trend 9: Management and organisational issues	0.50	0.25		
Trend 10: Domestic status	0.51	0.26		
**Trend 18: Wife's involvement	0.52	0.27	6.41	<0.000

* See chapter 9
** See chapter 8
– Denotes negative predictor

Whilst we defined 'intrinsic satisfaction' as being a function of actually doing the job, 'extrinsic satisfaction' is derived from issues that are external to the job or in the context in which it occurs such as job security, working facilities and so on.

All of the issues that predicted intrinsic job satisfaction also predicted extrinsic job satisfaction. Whilst it is surprising that this should be the case (since extrinsic and intrinsic job satisfaction are conceptually different aspects of job satisfaction as a whole), it certainly makes our discussion simpler and we can conclude that the issues already mentioned dominate the job satisfaction perceptions of the pilots. To these, however, we can add that the challenges of assuming responsibility and taking decisions are identified by pilots as positively contributing to their levels of extrinsic job satisfaction.

The predictors mentioned above crop up in the other aspects of job satisfaction, so there is little need to repeat our comments. However, we can add to the above discussion several interesting points. For example, the 'intrinsic satisfaction' derived specifically from the job itself is decreased by the disruptive aspects of the different schedules and rosters. Similarly the component of job satisfaction derived from working conditions is additionally (and negatively) predicted by 'fatigue and flying patterns'. From this we can conclude that as fatigue and tiredness increase due to environmental factors such as hotel, etc., job satisfaction will decrease.

By way of conclusion, we can state that the perceived level of job satisfaction derived by pilots is largely a function of the extent to which they can achieve career goals and ambitions, management style and organisational climate and morale. These, plus the overall status of the pilots' domestic scene, serve to reduce job satisfaction. However, such negative issues may well be offset by the welcomed experience of assuming responsibility and taking decisions.

Performance

The predictors of performance were fairly clear (Table 6.5).
Pilots who are at risk are those who identify working patterns, schedules and rosters as a primary source of stress. They tend to be older pilots who are anxious about completing courses and technical and medical checks. Also important is the impact of 'not flying as many hours as they would have liked'. The examination of performance is an interesting issue. The measure we used was designed by us for the study to try to achieve some quantitative assessment of performance as perceived and reported by the pilot himself. The above comments regarding the prediction of performance are interesting in that they at least give us some idea as to which issues pilots feel affect their performance. However, it would be an advantage to know

Table 6.5 *Predictors of self-rated pilot performance*

	Multiple R	R Square	Overall F	P
Trend 11: Fatigue and flying patterns	0.32	0.10		
Trend 3: Anxiety of courses and checks	0.39	0.15		
Trend 6: Insufficient flying	0.42	0.18		
Age	0.43	0.19	4.99	<0.000

explicitly how pilots would respond if asked to rank a list of factors that all might potentially affect pilot performance, on the basis of their perceived impact on performance. We in fact did this. We presented pilots with a list of twelve items and asked them to rank their top six in order of the perceived impact on their performance. Table 6.6 contains the ranks.

Table 6.6 *Rank order of factors that affect pilot performance*

	Rank						
	0 (Unranked)	*1*	*2*	*3*	*4*	*5*	*6*
Weather conditions	185	48	45	42	41	42	39
Inability to separate home from work life	358	4	15	19	14	13	19
Overfamiliarity (with type, routes)	270	13	17	23	32	39	48
Fatigue	29	231	68	40	40	24	10
Mind that is 'full of other things'	249	22	31	23	33	49	35
Relative time of day one is asked to fly	109	25	104	75	53	40	36
Interpersonal relations with aircrew	222	28	20	40	36	45	51
Upset pre-flight routine (at home)	371	2	4	11	16	20	18
Carryover effects of home life events	331	2	5	17	34	25	28
Health	134	25	68	66	52	44	53
Poor pre-flight preparation	209	15	29	36	40	52	61
Things not under direct control of pilot	198	27	36	47	50	43	41

One of the characteristics of ranked data is that it lends itself to analysis in different kinds of ways. Table 6.7 summarises some of the findings in Table 6.6, but from an array of different perspectives.

As the reader can see, it is possible to make such data say almost anything! There are, however, some basic deductions that most of us would agree with.

The most highly related items seem to be those that are operationally orientated, i.e. items such as weather conditions, other flight

Table 6.7 *Factors that affect pilot performance ranked in different ways*

(1) Most frequently identified items ranked as 1
    (i)   Fatigue
    (ii) Weather conditions
    (iii) Things not directly under the pilot's control
    (iv) Interpersonal relations with aircrew

(2) Most frequently identified items ranked as 2
    (i)   Relative time of day one is asked to fly
    (ii) Fatigue
         Health
    (iii) Weather conditions

(3) Least frequently identified items ranked as 1
    (i)   Carryover effects of home life
    (ii) Upset pre-flight routine
    (iii) Inability to separate home and work life

(4) Least frequently identified items that were ranked (i.e. items that were identified but ranked 6)
    (i)   Fatigue
    (ii) Upset pre-flight routine
    (iii) Inability to separate home and work life

(5) Most frequently identified items left unranked (0)
    (i)   Upset pre-flight routine
    (ii) Inability to separate home and work life
    (iii) Carryover effects of home life events
    (iv) Overfamiliarity (with routes, aircraft, etc.)

(6) Items most frequently ranked (irrespective of value), i.e. on the basis of number of mentions
    (i)   Fatigue
    (ii) Relative time of day one is asked to fly
    (iii) Health
    (iv) Things not under the direct control of the pilot
    (v) Weather conditions
    (vi) Poor pre-flight preparation

crew, or issues out of the pilot's control such as air traffic control and baggage handling, etc. By far most agreement centres around the pilot's perception of his current mental and physical state. Clearly, fatigue seems to be the key issue though we must remember that there is a tendency for pilots to use the term 'fatigue' to mean all sorts of issues, i.e. it is a 'buzz' word.

Interestingly, items of lesser importance tend to be those whose influence is indirect or modifying. Examples of such issues could include an upset routine at home before leaving to fly, an inability to separate home and work life or carryover effects of home life events.

The relative disregard for 'home-related' issues and those that do not directly have an impact on the pilot, when flying the aircraft, was not surprising. In many ways, this is a function of the nature of the task in our questionnaire. We asked pilots to rank items, hence all the positions are relative and not absolute positions. Second, the nature of 'performance' is itself difficult to tie down, so it is not surprising that pilots identify those issues that affect it most as being those that are probably the most obvious. Finally, if you give pilots this sort of task to do, they tend to either disregard or reduce their estimates of the impact of non-work issues, such as the effects of home life. In fact, we know from this and the preceding chapter and from the chapters that will follow that such issues *do* have an impact on pilot performance. However, in this sort of task, this effect is not registered. Additionally, of course, had the items been ranked in the context of another outcome variable, such as job satisfaction for example, then different results would have been found.

Some conclusions

In this chapter we have identified trends of pilot stress and related these to outcomes. The purpose of this concluding section however is not just to present a summary of these findings. It would be an advantage to do two things: first, to try to identify specific issues (i.e. particular items as opposed to trends) as predictors of outcomes, and second, to try to cross reference such items and outcomes to the material presented in earlier chapters, such as biographical variables.

Of the three outcome variables we have examined in this study, it is probably mental ill-health which is the outcome we would identify as having critical implications for the study as a whole. This is because conditions of mental ill-health pervade all aspects of the individual's 'psychology' and colour perceptions of the world and everyday life. In the above discussion we have examined those issues that predict mental ill-health, job satisfaction and job performance. The data we drew upon was that which concentrated upon establishing predictive relationships of the outcomes to the major *trends* previously identified in the earlier part of this chapter, plus a selection of biographical data. By using still further analyses, it was possible to examine the particular relationships between *specific* individual issues or items and the outcome variables. The objective of performing such an analysis is to reaffirm the points raised earlier. The statistical results are highly detailed. Whilst being of interest, they are too lengthy to present either here or in an appendix. We have therefore omitted them and report only the conclusions drawn from them. It will be seen that there are some interesting perspectives and it would seem appropriate to mention these further.

In terms of mental ill-health, it seems as though anxiety and overall mental ill-health are the most readily affected by the stresses that pilots perceive. It was for these two outcomes that the most significant associations were recorded. For mental ill-health (as compared to the other outcomes examined), the significant associations tended to be with domestic issues, confirming that it is generally home life that contributes most to neuroticism or lack of mental well-being. Although 'attaining one's own level of performance' and 'the build-up of things to do when flying' pervade much of the pilots' mental ill-health profile, in terms of anxiety, it is 'home life' in general and how compatible it is with the pilots' working life that seems to be critical. Practical issues such as the way in which home life is geared to flying are relatively major and important. Issues such as things that are left unresolved or minor domestic upsets also contribute to anxiety experienced by pilots.

Overall, however, there seems to be a close relationship between general mental ill-health and difficulties experienced in actually identifying the precise origin and nature of the source of concern. In other words, it is those sorts of worries that are not easily defined that tend to concern the pilot most. One plausible reason for this is that pilots tend to be able to cope with such issues, if they can somehow see a way out (which, of course, assumes that it is possible to identify the problem in the first place). Given that this is the case, a compatible goal as perceived by the pilots is the desire to try to fulfil one's own expectations and standards. This seems to be particularly true in terms of resolving problems to achieve a satisfactory solution.

There are some other trends that are also apparent in the mental health data. These are largely a function of the demographic characteristics. In terms of age, for example, previous trends have already been recorded and are well established in the psychiatric literature. The characteristics observed in the array of mental ill-health scales reflect those previously identified conventions. Generally, however, we found that the greatest differences lay in the comparison of the early career groups, the 21–30-year-olds, and the mid-life 41–50-year-olds. This was especially true in terms of somatic disorders and depression.

Depression is an interesting outcome which, in fact, cropped up regularly. The importance of home life has been noted here and will feature greatly in following chapters. One would have expected, therefore, those who have 'disrupted home lives' to exhibit elevated levels of mental ill-health. Consequently, divorced pilots were significantly more depressed than married pilots. Worst off were those who cohabited, for they were even more depressed than those who were divorced.

Of critical importance is how the pilot uses his leisure time. Three aspects are relevant. First, the pilot must take the opportunity to deliberately relax and wind down. If he does not, his level of anxiety and overall mental health may well be adversely affected. Similar results may ensue in the situation where the pilot does not have some sort of hobby or outside interest. In chapter 9, we will emphasise the third aspect which is important. This is the role of *social support from others* in helping to cope with any stress experienced. Mixing socially with colleagues seems to be a critical issue. The implications of not doing so seem to pervade all but a few aspects of mental ill-health.

Finally, there is an array of issues that define the pilot's working status. Elevated levels of obsessionality, depression and overall neuroticism seem more apparent in *long haul* as opposed to *short haul* pilots. A similar picture is portrayed in terms of rank, with captains exhibiting significantly higher levels of obsessionality, depression and somatic disorders. Consequently, similar trends were found in groups as defined by seniority. It must be remembered, of course, that an underlying factor will be age, as previously mentioned.

Recurrent trends characterise the underlying influences of job satisfaction. Strongest associations seem to be with macro rather than specific micro aspects of the job. Such large-scale issues may well underlie or moderate the perception of job satisfaction by the pilot as a whole. Interestingly, many of the issues that pilots perceive to be related to job satisfaction are not those that are exclusively domestic or occupational in their origin and effect. The job satisfaction experienced by the pilot would seem to be greatly determined by those issues that transcend the home/work interface. Dividing satisfaction into intrinsic and extrinsic, the former intrinsic satisfaction is largely a function of *career* issues, such as uncertainty or the absence of opportunities for potential advancement. On the other hand, extrinsic satisfaction was largely seen in terms of *organisation climate*, morale and so on.

The job satisfaction 'turning point' seems to be late thirties or early forties, since significant differences were apparent between the 31–40 and 41–50 age groups. Although the intrinsic satisfaction derived from the job, those who relaxed, for example, who had a hobby and who were satisfied with their home lives, also derived satisfaction from their work. In both instances, it is the extrinsic qualities that seem to be influenced.

Given the above comments, one might expect status defined by either rank or overall perceived seniority to be a fairly important determining feature of job satisfaction. Although there were some discrepancies in some sample sizes, captains clearly derive greater levels of job satisfaction from their jobs than do either first officers or senior first officers, particularly in terms of intrinsic satisfaction. Seniority generally was widely (and positively) related to job satisfaction. Interestingly, however, most differences lay not in levels of satisfaction of pilots from either end of the continuum but in groups that fell into the middle categories, thus reinforcing perhaps the idea of some sort of mid-career turning point in perceived job satisfaction.

Finally, there are four additional points we would like to raise about job performance. We previously identified that trying to attain one's own standard of performance greatly influenced subsequent *reported* performance (though to a certain extent, given the nature of the questionnaire task we asked pilots to perform, this is to be expected). Once again, the issues concerning the degree to which the pilot relaxes is a significant issue in the perception of performance. Other issues, however, have an impact on how well or badly the pilot rates his own performance. For example, senior first officers tend to rate their performance as being higher than do captains; long haul pilots do the same compared to short haul pilots.

Chapter 7
Life events

Life events, sometimes referred to as 'life changes', are simply incidents that occur in an individual's life, for example, marriage, divorce, birth of children, death of a relative, moving house and so on. Although these are generally identified by most of us as being important or significant, they need not be extraordinary events. The main issue is whether or not the person perceives them as important. Given our previous examination of the nature of stress, life events are perhaps the most 'obvious' life stressors. We all can identify times when we feel 'under pressure'. As we have said, this is a too simplistic conceptualisation of the effects of stress; however, we all can identify with the experience of some 'trauma' or other. Quite often it is the experience of such incidents that account for fluctuations in our health. This is especially true for mental health, as most events involve some sort of immediate effect followed by a period of recovery and adjustment.

Background research into 'life events'

It was observed earlier that the home/work relationship is an important facet in the perception of stress. Extending this, it will be fairly clear that 'life events' will play an integral part in the home/work/home feedback loop. A life event was defined, and in many ways this definition of an 'event' was unsatisfactory, since it is treated as being anything 'significant' that the person perceived to occur in his life or in which the individual is involved. Although it is tempting to think of these as negative 'events', positive events, however, can have adverse consequences.

Various attempts have been made (from as far back as the 1930s) to try to establish a link between the occurrence of life events and stress outcomes (Goldberg and Comstock, 1976). Most studies have employed some form of test or questionnaire, usually based upon the scale designed originally in the 1960s by Holmes and Rahe (Holmes and Rahe, 1967). Table 7.1 contains some of the items from this questionnaire. In general, events have some common characteristics:

- some events are indicative of the lifestyle of the person
- some events are indicative of occurrences involving the person
- they extend from the 'ordinary' to the 'extraordinary'
- some are negative (or are 'stressful' in the conventional sense), others are positive
- interpersonal events directly related to emotional life are especially important such as marital problems or frustration in love relations

It might be apparent by now that there are problems with this type of research. These are important and worth highlighting for they are the reasons why we undertook the approach we did in our study of pilots. For example, it is difficult to define the 'importance' of events, it is not clear how the impact of events on an individual is affected by time and, indeed, whilst different types of events have been found to be important, different types of *outcomes* have also been recorded.

Apart from these, there are other major points worth noting regarding the nature and extent of events experienced by individuals, and also certain methodological problems. Perhaps the most important factor is the interpretation of the event by the individual. For example, if one examines Table 7.1, it could be superficially concluded that these are mostly negative events. It is possible, however, that such events may be construed in a positive light – divorce after years of arguments is seen as relief or death after a long and painful

Table 7.1   *Selected life events from previous research*[1]

	Rank
Death of spouse	1
Divorce	2
Marital separation	3
Jail term	4
Death of close family member	5
Personal injury or illness	6
Marriage	7
Fired at work	8
Marital reconciliation	9
Retirement	10

[1] Selected from Holmes and Rahe, 1967.

illness is seen as a merciful release. The way the event is interpreted depends upon the idiosyncratic background or context of factors the individual finds him/herself in, and in which the event occurs. When one considers the endless catalogue of events that individuals might regard as 'significant', it will be readily apparent that the process of relating events to stress outcomes is not as easy as it would first appear.

To conclude, there are three messages from the background stress research into life events. The first is that although this research is fraught with problems, it would appear as though important events that occur in one's life do have significant effects on mental and physical health. Second, it is virtually impossible to establish the relative importance of life events to 'events' that occur elsewhere (such as at work). Third, more recent evidence has suggested that when attempting to study life events it is difficult to improve upon a simple summation of the number of events (as opposed to a more rigorous questionnaire method) (Rahe, 1978).

Life events in aviation research

Several points have already been noted in chapter 2 about the background to this research. One of these is that although there has not been much research performed in the past into the effects of life events on pilots, the research that has been completed has found that life events (on balance) are influential in accident and incident causation. The relationships between life events and subsequent pilot ill-health has also been demonstrated; however, the extent of the relationship is not quite so convincing. Despite the well-documented comments reiterated here, the examination of life events (also referred to as life 'changes') has persisted within aviation research. Over more recent years the term 'domestic stress' within aerospace psychology has virtually become synonymous with the life events/life changes data. Clearly the preceeding chapters in the present study illustrate this to be a gross underestimation of the impact of domestic and non-occupational factors on pilots.

What we did and why

We previously mentioned that the work of Alkov and his associates (Alkov, Borowsky and Gaynor, 1982) has attracted much attention. What Alkov has tried to do is to explore the relationship between accident 'proneness' and precipitating life events as a function of the pilot not being able to cope adequately or appropriately with the event. He devised a twenty-two-item questionnaire which is a mixture of events and personality characteristics. He then asked colleagues of pilots (usually flight surgeons) to rate a specific pilot on the scale. All of the pilots had been involved in some incident or other. However, Alkov knew beforehand which pilots had been 'at fault'. Therefore, he was able to analyse the results by comparing the profile of the 'at fault' and 'not at fault' groups.

There are, however, some problems with this approach. From this methodology one must assume that respondents perceive his range of items (twenty-two in all) as a complex web of inter-relationships. The further implication of this is that in the analysis of answers, one must assume a similarly complicated mental process on the part of the pilots (i.e. this is what Alkov identifies as the relationship between personality and stress coping as a function of 'acting out aggression' (Alkov, Borowsky and Gaynor, 1982)). Second, the method of analysis employed seemed rather too simple. One would have thought that in

Table 7.2   *Life events found to be significant in previous pilot research*[1]

Factor	Statistical significance
Aircrew members who were at fault in an aircraft accident were more likely than those not at fault to:	
Have marital problems	0.0202 *
Show signs of immaturity and instability	0.0324 *
Have recently become engaged or married	0.0411 *
Be making a major career decision (such as getting out of the service)	0.0017 **
Not be professional in their approach to flying	0.0007 **
Be having difficulty with interpersonal relationships	0.0047 **
Have recently had trouble with superiors or received disciplinary action	0.0029 **
Not be able to quickly assess potentially troublesome situations	0.0047 **
Recently have had trouble with peers or others	0.0279 *

* $p < 0.05$
** $p < 0.01$

[1] *Source*: Alkov, Borowsky and Gaynor, 1982.

the type of relationships being examined, a large number of variables would be interacting simultaneously to determine the end results – yet the analysis used in this previous research does not illustrate this. Finally, an examination of the items that Alkov found to be significant (Table 7.2) is sufficient to illustrate that items are of only limited practical utility. Second, we do not really know what is happening with those items which were not significant.

There are several important messages from our examination of life events research both inside and outside the context of aviation:

(a) the examination of life events in general is fraught with difficulties and hence a simple approach is recommended, but;
(b) the examination of life events in aviation has adopted the more complicated questionnaire-type approach which, if modified and expanded, looks potentially fruitful.

Hence, to conclude, our information is based upon two sources. The first is the simple analysis of whether or not pilots had recently experienced a life event, plus an analysis of the nature and impact of that event. The second is based on the expanded form of Alkov's twenty-two-item questionnaire approach. By their nature, the former will be primarily a qualitative analysis and include some direct interview material from pilots, whilst the latter will be primarily of a statistical nature.

Life events using a straightforward approach

On the basis of a relatively simple and straightforward approach to looking at life events, we divided our analysis into two parts. The first summarises how pilots feel generally about life events and their potential impact. The effects upon performance were felt to be more relevant given the fact that previous researchers have already attempted to look at the effects on other variables (such as ill-health).

Consistent with expectations, but contrary to the stereotype, pilots do feel that their performance can be affected by events occurring in their lives. The following statements are representative of the 96 per cent who feel this (Table 7.3).

'I can sum up my experience of life events in three ways. The first is the death of my first wife several years ago. I didn't go to work for several weeks. I then went back. I don't know how the effects

Table 7.3 *Pilots' reactions to life events generally*

	Yes Number	Yes Percentage	No Number	No Percentage
I think that performance could be affected by life events but not to the extent that is perceived by the pilot.	260	58.8%	180	40.7%
I think that performance could be affected by life events but not to the extent that is perceived by others (e.g. colleagues).	249	56.3%	191	43.2%
I think that performance could be affected by life events but not to the extent that safety and proper flight conduct are affected.	210	47.5%	230	52.0%
I think that performance could be affected by life events but not to the extent that minimum operational standards are not met.	218	49.3%	221	50.9%
Do you think generally that life events can affect pilot performance?	424	95.9%	18	4.1%
Do you think that negative events can affect pilot performance?	411	93.0%	31	7.0%
Do you think that positive events can affect pilot performance?	383	86.7%	59	13.3%
I think that performance could be affected by life events but not to an extent that is measurable.	244	55.2%	198	44.8%

crept into work. The company were marvellous. In some ways I suppose I found going to work a form of relief. The worst period was after she died. Before there was lots going on. Afterwards I went to pieces for a few weeks. I was struggling to do simple tasks like departure routines. I was given an office job which was a shock to the system and things gradually became resolved. My second experience of an event occurred before this. I had an affair with another woman. Worst was living a double life. Life was hectic, quite apart from lack of sleep! Thinking back there were times when my performance was lacking. I was in an unsettled state, not knowing where I was. Then the wife of a colleague told on me, that upset me immensely!! My first wife was super and didn't cause an upheaval. The affair finished and that was that. But the event was definitely reflected in my performance. My third experience was marriage. This was a positive event. I didn't enjoy my period of widowhood, I'm not sure how this had an effect on me but it is definitely a period that stands out.'

'I'm experiencing an event at the moment involving a major decision. I've got an interview with an overseas airline at the end of the month – surprising! The major point is that I like my home environment, Britain, and prefer short haul. But I'm going nowhere with my present company. In my previous marriage I had a poor home life and insecure job. It's better in the sense I'm happy but . . .! If I get past the interview, it's off to the Far East for formal interview, etc., but my wife is pregnant. You can see I have a major decision on my mind at the moment involving potentially massive changes. I'm very conscious of this even at work when flying. I hope, though, that they give me an offer so that the decision is mine and not theirs!'

The above two quotations provide interesting illustrations of the sort of material we are examining under the heading 'life events'. In the first situation, the pilot experienced what reads like a litany of events and experiences. There is a mixture of positive and negative events, producing varied and even ambiguous effects. This is a good quotation to start with, since it illustrates many of the qualities that we will now go on to mention.

The second quote above is interesting because the pilot was actually experiencing an event at the time of the interview, although the event was in fact 'work-related'. Two important features stand out. The first is that whilst the background context of the event is primarily negative (i.e. job dissatisfaction and job insecurity, etc.), the event itself is positive (i.e. an interview for a new and presumably better job). The second point is that an underlying factor important in the perception of events is the degree to which the pilot has *control* over the situation. We shall develop this theme later. Consider for the moment, however, an event which affects more than one individual. In this following situation, it is possible to illustrate that different people do seem to react in different ways.

'I suppose being hijacked in X several years ago could be called a
life event even though it was work-related! Many of the hours
spent at X airport were spent with a pistol pointed to my head. I
was captain. The co-pilot and engineer were both deeply affected.
In fact the engineer was never the same again. Within ten days I
was flying again. I had some disrupted sleep for about a week or so
but that's all, I'm not an introvert, I didn't analyse it.'

This event was work-related and provides an interesting comparison
in reactions between the captain, co-pilot and engineer. It is useful to
note the contrast between the pilot, who was back flying again after a
short period, while the engineer 'was never the same again'. The
situation described was, of course, an extraordinary event. However,
it does seem to illustrate that these things do occur and that individuals' reactions to them can be unpredictable.

Previously presented was a selection of the sort of events individuals have identified in other studies as being important (Table
7.1). It was noted that any might be construed in a positive or negative
way. Within the context of pilot performance, consider the following
contrasting accounts:

'Two positive events stand out that occurred simultaneously. At
the same time as getting a command I was elected a town
councillor. Being elected a town councillor had a beneficial effect
on the way I look at things, taught me a lot about handling people.
Again, the command – superb fun and a quantum leap mentally
and physically. Positive events make me more relaxed and sure of
myself at work.'

'Being "stood down" was the biggest event we had. It involved a
lot of arranging. I was under my wife's feet and so on. For the first
time I realised she needed time to herself. I got a job after five
months, but it was quite a shocking experience.'

There are two issues about which one can make inferences. The
first, as previously mentioned, is the background reason *why* such
events are interpreted by the individual in the first place as being
positive or negative. The second concerns the nature of the effect on
performance. It will be noted that we simply measured the 'effects on
performance.' We can make no inference about whether these events
positively or negatively affect performance. This is because both
positive and negative events may *both* either boost or hamper optimal

performance. Whether it does either is again 'context-specific', and outside the scope of this study. Whilst clearly the latter is of greater interest (i.e. a negative effect), really *any* effect is worth noting irrespective of direction.

The data collected would also tend to indicate that the experience of an event is a very *personal* one. For example,

'Events I would identify in chronological order would be births of children and death of my father. In terms of the first birth I had to fly the day before and the day after the birth. In the second birth the company were more flexible. In the death of my father, I wasn't away and also my brother lived close by. The effects were mixed. Not worry but tension and shock, I felt annoyance in the first instance, the second was OK and the third I handled fairly well – I think. One issue is the unpredictability of events and the chance you might be away. There is also the fact that often not only must you cope with your own emotions, but also the emotions and reactions of others. That can be lingering and ongoing. It's a very individualistic thing. It's tangible in one sense, something is happening but it's a very personal experience and difficult to describe.'

There would appear to be four lessons to be learned from this example. The first is that an individual's reactions to a life event can be a very personal and private experience, one that is difficult for the individual to pin down and even more difficult to articulate to a third party. The second point, which must be a function of the first, is that feelings are mixed. Whilst one or several reactions might predominate, the individuals will experience a range of outcomes. An interesting slant is that a key feature in the eventual outcome reaction and effect is the degree to which the pilot can *predict* the event that will occur. This is compatible with previous research, which indicates that pilots as individuals tend to thrive under conditions which involve a high amount of *control*. In fact, it is often said that much of the job of a pilot is keeping 'ahead of the game'. Not only are the reactions mixed, but they appear to be indeterminate in their duration. They tend to linger, which, no doubt, makes them all the more difficult to cope with. An additional factor that would appear to make coping with life events more difficult is the fact that the feelings of others have to be managed also.

Previous research into pilot personalities indicates that many pilots, although to the outside world they appear to be calm and confident, in actuality can be as emotional as anyone else. When one asks pilots about the privacy of experience that surrounds such events, opinion seems fairly evenly divided over whether there is an appearance of the effects of events at work, or whether events could *potentially* affect performance. The extent to which such effects were obvious to others was not thought to be so great as to be great enough to be spotted by the pilot's colleagues (56.3 per cent) or, indeed, to extend to a degree that was measurable (55.2 per cent). Of even greater interest is the fact that just under 60 per cent of pilots felt that whilst life events might well affect them, it might not do so to such an extent that they *themselves* will perceive the effect. In a slightly different context, consider the following:

> 'Death of father five years ago. It was a first bereavement and quite unexpected. I was very emotionally involved and often quite close to tears and felt depressed. The company were good – I got two days off! I'm sure I was below par at work, not concentrating wholly on my job. At the time I *was* aware of the effects on me. With hindsight it was the right decision to go back, but if anything had happened I would have been in trouble for going back. In contrast a positive event was getting a command totally unexpectedly. I remember the transition was pleasant but there simply wasn't time to monitor the pleasant effects of the event on me.'

The comparison between the two incidents this pilot experienced is interesting. In the latter, the pilot experienced a positive event, and already being able to identify his feelings as being positive or pleasant, but is unable really to be any more specific. One would have thought that the experience of the former, a negative event, would have evoked a different response, particularly in the sense that we tend to pay attention to things to a greater extent when we are experiencing something negative and tend to be comparatively complacent and pay not so much attention when we are feeling positive and satisfied. Indeed, this does seem to be the case. Here we have a pilot who clearly identified a negative experience and who is able to pin down the effects. We must remember, however, that these comments are reported with the experience of hindsight. Consider the following account which combines both hindsight and ambiguous effects.

'For at least two or three years prior to the incident, there had been a steady deterioration in the state of my marriage to the extent that I would get up in the morning, unnecessarily early, to get out of the house before my wife and child woke up. On this particular morning this did not occur and I was subjected to a non-violent but angry argument which left me emotionally boiling, a state I remained in throughout my drive to the airport, through the flight planning and indeed up to the incident itself. It was only when the cause of the conflict was removed, that I realised what a strain I had been under and how it made me entirely oblivious to what was going on. If anyone had suggested that I needed help I should have said it was completely unnecessary.'

Chirp no. 1, 1983

So there is some evidence that the effects of events are not always so tangible as they might be. Clearly it is only after the event, when the major impacts have started to recede, that the pilot can sit back and coolly analyse how he coped with the situation, an experience that many of us can identify with, if only to a more limited extent.

Still further, one final conclusion that must be drawn from this data is the insidious nature of such events and how they affect the individual. This has an important connection with 'readiness to fly'. Most companies operate a system whereby if the pilot deems himself unfit to fly, he can withdraw from a flight and (pending the result of an enquiry) later return to work. However, since the majority of pilots (and indeed others) may not be *aware* of stress (or its manifestations), clearly the process of deciding upon fitness for flying is not as accurate or reliable as it might be.

Pilots must perform to exacting standards on each and every occasion. Hence safety, 'proper flight conduct' and adherence to operational standards are key issues. There was a clear split between those who felt that any effects of life events would be so extensive as to impinge on to such key aspects of performance and those who did not feel this. The following is not atypical.

'Before our divorce, gradually disagreements got bigger and bigger and we started doing things separately. Such things can affect you when you're flying but only when you have time to think about them, i.e. during the cruise period – it's possible to miss something that might have occurred. We divorced after seven months of separation. But I'm delighted and very happy, so is my ex-wife. One of our sons has gone to live in America, the other lives with my ex-wife. She has accepted the situation. It's a positive event [the divorce]. I'm not *aware* of any change in my flying though I do feel more relaxed. I can put more attention into flying – or at least I *think* I can!'

Here is an example of a bad situation turned good. The interesting point to note is that the pilot has identified stress manifestations in both a negative and positive way. However, there appear to be definite limits or boundaries to these effects. They appear only to be active when the pilot is in the period of flight, when little (or relatively little) is occurring. For this we can infer that although effects are perceived, they're not so extensive as to threaten safety. Other pilots also reflect this opinion.

'When my wife was ill after our second child, that was recurring and in the back of my mind. It was so serious it broke the barrier between home and work. If you haven't got 100 per cent concentration on the job it's bound to affect performance. Whether it does so to an extent that is measurable I don't know. At work I do think about home events, but I don't worry about them . . . or at least I think that's the case!!'

Again the impression is that whilst stresses and strains are recorded, they appear to be limited. There are, however, some interesting issues. For example, both of these quotations finish with an after-thought: 'I *think* I can' and 'I *think* that is the case.' Given our previous comments regarding ambiguity and the fact that a pilot may not always be aware of what's going on, clearly one must take account of this in drawing conclusions about flight safety.

Just over half (52 per cent) felt that safety and proper flight conduct might well be affected, and that minimum operational standards might not be met (50.7 per cent). One can draw several conclusions from this data. We now know that the issue mentioned previously regarding magnitude of effects of life events is complicated still further by the fact that the pilots feel that it might not be assessed in quantifiable terms, i.e. there is an effect but it is difficult to measure. The latter information has told us that flight standards might well be affected by a pilot who has experienced an event. Since opinion was split, again, one must assume that the effects of life events are relatively difficult to quantify and even more difficult to analyse in terms of attributing causation.

All of this information is relevant, thought-provoking and interesting; however, its generality only lends itself to broad analyses. We therefore integrated into this section some questions not unlike those asked in the data presented above. Pilots were asked, however, to answer the same questions (i.e. about positive/negative effects, safety and standards) only with reference to particular events that the pilots *themselves* may have experienced (Table 7.4).

A large majority of pilots said that they could identify some sort of 'event' which had an effect on their lives. When one examines the nature of events identified by this majority (78 per cent), no trends are revealed which portray this population of pilots as being significantly different from other occupations. For example, the sort of events identified most often were moving house, changing job, birth of

Table 7.4 *Pilots' reactions to life events experienced by them*

	Yes		No	
	Number	Percentage	Number	Percentage
If you consider your life generally, would you say that you have experienced any event which has been important to you?	344	77.8%	97	21.9%
If yes,				
How long ago was it (months)?	Average 137.15			
How long did the period of the event last (months)?	22.54			
Do you think your performance was affected?	151	43.9%	165	48.0%
Do you think your performance could have been affected without your realising?	175	50.9%	143	41.6%
Do you think your performance could have been affected without others realising?	145	42.2%	169	49.1%

children and so on. Such events are not especially unusual. However, it is just this very quality that is important because it illustrates the fact that in many ways pilots are not an especially 'different' group. Pilots tended to identify events that had occurred a long time ago – an average of eleven years. Additionally, pilots were able to trace the effects of the events for two years after its occurrence,

'I was married before, divorce proceedings were not "traumatic" but it was certainly major, very unsettling and lasted one and a half years. Unsettling not so much from the flying point of view although there were times when it did intrude. I was the one who left, she was a survivor – I had to re-examine my thoughts and attitudes. During this period there were several times when it intruded. For example, I made an approach into Manchester once, which was too high and too fast. It was a time when I was experiencing particular problems at home. No specific point crossed my mind, it was just my general condition. It wasn't dangerous, just embarrassing. It occurred fairly near the middle period which was the worst stage.'

'About six years ago I came home from a trip and found my wife with her bags packed and literally on her way out of the house. It was very traumatic and upset me for about two years. Strangely the emotional effects were at their strongest after nine to twelve months when it all started to sink in. I felt a sense of loss, rejection, emptiness. I lost interest in my work. I don't think that my experience would have been any different had I not been a pilot. The event would have occurred anyway only perhaps in a different way. Having said that, being away so much didn't help our relationship.'

This pair of quotations illustrates several important features. The first pilot experienced a long period of distress characterised by introspection. It was not the fact that any specific thought crossed his mind, but rather a 'general condition'. In the second situation, the pilot experienced the 'after effects' of the event for a long period, but there appeared to be a time lag between the event and its consequences. Interestingly, in this latter situation, the pilot indicates that he and his wife would have split up even if he was not a pilot. He recognises that it may, however, have happened in a different way (though didn't identify which way). Neither does the pilot indicate how the *effects* might be different.

It should be noted, however, that there were wide variations in scores (Table 7.4), as one might expect. It is also apparent that although the effects of events were prolonged over a long period, the pilots concerned felt that their *performance* was not impaired. It must also be noted that their opinion was split, with 44 per cent indicating that their performance *was* adversely affected. Additionally, the majority of pilots felt that there might well have been an 'effect' without them realising. It is fairly clear, therefore, that life events do have some effect on pilots for a prolonged period and that such effects may have an impact on the pilots' performance. This influence, however, would probably be spotted by the colleagues, but not always.

Life events using a questionnaire approach

The reader will recall that some previous aero psychologists (Alkov and his associates) had conducted research into life events and life changes experienced by pilots, using a more complicated questionnaire-type approach. It was this method that was used to produce the significant items in Table 7.2. We modified the questionnaire slightly and, by applying the complex statistical procedures employed extensively throughout this research, uncovered three underlying trends in the twenty-two items. These are summarised in Table 7.5. We named the trends 'emotional losses', 'pilot characteristics' and 'emotional gains'. Pilots perceive the twenty-two items as simply comprising of three trends, and not a complex matrix of relationships.

From an investigative point of view, this was a 'good' result. The solution appeared to be relatively internally consistent in its meaning, since the items loaded on to a specific trend were not loaded on to any other trend. Additionally, throughout the whole solution, only one item, 'financial difficulties' (loaded on to Trend 1), provided any ambiguity. The solution can also be described as being extremely robust in that we were able to account for a high percentage of the variation in pilot scores that was due to the effects of events and pilot characteristics.

Of the twenty-two items originally presented, seventeen were significantly related to our three underlying trends. From the five remaining items, four loaded on to a fourth non-significant trend.

Table 7.5   *Underlying trends in pilots' life event data*

	Loadings
Trend 1 (58.30) Emotional losses	
Recently undergone a marital separation (for reasons other than duty location)	0.87
Recently undergone divorce	0.82
Have marital problems	0.80
Recently had a death in the family	0.68
Recently lost a close friend through death	0.67
Have financial difficulties	0.54
Trend 2 (26.50%) Pilot characteristics	
Exhibit professionalism in his approach to flying	0.82
Impress others as a good team member	0.79
Exhibit the characteristics of maturity and stability	0.77
Exhibit the ability to quickly assess potentially troublesome situations	0.77
Impress others as a good team leader	0.76
Exhibit mastery of his aircraft within operational parameters	0.73
Handle life difficulties well	0.65
Have a sense of humour and humility concerning himself	0.55
Trend 3 (9.90%) Emotional gains	
Recently got married	0.81
Recently became engaged	0.76
Recently have a new addition to the family	0.64
Trend 4 Not significant	
Make any recent major decisions regarding the future	
Have difficulty with interpersonal relationships	
Recently had trouble with superiors	
Recently had trouble with peers	

These items concerned 'making important decisions about the future', 'difficulty with interpersonal relationships', 'trouble with peers' and 'trouble with superiors'. It is interesting to contrast this finding with the fact that all four items had been previously identified by Alkov as being important (Table 7.2)!

Some further comments regarding the relationships of items to trends are pertinent. No judgment was made, when we presented the items to the pilots, about whether or not an item appeared to have a positive or negative connotation. As we have said, to some extent this was to be expected since the direction in which an item is interpreted depends largely on the background circumstances of the individual.

Another point that must be mentioned is that, unlike Alkov, we wanted pilots to rate all items (rather than just those they thought applied). Hence, in doing so, an item that is not important is given a low score and the obverse for important items. The ultimate implication of this, from the point of view of our analysis, is that the reason an item is not significantly related to a trend is because that is the way the pilots perceive it, and not due simply to the item not being rated. Given this background, we can now proceed with a clearer examination of the contents of the three trends.

'Emotional losses'

Compared to the other two trends, those issues collectively labelled as involving 'losses' are by far the most important. All of the items are highly related to the trend. Interestingly, all of the top three items are related to marriage. Examination of their order reveals them to be about marital separation, divorce and problems. One might have hypothesised 'divorce, separation, problems', a sequence whose order is based on magnitude. But, one can only assume that the inversion of separation and divorce is based on the fact that a marital separation is an ongoing phenomenon, whereas divorce is a final endpoint. (Consequently one can only assume that problems were rated third because their *lack* of magnitude was more important than the fact that they are ongoing.)

The next two items are about *death*, which could lend itself to the conclusion that marital issues are of greater importance than issues related to friends and other family members, even though the context is one of loss of life. Of course, one must remember that traditionally the greatest life stress is death of spouse.

'I was breaking up with the woman I lived with for eight years. Also at the same time I was doing a command course for X. The course went absolutely fine. I just didn't carry stress over from home to the job. But perhaps things worked the other way around. I'm sure she had a harder time because I was on the command course than otherwise would have been the case, in the same circumstances of a break-up.'

This is an interesting illustration, since it involves coping with a situation made more complicated by the simultaneous experience of several events. What is interesting is that the statement is a fairly typical illustration of the attitude many pilots adopt and their under-lying personality characteristics. Certainly this is comparable with the sort of characteristics that were extracted in this second trend. Emotional losses are compatible with the 'common sense' view of an 'event'. This finding particularly in terms of marital issues is certainly compatible with previously recorded findings in the life events literature.

'Pilot characteristics'

These items concerned characteristics of the pilot. Although all are superficially positive in their content, the rating scale gives the opportunity for a negative quality to be reflected where appropriate.

There was little disparity in magnitude of the loadings, indicating that little discrimination was made between most of the character-istics. Dominant issues were qualities such as 'professionalism', although the concise contents of such qualities, unfortunately, are not immediately clear. It is often said in aviation that if a pilot is experiencing trouble he should be sufficiently well-trained, experi-enced and 'professional' to be able to cope. Team leadership and integration are also important, confirming Alkov's previous com-ments regarding the quality of interpersonal relationships. Least important is a pilot's self-concept (though clearly such issues are difficult for other people to assess).

Ironically, the fact that the individual typically handled life diffi-culties well was (although significant) not a relatively strongly loaded item. One can conclude that the prediction of life event/incident relationships by pilot characteristics is even more difficult, and other factors (such as professionalism) tend to be regarded as clearer indices by pilots than more complex individual characteristics (such as esteem).

'Emotional gains'

This underlying trend was comprised of three items. However, it is clear from its examination that issues involving marriage appear yet again to be most important. Whilst all involve a high degree of emotional content and experience, one would have hypothesised that an addition to the family would be the most highly rated event, particularly since such an event contains the important quality of long-term commitment. This was not found to be the case. The previous qualities mentioned earlier such as open-endedness and *relative* magnitude (i.e. relative to the situation preceding the event) appear to explain the sequence of events relating to the trend. For example, 'becoming engaged' or the 'birth of a child' are events that are preceded by a period of expectation. They are also fairly open-ended events. These qualities are also true of marriage; however, this has the additional quality of degree magnitude of the event. There are other factors that one must remember which might help explain why birth of child was not rated by pilots as being the most important event. One simple factor is that many pilots may not have experienced the event itself. Another more complex factor is that the emotional consequences of 'additions to one's family' may well be mediated or reduced by the effects of age and maturity.

Discussion of the 'questionnaire' approach

At the onset it must be stated that differences in results must clearly be a function of the sample and adaptations in methodology involved. We feel, however, that even if the effects of methodological changes could be controlled for, the end result would not be substantially changed. In other words, although Alkov's approach is interesting, the relatively simple approach previously reported would seem just as worthwhile.

We previously noted that some items were not related to the four significant trends we discussed. We also noted that of the five items, four were issues that Alkov had found to be important! These were:

– Make any recent decision regarding the future.
– Have difficulty with interpersonal relationship.
– Recently had difficulty with superiors.
– Recently had difficulty with peers.

We know that relative to our three trends, those issues were not significant. However, since they are grouped together and since Alkov previously identified them to be important, there might well be something common between the items. There might well be some truth in Alkov's previous assertion that the effects of life events on pilots (with reference to accident proneness) are a function of inadequate coping and the acting out by the pilot of aggressive tendencies. We may have found some support for this theory, although clearly the quality of the finding is in the context of a non-significant result.

Life events

We confirm, by this questionnaire-type approach, that a life event experienced by a pilot will have a subsequent psychological and behavioural effect on him, and that the pilot's personality characteristics will themselves have an effect on the extent and impact the event will have. We must also add, however, that these findings are not really of much use in a practical sense, particularly for those who have to manage and assess pilots. If one adopts an academic approach, one must note that Alkov attempted to rationalise his findings by suggesting that he had uncovered a link between accident proneness and a set of personality/stress coping factors. He said it was all a function of 'acting out aggression'. We feel that our findings demonstrate that if one asks pilots to rate the items, they tend to do so in very simple terms (i.e. on three trends). Pilots simply tend to make fairly broad distinctions and do not make judgments that concentrate on any single specific issue.

Conclusions on life events

We stated at the beginning of this book that life events are major 'landmarks' that we all experience and can identify as being particularly emotional. Within 'pilot stress', the previous literature contains many (and some quite successful) attempts to isolate the nature of events that pilots experience and their possible effects. The study of these particular sources of stress seemed, therefore, potentially relevant and interesting.

In this chapter we have presented an array of material drawn from the same source, adopting several different approaches. One approach was of the qualitative interview type, backed up by simple statistical analyses. The other approach was based on the work of previous researchers and used a revised and expanded questionnaire, plus a complex statistical method of analysis. The role of this conclusion therefore is to try to draw together some of the points raised and to integrate the material into the wider home/work interface for pilots.

In terms of the events identified by the pilots themselves, some included events that were work-related. We previously defined 'events' to be more general in content, but to illustrate the point we want to make, work-related events have been included. Events are both positive and negative with most pilots citing events that occurred a relatively long time ago. There was, however, a fairly wide range in the length of time elapsed since the event cited. Analysis of the effects of these events revealed a wide range of outcomes. Clearly the reactions between the 'coping' and 'not coping' pilot are varied. The reactions tend to be private and rather personal, but certainly we found pilots well able and willing to articulate their feelings after the event. Again, feelings and reactions are mixed. There is a broad array of reactions, in fact, so much so that to a certain extent the pilot may experience a degree of ambiguity as to the form of his own reactions to an event. It would appear that the 'conscious' effects are apparent to the pilot at work in situations of low demand on his attention, i.e. during the periods of cruise. Additionally, however, the more insidious effects in an unconscious form will still be present. Clearly such effects are important, particularly during periods of relatively high pilot-attention demand, e.g. take-off and landing. Although *introspection* appears to be important, specific thoughts about the event do not really appear to be considered to be of greatest impact. It is the more generalised 'feelings' that would appear to be of greater relevance. All these factors contribute to the insidious nature of the effects, previously mentioned. These subtleties are further illustrated by the fact that pilots may themselves not perceive any effect. The effects may not necessarily be noted by others, and finally, what is an ordinary event to others may well be *important* to the individual, and hence others may not be on the look-out for any effects.

Some underlying issues or themes also seem to emerge in this qualitative-type data and approach. The first is the degree to which the pilot can predict (and measurably prepare for) the event, which will have an important bearing on the nature of the impact. In other words, the lower the ability to predict, the greater the perceived impact. Examination of the pilot personality reveals this need for mastery and control. Also, since a large part of the job of pilot involves the elements of prediction, it is not surprising that this should emerge with respect to broader life events. Second is the fact that *ongoing* situations are the worst. Clearly because no end is in sight (and no way or means of a resolution apparent), the situations in which events occur are complex. Indeed, the pilots identify the fact that it is not just their own reaction that they have to deal with, but also the reaction of others. Finally, hindsight can be illuminating. Hindsight is, of course, a great teacher, but it seems particularly interesting from the pilot's point of view, in terms of the perception of a life event and how he subsequently reacted and dealt with it.

To conclude, the pilots agree that life events do have an impact, but one that, on balance, they can cope with (or think they can cope with).

Life events

Whilst recognising the unconscious nature of effects, pilots also feel (again only on balance) that their colleagues would probably notice any effects. The long-term implication of these insidious effects is that the pilot may well be all right until *tested*. It is only when tested, such as in an emergency situation, that the effects of these events may be registered. In other words, most pilots have a capacity to cope, but one which may well break down.

The life events material was also examined using the more complex questionnaire approach. We need not reiterate the conclusions previously drawn above. It is sufficient to state that pilots appear to use a relatively broad scale on which to assess life events, i.e. 'emotional losses', 'pilot characteristics' and 'emotional gains'. Whilst the contents of the trends were interesting, on balance the findings seem to support the contentions of other researchers in the life events area that a more simplistic approach is better. Clearly, however, the combined implications of the findings indicate that life events do indeed play an important role in the lives of pilots, and potentially can form an integral part of the stress that arises from the interaction of work and home, as well as stress that arises from their lives in general.

Chapter 8
Pilots' wives

Introduction

There can be little doubt that factors affecting one's spouse, whether they be of domestic or work origin, will have inextricably interwoven and important implications for one's own life. With reference to the former, the domestic scene, the connections are fairly clear. However, the same is also true of the latter. Although many professional people hold an idealised belief that they manage to keep their work lives and home lives separate (if only at a mental level), in reality the two facets of their lives are interdependent. Like it or not, one's work will affect one's home life and all those other people that comprise one's family. In this case intuitive common sense and some introspection of our own lives are enough to tell us that this is the case.

Within a managerial setting, the role of executive's wife has been partly examined (Cooper, 1982). It was concluded that not only was the wife greatly affected as an individual, but was regarded as a type of 'asset', instrumental in the executive's career progression. No doubt this refers to the social side of things and, in fact, the absence of this 'asset' role is recorded in the reverse situation, where the executive is female and the asset is her male partner (Davidson and Cooper, 1983). (Incidentally, this apparent lack of an active role in spouse's career progression is something that the pilots' wives complained of and will be mentioned later.)

Probably the best way to begin talking about pilots' wives is to establish some fundamental facts about them. By far the most important is the fact that they tend mainly to come from aviation backgrounds themselves. This is clearly the case with present or former air hostesses as, indeed, many pilots' wives are. However, many have had some other previous contact with aviation and the world of flying either through family or friends. The important implication is clear. Many of them know what to expect before embarking on a long-term relationship with a pilot. This is an interesting point and one worth developing for the reality is contrary to most lay people's image of a 'pilot's wife'.

Superficially, a pilot and his work are seen by many people as glamorous. His wife is someone who is seen as being part of this glamorous lifestyle which, combined with a relatively high disposable income, is regarded as desirable. We shall see subsequently that this stereotype is not borne out. But what is interesting is that these women still embark on marriage in the knowledge that this glamorous lifestyle is a figment of the general public's imagination. We shall see, indeed, that the lifestyle does have important drawbacks and, in some instances, is contrary to several of *their own* basic expectations. To readopt our initial theme, it is worth pressing further the fact that a spouse's life and work will affect the other partner. In many ways this forms the basis of why pilot's wives were examined by us in the first place. There were several reasons for doing this. First, to see how *they* perceive the job of pilot, and second, to gain insights into the experiences of the wives themselves and later make inferences about the home lives of pilots. This latter option itself was chosen for a variety of reasons, not least of which was the fact that the wife would only be able to comment upon those work experiences of her husband with which she comes into contact. Another attraction is, of course, that not only can fresh insights be gained about another group that will be affected by aviation but also, in turn, as we have indicated, the wives themselves will also have an effect on the pilot. In other words, it would be of optimal benefit to look at *both* sides in the home/work equation, i.e. how the pilot and his work affect his wife and home life and in turn how his wife and home life affect the pilot. It was with this in mind that we undertook a large-scale study to run concurrently with our main study of the pilots themselves. Actual conduct of the wives research was, indeed, not unlike the main study – initial interviews with pilots' wives and a large-scale follow-up questionnaire survey. The purpose of the investigation was to empirically determine which sources of stress were important in the lives of pilots' wives and to see which characteristics would predict the psychological state of 'life satisfaction'. We can define this as being the same as 'job satisfaction', only it refers to one's outlook on life in general.

Measures used in wives study

In addition to a brief biographical measure, there were four main orientations the questions used. These may be divided into two halves:

(1) questions regarding the wife's stresses and strains;
(2) questions regarding the wife's impression of pilots' stresses and strains.

Questions regarding the wife's stresses and strains

There were two parts. The first part concerned *domestic* sources of stress.

There were three options available. The first was simply to use the domestic stressor measure from the pilots. The second was to use the pilots' domestic stressor and to add the stressors identified by pilots' wives. The big advantage of both of these methods is that they have some degree of cross-referencing of answers to common items between pilots and pilots' wives.

However, it was decided that the third option would be adopted. This was simply to use the domestic stressor questionnaire based solely on data derived from interviews with pilots' wives. The major objective of conducting a study on pilots' wives was for purely investigative reasons, and this latter option was more compatible with such underlying motives. Respondents were required to rate each item on its stressfulness.

The second part concerned the measurement of *life satisfaction*. This was a questionnaire previously published in the psychological literature (Warr, Cook and Wall, 1979). It yields five scales which were intended to be used as a measure of stress outcomes. Not only was this intended to reveal insights into the quality of pilots' home lives as perceived by their wives, but it was also intended to see how the above scales predicted its value, i.e. which stressors determined which aspects of life satisfaction.

Questions regarding the wife's impression of pilots' stresses and strains

Again there were two parts, the first concerning *factors at work that affect the pilot at home*.

It was found that many of the items identified by pilots were, in fact, reiterated by their wives. Hence it was decided to use the list of factors identified by pilots. However, the task was for the wives to assess the degree of stress they felt it caused their husbands.

Pilots' wives

The second part of this section concerned the *effects of work stressors on pilots at home.*
This list was produced by the pilots, but *not* administered to them. The aim of including it here was to measure the nature of effects of job stress on pilots at home, as perceived by their wives and expressed in non-clinical terms. Potentially this is quite interesting in the light of previous evidence that pilots tend to deny and suppress the effects of stress and are not always aware of (or at least do not report) how things really affect them.

The sample of pilots' wives

In addition to a sample of thirty-five or so pilots' wives who were interviewed (and who provide the qualitative data quoted here), the main sample was comprised of 282 wives of civil aviation pilots. The sample was chosen randomly and represents 56.4 per cent of the 500 wives to whom we originally wrote and asked to take part in the study. As indicated previously, response rates in postal surveys are notoriously low, so this rate of response from a sample who are not even the main subject group of the study clearly demonstrates awareness and the interest in the topics investigated! Some further basic facts about the wives are provided by the demographic characteristics summarised in Table 8.1. In many ways the characteristics observed in the sample of pilots' wives reflected the same trends as were found in the sample of pilots. This is not really surprising given the method of collecting the sample.

The age range of wives were: 17.2 per cent between 21 and 30, 56 per cent between 31 and 40, 22.4 per cent between 41 and 50. There were marginally more women who worked (54.4 per cent) than did not work (45.6 per cent), which is consistent with the fact that Britain has amongst the highest proportion of working women in Europe (Davidson and Cooper, 1984). Of those who worked outside the home, most of them were full-time (40.5 per cent), though the majority worked either part-time (33.8 per cent) or occasionally (25.7 per cent). We shall see later that such figures reflect the difficulty the pilot's wife has with reconciling the domestic routines, a pilot's home life and a full job (quite apart from any type of 'career').

In accordance with employment and company size, the main employer of pilot husbands was British Airways (60.8 per cent), although all the three other major independents together with a range of smaller commercial airlines were also represented. Just over half

Table 8.1 *Wives' biographical data*

		Number	Percentage
Age	21–30	43	17.2%
	31–40	140	56.0%
	41–50	56	22.4%
	51–60	11	4.4%
Do you work?			
	Yes	136	54.4%
	No	114	45.6%
If yes, do you work			
	Occasionally?	35	25.7%
	Part-time?	46	33.8%
	Full-time	55	40.5%
Husband's present employer			
	British Airways	152	60.8%
	British Caledonian	30	12.0%
	Britannia	24	9.6%
	Dan Air	19	7.6%
	Other	25	10.0%
Husband's rank			
	Captain	136	54.4%
	Senior first officer	95	38.0%
	First officer	19	7.6%
Husband's flying			
	Long haul	77	30.8%
	Short haul	151	64.4%
	Domestic only	12	4.8%

the husbands (54.3 per cent) were of the rank of captain, the remainder were of senior first officer (38 per cent) or first officers (7.6 per cent) in rank. Wives of 'long haul' pilots represented slightly over 30 per cent of the sample, with 'short haul' nearly 65 per cent and 'domestic only' just less than 5 per cent of the total.

The nature of stress experienced by pilots' wives

The items were generated by interviews, then summarised and presented to wives in the questionnaire. The wives were asked to regard the items as sources of stress. The task, therefore, was to rank them on their 'stressfulness'. The results were then analysed and statistical trends extracted. The content of each trend was examined and the following labels given:

(1) domestic role overload
(2) fear of husband's job loss
(3) threats to marital relationship
(4) distance from husband's career
(5) job's impact on social life.

Examination of the contents of each trend reveals how these labels are appropriate (Table 8.2).

Table 8.2 *Underlying trends in stresses perceived by pilots' wives*

	Loadings
Trend 1 (56.6%) Domestic role overload	
Husband who doesn't try to understand stresses he creates	0.61
At times feeling like a 'one-parent family'	0.59
Increased domestic workload	0.53
Husband's tiredness	0.50
Difficulty experienced in involving husband in things he has missed	0.48
Feelings of being rejected and upset when husband is tired	0.46
Having to deal with things as they occur and not letting them ferment	0.31
The fact that the job of pilot isn't 'social' (i.e. being isolated as a family unit)	0.30
Trend 2 (9.3) Fear of husband's job loss	
Threat of redundancy or early retirement	0.82
Health and potential job loss	0.79
Difficulty in career change over 40	0.75
Trend 3 (9.2%) Threats to marital relationship	
Conditions of work that almost 'foster' promiscuity	0.92
Tremendous need for trust in marital relationship	0.84
Trend 4 (7.3%) Distance from husband's career	
Absence of role in husband's career progression	0.83
General lack of involvement in husband's working life	0.69
Lack of recognition in contribution to pilot's quality of life	0.57
Employers who are oblivious to home life or regard wives as unimportant	0.42
Trend 5 (6.6%) Job's impact on social life	
Friends not being able to go out mid-week	0.58
People socially who consider your lifestyle as 'glamorous'	0.51
Not being able to mix socially at weekends	0.44
People don't drop in for fear of intruding	0.38
Responsibility of being married to a pilot	0.33

Trend 1: domestic role overload

Examination of the first trend reveals three overall broad themes: isolation, responsibility and tiredness. The first theme, *isolation*, itself operates at two levels. One is in terms of feeling like a 'one-parent family', indicating the feeling that wives have to take a very dominant role in child-rearing activities (no doubt a function of timetabling, and also of tiredness of pilot). At a second level, isolation is prevalent in the sense that the actual job of pilot does not lend itself readily to a wide circle of friends. The second broad theme in this first major trend concerns *responsibility*. Once again, this has two aspects: first, in terms of having a relatively high amount of work to do in the home to which the pilot can only make a limited contribution, and second, responsibility in an 'interpersonal sense' of having to talk through arguments and disagreements until some sort of resolution is reached, and also having to try to involve the pilot in things he has missed. Both are quite difficult processes in that they require the wife to behave in a way that is counter to what would be normal, i.e. both are unnatural and artificial processes. The third broad theme in this trend concerns *tiredness* of the pilot. Quite apart from the fairly expected implications of being married to someone who routinely experiences fatigue, an additional interesting consequence is that the wife experiences, to some extent, a feeling of being 'rejected' as a function of her husband's tiredness and consequent disinterest.

To conclude, the trend was given the label 'domestic role overload': the message behind this was relatively straightforward. A major component in the nature of stress experienced by pilots' wives concerns the domestic features of dominance and responsibility inherent in their roles, with a resultant burden that is too much to carry. The latter is exacerbated by the relative degree of isolation from others – a theme which we will pursue later.

Trend 2: fear of husband's job loss

This is an ongoing fear or threat that pilots and their families learn to live with. Quite apart from the usual difficulties experienced by people (especially over 40) in finding employment after working in a highly specialised field, the problem in the case of pilots centres around the fact that the skills of a pilot and knowledge achieved by training and experience are not readily transferred into occupations in other walks of life. One must remember also that generally speaking only around 50 per cent of pilots' wives work, and of these only 40 per cent work full-time. We shall note later that those who work (or have worked in the past) experience immense difficulties in doing so quite apart from the usual problems married women face in working. To summarise this, the fear of husband's job loss is made worse by the fact that pilots' wives find it difficult to pursue a career of their own, which in other circumstances would at least provide some form of potentially necessary future financial support.

Some comment ought to be made regarding the order of items in the trend. It is interesting to note that economic issues of redundancy and early retirement prevail over other causes. This is especially interesting when one considers the age of the sample. We indicated that around 50 per cent of the wives were in the 30–40-year-old category (it is reasonable to assume that their husbands more or less mainly fall into this category). The finding is interesting, for it shows (a) an awareness and fear of negative economic conditions in the aviation industry, and (b) that such a fear is not exclusively held by older wives and by pilots as one might expect. The final point is that this trend is the second trend extracted, so one may infer that these fears are quite prevalent in the wives' thoughts.

Trend 3: threats to marital relationship

The emergence of this issue as an identifiable trend is intriguing. The two items that comprise the trend emerged in the preliminary interviews with the wives and hence were included for that reason. The issue is made doubly interesting for two reasons. The first is that to the layperson, the lifestyle of a pilot is 'glamorous' and allegations of promiscuity are common, generally denied by pilots. Clearly, from the wives' point of view, this denial is inconsequential. The castigation by pilots of such issues as being groundless speculation and unfounded nonsense is at apparent variance with the opinion of their wives. Additionally, one must recall the fact that many wives are ex-hostesses or have had some contact with aviation other than through their husbands and, in reality, are in quite a good position to judge. A further extension of this logic is that there might well be some truth in the claims of promiscuity.

The second fact that makes this interesting is that this was the third trend extracted, therefore, one may infer that this trend, comprising of only two items, seemed to be quite dominant in their opinions. In an unsettled marriage there is the danger that all blame be attributed to pilots, whilst clearly important negative events that occur within marriage are the 'fault' of all parties concerned. Pilots seem to be seen, in this instance, as the guilty party. Consider the following two parts of a conversation between wives.

'Many of us have been air-hostesses ourselves and know full well what happens – how do you think I got to marry a pilot? We know about the sort of temptations that exist and also the opportunities there are to exploit them. It wouldn't be the first time that a pilot has been found working two rosters – one for the airline and one for the girlfriend! Given the fact that wives and families are generally kept in the cold, the pilot can more or less do what he wants. The major point is that there has to be a great deal of trust within the marriage to keep family life secure. Trust is important. Things can work both ways with pilots being away so much, it would be easy for a lot of us to have a boyfriend or at least a part-time boyfriend! Yes, I think trust is important.'

'Yes I agree. I don't know exactly what the divorce rates are for pilots. I get the impression it's quite high. Of course, you only know about divorces and things that go wrong – never about things going OK. I think though that if either partner, as has just been said, wants to play around they can – certainly more easily than some of our manager-type friends. Yes, trust is crucial.'

The underlying theme of trust is, of course, implicit in most marital relationships. The main theoretical issue is of lasting trust and whether or not positive or negative results occur as a consequence of whatever 'test' may occur. In other words, the concept of trust can be thought of positively as being a healthy component of marriage or negatively as a 'duty'. The direction obviously depends upon the state of the relationship. One can only assume, since explicit evidence was not forthcoming, that the wives see the process of trust and test in a *negative* way, with themselves presumably as the innocent parties.

Trend 4: distance from husband's career

This trend is neatly summed up by the following quotation,

'Employers, in general, don't really pay much attention to home life. You only have to look at the way rosters and schedules are arranged, then changed and changed, to see that. But it is also apparent in other ways. For example, the actual job and career of a pilot is very separate from home life. I mean that there isn't much we can do to help our husband's progress. We can only make indirect contributions by affecting (and in fact determining) the quality of the pilot's lifestyle and how difficult or easy it is for him to fly.'

Clearly the theme is one of an absence of an active role in the pilot's career progression. This may be examined at two levels. The first is the fairly obvious one where, due to the nature of the pilots' job, the wife cannot make the sort of contribution that would normally be expected in a traditional middle to upper managerial job – the role of the spouse as an asset, who furthers the spouse's career by the mechanism of social interpersonal relationships. Some wives contrast the existence of this role in the RAF and civilian flying. The implications of a lack of contribution of wife is not a subject that is apparent from the pilots' point of view; the 'spouse as an asset' concept is inactive.

The second level of examination is the idea that a secondary consequence of not being able to make a contribution is that *they* feel that they are seen by others (organisations) as not being able to make *any* sort of contribution to the quality of pilots' lives. This extends further than simply working life, for the wives clearly feel that employers generally do not recognise the contribution that wives can make to the pilots' quality of life. Additionally, such employers are seen by wives as trivialising the importance of home life. By this the wives mean that in a decision situation, for example, if employers had to choose between a decision that would affect work and one that would affect home (the obvious issue of rostering), then it would be home that would suffer. The impact is that this extends, still further, the distance of the wife from her husband's work. The overall culture is one of segregation and alienation.

Trend 5: job's impact on social life

This was the final overall trend that was prevalent in the nature of stressful experience reported by pilots' wives, and concerned yet again the theme of separation and isolation. We know from the previous trend that wives feel isolated in the occupational sense. We know also, from the first trend, that within the domestic scene the wives see themselves as playing a dominant role. It would appear from the existence of this fifth underlying trend that the feeling of isolation extends outside the home into general social life.

An item concerning 'the responsibility of being married to a pilot' is loaded in this factor. The inference one may draw is that social problems tend to hamper the social systems of support available to the wife to such an extent that she feels responsible for what might occur at work, to her husband.

At a simplistic level it might be surmised that this reflects the disruptive influence of flying schedules and rosters in the form of cancelled arrangements or restrictions in socialising. This is so given that the isolated role of the pilot's wife, the fact that the whole family unit is socially isolated, is not only seen as being stressful in itself, but will also have important limiting effects on the nature and degree of social support available to the wife in order to cope with whatever stresses she might be experiencing. In other words, her coping resources are greatly diminished.

Overall summary of underlying trends

To draw this data together, it is appropriate to direct some further comments towards the nature of the five underlying trends that were found in the perceived stress reported by pilots' wives. Two points are worth highlighting. The first concerns the means by which trends were extracted. This is done by sophisticated statistical means. Generally, the results one obtains depend largely on the statistical model one adopts and also the options that one uses in actually performing the analysis. The procedures we used were those that were psychologically as meaningful as possible to suit the situation, so as to minimise any bias in determining results. On this basis, one can state that the trends extracted were relatively 'clear' (in the sense that each covered different themes) and relatively 'pure' (in the sense that items loaded on to one trend were not found to re-occur or to load on to other trends).

The second point worth noting is the order of the trends. The first two trends are primarily practical in nature (i.e. too much work and dominance at home; fear of husband losing his job), whereas the remaining trends all contain some sort of emotive content (i.e. marital problems, lack of recognition, social isolation and so on). The overall inference must be, therefore, that pilots' wives are all fairly pragmatic in their approaches to life, since stresses of this kind feature uppermost in their responses.

Levels of life satisfaction reported by pilots' wives

We all understand what we mean by the term 'job' satisfaction. It is a psychological state, with various components, which embraces our general attitudes towards our jobs and work. A similar concept may be applied to life in general. In addition to an overall total composite scale, the measure we used has four subscales. The subscales are satisfaction with personal life (e.g. health, education, social life), satisfaction with standards and achievements (e.g. housing, neighbourhood, leisure time), satisfaction with lifestyle (e.g. what you have accomplished in life), plus a simply overall global view of life satisfaction (i.e. taking everything into account).

The data in Table 8.3 gives us an elementary idea of the wives' life satisfaction profile. Deductions from this are fairly self-evident. Greatest positive contributions were made to life satisfaction by satisfaction with lifestyle and personal life, and less by standards and achievements. Generally, as one might expect, these results reflect the trends extracted previously (above).

Such data is interesting, but is not really particularly informative until it is placed into some sort of perspective. This is done by comparing our observed results with another external yardstick. Yardsticks, such as other professionals' wives, unfortunately were not available for comparison. It was decided to use the population on which this measure was originally designed as a comparative group,

Table 8.3 *Wives' life satisfaction: comparisons of subscales*[1]

	Total marks available	Wives' score	Percentage
Satisfaction with:			
Personal life (4 items)	28	21.7	77.5%
Standards and achievements (7 items)	49	30.9	63.0%
Lifestyle (4 items)	28	22.1	78.8%
Total life satisfaction (15 items)	105	74.8	71.7%
Overall life satisfaction (1 item)	7	5.5	78.5%

[1] To facilitate comparisons between subscales, wives' scores were expressed as a percentage of total marks available for each particular subscale.

Table 8.4   *Life satisfaction: comparisons of wives' scores with expected norms*

	Wives' average	Norm	Decile	Range
Satisfaction with personal life	21.69	21.51	5	21.50
			6	22.20
Satisfaction with standards and achievements	30.98	25.44	8	30.80
			9	3370
Satisfaction with lifestyle	22.14	20.22	7	22.10
			8	23.00
Total life satisfaction	74.82	67.09	7	71.90
			8	76.00
Overall life satisfaction	5.51	5.13	7	5.27
			8	5.64

i.e. male blue-collar workers in the manufacturing industry. On the face of it such a comparison would appear meaningless. Whilst there is some validity in this, at least the comparison gives us some idea as to the relative scores of pilots' wives. As may be expected, Table 8.4 reflects the fact that the pilots' wives have higher levels of life satisfaction than the blue-collar workers.

In three out of five scales, wives' scores fell into the 7–8 decile, and for satisfaction with standards and achievements their scores fell into the 8–9 decile. It was only in terms of satisfaction with personal life that minus scores fell down into the 5–6 decile.

Having examined the wives' life satisfaction scores, and obtained some idea of their relative importance, the next step in examining this data was to see if there were any underlying trends in these women's opinions which might have either explained our results or which prevent us from addressing our remarks to the group in general. Since such sets of analyses involve the examination of several issues at once, they are referred to as bivariate analyses.

Bivariate analyses

Table 8.5 summarises some of the relevant results. They mainly concern the relationships between biographical characteristics and life satisfaction levels.

In terms of age, it is the 41–50-year-old group who seem to be

Table 8.5 *Significant differences in life satisfaction scores between different groups of wives*

	Groups	't'	Significance
	Age		
Overall life satisfaction	(21–30, 41–50)	2.23	0.028
Personal life satisfaction	(21–30, 41–50)	2.45	0.016
Overall life satisfaction	(31–40, 41–50)	3.26	0.002
Personal life satisfaction	(31–40, 41–50)	2.52	0.013
Standards and achievements satisfaction	(31–40, 41–50)	2.78	0.006
Total life satisfaction	(31–40, 41–50)	2.48	0.014
	Wife working		
Overall life satisfaction	(wife in work, wife not in work)	1.98	0.049
	Husband's employer		
Overall life satisfaction	(BA, BCal)	2.87	0.007
Personal life satisfaction	(BA, BCal)	3.33	0.001
Total life satisfaction	(BA, BCal)	2.49	0.018
Lifestyle satisfaction	(BA, small indep't)	2.04	0.043
Total life satisfaction	(BA, small indep't)	2.31	0.022
Overall life satisfaction	(BCal, small indep't)	−2.08	0.043
Personal life satisfaction	(BCal, small indep't)	−2.92	0.005
Personal life satisfaction	(Britannia, small indep't)	2.00	0.050
	Husband's rank		
Overall life satisfaction	(senior first officer, captain)	2.53	0.012
Personal life satisfaction	(senior first officer, captain)	2.41	0.017

different. Compared to the 21–30-year-old group, they are significantly less satisfied with their personal lives and with life overall. A
similar trend was revealed in comparison with the 31–40-year-old
group. Not only were the 41–50 group significantly less satisfied
overall and in terms of total life satisfaction, they were also significantly less satisfied with personal life and with standards and achievements.

Whether the wife was working or not seemed important. The data
confirm previous suspicions in that wives who do not work were
significantly more dissatisfied with life (overall) than wives who had a
job.

It would appear that husbands' employer is an important variable.
This was an interesting result which can best be summarised as
follows:

(1) Wives of BA pilots were significantly more satisfied overall and in
    terms of total life satisfaction and personal life than wives of BCal
    pilots.
(2) Wives of BA pilots were significantly more satisfied with lifestyle
    and in terms of total life satisfaction than wives of pilots who flew
    for smaller independent companies.
(3) Wives of Britannia pilots were significantly more satisfied in
    terms of overall life satisfaction and personal life than wives of
    BCal pilots.
(4) Wives of Britannia pilots were significantly more satisfied with
    personal life than wives of pilots who flew for smaller indepen-
    dent companies.

Finally, the results indicated that the pilots' rank was an important
factor. Wives of captains were significantly more satisfied overall and
with aspects of their personal lives than wives of senior first officers.
One must remember, however, that the rank of SFO does not exist in
all airlines, so one may expect some doubt over the reliability of the
finding. However, on face value the above conclusion that husband's
rank is important seems reasonable.

Conclusions on the wives' levels of satisfaction

We regard 'life satisfaction' as a psychological state, not unlike 'job
satisfaction'. The conclusions from the above data are relatively
straightforward. The wives appear to possess fairly high levels of life

satisfaction and in this respect conform to the layperson's simple view of pilots' wives and their lifestyles. There are several points, however that must be firmly stated. First, it is apparent that there is variation in scores between different aspects or component parts of 'life satisfaction'; second, there are variations in subgroups; and third, as we have seen from previous data, the wives can identify definite sources of stress in their lives. The next step in the analysis is to see how our previously identified underlying trends (i.e. domestic role overload, fear of husband's job loss and so on) are related to the different aspects of life satisfaction. In other words, we want to find out what determines, or statistically predicts, life satisfaction.

The prediction of life satisfaction

To see what predicted life satisfaction, we included the trends in wives' stresses plus a selection of biographical details. The results are presented in Table 8.6. For clarity of presentation the discussion of these results has been subdivided.

Personal life

It was found that dissatisfaction with personal life for pilots' wives was strongly associated with 'domestic role workload' and the 'job impact on social life'. It is not surprising that this is so. The position regarding the role that wives have to play at home is illustrated further by the following comments.

'Things which cause me stress in my life are in the main, I think, typical of most women of my age and position – things like coping with teenage children, the state of the world and domestic trivialities. These are generally unrelated to my husband's work now however. If I had been asked this question when our children were younger I would have included night-stops and tours of duty in this category, I used to dread having to cope and be alone with young children.'

'The untidiness of the family and the never-ending jobs to catch up on. My husband's bad crotchety moods make it worse. Sometimes I think I'd like a soundproof room to retreat for peace and quiet. Not managing to do things the way my husband wants them done. Possibly I'm not well organised enough.'

Table 8.6   *Predictors of wives' life dissatisfaction*

	Multiple R	R Square	Overall F	P
Dissatisfaction with personal life				
Trend 1: domestic role overload	0.50	0.25		
Trend 5: job's impact on social life	0.54	0.29		
Trend 2: fear of husband's job loss	0.56	0.30		
Husband working for smaller airlines	0.56	0.31	7.58	<0.001
Dissatisfaction with standards and achievements				
Trend 1: domestic role overload	0.28	0.08		
Husband working for smaller airlines	0.31	0.10		
Wife not working	0.34	0.11	3.02	<0.001
Dissatisfaction with lifestyle				
Trend 1: domestic role overload	0.26	0.07		
Trend 5: Job's impact on social life	0.30	0.09		
Younger wives	0.33	0.11	2.99	<0.001
Overall life dissatisfaction				
Trend 1: domestic role overload	0.30	0.14		
Trend 5: job's impact on social life	0.43	0.18		
Wife not working	0.45	0.20	4.83	<0.015
Total life dissatisfaction				
Trend 1: domestic role overload	0.43	0.18		
Trend 5: job's impact on social life	0.46	0.22		
Husband working for smaller airlines	0.49	0.24		
Trend 2: fear of husband's job loss	0.50	0.25	6.43	<0.001

'As I have always been married to a pilot, I can only look at other people's lives to imagine what it must be like to have a husband with a "normal" job . . . and having to gear the home life to flying in as much as meals and quiet times! One is organised with this in mind. So is our social life – no late nights before flying, no drinks before flying, etc. I take (or feel I have to take) the initiative at times, especially with family problems or stresses. These have to be dealt with as they occur and not allowed to ferment – 'never let the sun go down on a quarrel' I think has to be observed. I find that children turn to *me* with their queries or worries because I am the one who's there. The men know what they are taking on and how it will affect them since the company publish the schedules in advance. This is so you can (in theory) try to organise your life – much of this is up to me. I think my husband tries very much to understand the stresses his job causes and he's trying to minimise them. He's very tolerant when I explode after waiting an hour or so outside the airport or when we turn down yet another invitation because he is working. After nearly seventeen years of shifting patterns of work I accept most things automatically but certainly at times I really do find that I mind and that sometimes it's all too much.'

'My home life is rather chaotic as my husband's work has no definable pattern. . . . The only set routine is the children's school routine. Our house is quite old and needed rather a lot of work doing to it when we moved in twenty months ago or so. It still needs a lot of work. With the slack winter programme of my husband's company, he has been able to work on the house himself (and supervise my efforts). I still think I ought (or am expected) to make a significant contribution in working on the house.'

Social life was also adversely affected.

'Our social life – especially trying to organise events with friends – is frequently difficult. Things have to be arranged a long time in advance and may quite easily have to be cancelled at the last minute for a variety of reasons. One minute he's here, the next he's gone, so we have to be very flexible. It's not easy to build up a circle of friends of any form with such underlying uncertainty.'

'It's virtually impossible to reconcile our lifestyle with everyone else's "Monday to Friday, 9–5 plus weekend" type of schedule. It's very difficult to do anything on a regular basis. We do have patient and understanding friends, but I know that people don't just 'drop in' because they feel they might be intruding; weekend entertaining is usually out. On balance I think, apart from our good friends, that people are reticent in asking us out. Also others aren't usually that keen to come out mid-week when we can make it.'

In addition, the situation was compounded further by two factors. First, the 'husband working for smaller airline' was a significant underlying factor, indicating that wives of pilots working in larger organisations tend to have higher levels of personal life satisfaction. The second compounding issue was that concerning fear associated with the pilot losing his job, as the views of one wife clearly illustrate:

'My husband has just returned to work after a short period of sick leave. Recently I've got to thinking of what we could do if he lost his job – for whatever reason not just ill-health. He's in his late thirties so he might just be young enough to do something else. We have talked about this. I think he realises more than he would admit to me that if he had to give up flying not just for ill-health but for whatever reason he would find it difficult to find something in which he could use his skills. It's not so bad now but in ten or fifteen years' time it could be worrying.

Lifestyle

Yet again, the same picture emerges here, with role overload and problems with social life being prevalent. One may add, however, to the analysis that it is lifestyle that appears to carry greater importance for younger wives. In fact, one wife commented:

'One of my biggest regrets is the fact that I have not yet achieved all my life's ambitions. This is probably because for the last fifteen years I have been bringing up children. This isn't particularly special since many wives and mothers would say the same. However, I think things are definitely made worse by being married to a pilot. I just cannot see how I could arrange things *and* have a job. The picture may well be brighter for the future now that my children are older but I just hope that I'm not too late! I think that, given the state of affairs in the country, younger wives must feel very frustrated. I know I did when I was first married eighteen years ago. But, so much has changed. I'm sure *now* it's worse for younger wives.'

Global and total life satisfaction

Interesting contrasts are apparent in the comparison of global life satisfaction and total life satisfaction. It will be recalled that global life satisfaction was a single question based on 'taking everything into account'. Total life satisfaction was a complete score of all preceding scales. For global measure and total scores again the two foundation stones of dissatisfaction were significant, i.e. 'domestic role overload' and 'job's impact on social life'. Differences, however, were apparent, Global dissatisfaction was additionally found to be determined by the wife not having a job of her own. On the other hand, the total life dissatisfaction scale was additionally important for wives whose 'husbands worked for small airlines' and in a situation where there is a 'fear of husband's job loss'.

The interest in comparing and contrasting these two scales lies in the fact that taking everything into account (global view), the wives' perceptions were moderated by issues directly involving themselves such as having a job. The total life satisfaction score, however, was moderated by features of the husband's life (i.e. smaller employing airline, job loss, and so on).

Factors that did not predict

It will be noted from Table 8.6 that the factors concerning 'threats to marital relationship' and 'distance from husband's career' did not statistically determine the life satisfaction outcome variables. It must be firmly noted that the fact that they did not predict these particular outcomes does *not* mean that they are of no importance. These issues were firmly identified as underlying trends in the experience of the wives. The result simply means that *relative* to the other issues, their predictive value was not as important.

Discussion

It is fairly obvious from the above results that the main source of stress on wives of pilots are those to do with the *overload* experienced by them in their domestic role, which is a direct consequence of their husband's job. This is clearly exacerbated by being 'isolated'. This feeling of isolation appears to operate at different levels. At a micro level, wives feel in the same situation as the 'one-parent family'. They have to cope with the normal day-to-day running of the family, and in doing so are unable to share much of this responsibility with the absent (mentally or physically) husband. The additional worry within this micro level is that the wife is in many ways expected to provide support for the pilots. It is this sort of social support for her husband which she herself does not have available to her in her own life.

The second level is the macro level of the 'isolated family unit'. Close community ties cannot develop due to adverse effects of irregular work scheduling on family plans and social engagements with friends and relatives. One can only assume that pilots' wives rely on their own reservoir of resources to cope.

An interesting result was the overall influence of whether or not the wife had a job of her own. Those without jobs were found to be more vulnerable to the various forms of stress and life dissatisfaction than those engaged in employment (on either a full-time or a part-time basis). A personal and global source of satisfaction is derived from the sense of achievement associated with employment and the pursuit of a career. No doubt, also, the experience of employment carries with it all of the reasons why people work which, from previous evidence here, one can speculate would probably increase the opportunity for meeting people (and for social support).

The finding of increased satisfaction on the part of working wives is interesting in that it is confirmed by previous research, which indicates that women with husbands in white-collar and professional jobs are increasingly entering the labour market. Additionally, despite having initial problems in 'settling in', they report greater overall levels of life (and job) satisfaction.

Another trend worth highlighting was the fact that wives of husbands working for smaller airlines were experiencing more life dissatisfaction than wives of husbands working for larger airlines. It could be argued that this is because of the work demands and schedules of smaller airlines, which are not as regularised and more changeable and, therefore, potentially more disruptive to family life. If one adds the proviso of seasonal fluctuations this might well be the case. On balance the data indicate that such disruptive effects are present. Scheduling systems were again important. In fact, such issues and procedures, such as 'bidding' for flights, were seen as potentially more disruptive than other factors associated with smaller companies.

An alternative argument is one based on commercial pressures. Smaller airlines tend to employ staff to their fullest. This means flying staff to the extent permitted by legal limits. As indicated earlier, this is usually a function of seasonal factors, since a large part of the work of smaller airlines is chartered holiday flights. The other aspect of 'commecial pressures' is in the sense that smaller companies have to fight harder to survive, which has its effects throughout the whole company – especially at the sharp end. The implication of this is, of course, that the employment position is more precarious and the quality of employment more strained.

Although not a central factor, the wife's personal and overall life satisfaction can be affected by worry about her husband's job security, i.e. fear of job loss. We have noted elsewhere that, as expected, this is of concern to the pilots themselves. Whilst the major reason is general recession in the aviation business, other reasons were also identified. One may add, however, that since wives of pilots have taken on domestic responsibility for maintaining the home and lifestyle of the family unit, it is not surprising that they should be concerned about the threat to the family's standard of living that their husband's job loss would inevitably entail.

Postscript to the 'wives study'

As indicated earlier, the wives study played a dual role. The first was to reveal more about the nature of stresses they experience and determinants of mental well-being. The second role was to afford some insight into the experiences of pilots as perceived by their wives. These perceptions concerned two factors. First, the nature of stress experienced by pilots, and second, the ways which stress affects pilots.

Wives' perceptions of pilot stress

Clearly these are going to be limited. In fact, the wives will only really be able to comment on those stresses with which they are familiar and with which they came into contact indirectly through their husbands. This is reflected in Table 8.7.

The items presented were an array of stressors which the pilots themselves had identified as having 'carryover effects' into home life. The table reflects two underlying themes (statistically extracted). The first theme in this 'carryover stress' was labelled 'work pattern fatigue'. It was comprised of eight items similar in content. It could well be claimed that the result simply reflects the content of items in the questionnaire (i.e. you get out what you put in). However, from

Table 8.7   *Wives' perceptions of pilot stress*

	Loading
Trend 1: work pattern fatigue	
How time of working determines when to sleep	0.89
Tiredness and fatigue	0.70
Return back and time of arrival	0.69
Patterns of flying	0.64
Scheduling and rosters	0.55
Unpredictability of flying	0.39
Length of trips	0.36
Effects of minor day-to-day things	0.31
Trend 2: courses and checks	
Anxiety of courses and checks	0.92
Preparation necessary for courses and checks	0.91

the clear extraction of these issues as a single distinctive trend one can conclude that wives make only broad discriminations between job stressors generally based upon this major issue of whether it spills over into home life or not. It must be conceded, of course, that this was to be expected for reasons given earlier.

There were two items that comprised the second underlying trend. These clearly highlight the *anxiety* experienced by the pilot, manifested in a way that is visible or apparent to the wife. The main point to make is that this stress associated with both the 'preparation for and experience of courses and checks' is seen as a conceptually separate and different facet of pilot stress.

Of the sixteen items entered for examination, there were six that were not extracted as significantly related to either trend. These were:

– social problems associated with rosters
– seasonal effects
– changes in family life due to job (e.g. change of base)
– factors out of the pilot's control
– number of sectors flown
– carryover effects of personality clashes

The non-significance of the last three items at least serves the purpose of confirming that wives really only can comment on those issues with which they come into contact. Changes in family life due to the job, such as change in base, presumably were not significant because of the

Table 8.8   *Wives' perceptions of the effects of stress on pilots*

	Loading
Trend 1: irritability and tension	
Becomes annoyed and angry (on a 'short fuse')	0.84
Becomes short-tempered and finds it difficult to laugh things off	0.84
Becomes irritable	0.74
Becomes tense	0.44
Becomes aloof and withdrawn	0.32
Trend 2: decreased performance	
His efficiency decreases	0.84
His level of concentration decreases	0.77
Performs jobs and tasks unsatisfactorily or incompletely	0.65
Feels tired	0.39
Expresses an awareness of physical effects	0.35

infrequency of such events. Similarly, seasonal effects only apply to that part of the sample working for smaller airlines, whose business is primarily chartered holiday flights. Finally, one can only assume that the underlying implication of the item was already summarised by other issues, no doubt those in the first trend such as 'schedules and rosters' and 'unpredictability of flying'.

Wives' perceptions of the effects of pilot stress

The underlying assumption here is that if the pilot is experiencing stress at work, then his wife will be in an optimal position to observe any changes in the pilot's behaviour or attitude that might be apparent at home. This is interesting data, since most previous studies have relied on self-report data from the pilots themselves and have made little attempt to cross-reference this with the perceptions of other observers (such as wives).

It can be seen from Table 8.8 that, again, two trends were apparent. The first was named 'irritability and tension' and the second 'decreased performance'.

These trends require little further analysis since they are fairly self-explanatory. The former refers simply to mood and the latter to a relatively objective assessment of task performance. This clear duality emphasises the simple dichotomy in the wives' perceptions of the effects of pilot stress.

Chapter 9
Coping with stress

Introduction

Occupational stress, as a research theme, has been of recurrent interest from the late 1960s, and continues to be so now, well into the 1980s. Like any other area of interest, there are trends in research; for example, earlier studies concentrated on 'executive stress', whereas now concentration is focused on the lower end of the occupational hierarchy. With the rise and fall of research topics, a small yet steadily growing body of research has concentrated on coping mechanisms involved in stress. This is also referred to as 'stress management'. The research is wide-ranging in its quality, from that which scientifically examines the psychological and behavioural processes involved in coping to those studies which read like extracts from popular magazines.

An introductory issue relevant to the present study is that much of the material is of American origin. Traditionally, Americans seem more ready to accept the existence of their psychological character-istics and the fluctuations that can occur in them. Moreover, if they are experiencing 'problems' (e.g. stress), they seem more prepared to seek assistance or help themselves. There does not seem to be an apparent equivalent propensity in our own British culture. (Perhaps there is some truth in the 'stiff upper lip' element of our cultural stereotype!) Given our previous comments about pilots, we would expect the study of stress coping mechanisms to be doubly difficult. In other words, trying to examine coping in British individuals may well not be simple. Trying to examine coping in people who are British and pilots will be even more difficult, owing to the nature of the pilot personality.

The key word in the above sentence is 'apparent', for we *all* possess a repertoire of psychological mechanisms or defences which we evoke to combat stressful or unpleasant experiences. Some will be uncon-sciously triggered, and it is only if they are unsuccessful that we will really take any notice. Others may be conscious or may even simply be related to the way we deliberately behave and organise our lives.

To conclude: the subject matter of this chapter is those processes, strategies and behaviours that pilots identify as cushioning the impact of the stresses they experience. In fact the 'psychological prop' analogy (i.e. something to lean on) is not as flippant as it might at first appear to be.

Coping strategies

To set the scene further, consider the following list of issues:

'When under stress I . . .'

1 Overeat
2 Smoke
3 Drink alcohol
4 Take medication (i.e. tranquillisers, energisers, sleeping pills, etc.)
5 Fight/argue
6 Withdraw
7 Sleep
8 Drive around
9 Call in sick to work
10 Cry
11 Read
12 Become involved in a hobby
13 Socialise up
14 Talk to my spouse or friend about it
15 Exercise

Although these may be identified as possible symptoms or signs of stress, they are also the sort of things that people do to try to *cope* with the stresses they are experiencing. Incidentally, the list was extracted from a recently published American stress management book of the type referred to above (Farmer, Monahan and Hekeler, 1984). It is apparent that whilst being of some help at an individual level, checklists of this kind are of only limited practical utility, particularly in our investigative study. It was, therefore, decided to adopt a different strategy.

Underlying trends in coping

In our exploratory interviewing we simply asked pilots how they coped with any stresses they encounter. We uncovered thirty-three items which seemed to be important. We asked pilots to rate these items on their importance in helping to cope with problems or stresses they might experience. The data was then analysed again using the complex statistical procedure previously referred to. Four underlying trends were extracted, as may be seen from Table 9.1.

Trend 1: stability of relationships and home life

The first trend to be extracted was called 'stability of relationships and home life'. As with previous analyses of trends, this was the first to be

Table 9.1   *Underlying trends in pilot coping strategies*

	Loadings
Trend 1 (46%): stability of relationships and home life	
Stability of relationship with wife	0.77
Home life that is smooth and stable	0.70
Home life that provides a psychological platform	0.64
Home that is a refuge	0.63
Talking to an understanding wife	0.56
Wife who is efficient in 'looking after things'	0.49
Wife who modifies her own behaviours and demands to suit you	0.43
Wife who has known you through your flying career	0.36
Trend 2 (13.5%): reason and logic	
Unconsciously separating home and work	0.55
Deliberately suppressing emotion	0.55
Staying emotionally aloof or shrugging things off	0.53
Deliberately avoiding confrontation	0.52
Trend 3 (11.5%): social support	
Talking to understanding friends	−0.82
Talking to understanding colleagues	−0.71
Talking to an understanding wife	−0.34
Trend 4 (8.4%): wife's involvement	
Wife who involves herself and is interested	0.67
Home life that is geared to flying (in practical terms)	0.60
Wife who had prior knowledge of flying or who flies	0.58
Wife who has known you through your flying career	0.50

extracted and hence inferred to be the most important. In addition, since the trend was composed of eight items, it is also possible to infer that a similarly large component of pilots' coping strategies is a function of such issues.

The structure of the items may be divided on two criteria, content and quality. In terms of content, a fairly clear split is apparent between those items that relate to the pilot's wife and quality of marital relationship (such as stability of relationship, wife who is efficient and understanding), and those items that refer to home life in general (such as 'being smooth, stable' and a refuge from work).

In terms of quality, issues such as the provision of a psychological platform, a refuge, and stability, all have a high psychological content, that is, they act as a 'prop' or method of support. At the other end of this spectrum are those that are primarily practical in nature, for example, 'efficiency of wife in looking after the household', 'a wife who modifies her own behaviour and demands to suit the pilot' and so on. Such underlying criteria are apparent from discussion with the pilots themselves. For example, one senior captain remarked:

'Stable open relationship is important with both partners mutually accepting the situation. More so from the wife than from the husband. The consequences of failure to cope – I get more tired, more quickly. I find sleep important. Pre- and post-flight routines are important. You can't guarantee sleep in the pre-flight situation and you're still wound up in the post.'

It is clear that the coping features involved are a complex mix of both practical and emotional issues. Whilst the main point to note is that elements of the pilots' domestic scene are clearly identifiable as 'coping' issues, it can be a 'double-edged sword', as one first officer reflects:

'I can think of one instance in which home life was not supportive. My wife can get very jealous and . . . where the effects of the disruption carry on for a fairly long time then home life can definitely have the opposite effect. It can be disruptive – very much so probably because it is so important in coping.'

In other words, it is this very fundamental role which by its importance can in itself be a source of stress, in situations where the home life conditions are poor or where the relationship is not stable.

It will be noted that nearly all of the items are structural, descriptive features rather than underlying *processes* involved in coping. In fact, the only process involved here is 'talking to an understanding wife', which is identified elsewhere (Trend 3 below). When examining structures and processes involved in coping (in this case, in relation to marital relationship and home life), it is to be expected that one will be dependent upon the other. On balance, however, it is reasonable to conclude that structures, rather than processes, can be more readily organised and manipulated by the individual pilot to facilitate his coping needs. In turn, this will involve some conscious effort on the part of the pilot. However, consider the comments from the following two pilots.

'I deliberately organise my life to satisfy my needs – I'm a survivor. I cope well because I get this tremendous need to win. I like to do things well. I tend to prefer things that aren't complicated. I organise my life to be uncomplicated.'

'Home life is important in coping. I do a lot of different things. I'm not married and have got more time. I keep busy. I don't see it as "coping", just part of my day-to-day functioning. I suppose in some way it perhaps could be called "coping". I don't think about it much, I just stay busy and that's it.'

Although in the second instance the pilot is unmarried, the comments seem to illustrate the fact that this ideal structure of a well-organised and stable home life, a sound psychological prop and a desirable marital relationship may not necessarily be in existence due to any conscious deliberate effort by the pilot.

Trend 2: reason and logic

This trend was composed of four items and was called 'reason and logic'. This is particularly important for several reasons. Perhaps the most important is that it places our first trend in context. We shall see that closely related to 'logic and reason' is the overall separation of home and work lives. Given these two sets of factors, the important role of the pilot's home life, especially those issues identified above, can be placed in perspective. In other words, one of the reasons home life and marital relationship are so important is because pilots try to keep home and work lives separate. This tendency to try to separate home life and work life is a process often quoted by pilots, and these 747 captains suggest:

> 'I'm virtually schizoid! Work and home are totally separate. Role of home life is 100 per cent. My wife involves herself in whatever is happening to me or whatever is worrying me. She's supportive *always*.'

> 'When I'm flying I don't tend to think about home life. My home life has always been supportive. Everything is geared around my flying. I have a good life – I suppose I live two separate lives.'

In the normal course of events this separation applied to the day-to-day stresses that exist both at work and home. However, this role separation can also apply to more important issues.

> I had a minor motor accident a long time ago – "driving without due care and attention"! I was shaken up a bit – it was my fault. I had to fly to Manchester the next day though Scheduling took a dim view of it. I suppose the stress was by feeling like a criminal. I had had a few gin & tonics in the mess. This may have had an effect. Anyway, I got off with a fine. It didn't really affect my work because of this separation.'

In this situation, the pilot experienced a car accident not unlike the 'events' we discussed earlier (chapter 7). The impact of the 'event' was cushioned by the separation the pilot was able to achieve between his home and working lives.

Whilst the separation can be modest in effect, it may be quite dramatic. For example,

> 'I lead two completely separate lives. My family have a farm. When I'm there I tend to get involved in the agricultural scene and forget aviation. It wasn't deliberate. The family lifestyle and the family farm and the separation is something that has grown up with my career. I spend about half my time on the farm and the other half flying.'

In this instance, the first officer quite clearly identified leading two lives. One life is spent flying, the other is spent on the family farm. Even in this extreme example, however, it will be clear that the switch from home to work and vice versa is not a simple one. There may well be a hazy transition period. One would expect to find such a 'twilight zone' from a psychological point of view rather than from a real practical one. However, some pilots identify a period that correlates well with this mental transition, as these two pilots do:

'There is a transition period in actually travelling to work. I don't think I'm good at coping with stress in terms of reducing it but I'm good at coping in the sense of operating under it. I think it is bad often to try to reduce it because often under stress the adrenalin is going which is good.'

'There is a cooling off period in the car which is probably the main way I cope. There is a residue of time between home and work when one has time to think about things.'

It would appear that the period spent actually travelling to work helps in this mental change. Although the former quotation is in the context of from home to work (and not readily identifiable in the latter), one can assume that the 'cooling off' period also applies from work to home. Either way, however, this separation is just part of the story, because it is the application of 'reason and logic' that is important also. Two issues seem to be involved in the extent to which the pilot permits stressful issues to have an impact on him. These seem to be suppressing emotion and staying aloof. For example, consider these two brief comments:

'I don't get emotionally involved with people, except at home where I'm involved totally. A good home life can make you feel good at work. There are positive links.'

'I don't worry about things. I usually try to sit down and mentally work things out. I work out solutions to the problems by logic. It's very much a black or white situation. Either you are fit to do the job or not.'

Unlike the first trend, these are the coping issues or processes the pilots employ. Additionally, although they are of high psychological content, they may well involve a conscious process or effort on the part of the pilot. In fact since they are able to identify a transition period, and are able to identify that suppression and detachment are successful coping strategies, we can infer that inherently unconscious psychological processes may well have some conscious element. At one level, the conscious suppression of emotion and the application of reason and logic could well be claimed to be a dubious proposition (try it yourself and see if it works). At another level, the process does seem to contain a certain degree of logic and intuitive appeal. This is particularly so when placed in the context of the pilot personality, which is portrayed by characteristics such as denying internal emotional life.

Although of lesser importance, the process of *deliberately* avoiding stress is relevant. The implementation of such a strategy is not altogether clear, as illustrated by the following:

'There is an unconscious level of separation of things. The easy way [of coping] when I was in the Navy was by shouting commands at people! I tend to cope by *avoiding* stressful situations. How can you cope with stress on a plane? I've no idea whatsoever. This project generally has got me thinking. I concluded that if a situation occurred that involved stress [in others] on the flight deck, I don't know what I would do to cope.'

As an aside it is interesting to note that the view of stress mentioned by the pilot above is one that concerns more immediate short-lived stresses, such as those faced on the flight deck rather than the longer-term more pervasive stresses referred to in previous chapters. Interestingly, this pilot does not know how he would cope with stress (presumably within himself as well as within others) on the flight deck.

By way of conclusion, it is perhaps worth addressing an apparent paradox in that the existence of the first trend infers a link (albeit supportive) between home and work, whilst the second infers a separation. Examination of the contents reveals that in terms of the impact of issues, there is a separation. Coping, however, may be achieved by either a separation of lives or, where no separation occurs, where only a supportive relationship seems to predominate.

Trend 3: social support

This is the sort of trend one would expect to find in a psychological study on coping with stress. The literature quite clearly indicates the importance of social support in mediating the effects of stress and, in fact, a lack of social support has been identified as a stressor in its own right. Such issues have been recorded as taking various forms; however, 'talking' is probably social support in its simplest form. Indeed, given the nature of the pilot's job, one would have anticipated that talking would be an important coping mechanism, since the performance of the job of pilot does not readily permit a verbal release as might be afforded in other occupations.

The trend was composed of three items: talking to understanding friends, talking to colleagues and talking to one's wife. The order of these is interesting, since one could have expected colleagues or wives to be the primary source of support. In any case, the psychological literature does not indicate *either* to be most important in the work situation! Traditionally, it is the relationship with one's supervisor that seems to be critical in mediating stressful experiences. Also of note is the clustering of these items with respect to their relationship to the underlying trend, i.e. friends and colleagues and then wives. This can be interpreted in a number of different ways.

The importance of the role played by the pilot's wife has been recorded here and in other contexts (chapter 8). The apparent clustering of friends and colleagues might simply be a function of the fact that they are all the same individuals, i.e. pilots tend to have friends who are also pilots. Theoretically one would expect colleagues to be the richest source of social support, since the job is predominantly technically orientated. Alternatively, since friends are rated the highest of the three groups, we must assume that they are also the most highly valued. One other possible alternative is that these friends are not pilots and have little knowledge of flying. Given our previous comments about segregating home and work lives, this latter interpretation would give our findings an internal consistency.

One final point to note is that the relationships of the items to the trend were negative. This does not render our above comments invalid. The negative statistics simply mean that these features are *not* present. In other words, *not* being able to talk to friends, *not* being able to talk to colleagues and *not* being able to talk to one's wife will result in impaired coping with stress.

Trend 4: wife's involvement

The fourth trend to be extracted statistically concerned the importance of the pilot's wife in cushioning the impact of stress on the pilot.

Coping with stress

The qualities inherent in the trend may best be summarised by this comment from a long haul captain:

> 'I've got to have stability. My home life is geared to flying – towards me flying. My wife is good about meals, timekeeping, petrol in the car and so on. She doesn't raise important things before flying. Things that go wrong like meals, petrol and so on would put me on edge. My wife was a stewardess and knows what goes on. This helps. After flying it takes me time to wind down – I'm usually not very sympathetic to her or my child. She understands. Also I can come home and tell her things. She will appreciate what I'm saying.'

If one were to read this out of context, it might be concluded that pilots are a rather self-centred and almost conceited lot, with the endearing wife and domestic household all working towards the pilot fulfilling his job. To a certain extent this must be true. However, in identifying the need to have a wife who is interested, who is understanding and is involved in the work her husband does, there is a strange vulnerability on the part of the pilot in that so much seems to be in the hands of others.

The issues identified in this trend must be in some way related to the first trend extracted (though from a statistical point of view they are separate). Their separate identities in turn indicate that there are qualitative differences between the two. To reiterate, two themes previously identified in relation to the first trend, the psychological element (in the form of an involved, knowledgeable and interested wife) and the practical element (in the form of a home life geared to flying) are both present here.

The reasons for differences between the first and the last trends may well be related to those very same issues already identified earlier in this text (chapter 8) as being a source of stress to wives, i.e. taking a dominant domestic role at home.

Further trends involved in coping

The previous four trends were extracted by statistical means. However, our interview data also provided a miscellaneous array of issues not previously identified as being important in pilot coping strategies.

> 'I have an almost Eastern philosophy about life and what it has in store for us. I accept myself and my lot in life.'

Few pilots would go quite this far! However, there is a wide range of further topics which might serve to illustrate the wide and various elements pilots identify as important in coping.

Throughout this text, we have confined our attentions to those elements of stress that are negative, i.e. 'stressful' in the traditional sense. However, in the same way an actor or musician 'rises to the occasion', stress can have positive qualities that manage to get the adrenalin going. For example, one short haul first officer remarked:

'I indulge in competition bridge and union motor sport. This, especially the latter, is stressful – but deliberately so. I also find that keeping fit physically helps me mentally to cope with things.'

In this instance the pilot deliberately uses the stressful elements of a risky sport to release tension and cope. This pilot also identified sport as being important and also the need to stay relatively fit. Given the great increase in public awareness of fitness and participation in fitness-inducing sports, it is not surprising to find the trend apparent in pilots.

'I'm not a fitness fanatic but I do a lot of running which I find good for bringing my mind together. It's noticeable at work. Those who are fit are more balanced at the job – I don't tend to fly off the handle, am better at man management. From my knowledge the fittest ones tend to be the best all-round pilots.'

In a previous chapter on pilots' characteristics, we concluded that, surprisingly, pilots did not engage in 'fitness-inducing activities' to the extent that one might expect. Here, too, in terms of coping, it would appear that a deliberate effort must be made on the part of the pilot for this to be a successful coping strategy.

Those who pursue the career of pilot are highly selected and (more importantly) self-selected groups. Only those who take real delight in flying are those who will eventually succeed in their jobs. It might be hypothesised, therefore, that the process of flying itself may not only be a pleasurable experience but also a coping strategy in its own right. Consider the following comments by two pilots in our sample:

'I tend to sulk a bit; I used to play a lot of rugby – that helped to get away from things. It [stress] never gets so great that I've got to do something drastic. Flying relieves a lot.'

'If something major was worrying me I would involve myself in other things whilst still flying the aircraft, e.g. read about staff regulations, manuals. Also talk a lot with the rest of the crew.'

In the first situation, the pilot identifies the fact that 'flying itself' relieves tension. In other words, the whole process is so complicated and involves so much attention that the pilot concentrates on this alone and leaves other worries and stresses to one side. In the second situation, it should be noted that the pilot engaged in these activities during the period when the plane is usually on automatic pilot!

Finally, the following brief comment from an older, more senior captain raises an interesting issue:

'Expansion of outside interests is important. I've seen pilots age twenty years on retiring at 55. I have a small venture going. A small sports shop. Having something in the pipeline is important. Pilots are very mollycoddled. It's quite a cruel world outside.'

This is an interesting quotation, because it takes a wider perspective. The coping issues identified in this chapter are applicable to all those medium- to longer-term issues identified in the preceding pages of this text. In this instance, however, the pilot identifies a whole strategy for dealing with a massive change in his life (i.e. retirement). This, in some ways, brings us full circle back to the first quotation concerning oneself and one's plight, and serves to illustrate the full richness of coping issues identified by pilots.

Predictive powers of pilots' coping strategies

We used techniques previously employed to discover which of our four underlying statistical trends predicted each of our three outcome variables, i.e. mental ill-health, job satisfaction and job performance. However, before disclosing results, some introductory comments are worth noting since for coping strategies, the situation is slightly more complicated than for the predictions previously mentioned in other instances.

We might hypothesise that if a strategy is to be 'confirmed' by rigorous systematic means, we might in fact expect to find a *negative* relationship between the coping strategy and the outcome variable. This is the simple situation where as the coping strategy decreases the outcome variable (e.g. stress) increases. Clearly, however, the direc-

Table 9.2   *Significant coping trend predictors of mental ill-health,
job satisfaction and performance*[1]

	Predicting trend
Anxiety	– Social support
Obsessionality	Stability of relationships and home life
Overall neuroticism	– Social support
Intrinsic job satisfaction	– Social support
Job itself intrinsic satisfaction	– Social support
Overall satisfaction	Wife's involvement

[1] This table is based upon results fully displayed in Tables 6.3, 6.4 and 6.5.
– denotes a negative predictor.

tion of our outcome variable itself will be important – for example,
mental *ill*-health is negative, whereas job satisfaction is positive. To
summarise, therefore, one might expect the following: as the power of
our coping strategy increases, mental ill-health would decrease and
job satisfaction and performance would both increase.

Although this data has been previously presented, a summary table
of significant predictions is presented here (Table 9.2).

The trend concerning social support in the form of 'talking about
problems' and stresses to friends, colleagues and one's wife was a
positive predictor of anxiety and overall neuroticism. This is an
important result because it means that the findings of other studies
(with respect to the role of social support in coping) also apply to
pilots. Given our previous comments about the direction of the
prediction, the fact that the predictive power in this instance is
positive needs further explaining.

The positive predictive *power* of social support and anxiety/overall
neuroticism is a direct function of the composition of the trend itself.
It will be recalled that the three items concerning talking to one's
friends, colleagues and wife were themselves negatively related to the
trend. In other words, it was the *lack* of social support that was
important. Consequently, the predictive *power* in this instance is
positive, i.e. as the degree of social support becomes less and less, the
level of anxiety and overall neuroticism becomes greater and greater.
Clearly, then, this result does make sense.

Further surprises are also apparent in that the trend concerning
stability of home life and marital relationship was found to be a

positive predictor of obsessional behaviour! This positive relationship means that as stability of home life and marital relationship increases, so too does obsessionality.

To remind ourselves, the trend concerning home life consisted of such issues as 'stability of relationship with wife', 'home life that is smooth and stable' and 'wife who has known you through your flying career'. Characteristics of obsessionality are 'being too conscientious', 'being a perfectionist', 'feeling happiest when working', and 'feeling upset by a disruption in routine'. An explanation of the observed positive relationship could well lie in the fact that both the coping trend and the outcome involve recurrent, stable and routine elements. If viewed in this way, the positive relationships may well be intelligible.

Job satisfaction was positively predicted by the trend concerning the wife's involvement in helping the pilot prepare and recover from flying, and the interest she takes in his job. We have previously stated the important role the pilot perceives is played by domestic factors in both his working life and life in general. This finding means we can definitely state that the wife's interest and involvement will mediate the effects of stress in the form of an important psychological state, i.e. job dissatisfaction. Interestingly, however, the trend did not predict either mental ill-health or job performance.

Given the fact that the trends made logical sense, were supported by qualitative interview data and were consistent with the findings of this and other studies, one would have expected each of the four coping trends to have some significant degree of predictive power. It will be seen from Table 9.2 that only the trend 'reason and logic' (Trend 2) was not significant. In some respects this was to be expected in terms of job satisfaction and job performance. The former might not be predicted by logic and suppression of emotion because the issues involved, such as using skills and abilities, opportunities for advancement and so on, may not generate a sufficiently high degree of emotion to warrant this as an appropriate coping strategy. Performance variables too may not be appropriate. One would, however, have expected mental ill-health to be predicted by the trend. If one examines typical issues such as worrying for no reason, uneasiness and restlessness, feeling panicky and so on, then the application of an unemotional approach seems valid. One can only assume that the reason for this non-significance is that reason and logic was non-significant *relative* to the other issues included in the prediction process.

Conclusions on pilot coping

The subject matter investigated in this part of the study was novel in that we were unable to identify or locate any previous piece of research that had attempted to examine coping in pilots. This part of the study was included with the intention that it provide material of supplementary and background significance. It is apparent, however, that the data uncovered are considerably more interesting and illuminating than had at first been anticipated.

But where does this material on coping fit into our conception of stress and its impact on the pilot? The answer to this might well be demonstrated by the two representations shown in Figures 9.1 and 9.2.

Figure 9.1 is not intended to be a serious theoretical model. It is merely a visual aid to illustrate how coping fits into the stress/outcome relationship. A closer examination of the right-hand part of the

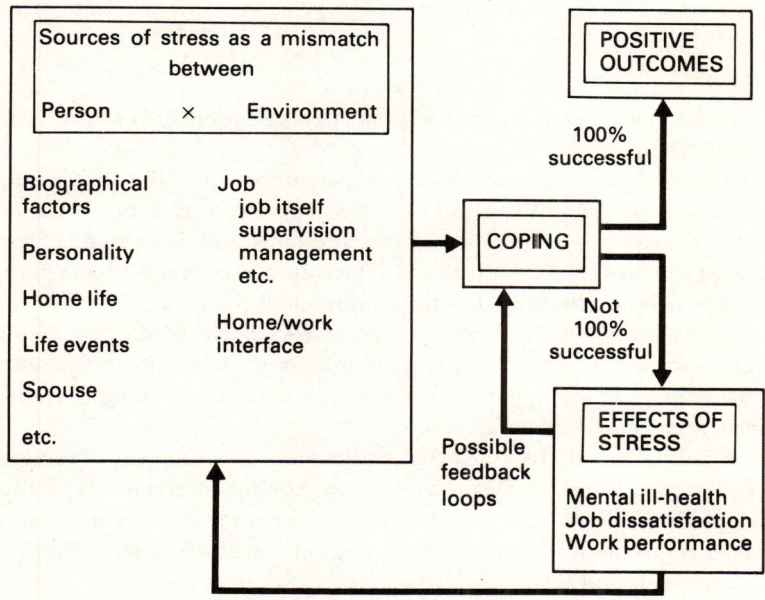

Figure 9.1 The position of coping variables relative to sources of stress and their effects

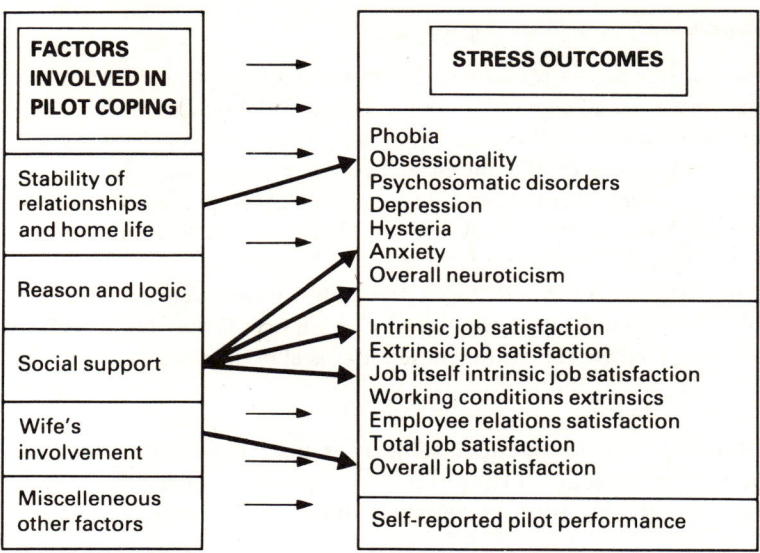

Figure 9.2 The relationship between pilot coping strategies and stress outcomes

diagram (i.e. the stress coping/stress effects relationship) is provided in Figure 9.2.

There are three primary issues that are important. The first is that there will be present in a pilot's coping repertoire an array of issues which have not been identified in the present study. Like many of the issues we have examined, some will be common to many pilots whilst others will be idiosyncratic to the individual pilot.

The second point to note is that arrows between the two parts represent the 'not 100% successful' relationship. Additionally, note that coping will be a function of a complex matrix of factors, in this case represented by small arrows.

Finally it should be noted that there seems to be an array of critical coping relationships that characterise coping strategies typically adapted by pilots (represented by large arrows). Can we draw any conclusions about these critical coping relationships as identified by the present study?

Our attention may be focused as an array of definitive statements that address underlying issues in pilot coping.

(i) Home life and the pilot's wife play a massive role. We already know this from the wives data; however the message is clearly repeated here. Interestingly, there is a marked absence of work-orientated or occupationally based coping issues.

(ii) There is emphasis on the practical rather than the emotional. Pilots tend to identify practical solutions to problems although ultimately the *impact* of those solutions must, of course, be psychological.

(iii) There is emphasis on structures rather than processes. This may well be related to point (ii) above. However, it is conceptually distinctive in that much of pilot coping seems to be a function of the way they organise things rather than the processes that exist within that organised structure. This in turn implies a deliberate, conscious effort on the part of the pilot.

(iv) There is little attention paid to coping with short-lived stresses or long-term stress. This may well be a function of the original questions asked throughout the study and the nature of the material on which we have concentrated. However, there is still a notable absence of coping with 'pilot stress' of the traditional kind (i.e. in the cockpit) or indeed of coping with large-scale stresses (i.e. those associated with life events, overall career, etc.).

(v) Psychological states seem to be a greater function of coping issues than do other indices. Again this will be a function of the approach we have taken and the way the variables were mentioned. However, mental ill-health and job satisfaction seem more closely determined by coping issues than does performance, for example.

Chapter 10
Some concluding comments

1985, when this book was written, was a bad year for aviation. The number of deaths through flying accidents was just under 2,000. Indeed, hardly a month elapsed without some media reference to flying-related accidents or incidents. In the examination of the causes of these incidents it must be recognised from the onset that these are not simple. One must adopt a realistic approach. Past data reveals that the causes are multitudinous. Whilst some 45 per cent or so of accidents may well be attributed to 'pilot error', such a simplistic statement conceals the fact that even if human error were the issue it should really only be regarded as being the major issue. In other words whilst mistakes made by pilots might well comprise this 45 per cent figure, other factors too will play a part.

In this text we have examined various issues within the broad umbrella of 'occupational stress'. The role of this chapter is not simply to reiterate the findings presented in previous chapters. Rather, we wish to draw together the material and highlight several important characteristics.

It will be recalled from chapter 1 that there were two basic issues of concern relating to the nature of occupational stress. The first was that it was possible to compose a model of occupational stress. This was based upon the analysis of many different occupations over a period of twenty years or so. The second issue, which was highlighted in chapter 2, was the apparent lack of a similar model for pilots. On the basis of our present studies, such a model for pilots is presented below in Figure 10.1.

In this basic model we present only specific parts of the data described earlier in the text. This is not to say that the features we include are more important. It is simply the case that inclusion of all or most of the results established would present a confusing picture even if based upon a relatively simple model.

We have included the eleven trends in sources of stress for pilots as comprising the main force in the stress equation. The other is the

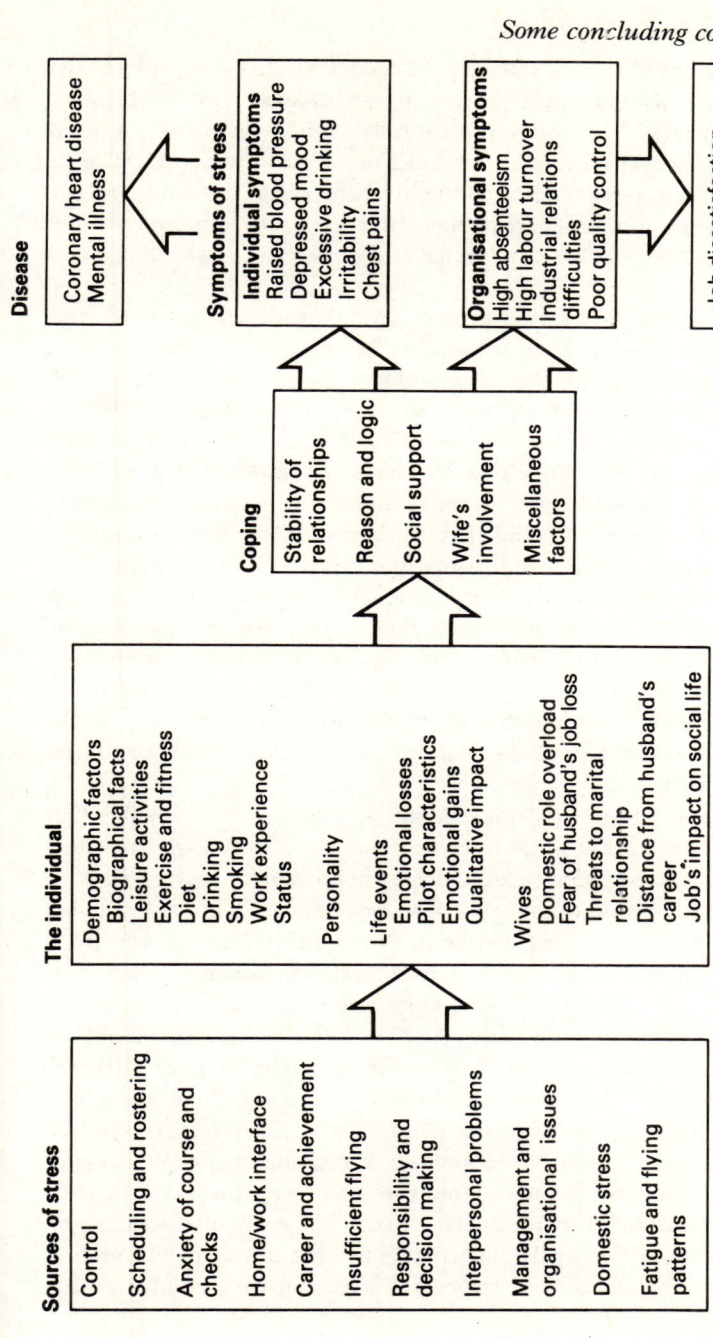

Figure 10.1 A basic model of pilot occupational psychosocial stress

individual. Within this category we have included biographical details, life events, the role played by spouse and finally the pilot personality which we did not specifically examine in our studies, but which we accept as well documented. On the assumption that stress is a result of a mismatch between the individual (e.g. the pilot) and the environment (e.g. job, career, home life, etc.), the situation will not always be negative. The individual may cope. We have included this in the basic model.

The data that is omitted from the model concerns the predictive relationships identified in the study. Were these to be included, the two broad arrows in the model would be replaced with a complex web of inter-relationships (too detailed to present here). The main issue to remember is that whilst we have identified a series of significant predicting variables, the causes of stress are *multivariate*. Rarely is any single cause to blame. It is usually some combination of sources of stress that interact with the characteristics of the person, combined with a failure of coping strategies, that leads to stress outcomes.

Since the role of this chapter is to record our lasting impressions of pilots and the nature of stresses they experience, an examination based simply upon individual and organisational consequences and implications would seem appropriate. Certainly by the nature of stress research it is possible to record both. On balance it is the implications for individuals that appear to be the more readily recorded.

What are our lasting impressions of pilots as individuals? Our overriding picture still has to be fairly positive. Although this study has concentrated upon stress, which by definition is negative, we have completed our studies in a good frame of mind. Pilots as a group are motivated, goal-directed, assertive individuals. They clearly think about the nature of the job they do. It seems reasonable to state that nearly all pilots enjoy their jobs. The major issue to note however is that such comments are most reliable when one defines 'the job' in terms of the actual process of flying aircraft. The terms of reference are important. This 'hands on' aspect of the job is one we would certainly identify as being the overriding positive feature. Indeed throughout our studies little mention was made of cockpit stress or negative features of flying. It would seem however that it is when 'the job' is defined in wider terms that other stress-related issues emerge. This can be verified by the identification of the eleven major trends in the study. Since this is important some further comments are pertinent.

Whilst it would be possible to devote this discussion to detailed examination of these stressors and their effects, there seems little point in repeating the conclusion already drawn. From a general viewpoint, however, the trends highlighted were consistent with those identified in other occupations. They were not however particularly well documented in the context of aviation. One lasting impression is that the relationship between work and home is as close as is the case for many other jobs. Whilst, from an individual's point of view, pilots report being able to separate the two, this seems a dubious process but one that is often quoted. Interestingly, there is a contrast between this dominant strong characteristic and the fact that many of the issues identified by pilots to cope with psychosocial stresses are in the hands of others, especially of the spouse. The pilots seem so strong and yet strangely vulnerable.

Whilst the process of actually flying the aircraft is a positive one and whilst other perceived psychosocial issues contribute negatively to the pilots' working experience, these are both in the context of the individual pilot's career prospects. The lasting picture here does seem rather gloomy. The pilots seemed to be frustrated, with little choice. Perhaps more important was that the prospect of being able to choose or being able potentially to change career was absent.

Clearly career issues will be more relevant for younger pilots at the beginning of the career cycle. Those at present in the mid- to late career stages have at least experienced some period of relatively 'good times' when the world of aviation was expanding and operating within a positive economic and psychological climate. Those presently in the early career stages are working in a relatively depressed situation. The prospects for individual pilots from a career perspective do not seem good. However, given other factors such as the massive positive contribution made by actually flying aircraft and the equally large developments in the application of new technology to flying, these may well offset dissatisfaction with career progression.

Finally, from an individual perspective, one is left with a surprising impression in terms of the stress outcomes measured in this study. We assume that pilot performance was satisfactory since all the pilots were flying at the time of examination though we did record fluctuations in how the pilot perceived the quality of performance. It was job satisfaction and mental ill-health that provided the more interesting individual data. Again, in terms of lasting impressions, it was surprising to record the fact that pilots did not conform to the satisfaction stereotype. Additionally, whilst all the pilots were medically fit, we did identify a group of pilots whom we feel would have provided interesting subjects for further study on the basis of their relatively elevated levels of mental ill-health.

The implications of our studies from an organisational perspective are much more difficult to tabulate. There seem to be three points most noteworthy. The first is that it seems reasonable to propose that a greater awareness be developed by organisations of the potential role played by psychosocial stresses and the implications that these can have at an individual level for pilots. We do not of course mean that pilots be psychologically tested wholesale. We simply mean that since no attempt appears to be made to measure or indeed manage for such issues, such a deficiency ought to be rectified. This is particularly valid given the relative percentage of contribution made by human error to accidents and incidents and also the relatively incomplete picture that is drawn in such a situation. An increased awareness would surely shed some light on a potentially important area that has otherwise been left untapped.

The second organisational implication related to this is that such issues are notoriously difficult to manage. Hence the lesson is that some attention should be focused upon the development of systems of identifying such stresses and the signs related to them with the overall goal of preventing the long-term outcomes of the sort measured by us.

The third implication is at this practical management systems end of airline operation. For example, stress management programmes have been installed and operate effectively in many other organisations outside aviation. These have been adapted by some American and European authorities with limited success. Such programs are to be welcomed. Whilst being only one solution they are indications of a responsible and intelligent industry coming to terms with a sensitive issue in a mature way.

It is with these thoughts in mind that our final overriding impression has to be as follows: 'Flying is a very safe way to travel.' We have looked throughout this text at a topic which is negative and controversial. It should be noted however that pilots are subject to regulation and control like no other profession. They are recruited, selected and trained to the highest standards practically possible. Additionally they and the quality and level of their performance are closely monitored. The standards to which they must adhere are clearly stated and indeed are themselves subject to monitoring and review. In other words, pilots operate within a complex system of checking, referencing and cross referencing. What we have done in our studies is to highlight the fact that despite all this, pilots are humans like the rest of us. It would seem appropriate therefore to finish with the following quotation. It is one that we used earlier to illustrate a point. It is worth repeating here for it sums up well how pilots feel about their work and their level of commitment to flying.

'. . . people are in it because it's their way of life. They are in it because they want to be; they don't want to be in anything else and would probably be happy to fly, even if they were not getting paid for it – well, almost! It's not really a job, it's part of one's life.'

Appendix: Questionnaire sent to pilot sample

UMIST
The University of Manchester Institute of Science and Technology
P.O. Box 88, Manchester M60 1QD, United Kingdom
Telephone: 061–236 3311
Telex 666094

Department of Management Sciences

SJS/LD

November 1983

Dear Sir

We hope you received the letter we sent to you some time ago, regarding the survey we are conducting by post. Please find attached a questionnaire booklet and a pre-paid envelope in which to return the completed questionnaire directly to us in Manchester.

Just to recap, we are trying to establish the nature of the relationship between home life and pilot health and performance. It is an important, large scale study and upon completion, the results will be carefully considered by all interested parties (e.g. BALPA). Certainly all of the pilots we have spoken to so far have been very enthusiastic to express an opinion.

You will find a number on the questionnaire. This is your reference number. Names and numbers are held for us by the Project Research Associate who is a medical doctor. They remain strictly confidential, however, if you wish, you may remove this number.

This booklet should take about 90 minutes of your time. We hope you will not find this over-burdensome. Once again may we take this opportunity to urge you to participate in the survey and thank you for your contribution.

Yours faithfully

Professor Cary L Cooper   Dr S J Sloan
Professor of Organisational Psychology   Research Fellow

THIS IS A CONFIDENTIAL SURVEY. *ALL INDIVIDUAL DATA* WILL REMAIN MEDICALLY IN CONFIDENCE AND WILL NOT BE DIVULGED TO ANY EXTERNAL BODY.

HERE IS YOUR REFERENCE NUMBER. YOU MAY REMOVE IT IF YOU WISH.

THE BASIS OF THE PROJECT

The identification of factors that affect the ways in which pilots behave have long been recognised as important topics for research. From such factors, the features of the pilot's home life have been identified as important, however their precise contribution to deviations from correct flight conduct has not been intensively researched.

A small amount of psychological literature does examine the influence of pilots' home lives on performance and on pilot health also. Several points are noteworthy. The characteristics of home lives examined almost exclusively concentrate on 'life events', i.e. very important things that have occurred in the pilot's life. Additionally, such effects are usually examined in the context of accidents and incidents. Whilst of tremendous importance and interest, we do not know the precise nature of the relationship between home and work for pilots. Secondly, we do not know the nature of such effects on a day to day basis. Other sources of data such as reporting systems do not permit the investigation into underlying causes beyond that which is volunteered in the report itself. Hence an investigation is necessary.

WHO WE ARE

UMIST is a technological university with an international reputation. The Department of Management Sciences has expertise in

Occupational Psychological research. We have been commissioned by an interested external body (United States Air Force) to undertake this research.

THE PROJECT OBJECTIVES

The objective of this project is to establish the domestic determinants of pilot health and performance and to examine the nature of the relationship between home and work for airline pilots.

DEFINITIONS

We would like you to interpret terms widely, for example 'domestic stress' and 'stressors'. These should simply be interpreted as referring to problems, events, occurrences which are important in home life. Remember too that stressors may be positive as well as negative.

The word 'stress' should also be interpreted in a wide sense, i.e. worry, tension, anxiety, anger or perhaps just mild irritation.

EXPLANATIONS

PLEASE READ THIS INFORMATION PAGE BEFORE YOU START TO COMPLETE THE QUESTIONNAIRES.

THIS BOOKLET

In this booklet you will find an array of questionnaires. They are mainly checklists and each has its own set of instructions and notes. Please read each set of instructions before starting to complete each checklist.

HOW TO ANSWER

For most of the questions please ring the number opposite your answer. If you make a mistake and ring the wrong number, cross it out and ring the correct number. For example,

The year is   1981   1

1982   ~~2~~

1983   ③

For other questions you simply write in your answer or complete as scheduled. If required to insert a number, please enter only one digit into each box provided.

WHAT YOU MUST REMEMBER

Since I shall not be present, I am depending on you to complete all questionnaires under 'scientific' conditions, so please note the following points.

1. Try to give your first and natural answer. This is best achieved by working quickly, but try also to be as honest and accurate as you can.
2. The questionnaires are to be completed by you and no one else.
3. The data must be given by you in private.
   Remember that I shall be keeping the data confidential, so must you.
4. Although some individual questions might seem unusual, remember that I shall be looking at groups of items, so please answer all questions.
5. Please remember the overall project basis, objectives and definitions.
6. Please ignore the numbers in brackets. These are for my own numbering scheme.

AFTER YOU HAVE READ THIS PAGE, PLEASE READ THE NEXT PAGE FOR FURTHER EXPLANATIONS

The scales which follow are a mixture of checklists and questions which relate to home life, work life and the relationships between the two.

You will find checklists compartmentalised into different sections. Within each section you may find more than one part. Please answer all questions.

The sections are separate and relate to the following issues:

Section		No of parts
1	Biographical data	1
2	Problems, occurrences, issues that are entirely domestically oriented	1
3	Problems, occurrences, issues that are entirely occupationally oriented	1
4	Relationship between work and home	1

PLEASE REMEMBER PREVIOUS INSTRUCTIONS AND TURN TO THE NEXT QUESTIONNAIRE WHICH EXAMINES BIOGRAPHICAL DATA

SECTION 1: BIOGRAPHICAL DATA

YOU AND YOUR FAMILY

(Card 1)

Sex?	Male	1	(1)	Age?	21–30	1	
	Female	2			31–40	2	(2)
					41–50	3	
					51–55	4	

Marital Status?		Married	1	
		Single	2	
		Divorced	3	(3)
		Widowed	4	
		Separated	5	
		Cohabiting	6	

If married *now* (either formally or in common law), does your partner work?

	Yes	2	
	No	1	(4)
	N/A	0	

If *yes*, do they work	Occasionally	1	
	Part-time	2	(5)
	Full-time	3	
	N/A	0	

Number of children ☐ ☐ (6) (7)
(number)

Age of children

Number of children over 18 years old ☐ ☐ (8) (9)
(number)

Number of children under 18 years old ☐ ☐ (10) (11)
(number)

Number of dependents (include partner and children where applicable)
☐ ☐ (12) (13)
(number)

YOUR INTERESTS

Do you find time to relax and 'wind down'	Always Sometimes Only when possible Not usually	1 2 3 (14) 4
Do you have an interest or hobby	Yes No	1 2 (15)
If yes, is it related to work	Yes No N/A	1 2 0 (16)
In general do you mix socially with other aviation colleagues outside work	Yes No	1 2 (17)

EXERCISE AND FITNESS

	Almost Always	Sometimes	Almost Never	
I maintain a desired weight avoiding overweight or underweight	0	1	3	(18)
I do vigorous exercises for 15–30 minutes at least 3 times a week	0	1	3	(19)
I do exercises that enhance my muscle tone for 15–30 minutes at least 3 times a week	0	1	2	(20)
I use part of my leisure time participating in individual family or team activities that increase my level of fitness	0	1	2	(21)
I do some type of gentle stretching exercises at least 3 times a week to improve flexibility	0	1	2	(22)

SMOKING

Do you smoke now?	Yes	2	
	No	1	(23)

If *yes*, how long have you smoked for (24) (25)
(enter oo if not applicable) ☐ ☐ years
(number)

If yes, please complete the following:

If you do not smoke, go to ⌐

	Almost Always	Sometimes	Almost Never	
I avoid smoking cigarettes	o	1	2	(26)
I smoke only low tar cigarettes or I smoke a pipe and cigars	o	1	2	(27)

	Yes	No	
I smoke less than half a pack daily	o	1	(28)
I smoke more than a pack daily	1	o	(29)
I have recently reduced my smoking	o	1	(30)
I have plans to quit smoking	o	1	(31)
I am currently attempting to quit	o	2	(32)

If you do *not* smoke now, have you ever smoked

	Yes	2	
	No	1	(33)
	N/A	o	

If yes, how long ago did you (34) (35)
stop (enter oo if not applicable) ☐ ☐ years
(number)

EATING HABITS

	Frequently	Sometimes	Almost Never	
Do you eat a variety of foods each day such as fruits, whole grain breads and cereals, lean meats, fish and poultry etc	0	1	2	(36)
Do you eat foods high in fat, saturated fat and cholesterol (e.g. fatty meats, eggs butter, cream, organ meats such as liver)	2	1	0	(37)
Do you eat salty foods, add salt at the table or use a lot of salt in cooking	2	1	0	(38)
Do you eat a large amount of sugar (especially sugary snacks, desserts and soft drinks)	2	1	0	(39)
Do you eat a high fibre diet including lots of whole grain bread and cereals, fresh fruits and vegetables	0	1	2	(40)

ALCOHOL

Do you drink alcohol	Yes	2	(41)
	No	1	

If *yes*; how many days per week do
you drink
    (42)
□
(number)

On those days on which you drink, on
average   how many drinks do you
have
    (43)
□ □
(number)

(Please complete the following)

Do you drink more than 2 drinks per day	2	1	0	(44)
Do you use alcohol as a way of dealing with stressful situations or life problems	2	1	0	(45)
Have you ever felt the need to cut down on drinking	2	1	0	(46)
Have you ever felt guilty feelings about drinking	2	1	0	(47)
Has anyone ever told you they think you drink too much	2	1	0	(48)

YOUR WORK HISTORY

Present employer, etc	B.A.	1	
	B. Cal.	2	
	Britannia	3	(49)
	Dan Air	4	
	Other (please state)	5	

....

(50) (51)

Number of years with present employer ☐ ☐ years

Current aircraft type ....

(Insert number as appropriate)
(one digit per box please)

(52) (53)

Number of years' experience on ☐ ☐ years
aircraft type

(54) (55) (56) (57)

Number of hours' experience on ☐ ☐ ☐ ☐ ☐ hours
aircraft type

(58) (59) (60) (61)

Total flying hours' experience ☐ ☐ ☐ ☐ ☐ hours

(62) (63) (64)

Average flying hours' experience ☐ ☐ ☐ hours
per month

Number of landings performed in aircraft type

(65) (66) (67)
☐ ☐ ☐

Total number of landings performed

(68) (69) (70)
☐ ☐ ☐ ☐

Average number of landings performed per month

(71) (72) (73)
☐ ☐ ☐

Average length of sector you fly
Do you generally fly . . .

☐ ☐ hours

Long haul	3
Short haul	2
Domestic only	1 (74)

Are you . . . .

Captain	4
Senior first officer	3
First officer	2 (75)
Other (please state)	1

....

Do you work at any other function (e.g. training captain)?

Yes 2
 0   1 (76)

If *yes*, please specify..

Seniority: This is difficult to assess since different companies use different methods. Please give us some indication of your seniority, using the scale below, on whichever basis is most important for you within the organisation for which you work.

Very High	5
High	4
Middle	3 (77)
Low	2 (Go to card 2)
Very Low	1

PLEASE CHECK THAT YOU HAVE ANSWERED *ALL* QUESTIONS AND PROCEED DIRECTLY TO THE NEXT SECTION

SECTION 2: DOMESTIC FACTORS

This section examines those issues that may be important in your home life. We are examining factors that are entirely domestic. We wish to measure their importance for you generally. Please ignore whether or not you feel these may or may not have carry over effects into your working life. Simply indicate whether something is stressful or not for you as an individual.

THE QUESTIONS

On the left hand side of the page is a list of factors which might be causing you stress.

THE ANSWERS

To answer the question of whether or not a certain factor may be causing you stress or not, circle the appropriate answer from the list on the right hand side.

PLEASE REMEMBER

There are no right or wrong answers
Give your first and natural answer by working quickly but be accurate?
There are six possible answers provided, remember that you may use any one to answer each question
Remember we interpreted the word 'stress' widely
Please ignore the numbers in brackets

	Causes me *much* stress	Causes me *some* stress	*Sometimes* causes me stress	Causes me *little* stress	Causes me *no* stress	(Card 2)
1. Disagreements, arguments, differences of opinion	5	4	3	2	1	(1)
2. Quality of marital relationship with partner	5	4	3	2	1	(2)
3. Degree to which household is 'geared to flying'	5	4	3	2	1	(3)
4. Family health	5	4	3	2	1	(4)
5. Nature of the home social environment	5	4	3	2	1	(5)
6. Lack of money	5	4	3	2	1	(6)
7. Dependability in, and competence of, spouse	5	4	3	2	1	(7)
8. Potential for extra marital relationships	5	4	3	2	1	(8)
9. Build up of tasks, duties and things to do	5	4	3	2	1	(9)
10. Issues associated with children (health, education etc)	5	4	3	2	1	(10)
11. Domestic situations that aren't clear cut	5	4	3	2	1	(11)
12. Worries on behalf of others	5	4	3	2	1	(12)
13. Conflicts of interests and resulting compromises	5	4	3	2	1	(13)
14. The 'good' use of time at home and how it is spent	5	4	3	2	1	(14)
15. Inability of spouse to fulfil their own abilities	5	4	3	2	1	(15)
16. Absence of calm, stability and dependability in home life	5	4	3	2	1	(16)
17. Constant, ongoing irritations	5	4	3	2	1	(17)

	Causes me *much* stress	Causes me *some* stress	*Sometimes* causes me stress	Causes me *little* stress	Causes me *no* stress	(Card 2)
18. Disappointment when others fail to meet expectations	5	4	3	2	1	(18)
19. Degree to which your personal goals and aims in life have been achieved	5	4	3	2	1	(19)
20. Success or failure of one's effort to achieve	5	4	3	2	1	(20)
21. Inability to identify problems (and hence solution)	5	4	3	2	1	(21)
22. New and unfamiliar experiences	5	4	3	2	1	(22)
23. Others not obeying or things that go wrong	5	4	3	2	1	(23)
24. Enforced or adapted roles at home	5	4	3	2	1	(24)
25. Responsibilities of home activities (e.g. PTA, Councillor etc)	5	4	3	2	1	(25)
26. Spouse's lack of understanding about the job	5	4	3	2	1	(26)
27. Not having someone to talk to about your work	5	4	3	2	1	(27)
28. Interpersonal relationships	5	4	3	2	1	(28)
29. The degree to which home life is the way you want it	5	4	3	2	1	(29)

PLEASE CHECK YOU HAVE ANSWERED *ALL* QUESTIONS
AND PROCEED DIRECTLY TO THE NEXT SECTION

SECTION 3: OCCUPATIONAL FACTORS

This section examines those issues that may be important in your life. We are examining factors that are entirely occupational. We wish to measure their importance for you generally. Please ignore whether or not you feel these may or may not be carryover effects into your home life. Simply indicate whether or not something is stressful or not for you as an individual.

	Causes me *much* stress	Causes me *some* stress	*Sometimes* causes me stress	Causes me *little* stress	Causes me *no* stress	(Card 2)
1. Career opportunities and lack of potential advancement	5	4	3	2	1	(30)
2. Seniority systems	5	4	3	2	1	(31)
3. Impending major career change or threat (e.g. redundancy, redeployment etc)	5	4	3	2	1	(32)
4. Not enough hours actually spent flying	5	4	3	2	1	(33)
5. Sharing of work evenly	5	4	3	2	1	(34)
6. Scheduling	5	4	3	2	1	(35)
7. Patterns of flying (i.e. relative times you are asked to fly)	5	4	3	2	1	(36)
8. Interpersonal problems with aircrew	5	4	3	2	1	(37)
9. Interpersonal problems with cabin staff	5	4	3	2	1	(38)
10. Style of management	5	4	3	2	1	(39)
11. Lack of management support	5	4	3	2	1	(40)
12. The whole experience (before & during) of medical checks	5	4	3	2	1	(41)
13. The whole experience (before & during) of checks on your flying ability	5	4	3	2	1	(42)

	Causes me *much* stress	Causes me *some* stress	*Sometimes* causes me stress	Causes me *little* stress	Causes me *no* stress	(Card 2)
14. Misuse of time (e.g. low amount of preparation time, delays)	5	4	3	2	1	(43)
15. The aggregate or cumulative effects of minor tasks when flying	5	4	3	2	1	(44)
16. Fulfilling role expectations (either as a captain or first officer)	5	4	3	2	1	(45)
17. Changes in your experience of flying (e.g. conversion course)	5	4	3	2	1	(46)
18. Impact of a lack of flying (practice effects)	5	4	3	2	1	(47)
19. Future career uncertainty	5	4	3	2	1	(48)
20. Tiredness and fatigue	5	4	3	2	1	(49)
21. Morale and organisational climate	5	4	3	2	1	(50)
22. Conditions of employment	5	4	3	2	1	(51)
23. Factors not under your direct control	5	4	3	2	1	(52)
24. Periods in flight of high workload	5	4	3	2	1	(53)
25. Inherent responsibility of your job	5	4	3	2	1	(54)
26. Making important decisions	5	4	3	2	1	(55)
27. Ambiguous factors or difficulties in problem identification	5	4	3	2	1	(56)
28. Attaining your own personal levels of performance	5	4	3	2	1	(57)
29. Situations that are ongoing	5	4	3	2	1	(58)
30. Effects of being over familiar (with routes, type, routines, etc.)	5	4	3	2	1	(59)

PLEASE CHECK YOU HAVE ANSWERED *ALL* QUESTIONS
AND PROCEED DIRECTLY TO THE NEXT SECTION

SECTION 4: FACTORS OF WORK THAT MAY AFFECT YOU AT HOME

This section examines those sources of pressure or problems that are associated with your job. In particular it concentrates on those that may have an effect on you outside work i.e. we are examining the relationship in the direction of work to home.

	This item is a source of pressure to me . . .	Always	Usually	Sometimes	Seldom	Never	
1. Tiredness and fatigue		5	4	3	2	1	(60)
2. How time of work determines when to sleep		5	4	3	2	1	(61)
3. Returning home and time of arrival		5	4	3	2	1	(62)
4. Scheduling and rosters		5	4	3	2	1	(63)
5. Patterns of flying		5	4	3	2	1	(64)
6. Unpredictability of when you are asked to fly		5	4	3	2	1	(65)
7. Social problems associated with rosters		5	4	3	2	1	(66)
8. Seasonal fluctuations in workload (& hence effects)		5	4	3	2	1	(67)
9. Changes in family life due to job (e.g. promotions, base change etc)		5	4	3	2	1	(68)
10. Anxiety of courses and checks		5	4	3	2	1	(69)
11. Preparation necessary for courses and checks		5	4	3	2	1	(70)
12. Factors out of your control (i.e. delays, cancellations).		5	4	3	2	1	(71)
13. How long a single period of flying lasts (trip)		5	4	3	2	1	(72)
14. How many sectors you are asked to fly		5	4	3	2	1	(73)
15. Effects of minor day to day things		5	4	3	2	1	(74)

This item is a source of
pressure to me . . . .

Always
Usually
Sometimes
Seldom
Never

16. Carry over effects of personality
    clashes or interpersonal issues   5 4 3 2 1   (75)

(Go to card 3)

PLEASE CHECK YOU HAVE ANSWERED *ALL* QUESTIONS
AND PROCEED DIRECTLY TO THE NEXT SECTION

This section contains 4 parts which examine the following

Part 1   Home factors which may affect you at work
Part 2   The nature of effects home factors might have for you at work
Part 3   Factors that might generally affect pilot performance
Part 4   The relative importance and difference between sources of stress from home and from work

SECTION 5: PART 1 HOME FACTORS WHICH MAY AFFECT ME AT WORK

	The size of effect I perceive is	Very high	High	Moderate	Low	Very low	(Card 3)
1. Efficiency of pre-flight preparation time (i.e. period at home *not* on duty)		5	4	3	2	1	(1)
2. Issues or situations that are ongoing or left unresolved		5	4	3	2	1	(2)
3. Overall satisfaction with home life		5	4	3	2	1	(3)
4. How satisfied one is on how things have been left		5	4	3	2	1	(4)
5. Lack of stability		5	4	3	2	1	(5)
6. Division of loyalties		5	4	3	2	1	(6)
7. Indirect results of home life activities		5	4	3	2	1	(7)
8. Marital problems		5	4	3	2	1	(8)
9. Spouse's attitude to flying		5	4	3	2	1	(9)
10. Length of time spent at home		5	4	3	2	1	(10)
11. Serious events that occur		5	4	3	2	1	(11)
12. Particular arrangements that have been disrupted		5	4	3	2	1	(12)

SECTION 5: PART 2 NATURE OF EFFECTS OF HOME FACTORS ON WORK

	I can usually tell when I'm experiencing stress because at work I react this way	Always	Usually	Sometimes	Seldom	Never	
1. Tendency to worry		5	4	3	2	1	(13)
2. Experience of tiredness due to disrupted sleep		5	4	3	2	1	(14)
3. Increased alcohol consumption when not flying		5	4	3	2	1	(15)
4. Decreased concentration		5	4	3	2	1	(16)
5. Reoccurrence of factors in thoughts of item during period of low workload (e.g. cruise)		5	4	3	2	1	(17)
6. Make errors without knowing why		5	4	3	2	1	(18)
7. Make errors of omission		5	4	3	2	1	(19)
8. Decreased quality of preflight preparation		5	4	3	2	1	(20)
9. Tendency to talk about the issue at work		5	4	3	2	1	(21)
10. Slows one down		5	4	3	2	1	(22)
11. Tendency not to listen as intently		5	4	3	2	1	(23)
12. One's mind becomes detached from tasks in hand		5	4	3	2	1	(24)

SECTION 5: PART 3 FACTORS WHICH MIGHT
GENERALLY AFFECT PILOT PERFORMANCE

Below you will find a list of factors that might determine a pilot's
performance. We would like you *to rank them*

(i) Rank your top 6 items in their order of importance in actually
determining pilot performance on the basis that you have perceived them in the past and know from your own experience how
important or unimportant each of them may be.

PLEASE RANK THE MOST IMPORTANT AS 1 AND THE
REMAINDER DOWN TO THE LEAST IMPORTANT AT 6.
PLEASE DO *NOT* RANK ITEMS AT EQUAL VALUES.

1. Weather conditions .... (25)
2. Inability to separate home events from work life (26)
3. Overfamiliarity (with type, routes etc) .... (27)
4. Fatigue .... (28)
5. Mind that is 'full of other things' .... (29)
6. Relative time of day one is asked to fly .... (30)
7. Interpersonal relations with aircrew .... (31)
8. Upset preflight routine (at home) .... (32)
9. Carry over effects of home life events .... (33)
10. Health (not 'ill', but being 'out of sorts') .... (34)
11. Poor preflight preparation .... (35)
12. Things that are not under direct control of pilot
    (e.g. ATC) (36)

SECTION 5: PART 4 RELATIVE IMPORTANCE AND
DIFFERENCES IN NATURE OR SOURCES OF STRESS

The aim of this part is to gain an insight into the relative importance
generally in your life, of sources of problems that are work oriented
and those sources that are home oriented. Additionally, we would like
some indication of the nature of the differences and similarities
between the two groups of sources.

1. Generally in my life sources of pressure that arise from work

account for the following percentage of that which I experience:

0–10%	0	51–60%	5	
11–20%	1	61–70%	6	
21–30%	2	71–80%	7	(49)
31–40%	3	81–90%	8	
41–50%	4	91–100%	9	

2. Generally in my life sources of pressure that arise from home account for the following percentage of that which I experience:

0–10%	0	51–60%	5	
11–20%	1	61–70%	6	
21–30%	2	71–80%	7	(50)
31–40%	3	81–90%	8	
41–50%	4	91–100%	9	

Below is a series of statements. Please indicate your feelings towards each statement by putting a circle around the number of the answer that best describes how you feel. It is the words of the answers that are important, not the numbers, so read the answers carefully.

3. One can become more emotional about domestic things than job factors

Strongly Disagree	5
Disagree	4
Uncertain	3
Agree	2
Strongly Agree	1
	(51)

4. Although problems may arise associated with work, it is always possible to leave and get away from them, which one cannot do with problems that occur in home life.

Strongly Disagree	1
Disagree	2
Uncertain	3
Agree	4
Strongly Agree	5
	(52)

205

5. Job stresses occur in short bursts

Strongly Disagree	1
Disagree	2
Uncertain	3
Agree	4
Strongly Agree	5
	(53)

6. Job stresses occur at an acute and intense level

Strongly Agree	1
Agree	2
Uncertain	3
Disagree	4
Strongly Disagree	5
	(54)

7. Home stresses tend to be ongoing

Strongly Agree	5
Agree	4
Uncertain	3
Disagree	2
Strongly Disagree	1
	(55)

8. Problems at work tend to be 'cut and dry' requiring 'yes or no' answers

Strongly Disagree	1
Disagree	2
Uncertain	3
Agree	4
Strongly Agree	5
	(56)

9. Chains of command, which do not exist at home, help to deal with job pressures

Strongly Agree	5
Agree	4
Uncertain	3
Disagree	2
Strongly Disagree	1

(57)

10. Home stress tends to occur at a less intense level

Strongly Disagree	5
Disagree	4
Uncertain	3
Agree	2
Strongly Agree	1

(58)

11. Home problems are more far-ranging in their implications for me

Strongly Agree	5
Agree	4
Uncertain	3
Disagree	2
Strongly Disagree	1

(59)

12. Relative importance of home versus work sources is seasonally dependent

Strongly Agree	1
Agree	2
Uncertain	3
Disagree	4
Strongly Disagree	5

(60)

13. Home factors have greater effects for me

Strongly Agree	1
Agree	2
Uncertain	3
Disagree	4
Strongly Disagree	5
	(61)

14. Stress at work is under my control, unlike things at home

Strongly Disagree	1
Disagree	2
Uncertain	3
Agree	4
Strongly Agree	5
	(62)

15. Stress (from either work or home) can ultimately only be overcome if I have a job

Strongly Disagree	5
Disagree	4
Uncertain	3
Agree	2
Strongly Agree	1
	(63)

16. Stresses tend to separate but affect me 100% in each

Strongly Disagree	1
Disagree	2
Uncertain	3
Agree	4
Strongly Agree	5
	(64)

PLEASE CHECK YOU HAVE ANSWERED *ALL* QUESTIONS
AND PROCEED DIRECTLY TO THE NEXT SECTION

SECTION 6: LIFE EVENTS AND LIFE CHANGES

We define a 'life event' quite simply as something that occurs in an individual's life, that has an important effect on that person. The event may be either positive or negative. For example, getting married, moving house, birth of child etc.

This section has 3 parts which examine the following:

Part 1   Life events you may have experienced personally
Part 2   Life events generally
Part 3   Life changes and experiences

SECTION 6: PART 1

If you consider your life generally, would you say that you have experienced any events which have been important to you
Yes   1
No   2   (65)

    If yes,
    What was the event (please explain briefly) ....
                              (66) (67) months
    How long ago was it   ☐   ☐

                                       (68) (69)
    How long did the period of the event last ☐   ☐   months

    Please sum up in a phrase your feelings *before the event* ....
    ....

    during the event ....
    ....

    after the event ....
    ....

Do you think your performance was affected

Yes	2
No	1
N/A	0
	(70)

Do you think your performance could have been affected without you realising

Yes	2
No	1
N/A	0
	(71)

Do you think your performance could have been affected without others (e.g. colleagues) realising

Yes	2
No	1
N/A	0
(72)	

SECTION 6:   PART 2

Do you think generally that life events can affect pilot performance

Yes	2
No	1
	(73)

Do you think that negative events can affect pilot performance

Yes	2
No	1
	(74)

Do you think that positive events can affect pilot performance

Yes	2
No	1
	(75)

Do you think that performance could be affected by life events but not to an extent that (answer all parts)

is measurable	Yes	2
	No	1
		(76)

is perceivable by the pilot	Yes	2
	No	1
		(77)

is perceivable by others	Yes	2
(e.g. colleagues)	No	1
		(78)

safety and proper flight	Yes	2
conduct are affected	No	1
		(79)

minimum operational standards	Yes	2
are not met	No	1
		(80)

(Go to card 4)

SECTION 6: PART 3

The list below is a series of descriptions of characteristics, life events and life changes which have been used by other researchers to describe pilots who have made errors or who have been involved in incidents. Some are positively oriented, others are negatively oriented, however all might have some effect on pilot performance.

TRY TO THINK OF A TYPICAL EXAMPLE OF A PILOT WHO IS PERFORMING BADLY OR WHO IS LIKELY TO MAKE AN ERROR. CAN YOU DRAW A PORTRAIT OF HIM?

We would like you to do so by weighting each of the items using your own experience and knowledge of other pilots. Use the scale below to rate each of the descriptions on the extent to which they are or are not influential in describing such a pilot.

EXAMINING YOUR PORTRAIT, DOES SUCH A PILOT . . .

	Very definitely Yes	Definitely Yes	Yes	No	Definitely No	Very definitely No	
1. Characteristically exhibit poor judgment	6	5	4	3	2	1	(1)
2. Recently undergo a marital separation (for reasons other than duty location)	6	5	4	3	2	1	(2)
3. Handle life difficulties well	6	5	4	3	2	1	(3)
4. Recently have a death in the family	6	5	4	3	2	1	(4)
5. Impress others as a good leader	6	5	4	3	2	1	(5)
6. Have marital problems	6	5	4	3	2	1	(6)
7. Exhibit the characteristics of maturity and stability ...	6	5	4	3	2	1	(7)
8. Recently lose a close friend through death	6	5	4	3	2	1	(8)
9. Exhibit mastery of his aircraft within operational parameters	6	5	4	3	2	1	(9)
10. Have financial difficulties	6	5	4	3	2	1	(10)
11. Recently have a new addition to the family (i.e. birth, adoption etc)	6	5	4	3	2	1	(11)
12. Impress others as a good team member	6	5	4	3	2	1	(12)
13. Recently undergo a divorce	6	5	4	3	2	1	(13)
14. Recently become engaged	6	5	4	3	2	1	(14)
15. Make any recent major decisions regarding the future	6	5	4	3	2	1	(15)

	Very definitely Yes	Definitely Yes	Yes	No	Definitely No	Very definitely No	
16. Exhibit professionalism in his approach to flying	6	5	4	3	2	1	(16)
17. Have difficulty with interpersonal relationships	6	5	4	3	2	1	(17)
18. Recently have trouble with superiors	6	5	4	3	2	1	(18)
19. Have a sense of humour and humility concerning himself	6	5	4	3	2	1	(19)
20. Exhibit the ability to quickly assess potentially troublesome situations	6	5	4	3	2	1	(20)
21. Recently get married	6	5	4	3	2	1	(21)
22. Recently have trouble with peers or others	6	5	4	3	2	1	(22)

PLEASE CHECK YOU HAVE ANSWERED *ALL* QUESTIONS
AND PROCEED DIRECTLY TO THE NEXT SECTION

SECTION 7: COPING

So far, we have examined problems that arise at work and at home, the relationships between the two and events that may have occurred in your life. This section is about 'coping' with such things.

When faced with various problems or situations we all try to deal with them by some form of 'coping'. Some ways are automatic or unconscious whilst others are things that we are aware of doing. Some might simply be the way you have organised things or indeed just particular personal preferences.

Examine the list of items. They are a list of characteristics, techniques or factors that might be important in helping you cope with problems or stresses. We would like you to assess each item on whether or not it is important for you using the following scale:

7   OF PARAMOUNT IMPORTANCE TO ME IN COPING
6   VERY IMPORTANT
5   IMPORTANT
4   NEITHER IMPORTANT NOR UNIMPORTANT
3   UNIMPORTANT
2   VERY UNIMPORTANT
1   OF NO IMPORTANCE WHATSOEVER TO ME IN COPING

Insert the number of your answer beside each item.

Please note: if you are unmarried, for the word 'wife', please substitute 'partner', 'girlfriend' etc, as appropriate

Please insert the
number of your answer:

1. Wife who had prior knowledge of flying or who flies .... (23)
2. Hobbies .... (24)
3. Home that is a 'refuge' .... (25)
4. Talking to understanding friends .... (26)
5. Unconsciously separating home and work (leading two lives) .... (27)
6. Deliberately avoiding confrontation .... (28)
7. Process of flying itself helps .... (29)

8. Home life that is 'geared to flying'
   (in practical terms) .... (30)
9. Wife who involves herself and is interested .... (31)
10. Involving oneself in physical pastimes and exercise .... (32)
11. Home life that provides a psychological platform .... (33)
12. Talking to an understanding wife .... (34)
13. Working things out by logic .... (35)
14. Wife who modifies her own behaviour and demands
    to suit you .... (36)
15. 'Staying busy' .... (37)
16. Home life that is smooth and stable .... (38)
17. Talking to understanding colleagues .... (39)
18. Living in a non-flying social environment .... (40)
19. Planning ahead .... (41)
20. Staying emotionally aloof or shrugging things off .... (42)
21. Not 'bottling things up' .... (43)
22. Wife who is 'efficient' in looking after things .... (44)
23. Expanding one's interests outside aviation
    (e.g. small business venture) .... (45)
24. Sleep .... (46)
25. Stability of relationship with wife .... (47)
26. Deliberately suppressing emotion .... (48)
27. Reversal of roles at home .... (49)
28. Smoking .... (50)
29. Selective attention (i.e. concentrating on single
    problems) .... (51)
30. Using distractions (to take your mind off things) .... (52)
31. Drinking alcohol .... (53)
32. Stability of relationships with colleagues .... (54)
33. Wife who has known you through your flying career .... (55)

PLEASE CHECK YOU HAVE ANSWERED *ALL* QUESTIONS
AND PROCEED DIRECTLY TO THE NEXT SECTION

SECTION 8:   JOB SATISFACTION

This set of items deals with various aspects of your job. We would like you to tell us how satisfied or dissatisfied you feel with each of these features of your present job.

Please use the scale below to indicate your feelings.

PLEASE REMEMBER

There are no right or wrong answers
Give your first and natural answer by working quickly but be accurate
Remember to answer all questions
Please ignore the numbers in brackets

Just indicate how satisfied or dissatisfied you are with each of the various aspects of your job by using this scale

1. I'm extremely dissatisfied
2. I'm very dissatisfied
3. I'm moderately dissatisfied
4. I'm not sure
5. I'm moderately satisfied
6. I'm very satisfied
7. I'm extremely satisfied

Simply write down in the box provided the number of your answer.

1. The physical work conditions	☐	(56)
2. The freedom to choose your own method of working	☐	(57)
3. Your fellow workers	☐	(58)
4. The recognition you get for good work	☐	(59)
5. Your immediate boss	☐	(60)
6. The amount of responsibility you are given	☐	(61)
7. Your rate of pay	☐	(62)
8. Your opportunity to use your abilities	☐	(63)
9. Industrial relations between management and workers in your firm	☐	(64)
10. Your chance of promotion	☐	(65)

11. The way your firm is managed · ☐ (66)

12. The attention paid to suggestions you make · ☐ (67)

13. Your hours of work · ☐ (68)

14. The amount of variety in your job · ☐ (69)

15. Your job security · ☐ (70)

16 Now, taking everything into consideration, how do you feel about your job as a whole · ☐ (71) (go to card 5)

PLEASE CHECK YOU HAVE ANSWERED *ALL* QUESTIONS
AND PROCEED DIRECTLY TO THE NEXT
QUESTIONNAIRE

SECTION 9: YOUR HEALTH

Since you are flying, you hold a valid Medical Certificate. Hence by definition you are 'fit'. As you will realise of course it is possible for your health to fluctuate yet still meet the minimum standards required. The implications for us as researchers is that we must resort to simply examining various degrees of fitness. In turn, this assumes that whatever measurement scale we use is fairly sensitive.

To induce some sensitivity into our measurements we are assuming that you can monitor fluctuations in your health. The questionnaire in this section is concerned simply with the way you feel or act.

To reply please circle your answer.

(Card 5)

1. Do you often feel upset for no obvious reason? Yes No (1)

2. Do you have an unreasonable fear of being in enclosed spaces such as shops, lifts etc? Often Sometimes Never (2)

3. Do people ever say you are too conscientious? No Yes (3)

4. Are you troubled by dizziness or shortness of breath? Never Often Sometimes (4)

5. Can you think as quickly as you used to? Yes No (5)

6. Are your opinions easily influenced? Yes No (6)

7. Have you felt as though you might faint? Frequently Occasionally Never (7)

8. Do you find yourself worrying about getting some incurable illness? Never Sometimes Often (8)

9. Do you think that 'cleanliness is next to godliness' No Yes (9)

10. Do you often feel sick or have indigestion? Yes No (10)

11. Do you feel that life is too   At times   Often   Never   (11)
    much effort?

12. Have you, at any time in your   Yes   No   (12)
    life, enjoyed acting?

13. Do you feel uneasy and   Frequently   Sometimes
    restless?   Never   (13)

14. Do you feel more relaxed   Definitely   Sometimes
    indoors?   Not particularly   (14)

15. Do you find that silly or   Frequently   Sometimes
    unreasonable thoughts keep   Never   (15)
    recurring in your mind?

16. Do you sometimes feel   Rarely   Frequently   Never
    tingling or pricking sensations   (16)
    in your body, arms or legs?

17. Do you regret much of your   Yes   No   (17)
    past behaviour?

18. Are you normally an   Yes   No   (18)
    excessively emotional person?

19. Do you sometimes feel really   No   Yes   (19)
    panicky?

20. Do you feel uneasy travelling   Very   A little   Not at all
    on buses or the Underground   (20)
    even if they are not crowded?

21. Are you happiest when you   Yes   No   (21)
    are working?

22. Has your appetite got less   No   Yes   (22)
    recently?

23. Do you wake unusually early   Yes   No   (23)
    in the morning?

24. Do you enjoy being the centre   No   Yes   (24)
    of attention?

25. Would you say you were a   Very   Fairly   Not at all   (25)
    worrying person?

26. Do you dislike going out   Yes   No   (26)
    alone?

27. Are you a perfectionist?   No   Yes   (27)

28. Do you feel unduly tired and exhausted?   Often   Sometimes   Never (28)

29. Do you experience long periods of sadness?   Never   Often   Sometimes (29)

30. Do you find that you take advantage of circumstances for your own ends?   Never   Sometimes   Often (30)

31. Do you often feel 'strung up' inside?   Yes   No   (31)

32. Do you worry unduly when relatives are late coming home?   No   Yes   (32)

33. Do you have to check things you do to an unnecessary extent?   Yes   No   (33)

34. Can you often get off to sleep all right at the moment?   No   Yes   (34)

35. Do you have to make a special effort to face up to a crisis or difficulty?   Very much so   Sometimes   Not more than anyone else (35)

36. Do you often spend a lot of money on clothes?   Yes   No   (36)

37. Have you ever had the feeling you are 'going to pieces'?   Yes   No   (37)

38. Are you scared of heights?   Very   Fairly   Not at all   (38)

39. Does it irritate you if your normal routine is disturbed?   Greatly   A little   Not at all (39)

40. Do you often suffer from excessive sweating or fluttering of the heart?   No   Yes   (40)

41. Do you find yourself needing to cry?   Frequently   Sometimes   Never   (41)

42. Do you enjoy dramatic situations?   Yes   No   (42)

43. Do you have bad dreams which upset you when you wake up?   Never   Sometimes   Frequently   (43)

44. Do you feel panicky in crowds?

Always   Sometimes   Never
(44)

45. Do you find yourself worrying unreasonably about things that do not really matter?

Never   Frequently
Sometimes   (46)

46. Has your sexual interest altered?

Less   The same or greater
(46)

47. Have you lost your ability to feel sympathy for other people?

No   Yes   (47)

48. Do you sometimes find yourself posing or pretending?

Yes   No   (48)

PLEASE CHECK THAT YOU HAVE ANSWERED *ALL* QUESTIONS AND PROCEED DIRECTLY TO THE NEXT SECTION

Appendix

SECTION 10: PERFORMANCE

As indicated in our objectives, we are trying to examine the determinants of performance. Since this is a survey by post, we must resort to indirect measures of performance by asking you to assess yourself. You will realise that this is, of course, somewhat unsatisfactory. Hence we are again relying on you to instill as high a degree of sensitivity as possible, by being as honest and as accurate as you can.

1. Think about your last few flights recently
2. Consider how well or badly you performed
3. Examine the list of elements below, they are different ways of assessing performance
4. Please rate yourself on the scales by circling the number of your answer. Remember, we are relying on you to make this as accurate a measure as possible.

1. Being ahead of the game:
   Ahead for 100% of flight   2 1 0 1 2   Behind for 100% of flight   (49)

2. Excess mental capacity:
   Plenty of excess capacity during flights   2 1 0 1 2   No excess capacity during flights   (50)

3. Coping with things that go wrong:
   Coped very satisfactorily   2 1 0 1 2   Coped very unsatisfactorily   (51)

4. Attaining self-set levels of performance:
   Attained self-set levels of performance for flights   2 1 0 1 2   Did not attain self-set levels of performance for flights   (52)

5. Smoothness and accuracy of approaches
   Very smooth & accurate approaches   2 1 0 1 2   Very unsmooth & inaccurate approaches   (53)

6. Smoothness and accuracy of landings:
   Very smooth & accurate landings   2 1 0 1 2   Very unsmooth & inaccurate landings   (54)

7. Degree of basic airmanship exhibited:

	2	1	0	1	2	
Very high degree of basic airmanship		——————		——degree of basic airmanship		Very low (55)

8. Overall smoothness of flights:

	2	1	0	1	2	
Very smooth		——————		——unsmooth		Very (56)

9. Quality of interpersonal relations with aircrew:

	2	1	0	1	2	
High and satis-factory quality		——————		——unsatis-factory quality		Low and (57)

10. Degree of mental and physical coordination:

	2	1	0	1	2	
Very high degree of coordination		——————		——degree of coordination		Very low (58)

11. Number of errors made:

	2	1	0	1	2	
Relatively high number		——————		——low number		Relatively (59)

12. Extent of errors made:

	2	1	0	1	2	
Relatively high importance		——————		——low importance		Relatively (60)

13. Satisfaction with flights generally:

	2	1	0	1	2	
Very high degree of satisfaction		——————		——degree of satisfaction		Very low (61)

14. Ability to divide attention:

	2	1	0	1	2	
Very high ability		——————		——ability		Very low (62)

15. Many pilots when asked to assess the quality of their performance reply that it is 'just at feeling' – can you assess yourself on a scale in this way?

	2	1	0	1	2	
Very good		——————		——poor		Very (63)

Appendix

WHAT HAPPENS NOW

Your questionnaire will be scored by me and after transfer to computer, will be statistically analysed with the questionnaires of other respondents. Together with the writing and submission of reports this process should be completed by May 1984.

You will have realised by now that this is an original and unique piece of work which examines issues of paramount interest. Your contribution is therefore gratefully acknowledged.

FINAL CHECK

There are several final points I would like you to make:

(i) Please check through the booklet again and ensure that you have answered all questions

(ii) Please send the completed questionnaire directly to me in the pre-paid envelope enclosed

ONCE AGAIN, THANK YOU FOR PARTICIPATING

Bibliography

Aitken, R.C.B. (1969), 'Prevalence of worry in normal aircrew', *British Journal of Psychology*, 42, 283–6.

Alkov, R.A. and Borowsky, M.S. (1980), 'A questionnaire study of psychological background factors in US Naval aircraft accidents', *Aviat. Space Environ. Med. 51*, 860–3.

Alkov, R.A., Borowsky, M.S. and Gaynor, J.A. (1982), 'Stress coping and the US Navy aircrew factor mishap', *Aviat. Space Environ. Med. 53*, 1112–15.

Beattie, R.T., Darlington, T.G. and Cripps, D.M. (1974), *The Management Threshold*, British Institute of Management Paper OPN 11.

Beehr, T.A., Walsh, J.T. and Taber, T.D. (1976), 'Relationship of stress to individually and organisationally valued states: higher order needs as a moderator', *Journal of Applied Psychology*, *61*, 41–7.

Canon, W.B. (1929), *Bodily Changes in Pain, Hunger, Fear and Rage*, New York, Appleton.

Caplan, T.D., Cobb, S., French, J.R.P. *et al.*, (1975), 'Job demands and workers' health', Washington, DC, US, Dept. of Health, Education and Welfare Publication no. (NIOSH) 75–160, US Government Printing Office.

Carrol, J.J. (1969), 'Emergency escape and survival factors', *Proceed. Aerospace Med. Assoc.*, Florida.

Chirp (1983), 'Confidential and anonymous reporting system', RAF IAM, Farnborough, Hants.

Christy, R.L. (1975), 'Personality factors in selection and flight proficiency', *Aviat. Space Environ. Med. 46*, 309–11.

Civil Aviation Authority (1975, 2nd edn 1982), 'The avoidance of excessive fatigue in aircrews. A guide to requirements', CAP 371.

Cobb, S. and Rose, R.H. (1973), 'Hypertension, peptic ulcer and diabetes in air traffic controllers', *Journal of Australian Medical Association*, *224*, 489–92.

Cooper, C.L. (1981), *The Stress Check*, New Jersey, Prentice-Hall.

Cooper, C.L. (1981), *Stress Research Issues for the 1980s*, Chichester, John Wiley & Sons.

Cooper, C.L. and Davidson, M.J. (1982), *High Pressure: Working Lives of Women Managers*, London, Fontana.

Bibliography

Cooper, C.L., Davidson, M.J. and Robinson, P. (1982), 'Stress in the police service', *Journal of Occupational Medicine*, *24*, 1, 30–6.

Cooper, C.L. and Kelly, M. (1984), 'Stress among crane operators', *Journal of Occupational Medicine*, *36*, August, 575–8.

Cooper, C.L., Mallinger, M. and Kahn, R. (1978), 'Identifying sources of occupational stress among dentists', *Journal of Occupational Psychology*, *51*, 227–34.

Cooper, C.L. and Marshall, J. (1976), 'Occupational sources of stress. A review of the literature relating to coronary heart disease and mental ill health', *Journal of Occupational Psychology, 49*, 11–28.

Cooper, C.L. and Melhuish, A. (1980), 'Occupational stress and managers', *J. Occ. Med., 22*, 588–92.

Cooper, C.L. and Payne, R. (1980), *Current Concerns in Occupational Stress*, Chichester, John Wiley & Sons.

Cooper, C.L. and Roden, J. (1985), 'Mental health and satisfaction among tax officers', *Social Sci. and Med., 21*, 7, 747–51.

Cooper, C.L. and Smith, M.J. (1985), *Job Stress and Blue Collar Work*, Chichester, John Wiley & Sons.

Cox, I. (1980), 'Repetitive work', in *Current Concerns in Occupational Stress*, C.L. Cooper and R. Payne (eds), Chichester, John Wiley & Sons.

Crisp, A.H., Ralph, P.C., McGuiness, B. and Harris, G. (1978), 'Psycho-neurotic profiles in the adult population', *Brit. J. Psychol., 51*, 293–301.

Crown, S. and Crisp, A.H. (1966), 'A short clinical diagnostic self rating scale', *Brit. J. Psychiat., 112*, 917–23.

Crown, S. and Crisp, A.H. (1979), 'Manual of the Crown-Crisp experiential index', London, Hodder & Stoughton.

Crump, J.H., Cooper, C.L. and Maxwell, V.B. (1981), 'Stress among air traffic controllers: occupational sources of coronary heart disease risk', *Journal of Occupational Behaviour, 2*, 293–303.

Crump, J.H., Cooper, C.L. and Smith, M. (1980), 'Investigating occupational stress: a methodological approach', *Journal of Occupational Behaviour, 1*, 191–224.

Davidson, M. and Cooper, C.L. (1983), *Stress and the Female Manager*, Oxford, Blackwell.

Davidson, M. and Cooper, C.L. (1984), *Working Women: An International Survey*, Chichester, John Wiley & Sons.

Erickson, J.M., Pugh, W.M. and Gunderson, K.E. (1972), 'Status congruency as a prediction of job satisfaction and life stress', *Journal of Applied Psychology, 56*, 523–5.

Farmer, R.G., Monahan, L.H. and Hekeler, R.W. (1984), *Stress Management for Human Services*, Beverly Hills, Cal., Sage.

Flack, M. (1918), *Flying Stress*, London, Medical Research Committee.

Fletcher, B.C. and Payne, R.L. (1980), 'Stress and work: a review and theoretical framework 1', *Personnel Review, 9*, 1, Winter.

Folkard, S. and Monk, T.H. (eds) (1985), *Hours of Work – Temporal Factors in Work Scheduling*, Chichester, John Wiley & Sons.

French, J. and Caplan, R. (1972), 'Organisational stress and individual strain', in *The Failure of Success*, A.J. Marrow (ed.), New York, Amacon, 31–66.

Goerres, H.P. (1977), 'Subjective stress assessment: a new simple method to determine pilot workload', *Aviat. Space Environ. Med.* 48, 558–64.

Goldberg, L. and Comstock, G.W. (1976), 'Life events and subsequent illness,' *Am. J. Epidemiol.*, *104*, 146–58.

Green, R.G. (1977), 'The psychologist and flying accidents', *Aviat. Space Environ. Med.* 48, 922–3.

Haward, L.R.C. (1977), 'Pilot beware!' *Nursing Mirror*, June.

Holmes, T.H. and Rahe, R.H. (1967), 'The Social Readjustment Rating Scale', *J. Psychosom. Res.*, *11*, 213–18.

Kasl, S.V. (1973), 'Mental health and work environment. An examination of the evidence', *Journal of Occupational Medicine*, 15, 506–15.

Kelly, M. and Cooper, C.L. (1981), 'Stress among blue collar workers', *Employee Relations*, *3*, 2, 6–9.

Kroes, W.H. (1976), *Society's Victim – The Policeman: An Analysis of Job Stress in Policing*, Springfield, Ill., Charles C. Thomas Publishers.

McCarron, P.M. and Haakonson, N.H. (1982), 'Recent life change measurement in Canadian forces pilots', *Aviat. Space Environ. Med. 53*, 6–13.

McMichael, A.J. (1978), 'Personality, behavioural and situational modifiers of work stressors', in *Stress at Work*, C.L. Cooper and R. Payne (eds), Chichester, John Wiley & Sons, 127–47.

Margolis, B.L., Kroes, W.H. and Quinn, R.P. (1974), 'Job stress: an unlisted occupational hazard', *Journal of Occupational Medicine*, *16*, 654–61.

Otway, H.J. and Misenta, R. (1980), 'The determinants of operator preparedness for emergency situations in nuclear power plants', paper presented at the Workshop on Procedural and Organisational Measures for Accident Management, Nuclear Reactors International Institute for Applied Systems Analysis, Laxenburg, Austria, 28–31 January.

Pahl, J.J. and Pahl, R.E. (1981), *Managers and Their Wives*, London, Allen Lane.

Parry, M. and Fokkeman (1958), *Aviation Psychology in Western Europe*, Amsterdam, Swet & Zeitlinger.

Payne, R. (1980), 'Organisational stress and social support', in *Current Concerns in Occupational Stress*, C.L. Cooper and R. Payne (eds), Chichester, John Wiley & Sons, 269–98.

Rahe, R.H., (1978) 'Life change measurement clarification', *Psychosom. Med., 40*, 1–2.

Reinhardt, R.G. (1966), 'Accident proneness in aviation', *Texas Med. 62*, 75–8.

Selye, H. (1946), 'The general adaptation syndrome and the disease of adaptation', *Journal of Clinical Endocrinology, 6*, 117.

Selye, H. (1976), *Stress in Health and Disease*, London, Butterworths.

Bibliography

Selye, H. (1983) 'The stress concept: past, present and future', in C.L. Cooper (ed.), *Stress Research: Issues for the Eighties*, Chichester, John Wiley & Sons.

Shirom, A., Eden, D., Silberwasser, S. and Kellermann, J.J. (1973), 'Job stresses and risk factors in coronary heart disease among five occupational categories in Kibbutzim', *Social Science and Medicine*, 7, 875–92.

Shuckburgh, J.S. (1975), 'Accident statistics and the human factor element', *Aviat. Space Environ. Med. 51*, 46–50.

Simonov, P.V., Frollov, M.V., Ivanov, E.A. (1980) 'Psychophysiological monitoring of operator's emotional stress in aviation and astronautics', *Aviat. Space Environ. Med. 51*, 46–50.

Ursano, R.J. (1980), 'Stress and adaptation: the interaction of the pilot personality and disease', *Aviat. Space Environ. Med. 51*, 1245–9.

Warr, P., Cook, J. and Wall, T. (1979), 'Scales for the measurement of some work attitudes and aspects of psychological well-being', *Journal of Occupational Psychology, 52*, 129–48.

Wilby, J. (1985), *Good Career Guide*, London, *Sunday Times*.

Yanowitch, R.E. (1977), 'Crew behaviour in accident causation', *Aviat Space Environ. Med. 48*, 918–21.

Index

Pilots Under Stress

S.J. Sloan and C.L. Cooper

Routledge & Kegan Paul
London and New York

First published in 1986 by
Routledge & Kegan Paul Ltd
11 New Fetter Lane, London EC4P 4EE

Published in the USA by
Routledge & Kegan Paul Inc.
in association with Methuen Inc.
29 West 35th Street, New York, NY 10001

Set in 10 on 12 pt Imprint
by Inforum Ltd, Portsmouth
and printed in Great Britain
by T J Press (Padstow) Ltd
Padstow, Cornwall

Library of Congress Cataloging in Publication Data

Sloan, S. J. (Stephen J.)
Pilots under stress.

Bibliography: p.
Includes index.
1. Aeronautics—Psychology.   2. Stress
(Psychology)   I. Cooper, Cary L.   II. Title.
TL555.S58   1986   629.132'5216'019   86–13923

British Library CIP Data also available

ISBN 0-7102-0479-5

Contents

Figures

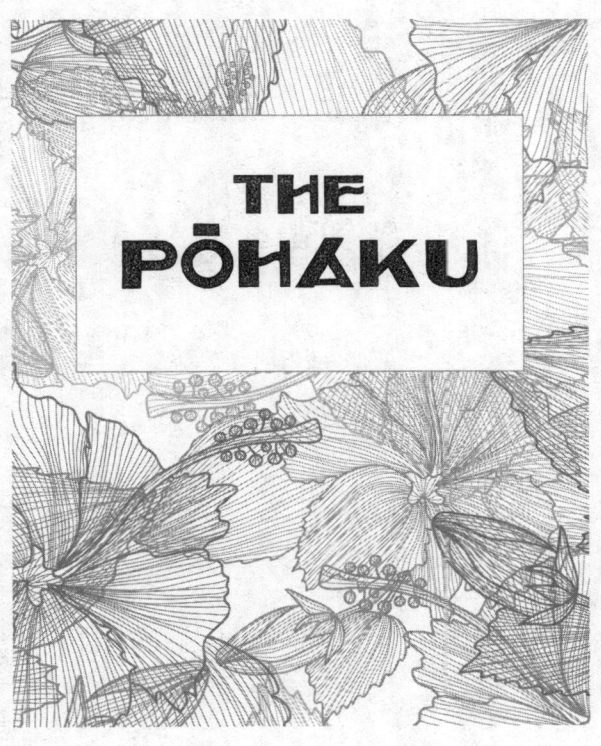

THE PŌHAKU

ALSO BY JASMIN `IOLANI HAKES

Hula

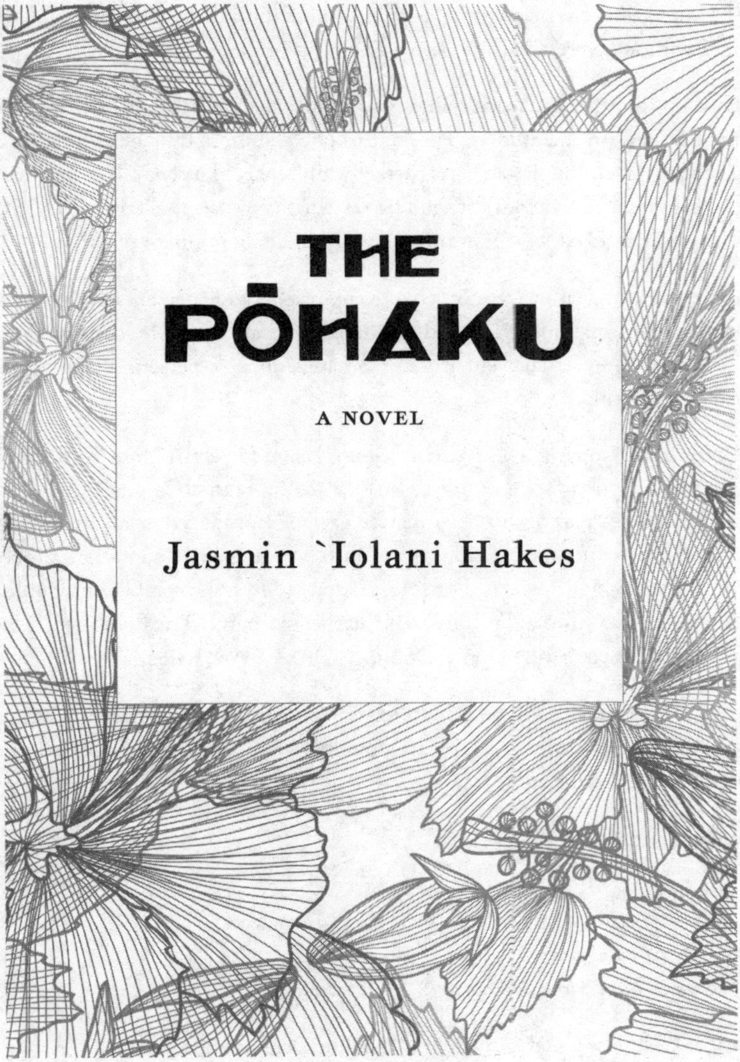

THE PŌHAKU

A NOVEL

Jasmin `Iolani Hakes

HarperVia

An Imprint of HarperCollinsPublishers

For Aunty Jamie, the keeper of our family stories

YOU HAVE TO KNOW WHO YOU ARE.

Henry Azbill, Mechoopda Maidu, Hawaiian

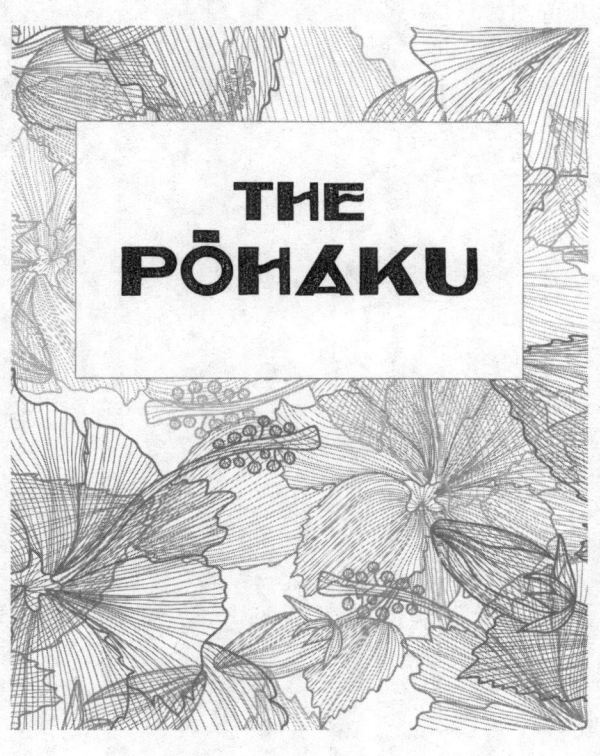

THE PŌHAKU

AUGUST 19, 1992

Mo‘opuna, the hospital says i can’t bring food or plants inside the building. What is their medicine, compared to ours? I brought a jar of homemade māmaki brew too, but with those tubes in your mouth I don’t know how to make you drink it. Good thing I brought the salt and ti. But now that I’m looking at you I don’t know how I’m going to lomi your body with them either. You’re covered in so many wires and tubes that it looks like you’ve turned into a very sad and broken marionette—Pinocchio in reverse. Even your legs are bound, can you feel that? Each wrapped hip to ankle, in something like the oversized cuffs they use to check your blood pressure. For circulation, the nurse tells me.

The nurse on duty is a small Filipina lady wearing scrubs printed with baby ducks. They remind me of the rubber ducky you used to chew on when you were small enough to bathe in the sink. When I think of how much of your life you’ve spent in water, I just don’t know how this could happen. Going to Queen’s Bath at night. I could maybe understand a tourist being so stupid. But a local kid from Kaua`i is supposed to know the island better than that. What could you possibly have been doing down there? The cops interviewed those ragamuffin friends of yours, as if they'd ever tell a cop anything. Wait until I find them. They're going to talk to me or else.

The nurse checked all your bells and whistles and plugged notes into your chart before noticing that the tape holding your breathing machine in place needed adjusting. How come they can invent all these fancy machines and give people new body parts but nurses have to carry around tape in their pockets to hold things together?

Why is your face still blue, your fingers still pruney? You've been out of the water for hours now. Give me your hands. We'll start there. Does that feel okay, is the salt too rough? Maybe I'll just go over your body with the ti today. I can put the salt in the corners of the room, try to clear it out that way.

Your cuticles are raw. Looks like you started biting them again. Something about them makes me feel old and helpless. I want to ask you why you'd hurt yourself like that. I want to clean your ow-ies and kiss them better as if you were small. Tomorrow I'll come with wipes and hand lotion. Maybe some polish for your nails. Give you a manicure and get you all fancy. Your feet are banged up in ways that can't just be from this. I heard you've been sleeping in your car, but your feet make me wonder if that's true. Maybe it's better I don't know. My heart hurts. I want to brush your hair and wipe you down so the nurses and doctors here don't think you are forgotten or lost, that you have someone praying for you, but they were the ones to put you in this bed and cover you with these blankets so it's already too late for that.

I keep a picture of you in my purse, from May Day the year you lost both your front teeth. You were always kalohe, but through that gap the rascal in you came out full throttle. Back then your mama was always asking me what she did that she was being punished for. (Why she bothered asking, I dunno, she knew just as much as all of us how good her snout was for rooting up trouble.) Then she would laugh and laugh. You know her, before everything happened she was always laughing. Underneath she was a good person, right to the end.

The nurse says everything is looking A-okay even though you look terrible. I'm sorry for saying so. What am I supposed to do with her words, celebrate? She only means nothing has gotten worse. She patted you on the shoulder before she left, as if you'd accomplished something.

Your security guard is the opposite of the nurse. A stone face that wouldn't laugh at a joke if I hit him over the head with it. He's informed me that he's under strict orders to never leave your side because they think you jumped on purpose. Since one of the walls of your tiny room is glass he's agreed to sit on the other side of it and watch you from there when I'm visiting. He doesn't need to know our business. At least he respects his elders. It's not like you're going anywhere. Who's he protecting anyway? Us from you, or you from yourself? You wouldn't have jumped on purpose. I can't ever believe that. I remember the last time we talked. It was when the hotel fired you.

"Because nobody wants to come to Hawai`i for vacation when there's a war going on," you explained, when I asked why. You were so excited to work there. Always telling me you couldn't come visit because you had to work. Didn't ever call in sick or come in late.

I wasn't good about paying attention to the news on TV, but you went on about Desert Storm, how a war in Iraq was to blame for The Coco firing your entire crew.

"The state is giving six million dollars to the Visitors Bureau to

3

run commercials on the mainland to try to get more tourists," you said. When I asked why the state didn't just put those six million dollars into helping everybody losing their jobs throughout the islands, you laughed. You were angry. Not suicidal.

You're so still. I don't trust the monitors. I watch them but I'm not sure what I'm hoping they will do. Maybe that they will tell me something God isn't.

From the other side of the window the guard is giving me funny kine looks. I'll slide the ti under your arms and tuck one under your lower back when he's not looking. That will have to do for now.

The ICU is on a floor of the hospital I have never been before. The rooms here are different from the ones downstairs, different from the open room with the big wide window facing the ocean where I gave birth to your mama. Less light, more whispers. This room has only one window and it's facing inside, imagine that, eh? From in here it's impossible to know which way is east, which way west. I wish I could bring in some fresh air for you, get some makani into your hair and sweep all these bad things out of your room. They were clustered thick when I first came in, this cloud I've been watching slowly accumulate around you in the years since your mama died. I wish I had done something then. I was so hardhead.

Yes, you're not the only one with a head too hard. Where do you think you got it from?

I thought you just needed time. I know I did. Needed to let the bleeding stop.

Your tutu needs a cigarette. I don't know if I'm supposed to be mad or sad right now. Forgive an old lady her pilau habits. I'll be right back.

. . .

Charlie Boy was downstairs, trying to convince the receptionist to tell him what room you were in. She wasn't budging on the only family rule. He looked destroyed. He's always been sweet on you, I think.

"Hey, aunty," he said after kissing my cheek. "Can I see her?"

"Come have a smoke with me," I told him.

He hadn't been at Queen's Bath, but he saw you at the party before. Didn't want to admit how trashed you were. He was worried he'd get you in trouble. I said we were past that.

"I heard the talk that she jumped on purpose. I wanted to come find you. I didn't want you to believe it. She didn't."

"Tell me what you know, Charlie Boy. Whatever we can tell the doctors about what happened the better they can help her."

I didn't think that was true, but I needed to know.

"She'd been drinking a lot," he said. "The cops came and broke up the party. Everybody scrambled as soon as we saw those blue flashing lights coming up the road. I thought I saw her jump into a guy's car with a couple of other girls. One of those girls, Iwa, lives down the street from me. I saw her when she got home. She said they had all pretty much sobered up by the time they got to Queen's Bath. They parked at the top of the trail to hang out and smoke a joint. No one planned to hike down to the water. They ended up going down the trail anyway. Iwa was slower than the rest of them, the moon was out but wasn't giving out enough light to navigate all the rocks. By the time she got to the cliff the group had already broken off into couples. From what they could see the water was calm. It was a freak wave. Came out of nowhere. There was a scream that sounded like it was coming from really far away. Iwa didn't think the scream and the wave were connected until they all found each other again and realized someone was missing. The guys dove in, managed to pull her out. Iwa couldn't say for sure how long she'd been in the water. The only thing she knew for certain was that it was an accident."

Oh Mo`opuna, what he said gutted me. That wave that swept you away. We've gotten so disconnected from the `āina. Of course you kids don't know how to read the ocean. Nobody grows their own food anymore, people want to dig into Pele's home on the Big Island to make a geothermal plant, pollution and garbage everywhere, hotels in Maui and O`ahu diverting all the water and drying up the islands, acid rain, pesticides . . . so much imbalance. Yes, you were stupid and when you wake up I'm going to kill you. But it's also not your fault. Something, well, something happened. Many, many years before you were born. So long ago that I thought it didn't matter anymore. But obviously it caused a disconnect that has only gotten more dangerous and deadly. For a long time I thought everything could heal itself, regenerate. Now I don't know. But that wave wasn't a wave. It was the truth that I tried to ignore. I can't ignore it anymore. There's only one thing that can save you. I know what I need to do. I think I've always known, though the duties we inherit don't always come with the courage they require, yeah? I've prayed and prayed and it keeps leading me back to the same answer, so I'm taking that to mean God is giving me the go-ahead. I will look to Him to guide both of us through this.

On second thought, maybe I should leave Him out of this. Maybe this doesn't have anything to do with Him at all. Oh Mo`opuna, the pain it causes me to say that. It's been a long time since your tutu left God out of anything.

Here's the thing. The doctors and nurses think they know what's wrong with you, but they say only time will tell. They tell me all they can do is help your body by keeping your parts going until your body can do that on its own again. They have no idea

6

how wrong they are. No idea how little they know about what is really going on here. But I know. I knew before the phone rang; the truth had been dripping into the bottom of my stomach for some time, turning it sour.

Have you ever heard your tutu apologize for anything, admit she makes mistakes? Now I am, okay? It's not too late. I was a coward and we all paid, all continue to pay, for it. Just stay alive long enough for me to make it right. I promise I will.

Mo`opuna, your tutu has been keeping a secret for a very long time. The Empty, that feeling deep inside you, that thing that's been eating away at you for so long, was born in the vacuum of that secret. This is not an excuse, but if you lock something away long enough there's a chance you forget where you put the key. I was supposed to tell it all to your mama but I never got the chance. Or maybe I stopped believing any of it meant anything. Maybe I thought it best to leave the past in the past. Just goes to show how wrong an old lady can be. Wake up and be mad at me, okay? The nurses say your ears still work and that I should talk to you. Aiyah, if that's not the universe right there giving me my marching orders then I don't know what would be!

What I have to tell you will take some time, so you're going to have to stick around, you hear me? I'll tell you everything, but you have to promise me that you will listen. That you will stay alive and fight and let your tutu give you the medicine.

• • •

It took from March 17, 1768, to sometime in 1777 for the princess Ka`ahumanu to be born. Nine years.

Dates. They like to think they're important.

Whatever the day and year, the date was not significant, in spite of it being what would change the world forever.

In the time before the islands were united into a kingdom—you knew this right, the islands weren't always united, your teachers taught you at least that much in school? Auwe, the fact that I don't even know the answer to that question fills me with shame—there was much fighting between them. Each island was ruled by local chiefs and its own king. Territory was defended, protected, and fought over. In this time there was a hill in the district of Hana on Maui called Pu`u Ka`uiki. This hill was part of the territory controlled by a Hawai`i Island chief who was a supporter of the efforts to conquer the island and make it part of a united Hawaiian Kingdom. This position had made enemies out of the other chiefs who wanted to maintain their independence, so when this chief's wife went into labor, he needed to keep their location a secret to ensure their safety. To the hill of Pu`u Ka`uiki they went. Within the hill was a series of lava tubes. The chief's wife ordered her husband's guards to remain outside, allowing only her female attendants to go with her to Ka`uiki Cave, deep into the heart of the hill.

· · ·

Aiyah. Mo`opuna, you're going to have to kala mai your tutu—I was so gung ho to get started and so upset about you and the wave and the thought of losing you that I jumped the gun. I started the story wrong. It's been so long time since I thought about any of it that it's gotten jumbled in my head.

At church, Pastor says that before God made the world there was only darkness. I won't disagree, but I will add this: Just because someplace is dark, don't mean it's empty. Darkness holds all kinds of things. My point is that sometimes it's hard to say when or where something started. From that darkness did the chicken come, or was it the egg? The prophecy of the pōhaku was like that. It rose from the lava core of Middle Earth, its mana so deep and

8

powerful that it was as if the knowledge of it had no beginning, no origin. But just like a game of Telephone, the truth behind the prophecy got so warped over time that eventually there was very little, if any, truth still in it at all.

In the schoolbooks everything always begins when the men of Europe enter the picture, so we gotta start there too. (Told you this was going to take a while! But stay with me. Please stay with me.) They say Captain Cook was searching for a northern passage between the Pacific and Atlantic Oceans and that's how he came to our homeland. This is true, but the sort of true that says Hawai`i is the fiftieth state of America.

• • •

At the heart of the real story of Captain Cook is the prophecy of the pōhaku and the story of a man named John Montague. You're not going to find this story in any history book. You stopped trusting me a long time ago, but for this you are going to have to give me another chance.

1740

JOHN MONTAGUE, THE 4TH EARL OF SANDWICH, SET OUT INTO the world around the year 1740, as was customary at the time for up-and-coming fancy pants to do after college. Montague's official destinations were Turkey, Greece, and Egypt, but there were other, unofficial stops too.

It was in Jerusalem that the true mission of his trip was revealed to his crew. You see, during his time at Cambridge, Montague had befriended an eccentric old history professor whom everyone dismissed as having lost his marbles. From him Montague learned about the Lost Tribes of Israel, in particular the group who were said to have escaped Jerusalem by ship, sailing around the world until they reached Oceania, driven by a prophecy that spoke of a special map, one that explained the connection between the earth and the cosmic corridors surrounding it, the planets and stars beyond what the eye could see. A user's manual of the universe, basically—and endorsed by God.

But Jerusalem was a dead end. If anyone there knew about a

lost tribe or a prophecy, they weren't talking. It was as though the people who had gone in search of it had never existed, which made him wonder if there was any legitimacy to this prophecy. Full of doubt, Montague returned to England and took the seat in the House of Lords that had been waiting for him since the day he was born.

But the prophecy continued to haunt him. A man like Montague was raised to think he deserved the keys to the whole universe, a world for him to conquer, just waiting. If there was a user's manual of the universe, he thought he deserved to have it. Over time, his doubts faded. Remember Uncle Jay's dog, the time it bit down on his favorite chicken and wouldn't unlock his jaw no matter how much Uncle Jay beat him? That was how Montague felt about Britain using all its resources to become the greatest naval power on earth. No matter what anyone else talked about in Parliament, every day the Earl of Sandwich came only with another long rambling speech about Navy this, Navy that. No one could understand his obsession. He wasn't the kind of guy who felt the need to justify himself, but the bottom line was that if he was going to discover what lay at the heart of the prophecy, he was going to need ships to search the seven seas in every direction.

Building the greatest fleet in the world was what would secure England's world power—Montague declared this not only to his wife who was, by then, slowly losing her mind, and then to his mistress, but also to the unfortunate men stuck with him on those long nights at the card table. His colleagues and friends suffered through this broken record monologue while he scarfed down the only meal he ever ordered: sliced meat flanked by slices of bread (a *sandwich* that he'd come up with specifically so he could eat without leaving his game and his audience). But he knew what people are only now beginning to realize: our wisdom runs deep. To learn more, to uncover whatever truth existed of it, he needed to follow the footsteps of the tribal peoples swallowed by Oceania. It was a watery trail. He bided his time and kept impressing upon anyone who crossed his path that boats were needed, for everyone's sake.

Chip away at a tree with a butter knife long enough and it'll eventually give up and fall over. By the year 1745, Montague was the commissioner of the Admiralty, responsible for the command of the British Royal Navy. But between the Jacobite rising and the threat of French invasion (historical incidents that don't have anything to do with our family so I can't say I'm an expert, and for the purposes of your kuleana, nothing we need to go into), and the murder of his opera singer mistress (and mother to nine of his ten children, the snake), his heart shattered in a way it had not during his wife's slow demise. He was forced to shelve his desire to conquer Oceania.

Still, Lord Sandwich allotted much of his admiration and affection to a certain Captain James Cook. It was to Cook that Lord Sandwich first revealed the true nature of his dedicated support of the navy, and to Cook he bestowed the duty of uncovering the truth of the prophecy.

The two men made certain to emphasize publicly that the first expedition was being undertaken in the name of science. They were not pirates. This was not a sunken treasure. Technically, they were not lying. What was science if not the earth, the stars, the planets? And after only a short time at sea, Cook understood that the world's waters were sprinkled with little sprays of land freckles, reflecting a night's sky full of stars. Anyone could find a continent. It was like looking for an elephant in a bathtub. Islands were harder. They moved, they swayed like trees in the wind. A fraction of a calculation and they would slip through your fingers, disappearing forever into the horizon. He wanted to collect them, to wear them around his wrist like charms.

On his first voyage, Cook rounded Cape Horn and arrived in Tahiti. There he discovered a different form of the story, understood by the peoples of Tahiti to be a long-standing myth, the kind you tell your children. The islanders he encountered revealed nothing, but this is never the way men learn. It was the shadows in the night, the curried favors, and the traded fabrics and goods that brought the good information forward.

Their version of the prophecy spoke not of a divine user's manual of the world but of a great power whose emergence was imminent. Not a map, but a human. One who would pulse with the understandings of the world and beyond. Who would hold the secrets of the cosmos and inner earth both, secrets that would not only command the high seas but also everything they touched. Cook considered how this might apply to him should this person become his ally, or better yet, his captive. With this he'd be more than a hero. He'd be a god.

He continued on until he reached New Zealand, where he stayed long enough to make contact with the Māori people and determine that they too believed in the existence of what he was searching for. He mapped the entire coastline of New Zealand and then set sail again, sure he was getting close to his target.

Every archipelago hummed with the undercurrent of anticipation for the coming of this great power. Stories varied how it would reveal itself. All Cook could do was continue onward toward the watery edge of the earth.

Finally, in Australia, he skirted its perimeter until he came upon the Gweagal people. If they had any knowledge of the prophecy, they admitted none of it to Cook. His pursuit had gone cold.

When he returned to Britain he was publicly lauded as a

scientific hero, but he privately implored Montague to set him out again the minute that his ships were repaired. On this first journey he had discovered what and where the prophecy wasn't. Given that the world could only be so large, this was almost the same as finding out where it was. All he needed was another go. So out he went again, and after finding his way to New Zealand, he gave the surrounding seas a good sweeping, instantly ruling out the Friendly Islands, Easter Island, Norfolk Island, New Caledonia, and Vanuatu.

This second tour catapulted him into celebrity. A star was born. Only he and Lord Sandwich considered his success premature, their secret mission not yet accomplished. The third time would be the charm.

It was cyclone season, but instead of hampering their journey to Tahiti, steady trade winds worked in Cook's favor, pushing his ship through doldrums and the calm belt of the equator. He took this as a blessing from above. Unlike the winds of his two previous voyages, when it felt like the entire world was resisting him, these winds felt like the encouraging tug of a hand urging him forward. Finally, after weeks of open waters, birds of the most vibrant reds and yellows he had ever seen began appearing in the sky above them. The air became laced with a sweet perfume that was noticed even below deck. His dreams became saturated with flavors that stayed on his tongue long after he woke. A rumor circulated among the crew that someone had overheard him humming to himself. They were in uncharted territory.

By the time Cook and his men arrived at the sweeping cliffs and unblemished white sand shores of Kaua`i, he knew without a doubt that he was where he had been trying to find all these years. *Sandwich's islands!* he shouted, voice shaking, to his men. He had found what he had dedicated his entire career to finding. He still had no clear idea exactly what this prophesied power would reveal itself to be, but Cook was sure that it was only a matter of time before he found it.

As it happened, he and his men had arrived at the onset of Makahiki, when working ceased and our festivals to celebrate life and the `āina's abundance began. Since these were still the days when akua freely walked and lived among humans, it was critical to determine if this white foreigner was an akua, akin to the highest ranking ali`i of the islands. If he was, he would need to be treated with a particular formality and protocol of respect. Keep in mind: Lono, being the god of fertility and agriculture and peace, is celebrated in the utmost during Makahiki times. Lono was always welcomed and invited to participate in these celebrations of the `āina's production, so it was not unreasonable to consider the possibility that this year he had accepted the invitation of the people and showed up. It needed to be determined whether this foreigner was in fact our god Lono, given that, in all their encounters with those of other lands, he came in a form no one had ever seen before.

Soon enough, locals in canoes circled the ships, quickly identifying Cook as the person they needed to focus their attention on. On board, they relayed the prayers of protection in this space where strangers meet, dictated by protocol, then grasped arms with the haole visitors. They bowed. All aboard agreed: this man was not Lono, but he was a chief among men.

Having learned some things about how tribes work during his travels to Tahiti and New Zealand, Cook signaled for his men to present these island representatives with gifts that the locals promptly brought to shore and presented to the mo`ī of the island. Those who had gone to determine the intentions of the ships told the mo`ī that the visitors meant no harm. They themselves were great explorers of the world, and although the white chief and his men looked like no one they had ever seen before, the islanders could recognize fellow explorers when they saw them.

The mo`ī, as with all the ali`i of all the islands, was a member of what would become Hale Nauā—I'll get to them, auwe, I can't believe I have told you nothing of anything, so many blanks to fill—and as such, prescribed to the active accumulation of knowledge of the outside world, of which they would add to the science and knowledge inherited from their grandparents, of which they would pass down to their children's children and their children's children's children. It is no accident that the word `ike means both knowledge and sight. To know is to see. To understand is to rule.

On the second day, with Jesus in mind, Cook asked: Would they take him to the women of the village who were with child?

The mouths of the islanders went thin and downward at these questions. Why he want to do such a thing?

Sensing a sensitivity, Cook withdrew. He was close, he could smell it. Before they left he gathered his ship's iron nails—locals were particularly interested in them, having access to no such metals on these lands—and told his men: trade these for the information that would aid their hunt. Any member of the crew who did so would be rewarded by none other than King George III himself.

Eventually one of his men reported: there were rumors of a birth a handful or so years before, one that had taken place somewhere secret, one that whispered of a prophesied delivery. They left Hawai`i sure they were close. For an entire year, Cook searched for yet another place that fit the description of the prophecy, but his search only solidified his certainty that what he was looking for was in Hawai`i.

When Cook's ships reappeared along the horizons of the

great island of Hawai`i, consideration was once again given to the chance that this was Lono, with his white sails come for Makahiki. Everyone knew that Kealakekua Bay was, after all, home of the Pu`uhonua o Honaunau and the sacred harbor of Lono.

This time, Cook did nothing to dispel the speculations that he was a god, nodding when he was bowed to as if he deserved it (by then he was a celebrity in Europe, like one of those guys you used to hang posters of in your room). For more than a month his white sails flapped in the offshore breeze of Kealakekua. He and his men participated in the merriment of festivals and celebrations. They ate local food without restraint. They engaged with local women without restraint. Cook got bold. He demanded to be taken to the special child of the prophecy he knew was on the island. A line in the sand was drawn. He had gone too far. He was ordered to leave. He refused. A crewman bashed in the head of a local chief's guard when he refused to talk. The mo`i of the island again ordered them to leave, this time swarming their ships and threatening to destroy them if they did not retreat immediately. With a tight grim look on his face, Cook ordered the anchor pulled.

From what I understand, rough waters were the reason the ship turned around only four days out, but I have come to believe that the only real storm that raged was the one inside Cook. It spilled out upon the decks. He had not found what he was looking for, but he had already laid claim to this immeasurable power. What you lay claim to, what you plant your flag upon, is as good as yours. That was the way his world worked. They turned the ship around.

This time, Cook was not greeted as the potential Lono. Instead, he was greeted with rocks. He was undeterred. He ordered his men to gather the wood of a structure upon a burial ground. By the light of this sacrilegious fire he took a young ali`i hostage, demanding the location of the prophecy. The people of the island commandeered one of the ship's small cutter vessels instead and demanded the release of the ali`i. But Cook's venom had seeped into his men. They lifted their jaws. They would not be told what to do by a group of half-naked islanders, they were on a mission for the King of England! By God, a mission for Christ Almighty! They fired.

News spread immediately. The ali`i was dead.

They continued firing their guns but were soon surrounded. There were only so many bullets. The men of the white sails scrambled from the shore, fighting each other onto the small boats that would return them to the relative safety of their ships. Cook was not among them.

The men of the ship licked their wounds for three days. They could not bring themselves to leave without their captain. They positioned their cannons and fired. Every Hawaiian on shore fell. The few surviving Englishmen rowed to shore and scavenged through the corpses. They collected what they could of Cook now in pieces and tucked tail on the first winds out.

The secret remained a secret because there was no one left to spill it, and no one still searching to uncover it.

AUGUST 20, 1992

THE HOSPITAL ONLY ALLOWS ONE VISITOR IN THE ICU ROOMS
at a time, mo`opuna. In case you were wondering if anybody
else is coming to see you. They're all down there in the parking
lot, all the nīele nancys who suddenly have all the stories about
seeing you around or hearing you were up to this or that but
never said a word to me all those weeks I was looking for you.
If you had a window facing out you'd be able to see the tailgate
party happening in your honor, though instead of beer and pu-
pus there's cigarettes and tears. I told them I'd share the visitor
pass when I'm good and ready. For now we have business, yeah?
Believe me: This might have started with Cook, but it does not
end with him.

Charlie Boy tracked down your car. He's driving it home to
Waimea today. He's a good boy. I don't know why you never liked
him. What were you doing all the way in Anahola? Is that where
you've been? If we turn on the TV maybe we can catch the news
talking about you. They used a picture of you that I never saw

before. I wonder who gave it to them. Your car, the picture, there is so much about your life I don't know.

There was a big story in the paper about what happened, too. It got a bunch of folks all gung-ho about closing Queen's Bath and blocking the trail. Little good that will do, I think. When people want to get into a place, they find a way.

Tutu is making a mess of your nails. I've never been good with stuff like this, you know that. The color is good though. Sparkly. I'd do your feet too but your toes are hooked up to beeping monitors the same as the rest of you. They glow red under your blanket. Your hands are so cold.

I don't recognize any of the nurses today. These ones come and check on you as if I'm not even here, moving you around without saying anything to me. There's a new doctor on duty this morning too. You want to know what he told me when he took me out into the hall? He said you slipped deeper into wherever you are. You're scaring me, honeygirl. Yesterday I talked too much about the white man, but I need to set this up right so you understand. Give you the whole picture, all the way from the beginning. You'll see where we come in soon enough, see why your tutu is ashamed. See my failure. But you'll also see why you have to stay alive, okay? Here we go, back to the cave.

1768

Outside the birthing cave, shadows moved across the glowing surface of the ocean. Constellations, `aumakua, sharks and owls, creatures large and small from above and below, and then vines and flowers and plants of all kinds, mountains and valleys and all the lands in between, a seismic ripple across it all. Inside the cave no one moved or spoke. Only the chanters kept chanting their prayers of protection and blessings, not daring to stop. For the various attendants, the midwife and her apprentice, kupuna, as well as the keepers of the sacred idols who watched over them all, a feeling of connection and communion filled every volcanic crevasse of the cave, permeating nostrils and seeping into mouths. This was pono, equilibrium, where mana flows smoothly and uninterrupted throughout.

The baby's father was ali`i, so the birth was always going to be notable, but that wasn't enough to explain the flicker of the torches or the way the cave went dark when the baby was expelled from its first, initial home. In all the time the chiefess had been pushing, she had not made a sound. Only squatted and set her jaw as if going into battle. But when the baby came out her strength went too. Her legs folded in on themselves. She collapsed onto the mat behind her. At the first cry, the others in the cave assumed it was the baby. But soon it was clear. This was not the sound of first breath.

The attendants scrambled in the dark to help her. They pulled on her arms and, when her legs proved unable, dangled her into a squatting position to expel the rest. The chiefess moaned. Kaluaua, the young midwife, pressed on her lower stomach. The womb was not empty. She pressed again. The scream that followed echoed deep into the island's innermost chambers.

What was assumed to be the afterbirth came out with the force of a surging tide. The attendants held out their hands to catch it but withdrew them as soon as they saw.

What had come from the womb of the chiefess glowed like freshly released lava. Kaluaua and her attendants took a step away. The chiefess stumbled backward. She pulled her legs into her body so as not to touch what had just emerged from it. Only the baby's tutu, who was there to make sure her son's child was welcomed into the world with the reverence it demanded, dared look directly at it. She moved from where she had been examining the baby and approached, switching places with Kaluaua. The glow intensified. Quick as if hunting for eel, the kupuna thrust her hand into the bloody mass, knees on the ground, her silver hair glinting as if holding the moon hostage. The glow from within the mass grew. The chiefess leaned in as if to protect it, but the kupuna held up her free hand and said nothing. After a minute her hand went still. She pulled her fist free and opened her hand, palm facing the sky. Blood dripped from her fingers.

This was a second child, hidden and birthed alongside the

afterbirth. It was roughly the size of a bulb of taro, hardly large enough to fill a man's palm. When the kupuna placed it with care upon a woven mat it curled into itself like the unfurled tip of a fern frond, stiff and unmoving yet pulsing with a heat that drew from the ground, from the walls, from the very air itself. Whether feeding from or giving to this ethereal epicenter it was impossible to say. The form glistened with inner birth waters and, although the inside of the cave was still dark, was seen by all. They watched in silent awe as it grew wider, unrolling, unfurling, expanding into impossibility. Solid then porous, the shape distending into a moʻo, a manō, a human, before shrinking, darkening, surface transitioning from dense to porous before settling into something in between. A stone roughly the shape of a pōhaku kuʻi, the kind to pound poi, a bottom-heavy oval, two depressions from which it was clear all was observed.

There have always been many kinds of pōhaku. It is what the islands are made of, what our existence rests its feet on. Pōhaku as tools, pōhaku to make our food. To build our walls and the protective boundaries of our ponds. There were already plenty sacred pōhaku too, kiʻi that held the mana of different akua, but none in the cave had ever witnessed or heard of anything like this.

From Kaluaua's arms the firstborn infant made a sound and reached out, whether for her tutu or for this new being no one was sure.

The kupuna lifted her arms toward both and began to chant a prayer that echoed into the deepest parts of the cave. The stone child pulsed along with the chant, the same glow of the placenta only stronger, like the sun growing from within and rising toward its surface, dimly lighting the entirety of the cave as if the torches resumed their burn. Only then did the baby open her mouth to announce herself to the world, revealing a tongue with a dark birthmark, already tattooed like that of a seasoned warrior. The stone child flickered. The kupuna motioned for the baby to be brought to its mother. Kaluaua, still early in her apprenticeship and unsure of her exact role in this extraordinary birth, approached with some hesitation and handed the child to her mother. As the baby nursed, it was impossible not to notice the glowing pulse of the stone child matched that of the human newborn's crown. The baby stopped mid-suck and seemed to search the room until it found what it was looking for. The kupuna, the only one who dared lift the stone child, brought it closer. The baby reached out a hand. Slowly, the kupuna leaned toward her mo`opuna, spreading its small hand. There was a sizzle when the baby and its stone sibling made contact, but the baby did not cry out pain. She did not pull her hand away for some time. When she did, the kupuna noticed a mark that matched the one on the baby's tongue, a tattoo that covered the entirety of her left palm.

"In my many years, I have seen many things," the kupuna began. "Kinolau in all forms, akua embodied in ferns and flowers, rains and winds, creatures of mauka and makai both. And we must also acknowledge the ki`i we have made, the images of our akua carved from rocks and coral and wood and shells. But to explain this, I believe we must first look to Ka Mele a Pakui.

"The first time Wākea, god of light and the heavens, came together with akua wahine Papa, goddess of earth and the underworld, they produced the island child Hawai`i Island. Maui and Kaho`olawe soon followed. It was the kōlea bird who told Papa about Wākea being with Ka`ula, with whom he had Lāna`i, and Hina, with whom he had Moloka`i. If not for that bird, we might never have had O`ahu, since it was Papa being furious with Wākea that propelled her into the arms of Lua. When Papa and Wākea finally return to each other, they birth Kaua`i, Ni`ihau, Lehua, and Ka`ula.

"Maui and Hawai`i are the first and second born of the union between Wākea and Papa, linking the two. But it is also here on this island of Maui and not far above ground from where we stand beneath that sits the hill Kaiwiopele—named the Bones of Pele because it is said that it is where Nāmakaokaha`i killed Pele when she was still in body form and had not yet seeped into the land."

She paused and gestured toward the human infant, who seemed to be watching. Shadows began to scroll across the stone child. Against its glowing surface the shadows had the effect of a pen.

"This child is the product of a union between a chief of Hawai`i and a chiefess of Maui. With her has come a message, one we cannot ignore. This pōhaku is kinolau and ki`i, image and body. Of Hawai`i Island but birthed in Maui, reminding us these islands are connected, as this child and this pōhaku are. They are bound together, united under single rule. As we shall be."

The kupuna placed the pōhaku carefully at the baby's feet and stood. At the mouth of the cave, she called out, "This child is Ka`ahumanu. She will rule, and all will bow in her presence!"

• • •

Later, no one felt qualified to put words to what happened. The unspoken consensus was that no one should ever speak a word of what had been born alongside the chief's child. Soon enough there were bigger things to worry about.

With his allegiance to Kamehameha and the efforts to conquer all other island rulers and unite the islands, Ka`ahumanu's father was an increasing threat to the king and other chiefs of Maui. There was no telling what would happen if what the kupuna had prophesied in the cave became known. Knowledge of the pōhaku's existence needed to remain indefinitely under the guardianship of the women who had been in Ka`uiki Cave. The political position of the chief was already putting the family in enough danger.

But with increased war between efforts to maintain individual rule versus unification, that was not enough to keep the family safe. When the family made plans to relocate to Hawai`i Island, the kupuna wrapped Ka`ahumanu in a thick tapa cloth and handed her to the midwife Kaluaua.

"Go with my mo`opuna," she ordered. "Watch over her. Keep her safe. She will one day be queen of all the islands. Mana is power, but also energy. Gods and goddesses have mana that is supernatural. The ocean has mana, taro has mana, you and I have mana, but they are all different forms, different levels. The mana of some things is so powerful they are worshipped and considered kapu. If there was a chief that was a close, direct descendant to an akua, his mana demanded no one look at him. If a ki`i had the mana of a god or goddess, that ki`i could offer the protection and strength of that akua."

The kupuna leaned toward Kaluaua and stroked the baby's cheek. She continued in a whisper.

"There's another kind of mana that is so powerful that, instead of protecting you, you need protecting from it. Take Pele. Her mana is powerful enough to create land. But go too close and it will destroy you. Do you understand what I'm saying?"

Kaluaua nodded. She promised to protect Ka`ahumanu with her life until the time came when Ka`ahumanu might take it.

• • •

Counter to the custom of the time, Ka`ahumanu's mother did not send her to be nursed by someone else. Still, Kaluaua was young and dedicated and made herself as necessary as a nursemaid. She served as second mother in every other way, giving in to Ka`ahumanu's every demand.

Since no one was allowed to acknowledge its existence, there was no one to say whether the pōhaku was in service to the princess or she was in service to it. But the power surrounding her was undeniable.

True to the kupuna's words, a mana surrounded the newborn queen and her pōhaku twin that was near Pele level. A concentrated mana unlike anything anyone in the cave had directly encountered before: a wellspring, a culmination of power summoned from deep within the womb of the earth, so deep it reached the outer stretches of space, the pull of the moon, the heat of the sun. The tides were all king tides. The sand trembled. Fish filled the ponds. The trees grew heavy with fruit. When she cried, the sky cried alongside her and the rain was felt miles away. When she was angry, waves pounded the cliffs. With the pōhaku, the princess had a bearing that went beyond royal. When others in the court asked Kaluaua about this new scent in the air, she feigned ignorance.

Inside their hale, Ka`ahumanu took her first steps. Outside, the battles to unite the islands raged on, keeping her father busy elsewhere. The man he supported, Kamehameha, was trying to make himself a kingdom, connecting all the islands under a solitary feather cape.

"I am not sure Kamehameha will succeed," the chief admitted to his wife when he finally returned.

"The battle of Kepaniwai took every ounce of strength we had. Fully conquering the islands will require more than that. It will require the help of the gods. But we must succeed. Unity is the only way Hawai`i survives."

The chief's wife pulled Ka`ahumanu into her chest and called to Kaluaua.

"I know what you are. I know who you serve. But there is too much fighting, too much death. One island with too much and another with too little. We need to be led as one. Husband, I am sorry I have withheld. We have what you say is needed. Kaluaua, tell him what happened in the cave."

Kaluaua was careful not to leave out a single detail, including the words spoken by his mother the kupuna. Fury turned the chief's eyes aglow with fire.

"How could I not know this," he roared. "To hide such a thing. I should kill you both."

His wife stood her ground.

"You fault a mother for wanting her child to be her child, even so briefly?"

The chief did not respond. To be a member of Hale Nauā was to never acknowledge even the society's existence. Like him, every member was ali`i, committed to the pursuit and perpetuation of the great body of knowledge Hawaiians had accumulated over the generations, although each had a different area of expertise. The chief was an expert in genealogy and could recite long hereditary poetry that unfurled for hours, but even he did not know when exactly the society was formed. All he knew was it was the society that had united the chiefs and kings of all regions in its single purpose of keeping alive all the science of the stars, the knowledge embedded deep in the ocean floor, and the secret meanings behind the great mysteries of the universe. So while the chief could not say exactly what this pōhaku was or what was the meaning behind its connection to his daughter, he understood that there were certain things that should be acknowledged and respected, even if they could not be understood. The mana emanating from the pōhaku was exactly what Kamehameha needed to win the battle for the islands. By telling him she knew he was a part of Hale Nauā, the chief's wife was telling him she knew what he would do the moment he learned of the pōhaku.

"The pōhaku is the strength and mana Kamehameha needs to bring in the remaining islands. We cannot think of our own desires in these times," he told her.

The chief's wife nodded solemnly.

"I know. But it is not ours to give him."

They all stared at the child squirming to be released from her mother's arms. The only way to give Kamehameha the pōhaku was to give him Ka`ahumanu.

The child remained firmly lodged between the two women until the girl's milk teeth fell out, which seemed to take much longer than usual because Ka`ahumanu was so big for her age. Indeed, everyone assumed she was older than she was. Glued to her mother regardless of her size, only when her teeth said she was ready was it time for the princess to be put on the road of her destiny, the one her markings did not let anyone—even for those who did not know why they were there—forget.

• • •

The mo`ī Kamehameha was thirty years old and already had sixteen wives. It didn't matter, the princess would not be ready to be his wife for many years. What mattered was that the people took note of the gift and the message it sent: Kamehameha had support. He was proclaiming himself the king of kings, the mo`ī of all, and the chief believed in him enough to give his firstborn.

Ka`ahumanu left her early childhood home on the day her final milk tooth fell from her mouth, moving to Kamehameha's protection and watch until she was of age to be married. She took with her the young nursemaid Kaluaua, whom she called Ua.

With the princess also went the pōhaku. Her parents prayed to the gods that this proximity to the pōhaku was enough for the mo`ī to forge eight islands into one, although after much deliberation it was decided that it was much more advantageous for them if Kamehameha believed it was the princess who brought him such divine blessings, as it would ensure their daughter was treated with the care and reverence that the seventeenth wife of a king might not get otherwise. Up until this point, knowledge of the pōhaku was still limited to those who had been there at its birth and to the chief who some might consider its grandfather, so Ua was put in charge of both the young princess and her secret stone sibling.

Ua was a dedicated caretaker of the princess, but in spite of her efforts to keep the pōhaku a secret, it did not take long for Kamehameha to learn the true value of the dowry brought by his wife-in-training headstrong girl. He too was a lifelong member of Hale Nauā, with eyes trained to see what could not be seen. He watched from a distance as Ka`ahumanu grew, Ua watching the two of them with a guarded wariness that she was becoming known for.

Growing old enough to be aware of the power she wielded but not old enough to know how to exploit it, Ka`ahumanu was given space to do what she pleased. Ua made sure every path before her was cleared, every view unobstructed. When the surf was up, Ka`ahumanu spent many hours gliding across the sparkling waves of the bay, Ua on shore, forever watching. When they bathed in Kiope Pond, a royal bathhouse built with lava rocks and coral mortar, Ka`ahumanu closed her eyes as Ua brushed and washed her hair. With Ua, she was a child. But once they were done at the bathhouse, Ka`ahumanu sent Ua away. Only then would she pull back her shoulders, return her chin to its elevated position, and spend the evenings drinking and smoking her pipe.

The pōhaku was at all times tucked within the knots of her tapa, its mana feeding off hers and vice versa until the unit grew so powerful it was nearly tangible, like the unseen energy of a dormant volcano coming undone, a giant sleeping with one eye open, a simmering force of nature.

When Ka`ahumanu turned thirteen Kamehameha took her as his wife. His wedding gift was his admission that he knew about the pōhaku and, if she agreed to continue to align her growing power with his, he would forever understand that her loyalty and service was and would remain exclusively with the pōhaku. There would be no subservience, no children.

Once they were married, Ua's position in Ka`ahumanu's life changed once again. As Kamehameha's wife, Ka`ahumanu had attendants hovering around her at all times. She had guards to keep her safe and lovers to make her laugh. Ua retreated, but not so far as to ever lose sight of her charge. Her new role allowed for a life of a different kind from most servants. She married and got pregnant, and instead of giving over her child, as any faithful and loving servant might, the child remained hers.

Ka`ahumanu's inability to have children did not prevent the king from distinguishing her in other ways. At every milestone he reached that brought him closer to a united Hawaiian Kingdom, there was an unspoken allegiance and understanding that his progress was in large part due to the support given him by the pōhaku, and Ka`ahumanu by default. She was the key, the critical element necessary for him to rule in entirety. To express his appreciation and acknowledgment of her role, he named her guardian of his heirs, the children sired by his other wives. To Ka`ahumanu, this was a mere token. The one who looks after the future monarch is not at all similar to holding the power of one. She brought her demand to Kamehameha, now the true moʻī: he would give her a seat on his high council. Kamehameha's eyes went round. He acknowledged the pōhaku's role in his trajectory, but this was too much. A wahine on high council!

Kaʻahumanu went red with rage. How dare he deny her what she asked. And how dare he call her a mere woman!

It was well known that the moʻī and the moʻī wahine butted heads like a pair of rams during mating season. Two steel flints clanging against each other, it was a wonder how their arguments didn't echo up the island chain. One day, though, in an early rainy season, when spears of light sliced through the sky and thunder roared in the distance, it got so bad that someone called for Ua.

Ua had seen them fight before, but this one was epic, even for them. Kaʻahumanu wanted him to appoint her to high council. Now he needed her and the pōhaku, the mana they produced together, and she was as responsible for uniting the kingdom as he was, saying that if not for her, it never would have happened. Kaʻahumanu pointed to Ua standing at the edge of the room, commanding her to stick up for her. Kamehameha looked at Ua, daring her to say something. Ua looked at her feet helplessly. He sneered. Kaʻahumanu was not the only one born to a destiny. What about the legend that prophesized a light in the sky with feathers like a bird that would signal the birth of a great chief? Kaʻahumanu was not the only one born within the folds of a legend. Kamehameha was the child heralded by a feathery light that foretold his greatness, which was why he had been hidden away in Waipio until threat of him being killed as an infant passed. And it was he who had overturned the Naha Stone, all three tons of it.

• • •

Ka`ahumanu could not be appeased with surfing or cigars, not this time. In the hush of night she left the compound, taking the pōhaku with her. Ua kissed her daughter softly on the cheek before slipping out the door to follow her queen. The whispering winds hid the sounds of their departure.

The year was 1793. Kamehameha had only just conquered the island of Maui—with Ka`ahumanu's help—but there was still all the others. With her departure, the tides deflated. The coconuts lost their milk. Even the mo`ī started wilting like a crop in a drought. His mana diminishing by the minute, he ordered all of his men out. Do whatever you need to do to get her back, he said.

• • •

The king's men left his compound in Kailua-Kona, Kamakahonu, via Kailua Bay and went up and down the Kona coast until someone finally pointed them in the direction of a village along Honaunau. They swarmed the bay and came upon a village keeping secrets. A guard recognized Ua and tracked her to where Ka`ahumanu was hiding in the Pu`uhonua o Honaunau, a sanctuary provided by the god Lono himself. Dragging her out of the Pu`uhonua would be desecrating the very nature of that holy ground, so they set fire to the village instead, hoping to draw her out to stop the destruction. But they underestimated her. In the light of the flames across the bay she made her way to Hale o Keawe, the holiest of temples, where none would dare enter, and buried the pōhaku as deep as she could, careful not to disturb any bones already resting there. This time, only she would know the whereabouts of the pōhaku. And Ua, of course, since it was Ua she ordered to dig the hole.

• • •

In spite of the missing pōhaku and the strain between him and Ka`ahumanu, by 1795 Kamehameha had conquered Oahu and Moloka`i. When he asked about the pōhaku, she said only that it was safe, and near enough. He begged her to bring it back. He put her on high council. He appointed her regent of his heirs. He made her kuhina nui, the highest position in all the islands, second only to himself, the king. Still, she would not return it.

After many battles, in April of 1810 Kamehameha negotiated a peaceful unification with Kaua`i. Hawai`i was now a kingdom.

1819

Ua insisted Palila was too young to go out to the boats. The girl only wanted to see the goods the white men brought to trade for the sandalwood they loved so much. Her mother said that wasn't all that was happening on those boats and forbade her from going anywhere near the canoes that made trips across the bay to the great sails floating on the water. But Ka`ahumanu felt it was her queenly right to engage at will with the men from the ships. Every ship captain came with gifts for the queen. And that was only the beginning of the things she saw.

Her mother was so old fashioned. She was constantly chanting and praying for protection as if the white men and everything they had brought was a direct threat to them both. Unlike her mother, Palila was thick, solid, as if the gods had built her with a layer of protection. Danger, if it ever got too close, would simply bounce off her. Ua didn't necessarily agree. Palila was tall for her age and her right eye sometimes strayed—at times she would be paying attention to a story her mother told and her eye would get bored and wander elsewhere. Ua would slap Palila's hand to tell her eye to return to its duty. Eyes worked in pairs, as partners. One did not do things without the other. But to nīele people Ua justified her daughter's wandering eye by coupling it with her height, always insisting her daughter was named after the honeycreeper because she had been sent into the world to observe from above, as a bird does.

Palila might have been too young for the boats, but her mother couldn't argue when Ka`ahumanu decided Palila was old enough to wave the large fan that kept the queen cool. Ka`ahumanu was a striking woman who had also been tall as a child. In Palila she found familiarity.

This gave Palila an even better view of all that went on onshore. From what she could see, the interactions looked harmless enough. Her favorite thing was watching the queen beat the men in chess. The men were so sure of themselves and always so shocked. Palila was always glad for her fan, where she could hide her laughter. It was as though they had never seen a woman with a head on her shoulders before.

Unlike most everyone else in the village, Ka`ahumanu did not reveal any interest in the geography books that had come with the Europeans. Instead, she wanted to know the words of the book, to learn how to read. Palila agreed. She secretly hoped someone would teach the queen, since then there would be a chance she could learn too. It all seemed harmless enough.

That changed the day a new ship arrived in the harbor, along with a new, bold captain. Captains had come before, but this one was different. Palila heard him before she saw him, followed by what sounded like animals, although none she was familiar with.

"What do you mean, I can't bring any of this into the queen's tent? These are gifts!"

Ka`ahumanu ordered Palila to peek outside to see what was going on. There were animals all right, not to mention chests that appeared heavy and baskets of colorful produce and other food. The captain caught sight of Palila.

"Tell your queen she is going to want to see what I have brought her," he said loud enough to make it unnecessary for Palila to relay the message.

"These seeds grow stone fruits," he explained of the final basket he had laid at the queen's feet.

When he was done, he ordered his attendant to lay out the delicacies brought all the way from England. He poured liquor into two of the daintiest glasses Palila had ever seen. To her surprise, the queen accepted the drink with a smile. Then the captain waved his hand toward the silver platter heavy with food.

Palila was not familiar with the feeling of fear, beyond watching her mother constantly live under some form of it. But her stomach dropped when she saw the queen reach out, her fingers grazing the various cakes and delicacies, considering. She brought something to her mouth and began to chew. The captain took his turn, taking something from the same tray and eating it. Palila's skin went cold. This could not be. It was kapu for women and men to eat together. Ever. Everyone knew that, even the queen. There was always a huge platter of fruits and fish laid out next to the chess board during the queen's long games with visiting sailors, but it was always understood this was food for them, not her.

Palila realized she was not breathing. She took a deep, jagged breath and went lightheaded. She waited. For lightning to strike, for the gods to come down. For the queen to fall over and die in front of her. But nothing happened. The queen and the captain continued eating and drinking until there was no food or liquor left. And still she did not get sick or die.

Ua was always worried. She was a steady stream of chants, prayers, and offerings. Palila did not want to even imagine her mother's reaction to this. When she returned home that evening she was still shaking. All night she dreamed of fire and teeth. In the morning, she burst into tears as soon as Ua looked at her.

"She, she ate," Palila gulped. "With the men. Yesterday. And nothing happened." She dropped her head in shame as if she were somehow at fault.

She waited for her mother to react, but when she didn't, Palila looked up. Ua sighed and offered her a tight smile.

"It is time for you to know the truth, Palila."

Ua took her hand and led her down to the water. Ua jumped in and motioned for Palila to follow. They swam out so far that Palila couldn't see the shore. Ua told Palila of Ka`ahumanu's pōhaku and all that had happened in the cave.

"But if the pōhaku is still buried under the temple in Honaunau on Hawai`i Island, how could it be what kept her safe from violating kapu yesterday?"

Ua kicked, turning her body to face the horizon to consider the question. Palila ducked under a wave and waited.

"To most of the questions you have, I have no answers. But this one I do. The pōhaku is no longer under the temple in Honaunau."

When Kamehameha decided to move the center of the kingdom to Lāhainā in Maui, the queen and Ua were of course expected to join him. The queen would not think of leaving it behind. But because the mo`ī watched her like a shark in the eternal hope of one day getting his hands on the pōhaku, it was Ua who had been tasked with making a secret journey to where it was hidden. She kept it with her during their move so that the king would never know it had returned.

Ua spoke so softly that the water had to carry the words to Palila's ears.

"My life is in service to Ka`ahumanu, but she is part of this pōhaku or this pōhaku is part of her, so I am in service to it too. I am not qualified to explain or interpret the ways of the akua or the ali`i and neither are you. All I know is that it draws from the earth things these white men deem valuable."

"Like the sandalwood?"

"Like the sandalwood."

Palila thought about it all for a moment, kicking her legs to stay above water as a wave made its way past them.

"Does Liholiho know about the pōhaku?"

Ua shook her head. "As far as I know, no."

"Then why does he do what Ka`ahumanu says?"

At that, her mother broke into a smile. She splashed Palila with water.

"Who wouldn't?" She laughed and began swimming to shore. Palila followed, comforted. If her mother believed everything was going to be okay, so did she. Plus, now she had a new thing to believe in, another layer of armor to fortify her. The pōhaku.

To Palila, kapu had always been more about sandalwood than anything else. Kapu meant only those allowed by the king could cut down the trees. He said who, he said when, he said where. He said how much. It was a system that ensured the trees would have time to replenish themselves and the groves were protected and things stayed in balance. The king was not just mo`ī of the people. More significantly, he was mo`ī of the `āina. Now her mother put this into new context: the cutting of the sandalwood caused the queen a pain deep in her stomach. She had long suspected it had something to do with the pōhaku and the queen's connection to it, although the only evidence she had of that was the knowledge that it had been Ka`ahumanu who had implored the king to institute these harvesting restrictions.

39

`Ai kapu was the world in balance. How things were supposed to be. She could not imagine eating with her father or the other men and boys of their village. She had no interest in knowing the taste of pork or banana or coconut. That was boy food.

By the time Kamehameha died, Palila had seen Ka`ahumanu eat with the men from the ships at least twice more. So it was no surprise when she heard Ka`ahumanu speak about the changing times. How Hawai`i needed to keep up with the rest of the world. What was a surprise, and what she struggled to understand, was how she got Liholiho, now King Kamehameha II, to go along with the breaking of kapu, to have a meal with her together. How could they trust that they would not be struck dead by the akua for such forbidden behavior? She would follow Ka`ahumanu and her mother to the edge of the ocean, but Palila wanted no part in that. Thankfully her mother didn't seem to have any interest either.

What her mother did have interest in, though, was Palila learning how to read. She couldn't believe it, she had assumed she would have to fight long and hard for that! But in this her mother agreed with Ka`ahumanu—the world was changing, and Hawai`i along with it.

For as long as Palila could remember, the majority of the ships came from New England. When America went to war, ships from that place called Boston were few and far between. But that didn't mean the demand for sandalwood decreased. The opposite.

With the end of kapu also came the end of the sandalwood restrictions. Trade quadrupled, as did the boats coming through. So much so, in fact, that Ka`ahumanu started planting the seed of moving the kingdom's headquarters once again, this time from Lāhainā to Honolulu, closer to the port. Palila cried when she heard this, but Ua reassured her there were two very large reasons they would not leave their home. The first was that Ka`ahumanu was kuhina nui but Liholiho was king, the eldest son from the great king's first wife, and Liholiho was adamantly against the idea. The second reason Palilia's mother offered her was even more reassuring: Here, the pōhaku remained safely hidden in the cave where it had been born into the world. There was no way Ka`ahumanu would move it.

1820

Hiram and Sybil Bingham were newlyweds, but their
journey on the *Thaddeus* was no honeymoon. Married in Hartford,
Connecticut, just eight days before they set sail, they were more
excited about what they were bringing to Hawai`i than what they
might find when they got there. The Lord their Savior!

Hawaiians had been coming to Massachusetts and Connecticut for some time by then, either sent by family to be educated or
passing by as a ship hand. The Binghams could hardly wait to get
to Hawai`i and start spreading the Gospel, something Hawaiian
savages were so obviously starving for.

It was evening when the boat pulled into the bay. Most of the
people on board were sick. They prayed and went to sleep. In the
morning, a welcome party was sent to greet the ship with milk,
eggs, and melon, the fuel they needed to get on their feet and on
dry land.

. . .

As soon as the *Thaddeus* arrived in Hawai`i, Palila knew there was something different about it. She turned to her mom to gauge her reaction, though she kept an eye fixed on the ship. Ua would not stop chanting, pleading for protection from the coming darkness. Palila shivered under her armor. This was not a boat looking to trade for sandalwood. When an audience with Queen Ka`ahu-manu was requested, she had closed her eyes and hoped the request would be denied. But the queen was bored and itching for a new chess partner.

The man who came bore no gifts and had no interest in chess. For reasons beyond what Palila could imagine, the queen seemed to take no offense. The man said he had the greatest gift anyone could ever give, though it would take time to share. His hands were empty, but the queen appeared satisfied with his words.

Palila kept the queen's fan at a steady wave, up and down as if she were an offshore breeze. When the man caught her staring, she unfocused her eyes and pretended to be gazing over his shoulder. It didn't matter. He focused almost solely on the queen, as if she were the sun.

The others who had come with the man, Bingham, had fanned out across what had once been a lush bay and was quickly becoming a busy harbor full of people and commerce. Trunks of books and silk dresses and ribbons were unloaded from the ship and discovered to have grown mold in the journey over the ocean. There was a great cry of disappointment and for a moment it was not quite certain if the women would revolt and demand a return to Boston. From Hawai`i's standards, the place was quickly growing in comfort and convenience. Storefronts and homes were taking the place of fishponds and farms. But it was a far cry from Boston. After a moment of tense silence, Sybil

Bingham secured her hat and motioned that she was ready to go ashore. It was a message to the other women; they followed obediently. They were all still a little green around the gills, but none had the energy to protest.

They'd already been there a week, but it was still all the village could talk about. These New Englanders had not come to take anything from the islands, they had come to give! To build schools that would teach them to read and write, to build places of medicine to bring the sick and wounded. They wanted to do all this in the name of an akua, one greater and more powerful than any they had ever known before. But it was one Palila had never heard of. The strangeness she felt when first meeting Bingham festered. Later that evening, she told her mother of her unease.

"They make me think of the men you told me about who came a long time ago looking for the pōhaku."

Once again Ua surprised her daughter by not flying into a hurricane of worry.

"Rest easy, child," she said. "I have listened to the talk. This akua they speak of is not connected to the pōhaku. The queen's pōhaku is bound to the `āina. When they refer to their kingdom, they point to the sky."

The next evening Palila brought a new worry.

"She told them to go to Oahu. To build their schools and their medicines there. That Honolulu would be the center of Hawai`i from now on."

Ua stopped stroking her daughter's hair and stood.

"I'll be right back."

Palila was asleep by the time her mother returned. She kept her eyes closed as Ua lay down alongside her and pulled her close.

She thought she heard her whisper that everything was going to be okay, but Palila might have been dreaming.

The next morning, the queen ordered all heiau, every single temple throughout the island, be destroyed and their things packed. They were moving. Whatever Ua had pleaded with the gods to do, they had not listened.

1821

THE QUEEN HELD OUT HER HAND, HER NAILS METICULOUSLY
cleaned and freshly trimmed, her skin glistening with oil. Palila
placed the pōhaku in her waiting palm, trying her best to keep her
wrist steady. She was not sure what or who was to blame for the
quiver of her bones, the queen or the pōhaku or both. She could
not shake the memory of the queen ordering the destruction of
all heiau. If the queen's newfound Christianity was as absolute
as she feared, the risk was high that it was only a matter of time
before she turned on the pōhaku and decided that it too must be
destroyed. If and when that time came, Palila wasn't sure she had
the courage to protect it. She felt as though she were staring at
two plants growing from the ground. They were nearly identical
except one was what would save them. The other would destroy
them all.

The stone was pocketed in a flash, tucked into the folds of
the heavy layered skirts the queen had taken to wearing since
their move to Honolulu. She brushed her lady's hair and placed

her favorite hulu manu on her head. Its yellow and red feathers gleamed a sacred bright against the queen's forehead.

It was still hard to believe this was her duty now. Her mother had gotten sick almost as soon as they touched the shore of Honolulu Harbor, as if the only thing that had been protecting her from the sickness and death growing all around them was her belief that what had happened was an impossibility. Once that was taken from her, she had not lasted long. It had been eight long months that felt like eight minutes but also somehow eight years. Palila told herself that at least Ua had died with hope in her eyes—the queen might have ordered the heiau of Lāhainā destroyed before their departure but at the same time she had quietly ordered Ua to once again retrieve the pōhaku, this time from the cave of its beginning. As long as the queen still believed in the mana of the pōhaku and her connection to it, Ua could believe that Hawai`i would survive. Some moments, Palila felt grateful that her mother had died believing that. She wished she still could.

Once Ua was gone, the bold and confident curiosity Palila had carried with her as a child was replaced by a skittish fear tempered only by the adherence to endless superstitions. Her world was no longer a familiar place. Before her mother died, Palila had assured her that she understood well the responsibilities she was inheriting, to the queen as well as to the pōhaku—keeping it safe and protecting its secrecy—but today, her understanding of those responsibilities felt stretched. Preparing the queen to be married—she had no idea what she was doing. There was no substitute for her mother.

This was the queen's second marriage in as many weeks, the first to Kāheawai, king of Kaua`i, and second to his son, Kahuelokū. The queen often collected lovers, but never husbands, so Palila understood and interpreted these marriages to be more of an accumulation of power, a political move. The queen was still playing chess, the only thing that had changed was the board she was playing on.

A feast was prepared. Palila chanted a silent prayer of thanks that she had become accustomed to eating with men. She was by now an expert in making herself invisible, but the queen had already made it clear that her new subject was to partake in the celebrations this time. In honor of her mother. There would be no standing behind her, holding court with the fan that kept both the queen comfortable and Palila feeling safe.

The ceremony of marriage to Kaumuali`i had been presided over by the missionaries and their god from New England, but at this second one there was not a one of them to be found. They didn't approve, apparently. They insisted marriage was between one man and one woman only. Palila secretly rejoiced in the queen's defiance. Everything was changing too fast. English schools and churches popped up overnight. Looking around, sometimes there were more foreigners on the growing-busier-by-the-day streets of Honolulu than anyone else. It was nice to know that the queen's feisty stubbornness remained the same, if nothing else did.

She made a small plate and took a seat on a mat within the queen's periphery where she could be summoned if needed. A young man plopped himself at her side almost immediately. She bristled. He smiled. She ignored him as she ate and tried to make it as obvious as possible that she was not interested. Her wandering eye did no such thing, dismissing her commands entirely, and focused its full attention on this new sight. He ignored the fact that she was ignoring him, smiling and watching the festivities as if they were watching them together. It irritated her more than it probably should have. She blamed it on missing her mother and refocused her attention on the queen, making it clear that she was there with a job to do and would not be distracted. That became difficult once the queen and her new young husband excused themselves. He introduced himself as Keawe. She sighed, resigned.

"Palila," she told him.

"Ah," he said with a bright smile. "Honeycreeper!" He belted out a whistle. *Dee-dee-dee dee-dee-dee-dee-dee.*

She couldn't help but laugh. His whistle did indeed sound exactly like the distinct song of the palila bird.

"I'm impressed. I thought palila was only on Hawai`i Island."

"Oh, no. Here in Oahu we have big mamane. Very big." He spread his arms wide like the tree that fed the bird and wagged his eyebrows as if to say something else. She laughed again and pushed her food aside. She was done pretending to eat.

He was easy to talk to, a good listener. They walked along the beach while he asked questions about her parents and her life in Kailua. She tried to describe how different everything in Honolulu was to Kailua, how far away that all felt now. When he asked her how she felt about these new churches and the god from New England, she changed the subject. He let the subject drop and for that she rewarded him by allowing their conversation to continue.

Time passed quicker than she had ever known it to before. They sat and talked long into the night, about everything and nothing.

Before long they were inseparable. Palila was thankful the queen was distracted with her new husbands and her new studies, since she'd finally decided that it would not be demeaning if she learned to read and write herself instead of having someone read to her and write for her. Palila was learning too, although she didn't care for the fact that learning to read meant studying the Bible and learning to write meant writing only things the god from

49

New England would approve of. He didn't approve of a lot, she was learning. Bible study was a new requirement of being in such close proximity to the queen, an assignment she endured in the way she imagined her mother would expect her to. She was bound to the promise she had made her mother to keep the pōhaku secret, so she could not explain to Keawe why she remained in such close service to the queen when it was no longer requested the way it once was. But being close to the queen meant being close to that fist-sized bulge in the queen's skirts that only Palila could see, since she was the only one who knew something was there to fill it. And then one day the queen's skirts held nothing. Palila swallowed her panic, waiting for the inevitable time of prayers when the queen's private quarters were usually empty. When the bells of the church chimed, she moved as fast as she dared.

In the room, she didn't have to search long. The pōhaku had a gravitational pull that everyone attributed to the queen. Only Palila knew its true source. Scattered across a table were various trinkets, the ever-growing collection of worldly and exotic gifts from visitors and admirers. Palila poked around in the disarray until she came to a large wooden bowl. There it was, among a collection of hair combs and necklaces as though it was nothing but an ornament. A mix of feelings surged through her body— horror, anger, fear. Could the pōhaku now mean so little to the queen that she didn't even feel the need to hide it and keep it safe? If she had even tried to hide it at all? Fearing she had taken too much time, Palila quickly wrapped the pōhaku in a cloth and shoved it into the waistband of her dress. She gasped. Its sudden proximity to her body made her heart beat fast like the wings of a bird. She stooped slightly and slipped out the door as quick as she was able without drawing attention to herself. Back in her own hale, she searched for a place to keep it. Nowhere felt safe enough. There was no way she could keep it on her as the queen had—surely the queen would notice. But there was also no way to leave it here—surely Keawe would stumble upon it eventually. Finally she settled on the only place not even Keawe would dare look without her permission. In between her legs.

Up until recently, that would not have been the case. But the god from New England didn't like nudity or sex or anything that made you think of sex. Now it wasn't just ali`i women wearing long Victorian skirts and high-collared shirts. Honolulu was beginning to look like how Palila imagined Boston to be, everyone bundled in layers of fabric, sweat dripping down every crevice. But that would be of use now. Keeping it wrapped in the cloth she had taken from the queen, she placed the pōhaku as high between her thighs as she could without inserting it inside herself. She secured the bundle with a long strip of fabric wound around her hips and through her legs repeatedly until it formed a sort of bloomers. Then she got dressed and tested it out, walking in circles, sitting, standing back up. Sitting was uncomfortable but not impossible if she tilted her hips to one side. She moved in every way she imagined having to move to do what she did for the queen every day. The pōhaku stayed where it was. There was the issue of undressing and Keawe, but for all her planning, in the end Keawe hardly noticed. He had taken to coming home from the docks smelling like drink from the ships, which made him silly and giggly. The first time she went through the process of making him face the wall until she undressed and slid into bed he laughed until he cried. Palila was amazed by how quickly things could become normalized. The bulge in her skirt, the pull of the pōhaku, the smell of his breath, the crowds of the streets. The church.

The first day Palila wore the pōhaku she was so nervous she nearly fainted. Surely the queen would sense its proximity. In those early days she dreaded the thought of being discovered, but as the days wore on and the queen said nothing, her dread was replaced with dismay. How could the queen not notice its absence? Had this new god really replaced everything?

She had been wearing the pōhaku for less than a moon cycle when Keawe came home clear-eyed and fresh-breathed one evening. He relayed what he had heard on the streets. A mass burning of ki`i in any form, whether it be image, statue, or likeness of any gods or goddesses but also ancestors and aumakua, had been ordered throughout the islands. Everything would be destroyed. Palila was slow to respond. There was no denying it a single day longer. Hawai`i was no longer the place they once knew. Their akua were no longer safe.

Outside, no one dared say a word in protest. But there were murmurs.

The next morning Keawe skipped work and joined others who planned to gather what ki`i they could and go deep into the forest to hide them somewhere they would not be found. Meanwhile, Palila made her way into the queen's quarters. The queen was sleeping when she entered, so Palila lay down at her feet to wait. When she woke, Palila smiled and brushed the queen's hair the way she had watched her mother do countless times when she was a child, the way she knew put the queen at ease.

"You have it," the queen said. It was not a question.

Palila placed the brush at her side and waited for her punishment. She nodded. When the queen did not respond, she asked her if it was to be destroyed along with the ki`i at the burning. The queen still said nothing. Palila's mind raced. If she tried to run and hide surely she would be caught. She could fling herself over a cliff, throw herself and the pōhaku into the ocean, where at least it could not be found and destroyed. If its destruction was even possible. But no. She was in service to the queen as much as she was in service to it. Whatever the queen demanded, she would obey.

She asked again. This time, the queen answered.

"No. Keep it. For now. There will be one more test. Kapi`olani will soon go to Pele and challenge her. If she comes out alive, then I will know without doubt there is only one God. Only then we destroy it, for it will mean nothing. Understand?"

Palila longed to ask her what she meant. Kapi`olani, challenge Pele? What did that even mean? But she knew who she was talking to. She had no choice but to agree.

• • •

The bonfire raged unlike any other, as if the flames knew what was being burned and all they symbolized could never be brought back again. To see the sacred ki`i collected from every corner of Oahu laying in a heap on the street made Palila feel sick. On Hawai`i, one hundred and two ki`i were destroyed. Palila felt the pain of every single one. The air dripped. Keawe returned late into the night, weary and silent. Palila did not ask him if they had been successful. There could be no success. Now there was only survival.

AUGUST 28, 1992

OH LITTLE BIRD. I SWEAR ON MY LIFE I SAW YOUR FINGER MOVE. How come the nurses aren't more excited? They look at me like I'm an antique. Your doctor, the one I like because I can tell he's telling me everything he knows, said that it might be a good sign but that it could also be nothing. He doesn't want me to get my hopes up yet. We're not out of the woods, he says. Nerves twitch.

Him saying that reminded me of a time when I was little. We were camping. Someone put a frog on the hibachi. Even though it was dead it started hopping around on the grill.

Do you feel me tapping your knee? If your legs weren't secured down would your reflexes react?

Okay, I'll stop crying, I'm not here to get hysterical. No more talking about dead frogs. This is not the same thing. Your finger moved and I know that means you are coming back to us.

Have you missed me? The last few days I've been acting in faith. I finally cleared out my dusty sewing room. I think you're going to like it. The Ishigos down the street, do you remember them? They had a garage sale. I scored a futon for you. They said it was hardly used at all. I pulled in the extra dresser from the garage. That old Waimea house is falling apart, but when you get out we'll fix it up together so you can one day raise a family there.

I cleaned out your car. Brought all your things into your room. I thought my heart couldn't break into any more pieces. Someone living in their car might stuff all their things in the trunk and sleep on a heap of trash, but not you. All your clothes and shoes organized in plastic bags, the glove compartment with its tiny things of shampoo and soap and toothpaste—you always did find comfort in small spaces. It made me think about the first time you ran away. You were six and tiny, the backpack hanging from your shoulders was bigger than you. Do you remember? It was a busted old thing. Looked straight off the back of one of those guys who were always hanging around your mama, the stringy haired ones who lived along the river and buzzed around her like flies. Back then I couldn't decide which was worse—the thought of her down at the river leaving you home alone or the possibility that she was inviting them over to reap the benefits of the county housing voucher she only got because she had you.

I wanted to come visit a million times. I wasn't allowed. Every time I drove by I got nothing but dark windows with screens full of holes and a bunch of rusted cars I didn't recognize cluttering the rutted gravel driveway. I drove by every day anyway, made it my regular route to the market. One time there was a dinky-looking tricycle in the yard that made me hope I might see you riding around outside, but the next day it was gone so it must have belonged to a neighbor's kid. All those old prefab houses were smashed together with no fences to split up the yards, so that was honestly the most likely explanation. Still, I prayed to see the bike again the same kind of hard that I prayed I'd see you. Prayers for deaf ears.

You shook your head when I asked what you and your mama were fighting about or how exactly you got all the way across town to my house by yourself but I was so happy to see you that I didn't push for you to tell me nothing you didn't want to tell. Frankly I wasn't even sure how you had remembered the way to me. It had been almost a year since I stopped babysitting you, and your mom hadn't brought you over a single time since. I thought you were too young for the kind of memories that stick but maybe you remembered more than I thought, since there you were at the door.

I don't know all what she told you about why you stopped coming to visit me. Believe me that it had nothing to do with you or with me not wanting to see you. Sometimes it was hard to know when I needed to be grandma and when I needed to be mom. I don't think I really understood back then that there might be a difference.

She was doing good for a time, got a job at the drive-in near the middle school that paid her in burgers that put some meat on her bones for a change. I was surprised she asked but was happy to help out. Every day you and I would plan what we'd do the next day, go to the beach or the park or the library for story time, and sometimes your mama would hang out for a bit when she was pau with her shift. We even ate dinner as a family a couple of times too. I can't tell you how hard I cried of happiness with you falling asleep on the couch and her eating and laughing with a mouthful of fries, telling me all the gossip about her coworkers. I started to think maybe the darkness was done with us, that our dues had been paid. I stopped lighting candles at church and started saying gratitude prayers instead. But then the so-called overtime started, the double shifts and you sleeping over and waking up in the morning asking for your mom and me not having any answers to your questions. Her cheeks started sinking in and there were no more burgers. I went to the drive-in already knowing. She'd stopped showing up weeks before. They hadn't even technically fired her because they hadn't been able to get ahold of her to tell her. They asked me to return her uniform if I ever managed to track it or her down. At the time I really believed that if I kept watching you it would be giving her a free pass. I thought giving you back would save her, that needing to take care of you would mean she needed to take care of herself. Now I wonder why the heck I thought that was a good idea. She's my kid. I didn't know how to be her mom and take care of her when she was little, how did I think she would know how to do that for you?

On the airplanes they tell parents to put your oxygen mask on first but I always rolled my eyes. I wasn't a good mom but I wasn't bad enough that I was going to watch my kid suffocate to death while I was sitting there breathing in all that good air. But I guess that's what I did. At least that's what she yelled at me later. She didn't say suffocate though, she said drown. That I was just dumping more problems on her. Standing by watching her drown.

I never could have imagined what was going to happen, to you or to her. The aunties at church all told me I should go to court and get custody of you but I knew if I did that my daughter wouldn't ever forgive or talk to me ever again, and as bad as our fights were she always eventually got over it and came back. I told myself it was only a matter of time and this too would pass. But then you show up sobbing with that big dirty backpack and dark circles under your eyes as if you hadn't had a single night's sleep in your entire life. Your hair was crawling with ukus and you smelled a ripe sort of sweet and I had to hide my tears and smile so you'd come in. You let me give you a bubble bath. I looked for marks on your body to tell me what happened but whatever marks there were, they were only on the inside. So I made you hot cocoa and you snuggled with me on the couch to watch a show, but when the show was over you got upset all over again and wanted to go somewhere no one could find you. I showed you the little cubby in the shed behind the lawnmower and even though it was dark and there were spiderwebs you rushed in as if it were a life-size dollhouse. You set up that backpack in the corner and pulled out blankets and pillows and your stuffed animals. I left you there with a plate of crackers and jelly when you said you wanted to be by yourself to "make your house." I went back inside and started making dinner. I will admit that I got it in my head that it would only be a matter of time before your mama showed up at the door.

I do think about what might have happened if I just called her and told her you were with me that day. We never know the thing that would make the ripples in the water go in a totally different direction, yeah?

This is my understanding of what happened. It was the middle of the day but supposedly she was sleeping when you snuck out and ran away. When she woke up she assumed you were in your room. At some point she realized you weren't. There'd been a party the night before, there were people littered around the house, passed out wherever they'd fallen. A cousin of the guy who she'd been going out with happened to be awake. She found him in the kitchen going through the cabinets, probably looking for something to eat. She asked him if he'd seen you. The conversation went south. People in the living room woke up to her screaming. The guy she was seeing got involved. She was hysterical. Went at both of them with a butcher knife, chasing them from room to room demanding they bring you back from wherever they'd taken you.

The people running in your mom's circles were the kind who believed everybody should mind only the business in their own backyards, so while all of this was happening they were all either pretending to be sleeping or slipping out the door. But your mama hadn't necessarily made friends of the neighbors, who by then had the police on speed dial and picked up the phone so often it was a wonder they got anything else done. The cops were familiar with the incoming number and the reported address so no one jumped up to rush over, given how many times the officer on duty had had to explain to the neighbors that, even if the house was full of druggies like they insisted, they had searched the house a number of times and never found a thing. Everybody was allowed to have friends. And yeah she had a kid that looked a little left to the wind, but what were they supposed to do about it?

But this time when the police cruiser finally pulled into the driveway the screams coming from inside the house were reason enough to flip on their lights and hurry on in.

The reports say no one was seriously injured. The cousin of the boyfriend was talked out of filing a report against her because everyone understood that her aggravated state was due to the fact that she had no idea where her child was. The police were willing to take a look around but they weren't too worried—so many times they'd seen how you were left on your own, playing in the dirt outside wearing nothing but a pair of saggy old panties. Although no one said it, everyone was sure you'd just wandered off when she'd been too high to notice. But whatever your mama was on was still flowing in her veins and, coupled with her mama worry, she wasn't seeing straight. She went from screaming at the men in the house to screaming at the men who had been called to stop her screaming. Without a knife in her hands the threats were less threatening, but it wasn't until she broke and started crying that they finally agreed to help her look for her missing child. Under beds and inside closets they went. You were nowhere to be found. Then they went outside to search the yard with their heavy police flashlights because by then it was twilight. Inside the house she fished out a cigarette but was too flustered to find her lighter, so on went the gas stove. Soon enough the windows weren't dark anymore and the cops were running back inside. It took a while for them to put out the fire.

Her hands were blistered and black when they found her, but she seemed not to register her injuries. By that time someone at the station gave me a call to let me know what was going on over there. You didn't want to go but I eventually got you in the car. By the time we got there she was lucid enough to see that her kid, the root cause of this cyclone of chaos, was alive and well and with the one person in the world she would have rather you not be with.

At the hospital they did a blood test that was later used to determine that the damage done to the house was hers to undo. She didn't have the money to fix it, of course, and on top of that she felt she should not have to pay rent in a place that was so clearly unlivable. It didn't take long for county housing to pull the plug on the entire situation.

I told her you both could stay with me for as long as she needed, but in her version of what happened I was entirely at fault. If I had just told her you were with me none of any of it would have happened. Which is probably true, but there was no rewinding or talking to her about what might have made you run away in the first place. I'd lost her her county housing. My punishment was not being allowed to know where the two of you went from there. Whether it was a four-star hotel or a couch in a crack house, I could only guess.

It's funny, now that I think about how I saw things back then. The no rewinding part. From that moment on things were like a slow spiral down a toilet drain that now I'd give anything to rewind.

I don't know what's worse, wondering where you went wrong or knowing down to the minute exactly when you did. I guess that's not important right now. What we should be celebrating is you moving that finger!

Your guard is a different one today. Was he the one who took the ti leaf out from under you? Couldn't have been the nurse who changes your sheets. I know her, she would know those needed to stay. Silly, yeah, a thing like that making me cry. All these years and all the hardhead things you and your mama have put me through and not a single tear but now they won't stop. This morning I was halfway through my coffee before I realized I was dropping tears all over the table.

You remember Uncle Kapono from down the road, the one with the koi pond in the yard? He brought over kahili ginger for me to bring but I got only as far as the elevator before someone stopped me. I made the ladies at the front reception promise they would find a pitcher or a vase or something and put the flowers somewhere they could be smelled. The thought of those flowers dying in the trash just about wrecked me. Nothing is going to die under my watch, you hear me? NOTHING.

Maybe I'm not holding it together so good.

Okay, I took a break, washed my face in the bathroom. Enough of me moaning and groaning yeah, we have business! Time to go back to the job that needs to be done. By now, mo`opuna, you must be thinking I am pulling one over on you, yeah, cuz how else would your tutu know all these things? Well I'm not. This is not a story. It is a memory.

1824

THE HIGH CHIEFESS KAPIOLANI REFUSED TO TAKE A CANOE
from Hilo to Pele's home in Halema`uma`u. For this journey only
her own two feet would do. A crowd gathered, unbelieving. Some
were eager for her to succeed. Some incredulous. Others followed
her along the long rough lava trail in mourning, her death all but
certain.

For thirty miles the chiefess labored. Bruises bloomed along
the bottom of her feet. Blood dried in rivulets along her toes.
At the foot of the crater, the protectors of Pele implored her to
stop. They begged her to chant her respects. To honor the goddess
that had birthed the island. They pointed down into the roiling
red lava within the cauldron of the crater. An obvious sign Pele
was disturbed. Please put her at ease, they pleaded. Relief flooded
their faces when she announced her intention to pray. It turned
to horror as the words of her prayer filled the air and echoed off
the crater's towering cliffs. To the Christian god she spoke. She
walked slow around the craggy ohia trees and towering hapu`u.

She brushed her fingers along the spines of the palapalai. She knelt along the sacred `ōhelo and gathered a large handful. The crowd waited for her to make the offering. They gasped when she ate the berries instead. She made a show of chewing, of swallowing. Within the murmurs she motioned for silence.

"To all of you who have joined me, stand as my witness. The Lord is my God. Pele cannot hurt me."

She pointed to the main vent leading into the depths of Halema`uma`u before continuing.

"I will go down. If—no, when—I return, you all must abandon this worship of idols and join me in acknowledging the one and only true God."

As was recounted later, she entered the volcano without hardly a hesitation.

• • •

When word arrived in Honolulu of what Kapiolani had done, Palila did not hear of it immediately. She had been experiencing piercing stomach pains for some time and now they had grown so bad that she was unable to get out of bed. Ka`ahumanu was so preoccupied with her Christian studies that she hardly noticed her absence. Unlike Keawe, who grew more worried by the day. Palila tried to reassure him that she was going to be okay, but it grew increasingly difficult to do. At first she was convinced it had something to do with the extended exposure of the pōhaku to where her body lay open, but she could not explain this to him without also explaining the pōhaku and why she had kept it there in the first place. Once it became clear that the queen knew it to be in her possession then she had no reason to keep it there, but by then she had gotten accustomed to its pull and felt empty without it. She stopped only when the pain became impossible to ignore.

By then there had been no reason to keep it hidden. Keawe was too concerned for her well-being to look around wondering about anything else.

After a night of tossing and bad dreams she woke up one morning drenched in sweat. The pain in her stomach had stretched its arms and was now holding her entire body hostage. Keawe wanted to take her to the building that offered the new medicine, but it was run by the women of the church. She wanted nothing to do with their medicine or the price they might demand of her for their assistance. Besides, she was too far gone for that. There was too much pain to imagine being carried. Keawe was finally done listening to her. He set out to get help, for someone to at least tell them what was wrong with her. He promised to be back as soon as possible.

He was away a long time. Palila wondered if she was dying. She felt weak. She couldn't remember the last time she had eaten. She took a sip of the water Keawe had left within her reach but had to fight to keep it down. It took all the energy she had. A few minutes later she sank back into a dark and disturbed sleep.

The shadows were long by the time Keawe returned. He shook her awake. Coming out of her fog, she noticed he looked even more worried than when he had left. Before her eyes cleared she thought maybe she had died, that she was a spirit witnessing him discover her body.

"Palila! Please wake up. There is something I must tell you."

He helped prop her up.

"What is it," she asked weakly.

"Someone is coming, I got someone, I mean, I did not forget."

His alarm cleared her head.

"Of course you didn't. What are you talking about?"

"News just came in from the harbor. All of Honolulu probably knows by now. Chiefess Kapiolani, she has done it."

"Slow down. I don't understand."

"She went to Halema`uma`u. Walked into the crater. People saw with their own eyes. She ate `ōhelo. She prayed to the Christian god. They warned her what would happen. She ignored them all. She went down into Pele's home and challenged her!"

"No. I can't believe it."

Keawe stood and shook his hands in the air.

"That's not all! She is alive. She came out."

"No."

He nodded before dropping his head. "Yes."

It took a moment for her to put the pieces together of all that would come of what the chiefess had done. She asked her next question already knowing the answer.

"Does Ka`ahumanu know?"

Keawe held her gaze.

"There is no way she doesn't."

Tears streamed down her face. She began to moan, the pain a mix of physical and of the heart.

Later, after she had rested and the spasms coursing through her body had finally subsided for a moment, she told Keawe all she had vowed to her mother she never tell anyone. But she needed his help. She knew what she needed to do, she had thought about it often since her final conversation with Ka`ahumanu about the pōhaku, and her body was too ill to carry through with her plan on her own now.

Keawe was adamant.

"There is no way I am doing that. You won't survive it."

"I am already dying. We have to face that. It is the pōhaku that matters, the pōhaku that needs protecting."

"But I can take it. I can hide it with the ki`i."

She cut him off.

"Absolutely not. This is my kuleana, and mine alone. There is only one way to keep the pōhaku safe now, and that is to disconnect it from the `āina in some way. It can't just be thrown into the waters off a cliff somewhere. Ka`ahumanu knows these waters, she would feel it. It would call out to her. I need to go far away to the other side of where the sun goes. I cannot return. If I do, the queen will be able to smell where I have gone. I will bury the pōhaku at the bottom of the deepest sea, and I will stay with it. Do not try to stop me. This is the purpose of my life."

He could not say he didn't understand. He said only that he had already called the kahuna lā`au lapa`au to heal her, although he was not certain how long it would take for him to arrive. Now that the old ways were forbidden so too were the ancient medicine practices of the kahuna lā`au lapa`au, so that meant the healers needed to come in secret to those who needed them. But even if he was coming in secret, it did not mean people did not know. If Keawe sent him away before allowing him to see Palila, there would be much talk. Besides, Keawe had insisted on his help, describing the situation as life or death. The kahuna was taking great risk to come in broad daylight.

When the kahuna finally arrived, Keawe went outside to allow Palila to be examined in private. He settled in to await instruction. Surely the kahuna lā`au lapa`au would need his assistance gathering whatever plants would be needed for the medicine he was sure would heal Palila and change her mind. If she did not think she was dying, there was no way she would go so readily to meet her death.

The kahuna was inside for only a few minutes. Keawe jumped to his feet. The kahuna smiled.

"Your wahine needs only to eat more! Your keiki li`ili`i is a hungry one, it's taking everything out of her."

She wasn't dying, she was hapai! Keawe shouted to the skies and rushed inside, jubilant.

His joy died as soon as he saw Palila's face. Instead of being happy, she was sobbing.

"What's wrong? Do you not want this child?"

That wasn't it. How could she do what she needed to, now that it would not be only her she was doing it to? And if she couldn't do that, how would she ever protect the pōhaku from the queen who was now surely on her way to destroy it?

Keawe asked her to bring it to him and was surprised when she pulled it out from under the thin blanket covering the bottom half of her body. She hesitated before placing it in his hands. There was no time to consider other options. He described the lava tube he had encountered once while diving deep under a reef shelf, following a series of underwater canals. He had discovered it accidentally while chasing a fish down a hole one day long before they had met. Only those with the deepest of breath could reach it and even still he had realized too late that he had been under rocks too long and there was no way to retrace his strokes and return to the surface without another breath. He had found the air pocket only by the grace of akua.

Palila told him to go. Despite the risk, it was the only chance they had.

1824

Palila had been right—Ka`ahumanu sent for her soon after the news about Kapiolani showed up in the newspapers and became official. Another announcement followed soon after. Ka`ahumanu would be baptized, she would become a Christian Protestant. Queen Ka`ahumanu would henceforth be known as Faith.

But before that she wanted the pōhaku. When Palila was summoned, Keawe tried to convince Palila to send word of her illness and pregnancy instead of going in person, but she knew it would make little difference to the queen. The only thing working in her favor was that she suspected the queen was trying to recover the pōhaku without her new Christian advisers knowing. She was still trying to figure out exactly how to use that hunch to her advantage when she found herself being ushered into a room full of long shadows and ink. The door closed behind her. It was the first time they had been alone together for a long time. There were no pleasantries.

"You know why I called you here."

"Yes."

"Then give it to me." The queen held out her hand.

"I don't have it."

The queen lowered her hand as if it were a bird feather drifting to the ground. She gathered her skirts and stood, all six feet of her coming to stand so close to Palila that, although she looked down on nearly everyone else, to the queen she had to look up to maintain eye contact. The queen burrowed into her with her eyes. Palila resisted the urge to blink and prayed her wandering eye behaved.

"You don't have it? You dare lie to me? I know you would never get rid of it."

Palila thought fast. Her wandering eye twitched.

"I've been ill. Let me tell you why. After the burning of the ki`i I began having terrible dreams. I thought I was being punished by your god for still believing in the pōhaku. I got scared. The dreams got worse. One night the dreams were so terrible I knew they were telling me I couldn't keep the pōhaku any longer. I offered it to Kanaloa. I took a boat far out into the ocean and threw it as far as I could. I have not seen it since."

The queen was so close that now she was bending over Palila. She looked ready to crush her.

"Where," she demanded. "Where in the ocean?"

"Beyond the cliffs of Hanauma."

The queen fumed. "Why should I believe you?"

"Because you would kill me if I lied," Palila said simply. She would kill her either way, she reasoned, silently imploring the spirit of her mother to give her the strength she needed to carry through with this untruth, to maintain the courage to lie. May akua strike her dead if she failed.

The queen considered her next move. Then she laid out to

Palila what was going to happen next. She was going to scour the beaches of all of O`ahu. She would search the island inside and out. It would not be hard to find it. The queen and the pōhaku were connected, or did she forget? Did she underestimate the queen or think she had grown weak? Far from it. She was a Christian now, but that did not mean she would accept being made a fool of. She would let Palila go because she was with child and out of the respect the queen held for Palila's mother, but that was not to be interpreted as the queen having changed her mind. Her mission remained the same. The pōhaku needed to be destroyed. The kingdom of Hawai`i was now a Christian nation.

"Bring it back to me, Palila. Do not serve it. You and your child will never join the kingdom of heaven if you continue to believe in such things. You will both be condemned to hell."

Palila put a protective hand on her belly and nodded to say she understood.

• • •

The queen died seven years later surrounded by her Protestant mentors and a Bible written in Hawaiian, a gift and the first of its kind. Palila, fearing the queen had put a curse on her, was surprised her child was born healthy. The next seven years, marked with relative peace and good health, were even more surprising. Palila and Keawe had withdrawn from Honolulu while Palila was still pregnant, in large part because she had taken to seeing ghosts everywhere and could no longer stand the smell of so many people crammed in such a small area.

In the last seven years of the queen's life, she moved further and further into the world of the Binghams and the god of New England while Palila moved further and further into her head. She and Keawe maintained a quiet farm far from the hustle and politics of the harbor, small enough for her to care for on her own whenever Keawe found work. But Palila was haunted, seeing menehune edging up to the borders of their fields or a message from the gods in every gust of wind and fallen leaf.

The closest neighbors knew them for keeping their heads down for the most part. Palila was too paranoid to ever let her guard down even long after the queen had passed. It was as though the queen's words had crawled inside her and lodged themselves there, her fear festering. Yet Palila took joy in their child, Kalehuna, despite their cloistered existence. Kalehuna, a wild inquisitive girl from her early years, was not allowed to leave her mother's sight, especially not to the harbor, which was of course exactly what Kalehuna especially wanted to do. Instead of going to school to learn how to read and write, Palila insisted on teaching her at home, placing emphasis more on teaching her daughter to care for their fishponds and fields than on what they would find in books from New England. In this way, Kalehuna spent the first thirteen years of her life safe, albeit from a threat she could not see or understand. But unlike her own mother's approach, Palila spoke freely to her daughter. Between the secret-keepers there would be no secrets, Palila reasoned.

As many things do, the pōhaku expanded in its absence, its power its own religion.

Kalehuna became well acquainted with the stories of the pōhaku. Palila spoke of the pōhaku with such vibrancy that it made the child conjure large unimaginable invisible forces at work, planting a seed of keen suspicion that they were living in a time when akua still walked among them. Whether there was fish in the pond to catch or if the pond was fishless, if there was too much rain or not enough, Palila had an explanation that without fail included an omen, an ʻamākua, or a message from the ʻāina.

The `āina was in constant communication. All you had to do was listen, she would say. This had the opposite effect on Kalehuna than she intended for her. The girl developed a deaf eye and a blind ear when it came to the daily rhythms of the land they were responsible for.

"Are you listening, Kalehuna?" her mother would ask.

"If you're going to tell me what everything means, why should I pay attention?"

For that she got a box on the ears.

• • •

Palila felt no need to hold back when it came to her suspicions of the direction their world was heading. While it was difficult for Kalehuna to fathom what her mother could possibly be so frightened of about the world beyond their farm, she did embrace her mother's distrust of Christianity and its growing influence, especially given that her father nearly always returned from the harbor stinking and stumbling as if his insides had been turned upside down.

Kalehuna learned how to read and write with ease, although she saw no need. She was content. Nearly thirteen, she knew where she belonged and felt secure in what the future held for her. Her mother would guide the way.

• • •

That all changed the night Keawe didn't come home. Palila fluttered from her bed to various rooms in the house and back again until dawn. Kalehuna woke every so often to the creaking sound of her mother's movements, a sound that told her she should be more worried than she had thought to be. The night's silence was loud.

The next morning Palila shook Kalehuna awake. Already fully dressed, she issued a series of chores for the girl to do while she would be away.

"I'll be back as soon as I track down your father," she promised.

Kalehuna assured her mother that she would be fine. The restless night had left her tired. She closed her eyes and tried to go back to sleep. She listened to the sound of her mother's footsteps as they faded away. Soon they were replaced by the chattering sounds of the forest. Eventually she got tired of tossing and turning and resigned herself to getting up. Doing her chores would at least force time to move faster.

She wasn't hungry until late in the day, the interrupted balance of her family interrupting the routine of her stomach. She sat down and was still mid-chew of her first bite when she heard a clatter. She held her mouth still to listen. After a moment the sound happened again. This time she realized what it was: the sound a rock made when it fell to the ground. Curious, she went out to the lanai. Nothing out of the ordinary that she could see, but soon there was the sound again. This time it was slightly louder. It drew closer. Then it was followed by the sharp footfalls of someone running. A moment later her father burst through the thick foliage, scrambling as fast as he could with his bare feet, stumbling and wild-eyed. He looked like he was being chased by a pig, which didn't make sense. Pigs didn't throw rocks. Kalehuna looked around quickly to see if there was something within reach that she could throw at her father's attacker. Before she could even reach out a hand to help, his pursuer entered the clearing. It was her mother.

Palila was not running. She stopped every foot or so to reach down and grab another stone, hurling it toward Keawe before continuing on. At the foot of the steps Keawe turned and dropped

73

to his knees, lifting his arms toward Palila in surrender. He struggled to catch his breath. His clothes were a mess. But he was in one piece and unharmed.

Palila threw another stone. It missed Keawe and bounced off the ground behind him. Kalehuna knew her mother had better aim than that. It was clear she was not trying to injure him, at least not seriously. But it was also clear that she was furious. Palila began to yell as the distance between the two closed. Kalehuna stayed as still as possible on the porch, not wanting to draw attention to herself or get embroiled in whatever might be going on.

"Out all night, passed out in a ditch! You could have been killed!"

She stopped a few feet from where Keawe was kneeling and folded her arms over chest, her face flushed with anger.

"So careless with your life. Do you ever think about what would happen if something happened to you? Not to us. We would be fine!" She made a sweeping gesture around the property. "Me and your daughter have done all of this while you are down there doing who knows what, bringing that pilau New England stink back with you every time. But the pōhaku. Oh," she pressed her hands to her chest and swayed. "You are the only one who knows where *it* is, the only one who can get it if we need to get it back. I trusted you to help me. And you are failing. You hear me? Failing! I cannot even look at you right now."

With that she stalked past him into the house and slammed the door. Kalehuna watched as her father covered his face with both of his hands and began to sob. He crumpled to the ground. She had never noticed how frail he was. How old. Yesterday she had thought of him as unflappable in even the strongest of winds. Now she wanted to run to him and throw her arms around him but was scared that even this might break him.

74

Keawe's sobs ebbed. He wiped his face with his shirtsleeve and dusted the dirt from the knees of his pants. His head hung heavy as he turned to enter the house. When he caught sight of Kalehuna he offered her a weak smile before going through the door. Their voices carried clear through the windows. Kalehuna pressed herself against the wall to listen.

"Pali, I'm sorry."

"I'm the one who's sorry. I never should have trusted you. I should have gone out on that boat and taken charge of it myself."

"Don't say that! We were given Kalehuna, it was a sign. Things happened the way they were supposed to."

"And this? Is this how things are supposed to be too? You drinking poison, slowly killing yourself? No. The pōhaku is no longer safe where it is. We are not safe. Kalehuna is not safe. There is a rot. It's spreading. Every day I feel it getting stronger. It is coming for us. I have decided: You must get the pōhaku back. Bring it to me. It needs to be under my care again."

Keawe let out a cry. "I cannot! My lungs. They are too bad now. It is too far down. I am too old. That is a journey for someone who is young, someone who could bring the sky with them when they cross the rocks from beneath. It is safe where it is. It was hidden well. Why move it now?"

"You don't want to go, fine. Instruct our daughter. Give her the directions, tell her where to go."

The quaver from Keawe's voice disappeared.

"No. Absolutely not. I will not allow my daughter's life to be put at risk. Please. Just let it be."

"Don't you see how things are getting worse? If we leave it where it is then it too will be forgotten or become rotten and then what hope can we hold for us? For Hawai`i?"

For a while there was silence and then whispers too small for the air to carry them out the window where Kalehuna sat waiting. When Keawe found his daughter on the porch, he pulled her into his arms and squeezed. He whispered into her ear. Things were going to be all right. Things would go back to the way they were.

• • •

But things never did. Keawe failed his first two attempts to get the pōhaku back. The first time, he followed the wrong rocky underbelly and nearly drowned. It took him almost a week to recover, Palila prodding him every day to go again. A clock in her mind had started to tick. Every day she woke up convinced it was their last, that they had finally run out of time. Six days later Keawe tried again, this time with Palila in tow. Kalehuna begged to go with them. Palila would not hear of it. After a couple of false starts, Keawe made it to the lava tube that led to the small underwater cave where he had left the pōhaku. But by this time the light was fading. The afternoon tides were approaching. He searched for as long as he could but at a certain point gave up. It was possible the tides had moved it.

Keawe was exhausted. Years of drinking too much and the hard physical labor of the harbor had depleted his body. Still, Palila pressed. This time they would not make the journey home for him to recover. Neither of them could keep doing this. It was decided between them that they would sleep on the beach instead. Keawe could go back to the cave as soon as they woke up.

The tide was low when dawn broke. Palila and Keawe embraced. Palila pressed her nose against his and chanted a blessing as he entered the water. He took a deep breath and disappeared.

After what seemed like forever, Keawe's fist broke the surface of the water, triumphant. Palila rushed to greet him. She hugged him with a strength that matched her relief and gratitude. Her hands shook so much she nearly dropped the pōhaku when he released it into her care. She pressed it to her cheeks. Yes, there. There was that pull she remembered.

She was quiet on the walk home, the pōhaku tucked securely between her breasts under her shirt. She was quieter than usual through dinner as well. The next day she told Keawe she was going with him to the harbor. Unwilling to hide the pōhaku away from her again but as equally unwilling to take it to the harbor—a place full of lurking dangers and threats—she left it in the care of Kalehuna, who was stunned by it. In a way, it was exactly as her mother had always described and at the same time was much smaller and more real.

That day she tended the fields and surveyed the ponds all while trying to adjust to the feeling of being slightly tilted.

"It was as though I were on a canoe making its way across a mountain of oceans, up and then down," she whispered reverently to her mother.

She couldn't help but notice how the land around her reacted too. As if a barrier of communication between her and the natural world around her had been broken down.

"Everything was talking to me. But the voice wasn't coming from outside, it was coming from inside. Telling me where it was thirsty. Where a section of the pond was crumbling. It's like how you said, except more. Way more."

Palili nodded. The next day Kalehuna easily identified a freshwater spring near where she and her father had searched endlessly for one. Nature was speaking to her and she could suddenly hear its call.

The next day her parents again left her to tend the fields alone while they went to the harbor. She was still outside exploring when her parents returned home. She ran toward them, eager to share what she had experienced that day. Their faces were somber. Neither smiled. They asked her to follow them inside, that they had something they needed to talk to her about. Kalehuna took a seat and placed the pōhaku in her lap. She covered it with her hands without realizing. Palila took note.

"Since the beginning, our people have been great explorers. I remember my childhood in Kailua when the first mo`i still walked among us. He would always talk about wanting many big ships for this new united kingdom. He knew that it was in our ability to navigate the skies that our strength lived, and in our ability to cross oceans that our power existed. We have many connections to different places across the great ocean. Not far from here is a very large land, larger than all our islands put together. Within this land is a valley made up of a sea of lakes. You know that when you see an uakoko it means there is an akua near, but I don't think I ever told you about the uakoko that stretches all the way from here to this land of the valley lakes. It is an ancient bridge, connecting our two peoples."

Kalehuna looked from her mother to her father as she tried to make sense of what she was being told. Palila continued.

"The pōhaku is not safe here. But I am too old to make the journey to the place where it can be kept safe until Hawai`i is ready for its return. No ship would have me, and we do not have money to buy passage. But there is a man, a European, who has come in search of a group to help him settle a new village in this very place on the other side of the uakoko. He is respectable. A former captain of the French Royal Swiss Guard. But he does not know how to work the land the way Hawaiians do. You will join his crew and help teach him."

It was all too much for Kalehuna to take in. Before she could say anything, her mother continued, "You would not be alone.

Your father and I met with one of the young men who will be going. You will travel together, you as his wife. He will keep you safe and protect you. Your face tells me you are not happy about this. But consider this: because of the uakoko it will not be difficult to return. When it is safe for the pōhaku to return so will you, and you will have gained great knowledge and wisdom."

"But when? When will it be safe for the pōhaku to return? Why is it not safe now? No one is after it, no one has come here looking."

Her mother glanced at the window as if to confirm. For the first time, Kalehuna wondered if her mother's head was clear of spirits.

"All your life I have tried to protect you from the changes. But they have only gotten worse. The mo'ī is creating written laws, but the man advising him is from the church. He is inviting another, different kind of god to rule Hawai'i. He is replacing our ways and turning the kingdom into something ruled by British ways and American laws. Auwe! Think of what would happen to the pōhaku then. How would we keep it safe? It was the pōhaku who brought attention to the sandalwood. Who knows what the pōhaku could draw to the wrong hands. Worse, if it was destroyed? There would be nothing sacred left. My mother's kuleana was to protect that pōhaku in your lap. I inherited that kuleana and have spent my life trying. I don't know if I have succeeded, but it is still under our care, which means it is now your burden to carry, your responsibility to bear. You must take it to this place called California and protect it with your life. You will know when it is time to come back to us."

Keawe asked his daughter if she had anything to say. What could she possibly say to change her fate? Her mother was pushing her out into the very world she had spent all these years warning her about. She was not given a choice.

SEPTEMBER 4, 1992

OH MO`OPUNA, WHY IS THIS NOT WORKING? EVERY DAY I LEAVE here thinking maybe I have told you enough, that your ears have heard enough of the story to feed off the mana inside it, yet the doctors are losing hope. They don't say it. They don't have to. The nurses aren't making eye contact with me anymore. Less life in your fingers. The air in here is stale. Look when I blow air in your face the way I used to when you was a babe and I would dunk you in the ocean. Now not even your hair moves. What am I doing wrong? I'm telling you everything as fast as I can. Do you feel me squeezing your hand?

Let me find the remote for the TV. I already told them they don't have to keep the news on anymore. I explained that we wanted it on only to see all the stories about you, to see how many people out there are concerned and want you to pull through. But we don't need to watch this. A storm in the middle of the ocean that is getting weaker. If it's getting weaker and not headed toward us why do we need to hear about it, yeah?

Okay, that's better. Now I can hear myself think.

Everybody's asking how they can help. People see me trying to work on the house to get it ready for you. Some uncles offered to fix up the roof, patch things up. You know I don't like people doing things for me but this time I let them. If it was just me the house could fall down around me and I'd be okay. But you need it. So they're up there showing me all the pukas and I swear I don't know how the house is still standing. If it were one of the little pig's houses the wolf wouldn't even have to blow that hard to turn it into nothing more than a pile of sticks.

Remember the empty plot behind the house? It's ours too. I don't know if you ever knew that. I wonder if it would have made any difference. I started clearing it but it's so much work, too much. I think I have to leave it until you can help me.

How does family stop knowing each other as family? I have started thinking about this a lot.

I always thought one day there'd be one or two houses more on our land, turn it into our very own compound. When your mom was little I had this idea that we'd turn the plot into a huge garden because we didn't need another house back then. The garden would have plenty of flowers for her and we'd grow so much food that we wouldn't have to go to the grocery store except when we wanted whipped cream for waffles.

Your mom was the sweetest kid. Always bringing flowers home for me to put behind my ear.

I decided to start small. She napped on a blanket in the shade while I sectioned off a little rectangle of land. Turned it over, cleared it. When she woke up she used her tiny little fingers to poke long trails of holes in the dirt. Putting the seeds in was her favorite part. She liked her hands dirty, liked pulling weeds. Sometimes I would find her squatting over the seedlings after they started poking their heads through. I'd ask her what she was doing and she always said she was watching everything grow, making sure nothing came and hurt them. Once she accidentally pulled one. She tried to replant it but it died. We were pulling weeds and she got carried away. She cried and cried.

It's my fault our garden got overgrown. She lost interest after I stopped working in it but I swear it wasn't that I lost interest, I just got busy. So many things felt more important back then. I let everything get in the way. She was fine until she wasn't and by the time I noticed, it was too late. The weeds had taken over.

I should have raised her in the dirt. Back in those days I wasn't thinking of anything except that it would mean laundry for me to do and stains on clothes I couldn't afford to replace. Here I've been lecturing you about the disconnect between us and our land and I'm to blame as much as anybody else.

I should have done opposite everything I did from the very beginning.

Okay, no. We not going down that road of tears today. Should have done this a long time ago but there's nothing to be done about that. All I can do now is finish telling you what you need to know.

The beeping, it's hiccuping. Mo`opuna. Stop. Stop shaking. What's happening? No, stop.

Nurse! Nurse!

1839

THE *CLEMENTINE* WAS NOT JUST A BRIG. IT WAS A WORLD UNTO itself. Kalehuna stood facing the shore as the wind lifted their sails. A hand pressed lightly on her back. She let out her breath when she saw it was only a woman who looked no older than her, wearing similar clothes and with a similar skin tone.

"Don't be scared, I mean no harm. I'm Liliha. Are you traveling alone?"

"I'm not sure. I don't think so? My parents," at their mention she could feel her eyes fill. She returned her gaze to the harbor and cleared her throat. "My parents made the arrangements."

"Say no more. My husband decided this for us and my brothers. But it's a grand adventure, isn't it? Think of all we'll learn while we help the European. My husband said that with the promised wages we'll be back in Hawai`i by this time next year. Honestly, if it's as good as the European promises, I'm going to tell him we should stay there."

Liliha chattered on, not seeming to mind Kalehuna's silence. A part of her was glad to have this sudden companion, but a different part of her wished she would disappear. Having a Hawaiian woman at her side reminded her of what she was losing.

The islands faded on the horizon.

• • •

The first few days passed in an unspoken shuffling and negotiation and the weighing of factors that ranked from each according to his ability, to each according to his needs. The most desirable aikane of any ship depended on who was doing the desiring and what that desire was, and this ship was no different. For the genuine explorer: value and priority was placed upon those who might have valuable `ike about the outside world to impart. For the sea-worn: crewmen rich with youth and beauty and other expirable assets. For the lonely not-built-for-life-at-sea: love, companionship. But for the ambitious, the only desirable quality of an aikane was the power they would gain from such a transaction. In this way the cards rearranged themselves and the dust settled on a new, unofficial but no less official, ranking order as they set out into the open sea.

The coupling was not unfamiliar for the crew already on board, but the open fluidity of it, the simple fact that there was even a word for this that was not derogatory or illegal, was refreshing, if not revolutionary. Kalehuna was not familiar with the mating rituals of seamen, but to her it seemed a not extraordinary extension of how matters of sex and sexuality were treated back home. She did the math.

Unlike some of his men, Sutter had no interest in an aikane of his own. He made no pretense of it, although he depended way too much on the debts and favors of others to ever consider standing in the way of the urges of man, especially those at sea.

Kalehuna quickly learned that everything her mother had said about John Augustus Sutter was wrong. For all her mother's visions and efforts to protect her from the evils of the world, her haste in getting the pōhaku and Kalehuna out of Hawai`i had blinded her to the true nature of this man.

The man from Switzerland might have impressed the impressionable in Honolulu because he had dressed like a mating bird in fine silks of vibrant colors and spoke of all the important things he had done and all the important men he had done them with. But as soon as the brig set sail it was clear John Augustus Sutter was all bluster. He had no military service. The only thing he had left in Switzerland was a wife and a handful of kids. And he had no money. The *Clementine* would not be heading to Alta California, at least not initially. They were going to Alaska. This was a fur trade deal, paid for by the Russian American Company.

Sutter was, however, a master of talk. He spoke loudly about their work contracts and how much they would be paid once they got to their destination, all while immediately reducing the rations and charging everyone on board for food, drink, and passage. By the first week every single person who had joined the crew in Hawai`i was in a debt impossible to conceive of ever paying off.

There were ten of them in all on the expedition—she and her "husband" George who she had just met, George's brother Kawika, Liliha and her husband Henry, Liliha's two brothers, plus three other young men who looked to be roughly her age. Though it was strange to be on this ship in this situation, Kalehuna liked George immediately. He observed his surroundings in a quiet way that reminded her of her father. He had an air of confidence around him that made her feel safe.

She was not the only one who sensed his reliability and usefulness. Kawika was the only one of them who had experience working on a whaler before, but it was George who soon became the one Sutter called for the most, asking him to fix this or address that. Kanaka George, he called him. Day and night, Kanaka George.

That is not to say Kanaka George and Sutter became friends. When Sutter took notice of Kalehuna, there was absolutely nothing Kanaka George could do. To the stated fact that Kalehuna and George were married, it had not escaped Sutter that Kalehuna spent her nights under the protective wing of Liliha instead of her supposed husband. The *Clementine* was not large enough for secrets.

Kalehuna had fashioned the pōhaku into a sort of amulet that she could wear around her neck for the voyage. She had thought it would protect her.

When the order came for Kalehuna to join Sutter in his quarters, she looked to Kanaka George. Kalehuna silently begged him to intervene, but he only dropped his eyes to the floor and continued doing what he was doing. Similar to most days, Sutter had spent that one drinking, and that meant everyone on board needed to walk with light feet. If Kanaka George voiced even the smallest of objections everyone would have to pay for it.

"Close the door," Sutter ordered.

She did as she was told but stayed at the entryway, bracing for him to lunge at her. Instead, he pointed to her chest.

"Is that some kind of Hawaiian thing? You people think rocks are diamonds? Let me try to guess. Black magic? Something your dear old mum gave you to protect yourself after she signed you over to me?"

She realized her mistake instantly. Instead of protecting it, she had done exactly what she had been warned never to do. She'd put a spotlight on it. After a beat she forced herself to smile and tucked it under her shirt to get it out of sight. The bulge it created was almost more noticeable than when it had been out in the open. She tried to wave off its significance.

"Just a silly thing. My parents have a farm. This is from our fields. A way to bring a bit of home with me to this new world."

He leaned back into his chair and made a face that told her he wasn't buying it. She did the only thing she could think of to distract him. Took a step closer. Smiled deeper. Put a hand on his shoulder. Let it linger.

His face changed. When he reached for her, she did not fight. In this, she would find her power.

• • •

Sutter was eager to get to California, so they stayed in Sitka not a second longer than necessary. By then Kalehuna had moved into Sutter's quarters. She took pride in the trust he seemed to have in her and did not attempt to be rescued or reclaimed by Kanaka George, who had allowed this to happen. She was fourteen and officially a woman. Through his various mutterings in the quiet moments between them, Kalehuna had gathered that Sutter needed to make a profit in Alaska in order to continue on with his plans for California. The journey was taking much longer than he had anticipated. Soon they would miss the window of weather when travel would be possible. They pulled into San Francisco all concaved cheeks and bellies, having completely run out of food a few days before.

They made to debark at a nearly empty military settlement and were told almost immediately by a group of local officers that they needed to vacate the surrounding waters under order of the presiding Mexican commander of the area. Undeterred, Sutter put on the regalia he had paraded in Honolulu. Then he gathered the letters of recommendation and introduction he had collected along his journeys—letters he described to Kalehuna as the only currency he needed—and went inland to meet the Mexican commander. The crew of the *Clementine* was ordered not to worry. Sutter was about to ensure they would be welcomed with the kind of pomp and ceremony reserved for presidents and kings.

Instead, what they got was an order to evacuate. They were given two days to turn around.

Kalehuna's first step on California soil took place in Monterey. Unlike San Francisco, the Mexican government there greeted Sutter with enthusiasm. In turn, Sutter professed his endless loyalty, offering his service to aid their efforts in expanding and solidifying their precarious rule over California. When they accepted his offer, he set out immediately to ingratiate himself with the local society, making his rounds collecting political friends the way a child might collect seashells. Kalehuna was often surprised by his success. It made her wonder if his luck was due to his proximity to her and therefore the pōhaku. She reasoned that it was her intimacy with him that would keep her and the pōhaku protected for as long as she needed to be in California, which she could only hope would not be long.

Sutter made no secret of his ambitions. Kalehuna quickly caught on—he had traded everything he had in Alaska. The trade had resulted in a storehouse of supplies so meager everyone on board barely survived the remainder of the trip. But he wasn't the only one who'd skedaddled it out to the land of opportunity. There were many other settlers on the coast exactly like him. Realizing this, he quickly switched tactics, redirecting his vision inland. Officials laughed. The interior of California was untamable as a wild horse! The natives there were inhospitable and downright hostile. They'd run him out in a day. He insisted that this was the very reason he wanted to go. He pointed to Kalehuna, Kanaka George, and the others. He had his own natives. His Hawaiians were his secret weapon. He appealed to the governor. Twelve months, he said. It was the basic requirement to become a Mexican citizen anyway. Give him that time to prove he could civilize California's interior. If he could do that, he wanted a grant of land. The governor agreed.

Kalehuna's mother had described a valley, not a coastline. They were indeed heading in the direction she was supposed to be heading, toward the other side of the uakoko that connected California to Hawai`i. It was here she would find the safe passage home she would need when it came time.

Kalehuna was no longer the naive child she had been when they had first set out on the *Clementine*. Sutter was protective of her, but more in the way a man is possessive of his horse than of what she had grown up seeing between her parents. This was especially true when he drank, and more than especially true since their arrival to California. If he noticed a man looking her way he exploded, but never toward the man. That was when he locked her in her room as if she had invited the attention. She knew better than to mistake this for love, since he never bothered to conceal his own roving eye.

They scouted the area in preparation of their voyage inland, familiarizing themselves with the landscape. Sutter soon got his hands on a flotilla, two schooners, and enough supplies to seed his vision. They left immediately.

The river soon turned into a rivulet that dead-ended into a muddy pool. The small fleet turned around, retracing their watery steps. They came to another fork, one of many tines. Kanaka George pointed to one that looked wide and promising. Kawika pointed to another one. Both were wrong.

For seven days the group went up and down the same stretch of river so many times that everyone lost the ability to determine whether the river flowed north to south or east to west or left to right. Frustration grew. They set up camp along the river's edge for the night. Sutter tucked into his cup.

"I'm going to scout along the shore a bit," she told him. He dismissed her with a wave.

She didn't stop to sit until she could no longer hear any sounds coming from the camp. Pulling off her shoes, she submerged her feet in the river and closed her eyes. She put her hands in the rush and let it speak to her. Her skin soaked in the cool liquid. It told her where it had come from and the scent she needed to follow to get there. She stayed there throughout the night, observing the reeds sprouting up from the water and the rocks along the river-bed. She followed the curve of the small bay until it opened at her feet. By morning, there was no doubt. This was the Sacramento, the river they had all been searching for.

She smiled. On the ship she had figured out how to be of use. The pōhaku had just showed her how she was going to be essential.

1839

"Shh," she said to Sutter under her breath. She pointed into the river, pretending to show him something in the water. "We're not alone. Don't do or say anything until I can figure out if they mean us harm."

"The men should be told to ready their guns," he told her just as quietly.

"It would look as though we came to attack. Please trust me."

Sutter turned from her, irritated. He did not like being told what to do. She waited for a moment to see if he was going to sound the alarm. When he didn't, she turned her attention back to the shoreline.

Her mother could not have known how apt her description of California and this valley of the lakes was. The river was wide and rushing. By now they had passed countless tiny islands and distributaries. Mountains stretched high across the distance.

They'd entered a land of bounty. Whenever possible, Kalehuna and Liliha had collected reeds and anything else that could be put to use. Liliha had started weaving rope for nets. In addition to the deer and elk Kalehuna had seen flashes of, the river was flush with large fish. As if sensing the pōhaku, the animals turned her way as they passed, their dark eyes watching from behind thick foliage and tree trunks. The fish followed in their wake.

As they moved deeper into the California valley Kalehuna noticed trees with white feathers tied to their branches. Eventually the others on the flotilla noticed. Kalehuna motioned for everyone to stay alert and calm. The party slowed its pace. The group became quiet and watchful, a dawning of unease.

For two days the shoreline remained silent. Kalehuna remained on watch. On the third day she turned in the direction of the rising sun and nearly screamed. Two hundred men stood along the bank. They were painted in yellow, black, and red. She did not recognize the markings but did not need to. Weapons at the ready, the intentions of the men were clear. She silently implored Sutter to let her orchestrate their reaction. Before he could respond, she ordered the men in their party not to shoot under any circumstances. If one of them initiated a fight, they would all die.

Encouraged by Kalehuna, Sutter went to shore alone. He made a show of his hands being without weapon and greeted them with all the diplomacy he had employed to get this far. He supplemented his broken Spanish with hand gestures. Pointing to Kanaka George and Kawika, he gestured back and forth from the men on his boat to the men who surrounded him. Same, his hands said.

Kalehuna pantomimed a message she prayed Sutter would understand. Nodding slightly, he turned back to those along the shore and invited them to join his camp the next day after they had time to settle. He had gifts for them.

The men onshore allowed the travelers to come ashore and unload. The Germans and Americans among them refused to move. There was no way they were sleeping under the blood-thirsty gaze of murderous Indians who would kill them the moment they closed their eyes. Back down the river they went. Only three remained with Sutter and the Hawaiians. Kanaka George and Kawika made a show of mounting the cannons while the others pitched the tents.

The night was uneventful. The next morning there was a long line for the promised gifts. Sutter and Kalehuna quickly depleted the supplies he had bought on credit, being sure not to release any of the artillery. But the line of those who wanted gifts stretched ever onward. Kalehuna sent for the stockpile of Hawaiian sugar that she had observed tucked within Sutter's personal supply of whiskey. Sutter did not argue. The ripple that coursed through the crowd was a happy one.

• • •

They moved upstream a few curves of the river more before Sutter sent Kalehuna to survey the area. She decided on ground that was slightly elevated compared to the river, sensing that this was a place of changing weather, and surely that meant the water levels changed with it.

Once decided, Sutter wanted to celebrate.

There was a narrow window between when Sutter's ears were open and when he was too deep in his cups. Kalehuna waited for an opening, but he was racing himself to oblivion. She sidled up next to him and offered him bread.

"You are an amazing visionary," she said into his ear. "I have the power to bring your vision to life. But perhaps it would be wise to make clear to your men that the group that came with you from Hawai`i came as advisors, not indentured servants. You know, so we can execute your plans."

High off the triumph that had come with following her lead with the local natives, he slammed his mug down and unsteadily got to his feet.

"Hush now, all of you. I have an announcement. Given their expertise, my Hawaiians will manage the logistical details of this New Helvetia," he declared. "My very own Switzerland," he said as if tasting the words for the first time. Then he gulped from his cup and passed out.

1839

"This coffee's bitter," Sutter spit into the dirt near Kalehuna's feet. "Grand plan, giving away all our sugar. And food. And the seeds! How are we supposed to plant a harvest if we have nothing to plant?"

He was quick to forget they had given those things away as an act of friendship that was probably the only reason they were all still alive. She ignored his whining.

"Rain is coming. If we don't build shelter, we'll have much bigger things to worry about than bitter coffee. What's coming will not be gentle."

By now Sutter knew to trust her when it came to their environment. She insisted that these would not be the gentle rains of Hawai`i, but the warning was accompanied by a reassurance. Surrounding them was an abundance of materials.

Kanaka George and the others quickly set to work. Before a single drop fell from the sky they had two modest Hawaiian-style thatch roof huts to sleep in, complete with a long main building with enough rooms to house a blacksmith, a kitchen, and living quarters for Sutter.

As the construction raced on, Sutter focused on replenishing their stores. He sent for livestock and any other supplies he could get on credit. He promised a fortune awaited in beaver pelts, and cattle awaited any interested investors. His was a valley flush with bountiful riches waiting to be plucked.

Kawika had seen enough while working the fur trade to know all too well the moral limitations of men in charge. For a white man, Sutter was generous and friendly. He paid them in credit and booze. There were worse hands to be under. He took Kawika and Kanaka George to a vista where they could see the expanse of the valley stretching out before them. He swept his arm across the view. Sutter needed them to go out among the people of the valley, he explained. Be his emissaries. Vouch for him, for what they were trying to do, which was make the area a more civilized place for all of them. Sure, there were profits to be had. But he insisted he was altruistic at his core. Kawika and Kanaka George were instructed to spread the message far and wide: The foundation of New Helvetia needed to be laid with the cooperation of those who were already rooted here. The two men set out wearing only their malo, sending a message they did not yet have words for.

Meanwhile, small gains were large ones. Kalehuna and Liliha plotted out areas best suited for crops and gardens. They continued sewing nets to catch the river fish in the same manner fish were caught along the shores of their islands. Sutter's American and German colleagues caught fish one by one. The nets caught more fish in an hour than the Americans and Germans caught in a day.

The diplomacy gamble paid off. Local women edged closer, observing the camp. Intrigued by the nets woven by Liliha and

Kalehuna, they brought baskets woven so tight they held water without leaking. Sewing techniques were traded. Eventually the women brought local seeds and acorns and showed the Hawaiians how to grind the mix into meal for porridge and bread.

Not everyone was so won over. At times Sutter's men woke up to horses stolen and structures destroyed. Thick hordes of mosquitos joined the forces determined to chase them from the valley. Some developed fevers and cold sweats that Kalehuna feared were messages from the akua of the area. She worried their efforts to civilize were interrupting the balance of the area.

"Maybe we are catching too many salmon at once," she said after spending an evening in Sutter's quarters. "Stripping acorns from the trees. No one is leaving anything behind."

"What are you talking about, woman. This is what's wrong with you people. You complain about others having too much but when it's our turn to take you say we shouldn't."

"Taking too much is not the antidote to others taking too much. If every fish is harvested, there are no fish to make more fish." She buttoned her coat and laced up her shoes.

"Nonsense. These are the ways of man. If we leave riches behind, we leave them for the next man to take for himself."

The next day Sutter ordered an expansion of the main building, adding a second story and fortifying the surrounding walls with cannons at every corner.

SEPTEMBER 5, 1992

Here, quick before da security guard come back. Okay
don't move, I going put this in your hand. Ha, your tutu so funny,
yeah? "Don't move."

Aiyah. I going burn in hell.

We just going pour you a sip here, and me a sip there, and
cheers. I rub em on your lips for you. Mmmm. Good yeah? Sherry's best Christmas kahlua. I was saving um. Hmm, maybe you
don't know Sherry from church. She make kahlua—I don't know
why Sherry no make sherry, haha—every year and give errybody
a bottle, even people who tell her every year they don't drink, cuz
she says I should have some in the cupboard for when she comes
over. I never tried any so I didn't know what da fuss was about but
gotta say dis is good stuff. Want some more? Da heck wit it yeah?
Here, I even pour one out on the floor for your maddah too. Today
we celebrating. Who says no can have Christmas in September.
Christmas we celebrate life, Easter we celebrate life, but why don't

we celebrate every day? Life too short, yeah? One day we could be here, next day gone. Poof.

Tutu is mad at you. Yesterday you scared me for reals, Mo`opuna. I was screaming when dey make me leave and don't tell me nutting so all night I was staying up listening fo da phone ring to tell me you gone. Dat's wen I remembered da kahlua and den I was tinking if you gone why I going stay but den look! Here you are. I came up to your room not even knowing if was going be your room still. But here you are.

How come dey get da TV on so loud in da waiting room? Can hear it from here. Dey not talking about you anymore, da people on da news, but dey still talking about the storm, you know da one I told you dey was saying here look guys we tracking a storm but no worry because it's not coming toward you guys, but just in case we're gonna keep watching it and reporting about it all dang day long. You keep telling somebody no worry enough times and even da guy dat already wasn't worried going get worried.

Even get one name now. Iniki.

Oops, dere da guard, inside da purse go da bottle and shot glasses.

Poof, all gone.

Whoa, da walls moving. Where dese damn hiccups came from. I need to drink water. Tutu gonna put her head in your lap and rest for a minute, k? Like wen you was little. You put your head in my lap and I played wit your hair and let you watch your cartoons and I just watched and watched you, mout'ing da words to the songs dey was singing on da TV. Dos' was da days.

. . .

99

Oof, what time is it? Look at me, drooling on your blanket like a baby! Owww, get drums pounding in my head. Let me go to the lua real quick, then we can start our visit over. Let the guard try to drag me out with that visiting hours nonsense. It's Christmas, for crying out loud. The Lord's day.

1840

Kᴀʟᴇʜᴜɴᴀ ꜰᴏᴜɴᴅ ꜱᴜᴛᴛᴇʀ ᴀᴛ ʜɪꜱ ᴅᴇꜱᴋ ᴏɪʟɪɴɢ ᴏɴᴇ ᴏꜰ ᴛʜᴇ many pistols that were his constant companions.

"You called for me?"

He lifted the gun, examining it in the lantern light.

"Know how to shoot one of these?"

She nodded. "Well enough."

He offered it to her. She hesitated.

"Go on. These parts are still wild. A woman in your condition should be able to defend herself."

He waved the gun again. She knew him well enough to know he was losing patience with her. His magnanimity was always a quick flash in the pan. She took it and left without another word.

Outside, Kanaka George was walking toward the blacksmith's station when he noticed Kalehuna standing in the sun staring at the gun in her hands. He changed direction.

"Everything all right? Sorry, I didn't mean to startle you. Where'd the gun come from?"

"Sutter."

"He gave it to you?"

"I'm not sure why. But I got the sense I'm not going to be sharing his bed anymore."

Kanaka George led her by the elbow away from the entrance of Sutter's private quarters.

"Because you are hapai?"

"You too? How does everyone know I'm pregnant?"

"It's a small camp. Besides, you may have taken Sutter as your lover, but I'm still your husband. I haven't forgotten that," he said almost shyly.

There was no question whose child she was carrying, but if Kanaka George was jealous he never said. A fierce protectiveness surfaced that communicated to all that this was, in a way that only made sense to Kanaka George, his child. No matter what he was tasked to do, he made sure Kalehuna remained within his sight.

She'd been correct in her impression that Sutter would stop sending for her at night. She started sleeping exclusively in the common quarters. Kanaka George slept at her feet and ate only half his portions, insisting she take what remained. It almost made her wish the baby really were his. Almost.

The baby came early. Too early.

Kalehuna mourned but was also relieved. She had come to this new world to keep the pōhaku protected long enough to ensure its safe return to Hawai`i. A child would complicate that.

Kanaka George did not share her relief. After a long night of drinking, he left the boundaries of the fort and disappeared into the valley. He returned a few days later, sullen and reserved. Kalehuna clung to the pōhaku and asked for its protection, although she could not say from what.

"Kanaka George brought a cloud back with him," was what she told Liliha. Liliha did not see it.

The two women woke to Kawika shouting. They joined the others in the courtyard who were just waking up. The news rippled quick.

The camp had been attacked. Slaughtered cattle littered the periphery of the fort. All of their horses had been stolen.

There was no mystery as to who had committed these crimes. The perpetrators had them surrounded. They made no effort to hide, no effort to soften their message. Henry and Kawika agreed with Sutter on the situation. There was a time and place for diplomacy, but this was not one of them. That evening a small group snuck out of the fort. Sutter went. Kanaka George did not.

At dawn Kalehuna rose to the sound of gunfire. Her body was still weak from her miscarriage. Lightheaded, she rushed outside as quickly as she was able but saw nothing to explain the noise. She searched for Kanaka George and was relieved to find him still there. He was still not himself, but he was not the sort of man who could turn his back on a woman asking for his help. Especially if that woman was Kalehuna. Together they went to try calming Liliha, who was worried about Henry.

The men returned a few hours later triumphant and un-harmed, Sutter with Henry and the others behind. Sutter was eager to recount what they had done. Thirty natives dead, but he'd shown mercy. He could have killed so many more! He'd done only what was necessary to send a message. These attackers needed to understand that it was he who was the magnanimous, gracious one here, defending his home. The natives were the thieves, the trespassers.

Done with his storytelling, he motioned for a drink and called to Kanaka George to join him in his office. Sutter closed the door behind them.

What exactly was discussed between the two neither Sutter nor Kanaka George would ever say. The public announcement was that Kanaka George would be relocating to Hock Farm, an annex of the fort that was thirty-five miles north. Kawika sent Kalehuna a look that told her clearly how much he blamed her for his brother being sent away. Henry was appointed Kanaka George's position at the fort, although as overseer of labor, not head manager. That position would go to Baudet Genest, a Frenchman who had recently left Monterey and who claimed to be an expert agriculturalist who had the unheard-of ability to make sugar from beets. Kawika would remain in charge of all comings and goings on the river. Kanaka George would be leaving immediately.

Kalehuna followed him inside as he collected his things. As soon as they were away from the door he grabbed her arm.

"Come with me to Hock."

She wanted to tell him how much she wanted to. That if it was about her, she gladly would, regardless of how Sutter might react. But she could not lose sight of why she had come to California in the first place. Her duty was to the pōhaku. She needed to remain in this valley of the lakes connected by the uakoko. There was no way to explain. Instead, she put a hand over his callused fingers.

"He would never allow it. It would only make things worse."

Kanaka George didn't ask who "he" was. He gathered the last of his belongings and asked her to watch over his brother.

1841

THE FORT WAS CHANGING RAPIDLY. WITH HENRY AS OVERSEER, the camp had a new rhythm, one that left Kalehuna feeling out of step.

Henry had noticed that many of the local natives working for them were attracted to the blankets and goods they were now selling through the fort's store. With Sutter's blessing he developed a system of currency that turned these workers into customers. The blacksmith designed a metal disk that could be worn as a necklace. Whenever a certain amount of labor had been done, Sutter or Henry stamped a hole the shape of a star in the disk. The worker could then bring the disk and exchange it for anything they wanted at the store. Workers celebrated this tangible proof of their labor. It was something they could track and stockpile without having to rely on Sutter's sometimes spotty ledgers. If not for Baudet Genest, one might have said things at the fort were improving.

For all his bluster, Sutter was not malicious. Genest, however,

Kalehuna was not so sure about. She suspected Sutter had given him the appointment of head manager based solely on Genest's claim that he could get the fort producing sugar within the year—a claim Kalehuna found highly suspicious. For one thing, the man had brought only a bit of sugar with him as proof of what he could produce, saying he had made it during his time as chief clerk in Monterey. This, without bringing a single beet seed. The natives who had accompanied Genest on his move into the valley swore to Kalehuna that they saw him steal the sugar from the kitchen, but there was not enough evidence to bring to Sutter. She wasn't sure Sutter would care. They were both men who made impossible promises and outlandish claims.

She decided to keep an eye on him the only way she could, offering to help him get started on the production of his beet sugar. He responded with a sour look and continued to go through great lengths to make it clear that he did not share in his employer's view that the Hawaiians were an integral part of the fort's mission.

The beet sugar was not the only thing that rubbed her the wrong way about Genest. His arrival coincided with a curious up-tick in recovered stolen horses. Sutter had managed to curtail the theft of his horses and livestock, but the same could not be said for anyone outside the fort. To help encourage visitors to continue passing through, Sutter established a reward for any stolen horses recovered. Kalehuna was convinced that Genest was double dipping—conspiring with horse thieves to steal horses and then splitting the reward monies with the ones who'd done the thieving. It was yet another thing she lacked evidence to bring to Sutter, but she suspected he was far too happy with the profits of this steady stream of visitors to look too closely at anything that might disrupt it. He'd resumed calling for her in the evenings, but now that his fort was established and his fields planted, she'd lost what little leverage she'd had.

A mutual enemy is a powerful unifying force. Kawika loathed Genest even more than Kalehuna did. When Kalehuna found out she was pregnant again, her body rejected the child almost immediately, but it was Kawika who made sure she was watched over.

. . .

The smell of rotting fish filled the courtyard. The Englishman who had sold Sutter the "system for preserving salmon" blamed the failure on California.

"My methodology is foolproof," the man insisted. "But it does not work in Godless territory."

"God didn't rot this fish, you did," Sutter boomed.

Kalehuna had been bathing regularly in the river near the fort since the beginning. Salmon had always flooded its waters, so much so that at times it was impossible to see any water at all.

"Let me take over the salmon harvest," she told Sutter as soon as the Englishman had left the fort. "There are ways to do this thing you are trying to do. I need only smoke and salt."

Kalehuna had always found solace in the fact that Sutter was not a man of God. It made him less of a threat to the pōhaku, and less of a bore in his better moments. She had grown up under the shadow of her mother's paranoia about the god of New England and, while she didn't share in that level of concern, she had initially worried about her ability to deflect the kind of moral persistence missionaries were known for.

Liliha agreed they should consider themselves lucky. Part of why Sutter's Fort had survived those tenuous early days was that he had not tried to convert the locals into any sort of model of civilization in service to the one true God—a practice Kalehuna was all too familiar with given what was happening in Honolulu when she left. But after she lost that second child, doubt crept in. Her mind traveled down winding paths that led her in circles. Was it the pōhaku that was causing her belly to be ripe for seed? Or was it the pōhaku that was causing it to reject it?

• • •

Her salted salmon quickly filled the storerooms. Once again, she was in Sutter's good graces. She interpreted the success as a gift from the land. How could this thing, so clearly the kinolau of an akua, bring her such connection to her surroundings while also serving to disconnect her so completely from her own body? These were questions she could not share with anyone, even Liliha.

Months went by. She no longer kept the pōhaku caged within her shirt, but it was always on her person. She began offering the pōhaku prayers and spoke to it as if it could somehow transmit messages. She missed her parents desperately. She had stayed in California far longer than she ever could have imagined. But nothing told her it was time to go back.

• • •

The riot in the courtyard started with a skirmish over a currency disk. A circle formed as two squabbling men locked arms and fell to the ground. In trying to gain a good view of the tussle, spectators nudged up against each other. Nudges turned into shoves, shoves into punches. The mob grew until the noise became loud enough to reach Sutter's inner office. He ran outside to see what was going on, pistol in hand. He raised it and shot into the air. The sound barely penetrated the outer edges of the scrum. Sutter motioned to his managers. They reached for their cowhide whips.

Down rained the lashings. The whips whistled as they sliced through the dusty hot air.

The creditor Sutter had been meeting with in his office emerged from the building, shielding his eyes from the sun as he tried to make sense of the scene. The stocky fellow scribbled a note to himself into a folded notepad he extracted from his pocket. Sutter caught the movement in his periphery. He ordered his men to keep up their assault, then shouted at the creditor, pointing his finger at him as if he wished it were a gun.

"You tell that no good son of a bitch you work for that my cow-hides are worth more than any silver. He should be happy to have them. If he wants beaver pelts then he should have the goddamn sense to send someone for them after the hunt. Can't very well give you what hasn't been brought now can I. And if it's you that comes back collecting, make yourself useful and bring back some corn and beans while you're at it!"

Above the din, Kawika was the only one who caught the exchange.

· · ·

He reported to Kalehuna all he'd overheard after the mob dispersed to lick their wounds. She was usually up before the sun, but it had been days and she had not stopped bleeding and, unlike the first two times she had miscarried, Sutter had not stopped calling for her to join him in his bed. Nights were filled with exhaustion and pain. Her relief came on the days Sutter started drinking in the morning, which had grown in frequency. On those days she could rely on the fact that he was usually snoring before the sun had a chance to lower itself from the sky.

Once Kawika told her about Sutter's conversation with the creditor, she understood. Genest had so far not produced an ounce

of sugar, from beets or anything else. He'd spent his time overseeing the procurement of distillery equipment that had gone into
overdrive the moment it was assembled, taking advantage of a
sudden abundance of wild grapes. Which explained why Sutter
hadn't turned on Genest's beet sugar promises the way he had the
fish preserves man.

Kalehuna hoped the pōhaku wasn't to blame for the prolific
grapes. She was already worried they were upsetting the balance
of the lands here. With his distillery Genest could brew enough
brandy for Sutter's supplies to never run dry, no matter how hard
the man tested their production capacity.

Kawika barely finished his recount of the exchange with the
debt collector when the entire population of the fort was rounded
up. Everyone gathered in the courtyard. Genest had an announcement to make.

Leaning against Kawika to stay upright, Kalehuna couldn't
imagine what Genest had to say. She braced for the worst. To her
surprise, Genest was patient and reserved. He waited for quiet to
settle before speaking.

"Things are getting out of hand," he said. He pulled Aleka
from the crowd and motioned for him to speak.

Aleka had been with the group since Honolulu but had kept
to himself. Built like a palm tree, he was tall and skinny with a
hunched lean. His features were sharp, all angles, except for his
wild, every-which-way eyebrows and his cheeks, which bulged in
a way that made him appear childlike and unthreatening. As the
attention of the fort's various residents shifted to him, he removed
his hat and smoothed his hair with his hand. He cleared his throat.

"For a long time, I was not sure why I was called to leave Ha-
wai`i. At home I was only beginning to learn about God. I had
only just dedicated my life to him. Through him I learned to read

and write and see the world in a new way. Since our journey on the *Clementine* I have witnessed my fellow Hawaiians sink further into the dark ways of our past and those of `hewa Hawai`i. I have seen lives lost to temptation and sin."

Here he glared pointedly at Kalehuna before continuing.

"While others die in the name of greed. This vacuum of morality is getting stronger. There is a great evil here. I have been called upon to drive it out. What we are doing is the Lord's work. I have been given the blessing," he nodded at Genest, "to start a church here where we will observe the Sabbath and meet regularly for prayer and strength to resist temptation. We might have lost our way but there is hope yet. We can be forgiven. May the Heavenly Lord have mercy on our souls."

Finished, he bent his head as if to pray. Whispers of agreement and nodding heads shot a hot current of alarm up Kalehuna's feet. Liliha and Henry were in her periphery. She looked to them to gauge their reaction, but they did not look surprised. She even thought she saw Henry offer Aleka an approving nod. She pushed them from her mind. She searched her body for the pōhaku without realizing she was doing so, exhaling only when her hand located the familiar weight still safely lodged within the folds of her clothing.

So this is why she did not know Aleka—all this time, a Christian! Judging her, condemning her. She glanced around for Sutter. He was the only one not there.

. . .

The impact was immediate. News spread. Soon people were streaming into the valley not only from the direction of San Francisco and the ocean beyond but also the mountains. Aleka was quick to call it a miracle of God.

111

1843

Aleka stood at the door of his small chapel, greeting the people waiting to get in. From what Kalehuna could tell, every Hawaiian in the fort had laced up their best pair of shoes to stand in that line. She didn't recognize half the faces, so it could have been every Hawaiian in the region. Every Hawaiian in the fort except Kalehuna and Liliha, technically. So he preached in Hawaiian. Did that really explain his popularity? She asked Liliha.

"I listened to one of his sermons a few Sundays back. He presented God in a way that lifted the back-breaking work everyone's doing for the fort. Almost as though our labor is a righteous sacrifice for our families back home."

"Did you believe him?"

"I don't know. His message made me feel good and proud. So I guess since I liked what he was saying there's no real way to know if I believe him. I might just want to."

She kept her mouth shut regarding her opinion on the matter. The shysters who swarmed the fort with all its new economic opportunities were the only ones who openly and vocally opposed it, protesting the attempts to prohibit alcohol and gambling and general sin-committing that was oh so good for business. But the religious observances and their accompanying regulations were not strong enough to cause major upset, especially since none of the regulations were enforced.

Aleka maintained a safe distance from Kalehuna. She suspected it was an act of preservation, not mercy, although she was sure she'd never openly expressed her doubts about his profession. She tried to remember the first time they'd crossed paths.

The *Clementine* crossed the halfway mark between Hawai'i and their first stop that day. Kalehuna had begun to worry that Sutter would soon tire of her. She imagined him passing her off to one of his men, declaring her a resource to be used at will.

That evening she lingered in his bed instead of scrambling for her clothes and running for the door the minute he loosened his grip on her body. Before he could get up she offered to get him his clay pipe and tobacco box—smoking was strictly prohibited on the ship due to risk of fire, but since Sutter was the one to have made that rule, he was the one allowed to break it. As he packed his pipe, she asked him if he had a vision of what he was hoping to find in the new world. He shook his head.

"That's the wrong question," he said, taking a puff and staring wistfully through the cloudy porthole of his tight quarters. "It's what I'm going to make that matters."

"And what's that?"

"Let me show you."

He opened the cabin door and led her down the narrow hall, all the way to the edge of deck. He pointed toward the hazy horizon.

"Somewhere in that direction is where I'm going to erect an empire, fully self-sufficient. Where food will never be in shortage, where hard work will be rewarded with a wealth you wouldn't be able to imagine without having been where I've been and lived the lives I've lived."

As he went on and on describing all the great riches the crew would all soon have if they stuck to his plan, she tried to hide her surprise at such an egalitarian idea. And then it struck her. Here was the way to secure both her position and protection, both on the boat and wherever they were going. She told him she had a secret she wanted to share with him. He raised an eyebrow.

"I understand dirt."

His laugh was like a bark. "Dirt? Whatever do you mean by that?"

"I mean I understand land. I hear it, I read it, it tells me things."

Sutter didn't respond, at least not until they made it across the ocean. Kalehuna couldn't remember how exactly that first conversation had ended, but she did remember bumping into someone nearby as she made her way back to her room. Aleka. She'd always assumed he hadn't heard anything, and little could have been deciphered in her words even if he had. Still, his current wariness of her hinted that he may have come to his own conclusions. He may simply think she was a witch.

It couldn't have helped that when they landed in the new world, Sutter finally responded by tasking her with mapping out fields where he would build his agricultural empire, as well as where the cattle would fatten and thrive as if they were his own children. Once they got the basic buildings of the fort secure, he set her out to make good on her claims. So as to not give her too much control, he assigned one of his men to do the same in other parts of the area surrounding the fort. The fields designed by Kalehuna exploded with crops while the fields not under her care produced row after row of nothing at all, where in some places the ground remained perplexedly flooded in the hottest parts of summer and others stayed parched and cracked throughout the wettest months of winter.

Her success was as undeniable as it was visible, and she grew proud of the thick swaying sea of wheat and endless rows of fruit trees with branches that hung low and heavy like unnursed breasts full of milk. It might have been more than enough for a different sort, but Sutter was a man whose vision always outpaced his profits, so no matter how much they were able to grow, raise, and hunt, it was never enough. The cattle were moved to untilled grasslands so the crop fields could be expanded. Which was exactly how the next bit of trouble started.

• • •

When Kalehuna was a child, Palila taught her about all the many different moons and how those moons dictated what was to be done where. From the moons you could read the signs that would tell you what the water and fish were going to do, what the lo`i needed, what the misty rain was prescribing. And although those moons cycled, the weather in Hawai`i flowed linearly. This was not the case in the valley of the lakes. Here, there was a time when the plants and trees shed their leaves and went inward. When the air bit and the rivers bulged. This was followed by a period of glorious, saturated color, when the rains smelled of new life and the birds came back from wherever they had gone. And then came the time of no rain. The air dried up and went still. The waistlines of the rivers narrowed. Green turned brown. The grasses underfoot developed a crunch.

One early hazy morning during this period of dryness, Sutter sounded the alarm. There was smoke in the air. Sutter and his European workers rushed to the fields but found them untouched. Kalehuna noted that the natives who had joined their crews returned to their regular routine of starting the day, unconcerned. It was as if they had expected this. Kalehuna asked the women if they knew what was happening. The women nodded. She asked them to explain. When they were done, she sent someone to bring Sutter back so she could tell him that they had nothing to worry about.

The fires were lit on purpose. The flames renewed the oaks and brought down the acorns and regulated the temperature of the rivers so the fish would be happy. The fires brought medicine and sunshine and drove away the mosses that leached to the bark of trees and drained them of their nutrients. This is good fire, she insisted. Sutter's face grew a shade redder with every rationale Kalehuna offered.

He denounced the fires in their entirety. Fire was their biggest enemy. The notion that they had anything to do with the temperature of the river was the most ridiculous thing he had ever heard. Who cared about acorns? It was wheat that was in demand, wheat that offered value. And once started, fire was impossible to control. It threatened their crops, it scared the cattle to death.

He sent a message across the valley saying the burnings had to stop without exception. When the fires burned on he took it as an act of war. The next time he saw a tendril of smoke slithering into the sky, he set out with his men in its direction. When they still didn't stop, he ordered acorn trees cut down and declared a valley-wide prohibition of good fires. Still the good fires continued. Whenever a nearby settlement was found lighting them, he seized its entire population. Labor sentences were issued. The fort became full in a different way.

Kalehuna was horrified. Aleka's congregation continued to grow, as did the number of people who were there not of their own free will. The fort was headed in two directions, both bad. Worse, there was no longer any question of whether they had upset the balance of the valley. When the days began to cool, the summer wheat was harvested and the soil readied for its winter rotation. But the soil had changed. The wheat was foreign and drank of the ground in ways that the natural plants of the area did not. The fruit trees required water in the times of year when the region had none.

Sutter was unconcerned. If building an empire was easy, everyone would do it. Nature did not dictate things, man did. They would turn the valley into what they needed it to be.

Felled acorn trees piled up. Kalehuna mourned with the local tribe and tried to make amends as much as possible, assuring them this would be the last tree, then this one. She moved her hand along the amputated surfaces of what was left behind. Pearls of sap dotted the rings. Tree blood. She pressed the stickiness into her palms and cried.

This display of shared grief did little to soothe the anger that was starting to grow like an invasive species, infiltrating inward from the outer regions of the fort. Aleka accused Kalehuna of stoking the flame. He set out for the distant reaches of the valley and told anyone who would listen: This was the work of God, not man. Life revolving around acorns was for the primitive. If they embraced God and joined them at the fort, if they worked hard and rejected the sinful misguided ways of the past, they would find mercy and forgiveness and a clear path to heaven. The time of living in darkness was coming to an end. Come to the side of the light or burn in hell for an eternity. Their choice.

Kalehuna turned to Liliha and Henry for help to save what

trees remained, but the married couple only looked at her with pity. As if an unspoken message passed between them, Liliha put a hand on Kalehuna's arm as Henry stood and disappeared with the excuse he had fort business to attend to.

"But this *is* fort business!" Kalehuna shouted at his back. He didn't turn around or stop. A moment later Liliha and Kalehuna were alone. Liliha spoke to her soft and slow as if she were a child.

"The first time Henry went out into the valley with Sutter to confront the tribes lighting the fires they came back triumphant as a group. Privately, my husband was deeply conflicted. The peoples here look like our people. He had taken up arms against them, and in service to a white man. I think he may have even killed one or more, although he would not go into those details. He stopped being able to sleep, too haunted by the idea he might be walking a path in the wrong direction. This was right around the time when you and I started talking about Genest, and I promised you I would speak to Henry to see if there was something he could do. Well, I did talk to him, although I never told you what resulted from that conversation. Instead of confronting Genest, he ended up talking to Aleka. About where things might be headed but also about the work we are all doing here. Aleka says civilization and progress require sacrifice, and that the resistance we see is because we have a natural fear of change. Even if things are primitive, our first instinct is to protect and keep things the same because they are familiar. Even in hardship, if it is a familiar hardship, it is one we have learned to survive, so it is preferable over the possibility of even better things, because those might come with challenges we don't yet know. Kalehuna, trust me," she squeezed her hand. "I have shared in your concerns, but if you would only come to hear what Aleka has to say, you will come to trust that we are on the right side of God. The past was darkness. The work we are doing here, the work and the sacrifice, is to bring us all into the light of tomorrow. And you know, after Henry and Aleka talked, Henry sat for a long time thinking about things. He looked up to the sky and asked God for a sign that he exists. A minute later, do you know what happened? Genest finds him and asks him to be in charge of overseeing the continued expansion of the distillery. Can you imagine? God had sent a white man to deliver a message directly to him precisely when he needed it most, asking for his help. Of course he accepted on the spot. Aleka says fire is the work of the devil. It destroys and kills the food we grow to eat and the animals that keep us alive."

Kalehuna pulled her hand free.

"But what if Genest was not sent to him by God?" she said. "Alcohol is what is killing the men around here. How is making it an act of God?"

"White men have come to civilize this land, and they needed our help to do that. One civilized people helping another for the betterment of us all. You don't have to take my word for it. Go speak to Aleka. Take your questions to him. He is not one of these white men. When he goes out into the valley, tribes put down their weapons and listen to him, in part because they share the same skin. He works not for Sutter or the other Europeans but for us, for the salvation of our souls."

Kalehuna did not seek out Aleka or bring him her questions. She had no interest in what the man had to say. Liliha must have been right about the reaction Aleka was getting from the tribes though, because Aleka's congregation continued to grow while the protests surrounding the good fires began to wane.

Sutter turned his attention to his newest prize, the cut acorn trees that were worth so much when traded and milled nearby. The leaves on the trees turned orange and yellow. The silty shoreline of the river disappeared as its waterline rose. In areas where you had once been able to stand with ease, a few weeks later you would be swept away. Now, Sutter joined Kalehuna in her studies of the riverbanks in the mornings. In the afternoons he buried himself in his office, sketching his thoughts. The force of the current had got Sutter thinking. If he could harness it somehow, it would be more useful than a hundred men. What if he built a sawmill on the river? Just because something had never been done didn't mean it couldn't be.

SEPTEMBER 6, 1992

I DON'T KNOW WHAT'S WORSE—WHEN YOU WEREN'T MOVING AT all or this nonstop jerking spasming. I've been praying for movement, but this seems more violent, more harmful.

The nurses insist these convulsions are not causing you pain. How can that be true? At least when you were still I could imagine you as a Disney princess, sleeping until the right words lifted the spell. But these seizures are so violent. I can tell when one is coming on. A twitch in your fingers. Then your hands start to shiver as if you were naked in the snow. Before you know it there's an entire earthquake ripping through you. Your eyes opened in the middle of that first one when I was calling for help. I tried to get you to look at me but wherever your eyes were looking it wasn't in this world. I couldn't go to sleep last night, thinking about that face. All the pain and surprise. I remember taking your mama to get her shots when she was a baby and she only looked at the doctor when he stuck her with the needle, too surprised to cry for a second. Then she looked over at me and started wailing, as though she were crying not because of the pain but because I'd failed to protect her from it.

Tutu isn't proud of how she acted in here during her last visit. Sneaking liquor in. Shame on me. Plus, all this dredging up the past filled the air with ancient dust particles that entered my lungs and wouldn't come out. When I was leaving yesterday, I started coughing in the elevator and didn't stop until this morning. Even in my sleep, coughing until it felt like my toes were going to come out of my throat and I was turning inside out. At first I thought maybe it was God punishing me but I went to the doctor this morning and turns out I needed antibiotics.

When I first got here the nurse said the doctor wanted to talk to me but that she was on her rounds and that I should stick around until she makes her way over, as if I was going anywhere. They should know by now. Also, why would they do that to an old lady? All the places my mind went while I waited for her, auwe.

You used to roll your eyes when I'd say I could feel you in my na`au, that you and me and your mama was connected, that no matter how far away you might be from me I would know when you was in trouble, but I'm going to tell you something now that will finally make you believe.

The doctor said that there have been complications. You have pneumonia, which is apparently a common risk that comes with being on a ventilator because the tube going down your throat can trap the germs that we usually clear out by *coughing*. You hear that, mo`opuna? Your old tutu has been trying to cough for you.

The pneumonia is not the biggest problem. Like a ripple in the ocean that becomes a wave, the pneumonia caused a fever which made your brain hot and that's why you had that first seizure. They're trying to figure out how much damage it caused and if that damage might be permanent. I need them to listen to me. I know what is wrong with you. This tube that gave you pneumonia needs to come out of your mouth right now. Too long like this and your body is going to forget how to work on its own like how when cars were invented suddenly no one remembered there was any other way to get yourself where you needed to be. When I go home tonight and they start whispering to you that I'm trying to kill you don't believe them. I'm not trying to kill you, I'm *saving* you. This medicine isn't the medicine for the kind of sick you have. You trust me, don't you?

The doctor sent the hospital lawyers who said I need to go to the courts to make them unplug you from the wall. They think I won't but they don't know me. Court doesn't scare me. I'd rip these things off you right now if I didn't know they'd use that as an excuse to keep me from you and that can't happen because I have to tell you a different sort of story today.

I was young and working at the gas station when I was hapai with your mom. The hours were hard. Sometimes if the manager was in one of his moods he'd schedule me for the evening shift and then put me on the o-dark-thirty shift the very next day. It wasn't a fancy job but it paid some of the bills and it was what I could get without a high school diploma. You didn't know that, but neither did anyone else. I wasn't shame. It was just that when your mom was growing up I wanted her to respect me and get an education and I thought maybe knowing I didn't finish school would give her excuses to not listen to whatever I told her to do. School wasn't for me, even sitting in the front row I couldn't ever hear what the teacher was saying. It was like sitting in a steel drum. It never occurred to any of us that I might be deaf in one ear. Which makes me laugh now to think about, because my mom always used to tell me that I never listened. Little did she know!

She was wrong, though. I did listen. Everything I'm telling

you, she told me. I was young, must have been around ten because it was before we moved to the big house on the corner. For a long time I wondered why she told me so early. I was too young. It's a dangerous story. I'll get to why in a minute but let me explain this first. It wasn't until she passed that I understood it was because of the cancer. She must have felt herself get sick a long time before the doctors found anything. She was like that.

Death has been chasing this story for generations, trying to kill it. So I don't blame her for telling me before I was old enough. She had good reason, but that doesn't make the damage it caused any less.

I started having these bad dreams. Every night I found myself hiding under a table in a dark house. Bad men or bad things were hunting me. I tried as hard as I could not to cry or move or make a sound, but at the same time I was praying that they hurried up and found me, that the table I was hiding under would flip over and whatever was meant to happen would happen already because the waiting and the fear was more painful than anything else my brain could imagine. But they never found me. Night after night it was just me under the table, so scared I was shaking and always about to piss my panties.

I was twelve years old the first time I stole a beer from my dad's stash in the garage. People might have told you I liked to party back in the day but it wasn't about that, not in the beginning. I was only looking for a way to sleep.

By the time I was fifteen my mom was dead, my dad was checked out, and I was sober about an hour a day, which was about the time it took to wake up and vow that today was going to be the day I was going to go on the straight and narrow. I'd get out of bed, change my clothes and smoke a cigarette. I'd try to eat some breakfast but my head would be pounding so hard that the thought of food made me want to throw up. I'd start getting the shakes and the only thing that would make everything better was a beer. Just one. It always was only ever going to be one. The gas station didn't care if I showed up to work smelling like beer because that was probably one of the least foul smells of that place. As long as I didn't steal booze and my register never came up short there was no problem. Which is funny because whoever was in charge of inventory never noticed or bothered to compare what was being bought with what was being sold and there was no explaining how we was always running low of all the little bottles of vodka and whiskey behind the counter even though not a single soul ever came in and purchased one, at least from me, in all the time I was working there.

At the gas station I'd see all sorts. You can usually match a person to their car, but you can't ever match a person to the snacks they buy inside the market while their tank is being filled. The tourist with the raccoon eye sunburn and the diamond as big as an ice cube on her finger carrying a purse that cost more than my car buys a soda and throws a fit when I charge her ten cents more because I got the sizes mixed up on accident. A guy in a suit and tie looking like he just met with the president buys a case of Boone's Farm and a bucket of twizzlers and makes you wonder if he's on his way to a dorm party with a bunch of barely legals. The kid who rolls up in a rusted out hooptie and walks in barefoot with no shirt pays for his gas with a wad of cash in his back pocket but is the sweetest most respectful customer I'd have in a week.

Then came your grandpa, who must have spent every day driving from one side of the island to another because he started showing up almost every day, filling his tank, buying a hot dog that he'd take his time drowning in ketchup and mustard and relish while trying to make conversation. Once he figured out my schedule he'd pop in right when I'd be clocking out and by then I'd have put away enough of those little bottles behind the counter to want to make a party of it. We'd stay out all night singing karaoke at Uncle Dino's bar or playing pool at a place you wouldn't know because it closed before you were born. Eventually going out drinking got expensive. It seemed a smarter, less pricey option to hang out in a lifeguard shack with a bottle or two. Then presto bammo, wouldn't you know it but here your mom comes. Your grandpa didn't want me drinking or smoking with her in my belly but by then my body needed those things the same way it needed food and air. I'd had a lot of practice hiding the drinking at the gas station so it was easy enough to hide it from him too, which was why fast-forward to when your mom is a teenager and starts coming home with trouble on her breath and I feel like a hypocrite telling her to stay away from all the things. I'm not saying I'm proud of it, I'm just telling you like it is because I'm only doing this once and if I'm telling you the story then I'm telling you all of it. At the time, my thinking was that she got a taste for those things when my body was making her so there was a part of me that thought there was no use fighting it. When I told you earlier about the garden we planted when she was a girl I said I got busy and let the weeds take over. I said it as if I was some kind of career lady who'd gotten busy staying late at the office digging herself out from a mountain of paperwork and for that I'm sorry. There's a part of me that still wants to be a grandma sort of grandma and I didn't want you to think less of me. But we're too far down the road to turn around now. I was drunk. That is what I was busy being.

Of course, I couldn't tell your grandpa all this so I hid it as long as I could. Eventually he found out I was the one paying for the trouble she was getting into. That's the truth of his leaving. He didn't abandon your mom, or me. He just couldn't sit around and watch me slowly kill myself, dragging our daughter down with me. He wanted to take her with him, but in those days you could kill somebody with an axe and the courts would still flip the coin in your favor if you were the mom. Besides, with him she would have had to clean up. None of us said it, but we all knew that was the reason she didn't fight to go with him. It wasn't because she wanted to stay out of love for me, that's for sure. In fact, she resented me for it. Blamed me for him leaving, I think. During one of our many fights in the years that followed she said I was everything that was wrong with her life.

Huh.

She was probably right.

Did you know I went to jail? For almost three entire months. It's not illegal for me to let my kid drink as long as we're on our private property, but it is apparently illegal to give your kid the keys to the car and go to the store when you run out of what you're drinking and need to restock. How does that make sense? I was with her. I was being responsible, not driving drunk. The only reason she didn't have her license yet was because the car you drive during your driving test has to have car insurance and we were in between things.

Man, it was like the judge was friends with your grandpa or something. He threw the book at me. Child endangerment, DUI, plus DUI with Child Endangerment, which are stupidly three completely separate charges. But I guess three months isn't that long for all that. I got out early for good behavior. I can follow the rules when I want to.

Jail ended up being a good thing. Forced me to sober up, to get a good hard look at where my life was headed. By the time I got out I was ready to turn my life around. It felt like a second chance, like anything was possible. Like I could suddenly be the mom I knew I could be, maybe even go back to school and get a degree.

Three months is a long time to spend behind bars. It's an even longer time when you're on the other side of them. I didn't have a clue your mom was into hard drugs before I got arrested but there was no denying it when I got out. It's a long enough time to get yourself emancipated too, so there wasn't really anything I could do about it, not that I had any idea what I could do. That was also right around the time she got pregnant with you.

I didn't find out until much later, after everything was different.

1844

THE AMERICAN RIVER RAN THICK AS AN ARTERY. FROM THE mountains it poured into the valley and coursed along the fort before eventually joining the river that had originally carried them there. Where Sutter saw the possibility of power for a sawmill unlike any other, Kalehuna saw an answer to their need for water in the dusty dry months. Sutter sketched his plans. Kalehuna spoke with the men about diverting water from the river to the thirsty fields.

Her middle widened as the river did. Liliha, who Kalehuna was seeing less and less of since she and Henry had joined Aleka's ministries, pleaded with her to spend more time in the camp with the other women. Kalehuna refused. Only after Liliha's insistence did she even acknowledge the state of her body. She had lost two children. She was sure it was only a matter of time before her body rejected this one too. At one time she might have been sad about it. Now her eyes were open to the darkness she was in part responsible for.

There was no turning a blind eye anymore to the troughs dug around the trees of the fort, where natives were fed a daily allotment of thin porridge, forcing them on their hands and knees to eat like pigs. There was no pretense anymore of where the native children went when they were taken from their mothers. Sutter did not bother hiding the fact that he gifted them to creditors and other visitors he wanted to impress or sold them to those with fat pockets to fatten his own. There was no ignoring the blood in the river from the villages he continued to attack, no longer under the guise of enforcing the prohibition against good fires. He did not treat Hawaiian citizens that way, he was too smart to ever forget that they were not citizens of this nation, that they had a kingdom to come to their aid if necessary, but she did not want to bring a child into what this world was becoming. She returned to the river's edge and her work, waiting for the blood to come.

It never did. Her body shifted shape and a new mana awakened in her bones. She had heard others complain of sickness and fatigue, as she had before, but this seed made her feel strong and powerful. Fruitful. Sutter graced her with his absence. Kawika was working long hours managing the movement of goods and people up and down the river but checked in on her when he could, bringing her small gifts from his stops near and far. Liliha and Aleka pleaded with her to come to the Lord before the baby was born but she refused. Aleka finally gave up.

It was too painful to think about Hawai`i, so she did not allow the thoughts. She refused to think about her parents or wonder if they were still alive. There were more Hawaiians joining them in the valley every day, and they brought with them news that did not bode well. She no longer saw her return an inevitability. She had spilled blood on this soil. The waters of the valley were now deep in her skin.

She could never call this home, but it was time to accept that to this child it would be. On the night of the next new moon, she waited until the fort became still and every last one of the drunkards was snoring. The night was not clear but the moon was full and bright enough to cast a silvery sheen through the clouds. It offered enough light for her to make her way out of the fort and to the river. She climbed over the rocks along the shore, not stopping until she came to a sharp curve in the water. It was a ways from the fort but a place she would recognize easily if she needed to come back, although she could not imagine any reason she would. She moved from boulder to boulder until she found a rock small enough for her to dislodge but large enough to remain steadfast during the seasons when the river waxed and waned. When she finally found it, only the moon saw.

It took much longer to shift the boulder from its muddy seat than she had anticipated. By the time she moved it as much as she needed, she was panting. Her shoulders burned and her stomach was beginning to cramp. She buckled down. There was still much work to be done and she needed to be back at the fort before sunrise. She got down on her knees and began to dig. When the hole was calf-deep she sat back and released the sling lashed to her chest. She unwrapped the cloth bundle tucked inside and folded her hands around the familiar surface of the pōhaku. The moon flickered. She closed her eyes. The sounds and smells of the valley blurred. The air around her grew warm and moist in a way she had not felt for years. The cold slush of river mud caked to her feet was replaced with the warm silky softness of her family lo`i. Tears came in waves. She stayed that way for as long as she dared, a foot in two worlds. Then she smoothed the cloth and carefully folded it around the pōhaku. She placed it in the hole and covered it, first in sand and then a layer of pebbles. Last, she returned the boulder to its original position, sealing the pōhaku into the riverbed where it could stay safe and hidden.

She could not be its protector and also be a mother. For many years she had chosen one path over the other but now she had arrived at another fork. It was time to change direction.

By now she was confident in her understanding of the river. It would be safe here.

In a daze, she retraced her steps. Her body was off kilter, her footing unsteady. Her ankle twisted. She slipped, jerking her body as quickly as she could to land on anything other than her stomach. Her hip crashed into the ground, her elbow absorbing the brunt of the fall. It took a few moments to get her bearings. She took stock. Her arm hurt but she could still move it. Her ankle rested in an unnatural angle but did not appear broken. When she tried to catch her breath and rise a stabbing sensation shot through her leg and exploded deep in her lower back. Dragging herself to where the ground was flat and dry, she laid back and waited for the pain to subside. But it only grew and grew until all she could do was wiggle out of her bloomers and haul herself into a squat, twisted ankle be damned.

A searing pain lifted her out of her body, hovering slightly above in a place of observance. She watched from a distance as her physical form grunted and clawed at the ground, as she gasped and cried out and then grew oh so silent. She fell onto her back and curled her body upward, hands in position to serve its body. She unfolded. First a head, then a body. She pulled it to her breast and wrapped it tightly in the sling that she had constructed for the pōhaku.

The moon poured from the sky to meet the babe. Kalehuna unwrapped the child and let it bathe in the light. Her English name would be Beth, but to her mother, throughout all that was to come, she would be Māhinahina. The silvery haze of moonlight.

1844

Mahina toddled around the courtyard, oblivious to the change in the air. Rising tensions along the coast of California were beginning to move inland. The valley rippled with talk. Sutter was so preoccupied with the development of his sawmill that he hardly noticed. At least until it became impossible to ignore.

The heavy sound of horses stampeded its way through the fort's entrance. Kalehuna raced to the courtyard. She quickly located Mahina poking around an anthill in the grass and swooped her up just in time. A second later, the area was swarming with soldiers on horseback. It was clear from their uniforms that they were American, proving the rumors true. The relationship between Mexico and the United States was no longer cordial. A war was brewing, if it hadn't already started. Kalehuna was not surprised. Over the last few years, the population of the fort had become more and more American, the Mexican government hardly visible at all.

The officer's men dismounted and fanned out. Kalehuna picked Mahina up, holding the child to her chest like armor. She knew from overheard conversations that Sutter only cared who won the war because of how it might affect his interests. She looked to Sutter to gauge what kind of threat they were under, but he opened his arms cordially and welcomed the officer into his inner private office.

They were in there so long that the officer's men grew bored and restless. Kalehuna took Mahina into the lodge to try to keep her occupied and out of sight.

The men eventually emerged. Sutter was uncharacteristically subdued. He quietly sent for a half dozen of his most loyal men and ordered them to pack up. The Americans left with as little warning as they had arrived. Sutter and his men went with them, leaving Genest in charge.

Genest might have won Henry's favor, but Kalehuna wouldn't have trusted the man with a tomato plant. He was smart as a raccoon and just as determined to come out on top. The dust hadn't yet settled from Sutter's departure when Genest ordered the day's porridge watered down to the point that it was difficult to scoop a mouthful with two hands cupped together. Since the gruel was hard to scoop, there was still some left in the troughs the next day. Genest didn't bother refilling them.

"I'm not feeding anyone too picky to not eat what has already been served," he said.

Sutter was bad. Genest was worse. This was not the first time Genest had demonstrated a capacity for cruelty that reached beyond strict management, but Kalehuna couldn't say with certainty that having Sutter there would have made much difference. She was beginning to suspect that maybe Sutter wasn't as blind to his manager's tendencies as she first assumed.

Regardless of the reasons, it was clear she could not stay. She was not a slave; she had a say in where she lived. And she was a mother now. She could not take the risk her daughter would somehow come to harm.

. . .

Kalehuna had no idea if Hock Farm was in the direction of where the sun set or where it rose, if it was toward the ocean or in the opposite direction toward the blue mountains over her shoulder.

"It's near a river," Kawika told her when she admitted her thoughts of leaving.

"You've been there?"

"A few times. Was up there just a couple months ago."

"Why didn't you tell me? Did you see Kanaka George?"

Kawika frowned. Yes, of course he had.

"Can you take me?"

Kawika looked around. The lodge was otherwise empty. He nodded slowly.

"If I can figure out a way to convince Genest that I need to make another trip so soon after the last one, I'll take you."

Liliha had by then committed herself fully to Aleka's church, so Kalehuna knew getting her friend to go with her was unlikely. Still, she waited for a private moment to tell Liliha her plans. She implored her to join her. But Liliha had a secret of her own. She and Henry were leaving. He was heading out into the foothills to start his own church, an offshoot of Aleka's ministry. Liliha was going with him.

Sutter's Hawaiians, as he liked to call them, were Sutter's no longer.

• • •

By this time, Sutter had a small fleet of flotillas and flatboats used to transport cargo up and down the river. The largest was fully covered and designed to withstand shoreline attacks. With store-rooms and a small bunkhouse, the flatboat could spend weeks on the river without ever needing to bank. But it required a full crew, and Kawika needed to navigate this trip on his own.

To go upriver without a crew the smallest of their riverboats would have been best. But that vessel was hardly more than a flat sheet of wood just shy of sixteen feet long, which would have left them uncovered and without a cooking area. At that size they could navigate the narrow windings of the river with more speed and agility, but Kawika's professed reason for this unplanned trip to Hock Farm was an unexpected bumper crop.

135

Genest was expecting a load of goods out of this, so they took the largest they could manage on their own, a medium flat-bottomed boat with long oars on either side and a cargo area that would need to be filled.

Kalehuna spent the entire trip trying to keep Mahina from throwing herself off the side while manning one of the oars. Hardly two years old and already full of conviction and opinion, she was so feral that sometimes Kalehuna wondered if she or the wild had given birth to Mahina. The child was always moving, always observing, soaking in every detail like a dry cloth over a puddle. She hated wearing clothes even when frost covered the ground and her lips turned blue. She refused anything on her feet. Long before she could form a single intelligible word, she considered all things equally able to participate in conversations, or at the very least able to pay attention and listen to what she was telling them. She babbled incoherently to piles of leaves, chuckled at private jokes shared with mosquitos, and got into heated debates with her toes.

When Mahina first started crawling, Kalehuna was forced to confine herself to the fort after Mahina tried to follow her mother into the water, not stopping even after she was completely submerged. Kalehuna dove in and pulled her from the bottom of the river. As soon as they broke the surface Mahina wiggled in her mother's arms and tried to return to the watery underworld from which she had just been scooped. The child was full to bursting with curiosity and eagerness. There was no room for fear or hesitation. She played too close to the fire near the kitchens but did not care when she got burned. She put things in her mouth and swallowed before bothering to find out if it was food.

On the flotilla, she dangled the top half of her body from the side of the craft and let her arms move with the current. Kalehuna could only hold the child by the ankles and hope they did not hit rough waters.

The trio made their way up the river until it transitioned into another one. The land flattened; the trees thinned. They passed sprawling meadows full of birds circling above and taking turns plunging into the sweeping grasses.

Kalehuna did not know what she was expecting Hock Farm to be, but when they neared, she was shocked to see a scattering of orderly buildings surrounded by a picket fence and a grid of fields stretching out in the distance beyond. A handful of men was stationed along the river with guns and cannons. They waved when they saw Kawika. In comparison to the fort, Hock Farm was quiet, peaceful. Why had it taken her this long to come here?

A nervous jolt hit her stomach when she spotted Kanaka George. The last time they had seen each other, she had rejected his proposal to come here with him. There was no way to know for sure if that offer had expired.

Kawika tied off, securing the flotilla to the dock. Kanaka George stepped to the edge and held out his hand. Kalehuna picked Mahina up and offered the child to him, expecting her to resist. But Mahina jumped into his arms without hesitation. Kanaka George laughed, dispelling any awkwardness that might have grown in the time and distance that stood between them. Kalehuna and Kawika followed Kanaka George into the main house. Mahina climbed onto his shoulders and dug her hands into his hair, bouncing as if she were riding a horse. Kanaka George broke into a gallop. The girl's laughter echoed across the property. Kalehuna exhaled a breath she had not known she was holding. This had been the right thing to do.

For three days Kawika and Kanaka George burned the midnight oil, talking in low murmurs, trying to figure out how to gather enough goods around the farm to keep Genest from getting suspicious. During the day, Kawika accelerated the harvest while Kanaka George and Kalehuna took Mahina along the river's edge to explore. Mahina put her knees in the dirt, pinching the soil the way a distant aunt does to all the young cousins at a family reunion, the way she had seen her mother do countless times before.

Too soon it came time for Kawika to return to Sutter's Fort. The brothers embraced. Kawika returned to the boat, this time two passengers lighter.

· · ·

Kanaka George told Kalehuna that an environment makes a child. During their time at Hock Farm, Mahina showed Kalehuna all the ways in which this was true.

At first Kalehuna tried to keep Mahina contained as much as she was able, but it soon became clear that Kanaka George genuinely enjoyed the girl's company as much as the girl enjoyed his. The entire farm was run with a meticulous order that did not happen on accident. It demanded most of Kanaka George's time. Kalehuna didn't want Mahina becoming a bother or disrupting his routine. Kanaka George assured her that it was okay if the girl wanted to tag along, although neither of them were fully expecting the girl to wake up before the chickens, ready and waiting by the door before Kanaka George could even splash water on his face. She chirped and chattered while he went through his daily list of chores, eventually becoming familiar enough with the routine to help where she was able. She didn't return to the house until it was time for dinner, allowing Kalehuna to scrub the dirt from her fingers and eating whatever was put in front of her without protest.

It wasn't that the farm tamed Mahina. She was still wild as a vine. There was just something about its rhythm that matched the one that beat within her. She was stimulated and happily exhausted at the end of every day instead of vibrating from within, ready to burst. Protesting dinner or the sweater her mother forced on her when the afternoons grew a bite seemed now a silly bother. She had better things to do.

Mahina and Kanaka George were rarely apart. Watching them, it was easy for Kalehuna to see that they belonged together. Sutter might have put his seed in her body, but Mahina had never been his daughter and never would be.

Without the pōhaku, Kalehuna now felt unnatural in what had for so long felt natural to her. She had lost all sense of direction and now required a detailed map to go even a few steps beyond the picket fence. The truth was that Kalehuna had forgotten what it was like to be without the pōhaku. She realized now that burying it at the fork of the river had not severed her relationship with it, at least while she remained at the fort. It had still been close enough to be connected. Here, there was a clear and distinguishable void. The air was thinner. The stems of plants did not reach out their necks in greeting when she passed. Pollen made her nose itch. Mosquitos attacked her arms. The river told her nothing. Like the aftermath of a falling out with a dear friend, the land was no longer talking to her.

She was mildly surprised not to be more bothered by the disconnect. For the first time since arriving in California she was faced with its magnitude and all the millions of ways it was different from Hawai`i. But instead of feeling isolated and homesick, there was the calm realization that, somewhere along the way, this place had become home.

1846

"MEXICO AND THE UNITED STATES ARE OFFICIALLY AT WAR,"
Kawika reported. Much had happened since his last visit to
Hock Farm.

"That American officer who came to the fort was recruiting
for a militia that planned to lead a rebellion against the existing
government and establish California as an independent republic."

He could also say confidently what happened next because he
was one of the few who knew how to get in and out of the valley
with relative ease. Indeed, once Sutter realized he would need a few
more of his men, he had sent for him. They traveled by horseback
in the direction of San Francisco. As they neared the bay, Kawika recognized the area where they had first arrived in California,
where they had been given two days to leave. A few hours out from
their destination, Sutter got louder, cockier.

"The man in charge is a fool. He's no threat, trust me. He was
the first man I dealt with when I hit the California coast. He's all
bark. No bite. No bite at all. I know him, know how he thinks.

Just let me talk to him. We'll get what we need out of him. I'm the only man who knows how to talk to him—all these men of yours will only confuse him, get him on the defense. I tell you, good sir, diplomacy is the way things work around here. Give me an hour. That's all I need."

The American officer listened to Sutter with a bemused smile. Kawika nearly laughed out loud as Sutter described his version of the relationship that existed between Sutter and the man in charge of the area, the very one who had demanded their immediate departure. Sutter and his mouth! Business together and diplomatic connections indeed. But the officer seemed to believe him, and Kawika was not itching for a fight. If Sutter could strike some kind of peaceful deal, they would all stand to benefit.

When they reached the settlement, the officer held up his hand and ordered his men to stand down. Kawika remained with them, watching in silence as Sutter slowly approached the entrance and was swallowed whole.

The horses tapped their hooves and huffed out their noses, jittery with unspent energy. When it became clear there would be no immediate call for them, the soldiers slowly dismounted, smoking and pacing ruts in the grasses. Kawika got hungry and assumed everyone else shared in his wish for some declarative action—if they weren't going to be starting a revolution at what point could they eat a meal?

The sun climbed as high as it would go and then began to sink. By the time the foothills in the distance grew an orangey glow, the officer in charge had had enough. He ordered everyone to saddle up. They were going in. The troops followed his command without a word. No one had ever seen him this furious.

Kawika half expected to see Sutter chained to a chair, but no. There he was, stooped over and full of hiccups in a book-lined study with the man in charge, the same one he had been tasked to negotiate a surrender. Drunk! All that time they had been waiting outside and the only diplomatic agreement Sutter had fleshed out was the one between him and a bottle of brandy.

The American officer took control of the room the instant he entered it.

"Sober up," he ordered, grabbing Sutter by the elbow and yanking on him with such force that Sutter nearly fell on his face. The officer handed him off to Kawika and declared the man in charge a prisoner of war. Then he took command of the man's desk and wrote out a letter that was dispatched immediately to San Francisco. They were now in charge, it said. From that moment forward California would be a republic.

Kalehuna couldn't believe it. A republic? Kawika paused in his storytelling to give her the same look he had given Kanaka George only a few moments earlier. Stop interrupting. He wasn't done.

"Once Sutter conquered his hiccups," Kawika continued, "he told the American soldier that if they truly wanted to declare their own country, they need a flag. The natives didn't have a flag, and look what was happening to them. So they spent the night arguing over what the flag should have been. Sutter wanted a bear because you didn't mess with bears, but the American wanted a big star and launched into a long story about a place called Texas. By morning they decided that the republic needed to start in a spirit of camaraderie and cooperation. When the piece of fabric was hoisted up as the new flag of California, it had a star and a bear both."

SEPTEMBER 7, 1992

THREE MONTHS IN JAIL IS NOT LONG ENOUGH FOR OLD HABITS to die completely. It's only long enough for them to go dormant. A pity party wakes them right back up.

This time my daughter wasn't on my downward spiral with me, she was on a downward spiral all her own. We'd jumped out of an airplane together but once we were in the sky it was every woman for herself. I fell hard. Harder than I had ever fallen before. I wanted to block everything out and forget every bad decision I had ever made, forget that I had thrown away any chance I had ever had at a decent life. This went on for a blur of a time. Long enough for you to be born and me not to be around to see it. I think she was already pretty gone by then. It's nothing short of a miracle that you came out alive and healthy. I came to the hospital not knowing if I was going to be allowed in. At that point I couldn't keep track of the days we were fighting or all made up. I cried with joy when the nurse told me right this way and pointed you out to me. They were getting you changed and reswaddled so

I went to see your mom to ask her for permission to hold you. She was drowsy but alert enough to smile and nod. I want to say she was happy to see me.

The nurse brought you into the room in a little plastic cradle on wheels. She smiled a big smile and congratulated me, as if I had had anything to do with it. She wheeled the cradle to where I was standing. I nodded when she asked if I wanted to hold you.

I really did want to.

You were wrapped up tight as a burrito, your face all squished between the swaddling blanket and the hat on your head that was probably as tiny as a hummingbird nest but on you it was huge and drooped like a deflated balloon. She picked you up and held you out to me. I reached for you.

A beer for breakfast was always the only way I kept the shakes away. But I had known I was going to the hospital that day so I had forced myself not to. I couldn't meet my mo`opuna with beer breath.

A ruler distance away from you and my hands started to shake. The shake was big enough for the nurse to notice. She paused in the middle of handing you over. She kept the smile on her face but it turned into a mask. Her eyes filled with questions. I quickly looked away and said something about being tired and overwhelmed with happiness and that maybe my daughter needed some time with her daughter before I got involved. Your mom had fallen asleep by then. The nurse didn't say anything. She nodded and tucked you back into your little cart. Before she rolled you back to the nursery I asked if we could keep you there with us. Of course we could. But also of course the nurse looked at me as though she wished I hadn't asked. The rest of those visiting hours I stayed right like that, standing over you as you slept and grunted and made little shapes with your tiny perfect mouth, all the while trying to work up the courage to try again to pick you up. I was so scared of dropping you. I could barely bring a tissue to my nose without it falling out of my hand and onto the floor.

It was a different nurse who came in later to help you learn how to latch on. I am ashamed to say I used that as an excuse to get the hell out of there. You were barely a day old and I had already failed you. In the car my hands were shaking so hard it was a miracle I could get the car out of the parking lot. I drove thinking I was going to the liquor store but that's not where I went at all. There's a big church at the corner and my car pulled straight into its parking lot as if someone else were driving. I had never been to church before but I knew I needed help—the big enormous kind. I walked in with my face covered in tears and my nose running. I found the priest in his office and almost turned around because he was in a t-shirt and jeans and I thought maybe he was like a waitress or a doctor or something who were only that job when they had their uniform on. But he sat me right down and brought me a glass of water. I swear that was the first glass of water I remember drinking in maybe fifteen years, same as I swear I haven't had a drop of booze since. Except here. That was my only slip-up, I promise on my life. I feel sick thinking about it.

In America they say the past is behind you, the future in front of you. I think it's really the opposite. We're sitting in the back of a truck that's plowing toward the future but all we can see in front of us is what's already happened. We don't know anything about where the truck is going. Everything we know is behind us.

I'm telling you all of this because there was a time after you were born that I could have saved your mom. I didn't know it then, but I sure know it now.

Walking with Jesus finally gave me the strength to get sober. It didn't take me long to get solid enough to be able to hold you without being scared or nervous. The church helped me find a way to love myself, maybe for the first time. But I couldn't believe in God and this story both. I should have and could have given your mom what I am now trying to give you, but I had to reject it along with the booze and everything else. With the support of the church I committed to letting the story end with me. Without the church I was as good as ruined anyway.

All that I did, I always believed I was doing for us.

Do you believe we can make up for the past? I'm trying, Mo`opuna. The house is nearly all ready for you. The guys are almost finished with the roof—just in time too. The news is still talking about Iniki and they say it's not going to hit us but we're probably going to see some rain from it. No more puddles in the kitchen for us, no sir. All they have left to do is over the front door where the porch overhang meets the roof. That section never did do good in storms. Hopefully it's not too windy though. Like I said before: the wolf wouldn't need to even take a deep breath to blow that house in and get that little pig.

When you come home everything is going to be better. I promise.

1846

THE NEXT TIME KAWIKA CAME TO VISIT HOCK FARM HE HAD A different kind of update. In the first few days after declaring themselves the leaders of a new republic, Sutter and the men had descended on San Francisco like locusts. But they had been there hardly a week when Sutter stopped being able to talk about anything other than the fort. His worry that the fort would not run to his liking unless he was there to oversee it grew by the minute. The American-turned-California-Republic officer got an offer from America to become more than an officer, if only they ceded California. And so it was. California became part of the wider United States and Sutter returned to the fort. Kawika went with him.

When Kawika was finally able to make his way to Hock Farm to fill them in on all the hands the territory had just passed through, Kanaka George arranged a feast in Kawika's honor. They had come from the great kingdom of Hawai`i to help civilize a place and that's exactly what Kawika had done.

Kalehuna wasn't sure what she'd considered civilization anymore, but she kept those thoughts to herself.

. . .

Back at the fort, Sutter returned to his mill planning. He wanted to build a great big flouring mill, but he needed lumber, which meant he needed to set his sights on a sawmill first. (This is how the world works: white men sit in chairs and come up with crazy ideas that everyone else has to pull the smarts together to make it happen.) It was too big to build on dry land and then move into the river once it was complete, so they made a temporary dam that blocked the water just enough for them to get the ground good and ready for the mill. Because it was only a temporary structure it couldn't hold that much water for long. When the water built up it got heavy and, left too long, the entire thing would come down. So they took to blocking it off and opening it back up every day. It was quickly realized that building the pressure in the water was helping the men to do their work. The dammed water, when released, had force enough to wash away the rocks that would have taken the men months and a load of dynamite to get out of the way.

Soon enough Sutter was going to have the sawmill everyone has been calling him crazy for dreaming up. Kanaka George couldn't believe it and wanted to see it, but Kalehuna had her own questions. How far down the river were rocks moving? How much was the riverbed changing?

"I need to return with you," she told Kawika.

Kawika and Kanaka George met eyes. Neither of the men understood. Why did she care so much about the sawmill?

"What about Mahina?"

Kalehuna placed her hand on Kanaka George's forearm. While

Sutter had seen Mahina as a bastard he had no interest in, she could not risk his ambivalence by bringing her along. Besides, she had business with Sutter that she needed to take care of and that she was not at liberty to explain to them.

"This is no business for a child."

It was all she could say.

1848

Kalehuna passed through the fortified walls of the fort with her mouth hanging open. Two years away from the fort was a lifetime in Sacramento years. Nothing was like it had once been. The sparse thatched-roof huts were now a thick row of storefronts selling anything you could possibly need or want. The radius of the fort had grown thick with campground clusters of makeshift living arrangements full of people who had come from far and wide to do business. The sweet smell of grass and river was now a heady aroma of sweat, meat on the spit, tobacco, and the general sour of people packed together. Kalehuna was sickened and impressed despite herself. Sutter had pulled it off. She would never have put money on the man, but she had to give it to him. He'd done everything he said he was going to do.

She knocked on the door of his private office.

"Look who's back," he said when she peeked her head in.

"I," Kalehuna started. He held up a hand to stop her.

"Please don't insult me with excuses of why you left. I know where you've been. The only thing I want to know is why are you back?"

His icy blue eyes did not glint for her the way they once had.

"I left something, I need it back," she stammered.

He dismissed her with a flick of his cigar. She would be allowed to stay one night. That was fine by her. She had no plans to stay anyway.

Under the cover of night, she found the river as changed as the fort. The difference had nothing to do with the time of day or difference in season, but she couldn't quite say exactly what it was. There was no moon that night so she relied on instinct to draw her to her destination. She passed a fork she was sure had to be the spot she buried the pōhaku, but after scanning both sides of the shore she decided it didn't jog her memory at all. She continued on through the tall grasses and slippery rocks. Only when she passed the unmistakable flat slip of sandy earth where she had birthed Mahina did she feel her confidence surge. There was where she had fallen. A few meters from there, a fork. No, not a fork. The Fork.

The boulder was easy to identify not because it looked the same but because it had changed color. The hiding spot she had chosen for the pōhaku had betrayed her. The entire thing was now white as whale bone, its marrow sucked dry. When she rolled up her sleeves to dislodge it, she found that it was as light as a piece of coral. A small miracle it had not been washed away in the river's seasonal weight fluctuations. She pushed it aside with ease and gasped at what was revealed underneath.

The pōhaku was exactly where she'd left it, but the blanket of stones she had made to cover it had moved into the shape of a circle, the pōhaku centered perfectly in its middle. Unlike the boulder, the stones were still stone colored, except now they were veined with a bright gleaming yellow. Under them, the sand glittered the same color. The darkness of night did not diminish its glow. She took a pinch and rubbed it between her fingers.

She remembered what her mother had told her about the sandalwood trees, how the pōhaku turned certain elements of the natural world into a commodity that men would kill for.

She waded through the water toward another part of the shoreline to inspect the sand there. After a few moments she spotted another golden flake, and another. Her body became sick with dread. What kind of chaos would follow if there was more gold to be found? If what Kawika had told her was true, the construction of Sutter's new sawmill was upsetting the riverbed. She returned to the white boulder and folded a small pile of gold dust in a handkerchief before covering the rest with as many rocks as she was able.

. . .

She sewed the folded edges of the handkerchief together so the gold would not escape, and then posted it to her parents in Hawai`i, not allowing herself to consider the possibility they might not be alive. If one of them had died, she was certain she would have felt it.

Even by horseback the journey into the foothills took nearly an entire day. During that time Kalehuna told Kawika only what she thought necessary.

"So this rock you brought over with us is akua?"

She shook her head.

"Not really. It's more like a connection. To the land."

"So a dowsing stick. Except instead of finding water, it finds gold."

She let go of trying to explain. He didn't need to know anything more than what she had recovered needed protecting.

When they drew closer to their destination, they agreed that it was probably best if they surveyed the construction site without making themselves or their mission known to the men working there. They pitched camp just far enough away to not announce themselves and waited for the Sabbath, when the site would stay untouched for the day. It was the first time Kalehuna was grateful for Christianity, and to Aleka for spreading it.

On Sunday they were finally able to inspect the riverbed. It was worse than Kalehuna had imagined. From what she could see, the mill was complete. Behind the large waterwheel the river snaked through a narrow valley of rocks and sandy inlets before disappearing into the mountains beyond. The river backed up against the wheel and formed a pool. From there, the water moved through the wheel, turning it. At the point where the water was spit out the other side, it came in a gush that had rinsed the riverbed below of silt, leaving behind a cleaned bedrock. Although the gold was small, the largest the size of a grain of rice, it was easy to see once you knew what you were looking for. Kalehuna picked up a few pieces and held them in her palm. Kawika dropped to his knees and began collecting every speck he could. Kalehuna watched a glaze grow over his eyes. She put the bits of gold from her palm into her pocket and turned her back on the river so as to protect herself from its spell.

On Monday, they approached the men and asked if anyone had noticed anything in the water near the mill. Those who had already noticed the gold were all native to the area, and out of a long tradition of ignoring it they had continued to do so, though now some were beginning to question this. They could not eat it when they were starving or wear it when they were cold. Historically, it served little purpose. But the land was now being overrun by white men. What was considered valuable had shifted. The next day no work was done on the mill. Instead, they all set out in different directions and put down what they needed to. By evening they were all satisfied with the land claims they had chosen. They spoke long into the night of the new life that awaited them. Kalehuna asked Kawika to escort her back to Hock Farm as soon as possible, as she knew this river would soon become a dangerous place. That night the camp pooled their brandy and drank the whole lot in a long and extended toast to their great fortune. Even Kalehuna took a swig.

The brandy did her in. She overslept. By the time she woke up breakfast was over and the coffee was cold. Kawika was not in the tent they were sharing. She assumed he was already knee deep in the river, panning. She folded her blanket and went to tie it to her horse but found her horse was gone. So was Kawika's. Her heart began to race with dread.

She ran to the spots along the river where the other men were hunched in the water. None had seen him.

"Why would he leave a plot of land that was going to make him richer than a king," one of the men mumbled without taking his eyes off the riverbed. There was no reason for her to be so alarmed. He hadn't even stolen any of their findings.

She couldn't tell them that he had in fact stolen something. Something far more valuable than the gold in the river, something that he believed would take him to all the gold he would ever want.

Kalehuna spent the rest of the day numb. A callus grew over her heart. Her vision narrowed. For the years she had been charged with watching over the pōhaku, she had always imagined its biggest threat was Sutter or another of his Europeans. The fact that it was Kawika, the one person she had allowed herself to trust, stung far more than the pōhaku's absence. She spoke to no one. The men made sure she had food to eat but she never thanked them. She spent every waking minute plotting how she was going to hunt Kawika down and imagining what she would do with him when she did.

• • •

When Genest arrived at the camp Sutter was with him. They already knew about the gold. Sutter laughed when the men showed him their claims. Who did they think they were? Where did they think they were? As of the second of February that year the Mexican-American War was over. The Treaty of Guadalupe was signed and sealed. California was now part of the United States. And given the Naturalization Act of 1790, Sutter had been in California long enough to be a naturalized citizen of the United States, which gave that right to "free white persons of good character," which there was no arguing he wasn't. Also given the Naturalization Act of 1790, that law did not extend to the likes of Natives, indentured servants, enslaved people, free black people, or Asians. Certainly not Hawaiians, who were already citizens of an established country. That granted him claim rights and excluded just about everyone else in present company. The only person who those claims belonged to was him. He and Genest hurried back to the fort, pockets heavy. Kalehuna asked Sutter to return with them to the fort on one of their horses, since hers was missing.

"Give them a finger they ask for a hand," Sutter told Genest. They turned their horses in the direction of the valley and trotted away.

Kalehuna lost track of the days it took to get to the fort on foot. Two of those days were marked by heavy, unrelenting rains that forced her to leave the river's edge and take shelter under a thicket of trees. She ate berries and drank from the streams she followed downhill. When she finally arrived, the news that Sutter and Genest had hoped to keep secret was spreading faster than a summer fire in a haybarn. Kawika had, it would seem, bought a bottle of liquor with gold flecks on his way out of town. The merchant who sold him the liquor told the gunsmith who told the carpenter who told the blacksmith who told everybody else. Men were quitting their jobs on the spot, stopping mid-task, dropping their tools in the dust. In the middle of the exodus Kalehuna tried to gather as much information as she could about Kawika, his whereabouts, his plan, with little luck.

She would not, could not, return to Hock Farm without the pōhaku.

She allowed herself only to think of Kawika and the pōhaku, using her growing resentment of him to push out any thoughts of Mahina or Kanaka George or what they might be doing, how her child and the man she had left her with might be faring in this fever taking over California.

Her single-minded determination eventually paid off. A white man from San Francisco stopped by the fort for supplies and told her he had come across someone who matched her description of Kawika. She could not fathom why Kawika would be moving in a direction opposite the gold, but she didn't have anything else to go on. She headed toward San Francisco.

A year passed. She scoured cities along the coast, combing through the many inland towns that were suddenly popping up throughout California. After a while she began to hope that, in a strange twist of destiny, Kawika had returned with the pōhaku to Hawai`i. And yet she knew that was an impossibility. A number of Hawaiians had relocated to California in the years since the *Clementine* and they were more than happy to share news from home, even though none of the news was at all happy. France had attacked Honolulu Harbor, capturing the Honolulu Fort and raiding government buildings. The kingdom had just entered into a Treaty of Friendship, Commerce, and Navigation with the United States. Now the islands were being hit with an onslaught of measles, whooping cough, dysentery, and other illnesses. Death was everywhere. She no longer wondered why her parents had not asked her to return.

Her search for Kawika ended in a small fishing village south of San Francisco populated almost entirely of Hawaiians, where she was brought to a small clapboard church with whitewashed walls and a modest cemetery in the back. She traced the carving of his name on the gravestone with her fingertips. The priest told her he drowned. But how was that possible? Kawika had spent years on ships and boats and then river rafts. He was the best swimmer she knew. She'd once watched him rescue two Englishmen who had managed to both fall off a flotilla at the same time. He'd dived in without hesitation, hauling one and then the other to safety without hardly needing to catch his breath. The priest shrugged. Knowing how to swim only lessened the chance of drowning. It didn't make it impossible. And no, if Kawika had left any earthly possessions behind, the priest surely didn't know about it.

There was no more trail to chase.

• • •

Many things bothered Kalehuna about Kawika's death, but the one that plagued her most was that Kawika was buried at a church, as a Christian. Had it been part of an elaborate disguise to ensure she never found him? There was no way to reconcile a belief in the pōhaku with a belief in the Christian god. If his conversion had been genuine, had he simply discarded it? She was familiar enough with Christianity by now to know that there were tenets around theft and stealing. Wouldn't he have felt compelled to return what he stole from her? The notion of him lying righteously there basking in God's glory made her grind her teeth. She had half a mind to dig up his grave to make sure the pōhaku wasn't buried alongside him. The only thing that kept her from doing it was the understanding, deep within her, that if the pōhaku was indeed buried in the backyard of a little white clapboard church in the middle of California, surely earthworms would have danced a calligraphy of lace patterns atop the dirt.

She wandered farther south for a long time, aimless and angry and on the lookout in the sky and in the newspapers for unexplainable natural phenomena, a sudden swell of birds, the discovery of a trove of emeralds in someone's backyard, anything that might offer hope of being reunited with what she had committed her life to care for. She tried to imagine where it could possibly be or whose hands might be holding it, if it would cause the earth to split in two or the ocean to rise up and wash the entirety of California clean. She had let down her mother, let down Hawai`i. And in such a stupid, shameful way. If the rainbow bridge called to her now, she would have to ignore it. She felt sick considering the possibility that the catastrophes currently crippling her nation were because of the loss of the pōhaku. That it had been needed to protect and sustain Hawai`i, not the other way around.

She cursed Kawika, cursed his God, cursed the gold that had infected his brain and caused him to betray her in such a hideous way. She traveled along the ocean. Considered throwing herself in. When she couldn't bring herself to, she turned and went north.

...

Sutter's Fort had again transformed. Now she found a shell of what it had been only a short time before. The shelves of the merchant stores had been pilfered. Workers had since abandoned their posts, trading in the certainty of a small amount of money with the gamble of a much bigger one. The ash in the firepit of the kitchen was cold. The walls were full of ghosts. The courtyard was haunted with smells of yesterday.

She stayed only as long as she had to.

1857

Before sutter and his wife came to hock farm, mahina was the happiest she would ever be. For everyone else in Northern California it was a time of implosion, but for Mahina it was a time to run through the grass barefoot and catch fish with her hands and have long conversations with cows. She loved her father and who needed a mother. She and Kanaka George had jokes that never got old and a table with sometimes not a lot but always enough. She was his shadow. He taught her everything he knew about the land and farming. She took pride in her growing list of responsibilities around the farm.

There was a constant rotation of people who stopped by on their way to the mines, all of whom were willing to pay good money for a hot meal and a load of laundry done. Although Sutter's Fort had combusted due to the gold-induced exodus, there was still a demand for goods. With food from the fields and enough money coming in from laundry and food sales, there was always enough and never too much. Most of the men who had worked at the fort were now mining in the foothills, alongside men from around the world who had dropped lives and wives wherever they were from to join the race for riches awaiting in the icy cold waters of Northern California. Thankfully, there were also still families who worked and lived around the farm with plenty of children for Mahina to play with, even after she stopped being considered a child and the word playing went from climbing trees and digging up earthworms to kissing and touching in the stables.

Finally fourteen, she flitted like a butterfly in a field of wildflowers, moving from crush to crush, one month a girl who tickled her neck with feathers, the next a boy who could whistle entire songs and always knew where to find the juiciest blackberries. Of her many talents, her ability to weave stories from dreams was the one those who knew her found most entertaining.

Besides Kanaka George, her two favorite humans were Soyo and John. Soyo was a girl her age from the meadow tribe who had lived at Hock Farm for the many years before it became Hock Farm. John was the youngest son of a Hawaiian man who had brought his sons with him when he moved to California to join the miners. Gold fever by then had spread clear across the Pacific. Everyone knew there was not a poor man to be found in California. So many people in Hawai`i wanted to go east toward that wild west that men were offering money to shipowners to drag them along. John's father was determined for him and his sons to make their fortune. He packed them onto a boat bound for Oregon, which he hoped was close enough. Their only possessions were a compass, a watch, and a trunk full of hope.

They were a lucky sort. Halfway to Oregon the boat announced it was changing course and going to San Francisco, where the captain had just discovered he could sell the brandy on board for twenty dollars a shot to miners, a hell of a lot more than what he would be getting for it in Oregon. Lucky them but poor John spent the entire voyage across the Pacific quarantined in a tiny closet with a case of measles that no one expected him to survive. He did, barely, but it left him so weak that the family only got as far as Hock Farm before being forced to admit that John could go no further. Even if he did cheat death somehow, he would have been of no use to them in the mining camps.

The one thing John's father had going for him was his own set of brothers back home, one being a potato farmer in Kula, Maui. It didn't take much sniffing around to learn that all the men who had once farmed the food needed to sustain this rapidly growing population were no longer in the business of farming. In California there was only one industry, and that was digging.

With no one planting or harvesting crops, potatoes became a prized commodity. At Hock Farm, John's father leveraged his potato supplier in exchange for boarding for John. It was info Kanaka George could sell if he wanted, plus a boy who could help around the farm once he grew some muscle.

It didn't take long for Oregon potato farmers to point out how much easier and more economical it was to get potatoes from Oregon versus Hawaiʻi, so Kanaka George never profited from that potato supplier connection, but by then John's father and brothers were gone.

It was impossible to say if he would have been any other way if not for the measles, but John was indecisive and unassertive, as willing to go with the flow as the water that made up the river. No matter how many hours he was forced to work in the fields, his hands remained soft.

Soyo was the opposite. If John was the water and Mahina a butterfly, Soyo was the wind. She determined what direction everything bent. On any particular day, Soyo decided where they were headed, Mahina made up a story of why, and John went along for the ride. In the combination of their elements existed a harmony that kept the fruit trees pollinated. John, Soyo, and Mahina were a unit that made the elders smile and dredge up memories of simpler times. Their laughter made the wheat grow.

That all changed when Sutter arrived to claim his farm. Soyo and John didn't know to be concerned. Mahina didn't remember him. Kanaka George remembered him all too well.

. . .

The message arrived early one morning via riverboat. The main building of the farm was to be readied for Sutter, who had until that moment treated Hock Farm as an afterthought. With the fort no longer functional, he was turning Hock Farm into his permanent residence. The note did not bother to include the fact that he would be accompanied by his wife and three children who he had finally remembered to retrieve from Europe, where they had been packed and waiting for his letter all those years.

Up until that point, Kanaka George had not known Sutter was even married, much less a father. At least to children he acknowledged. He hid his surprise and the others followed his lead. There was nothing any of the Hock Farm workers could do about it anyway. Sutter owned the land. In this world that meant he was on top.

The Sutter family arrived before the farm formed an official opinion regarding this new development, so they were met and greeted warmly by some while others found the chores of that day much too demanding to allow for interruptions. Sutter returned to Hock Farm flanked by two sons who were children in adult bodies, an unsmiling daughter, plus a wife who appeared to have left her humor in Switzerland. Their first week at Hock Farm the sky cried the entire time.

His sons, having spent much of their formative years in the shadow of their father's absence, came to California less out of family loyalty and more out of a desire to show all who had expressed doubt that their father would ever send for them that they had been wrong. They launched immediately into a carousel of schemes. It quickly became obvious that while one had a head for business, the other didn't have much of a head for anything at all. Sutter's daughter, on the other hand, spent most of her time in the library reading.

"She just wants to show off," Soyo whispered as they peeked through the window into the library. Mahina agreed.

"Yeah. The country's not good enough for her."

John, typically, had no opinion.

Although now thicker in both waist and beard, the icy blue of Sutter's gaze had not diminished, nor had his appetite for liquor and accumulating debt. The once-steady prosperity of Hock Farm slid quickly into a seesaw economy as volatile as the whole of Northern California had become. Kanaka George never indicated to Mahina that she might have a connection to Sutter, but it would have made little difference. For the first time she started to consider a life beyond the farm.

Everyone else in the world seemed to be tossing one life over their shoulder to replace it with a different one, why not her? The only anecdote to this melancholia was time with John and Soyo, when she could not imagine being anywhere else.

1858

On a damp day in December when the harvesting was done for the season, Mahina, John, and Soyo were released into the wilds. Kanaka George usually used the winter months to repair what needed repairing, and that did not change now that Sutter was in residence. Nor were they anticipating visitors. Their chores could, for once, wait. Although they were getting older, they were still in the brackish years between adulthood and childhood. On this day they were children. They were allowed the afternoon to be them.

Soyo led them toward the river, which had taken on its winter shape. A trail of fallen leaves spilled through groves of naked trees and bare blackberry brambles that eventually opened to reveal a craggy, muddy shoreline unrecognizable from its spring counterpart. The water ran smooth and silent, as cold as it was blue. In spite of the sky hanging low, dark, and heavy, their mood was light. They stopped at the river's edge. The water appeared frigid, inhospitable. John looked at Mahina.

"Are we really doing this today?"

"You can surrender now, I always beat you anyway," Mahina said, laughing.

"And I always beat both of you," Soyo giggled as she sat on a rock and began untying her shoes. She pulled her socks off and stuffed them into her shoes. Wiggled her toes to show them she was ready.

"Well, not today, girls. Come on Mahina, hurry up."

They counted to three. There was a collective gasp when flesh hit liquid. Their breath came like fog out of their mouths. The water was so cold it burned. The icy water turned Mahina's feet bright pink. Not even thirty seconds and her toes were already numb.

She usually couldn't outlast Soyo, but she always outlasted John. He began to squirm. This was when he usually raised his white flag. All she needed was to last a second longer than him, but instead of surrendering, John got very still. Mahina could feel her insides turning to ice. Her feet tingled. They were the only part of her that existed, the rest of her body was merely an extension. She was suddenly sure she could stay in that water for as long as she needed to win, forever if necessary.

A trio of ducks floated by. She began to laugh. John and Soyo looked her way. Then they were all laughing, clutching their stomachs. Soyo, the reigning champion, finally put up her hands.

"Okay both of you! One more minute in here and you'll have to wait til spring before I can come out."

"Ha," John squealed. "Mahina, you're next!"

Mahina's name echoed faintly from elsewhere. It filled the air and got louder. The trio looked at each other.

"Mahina, where are you!"

"Are you in trouble?" Soyo whispered.

"Kanaka George doesn't like us in the water this time of year, says it's too unpredictable," she answered.

"That's it," John said, making his way to shore. "We're pausing the game."

He began pulling on his socks and lacing up his boots. Soyo and Mahina inched their way to dry land. Soyo winced. Thousands of invisible needles pulsed from heel to toe.

Kanaka George's head appeared through the reeds. The children braced for a scolding. None came. Kanaka George took in the scene with a blank face, motioned for Mahina to follow him, and turned back in the direction he had just come. They traded glances. Mahina fumbled with her socks, her fingers now just as numb as her feet. John and Soyo moved in unison, kneeling at her feet to help put on her stockings and shoes. When she got up she stumbled, her legs an aching combination of poking needles and blinding numbness. They threaded elbows and together set out in the direction of Kanaka George, none sure if they really wanted to get to where they were going or find out what was waiting for them.

• • •

Kalehuna appeared at the foot of Feather River in front of Hock Farm with no invitation or advance warning. In something resembling a trance, it was as though she was not aware of how long she had been away.

Mahina had not formed an opinion of Sutter before he arrived. Similarly, she never thought about her mother, who by then had been missing for nearly ten years. But suddenly there she was, forcing Mahina to decide how to feel about her. She looked to Kanaka George for guidance. He refused to meet her eye.

Kalehuna had not bothered explaining the details to Kanaka

George of what she was chasing after or why before she disappeared. At the time, he'd assumed there was something human going on between Kawika and Kalehuna, a lover's quarrel of some kind that they'd needed to sort out without a child around. He'd been so taken by Mahina, the daughter he had yearned for without realizing or expecting, that her leaving hadn't mattered. To him, it still didn't. What did matter was why she was back now, and how that might affect Mahina. Time had turned Mahina into his daughter and Kalehuna into a stranger.

As Kalehuna settled in, the walls grew eyes.

Over the next few days, it became obvious that Kalehuna had developed an itch, or maybe it was a twitch. Where she had before been in tune with the winds and the rains, now she was restless and agitated and so vastly out of sync with her surroundings that Kanaka George came to suspect she was haunted. She trailed off in the middle of sentences and paced deep ruts in the dirt while chewing a lock of her hair, so deep in thought it was difficult to engage her. When he asked after his brother, Kalehuna muttered something about Kawika having gone somewhere with the god from Boston, which is as much as he knew anyway. The only thing that prevented him from thinking that she had left her mind on the road during her travels was the vigilance with which she pored over newspapers and interrogated every stranger who passed through the property. The questions she asked were peculiar and always the same. Had they observed any unusual activity in their travels, any extraordinary animal behaviors or unexplainable weather? Had they come upon any town or city where fountains of gold had suddenly erupted from the ground? The strangers always laughed at that one, if they had come across that then they wouldn't very well have left it, so they would not have been at Hock Farm to answer her questions in the first place!

Despite this new oddness surrounding Kalehuna, Kanaka George was consoled by the fact that Mahina seemed to have little interest in her mother and her mother was hardly focused on Mahina. The girl withdrew into the safety and comfort of her friends, preferring to sleep and take her meals with Soyo's family instead of in the main house with him. She whispered her stories now instead of shouting them for the world to hear. It pained him, but he recognized that she was also nearing the time when children stretch their wings to test their strength. He began taking a bundle of food and tracking Kalehuna down in the evenings to share a meal with her instead of eating alone. In this way they became familiar to each other again. She said little in the beginning, so he filled the silence with what had happened in her absence.

"The Gold Rush changed everything. People full of big dreams were flooding California from all over the world. But it was the worst thing that could have happened for Sutter. It was as though his whole New Switzerland vision had been built on an ant hill. The men who stayed with him in the final days said as soon they chased one squatter away five more arrived in his place."

Kalehuna only fiddled with her food. He continued.

"He went down to Sacramento a few months ago. Said if the government wasn't going to get involved on their own he was going to make them do something. His mood was even fouler when he returned than when he left. Apparently they thanked him for his service, wrote him out a check he says is a toenail compared to what they owe him, and then sent him on his way."

"Thank you," Kalehuna said in a small voice. "For taking care of Mahina."

"She's my daughter. We've enjoyed each other." He didn't tell her how much simpler things had been with her gone.

"Do you think I could talk to her?"

Kanaka George told her it couldn't hurt to try.

• • •

Kalehuna followed the sneezes and found Mahina hidden in the reeds doing laundry. The river was swollen with spring and the current did much of the work for her.

The air hummed with bees feasting on the brilliant blanket of wildflowers that had appeared overnight, along with the pollen that tickled Mahina's nose and made her eyes water. She had long been waiting for the opportunity to show her mother that she was blatantly uninterested in her, so although she noticed Kalehuna's approach, she continued beating the clothes against the rocks without looking up.

Kalehuna was not impressed. She'd birthed this girl. She understood her. It would take more than the silent treatment to discourage her. Besides, she had come to tell Mahina the story that was her birthright. That required only ears.

Mahina offered no indication that she was listening, but when the laundry was done she stayed seated. Kalehuna got carried away with the beginning and lingered over the details. The hour grew late. She was only at the part where the queen and king were in a fight so gigantic the whole island could hear it.

"I hear Kanaka George calling us for supper. There is still much to tell you. Tomorrow?"

Mahina kept her face neutral. She picked up the laundry basket and headed toward the house. But Kalehuna understood Mahina more than she could ever know. To her, it was clear Mahina was paying attention.

She was impatient to finish the story, to give her daughter it in its entirety. But what was one more day when so much time had already passed? She assumed tomorrow would come soon enough.

But that tomorrow never came.

• • •

Spring was when they let the cattle out, so with the new season came the need to inspect and repair all the fencing of the property. The work kept John, Soyo, and Mahina busy for the next few days. The list of John's responsibilities on the farm had grown with his height and filled out with his shoulders. This was technically his job, but because their triangle worked more efficiently and effectively when whole, what was taken on by one was taken on by all.

The acreage of Hock Farm was not as imposing as that of Sutter's Fort—at least what it had been before the rush—but it was still a lot of ground to cover. Day after day they set out at first light and did not return until long after the lanterns had been lit. There was also planting and a million other animals and things needing attention that suddenly had no people to tend them.

Every day their workforce dwindled. The gold in the mountain sang to men in their sleep, luring them from their beds, leaving the fields with no one to care for them in the morning. But the resilience of the trio was undaunted by the amount of work or the dwindling hands to do it. They laughed and hugged and danced when it drizzled. The seeds they planted were happy for it, as were the cows. Three is usually a hard number, but it seemed to all on the outside looking in that they had found harmony in it.

When Kalehuna got word of an unusual storm in the east, she disappeared in that direction, although this time she took the time to tell Mahina goodbye. She promised she'd be back but offered no real explanation. Mahina reassured Kanaka George that she was not upset.

"It had been her being here that took getting used to, not her being gone," she insisted when he pushed.

It was a time when the winds could not decide in what direction to blow. Spring that year was particularly temperamental. A week of heat brought fear of a too-early summer, only to be broken by a drop in temperature so dramatic it left the grasses stiff with frost. It was as though winter had snuck in the back door and was refusing to leave. This was followed by a persistent rain that soon made outside work impossible.

Sequestered in the big house, the trio was no longer able to do their chores collectively. John was kept busy helping Kanaka George follow any leaks that appeared and keeping the house patched to the best of their ability. Mahina and Soyo were assigned to the kitchen, but their constant giggling and singing infuriated Mrs. Sutter, who had grown even more sour with the weather. They were soon separated.

Soyo remained in the kitchen while Mahina was to be Mrs. Sutter's personal maid. She could not imagine a worse job. The woman was demanding and impossible, often asking Mahina to do multiple things at once. She wanted Mahina to ball a nest of knotted yarn so she could continue her knitting, but when Mahina sat down to do that she screamed, wanting to know why she wasn't dusting the library. Which would have been tolerable if there had been some way to preserve the protective bubble of her friends that normally kept her shielded from things she didn't want to think about. Without them, the situation was misery.

Her mother and the story she had not finished telling her crept into her head as she moved from one task to the next. She took great pride in her imagination and its knack for weaving tales, but in this instance her imagination let her down. Her mother had introduced the story as her birthright. She'd said it was the reason why everything was increasingly strange and complicated, why she had been missing for so long, and why they had ever left Hawai`i. Then she disappeared again without finishing what she started. In spite of all that she had done or not done for Mahina throughout her life so far, this was what made Mahina finally resent her mother. The blank space of the unfinished story stretched and grew a little larger every day.

Mrs. Sutter cut into her thoughts. It was time for Mahina to sweep the floors and build up the fire. She could only hope that Soyo had been dealt a better hand. She'd try to slip away and see her that evening.

In the summer the kitchen was an unpleasant, sticky place swarming with flies. Being there was a punishment. But during the cold of winter it transformed into a sanctuary of light and abundance filled with the warm cozy smells of bread rising in the oven and soup bubbling on the stove, the functions of an axis upon which an entire house turns. In those times it was a privilege to be there. The same usually applied during long weeks of cold spring rains. Soyo had worked in the kitchen before. She found those in-between seasons equally pleasant. Peeling potatoes and making rainbows of sliced vegetables brought her joy. But this year the rhythm of Hock Farm was, like much of California, suffering from an arrhythmia.

She had only ever been invited to help when the cooks needed an extra pair of hands, when Sutter was entertaining a large group, or when the family was celebrating a holiday, but John was not the only one of them filling out and growing up. With age came dexterity. For the cooks, this meant they had someone to clean the kitchen when the day was finally over and someone to come in very early to put the bread in and prep ingredients. Because of the extreme range of her working hours and the lake of mud that had emerged between the main house and where she normally slept with her family nearby, Soyo spent that prolonged deluge sleeping at the house. Mahina, who Kanaka George had given a room in the main house long before the Sutters showed up, was thrilled to share her bed. Soyo was too, at first. The first night, they pulled the covers over their heads and curled their bodies into each other, knees and foreheads touching. From above, their outline made the shape of a heart.

"I've never spent a night away from my parents," Soyo whispered.

"We could sneak out right now and try to find a way there through the mud."

Soyo was quiet. She sniffled. Mahina covered her hands with her own and held them to her chest.

"You know, you're still with family. I'm your sister. You're safe with me."

Soyo smiled and nodded. Mahina started singing a song about the moon. Soyo soon joined in. They whisper-sang as the rain roared outside. Soon Soyo's breath evened and Mahina could tell she was asleep. Only then did she allow herself to feel unnerved. Of the three of them, Soyo was the strong one. With Soyo seeking shelter under her wing instead of the other way around, she felt exposed to the elements, vulnerable in a way that made sleep impossible.

• • •

As the clouds continued to empty themselves, Sutter dedicated himself to drinking what remained of winter's liquor. Because he wanted to avoid his wife as much as possible but also because he was almost always too drunk to climb the stairs to go to bed, he began spending most of his time in his study. He took to sleeping in his chair, if

174

he slept at all. If a meal was served while the door to his study was closed, everyone knew to let him alone. That his belly would be full of liquid, and that's how he wanted it. At times he went days without eating, forgetting that the human body required water and food. This was always followed by a period of erratic, manic snacking at odd hours. It was during one of these periods of voracious, insatiable hunger that Sutter stumbled into the kitchen late one night looking for some leftover custard pie from dinner and instead found Soyo, shirtsleeves rolled up and arms elbow-deep in a sink of dirty dishes. Soyo was on the edge of childhood, a toe just beginning to touch the threshold of maidenhood. It was precisely the age he found most beautiful. He suddenly remembered his body had other needs.

• • •

Soyo stopped sleeping. The rains stopped too, although her responsibilities in the kitchen did not change with the weather. Mahina, already burdened with her own nightmares and sleeping fitfully, woke a number of times to the sound of her best friend sobbing into her pillow. No matter how much she begged Soyo to tell her what was wrong, her friend only ever shook her head. But Mahina was not blind. All it took was a pair of eyes to connect those dots. The bigger problem was figuring out how to make it stop.

• • •

"Please don't tell John," Soyo pleaded when Mahina finally confronted her with what she suspected.

Mahina couldn't understand why Soyo was suddenly so passive.

"It's Sutter," Soyo reasoned. "No one can stop him."

She didn't care who he was. The pain he was causing Soyo was impossible to ignore. Mahina couldn't stand by as her friend slowly fell apart.

She found John in the basement, stacking boxes.

"How long have you known? Why didn't Soyo tell me?"

"She didn't tell me either. But she also didn't deny it."

His fists curled.

"Let's put rat poison in his liquor. Better yet, I'll introduce him to the wrong end of a pitchfork."

"I think we should start with telling Kanaka George. He can help us."

"Help us what, Mahina? He's never stood up to Sutter and never will. He might be older than us, but he's just as powerless."

They mulled over one impossible scenario after another. Eventually they begrudgingly acknowledged their mutual impotence and went to bed. Mahina was sure she would not be able to fall asleep, which was just as well. Soyo hadn't yet come to bed. Mahina decided to wait up for her. They could fall asleep together.

A few minutes or a few hours later, Mahina started. She must have drifted off. The noise that had jolted her awake came again. It sounded like shouting. She did not hesitate.

As she made her way down the stairs the sound took on new urgency. She rushed into the kitchen and gasped.

• • •

The belt zinged through the air, cracking sharply against its target. John huddled on the floor beneath it, his arms over his head to protect it. Sutter loomed over him, lifting the belt to whip again as soon as it met its mark. His roar was incoherent. Mahina's eyes jumped to various parts of the room as her brain raced to put the pieces of the puzzle of what had happened together.

Carrot scrapings scattered across the large wooden table that anchored the middle of the room.

Soyo's ripped apron on the floor.

Shattered fragments of a large pot.

The smattering of blood dripping off Sutter's head.

If Soyo was in the room, she was hiding where Mahina couldn't see her. John and Sutter were too engaged to even notice she was there. Mahina ran to wake up Kanaka George.

It took three men to pull Sutter off John. By the time they did, John had passed out from the pain. The boy lay among the wreckage of the kitchen, his shredded arms bleeding through what remained of his shirt. Mahina rushed to his side. Up close she could see bits of fabric embedded in the pulpy mess of his skin.

"Get the hell off of me!" Sutter shouted. He shook the men off.

"Take the boy to the barn," he seethed. He reached for the back of his neck and fingered the trail of dried blood that ran from his ear to his shoulder before threading his belt back into the loops of his pants. The men took John without a word, carrying him as carefully as they could.

Mahina watched from the doorway as long as she dared. Then it was time to find Soyo.

There, a small trembling lump under the thin coverlet they shared. Mahina crawled onto the bed and put her arms around it.

"John was hiding in the kitchen. I didn't know. I swear. He came out of nowhere."

"Shhh now," Mahina said. "It's all over. None of this is your fault."

Soyo sat up. She shook her head. The fact that it wasn't her fault absolved her of nothing.

"Is John going to be okay?"

Mahina tried to hide the shiver that ran through her body as the vision of John laying on the kitchen floor returned to her.

"We can't stay here," Mahina said, to herself as much as to Soyo.

Soyo wouldn't hear of leaving the farm, not without her family.

. . .

The next morning the cook dragged Soyo into the kitchen to clean up the mess. Mrs. Sutter took a whip to Mahina for her involvement in the previous night's events. The three of them, attacking her husband like that. Deplorable. And after all her family had done for them! Mahina knew there was no point in arguing with her. The woman saw what she wanted to see.

John stayed tied up in the barn for that entire day and deep into the following night. Sutter beat him until his arms got tired. Usually he ordered his men to pick up where he left off when there was a disciplining needing doing, but this was personal. He wanted John all to himself. He ordered everyone out of the barn. The sound of screaming followed.

Hours later Sutter returned to the house for a drink and a cigar. He closed his eyes for a short nap. When he was properly revived, he went back to the barn.

. . .

When John was eventually released, he withered to the ground, broken. Kanaka George advised him to head toward one of the mining towns where his father and brothers had most likely settled. The boy was no longer welcome at Hock Farm, and Kanaka George no longer had any power to protect him. But he needed time to heal before going anywhere. Just outside the boundary of the property, Soyo's parents took him in, eager to show their gratitude for his standing between Soyo and that man. They tended his wounds and kept him out of sight in case Sutter's men ever showed up looking for him.

As he healed, Mahina and Soyo struggled to return to life on Hock Farm. Soyo went rigid any time she thought she heard Sutter's voice, and Mahina's blood was always boiling too loud for her to take orders from Mrs. Sutter. What had happened to Soyo and then to John added to the growing storm that had been building inside her since Kalehuna left, taking her half-told story with her. Mahina's body shook with anger. Crystal vases and hand-painted China plates slipped through her shaking fingers. Eventually Mrs. Sutter had enough. She ordered Mahina and Soyo out of the house. They had both proved themselves worthless traitors. Meant as punishment, Mrs. Sutter had no idea the gift she was giving them. They held their smiles until they were a safe distance. Miraculously, they'd been set free.

SEPTEMBER 8, 1992

Every day your shakes are a little less, I can tell even with this wall of glass between us. It's to help protect you from germs, supposedly, but it's probably because I made such a stink about unplugging the ventilator that they are scared I'm just going to do it myself when no one is looking. I think they also don't like me sneaking things in. People here have a different definition of medicine.

They better be taking care of you. I keep telling them you aren't a danger to yourself, that there's a difference between being a stupid kid and being a suicidal one. The doctors say there's no proof that it was a wave that swept you away, so until you can wake up and tell them yourself, they are going to take all the precautions. Even if they are right and you did do this on purpose, it was only because you were empty because you didn't know your kuleana because your tutu is hardhead—but now you aren't empty because now you do know your kuleana and that your tutu is still hardhead but is changing, and that means you would never go try again to do what you did, even if it was on purpose, which I'll believe when you snap out of it and tell me yourself.

Speaking of which. If it *was* on purpose and you *did* try again, I will straight up kill you. Got it?

And also another speaking of which.

Mo`opuna, I am so glad you're alive. I knew you were going to come back. Knew it the entire time. But I wish you would say something. Your tutu needs to know you've been hearing me. You don't even have to talk. Just bat an eye or tap a finger or something.

You know, as much as your tutu always said therapists and mental health stuff were all cockamamie, it's true, your tutu is pretending you're in a padded room on a 5150. Not to anybody outside, trust me, I don't tell those hens out there nothing anymore. Just pretending with me myself and I.

It was a long journey from when they took out the breathing machine. Your doctor says there's still a chance. We're still in the running.

You're progressing. You regained your sleep-wake cycle thing. Now you're in a "post-coma unresponsiveness."

The next stage of consciousness.

Ha. Now your tutu is talking like one of those New Age gurus living Puna side.

This is like you're in preschool again, visiting me with your flashcards so we could practice letters. We're really working our way up the alphabet now, kiddo. From Post-Coma Unresponsive heading toward Minimally Responsive.

Which means we're back to you and me working on our communicating. If you aren't going to use your words then at least bat an eye. Tap a finger. Minimally Responsive means you respond, even if just minimally. Heck, in that stage they tell me you might even be able to say a word or three.

I've been trying to read up on it when I'm not fixing up the house so it's ready for you. The next stage after Minimally Responsive would be you saying Inappropriate Stuff. Aye you, get your head out of the gutter. Not *that* kind of inappropriate. Doesn't even have to be swearing. But Inappropriate like yelling inside a church in the middle of service or something. Like you're saying stuff but it's not exactly a conversation.

From there we go up to Confused, which I promise no matter how Confused you get I won't lose patience. We'll stay calm and unfocused. Ha. Okay, no more bad jokes from me, I promise. Maybe I got stuck back in stage Inappropriate, ha. Okay, okay, sorry, sorry. Final final last joke.

We won't stay in Confused long though, so no worries. Because before we know it you're going to be cruising easy street into Stage Oriented. Boom, the *you* that makes you you is awake. That's when you know who you are (now also in the ways that matter), you know why you're here, things like that. We walk on outta here and say goodbye forever to that damn bed and this damn place.

Man. I'm starting to sound like I haven't been to church in a long time. Sorry, but if I never have to spend another minute in this hospital room again it would be too soon.

I'm ready to start, mo`opuna. With whatever our life looks like now, with all the truth being out in the open and me coming to peace that the pōhaku is real and the story is real but also God is real and the Church is real and all that's helped give me strength and courage to quit drinking is real, and we go get the pōhaku and bring it home and live Happy Ever After. The Kingdom rises like the sun, America no more, the `āina is healed and pono is the name of the game. Who knows what happens then, right?

I know I know, I have to finish the story for any of that to happen. Well, peeling an onion takes a minute. Every chapter I've told has been like taking off a bandage I forgot I was wearing. Like wiping the dirt from a mirror and seeing what my real face looks like for the first time. It was hard at first but now it feels the opposite of whatever that is.

1860

THE PLAN WAS TO JOIN SOYO'S PEOPLE. THEY SET OUT AS SOON
as they could gather what the journey would require.

In spite of their name, the meadow people did not reside in the
meadowlands, at least not year-round. During the warm months
they moved like fluid through the valleys and the mountains, carrying temporary living structures with them as they hunted game
and gathered acorns and pine nuts. In the winter they stilled the
way the trees and shrubs around them did, retreating underground
to wait out the cold. The spring had been long and wet, but summer that year had passed quickly and it was already nearing mid-
autumn. Soyo would only be able to find them once they settled
for the winter, but winter made for hard travel. The window was
narrow. Even if they left now, their chances of finding them before
the snow fell were slim.

When Soyo's father was finally able to procure a horse for
them, it was time to go. The trio would have to take turns riding,
but at least it could carry the burden of their food and belongings.

On their final day before setting out, John announced he was going back to the farmhouse. To say goodbye, he said.

Soyo wouldn't hear of it. It was too dangerous. John insisted. He'd go in the middle of the night when no one would see.

"If you're going, I'm going with you," Mahina said. Soyo wouldn't hear of them leaving her behind. What one did, they would all do.

Mahina and Soyo did not try to stop him when he lit that match and set those delicate lace curtains in the living room on fire. By the time they got to where the horse was tied up waiting for them, the roof was beginning to cave. Only then was John ready to leave.

. . .

Their bodies were young and could absorb a certain amount of abuse. Still, the terrain was rocky and difficult. It became even more so the farther north they went. The scars of Sutter's rage lingered in John's right hip and caused it to lock every so many miles, even though the girls insisted he be the one to ride the horse. It required frequent breaks. Their going was slow. Gilded leaves shivered in the winds and rain of the falling season until they succumbed to their cycle and drifted to the ground. The balance between day and night tipped in night's favor.

The weather was not all that was changing. By the time Mahina, Soyo, and John neared the area where the meadow people spent their winter, what was happening among the three of them was undeniable. Their triangle had changed shape. At first Soyo and John tried to hide it, concerned how Mahina might react and reluctant to disrupt the harmony of their shape. But what had happened to Soyo and John at Hock Farm had bound them together in a way they hadn't been before. Mahina was tempted to make a show of being oblivious, though there was no denying the gravitational force developing between the two people she loved most in the world. The awkward, intentional distance they put between themselves, the lingering looks, the way John accepted Soyo's assistance even when his hip didn't demand it—Mahina put them out of their misery and gave them her blessing. The truth was she didn't blame John for loving Soyo. She was magnetic. A force of nature in a way free of sharp edges. She guided them all with a calm steadiness across a path only she could see. Mahina trusted her completely. If Soyo's heart had turned in Mahina's direction instead of John's, Mahina would have been honored to receive it.

The village was composed of eight houses that centered on what Mahina assumed was the village ceremonial house. Each house was round in shape and roughly uniform in size. The singular exception was the ceremonial one at the center, which was a much larger version of the others. From the outside, the units looked small and compact with walls made of bark and packed earth. Mahina was surprised to find herself in a wide open space when she stepped into the one she was invited to share. The walls were supported by a web of poles that captured the sturdy symmetry of a spider's work. The floor was sunken into the ground about four feet, keeping the space warm on the coldest of days. A ladder at the entrance brought her down. She was offered a sleeping mat and a fur blanket and space to lay both.

She tried to make herself useful in the preparations for their welcome feast, shadowing Soyo and taking note of how they did things here. She could not see where John might have gone but assumed he was being taken care of because Soyo was calm and laughing. Off to the side, a group of men tended a pile of rocks in a fire. When the rocks began to glow red, they were moved with sticks into a hole in the ground. Steaks of venison and rabbit were wrapped in large leaves and placed on the rocks. The hole was then filled back up. The earth oven was remarkably similar to what Kanaka George had described to her as a Hawaiian imu. The sight made her uncomfortable. She had been born in California. Her entire life had not stretched beyond Sutter's Fort and then Hock Farm. She felt a twinge of homesickness for Hawai`i, a place she had never visited. If only her mother had finished telling her the story. She didn't know anything about the place or family she had come from. Somehow, that story felt like the answer to all of her questions. She pushed the thoughts as far into the corner of her mind as she could and tried to focus on where she was. What was behind her was gone. There was only now.

But the feeling persisted and was only made worse once dinner started. They were invited to introduce themselves. John went first. His name song was long. She learned many things. The place where his family started. His mother's family and the island village of her birth. His grandfather, a chief. His father's service to the king. His brothers and where they were all born. When he was done, he was acknowledged and welcomed.

Mahina was invited to share next. She stood having no idea where to begin. She did not know the place name of her family or who her grandparents were. Her mother had left her at Hock Farm when she was barely a handful of years old. She knew nothing of her, which meant she knew nothing of herself. She assumed Kanaka George was her father, but he too had shared nothing. Of all the things they'd always talked about, he had never delved into the past. For the first time she wondered why.

Normally she could spin a tale as beautiful and intricate as a spider's web, but knowing a story already existed of her family, even if she didn't know what that story was, canceled out her ability to make something up.

Soyo came to her rescue. She introduced Mahina as her friend who was a gifted storyteller who could speak to animals. Her grandmother was the trees, her grandfather the water. This earned her a few understanding head nods and a handful of smiles. That was Soyo for you.

The meadow people did not have a marriage ceremony, but there was still a formality surrounding the matter. There would be an exchanging of gifts between John and Soyo. Then John would live with Soyo and her family for a period before they moved into a place of their own. Mahina was the closest John had to family there, so she set about learning to weave from other women to make baskets that they could then fill with gifts to demonstrate his good intentions toward his bride. The village women were master weavers, their braids tight enough to stop water from passing through. She learned quickly. It came easy to her. One of the elders observed and thought maybe she had done this work before, in another life. She did not argue. It was possible she had.

The women chattered as they wove. Through them, Mahina learned of the California beyond Hock Farm. The gold rush had displaced many and brought sickness and death to many others. Their community was less than half the size it was only a few years before. Mahina understood: As many Hawaiians continued to come to California to find work and fortune, just as many died of fever and illness. Stories swirled in her head about why it might be happening and where it all might lead, but she did not let the stories complete themselves, not even inside her head.

John moved in with Soyo. It was official. Over the next two years the strength and durability of their triangle survived despite the odds. Mahina lived in the collective home with them and Soyo's extended family, and when it came time for the couple to move into their own home, the fact of Mahina going with them was not up for debate or discussion. Although it was not common, the meadow people did not look down on a man having multiple wives. Once it was understood that John and Mahina were not siblings, the people assumed there was an arrangement between the three. And there was, sort of. Although it was Soyo in the middle, not John, who was gone more and more often, for increasingly longer stretches. He had kept his promise to Mahina to never talk about gold or her mother, but he had not promised to avoid thinking about gold, or to talk to Soyo about it. His years of working the land at Hock Farm, followed by his time shadowing the work of the meadow people, he felt certain he was developing the ability to read the signs that would lead him to gold. Gold that would purchase him the land and power to return his family to a place of rightful honor. To Soyo he admitted the recurring fantasy of sending for his brothers and erecting a huge family compound where they would want for nothing and have no need to work for increasingly erratic and untrustworthy white men. An island of their own somewhere in the middle of the California foothills.

Mahina was blissfully unaware. When John was away it was she and Soyo, laughing until tears rolled down their cheeks as they washed clothes in the stream, foreheads touching as they lay in the grasses and named shapes in the clouds, a space for two when they were in ceremony. They spent their first winter adjusting to their new life with the meadow people and, in spite of the awareness that the times of challenge and change were not over, they were happy. The wildness Mahina had been born with slowly bubbled once again to the surface. She felt in deep kinship with the meadow people.

Eventually Soyo's moon time stopped. Mahina knew why

before she did. A child, a boy, came in a dream. In it, he was encased in a thick fog. It was impossible to say if it were day or night. His face was unclear. But Mahina considered herself a storyteller, not a dream reader, and so told Soyo none of the surrounding details of the dream other than that there would be a child. Together they told John. If any of them wondered how this might change things between them, no one said.

By late spring Soyo's belly filled her smock. Disturbing reports of attacks and displacements of entire villages in other parts of the foothills prompted the meadow people to do something they had never done before. This year they would stay near their winter home instead of moving like fluid through the valleys and mountains as they usually did.

That summer John worked the salmon harvest with new purpose, leaving Soyo and Mahina alone for many hours and days. That spring had been an especially rainy one, and the rivers were still flush. One mid-July morning, Mahina awoke to find Soyo curled against her. A quick glance around told her John was already gone. It had been a full moon the night before, and both Soyo and Mahina had tossed and turned long after John had started snoring, so she was glad to see Soyo sleeping in.

The sun was standing heavy upon their roof, but inside the belly of their below-ground roundhouse the air was cool and light. Mahina pressed herself against Soyo's curled back and inhaled the scent of her skin, sunshine with an undertone of smoke. Her entire body exhaled as if it were smiling. This new life—away from Hock Farm, away from the mysterious gap left behind by her mother, away especially from Sutter and his volatile family—held all she never could have imagined possible. Stability. Peace. A calm that resonated as much internally as it did externally. In her time with Soyo's people she had found herself in relationship with her surroundings in a way that Kanaka George, for all he had taught her, had not told her was possible. Still, if she allowed herself to think about it, he was the one thing about her past that she missed, and missed deeply. Soyo moved slightly in her arms, bringing Mahina back to the present. Her eyelids fluttered before settling once again into stillness. Mahina smiled and nudged her.

"Come on sleepy, the day is nearly over. Time to prepare for dinner."

The rhythm of Soyo's breath changed. Her eyes burst open.

"How is it not still morning?"

Mahina chuckled and pressed her lips against Soyo's warm shoulder.

Outside, they shared the last of the small cakes of roots and berries they had ground and baked the day before. Mahina preferred the sweeter ones made with acorn flour, but the black oaks would not be dropping until the forest trees started to change color and the days got shorter. Even then there was no guarantee. Ever since the men of the fort had replaced their forests with fields for crops and cattle, there were fewer oaks than ever before. Plus, the rains had been heavy this spring. The trees tended to hold on to things after seasons like that. But overall she would take a root and berry bread over grasshoppers and worms. When they were kids Soyo had showed Mahina and John how to dig holes in fields and set fire to the grasses surrounding them, driving all the hopping creatures into their trap, but Kanaka George had always made sure the Hock Farm kitchen had more than enough to eat for Mahina to ever have been tempted to consider insects food. Now, she viewed many things in their immediate surroundings to be part of their diet in a way she never had back then. Refusing to eat grubs was the last holdout of her former life. Sure, she would eat grasshoppers when they were ground up and mix into the bread, but only because she could pretend they weren't there.

• • •

The sun was high and shadows small when the men returned. It was many hours earlier than usual. The women paused in their work, the children paused in their play. Everyone gathered outside the main roundhouse without a word. No one bothered explaining to the children that something was wrong. Even they could feel the chill that had filled the warm summer air.

Her body swollen with baby, Soyo grunted as she tried to stand. Mahina pulled her up and put a protective arm around her shoulders. Her free hand held tight to the knife she had been using to strip salmon.

The men emerged from the trees with a white man on horseback. The white man did not seem concerned that his horse was dangerously close to trampling a villager. Mahina caught sight of John and tried to make eye contact with him, but his face was turned toward the ground. His back was stiff. Warning prickled deep in her bowels. It was impossible to ignore the placement of the stranger's hand on his gun, a clear message. Suddenly, the white man's booted foot shot out to his left. The boot made contact with the head of an elderly man. The man crumpled. When his wife moved to assist him, the other women held her firm. It would not be wise to move until they understood what might be going on. The stranger addressed the fallen man without looking at him.

"Is this every one of yous thiefs and murderers? And don't you dare be lying now."

How to answer? Yes, this was everyone in the village, but not a single one among them was a thief or a murderer. The stranger did not wait for an answer.

"No matter. You all can spread the word: every male Indian living in these hills is to report to Bidwell Rancheria on July 21st. Anyone in violation of this order is in violation of the law. You hear me? The *law* of the great United States of America, of which you filthy animals are trespassing."

He yanked sharply on the reins of his horse. The men scattered to get out of the way without being trampled. A moment later the man was gone, his sour stench the only evidence he had been there at all.

The incident filtered through the foothills. It did not take long to learn what had prompted the roundup order.

At least the settlers' version of it all.

1863

Sᴀᴍ ʟᴇᴡɪꜱ ᴀɴᴅ ʜɪꜱ ꜰᴀᴍɪʟʏ ʟɪᴠᴇᴅ ᴊᴜꜱᴛ ꜱʜʏ ᴏꜰ ᴛᴡᴏ ᴍɪʟᴇꜱ ꜰʀᴏᴍ where his children attended school. Johnny, who was six, wasn't old enough to have to go to school on the regular, but on this particular day he'd begged his mom to go with his sister, Thankful, who was nine, and his older brother Jimmy, who was eleven. So the three went off to school. The sunshine hours passed in normal fashion. In the evening, the children began their journey home.

Little Johnny was thirsty. There was a creek just off the road. The three detoured toward the bank to let Little Johnny have his sip of water. Johnny rushed toward the creek, dropping to his knees and kneeling forward, his mouth open in anticipation. Thankful and Jimmy were a second behind him. They kneeled on either side of their brother. Before either of them could take a drink they heard a loud pop.

Jimmy fell straight face forward into the water. A bullet hole in his back.

There was no time for his siblings to scream. Four dark men

jumped out from the shadows. They pelted Jimmy with rocks. It was unnecessary. It was clear with his face in the water like that that he was gone.

More men, all dark and barefoot, emerged. They forced Thankful and Johnny up into the hills to a campout. The next morning they continued along. Thankful held on to her brother as best and as long as she could. When she couldn't hold on to him any longer the men relieved her of her charge. She begged them not to, told them she knew they were going to kill him. They smiled. They weren't going to do a thing to him, they told her. But when they came back she could see sure enough they were carrying only the boy's hat and his clothes.

After that they marched her over hill after hill and creek after creek. One night one of the men tried to rip the little gold hoops from her ears. Another wanted an earring too. They started fighting. All Thankful wanted was for the men to stop fighting, so she took her earrings out herself and gave them one each. Then came a time when they were all so busy walking it wasn't clear if any of them were paying attention to her. That's when she made a run for it.

She ran back toward the last creek that they had crossed and hid under a pile of driftwood. The men came looking for her but walked right past her. She waited and waited. Finally when she was sure they were gone she made a run for it. Soon enough there she was at the door of Mrs. Thomasson.

This all according to Thankful, of course.

How could anyone doubt or question that teary, trembling little girl? There was no question about poor Jimmy, shot in the back clear as day. Mr. Thomasson placed a pillow on the saddle of his horse and sat Thankful atop it. She took him to the last place she could remember being with her baby brother. Soon enough there Johnny was, naked as the day he was born, his life beaten out of him.

1863

LONG BEFORE THE DISCOVERY OF GOLD, RELATIONS BETWEEN those who had come and those who were already there had taken a hostile turn. The gold just made things worse.

Death was no stranger to anyone willing to risk it for the sake of land and the pursuit of prosperity, nor were the rumors of death anything new for those who'd stayed where they were. But these here were children. The horror of the crime mixed like gasoline with the reports and rumors coming out of California, the scalpings and attacks and general savagery of the people standing in the way of civilization and the pursuit of prosperity.

The outcry was more contagious than smallpox. It spread across California, spilling over its borders and seeping into the rest of the country until all its purple mountain majesties and amber waves of grain turned red with rage. People wrote into newspapers. They made buttons and signs. They tracked down their local electeds and demanded justice.

Not just for that poor Lewis family. For the entire civilized Christian nation still growing.

...

Bidwell Rancheria was only a few days' walk, but John wouldn't hear of Soyo joining them no matter what the settlers had threatened to do. Mahina didn't want to leave Soyo any more than she wanted Soyo to go with them.

In the days leading up to their departure, Mahina experienced a numbness in her mind and heart that was unlike anything she had felt before, not even after her mother left. Her heart was quiet, mourning a loss she could not articulate. She kept herself busy by loading their roundhouse with food and water. The salmon harvest was forgotten. All the men of the village could think about was who might be to blame, and who would be blamed.

None were under the impression that those two things were the same.

The speculative talk continued while everyone made their way to where they had been ordered, with the exception of Soyo and a few elders who refused. The meadow people were joined by other communities along the way. In one, a young man professed to have once worked where they were headed, what had been known as Rancho Arroyo Chico before the name settled into Bidwell Rancheria.

A pioneer before the gold madness, John Bidwell came to the area to graze cattle. He had come in relative peace, minding his animals and land. In general he showed respect to the people of that land, who in turn helped him with their labor. He was not long there when he met and married the daughter of a headman. No one disapproved of the match.

"His footprint was light," the former Bidwell employee explained.

With native help John Bidwell's wealth grew. John eventually built houses for those who had helped him. He did what he could to protect his workers from those who wished them harm. But he didn't become truly wealthy until he took his laborers and put them to work mining. With that money he was able to expand the ranch, adding more livestock, fruit orchards, and wheat and oat fields. Unlike most white men, John Bidwell paid natives the same as his nonnative workers. He was a fair man.

"Then why are you not still there working for him," someone in the crowd asked.

"I left after harvest, when less hands were needed. I could have stayed. But it is not in my nature to stay in one place season after season."

Those around Mahina and John nodded. They understood this.

"He is a good man," he continued. "There is no reason to worry. Bidwell believes in justice, but not what other settlers call justice. He will protect us in the same way he protects his workers."

For the rest of the journey, everyone tried their best to believe him.

· · ·

Mahina had never seen so many native peoples gathered in the same spot. It was clear that the order to report had traveled through much of the valley, foothills, and the mountains beyond.

"Which one is Bidwell," she whispered to John. A few yards away, men with rifles tucked in their elbows shouted orders. Using their horses, the armed men separated the women and children, corralling the men into rows.

John shrugged and shook his head. Until this order, he'd never even heard of Bidwell before. He had no idea which one of these men he could be.

And then it was time, although Mahina was not clear what it was time for. The men with rifles turned their horses in salute toward the front of what was clearly the main house. The crowd went silent. The screen door opened without a squeal. Four men emerged. One held the hand of a barefoot girl with two blond braids framing her face. Mahina supposed the girl was Thankful and the man Mr. Lewis. The other two men stood a step behind them and remained there as Mr. Lewis began to speak. His words were addressed to his daughter, but by the volume and sharpness in his voice it was clear he was talking to everyone but her.

"Don't you worry, Thankful. None of them will look at you or touch you or try to talk to you. If any of them even blink that's going to be the last thing they ever do, you hear? Now you come with me and show your old pa who done did kilt your brothers."

He moved toward the steps leading to the lawn. Thankful refused to move. Mr. Lewis let go of her hand and grabbed her arm.

"Come on now."

He pulled, not gently. Her shoulders sagged.

The first row took the longest, as Thankful was still resisting and needed her chin tilted up for her so as to look the men in the face. Later, John would tell Mahina he had held his breath when they passed through his row, trying to stand as straight and still as possible even though the journey there had inflamed his bad hip. By then the little girl was skimming so he hadn't had to hold still for too long. The horses whinnied. The settler spectators who had come to watch were growing impatient. The sighs of Mr. Lewis grew louder and longer with each row that ended without an identification. Then Thankful stopped.

"This one here," she said, pointing vaguely at a spot between two men. "He looks like one of them who took Johnny. Who took him away and stole his hat."

A woman standing next to Mahina started to cry. Mahina had never seen her before, but when she reached out her hand the woman took it and held on so tight her fingers tingled. It was hard to hear with her heart pounding all the way up in her ears. From where Mahina stood there was no clear view of the man who had been singled out. She didn't need to see. The cry and the shuffle of him dropping to the ground made it clear he wasn't admitting guilt.

The settlers had come to watch the show. They began to jeer and shout. Mr. Lewis pushed the man to the ground and kneeled on his lower back, pressing the accused man's face into the grass. The man made no move to resist. Still, Mr. Lewis and one of his men hogtied his hands and feet. They dragged him toward the far end of the lawn.

No one called for the sheriff. He was already there on the porch watching.

The sheriff whistled. Mr. Lewis looked his way.

"Over that way there, third row to the left. Him too. He's been making trouble and been a rock in my boot, I'd put up a gold nugget that says he's as guilty as the one you already got."

Mahina fought the urge to close her eyes and forced herself to be witness. The men deserved a trial. But that wasn't what this was looking to turn into. The accused were dragged off, two sets of boots leaving a trail that led to eternity. They were strung up to a tree.

Mr. Lewis, having the biggest grievance, was allowed the first shot. After that they were free game for anyone who felt like shooting.

• • •

Since Thankful wasn't sure exactly how many men had been involved in the kidnapping and murders, the rest of the group was told they were still under suspicion. They were ordered to remain where they were. Some took it literally, laying down blankets right where they were standing. Others moved toward the edge of the large lawn where there was shade.

"Mahina!"

She turned. It wasn't until she saw John that something inside her cracked.

"Do you think, do you think they were the ones who killed those boys?" she managed to ask through her sobs.

"Shhh now," John said, folding her into his chest. "You know well as me that that don't matter."

"We have to get back to Soyo."

"Yes."

"As soon as possible. What if the baby is coming?"

John grabbed her shoulders and pulled far enough away from her to be able to look her in the face. He pressed his forehead against hers. From that close she could see nothing but his eyes. Bloodshot, red.

"Look, I don't care what they say. They can't keep us here. We haven't done anything wrong. If they don't release us by sundown I'm leaving anyway."

Mahina could only nod. Now there was little to do other than wait.

• • •

Although it was indeed John Bidwell's rancheria, he was not the one to make the announcement. But as attention was being called for, he motioned for his workers to set out large baskets of bread, pots of corn mush and beans with sausage as well as enough hot coffee for all two times over, as if he hoped the meal would lessen the blow.

"The time of lawlessness is over. This is God's country now and we've been charged with cleaning out the vermin. There will be no tolerating violence and murder and savage behaviors. You are all hereby ordered to return on August 28th. You have until that time to round up your families and belongings. Bring everything with you. The federal government has generously reserved a large portion of land that will be your new home, where you will learn how to live like civilized people."

The ripple in the crowd was silenced by a rifle shot into the air.

"If you do not move, you will be moved. Dead or alive, don't matter none."

• • •

It was dusk by then. John headed for the line beginning to develop near the bread.

"I don't want that white man's food," Mahina said, folding her arms and glaring at him for putting his stomach before Soyo.

"Being stubborn isn't going to hurt them. And being hungry is only going to slow us down."

Still, when it came his turn, John turned down the beans and corn mush, grabbing only a handful of biscuits. Fuel enough to get them home, or what was home for as long as they were allowed to call it that.

• • •

The baby was born on August 12, 1863. Soyo took charge of the birth the way she did everything, instructing those in attendance with what she needed and where she wanted them at any given moment. Mahina was by her side every step of the way. The birth itself was smooth, the baby strong and healthy. After the first night with his mother, he was handed over to Mahina, who wrapped him to her chest and took him a short distance away to see if his name found them. When she returned, the baby was properly introduced to his father. He was Pono. Goodness, but also hope.

By the end of his first week of living in the physical world, Pono could hold his head up without assistance. None of his parents knew he was preparing for what was to come. That September was on its way, and with it the end of everything.

1863

THE BRIEF PERIOD OF IGNORING THE RELOCATION ORDERS
ended before it began. The newspapers and notices posted on every corner of every town made it clear. Anyone who remained
would be shot on sight.

"For our own good, ha," John snorted, tossing the newspaper
into the fire. "They say this is to protect us from settlers, but if that
were true they should be rounding up the settlers and removing
them since they are the problem."

They watched as a small spark jumped onto the paper and
took hold. It doubled instantly, becoming a claw of flame eating it
whole, a snake to its prey.

· · ·

At first Mahina assumed the armed soldiers posted around the
Bidwell Rancheria had been put there to protect the white men
since they were clearly in the minority, but word spread quickly
that John Bidwell, being a minority in his allegiances and
sympathies, had hired the men from San Francisco to protect

his native workers from the American cavalry there to enforce and oversee the roundup. Mahina pointed out one of the hires to Soyo.

"The young one over there toward the fields. He looks ready to cry. That gun is nearly bigger than he is."

Soyo, still as gray as she had been since giving birth, shook her head without looking up. In the short time they'd been allotted to empty the roundhouses and bundle as much as they hoped to be able to carry, the women of the village had not forgotten the new mother. They provided her a steady stream of hot medicines to drink, covering her body with poultices and prayers. Nothing helped. Mahina feared that what ailed Soyo had more to do with a sickness of spirit than of physical body. It was possible Pono had absorbed all of his mother's mana while inside her body and taken it with him when he exited. John insisted that it was only the looming relocation and all the uncertainty that came with that.

Whatever it was, the wind that was Soyo had gone quiet.

"Please drink," Mahina said, pressing the flask filled with medicine into Soyo's hands. Soyo sipped obediently. Mahina willed the liquid to take hold, replenish what had been lost.

People continued streaming in from every direction, rivulets of water becoming a pool. No one seemed to know anything more than anyone else, where they were to go or what was to happen to them. John made his way through the murmuring crowd to the tiny corner his family had staked out for themselves. He checked the baby before addressing the women. Pono was tightly wrapped against Mahina's back (a hard-won concession from Soyo, only after her occasional dizziness graduated into a nearly continuous sensation of being on a riverboat) and sleeping soundly, oblivious to the chaos swirling around him. John grazed the tip of his finger over the baby's cheek. His shoulders deflated as he turned to his wife.

"They say the boss pays all his workers alike, brown and white, so of course everyone had the same idea as me. I was one of the first who found the foreman though. I told him about my experience at Hock Farm. I told him I would and could do any work. I thought that might count for something. But no. They aren't hiring a single person. When I was leaving he told me I better mind the Bidwell guards—they had been hired to protect their workers from us just as much as from everybody else in this mess. Something doesn't smell right. Where is this Round Valley they are taking us to, this place they are saying will be our new home?"

His glazed eyes shot behind him toward the cluster of cavalrymen that would be escorting the group on their trip. It was a question they had all asked in every way there is to ask a question, continuously, to each other and internally, mournfully and angrily and with futility. He did not expect an answer. The only thing anyone knew was that there was no answer.

Soyo suddenly clutched her stomach and fell to her knees. She dry heaved once before her body rejected the last of the medicine from the flask, returning it to the earth from which it had been made. Mahina's chest tightened.

"John, stop! This is hard enough on her without you pointing out the obvious. You're making it worse. You would rather live here, another version of Hock Farm? Did you forget everything that happened with Sutter?"

"At least here we would still be free."

Soyo raised a shaky hand. They both rushed to help her up. John reached her first. Mahina slapped at the dirt on Soyo's skirts.

"Please," Soyo whispered, barely audible over the din of the overcrowded lawn. She reached for John's hand with her left, Mahina's with her right. Pressed them together as hard as she could.

"Please."

. . .

Captain Augustus Starr was not a large man. It was his eyes that marked him as the person in charge. Clear and piercing, his was a gaze that was either warm or icy, depending on the receiver. A respected merchant before the Civil War, two years of service had transformed him into a commanding leader who saw the roundup as a battle to win, who saw the people he was leading as a herd of animals, the kind of riffraff he had never allowed in his store. They were no better than raccoons, scavenging stealing nuisances. In spite of this, there was a line of people trying to appeal to him, to let this elder who is too elderly to march stay behind, to allow this one who had lost a foot be on a wagon, to please let this infant with the sunken cheeks and dead stare see a doctor before being forced to move.

Captain Augustus Starr barely blinked his shiny blues. He motioned to one of his men. Clear these people away.

A rifle shot into a cloud announced the beginning of the march. Starr's men nudged those nearest them to move along. A ripple raced ahead as people scrambled to readjust their loads and fold the blankets they had put down assuming they would be there at least one night.

The August sun was high and the shadows slim. It was clear to Mahina that Soyo was far from being the only one in poor health. Scattered throughout the crowd were dozens of versions of her. Male and female, young and old, all hunched under layers of heavy blankets, shivering and sweating simultaneously, the same gray cast, the cracked lips. Some clutched their stomachs in pain.

"How can we be a threat to them? Here, try to feed him a little before we start moving. It's going to take them a bit to get everyone cleared out."

Mahina handed Pono to Soyo and turned to help John secure their belongings. He spoke under his breath, loud enough for only her to hear.

"When I was a boy, I once asked my father why we left Hawai`i. He was always comparing everything to Hawai`i, finding ways to mention it or refer to it in any conversation. I didn't understand why he would leave a place he declared such love for. Then he told me something I've been thinking about more and more. White men, he said, see land as something to possess. Something to extract from. They do not understand the concept of being its steward. Of being rich with the land. They can only see being rich from the land."

Pono fussed, clearly still hungry and not getting what he needed from his mother's breast. Mahina lifted him out of her arms and began swaddling him into his basket, made complicated by her having use of only one hand, Pono sucking on the knuckle of the other.

Soyo let John lift her, too weak to protest. Her knees buckled at first; he held her until she patted his shoulder, a signal that she had found her footing. Only then did he continue his story.

"My father insisted he'd always known the white men who'd come to Hawai`i had found so much wealth in the `āina that they would hold on to the land like a hungry wolf would a rabbit with a broken leg. And the moment that happens, he said, that's it. It's over. The people who were there first are no longer hosts. Now they are unwanted guests. Impediments.

"We're not being rounded up because anyone cares a hoot about our safety and survival. No, we're being rounded up to turn occupied lands—lands still flush with gold—into empty, unoccupied land open for the taking."

Anger and the sun turned his skin the color of rust.

1863

THE MARCH DID WHAT NOTHING ELSE HAD. THREE WAS BECOM-
ing two plus one of two revolving components. When he wasn't
fuming, John focused his attentions on Soyo. Mahina on Pono.
Soyo was too ill to notice.

The pooled peoples now suddenly a tribe—as arbitrary a
union as joining all animals with the same color fur together regardless of habitat—were dragged across the floor of California.
It did not take long for Mahina to find Pono a nursemaid. Those
competing for last place in the march were the sickest, the eldest, and the youngest of them, but at least it meant there were a
number of women willing to feed Pono what they could implore
from their bodies. Ever since his mother took a turn for the worse,
he'd developed a listlessness that gave Mahina a whole new set of
nightmares.

From what the birds saw, the march was a slow rumble, a yellow dusty cloud, a steady stream of harmonious movement. From
what the dirt saw, chaos. Growing gaps stretched the march into

a dotted line rolling across the dry foothills. The weeping, pleading, and moaning of the sick and tired were enough to drive anyone mad, but Mahina far preferred that to the silence of the dead and dying. The growing fevers, chattering chills, and weakening coughs were all too similar to Soyo. John put distance between them and the others, not wanting to be slapped in the face with the fading spirit of the group at every whimpering step. They were allowed to pause, but the breaks were never long enough to rest. There was barely time to sit before they started up again.

Others did not. Others sat never to stand again. Still the group marched. Through endless brush and dry grasses whispering and scratching at their ankles. Past a woman lying face down in the dirt, a bayonet bloom across the back of her dress. In the meadow a few feet away, an infant not much older than Pono. Crows pecked where his skull had been bashed against a tree. John shielded Soyo's eyes. Mahina could not look away. She wondered what might have happened, praying that whatever it was, the mother had been killed first. The smallest of mercies.

John became the lookout for what he did not want them to see, though there was no closing eyes to sound. Of being followed by things with sharp teeth. Of the growing resignation. Of bodies of the sick in heaps, communing in death, the grunts and squeals of wild boar feasting on their bones. The flap of carnivorous wings. Worse were the almost but not quite dead. Mahina threw rocks at the hogs not willing to wait. They ignored her the same way they ignored the moans of those they ate. Carried on. All the more reason why Soyo had to get better.

For a while and against all odds, it seemed she would. While others who appeared afflicted with the same grayness fell into seizing fits that held up the line, while their stomachs turned inside out, their flesh beaded with cold sweats despite the dry heat that came from above and below, Soyo remained upright. At night there were sanctioned rest periods where she shivered herself into a ball, but once she was burrowed into herself she would sleep with a childlike soundness that reminded Mahina of their earliest days sharing a bed at Hock Farm.

John looked in the direction of the hills for salvation, Mahina to Round Valley. She stoked the small ember of life sputtering within Soyo.

"Round Valley will have doctors and blankets. Remember when we left Hock Farm? We had less than now. This will be no different. Pono will have everything he could possibly want."

John glared. Mahina put on her brightest smile and hoped her words held at least a small amount of truth in them. She took Soyo's hand and was rewarded with a faint squeeze, repeating her words in the coming days so many times that she began to believe them herself, that this would be a homecoming of sorts. That Round Valley would be like Hock Farm. A homestead, only this time it would not be populated with white rapists and European wives as bitter for being left behind as they were for being finally sent for.

"There's talk of a few men planning to make a break for it and take their chances in the hills," John told Mahina that night after Soyo and Pono had both fallen asleep.

"Don't tell me you're thinking of joining them! That's asking to be shot. We're slow as it is. With a baby they'd be able to track us with their eyes closed. And say we did manage to somehow disappear without them noticing. What then? There's no meadow people to return to, they are all here with us. To be alone and survive is its own kind of death."

"I'm not talking about going backward. This would be going forward, into the hills, where the land will provide. Where I know there is still gold. I know I can find it. With that, power and protection and peace."

"If gold ever found anyone any peace, I'll eat a rock."

John didn't bother to respond. In the morning they continued putting one foot in front of the other, the same distance between them but farther apart.

The climate changed as the days wore on, and with it the terrain. What was dry was now wet. The air grew heavy with moisture. Golden rolling hills blanketed with brush and golden grasses gave way to thickets of towering trees skirted with ferns and moss. Fog replaced the morning sun, dew so thick it resembled rain.

. . .

For fifteen of the longest days of her life, hope held Mahina's spirit together. In spite of Pono's blistered sunburned skin and exhausted whimpers, in spite of the growing animosity between John and herself and the distance that inevitably put between her and Soyo, in spite of the horrors, in spite of the death and the thirst and the heat and the hunger, in spite of all she would never be able to unsee of the trail, in the end she knew there would be Round Valley. *Once they got there all could be repaired* became the chant in her head that fueled her legs and back and shoulders.

Ahead, the marchers stopped. Voices rose from the diminished group but they were too far away to make out the words. John locked eyes with Mahina. She pressed Pono tighter to her body. John pulled Soyo into his side. Whatever they were arriving to, they would be united in it.

As they drew closer the voices turned into a wailing. This time it was not a wail of despair. This was a wail of mourning.

John saw it first. Or rather, he was first to see the absence of it.

"What do you mean we're here? How can we be here? There's

nothing here. You can't arrive at a place when there's no place to arrive to. Where are the doctors, the homes, the plots to grow food? Where are the streams to fish in?"

When the glue holding Mahina's spirit together disappeared it broke into as many pieces as there were stars in the sky. Round Valley was nothing but an utterance. There were no homes, no structures, no tents with medical care, no food, not even a single blanket. As they had done on the trail, the calvary ignored their questions, confirming only that they were indeed at Round Valley Reservation and reading with unjustified ceremony the order that this land, a gift from the federal government, was to be their new home. It was announced that the rounded-up should be grateful. They could do with the place as they pleased.

"Do what with what?" John asked loudly. Soyo tugged on his sleeve as he repeated what he'd said. A uniformed soldier adjusted the rifle tucked in the crook of his elbow and pulled his horse in their direction. His horse came dangerously close to stomping on John's feet. John didn't flinch.

"Care to repeat that?"

He cleared his throat theatrically.

"I said: 'do what with what.' As in, what is there here to do anything with? We were promised protection, a new home."

The soldier jeered.

"Got a handout from the government and all it gets back is complaints. Just like you people. Maybe if your kind weren't running around stealing and killing good civilized Christian folk you'd have more to show for it. If I were you, I'd shut my yap. Besides, aren't you injuns supposed to know how to live off the land?"

He laughed. The horse neighed as if for emphasis. John's back got a little taller, his chin a little more forward. Mahina knew him as well as she knew herself, and knew well when his fury was steering the ship. If he were facing her, she knew she'd see the outer edge of his left eye twitch.

"I don't see what is funny. No one here has killed a soul or stolen a thing. If you want to talk about being civilized, how about getting my wife a doctor? If you want to talk about being civilized, let me tell you about the kingdom where I am from."

The soldier was already bored and looking elsewhere. He shrugged as he moved on, saying something about that not being his jurisdiction.

With Soyo resting with Pono under a nearby tree, Mahina pulled John a few feet farther so she wouldn't hear them.

"Do you want to get us all killed, poking the bear like that? What were you thinking?"

"I was thinking that we don't belong here."

"Of course we don't belong here, none of us do."

"No, I mean us, you and me. We're not American. We're Hawaiian. That means they don't have jurisdiction over us."

Mahina fell silent as John's words hit her. She opened her mouth to respond and then closed it again. John smiled triumphantly.

"Before you celebrate, have you forgotten that I was born here? In California? I don't know where my mom was born, and I don't even know who my father is. I can't believe I have to explain this to you of all people."

His face went soft. He reached for her in apology. She pulled out of reach.

"I'm sorry," he called after her. She waved a dismissive hand without looking back. She spent the rest of that first day exploring the area. Other than the high picket fence cordoning off the reservation (as if they were cattle, John would say later), it was hard to tell exactly where Round Valley started and where it ended. The imagination did not serve the dusty basin well. Still, Mahina clung to all she had pictured the reservation would be during their long trek to it for as long as she could, which wasn't long. It did not take more than a good long glance to see that what John said was true. There was nothing here. Still. There was Soyo to consider, Pono to care for. Neither could continue on the way they had.

People continued slowly trickling in. Relief turned quickly to disbelief once the marchers realized they had arrived at their destination. Piled onto their preexisting exhaustion came wariness, frustration, anger. There was a natural sectioning off as people migrated toward the people they recognized, toward the spoken language their ear knew. Invisible divides reappeared between the tribes now supposed to be one.

Some began clearing spaces for cooking areas while others rummaged through belongings to forage what they could for temporary sleeping areas. Having grown up in a farmhouse and spending too short a time with the meadow people, Mahina took notes. She saw immediately that the biggest challenge wasn't going to be the land. It was going to be the people. The government had ordered people of many tribes and Nations to make a home together, to be family. But family didn't work that way. She might not have ever had one, beyond Kanaka George at least, but she knew that much.

Still furious with John, she was tempted to find somewhere she could sleep alone. But this was not the time for stubbornness. Like it or not, the three of them (now four) needed each other. They were all they had. Which, she reckoned, was what made them a family.

1863

Soyo insisted she was recovered enough to resume feeding Pono, but her body said otherwise. Mahina tried to assure her that this didn't mean she was failing as a mother, but that only made Soyo deflate more. She lay down and closed her eyes. Mahina took Pono to be fed. When they returned, John was back from his hunt, looking like flint about to catch fire. She put a finger to her lips and motioned to the baby, who had just fallen asleep. He paced as she tucked Pono into Soyo's arms. As soon as she was free, John yanked her to an area where they could speak privately.

"What is it? If you're only going to fuss more, I don't want to hear it."

"I'm sorry about what I said earlier. But that's not it. Listen," he said, looking over his shoulder.

She followed the direction of his gaze but saw nothing beyond more of the same chaos that had surrounded them for weeks now.

"John, you're starting to scare me. What is it?"

"I think I saw your mother. I mean . . . Kalehuna. Kalehuna. I think she's here."

It was the last thing she was expecting. She stared at him blankly as her brain raced to process this information.

"Are you sure? It's been years, and you hardly knew her."

He gave her a look.

"Don't take offense at this, but I know her about as well as you do. I was with you at Hock, remember?"

It all rushed back. She swayed. Pressed her hands to her eyes, which were suddenly stinging.

"Did she see you? What did she say? Did she recognize you? What did you say to her? Did you tell her I was here?"

"Whoa, whoa, slow down. Yes, she saw me. Saw me before I saw her. Took me a minute to recognize her, to be honest. From what it looks like she was staying with another tribe, I'm not sure where, when she got rounded up and brought here. Yes, she asked about you."

"What did you say? Did you tell her I was here?" she repeated.

He started to shake his head and then stopped.

"Technically no. I did tell her I knew where you were. She asked if I'd bring you to her if I wasn't willing to bring her to you."

She went rigid. Shook her head with a force that pinched her neck.

"Absolutely not. I don't have anything to say to that woman, and I don't want to hear anything she has to say to me. She's a liar. As fickle as a feather."

"I thought you might say something along those lines. But consider this: she asked about Soyo too. And when I told your mo—, I mean, Kalehuna, about her sufferings, she said she'd learned a thing here and there during her travels that might be able to help her."

"She's lying. She has no scruples. That woman would say any-thing. She's just trying to see if we have anything she needs. Trust me John. Some people are givers, others are takers. Time does not turn one into the other."

"But Mahina, think of Soyo. There are no doctors here. It can't hurt to try."

"She's getting better."

"She's still far from healed and you know it."

He grabbed her hand and pressed it between both of his. His fingers were thick with calluses.

"She's harmless."

Mahina wasn't so sure.

• • •

The march might have been over, but that did not mean the suffering was. September conceded into October. Night came sooner. The shadows got longer. The leaves on the trees changed color, then fell. Still they waited for doctors, provisions, blankets, anything to help them through that first winter. Still there was too much death and not enough of anything else.

In that time Kalehuna made herself necessary to her daughter and her friends. Mostly because Soyo had turned with the weather and, regardless of whether her elixirs and chantings did anything to help, they indisputably gave Soyo some kind of comfort. Which left Mahina agitated and irritated most of the time. Pono, already proving himself an empath, had picked up on this frequency and developed a colic sort of fussiness that left them all feeling freshly grated. But if Mahina wasn't so far from ready to start being truly honest with herself, if she was ready to be truly truly honest, she would have had to admit to the guilt that was baldly evident in the heaviness weighing her down. This wasn't just fatigue and worry caused by a sick friend and a perpetually screaming baby. This was guilt that her friend, no, more than that, the person she loved more than anyone in the world, was severely ill and that, for as long as she remained ill, Mahina was better off for it.

It wasn't that she wanted Soyo to stay sick. She would without hesitation give her life to save her dearest friend's. It was just that Soyo being sick meant Mahina needed to be by her side, which made it easier to rebuff Kalehuna's many attempts to spend time with her daughter alone.

The October day Soyo died was colder than it should have been, the sky thick with an impending rain that never quite broke. A wail went up and spread through camp quick as a tsunami. There had been too much death overall for anyone other than Soyo's immediates to feel the acute pain of this one, but the sound of Mahina's cry was so full of undiluted grief that it pulled the scabs off the hearts of strangers and made their own buried wounds fresh as new.

Dying is personal and death is universal, but grief is nonconforming. John, Mahina, and Kalehuna each processed Soyo's passing in their own ways. Mahina frantically covered Soyo's body with her own, as if trying to absorb her. She was so focused on this attempt to merge that she didn't register Kalehuna pulling Pono tight to her chest and heading for the bare hills, half running as if she could protect him from the death of his mother by taking him far enough away.

Later, when they were organizing a search party to help hunt Pono and her mother down, it would not be lost on Mahina that of course her mother had run, that that was what she excelled at most in life.

Unlike either of the women, John did not attempt to get near his wife, nor did he make a break for it. He sat as far as he could

without physically moving, going deeper and deeper inside himself until he was all but gone anyway.

After Pono and Kalehuna had been tracked down and yelled
at by Mahina—that Soyo being dead didn't change the fact
that Mahina had nothing to say to her and had nothing she
wanted to hear from her either—John told Mahina they needed
to talk. Hardly two words had passed between them since the
terse conversation they'd had when Kalehuna appeared out of
nowhere, John spending entire days talking in hushed angry
whispers with other men who were also full of hushed angry
whispers and Mahina filling her every moment caring for Pono.
Now he motioned for her to hand Pono off to Kalehuna, which
she did with theatrical begrudgingness.

"What's this about? Come on, out with it. Or did you drag me
here and make me leave the baby with That Woman just to glare
at the stars?"

Three should have been a hard number but it hadn't been.
Two, two was going to be impossible.

"Stop calling her That Woman. She's your mother. She gave
you life. She had her reason for leaving, one she's been trying to
explain to you I might add, but now she's here. Forget her for a
minute, that's not what I need to talk to you about."

His voice cracked. For a second Mahina thought she saw the
John she had once known, that sweet boy she missed almost as
much as she did Soyo, pass across his face, the sharp edges he had
recently acquired momentarily softened. She told herself it was
the moon playing tricks. He crossed his arms and then uncrossed
them. It occurred to her that he was nervous.

"Let me try to say this without you interrupting. I did something. Right after we got here. Nearly immediately, actually. I
wrote to my, our, *the* king. Reported to him what had happened,

how we had been rounded up. He finally responded. I am free to go. The guards have their orders to let me leave. I wanted to tell you as soon as it happened, but there was Kalehuna and then Soyo got worse and then there was no possibility of going anywhere so there was no sense saying anything about it anyway."

"Wait. You wrote to the king? After it was clear that if you did so you would be leaving me behind?"

"I asked you not to interrupt. I wasn't done. And yes. I did it because it was clear there would be no doctors, no medicine here for her, and at least if we went back to the hills of the meadow people we would be living on our own terms. I was going to tell you. Besides, you seemed fine with the idea of staying here with or without me."

"Resigned is not the same thing as being fine. And no, I would not have been fine with it. Wherever Soyo was was where I planned to make my home, you of all people should know that."

"Yes. And she's gone now."

"How quick you are to move on!"

"It's not about moving on, it's about accepting that I can no longer do anything for her, but I can still help my son. And I will not have him here, growing up surrounded by a fence as if he were livestock. She made me promise her I would find a way to get him out of here. I aim to keep that promise."

"You're leaving."

"Yes."

She turned to go back to camp. He grabbed her arm and pulled her to face him. She was too stunned by the force of his grip for her to resist.

"That wasn't all I was saying, woman. Close your mouth and open your ears. In my letter to the king I wrote that me and my family had been rounded up. My wife and son. I'm not looking to leave you behind. Pono needs you. I need you. Half the soldiers posted here think you're my wife anyway. They might as well be right."

Mahina started crying in spite of herself. It had been a too-long too-hard day.

"If that's you proposing marriage the day your wife died may the gods strike you dead."

"Come now. Death was her mercy. I'm not proposing marriage, I'm proposing survival. I can't take care of Pono without you. What are you going to do here alone? The only way we make it is together. That's how it was when we were kids and that's how it's always going to be."

As much as she resented his seeming ability to function so soon after Soyo's passing, her silence answered for her. She would go.

THEY DID EVENTUALLY BECOME MAN AND WIFE, AND NOT ONLY just to escape Round Valley. That didn't mean either of them ever forgot Soyo. She remained what glued them together right until the very end. The only reason this story stayed alive for me to be here telling it to you is because Kalehuna and Mahina were reunited. She didn't protest when John insisted to the authorities that, Mahina being his wife, Kalehuna was his mother-in-law and a Hawaiian citizen in her own right—which meant she had every bit the freedom to leave as he did. She didn't protest when he re-traced their footsteps and led their small party northeast back to the Sierra foothills either. I think she was too brokenhearted to care much where he took her or who came with them.

It wasn't until she and John had a daughter (they would go on to have four children, one girl and two more boys not including Pono, although those boys died young) that Kalehuna finally got her daughter to listen to her long enough to finish telling her their mo`okū`auhau, including everything about the pōhaku. That her trying to track it down was the only reason she left Mahina with Kanaka George in the first place. But mo`opuna, she was stubborn, probably as stubborn as you, so she didn't ever forgive her mother. But she did believe her, and we know that because as soon as Mahina's daughter was old enough she passed down the story just the way her mother had told it to her. In that way she tried to give her child what she wasn't given. Which, I have to say, is what all us grownups are ever trying to do.

If you can hear me, I hope you hear that.

My mom died young so for a long time I thought being alive was enough to give your mom what I didn't have. It wasn't. I know that now and I'm sorry. Anyway. You must be tired of listening to me tell you how sorry I am for everything.

I said a few days ago that this was a dangerous story, but I've also been telling you that you not knowing about the pōhaku was the reason your mama ain't with us anymore. It's why you felt the Empty that grew and grew until here we are. But in the real world the big things are rarely either or. They're usually *and*. It's a dangerous story *and* it's the one that is going to bring you back to life.

Anything that's truly powerful runs the risk of falling into the wrong hands. An arrow that runs true hits its target, but one that misses is a liability. The pōhaku was so full of mana that even the story of it contained mana that was meant only for its protectors, and even then. Even then. There is such a thing as too young. My shoulders weren't ready for it. But that's enough about me. The point is that you didn't have to know about the pōhaku to be destroyed by it. Take Sutter. In Hawai`i, the pōhaku balanced the energies between the `āina and its people. But in California, it upset the balance. Destroyed him in its proximity. And if you did know about it? Auwe. Say something to a room full of one hundred people and you'll get one hundred people hearing something that means a different thing to each of them.

Mahina and John might have ended up married with children but that don't always mean happily ever after. The anger that planted itself inside of John was there before the roundup. If I had to put money on it I would say you could trace it right back to that kitchen at Hock Farm where Sutter did what he never should have done to Soyo, but it grew during the roundup and grew more once she died, when he took Mahina and Kalehuna and Pono off the reservation at Round Valley and back up the way they came. Mahina and Kalehuna both hoped their departure would be enough to cool down those flames burning inside him. And it was, for a time. But just because a firepit is full of ash doesn't mean there's no fire. Poke at it and you'll see sparks fly.

During the march, John fell in with a few men who ran away before they arrived at Round Valley. Most of the group ultimately died. Only two survived. Those two eventually sent word to their families in Round Valley that they were well and good hiding out in the hills. John admitted later that the reason the men had run away and risked retribution toward their families was because they were all men who'd worked the mines. They insisted that the moment they got wise to the fact that the American government wasn't going to honor their stakes or their land claims they swallowed what they needed to swallow and stayed on as hired hands for those very claims. Their families needed to eat either way, they'd said to keep their jobs. But then they'd mismanage things or "discover" gold a bit down the stream or one valley over. Like all gluttons, some of those white vultures were so blinded by greed that they forgot they were meant to work the land. They'd skim their claim for gold and use those early proceeds to move on to the next claim, always thinking if only they found that perfect spot of land, pools of gold would be there waiting for them.

Never underestimate the wisdom of the one perched on the lowest rung, mo`opuna. They are the ones closest to everything, the ones with their ears pressed to the chest. The hungriest, listening to the heartbeat.

Those men on the march knew there was still gold waiting to be found in the hills, so that's where John headed the minute Mahina agreed to leave Round Valley with him. On foot it's hard to head toward a notion though, especially when instead of a map you have a somewhat skeptical subbed-in wife who won't stop asking where exactly was he leading them plus an older woman who won't stop talking like a mosquito in your ear on top of a baby who would have stopped crying if only they'd stop for a minute so he could use his hands and knees to explore the ground a bit.

Kalehuna knew she had to tell Mahina about the pōhaku one way or another, and she didn't feel she had the luxury to wait until they were alone. When she told Mahina about the pōhaku, she told the both of them.

. . .

All these things were like glowing embers in that ash-filled firepit of John's mind. They're running low on supplies and running even lower on patience for each other and he's trying to concentrate on where his na`au is telling him their wealth and security and new home is waiting for them and the horizon is cloudy and he's beginning to think he's lost hold of the trail leading him when suddenly there's the story of the pōhaku. Remember what I said about the hundred people in the room? Mahina heard a story about her family and the responsibility she was inheriting. That's not at all what John heard. What John heard was the same thing Kawika heard the night he decided to steal the pōhaku from Kalehuna. Where they settled no longer mattered. In his mind, once Mahina did what her mother had failed to do (that being track down the pōhaku and take up the mantle once again of being its protector instead of its seeker) gold would come to them. He stopped at a spot along the Feather River not far from where they had lived with Soyo's people and declared it their new camp.

A rotation around the sun brought Lucia `Alohi. Meaning to glitter, to shine, named that by her mother in the hope that her father would find in her the gold that he believed so deeply was their destiny.

Ideas fester and get infected same like paper cuts can. I'm sorry to say that's what happened with John. The deaths of their two sons one right after another couldn't have helped. But this was more than grief. The story of the pōhaku had lodged itself inside him. He slowly lost his mind until one day it was all the way gone. John got out his gun and shot Mahina in the stomach, then turned it on himself. The last conversation Lucia `Alohi and Pono heard between their parents was their father demanding their mother turn the pōhaku over to him. Mahina begged him to believe her that she didn't have it and didn't know where it could possibly be.

He didn't believe her.

Like most tragedies, this one rippled outward and onward. The siblings started arguing over the bigger implication of that deadly argument. Lucia `Alohi believed her mother and Pono his father, at least to the extent that he prescribed to the idea that there was gold that was theirs to claim, and somehow the women in their family were responsible for the delay in claiming it.

Lucia `Alohi was only twelve years old and Pono fourteen, but the siblings soon parted ways. Pono went off into the hills, where legends sprung up years later about a Hawaiian man who did indeed find a large stash of gold only to bury it where it could never be found by another. But the truth of that fortune, if there was one, died with him.

There are treasure hunters out there still looking for it if you can believe it.

As for Lucia `Alohi, she stayed for a time where her parents had built their home, where a small village of others who had evaded the roundup was beginning to grow again.

Her time there wasn't long. A riptide was coming through, one that would sweep her up and take her places she never could have dreamed.

1881

Aɴɴɪᴇ ᴀʟᴡᴀʏs sᴀɪᴅ ꜰʟᴏᴡᴇʀs ᴡᴇʀᴇ ᴛʜᴇ ᴇᴀʀᴛʜ's ᴡᴀʏ ᴏꜰ ʟᴀᴜɢʜ-
ing. Lucia `Alohi disagreed. Flowers were how it smiled. It laughed
in water, bubbling, splashing, gurgling down the rocky mountains
and through the valley. But now the river wasn't laughing. The
creek beds were dry. It hadn't rained in months. The Bidwell estate
had many farmers to manage the property and their many acres
of flora, but before Annie headed for a trip back to Washington,
DC, she left `Alohi in charge of keeping an eye on her precious
plot of knotweed, which `Alohi had always secretly found a curious
choice for a favorite. Knotweed looked more like an overgrown
bush of heart-shaped leaves. If someone were to have asked her to
guess what Annie Bidwell's favorite flower was before they met,
she probably would have said roses. She seemed a roses sort of lady.
But that was just one of the many ways Annie defied expectation.
Which was partially why this droop was a problem. `Alohi loathed
the idea of letting her down. She swore her heart would break right
in half if even a single bit of knotweed died under her watch.

To get through the hottest parts of the day, she started placing parasols over the plants to offer them shade. The rest of the day she spent praying her benefactress return from her trip to Washington, DC, sooner than not. The heat was unrelenting. The flowers continued to wither and still there was no sign of Annie. When the plants reached a point of apparent no return, her prayers switched gears. Instead of hoping Annie would be back as soon as possible, she started to dread the day she would.

Annie was an anomaly of a white woman in more ways than knotweed. From a prominent political family back east, she was born with the opportunity to spend the rest of her life living a life `Alohi liked to dream about: twirling voluptuous velvet skirts in frilled social circles, summers in Europe, the rest of the year attending grand galas and high teas. Instead, serious Annie married practical John Bidwell and, after a wedding attended by famous figures and presidents both current and future, she followed her new husband across the country to settle at Bidwell Ranch in the growing town of Chico. She quickly became known for being as empathetic, opinionated, and outspoken as her husband.

Her first test came as soon as she arrived. When she discovered her husband was still married to the daughter of a local tribe's headman, she didn't throw a fit or run for the hills. No, Annie went right up to the woman and demanded the key to the manor. In Annie's eyes, they hadn't married in a Christian way, and if God wasn't involved the marriage didn't count.

When the first wife refused, Annie simply changed the locks.

"Darling girl, be proud of your name. It's your heritage," Annie declared the day Lucia `Alohi gathered up the courage to introduce herself to the woman who'd quickly become a local legend. "Hold your head up high, `Alohi."

She protested as politely as she knew how. Heritage? She knew virtually nothing about Hawai`i and had never been there. Neither had her mother, possibly even her mother's mother. Besides, both of her parents were dead. If there had been anything connecting her to Hawai`i, it was gone now. But Annie only smiled and shook her head.

"It takes a lot more than that for our souls to lose hold of where we are from. Back in Washington I knew some Hawaiians. Out of Boston, if I remember correctly. Fine, godly folk. Voyagers and world explorers the lot of them. Their faces were full of warm sunshine just like yours. You know what I think? I think you are more Hawaiian than you realize. I will pray that one day you come to know that."

And just like that, regardless of Annie being twenty-five years her senior, they were friends.

After her split with her brother, `Alohi had taken a chance, making the long, hot journey to the ranch, the place from which her family had been expelled years before, in the hopes of a job in the kitchen. With such a welcoming beginning, she let hope bloom. She and Annie shared a passion for doing God's work and, much to `Alohi's surprise, they also shared a mutual definition of what that entailed. Deeply concerned for the welfare of the Maidu peoples who neighbored their property, Annie served as vice president of the National Women's Indian Association. When that wasn't enough, she expanded the umbrella of Bidwell protection by building a mission. Surely a woman with that kind of heart would give an orphaned young woman employment. Once she found an opening, she told Annie why she was there and asked her for a job. To her surprise, Annie shook her head and said no. Instead, she offered her an education.

"Mrs. Bidwell is back, Ms. `Alohi. She axed me to fetch you."

`Alohi shook herself back into the present.

"The word is asked, Charlie, not axed. You say it."

"Asked."

"Very good. Please go back to the main house to let her know I'll be there in a minute, I have to finish this up."

The boy nodded and took off running out the door of the school, where `Alohi had been working with Annie to expand its reach. So far, most of their students were children of Bidwell workers. Their hope was to grow their classroom to include children beyond the ranch, where they were more at risk of falling to the perils of alcoholism and violence. She finished stacking the new supply of Bibles that had arrived, taking them out of the box and placing them on the new-student supply shelf. She took her time, tracing the wording stamped on the leather covers, opening them wide to hear the delicious pop of their spines, placing each one neatly next to the others. She dragged the chore out for as long as she could, but still finished way too soon. She sighed. No doubt Annie had headed straight to her patch of knotweed and sent for her the minute she realized `Alohi had let it all die. She could already see the disappointment in Annie's eyes. She pressed her palms together and sent up a prayer for forgiveness. When she could think of nothing more to hold her back, she headed to the house.

• • •

Annie was sitting in the main hall. Vases bursting with fresh flowers filled every corner, celebrating her return. Annie paid them no mind, jumping from her seat and crossing the room in five long strides the moment she saw `Alohi on the porch.

They spoke at the same time.

"Come in, oh my, where have you been! I've had ants in my pants waiting."

233

"I'm so sorry, I don't know what happened. I tried . . ."

Annie put a hand on `Alohi's arm to silence her.

"What are you apologizing for? Shh now, I have something very exciting to tell you!"

"But your knotweed."

Annie waved her hand.

"My knotweed? Oh bother, that can wait. Come, let's sit. You won't believe what happened."

`Alohi couldn't fathom what could have possibly happened, but for all her seeming desire to tell her, the woman then got coy, ordering up biscuits and blackberries from the kitchen, plumping the pillows on the couch, shaking out her skirts as she sat. She took hold of `Alohi's hands and squeezed.

"You're starting to worry me."

Annie laughed.

"No need to be worried. It's only that I've been waiting forever for your ear."

"Well, you have it now."

"Indeed I do. I'll get right to it. While we were in Washington we got an invitation from the State Department. There was to be a dinner, followed by a ball. The White House had never hosted such a thing. Although I admit I wasn't necessarily excited about this development. I was distracted by a few personal matters. In any case, I thought I might be able to get a word in with the right people about a few of my causes. I set about preparing my points without asking John for more details. Turns out, it was a very official international affair. And the guest of honor was your king! King Kalākaua is on a world tour, traveling to Japan and China and across Europe. Now, keep in mind when I tell you what happened next that I met him once before, not long after I'd moved to California. There was a grand welcome in San Francisco, 1874 or '75 I believe. Everyone made a fuss, he was the first king to ever come to the United States, so the fuss was well deserved. And he was an absolutely lovely man. Very smart. Anyway, fast forward to this visit. After dinner there was a Grand March, but since he'd made the trip without his queen he asked if I would be so kind as to be his partner. Of course John was more than proud to lend his wife, and the highlight was it gave me a minute of the king's time, which I took full advantage of by telling him how much I admire him and about you and all the work we are doing here together. He smiled and nodded when I asked him if he'd like to meet you, since he's going to stop in Sacramento for a brief visit before he returns to the islands."

"Annie! How embarrassing. What do I have to offer a king? Why in heavens would he want to meet me? What would I even say?"

Annie nibbled the edge of a biscuit and thought for a moment.

"Let's not get ahead of ourselves. He didn't actually agree. No plans were made. I am sure his advisors have packed his schedule with appearances and meetings. All I am saying is there is a chance. And what nonsense! He may be a king but he is also a human being who shares your homeland. You are kin."

`Alohi admired Annie's confidence, but she wasn't so sure she agreed. That night she went to bed certain that would be the end of it. They would not hear from the king and life would go on.

Why then did she feel the pinprick of disappointment? How can one feel sad about something that hasn't happened, or yearn for something that an hour ago she hadn't known was possible?

• • •

The order came the next day, but it wasn't from the king. Word had spread. The Hawaiian representative to California happened

235

to live in Chico, and he happened to think Mrs. Bidwell's idea was a fabulous one. He even wanted to take it one step further. Hawaiians had been an integral part of California since the beginning of recorded history. If the president of the United States could roll out such a grand welcome for King Kalākaua, shouldn't his own constituents do the same? By constituents, he was careful to qualify, he meant Hawaiians. All of them. Those who had moved to California yesterday, those who had moved to California generations ago, and every Hawaiian in between.

The qualification gave `Alohi no wiggle room. She would be going to Sacramento to meet the king. Her stomach turned over.

An idea ballooned. This was (had been? still was?) native land long before it was considered America. If anything, they should be the ones welcoming the visiting king. Besides, the peoples of this land were now her family. Hawaiian, native—the lines blurred. If she was going to go to Sacramento, they would all go. Excitement whipped through Chico like a November wind.

Unlike the forced march of 1863, the crowd that made its way south was a cheerful, proud one. With all the singing and chanting, `Alohi forgot her nerves. Surrounded by so many people she considered family, greeting the king would be easy. A group effort. They moved, a cloud across the sky.

When they arrived they found the entirety of Northern California squeezed into dusty downtown Sacramento. The king's train stretched across the horizon alongside the American River. His car stood tall and regal, dwarfed only by the size of its audience. Mountain folk, valley folk, city folk, country folk, fishermen, farmers, politicians, children—the finely dressed covered in silk ribbons and lace, and the poorly dressed in faded fabrics, the only clothes they had—flooded the wooden sidewalks and cobbled streets. They packed the saloons and leaned out the windows of brick storefronts, all hoping for a peek at the Hawaiian king. No one seemed to know if he was going to make an appearance. Whispers erupted into cheers every time the door of a train car opened. The cheers turned quickly into disappointed sighs when it proved to be yet another of the king's entourage or uniformed train attendant. Someone in the crowd noticed a curtain in the main car move. A collective question rose from the crowd. Was the king watching them?

The representative that had instigated this gathering perched himself on the outer step of a car and waved his arms. Eventually people took notice and fell silent. `Alohi wondered why his cheeks were a curious shade of purple. She thought she'd been the only nervous one.

"The king doesn't have much time. They'd thought this was simply a brief stop, not an event. Since there's no reception area set up for him, he won't be coming out. He, they, I don't think his people were expecting such a showing here. You've all come such a long way. I told them to tell the king. But there's just no way to bring you all into his car."

An elderly woman `Alohi recognized from the kitchens of the Bidwell Ranch let out a loud noise resembling a goat's baa.

"Our voices don't take up room," she called.

With that, she started to sing. As others around her joined in, `Alohi recognized the melody. It was a creation song of their local tribe. It ended seamlessly as another group started theirs, and another. Finally, a Hawaiian woman `Alohi could not see began to chant. She shivered. A man in a stiff uniform appeared at the door of the car. He pointed at her and motioned for her to follow him.

• • •

`Alohi looked around to make sure the man had addressed her and not someone standing behind or beside her. Then she looked down. She was wearing what she usually did, layered petticoats filling a modest, dark skirt that hit below the ankles accompanied by a stiff, long-sleeved matching bodice pulled tight at the waist. Nothing remarkable, nothing that made her stand out. Had she done something wrong? She adjusted her sleeves where they grabbed her wrists and patted her hat to confirm it sat straight on the bun of her hair before climbing aboard the train. Her legs wobbly, she took each step slow and precise, making sure her foot was firmly on the platform before allowing its mate to join it. She repeated the process until she reached the top. Gripped the rail with her hands. She did not trust her legs. An usher greeted her before opening the door and stepping aside. A moment later she was staring right into the gaze of the king.

She dropped her eyes to the thick carpeted floor. There had been no time to ask anyone for the proper protocol. Did she bow? Curtsy? Kneel and kiss his hand? Seconds ticked by and still she stood like a naughty child awaiting punishment. Her periphery was lined with stacks of books she assumed he had collected on his European travels, given the titles. She focused on them without turning her head; the only escape she allowed herself.

When he finally spoke his voice was friendly and inviting. He told her she looked like his sister, which was enough to clear the cobwebs from her brain. Of course! She suddenly knew what was required of her. With a sharp pang she thought of her mother. All those years she had been training `Alohi and until this moment she had not understood. Her mother had given her everything she had ever needed. She took a deep breath and closed her eyes. It would be easier that way.

As she chanted she began to worry her genealogy chant might have gone on too long. Hadn't the representative said the king's stop was meant to be a brief one? She made a quick mental edit, cutting out most of the references to the pōhaku, since she couldn't imagine the king would have any interest in a story she wasn't sure had any basis in truth. The king's attention never strayed. When she was done, he clapped once and declared them kin. Her great-great-grandfather was a half brother to one of his ancestors. His advisors nodded with delight. Then he asked the room be cleared so he could speak to `Alohi in private, as family.

"How much do you know about freemasons?" he asked as soon as they were alone.

She startled. Of course she had heard he was one, everyone had, but she couldn't recall if that was knowledge she should pretend not to have. She scrambled for a not untrue answer.

"Freemasons? Nothing, your highness." Technically, knowing he was one and knowing what one was were not the same thing.

The king waved a hand.

"No matter. Tell me more about the pōhaku."

She asked for a glass of water and sipped it to buy time. It had been her understanding that her mother was the only person who had known about the existence of the pōhaku, and then when she passed on, `Alohi had inherited that title. But the king had clearly heard about it before. She wished she had the words to ask him how. Instead, she told him all she knew. He asked her once or twice to slow down and leave nothing out, but he never questioned her account. When she was done, he made himself a drink from a gold gilded bar cart in the corner of the cabin (she kept the opinion she and Annie held regarding alcohol to herself, but firmly declined when he offered her one). He sipped it slowly at the window, pulling back the curtain and offering the people outside a wave. `Alohi heard a muted cheer.

1881

"Leaving!" Annie watched from the doorway as `Alohi gathered her few belongings.

"Yes."

"For how long? To do what?"

"I don't know for how long, we didn't really discuss that. And do? I'm not sure, exactly. I assume he'll go into detail on the trip back."

"So you don't know how long you'll be there or what you'll be doing there."

"No. But you were the one who planted the seed. If I'm Hawaiian shouldn't I know Hawai`i? He has made me Guardian of the King's Kahili. An attendant of sorts. I'm not sure exactly what that duty entails, but I was made to understand that it is a huge honor."

She left out the part about the king offering her the position only after she told him about the pōhaku. Her mother had repeatedly told her about the pōhaku and its theft, emphasizing `Alohi's grandmother's inability to find it. It was clearly a story her mother deemed critical for her to know, but she had always assumed it was an allegorical one, meant to impart wisdom of some kind that she would figure out eventually. Like the Bible, but one read and prescribed to exclusively by the women of her family.

The last thing the king asked her before summoning his chief adviser and announcing she would be traveling with them back to the kingdom was if she herself had found anything new regarding the pōhaku's whereabouts. His face was so open and full of naked hope that she didn't dare admit it hadn't occurred to her to look.

• • •

Whatever `Alohi had been expecting when she agreed to travel with the king, this wasn't it. On the ship, she spent most of her days in the king's company. It took her half a day to realize he was easily the smartest man she had ever met, complete with a quick humor that complemented her own. She realized she was developing a crush the kind a little girl might have for her father if he was a certain kind of father (not her father's kind) and did not feel ashamed for it. Why not love and admire your king?

By day he taught her how to read the water and the wind, by night he taught her to read the stars. As soon as he learned how embarrassingly unversed in all things of their culture she was, he happily launched into lesson after lesson. She inhaled these lessons with matched enthusiasm, trading him entertaining stories of the wild happenings of the California foothills, most of them true. Over elaborate dinners he laid out his plans for putting Hawaiians back in the center of power. She didn't understand but didn't dare ask for clarification. What did he mean "back" in power? Wasn't he king? If he wasn't ruling, who was?

He was reclaiming his power, he said. She was too embarrassed by her own ignorance to ask him what he meant. The more he explained, the less she understood.

. . .

In the same way trees become a forest only if you back away far enough to see it, the particulars of home only make themselves known in their absence. `Alohi's bit of California was wild but familiar, a chart-your-own-course chunk of normalized chaos full of get-rich-quick schemes and lose-your-fortune-overnight realities. She had never seen a horizon without hills nor an ocean so brightly blue. Never before seen these curiously long trees with thin trunks that leaned to evoke a woman's hip. Her experience with cities was limited to San Francisco the day she boarded the ship bound for Hawai`i, a fog-shrouded glimpse of a forest of buildings.

Arriving in Honolulu, she was faced with her own ill-informed assumptions. An island, yes. But also a sprawling, busy city that grew as they pulled into the harbor. She disembarked. When her feet made contact with the land her body swayed as if she were still on the ship. The pier before her tilted. Her skin drank in the sweet salty air with a thirst she couldn't explain.

People milled around her. She reined in a sudden desire to reach out and touch them to make sure they were real. Her life had taken on a dreamlike quality. If she didn't know for certain she had never been there before, she would have sworn she had. There was an instant, innate sense of direction and place. Even the faces looked familiar. Familial.

Why did she feel like that here but not in California, where she felt so other?

How could someone feel at home in a place they had never been?

SEPTEMBER 10, 1992

I TOLD YOU THIS IS BOTH A DANGEROUS STORY AND THE ONE that is going to save your life. I need to tell you the other half of what that means.

Hurricane Iniki is getting stronger. It's still way south of us but if we get plenty rain tomorrow I'm not sure the roads will be clear enough for Tutu to come and I need to finish because tomorrow they're going to finally do it. I know I've been fighting for them to unplug you from the wall but now I'm scared and I need to be sure you have everything you need. I've been up all night and about to lose my voice with all this jabbering trying to tell you as much as I can before morning but don't think I haven't noticed those eye-lids of yours, fluttering away like you're having a whopping roller-coaster of a dream. Honey Girl, if you open your eyes right now you'd see your tutu sobbing like a little girl on Christmas morning after Santa's paid her a visit and given her everything on her list. I love you so much.

Thank God you're out of that room with the glass walls so I

can touch you again. Can you feel my hands on your face? You're so beautiful. The most beautiful thing. Some say nature is the art of God and that might be true but so is this: you are His masterpiece. You were stuck in our ohana's overgrown garden, weeds blocking your sun, drinking your water. I don't know how to, I don't know, uh, how to talk to God right now the way I usually do because I don't know how to live in a world where both He and this story are true and the nurses are finally kicking me out but I'm going to be right there in the waiting room praying that they both are as true as beans. I'm going to pray this works and pray He doesn't take my praying as a lack of faith because I know you're going to go right on breathing when they take you off this ventilator tomorrow. You hear me? *You are going to go right on breathing.* Even though I was up all night and I still didn't tell you the entire story. I'm trying but there's so much, too much. I'm going to pray that it was at least enough. And I'm going to be right on the other side of this wall yelling out the rest and I'll keep yelling until I've told you every last bit of it or until they let me back in this room with you, whatever comes first.

Those are the only two acceptable options, mo`opuna. You be good for the doctors, okay? You be good and keep listening and keep breathing.

1881

For two years the construction of `Iolani Palace faced a steady stream of delays and one thing after anothers, but as the year 1881 drew to an end it began to look like 1882 would be the year the king finally got his castle.

The days following `Alohi's arrival to Honolulu were marked with a scurry of household staff, a flurry of wooden crates, and a hurry of everyone other than her. She saw the king in passing a single time, if that half nod of acknowledgment could be counted as an encounter. With the busyness came a sense that she should be doing more, but with no clear order to do anything, she assigned herself the task of staying out of the way. She explored, taking long walks around the palace grounds. The walks got longer as her radius expanded into the city. In this way she came to know Hawai`i.

Guardian of the King's Kahili had sounded like an important role back in Sacramento, but from the looks of it, what really needed guarding was everything else. The white businessmen and plantation owners held themselves in a way that made her think of the white men whom her brother had pointed out as having stolen their father's land claims and gotten rich. A little above everyone else, authority figures who opened closed doors without knocking. It was something she hardly noticed in California, or maybe it was just that she had so normalized that there, the skewed dynamic so firmly entrenched in the soil. Here it was unsettling. Off-kilter. She recalled what the king had told her about his plans to reclaim. It was beginning to make sense.

Following the sound of a piano one evening, she found herself in a drawing room with the king's sister, Princess Lili`u. The princess insisted she didn't mind the interruption. `Alohi hesitated. She'd seen the princess a few times but only in passing, which had been a relief. The royal emanated a keen intelligence and an ability to see right into the heart of people. It was intimidating. `Alohi forced herself to take a step forward.

"I've been meaning to speak with you alone for some time," the princess said, motioning for `Alohi to sit.

Lili`u used her hands in a way nearly identical to what her brother had done on the train the day they met, but in all else they were different. Where he was jovial, she was subdued. Sad, even. She was around the same age as Annie, but in her eyes lived someone much older. She went to her desk and extracted something from deep within a drawer. Whatever it was slipped into the pocket of her skirts before `Alohi could see what it was.

"My brother tells me you are the keeper."

`Alohi was confused. Had she been appointed a duty she was not aware of? The princess noted her look of confusion.

"Were you under the impression that no one in Hawai`i knows about the pōhaku? You look surprised. Let me explain something. We too have our own ancient knowledge, our own societies to ensure the preservation of such things. Men like my brother are focused on mathematics. Stars, the tides. But there is another layer. Long ago there was a prophecy that has been forgotten by most—but that is not the same as being forgotten. There were other attendants in that cave when the pōhaku was born alongside Ka`ahumanu. It was kapu to speak of it later, of course, but that did not mean the pōhaku could not be worshipped, and worship it they did. It was what was going to protect us, hold us intact. For years the women believed it was still here, hidden. And then Palila, your ancestor, came forward. She had sent it away with her daughter. For safekeeping, she explained. But she knew as soon as it was gone that that had been a mistake. I don't know why she never made any move to send word to her daughter to bring it back. Perhaps she did and the message never made it. And perhaps she was not entirely wrong, there is the chance that Ka`ahumanu would have tried to destroy it when she converted. But it never should have left the islands. If that wasn't clear then, it is clear now. Here."

She took what had been in her pocket and pressed it into `Alohi's palm. A small handkerchief that looked to have been sewn into a pouch.

"Your grandmother sent it home from the hills of gold. Both of her parents were gone by the time it came, but the person who intercepted it is loyal to me."

She moved toward the window, pushing the lace curtain aside for a view of the busy square below. The curtain dropped back into place. The princess returned to her seat.

"The sharks are circling. The seats of power have been infiltrated. My brother has been public with his opinions and his desires for the kingdom, and they are in direct conflict with the interests of a certain community here who see him as a threat. Not openly, of course, but there is talk. He still considers many of them allies. They might have been, once, but no longer. A stomach might be satisfied by a meal, but that satisfaction is temporary. What is once considered a feast is soon considered scraps. It demands more, needs more. As the stomach grows so too does its needs. What was once hunger becomes gluttony. Anything shy of everything is not enough."

`Alohi nodded. The princess had unwittingly told the story of California's Gold Rush.

"And this is connected to the pōhaku?"

The princess stared at her as if trying to assess her. `Alohi's head was spinning.

"When you came to my brother it was understood that you were introducing yourself as the keeper of the pōhaku. It is critical that the pōhaku be returned to its rightful place as soon as possible. Our walls grow weaker by the day. The pōhaku is like an umbilical cord, connecting us to our life force, our sovereignty. Severed, it is a wound that cannot be cauterized. Without the umbilical cord, the mother and the child both die."

The princess paused, motioning toward a service of tea on a tray. `Alohi shook her head. The princess continued.

"I wish that were all. Look at the disruption it has caused in California. Do you really think it has not or will not continue to wreak havoc wherever it is? But here you are and it is clear to me that it is not with you. Are you saying you are not its keeper? If you are not, who is? We must get a message to them immediately. Balance must be restored."

`Alohi wasn't sure what she was saying.

"No. I mean yes. I mean, I don't know. Its keeper? You have been frank with me so I will be honest: my mother told me the story of the pōhaku before she died, but at the time I understood it only as something like a bad luck curse on our family."

"Bad luck curse?"

"Yes, my great-grandmother went crazy trying to protect it. My grandmother's life was dominated by a failed attempt to recover it. And when my father found out about it, he, excuse me, I'm sorry."

She cleared her throat delicately before continuing, blinking tears away.

"When my father found out about it he killed my mother because he thought she was hiding it from him."

Princess Lili`u was not surprised.

"Men have a different energy," she said as if that explained everything. "The pōhaku would interact with people in different ways. But what do you mean 'thought.' She wasn't hiding it from him?"

It was then that `Alohi understood what her mother meant by the secret. She'd assumed the secret was the existence of the pōhaku, but no. The secret that she had been tasked to keep (and had clearly failed to do) was its theft. The princess and whoever else she was alluding to in this unknown society of hers had believed her to be in possession of it. She lowered her head in shame.

"No. My mother didn't have it, never did. Someone stole it from her mother. She eventually tracked the man down but by then the man was dead and the pōhaku was gone."

The princess gasped. "Gone! Where did it go? Where is it now?"

"I am sorry to say I don't know. When my brother and I parted ways he took up the search, but only because he believed it would lead him to a fortune, as my father had believed. At the time I didn't care how much truth resided within the story. It didn't matter. It was the story that killed my mother."

1887: JANUARY

THE THIRTEEN MEN GATHERED IN THE BOOK-LINED DEN OF DR. Tucker, who set his tumbler of whiskey on his desk with the grandeur and gravitas of a courtroom judge pounding a gavel. The room fell as silent as the rest of the house, given all the staff had been dismissed for the night, lights off everywhere minus the lamp on the table. The lightlessness signaled the urgency. Coupled with the drawn heavy curtains there was an impression of invisibility.

"Hello, gentlemen. Welcome. You have all been invited to this meeting by either me or Mr. Thurston here."

To this Mr. Thurston raised his glass in salute.

"I applaud you in your discretion in accepting the invitation without any accompanying details and appreciate indulging in all the preliminaries. What we are here to discuss is a highly sensitive matter and can never leave this room."

He paused, waiting for verbal confirmation from each man before continuing.

"You are all successful men. All perfect specimens of the strength of our stock. All called here by no less than destiny. You have unequaled vision and energy. You have turned this savage country into one of Christian values, sound structure, and limitless opportunity. You have created an industry full of jobs for the natives here. You have showed them the civilized way to do things."

He lowered the volume of his voice.

"You have helped this floundering, wayward kingdom realize the untapped resources it was sitting on and letting go to waste. And how does the king express his gratitude? With public declarations about replacing us with Hawaiians who have no experience in public office. By limiting our ability to defend ourselves and the businesses we built from nothing. Look at his palace—the king has gotten rich off the sugar we grow, the sugar we produce and refine and ship. He's gotten rich off our backs, our connections, our sweat. He wants to take the food off our tables, to empty our children's pockets of their coins. Then what? We just sit on our laurels and wait for Britain or France or Russia to swoop in and claim the kingdom? I say enough."

He pounded a fist on his wooden desk to punctuate his point.

"Thurston here has a ledger. What he and I are proposing is the formation of a league—The Hawaiian League, if you will—that would work to protect our community and all we are building, now and into the future. By signing your name to this you are vowing to never reveal the existence and purpose of our society unless it is to expand our numbers. You vow to stand with us to protect the white community of the kingdom at any cost."

Not a single man dissented. The ledger filled in a blink. It was decided the league needed a charter, a constitution of sorts. They immediately set about drafting one.

"This gives me an idea," Thurston said when they were through. The men waited for him to continue. He held up a hand.

"All in good time. For now, we need to focus on building a military force."

"If only the Honolulu Rifles were still around," someone muttered.

"They are," said a Boston-born sugar man with a long neck and a deep voice. "Not officially, of course. But the men themselves live on. They would at least know which supply stores would help us with rifles and ammunition."

"Yes, perfect," chimed Dr. Tucker.

"Isn't that a big jump," asked a newly sworn-in member of the House of Representatives.

"Hardly," the son of a prominent missionary replied. "The king is planning to remove our entire block from Cabinet. We can't very well let that happen. Gentlemen, I dare say this might get ugly, but if executed correctly, it will be quick and painless. You," he said, pointing to a cluster of men to his left, "do whatever needs to be done to get as many guns under our belts as possible. Meanwhile, I believe the rest of us should get to work drafting a constitution. We drafted one for this committee, but wouldn't you agree the kingdom is in need of a new one, one that has its priorities clear?"

"Hardly enough," Thurston cried. "You might as well go to the king right now and ask his permission. Let's not fool ourselves. I'll say it. We are talking about an overthrow. Possibly violent. A fly doesn't leave a dish just because it gets shooed away. Kalākaua became king by promising the masses a heathen resurrection. As long as he is alive he will fight this and fight us. The king needs to be silenced. Permanently. Then we declare this a republic, governed by us."

The men mulled this proposed direction in silence.

"Careful," the Mission Son said eventually. "The king has made many friends and allies during his travels abroad. There are diplomatic representatives here from across the world. Treaties to consider. A hostile takeover the kind you're proposing risks rebellion and international involvement."

"Diplomats who I have no doubt could be made to understand that someone is going to take over this kingdom, so it might as well be us. I say we vote," Thurston said, his knee bobbing up and down in agitation. "A full coup, or a new constitution."

The constitution route won by a narrow margin, but the vote would be honored by all. They were God-loving protectors of democracy, after all.

"Constitution it is," said Thurston, getting to his feet. He folded his hands behind his back and began to pace. The men, hyped on their own superiority, started a list of all that needed to be amended.

"We, the European and American residents of Hawai`i, need to be given the right to vote regardless of citizenship."

"Our appointed executive cabinet needs to be able to make decisions without interference."

"And the legislature should be able to draw from the public treasury—for emergencies, of course—without consent of the king, especially one with such limited experience in complex matters such as the ones we are challenged with. We are not children, we should not have to seek permission to do our jobs!"

"And by god, the people who should vote on matters of utmost importance are primarily landowners. They have a vested interest. Not the commoner who owns nothing and gets paid nothing and pays no taxes in return. Owning taxable land should be a requirement, to vote and to be a member of the Legislature."

A round of ayes. Thurston stopped pacing.

"A hiccup, gentlemen. The king would need to ratify this new constitution with a signature. What if he refuses?"

"I am sure we will figure that part out," the Mission Son replied.

He smiled. Then they all did.

1887: FEBRUARY

For five years `Alohi served as princess royal Lili`uoka-
lani's Lady-in-Waiting. During that time the princess gave her
unrestricted access to the kingdom's resources in the hopes that it
would help `Alohi find the pōhaku, or at least a clue as to where
it might be. Every morning the princess summoned `Alohi to her
piano room where they were assured privacy. Every morning she
asked her if anything new had been discovered. Every morning
`Alohi wished she could answer in the positive, because in truth
she had learned more in her service there than she could have ever
imagined, although as much of the knowledge she had gained had
been residing within the people she met as what she gathered in
the library she was given unlimited access to.

Her first language had been Hawaiian, but being in Hawai`i
had allowed her to experience the language's full depth and lyrical
qualities. She'd learned about the work and workings of the gods
Kū, Kanaloa, Kāne, and Lono, not to mention Pele and Hi`iaka.
She'd studied complex star charts that had forever altered how

she viewed herself in relation to the natural world. She'd familiarized herself with the local flora and fauna—so different from what she'd grown up with in California—as well as the medicinal powers many of them held. She'd come to be able to distinguish the sounds of a honeycreeper versus a plover versus an `elepaio. Her eyes had developed the ability to see what the moon did to the tides, her ears the ability to listen to the winds and know what they were saying. Most of all, she'd learned about Ka`ahumanu. Ka`ahumanu, intrinsically connected to the pōhaku. Its other half. Her spirit disengaged from her body. She observed herself from a distance. This family history, family lore, the one thing her mother had been able to pass down to her, what `Alohi had dismissed for so long as merely a story shifted shape, grew larger, deeper, until all the unlikely things that had happened—meeting the king, joining his court, moving to Hawai`i—no longer felt like they had happened by chance. Her kuleana had called her home.

She did not share in the princess's notion that there would be anything to be discovered in her search here—the pōhaku had been stolen in California, not Hawai`i—but that was not something she would have dared argue with the princess about. Instead, she prayed the past might give her guidance for the future.

. . .

`Alohi knew that when Lili`u asked her that morning if she'd learned anything, she was referring to only one thing. The princess wanted to know about the pōhaku, and to that, `Alohi answered the way she did every day, the only way she could. She had not.

1887: APRIL

Honolulu Star-Bulletin

Hawaiians Abroad! The 1887 Royal Tour

Hawaiian Gazette

Queen Kapiolani and Suite Are Booked for the Outgoing Steamer Today, En Route for San Francisco

Boston Herald

President and Mrs. Grover Cleveland Honored Hawaiian Queen Kapiolani and Princess Lili`uokalani at White House State Dinner

Boston Globe

Queen Kapiolani and Members of Her Party Will Soon Be Visitors in the Athens of America

"Aloha!"

That's what Her Royal Highness Queen Kapiolani of the Sandwich Islands said to her sister-in-law, Princess Lilino-Kilani [Lili`uokalani] when they met in the Baltimore & Potomac railroad depot in Washington yesterday afternoon.

When the queen said "Aloha!" the princess politely responded: "Aloha Oe!"

Now it would have been a very serious breach of courtesy indeed had the princess neglected to add on the "Oe!" for while the Sandwich Islands expression for "How do you do?" "Aloha Oe!" is used only when the sovereign is addressed.

It would be well, therefore, if every Bostonian who desires to do the "proper thing" should practice the salutation industriously during the coming week, so that he may fittingly demonstrate the culture of the Hub by welcoming her Hawaiian majesty in a becoming manner.

1887: JUNE

THE ATLANTIC OCEAN DID NOT MOVE THE WAY THE PACIFIC did. At least that was the excuse the princess used when she did not leave her quarters for the entirety of their trip aboard the steamship *City of Rome*.

The fib was not far-fetched. The weather was so dreary and thick with fog that it was nearly as dark during the day as it was at night. The foghorn sounded at regular intervals all the way to London, which `Alohi thought was enough to drive anyone to retreat to the relative quiet of their room. But when the other concerned attendants finally left the room, leaving her alone with Lili`u for the first time since leaving Hawai`i, the princess confessed. Her retreat had nothing to do with the ocean or the weather or even the blasted foghorn.

The dreams that started in San Francisco had grown steadily worse the farther they moved from the islands, the princess explained. Or rather, the dreams themselves had not gotten worse, but the effect they had on her had. While they all differed

slightly, on the surface they were relatively innocent: set in Ha-
wai`i, sometimes in the palace, where Lili`u is in her drawing
room at her piano, except the piano makes no sound when she
tries to play it. Or she is in a room that normally allows her a
panoramic view of the busy streets of the city, but when she
looks out the window all she sees is white, like the backside of
a curtain.

Nothing overtly wrong, but every morning she woke with her
sheets dripping of her sweat and her stomach full of an unease
that was not put to rest when she inquired if news had come from
Hawai`i. None had.

The queen insisted no news was good news, but no reassurances could shake the feeling Lili`u had that there was danger on
the horizon. Danger that they were either heading toward or, by
being gone, were absent from.

. . .

When their tour had docked in California, `Alohi experienced a
wave of vertigo. Her two worlds collided. Seven years had passed.
So much had both changed and stayed the same. She too was
different. She thought of Annie and all she would have liked to
tell her, allowing herself a brief daydream that involved bridging
the distance that remained between her and her previous life. Of
anyone, she thought Annie would appreciate how much more
herself she now felt, in ways not measured by years or age. Before
she'd met the king, she'd considered herself Hawaiian because it
was how she could keep the memory of her parents alive. But at
some point her time in the kingdom, and not inconsequentially,
on the `āina, had seeped into her skin and embedded itself in her
fingerprints. Now she was Hawaiian in a way that announced
itself with every move she made. There was no time to indulge

in a journey to Chico. She was part of a royal court, and while that indeed came with many liberties, none of them necessarily equated into freedom to do whatever she wanted.

They had not landed in California to be in California. Their demanding schedule quickly put the state in their rear view. By the time they arrived in Boston, `Alohi had recalibrated, all thoughts of California behind her. And because Lili`u did not reveal her growing internal turmoil until they were out to sea on the *City of Rome*, there she was free to marvel. The Boston city council arranged for them a suite of apartments at the Parker House and, not to be outdone by Washington, DC, or New York, engaged the city's finest florist to make an indelible impression on the royals, which he did.

Every ceiling in every room was draped in bowers bursting with roses, hydrangeas, and lilies. Every table was heavy with flowers `Alohi had never seen, the arrangements accessorized with stuffed hummingbirds, bumblebees, and butterflies. But nothing topped the banana tree featured in Queen Kapiolani's private reception room. A banana tree in Boston! The city's finest florist was the city's finest for good reason: the hunchbacked tree was so full of ripe bananas that the queen took to eating one every time she passed through the room, just because.

That was not to say there was any lack of food. Each meal was a parade of pageantry, every fruit crowned with cream, platters so full of fish and crab `Alohi wondered if Boston had somehow depleted the ocean's supply, lamb jeweled with tomato rubies and English pea emeralds, omelets fitted with ruffled skirts of truffles and only the wildest of mushrooms. She ate until her seams groaned, their protest drowned by the constant serenade of cadet bands in rooms filled with Boston's Who's Whos. She properly clapped her gloved hands when the girls at Wellesley College sang the Hawaiian national anthem and waved at the reported twelve thousand people who had crammed themselves into Mechanics' Hall in the hopes of a glimpse of the Hawaiian Royals, having such a grand time that she nearly forgot she was not one.

But her illusions of grandiosity that sprouted in Boston were shattered as soon as she set foot on *City of Rome*. `Alohi was all too quick to learn that the tour may have temporarily interrupted Lili`u's daily inquiry regarding the pōhaku, but that did not mean it had drifted far from her mind. As the princess's affliction grew so too had her conviction that, whatever impending doom her na`au was trying to warn her about, its antidote was the discovery and subsequent returning of the pōhaku to its rightful place in Hawai`i.

`Alohi's balloon popped. While she had been dazzled by the festivities and royal welcomes at every stop, she was by no means a royal. She was there only because hers was what the princess considered a critical mission, one she had been failing to achieve in her nearly seven years as part of the Queen's Court. It was a failure she had not until that moment felt so acutely. No matter how many times she tried to console herself with the knowledge that it was impossible to inquire or search for something when that thing's existence was a secret, her shame would not be silenced.

In England they were greeted by a guard of honor a hundred men thick and an invitation to call on Queen Victoria at Buckingham Palace, which they did as soon as they refreshed their appearance. `Alohi could not muster up her previous excitement and delight, not even when Queen Victoria herself was a mere step away, kissing Queen Kapiolani on the cheek and Princess Lili`u on the forehead. It was impossible to celebrate the novelty of the two queens chatting intimately together on a sofa while the pōhaku was somewhere out there, either lost and forgotten or being misused and abused. She observed as if from a great distance the British royal presenting all eight of her children; later she could not recall if she had properly curtsied to the British queen once the audience was ended, but she was sure she would have been informed if she had not.

It was not until Jubilee Day itself, Queen Kapiolani and Princess Lili`u busy being whisked away to Westminster Abbey for a Thanksgiving Service with a banquet to follow, that `Alohi found herself alone. The rest of the entourage had so excitedly rushed to take in the many celebrations the day promised that no one noticed her absence.

The entirety of Britain and its Commonwealth turned out to celebrate its queen. As `Alohi got lost down the cobbled streets of London she passed people who, by their mannerisms and dress, seemed far more foreign than she. Hawai`i felt about as far from London as one could get. It occurred to her that perhaps it was more than simple geography that put distance between places.

Quite by accident she found herself on Great Russell Street, where a stately manor commanded attention. Boys called out at the doors, promising free entry to the greatest collection of antiquities in the world. A flyer made its way into her hands.

She changed course. To the British Museum she would go.

A mustached man in a top hat welcomed her into what he described as a "portal into the history of art, culture, and humankind."

"A massive collection of various curiosities," he boasted. "One could spend an entire week perusing our many collections."

"And what if one only has a bit of one day to spare?"

"Ah, then you must start in the Museum of Natural History, resting place of the famed Rosetta Stone." he insisted while holding a tip of his mustache as one might a cigarette. "Of course," he continued, "the Rosetta Stone is indeed *the* Rosetta Stone, but there are many other notable items deserving of the attention of a lady like you. Do you have an interest in Greece? We have a fabulous collection of sculptures from none other than the Parthenon."

At this he paused to gauge her level of enthusiasm. When she did not appear satisfactorily impressed, he rattled off the names of cultural objects, prints, and antiquities. Each made less of an impression on her than the last.

"And then, of course, given your, ahem, well—" Here he gestured with a backward wave of his hand toward her hair and then her face in what she could only guess was meant to reference the obvious fact that she was neither British nor white. "You might find interest in our exotic display of the South Seas, collected by the Captain James Cook himself."

Her jaw dropped. With a smug, satisfied snort the man turned his attention to an incoming trio, all of whom appeared to be drunk or bewildered or a bit of both.

She should not have been surprised to find Cook there. The man had been British. But it had never occurred to her that in some places the man was a figure to be celebrated. What she had learned about him from her mother and then various kupuna in Hawai`i had framed him as a bellwether of doom. Hawai`i's turning point. Now she was on the other side of the world, where the story inverted. The man in the top hat had not said Captain James Cook. He'd said *the* Captain James Cook, same as the way he'd said *the* Rosetta Stone, as if the two were of equal marvel. With dread souring her stomach she headed toward the South Seas exhibit.

She half expected to see the pōhaku, a spotlighted star of the exhibit sitting on a marble pedestal, reduced to circus attraction, days spent being poked and prodded by aristocrats. But there was little beyond a disappointing collection of books, drawings, and coins. Not that she was disappointed the pōhaku wasn't in the museum. That would have been horrible, but at least that would have been a horrible that offered direction and decisions and action.

No, that wasn't the root of the disappointment. This was disappointment of the other kind: the lost trail, the failed gamble of the man who puts his entire pot into a hand of cards that do not align, the flailing fish that finally realizes he has been taken out of water and brought inland. For a moment she had felt something akin to traction, that she'd finally caught the scent, but the scent had been only tobacco and body odor. She was back to square one, directionless once again, on the hunt for an object no one in the world could help her find, because none of them knew it existed.

1887: JULY

"I T IS IMPERATIVE TO RESTORE ORDER, TRANQUILITY, AND THE confidence necessary to a further maintenance of the present Government that a new Constitution should be promulgated," Thurston read from the working draft freshly pulled from the typewriter, a pencil tucked in his ear that he took down every few sentences to make notes in the margins.

"Should be *at once* promulgated," Mission Son inserted.

Thurston's pencil scritch-scratched the document before he continued through the eighty-two articles. When he was done, he handed the paper to the quiet man with the long handlebar mustache sitting in the corner before returning to his seat, crossing and then uncrossing his legs. The other men in the room were quiet, the only sounds in the room the ticking of the clock and the occasional crinkle of burning tobacco being pulled through the stem of a pipe. All the planning, all the secret meetings and hidden shipments and clandestine overseas communications boiled down to this moment. Finally, the mustached man looked toward Thurston and nodded.

"Look at that, gentlemen: the Chief Justice of the Hawaiian Supreme Court approves."

The room erupted in applause for itself. Thurston retrieved the document and handed it off to his secretary to retype with the incorporated edits.

That evening they met in front of the Armory building. Like an invasive species in a land with no predators, the Hawaiian League had doubled in numbers since the beginning of the year, as had its arms cache. The guns were distributed in organized silence.

The king was in his study. The League entered without knocking. The draft constitution was presented. The king handled it as one might a soiled napkin. He read for a long while.

"Should I carry through with my signature?"

He directed the question to the Chief Justice, who answered without hesitation.

"You must follow the advice of your responsible ministers."

"And these in front of me would be my responsible ministers?"

Thurston tilted his head toward the window, where the tips of rifles pointed toward the moon, making long shadows. The king reached for his pen.

• • •

The news rippled, a boulder crashing into a pond, reaching Queen Kapiolani and their traveling court as they were on their return voyage to Honolulu, somewhere between New York and San Francisco. No one knew what to say so nothing was said, as if by not acknowledging what had just been forced upon the king they could delay its truth.

ʻAlohi took care to find solid footing in the unsteady, terrifying space between train cars and knocked again at Liliʻu's door, unsure if she had knocked loud enough the first time, unsure too if the princess had in fact answered but had been drowned out by the clacking rumbling of the train rolling relentlessly across its tracks. She weighed the possible consequences she faced by barging into royal quarters without approval. Over the years she had taken for granted her unrestricted access to the princess. She was not used to being on this side of the closed door.

"Please, my lady."

She took a deep breath and pressed hard on the metal handle. The door opened without resistance. She slipped inside before it slid back into place, once again cutting off the rest of the world.

It was remarkably quiet in the car. Liliʻu was at her desk. She looked up from her writing but said nothing. She had aged ten years overnight. In that short time her chin and shoulders had retained their regal carriage, but her eyes had melted into pools of sadness. Every encounter between them when they were alone had always started with her asking ʻAlohi about the pōhaku, but now there were no questions. For all the princess knew, the worst had already happened. It was too late.

ʻAlohi launched into the speech she had prepared in her head, one that had assumed Liliʻu would still be filled with the urgency she always had when it came to the pōhaku. She described the Cook collection and the other exhibits in the British Natural History Museum full of gems and stones and butterflies and shells. And rocks. From there she had spent the rest of their time in London learning all she could about the Collectors of Curiosities, as she had come to think of them. These things in the museum had not collected themselves. Men had. Men who, given the track record so far of European and American men, no doubt did not hand over everything to museums for public consumption. Surely the most valuable, the most exotic, the true treasures were kept private. To those who already have money and power, the only commodity left to pursue is exclusivity.

Pretending to be on assignment for the royal family, she'd paid a visit to every shop of collectibles she could find, inquiring about a rock of a particular color and size—or a gemstone or a mineral deposit, she kept it vague—that was to be a gift to the king from his queen. A souvenir of sorts. When the shopkeepers began pulling out their treasures she would then say she had no time, they were due to leave that very moment, but there would be time in California, if they knew of anyone there who might be able to help her find the perfect gift for a king.

Three names, she told the princess. Of collectors in California. One, a man who coined himself a scientist. Two who had grown so rich so fast that they were bored by the unaccustomed padded ease their newfound wealth provided and had turned to piracy, drawn not by the thought of treasure but by the danger and intrigue of obtaining the unobtainable.

When the royal party arrived in San Francisco she would stay behind, urged by this horrible new development in Hawai`i but also because now she had a plan, now she knew with (near) certainty where the pōhaku would be found. The pōhaku, much like their kingdom's sovereignty, might currently be in the hands of white men, but she implored the princess to share in her belief that they had not yet come to a point of no return. She could undo this, could undo what had been done to the king, she alone could right this wrong. It was not too late.

The princess rose from her chair and approached `Alohi. Only then did `Alohi notice the tears falling from Lili`u's pools of sadness, only then did she become aware that she too was crying. Lili`u placed her hands on `Alohi's shoulders and leaned forward. Their foreheads and noses met. They stayed that way for a time, swaying with the train as it twisted around bends. If one hadn't read the newspapers, they might have appeared to be dancing.

. . .

She hadn't necessarily lied. The pōhaku might very well have been in the hands of a collector, hidden in some back room somewhere, brought out at fancy parties as the entertainment. But the names she'd presented to the princess weren't necessarily ones she felt confident about.

After the royal party left for the final leg of their journey back to Honolulu, `Alohi stayed a night in San Francisco. At the time she had every intention of hunting down each of those three men and not stopping until the pōhaku was safely in her possession.

She found San Francisco a changed place, a city full of busy strangers. With fading optimism she'd eaten a light dinner and retired early, telling herself she would wake up early and begin. But when the chilled foggy morning arrived she didn't ask anyone about the names tucked in her belongings, barring the attendant at the hotel who shook his head and said he'd never heard of anyone on her list.

As she made her way northeast to Chico she rationalized what she was doing. Collectors were presumably wealthy, rotating in circles that existed on a plane that ran parallel to hers. Her ties to the Hawaiian royal family notwithstanding, she was not blind to the way she was dismissed, ignored, and overlooked. Not just because she was a woman. People ignored the fine stitching of her dress, the intricate lace at her throat, the perfect fit of her hat. They saw only the brown of her skin, the black of her hair. The fact was, yes, she couldn't wait to see Annie, but this was more than that. She needed her. Annie could penetrate the upper echelons of society. Hell, Annie was the upper echelons of society, in a way `Alohi would never be. This was not her abandoning her mission. This was her being realistic.

Then she got there and there was all the excitement and welcome back gatherings and the tour of the newly expanded school and all the projects happening within the Bidwell home that were in dire need of `Alohi's help and it was as though urgency was a seesaw that could only hold for one or the other. Annie was educating the Mechoopda, she said. Rooting out their old unchristian ways and replacing them with sewing and English language courses. She'd even finally managed to demolish the last of the dome homes. Every single one of Annie's students now lived in proper wooden structures. There wasn't a woven basket in sight.

The first day sped into the next and the next. There was always something needing to be done that moment, things that were easy to do because they were right there in front of her. She told herself she would wait until things calmed down. When she found some quiet time with Annie she would ask for her help in inquiring about the people behind the names on the slip of paper now tucked in a drawer in her room. What was the harm in waiting a day or two? The horrible new constitution had already been signed. Besides, did she really believe finding the pōhaku and bringing it back to Hawai`i would erase his signature?

For three and a half years a quiet moment to explain to Annie what she needed never came. The suffrage movement, long near and dear to Annie's heart, was finally beginning to pick up steam. The National Woman Suffrage Association and the American Woman Suffrage Association were on the verge of merging. Things were changing, and Annie was convinced it was in large part due to the groundwork they were putting in. Her advocacy against alcohol was gaining traction. Her husband had been meeting with men from the National Prohibition Party. There was talk that they were going to nominate him for president.

Annie told `Alohi every day that she couldn't do what she was doing without her.

1890

THE DAY BEFORE NOVEMBER FLIPPED INTO DECEMBER, ALL that noise, all that busy bluster, came to a screeching stop.

...

Annie rushed through the mansion calling for `Alohi. It was midmorning, not yet time for their daily chat over tea, which was something like a review of all they had accomplished that morning as well as all that would demand their attention for the rest of the day. It was a time they both looked forward to, reserved not just for logistics and certainly not for gossip, which Annie considered a sin, but a comparing of happenings and observations, so that between the two women a temperature was kept of the wider estate and its workings.

As `Alohi made her way back from the storerooms to the main house she heard her name being called. She checked the time, wondering if somehow it was later than she'd thought. But no, there were hours still before their unofficially official break time. Her thoughts ran quickly through the possibilities. Had someone gotten hurt, was there some kind emergency? There was no smoke coming from the windows, no servants rushing in or out. She climbed the porch steps in mild irritation. Annie surely knew how busy she was that day. Whatever she had to say could no doubt wait an hour.

But it couldn't. Annie pressed the newspaper into her hands and opened it, stabbing her finger at the headline.

"Your king! He's on his way!"

Too shocked to know how to respond, she looked down at the newspaper in her hands. Her vision blurry, it took a few furious blinks to be able to make out the words.

Sure enough, the newspaper confirmed. King Kalākaua had departed Honolulu on November 25 on the American naval ship the USS *Charleston* and would be arriving in San Francisco on the fifth of December. The article failed to explain a reason for the king's visit, saying only that this was not going to be like his other tours of the country. This time he would only be visiting California. A private visit, in order to improve his health.

"Do you think he is ill?"

Annie had obviously already read it. `Alohi shook her head.

"It says right here that he's coming simply for 'rest and recreation.' It can't be serious. Surely the newspapers would say so."

"Then we absolutely must go to San Francisco to greet him! Oh my, think of all we could share with him about the schooling practices we've developed to help you natives. I think he'd be thrilled. Maybe he'd even take some advice. `Alohi? Are you all right? You look flush, child. I completely understand. This is very exciting news. Quite overwhelming. Come, let's sit."

She reclaimed the newspaper and placed it on an end table as she led `Alohi to a chair. `Alohi accepted the offered handkerchief and pressed it to her face. A private, unannounced trip? It could mean only one thing. He had come for the pōhaku.

She piled on excuse after excuse as to why they shouldn't surprise him as soon as he arrived in San Francisco. Annie finally relented when `Alohi agreed to make the trip after Christmas. For a time she was hopeful that by then Annie would have forgotten about it. It proved an impossible wish, given the constant reports in the newspaper about the king having toured this facility and attended the banquet, went to this service and this cruise and this baseball game. Every post delivery was an exercise in keeping her panic hidden, convinced as she was that that day would be the day she would get the king's summons, followed by the relief that no such thing had arrived, which was mixed with a disappointment that left her feeling like her heart was a dog chasing his tail. If the king had not come for the pōhaku, was his silence purposeful? Or worse, was she that insignificant that she had disappeared from his memory? Maybe the king and his sister had given up on the pōhaku, and more so her, altogether.

The thought provided no consolation.

She agonized and prayed for help in whether she was doing the right thing. Or if she wasn't, what the right thing to do was.

God provided no answers. She had never mentioned the pōhaku to Annie and was sure that, at best, Annie would not understand. At worse, she'd accuse her of worshipping idols. Which was not at all what this was.

Christmas came and with it reports the king had left San Francisco for Los Angeles, San Diego, then Santa Barbara. `Alohi stockpiled the newspapers in her room, reading them in private in the hope they said something that would tell her anything. In the mornings she felt groggy and clammy and unable to get out of bed. In the evenings soup was delivered to her room that she did not eat.

1891

THE FIRST WEEK OF JANUARY, ANNIE GOT WORD THAT THE KING would be traveling to Sacramento on the tenth. By then `Alohi was too worn down by her continuous tumble down a rabbit hole to put up a fight. She allowed herself to be packed up and taken to meet her fate. But by the time they arrived in Sacramento everyone there already knew: he had not gotten on the train modified and outfitted to bring him there.

The train depot ran a few feet from a portion of the American River. Now suddenly without a reason to be where they were, the women walked to its edge. The river glided past them. Annie looked to `Alohi to decide what they should do next. `Alohi stared into the river's flow, the path it was telling her she needed to return to.

"You go home," she instructed Annie. Her voice left no room for discussion. "This is my kuleana. I will go to San Francisco. He will end up back there eventually."

Annie was politically savvy. She knew when to pick her battles and when to let things go. She kissed `Alohi on the cheek and went back the way they had come without protest.

Once `Alohi was alone, she started in the opposite direction, toward San Francisco. She would not let fear or shame or anxiety determine her actions, not this time. The king only ever stayed at the Palace Hotel so he was easy to find.

She learned quickly that finding a king is not the same as gaining an audience with one.

• • •

The Palace Hotel was named the Palace for good reason. At the doorway of the circular Grand Court, an open-air inner courtyard where carriages were on a steady rotation of depositing and retrieving guests, crystal chandeliers offered a glimpse of tier after tier of thickly carpeted floors and velvet curtains and thick towering white columns that went beyond the clouds. Gold leaf gleamed from doorknobs and framed art. Even after her stint living at `Iolani Palace it was intimidatingly impressive.

From the sidewalk she caught a blast of warm, fragrant air that escaped every time someone passed through the entrance. She caught the doorman studying her. She returned his stare. He must have been at least ten years younger than her, a kid drunk on the minuscule amount of power his position allotted him. The cold had turned his nose the color of a summer berry. He rubbed his gloved hands together and cupped them over his mouth, blowing to warm them. The move made him look even younger, a child playing grownup. She nodded as if she were expecting to be recognized as someone who belonged there.

There was hardly three feet between them. It could have been worlds.

He gave no impression that she was a person worth acknowledging. She moved a step forward. He frowned and repositioned himself so that, instead of standing beside the door, he was standing in front of it.

"Pardon me, I'm here to see the king."

His eyebrows rose slightly but otherwise his gaze and stance remained unchanged. Through the door, she thought she recognized a face. She muttered as she passed the doorboy. If he wasn't going to open the door for her she would open it herself. She wasn't sure why she had felt the need to tell him who she was there to see anyway. Screening was not his job.

In the lobby she caught sight of the king's chamberlain. All thoughts of the doorman vanished. She pushed through the glass partition and rushed across the lobby.

The chamberlain was distracted and offered her little beyond an acknowledgment that the king had returned to the hotel late last night from an installation. From which he was currently resting. A servant appeared carrying a large phonograph. The chamberlain ordered it be taken to the king's quarters immediately.

"A recording of his Majesty, for future generations," he explained almost apologetically. His voice dropped to a whisper. "He is ill. His kidney again. But I know he will be very glad to see you. Perhaps tomorrow?"

She nodded and returned to her rented room on the other side of the city to wait. She did not bother going to bed. She knew sleep would not be coming.

The next day she arrived early, but it was too late.

The king was dead.

She forced herself to stay on her feet, helping turn the king's bed so his feet faced Hawai`i. Flowers covered his chest. It was later reported that the flags of every public building in San

Francisco were lowered to half-staff, but she saw only her own failure.

She refused to leave his side. Sat numb through the funeral service at Trinity Episcopal. Hardly blinked at the endless procession of dignitaries and officials, Masons and envoys. Cried as the American army and navy fittingly treated the king with full military honors. Walked behind his coffin to the harbor, followed by more than one hundred thousand people. She wondered blankly where they had all come from. There was no question whether she would accompany the king back to Hawai`i. She boarded the *Charleston* without bothering to pack her things. What did any of that matter anyway.

· · ·

News traveled slower than the funeral ship. In Honolulu, a crowd gathered to welcome an alive king home. `Alohi delivered the news to Queen Kapiolani and Lili`u.

They did not, could not believe her. She did not fight them. They went to see for themselves. The welcome decorations were ordered torn down and destroyed.

1894

THE LAGOON IN GOLDEN GATE PARK DID NOT HAVE WAVES OR currents or tides, so `Alohi drifted in slow circles in the single-hulled canoe. Visitors leaned against the thin wooden railing ringing the shore. Sometimes a child pointed, but for the most part people just stood and stared. She wondered not for the first time what she was doing there.

Lili`u, now queen, had appointed her in charge of Hawai`i's exhibit at the Midwinter Fair in San Francisco. She'd considered the appointment an honor until she learned about all the other countries developing exhibits to showcase. For months they brainstormed what plants could survive being shipped over and, minus magically transporting a beach or a valley or Diamond Head, what kinds of structures they could possibly build that would accurately capture all their kingdom had to offer. Given that many countries had already designed their displays for the World Fair in Chicago and were merely transferring them from one state to another, she'd been behind schedule since day one. She'd brought canoes and supervised the construction of a sequence of thatched-roof buildings and even erected a model of Kīlauea volcano right there in the middle of Golden Gate Park, but sitting in that canoe in the middle of that murky lagoon gave her a depressing view of their "Hawaiian Village." It was as sparsely populated of Hawaiians as Hawai`i had started to feel these last few years.

Minus the serial killer who had very nearly managed to discourage visitors from attending, the World Fair in Chicago had been the kind of financial success that inspired imitation, and it appeared the Midwinter Fair was on track to meet its goal of boosting California's economy. Thousands poured into the park daily.

`Alohi needed her spirits back. The following day she took a break from the Hawaiian Village to walk the Japanese Gardens everybody seemed to be talking about. She found them surprisingly tranquil, despite being situated among such chaos.

She normally didn't care for crowds, but she found comfort in this one. People were beginning to gather around the Daniel Boone Area to see Parnell the Lion, who hadn't yet killed his trainer and two others. But there was too much to take in for her to want to see a large cat, although she did have to admit that the goats rolling barrels made her laugh. A wedding procession marched down Cairo Street. There were Turkish dancers and a giant horse and rider made entirely of prunes. She turned in the direction of the Egyptian-style Palace of Fine Arts building. A sign appeared. The California Mines Exhibit. She was tempted to pass it by. Why would there be anything entertaining about seeing a display of all that had ruined her life before it had even started?

Then again, why not see what her father had killed her mother, not to mention himself, in pursuit of?

Inside, shiny gold nuggets, coins, and flakes were laid out on black velvet in long glass displays. A gilded globe held court in the center of the hall. A bronze statue of a trio of panners sat high on a thick marble pedestal. She moved from display to display, realizing only after she had gone full circle that she had been looking for some kind of mention of her family and all they'd sacrificed. Or any Hawaiian, for that matter. Her mood soured once again. This metal, this silly, stupid metal was why she no longer had parents? But no, she was not being honest with herself. It was not the gold her father had been looking for. It had been the pōhaku. The pōhaku that she had vowed to find, that she was no closer to finding than the day she started looking.

Unable to fill her lungs, she pushed her way toward the exit. A man protested. She mumbled an excuse me and continued pushing her way forward until she was back in the sunshine.

"Not impressed?"

"Should I have been?"

The man who asked shrugged and asked her if she wanted to join him for a drink. She said she didn't drink and he shouldn't either. He laughed a sort of bitter laugh that was so full of pain that she felt he should not be left alone. And he wouldn't be, not ever again.

• • •

Henry was from Round Valley too, although she didn't admit to having any connection or familiarity with it. He'd been sent to boarding school as a young boy, so he didn't follow the "old ways" was how he put it. He'd returned to Round Valley only when he'd aged out. There he found his older brother Frank deep in a poker game, as drunk as he was losing. The next morning Frank confessed to selling the deed to their father's property. Henry demanded Frank get the deed back. Frank told him it was already in the mail. So Henry did what anybody would have done. He tracked down the stagecoach and held it up, which didn't get him the deed but did earn him eight years in San Quentin.

"When I got out I started wandering," he said. "I figured I'd let the world lead me where it wanted me to go. It brought me here."

To her.

Her Christian heart went out to him. It might have even been possible that he was right, that God sent him to her for help. She decided right then and there that she wouldn't be returning to Hawai`i when the fair was over. All his sharp edges might as well have been gushing wounds. She felt compelled to drop everything to soothe the hurt.

They spent the rest of that day taking in as many of the demonstrations and showcases as they could during the gaps in her shifts at the village, where they were both surprised to discover one of the men who'd come with her from Hawai`i was his distant cousin. She took it as another sign.

Feet hurting and ready for dinner, they agreed that the Manufacturers and Liberal Arts Building would be their final stop for the day. She regretted that decision as soon as they entered the building. The place was huge, containing products and inventions made all over the world. Every booth inspired in Henry a different idea for what he could do to support a family (it was already clear to them both where this was going). She followed as he hopped from one contraption to the next. He was gloriously handsome when he smiled. Caught up in his profile, she almost forgot how much her feet were begging to sit down. Finally, she gave his arm a gentle tug. He nodded, pointed to one more section. She gave in. The section was at least in the direction of the exit.

Halfway down the row they came upon a man who introduced

himself as the German petrologist Philip Alfredson. It was a name
`Alohi would never forget.

A petrologist, the small man told the growing crowd peering
through his display cases, is a scientist who studies rocks and
minerals. He referenced a few of his specimens.

"Collected from my voyages all over the world," he explained,
a peacock on parade.

"But you're in the manufacturing building," Henry said loud
enough to carry across the noisy air. "What does your rock collection have to do with anything?"

It was as though the man had been hoping someone would ask
him that. He removed his spectacles and pinched the delicate wire
frame with his fingers before cleaning them with a handkerchief
produced from his pocket. He replaced the glasses carefully on the
bridge of his nose and twisted either end of his mustache into fine
points. `Alohi suspected it was part of an act, a way to build sus-
pense. Whatever it was, it worked. Instead of moving on, everyone
in their near radius stopped, a crowd that wanted answers without
having to ask questions.

"A fine question indeed, good sir. I am no mere collector. I am
a scientist, conducting a highly sophisticated study of the phenomena of the earth's surface. The relationship between its surface's properties as well as their influence on the human race. On
a geological basis, of course."

He went on, but `Alohi had stopped listening. There, in a small
case on display by itself, was a black rock. Volcanic. She did not
have to ask. There was no doubt about it.

The room began to spin. Hundreds of voices all talking at once
vaulted into her ears, rattling them. She had never been quick on
her feet but she demanded her brain come up with something to
say, something to ask to slow this down and buy her time to figure

out what to do. Should she ask where he got it? What he thought it was? Where he planned to take it once the fair was over? The entire situation sounded ludicrous even to her. She considered grabbing the case and making a run for it, but she would not have even made it to the door. Every building was heavily patrolled by the city's entire police force. How could she explain?

"`Alohi."

Henry waved a hand in front of her face.

"Where'd you go?"

She blinked and tried to focus.

"Come on, you must be as hungry as I am. I don't have to listen to any more of this nonsense. Rocks and their relationships. Ha."

With that Henry circled his arm around her waist.

"This is the only relationship I'm interested in."

She blushed. With a quick glance over her shoulder, she allowed herself to be led toward the food tent.

Philip Alfredson.

She tried to memorize his face, his hair, the exact location of his booth. She would go with Henry and then say she needed to freshen up in the ladies room. She'd find the man and tell him she needed to speak to him as soon as possible. She would make him agree. By then she would figure out how to get him to give it to her. Heck, Henry had held up a postal carrier for his father. Surely he would help her steal it, if it came to that. The Lord worked in mysterious ways.

First Henry wanted a beer to give them time to discuss what they should eat for their first dinner together and to celebrate finding each other. She barely registered as he burrowed a hole through the crowd and steered them to the refreshments.

What luck. A scientist, of all things! She wondered if the names of collectors she'd gotten all those years ago had had anything to do with the pōhaku finding its way to the German. Maybe none of them had ever had it after all, which would mean that it wouldn't have mattered that she'd allowed herself to get distracted with Annie's causes and put her own on indefinite hiatus. Now that the pōhaku was within reach she could finally absolve herself of that guilt. She plotted as Henry drank. Once she finally had it in her possession at long last, would she set off immediately for Hawai`i, or could she allow herself a few days to convince Henry to go with her? She could imagine Lili`u's face as she presented it to her. What would happen then? Would the changes in Hawai`i be immediate? Would the kingdom rise like a new sun?

She wished her mother was alive so she could share in this triumph with her. She hoped her ancestors could now rest easy. The pōhaku would soon be where it belonged once again.

SEPTEMBER 11, 1992

THIS STORM, MO`OPUNA. IT'S LIKE I'VE BEEN SAYING: THE
world is out of balance. Civil Defense is showing Iniki is going
to pass by, so we'll be lucky this time. But the longer the pōhaku
stays missing, the bigger the imbalance. Who says we'll be lucky
next time, or the next? How big are the fires and floods and
hurricanes and tsunamis going to get? How many people die?
Melting glaciers, polluted water, how much longer can we go on
like that?

The nurses are feeling hopeful. One said we aren't completely
out of the woods yet—no counting those chickens—but that she
was "growing more positive every minute that passes." I asked her
if you were going to remember.

"Remember what?" the nurse asked me.

"You know, remember all the stuff we told her while she's been
asleep. I've been telling her about our family. Things I should have
told her a long time ago."

No nurse ever looked at me with as much pity as that one did right then. She shook her head and hugged the chart in her hands. You aren't going to remember a thing.

I didn't believe her. I don't know why I got so upset about it. Once you're home you and I will be together and I can tell you the entire story again from the beginning, except now I'll have already put words to it once, I already worked through all the hard stuff. There's time. We have time.

. . .

I'm so tired I can barely think straight. Where were we? Oh yes: `Alohi.

By the time she finally got back to the hall in San Francisco the man was gone, of course. Tell you a secret, though: I always kind of wondered if she imagined him. If it did happen, it was the closest anyone ever got to getting the pōhaku back, but if so I got questions. How does a guy with an entire display of rocks and stuff just "disappear"? One possibility is she took a mighty long time getting back there. You know, time sometimes blurs details. My mom told me what her mom had told her, that `Alohi supposedly went back to the hall as promised, as soon as he finished that beer and they got something to eat. But how do we know that was exactly how it went down? Maybe your tutu has a dirty mind, but eh, what if she was all lovey dovey with this new guy and her eyes were full of hearts and stars and she didn't go back until the next day, and maybe that scientist guy was scheduled to be there only that one day? Or what if she wanted to have come across it so bad that she *does* go to the fair, she *does* come across a rock specialist or scientist or whatever but then all those years of guilt and blaming herself for the Bayonet Constitution and the death of the king et cetera et cetera kick in and her imagination takes over and she thinks maybe she saw the pōhaku and all through that first dinner with

Henry she's thinking about it, convincing herself that what she saw must have been the pōhaku, and because he was gone when she got back her brain latched on to that inkling of a thought that was never really certain but became certain in the face of the inability to prove otherwise?

Just stirring the pot. It was the pōhaku, I know it was. If you'd have met my grandma you'd know she wasn't the speculating type. My mom didn't tell me the story until both Gramma `Alohi and Papa Henry had both passed so it took me some time to connect the story to the grandparents I'd spent time with when I was little. Gramma A wasn't a joker but she always made people smile. She was sunshine, honest as daylight. Papa as a young man was not hard to picture, even as old grandpa it was easy to see he would have been half the man he was if he hadn't had her to fill what he was missing.

I did try, you know. To find it. He was a real guy, that scientist. Before he died he published a few essays about his studies. Those papers include multiple references to what can only be the pōhaku. But that was as far as I got, because turns out the guy was Jewish and when Hitler took over Germany he either didn't like what this guy was writing about or he read between the lines same way I did and realized the pōhaku was a magnetic anomaly or something technical or even something otherworldly, but at the end of the day whatever it was would bring power to the Nazis in their quest to take over the world. Boom, scientist guy ends up in a concentration camp and any trace of the pōhaku disappears. I'm not saying the Nazis took it, but the first time I watched Indiana Jones I thought oh yeah, maybe they didn't just put this guy in a concentration camp because he was Jewish. Maybe Hitler was picking up where Cook left off, driven by a prophecy.

But now you get why I need your help, yeah? Tutu can barely

fly interisland anymore. Can you imagine me going to Europe by myself? And then what would I do, ask the first person on the street where I can find old Nazi loot, specifically a rock? Or the collection of a long-gone scientist that was overlooked by the Nazis that maybe is sitting in a warehouse or a dusty attic somewhere?

My hip hurts when I walk to the end of the driveway to check the mailbox. No. I was never going to be able to do that. All anyone in our family has been able to do since that fair is keep the story alive in the hopes that one day something or someone is going to blow the dust off this mystery puppy and we get it back. I almost failed to even do that.

But I'm so proud, SO proud, that that something or someone is you, Mo`opuna. The buck stops with you and thank God because I don't know how much longer we can go like this. Because you're going to survive, you hear me?

You will survive and you will move up the alphabet and then we will move on and then we will live forever.

For now let me kiss your forehead. I gotta get going. The TV says Iniki is coming in strong and headed toward us. I've gotta go back Waimea-side and tape up the windows. Maybe staple down the roof somehow. Your tutu is going to make sure you have a home to go home to when you're ready to bust out of this joint.

This is not the end. This is the beginning.

THE NEW YORK TIMES, SEPT. 13, 1992, SECTION 1, PAGE 1

Hurricane Hits Hawaii; 2 Reported Dead in Kauai

BY SETH MYDANS

Sugar cane fields were flattened, trees and telephone poles were flung through the air and hundreds of homes were ripped apart Friday as a monster hurricane roared across the lush island of Kauai, but remarkably few deaths and injuries were reported.

Today, after a helicopter tour of the island, Kauai Mayor JoAnn Yakimura estimated the damage at $1 billion and described "an incredible alteration of the landscape" in which miles of vegetation were leveled and huge chunks of the coastline had been swallowed up by the ocean.

Gov. John Waihee called the hurricane that roared across the western edge of Hawaii early Friday afternoon "probably the worst disaster we have ever had in the state of Hawaii."

Across Kauai (pronounced kuh-WHY-ee), which is home to 55,000 people, residents stood in the shells of their small houses and spoke with awe of 160-mile-an-hour winds that screamed like a person in pain and whipped roofs from above their heads.

Many tourists, evacuated from hotels and living today on rationed food and water in others, reacted more angrily.

"Sure, I'm having fun," said Susan Bernstein of Los Angeles as she stood in a two-hour line to buy supplies at one of the only operating grocery stores in Lihue, the main city of Kauai. "No water, no electricity, no phones, nobody knows I'm here. You can't go in the water. What are you supposed to do?"

Power, water and communications remained out of order in most parts of the island today, but the National Guard began at dawn to fly in supplies and help to get public services operating. Guardsmen took up posts at the port and at some warehouses in retail areas to prevent looting.

Kauai, a circular island about 30 miles wide, may still be the most pristine of all the Hawaiian Islands. Its spectacular vistas include jungle-draped hillsides festooned with wild orchids and red bougain-villea, and the rugged Waimea Canyon, described by Mark Twain as the Grand Canyon of the Pacific.

Until the tourist boom, beginning in the 1970's, brought chain-owned hotels like the Sheraton, Hyatt Regency and Hilton, sugar cane was Kauai's principal industry, with fields of bananas, papayas and taro stretched along its roads.

But much of the landscape familiar to tourists, and residents, had changed drastically once the hurricane passed. In every direction, the tops were blown off palm trees, telephone poles were snapped in half, sheets of metal siding were wrapped around tree branches and electrical wires hung in webs over roadways.

Along the coastline, boats were tossed onto the shore. At the airport, small airplanes and helicopters were flipped onto their backs.

"My house was destroyed. My mother's house was destroyed. My neighbor's house was destroyed," said Jeff Callejo, a policeman. "Everywhere on the island you go you see destruction."

Standing inside the shell of her house, Nina Dacay said the wind was "scary, scary." She added:

"The walls and the partitions were all vibrating. The only safe place we had was by the stove where there is a brick wall. Outside we saw iron roofs, plaster roofs, just floating into the air. I can imagine what the people in Florida went through. It was horrible."

In the face of such horrors, Governor Waihee said, "It is extraordinary" that so few deaths or injuries had been reported.

He said one man in his 90's died from flying debris and one woman in her 70's died, apparently when her house collapsed.

Tourists, facing their fears not at home but in the midst of a tropical idyll, seemed bewildered, lost and angry today.

Philippe Besson, 28, visiting from Switzerland, stood forlornly with his girlfriend at the deserted airport, holding tickets for a 10 A.M. flight. Because of damage to the control tower, all commercial flights had been canceled.

During the storm, Mr. Besson said, he joined 500 other hotel guests in a shelter at an elementary school where the wind was so strong that people took turns pushing against a door to keep it closed.

Standing in line for groceries in the hot sun today, Lorraine and David Friedlander, newlyweds from Rochester, said they were shopping for anything that would tide them from the 7 A.M. meal to the 4 P.M. meal at their hotel.

A perfect end to their honeymoon, they said, would be if a military transport would evacuate them from the island.

"It's surreal," said Andy Brauer, who is on vacation from Los Angeles. "One day we are at a luxury hotel and the next day we are in this war zone with no electricity and no water. We can't use the toilet. We don't have any candles. We're on a candle hunt now."

The director of the National Hurricane Center in Coral Gables, Fla., Bob Sheets, said Hurricane Iniki was about as powerful as Hurricane Andrew, which roared through South Florida and Louisiana last month. (The name, from a Hawaiian word for "piercing as pangs of wind or love," is pronounced ee-NEE-kee.)

Hurricane Iniki has disrupted the filming of Steven Spielberg's big-budget dinosaur movie, "Jurassic Park," which is being made on Kauai.

Thomas Pollock, chairman of the MCA Motion Picture Group, the producer, said the storm led to the cancellation of the final day's shooting. Mr. Pollock said Mr. Spielberg and the entire cast and crew were safe, and the expensive sets were intact.

"None of the dinosaurs were injured," Mr. Pollock said.

ACKNOWLEDGMENTS

Many of my ancestors were immigrants, which meant what there was to pass down to future generations were primarily things they could carry with them: stories and recipes. For the writing of much of this book, I began my workdays lighting a candle and facing the small collection of pictures of a few of them that have survived the test of time. It is on the shoulders of these people I, and now my daughters, stand, and I hope in some small way this honors everything they overcame, sacrificed, worked for, and kept alive for us.

When a book takes this long to write, it often comes with a story and a long list of people who have helped nurse it into being. This one is no different.

I am enormously and eternally grateful to Sarah Bowlin for her boundless support, brilliance, and friendship. There aren't enough adjectives to fully express your awesomeness.

Thank you to Judith Curr and the entire team at HarperVia for your continued faith in me and for believing in this story long before it was one. Rakesh Satyal, you are a dream to work with. Thank you for turning my jumble of ideas into something readable.

Shortly after I moved to Sacramento in 2003, I took my daughter on a field trip to Sutter's Fort, where I learned about the ten Hawaiians who played a critical role in the development of the region, but who had been left out of the history books. For the next twenty years, I went down many a rabbit hole in search of any details I could find about them. Ronald Williams at the Hawai`i State Archives, Michael F. Magliari, Ryan Browar and the Special Collections department at Chico State, Nancy Jenner at Sutter's Fort State Historic Park, Benjamin Madley at UCLA, and the `Iolani Palace Collections and Archives were instrumental in this hunt, generously lending guidance and research that grew the backbone of history this story needed to stand.

Eternal gratitude also goes to the organizations dedicated to supporting writers, especially the people who work so hard to make it all happen. VCCA and Storyknife—this book would simply not exist if not for the time and grace you gifted me. Thank you also to Ann Hood and my Writers in Paradise cohort for the insightful (and helpful!) feedback.

The seed of this story may have begun long before I had any idea how to throw words on a page, but I was lucky to have friends and family who carried me through countless drafts, making sure I was housed and fed, and watering me when I was wilting.

Laura Lynne Powell, the book might have changed, but that day at the Sugar Mill all those years ago when I read you one of the very first chapters will live forever in my heart. Our friendship is my lifeblood. Dan and Anne Hakes, I am forever grateful for your love (and the door you continue to hold open for me). Brian Hakes, I love you brother cousin. Jennifer Basye, thank you for always having such rich ideas and pointing out rocks I should turn over. Farnaz Fatemi, Christina Berke, Su Hwang, Vida James, and Joy Huntington—everyone should be so lucky. Otters and moose and double rainbows forever. Erin Hollowell, there is no quantifying the good you do in the world, but I sure am grateful for it.

Every day I am astonished at the Los Angeles literary community I am by some miracle a part of. I can't remember how I existed here without the Slackers and Group Chat and I don't want to. On that note, Celia Laskey, Laura Warrell, Diane Marie Brown, Ashley Coleman, Greg Mania, Hannah Sawyer, Carolyn Huynh, Liz Silver, Gabrielle Korn, and Ilana Masad: Surrounded by all of you, anything and everything feels possible. Tommy Pico, thank you for giving me the one word I needed most (a poet's superpower, I know, but still).

My family is my reason for being, and for this they got to put up with me disappearing for weeks at a time and obsessing about this story for decades. Mom and Nina, for all the many iterations you've read of this, I hope this one rises to the top. Riana and Mila, my triangle, I love you both with everything I am. Steve, you are my rock. Isabella Puamōhala, you can interrupt my writing any time.

Above all, a huge mahalo to all the booksellers, librarians, teachers, reviewers, podcasters, and readers who have read and supported my work. You make this life possible.

1 MONTH OF
FREE
READING

at
www.ForgottenBooks.com

By purchasing this book you are eligible for one month membership to ForgottenBooks.com, giving you unlimited access to our entire collection of over 1,000,000 titles via our web site and mobile apps.

To claim your free month visit: www.forgottenbooks.com/free629883

ISBN 978-0-483-52300-5
PIBN 10629883

REPLY

TO THE

EDINBURGH REVIEWERS.

BY THE

Author of "the Resources of Russia," &c.

BOSTON :
PUBLISHED BY MUNROE & FRANCIS,
NO. 4 CORNHILL.

1813.

HRus

E917re

589716
4.8.54

VARIOUS reasons urge me, against my original intention, and much against my wishes to offer to the impartial public, some remarks in reply to the Edinburgh Reviewers, who, in reviewing my work, " The Resources of Russia," have, at length, condescended to place me on their list of proscription ; a condescension, which I should deem highly flattering to me, were it not for the extreme prodigality with which they multiply their favours, and thereby render less valuable, and less to be envied, the honour and distinction thus conferred.

When first I heard of the awful summons being issued, to drag me forth from my obscurity and insignificance, into the presence of that dreaded, scribbling Queen of the north, whose goose-quill-sceptre reaches across the Atlantic, and from whose decision there is no appeal ; I felt a certain unpleasant and undefinable sensation, such, I ween, as agitated the breast of the hapless Anne Boleyn, brought for trial before the tyrant, wedded to her only for her destruction ; or such as the guiltless victims of the inquisition must have often experienced ; or such, to come nearer home, as have been excited in many an American citizen, when forced to appear before Napoleon's *council of prizes*, where, as in the two former cases, innocence is the worst possible defence, being the principal and the only crime. My apprehensions were the greater, as, besides the proscribed defence of innocence, and the assurance of being condemned by that which should acquit me, I was but too conscious of other no less heinous and unpardonable offences against the

Scottish Majesty. *I knew, and could not deny, that I was born and educated in Russia;* for this first offence, though it is involuntary, I expected to be doomed to suffer, as mankind have suffered for the original sin, without, however, their hopes, already fulfilled, of final redemption; that is, to wander an outcast through the wide world, like a second Cain, with a heavy curse for my burthen, and with a placard stuck to my back, as a speaking memento of my excommunication, and a warning to all not to relieve me. *I dared, barbarous Russian that I was, to use, without even asking her permission, the language in which her Majesty sends forth her anathemas, and thus, with her own weapons, and on her own consecrated ground, to invade her temple, and with unhallowed touch profane the skirts of her garments;* for this I expected to be chained to the Caucasian rock, like Prometheus, for stealing Jupiter's fire, or to be buried under mountains, like the Titanian race, who impiously attempted to invade the seat of heaven. *I loved my country, and have endeavoured to vindicate her character, evincing therein my just detestation of the great slanderer Dr. Clarke; my anger, equally just, towards "all the talents;" my unrepented rebellion towards the Edinburgh Reviewers; and my deep indignation towards all sworn and declared enemies of Russia;* for this I could scarcely realise in anticipation any adequate punishment. *I told that which my countrymen have proved true; and which, by destroying the infallibility of her Majesty's wisdom, made her blush for her human infirmities;* for this last and the worst of all offences, my death would be insufficient to appease her vengeance, and to be broken alive upon the wheel would be considered only as an act of mercy.

My fears, however were groundless, and I soon felt, that, had the Edinburgh Reviewers praised me, I should have experienced a severe disappointment; for not to be abused by them, in a cause like mine, would have deprived me of my greatest triumph. It was I that first made war upon them; a war, the principle of which, as a Russian, I shall maintain, so long as I live as I feel their wanton hostility to my country; and it was a sore grievance to them, to find themselves thrown upon the

defensive, contrary to their usual career, in which they have seldom been attacked, but almost always appeared the assailants. My glove of defiance remained a long time suspended in their sight; and it was not till they were absolutely forced by the public opinion, that they dared to take it up; and the weak manner in which they combatted, corresponds exactly with their strong reluctance to the combat; so that, to show their manner, for the edification of their American admirers, is a greater object with me, than to boast of a victory which I am conscious of having obtained. I say *American admirers*, because in England the Edinburgh Reviewers pass exactly for what they are worth; great only with their own party, and no conjurers with the rest; and I claim victory, because to my work on national topics, they have replied chiefly by personal invective, and defence of their party. What? Have I, an insignificant individual, as they affect to consider me in the beginning, forced them to a long, sophisticated, elaborate defence of the Foxite administration, and even to an occasional appeal to me—to me, the chorister in count Worontzow's chapel? Have I, a barbarian, a Cossack, a Calmuck, forced them to give up their favourites, the sullen Hutchinsons,* the beardless Erskines, the frothy Whitbreads, and other worthies, without even a word to exculpate them? This is a triumph indeed! But I will recur to their personalities, in using which, they complain of my own disrespectful language towards their friends; a complaint, which in the mouth of an Edinburgh Reviewer, the abettor of Dr. Clarke, and against a Russian who is belaboured even while the complaint is making, sounds very much like religion preached by an atheist, temperance

* The only thing they say about Lord Hutchinson is, that I charged him with *false representation of the state of affairs.* Most certainly I did make this charge; and what is worse, I renew it now, and will repeat it again and again, until the Edinburgh Reviewers have disproved it, which they seem loth to attempt. They said, I assigned as the reason, that he *was disappointed of being the commander in chief of the Russian forces.* I assigned no such reason, but merely noticed a report to that purpose, spread for aught I know, by their own friends : therefore the Edinburgh Reviewers must share, with his lordship, the danger of *false representation.*

commended by a drunkard, the privilege of chastity claimed by a common prostitute; or, what is still worse, the universal justice and universal rights, by sea and land, coming from the lips of a Buonaparte ! !

The mighty Edinburgh Reviewers, at the first onset, are
compelled to acknowledge that, " some how or other, the
invader lost his army in Russia," therefore, with *prodigious
kindness*, they concede what it is not in their power to
withhold, and with *astonishing liberality* they decline dispu-
ting with me about the Russian prowess, when it no longer
can be disputed : but then they say, that my statisticks
must have been taken from some common school book or
gazetteer, which is an insinuation, if it be any thing, that
what is common and accessible to all, cannot be true, or, if
true, loses its value, and must be spurned as a thing unworthy of those who pretend to distinction in learning.
In truth, this is the only ground on which I can account for
the extraordinary ignorance of Dr. Clarke and the Edinburgh Reviewers about the Russian affairs ; a subject on
which, almost in every thing they have said, they have been
grossly mistaken : and if the book or gazetteer which, they
say, is known to every school boy, but, I confess, totally
unknown to me, really exist somewhere ; then their ignorance, while in possession of so common a source of information, is not only extraordinary, but highly criminal ;
or if they cannot tell where, when, and by whom this book
was published, which, I suspect, exists only in their imagination, then they have made use of a shallow artifice
which recoils upon themselves, and renders their ignorance glaring, if real, and disgusting, if pretended. Thus
the important discovery, by which falsehood had obtained a ~~had obtained~~
circuitous and difficult path, is to be exalted above all truth
that is easy of access, and can be reached by a plain open
road, belongs of right to the Edinburgh Reviewers: and
such is their bold proficiency in this novel doctrine, that I
suspected it, long before they confessed it, by implication.
It is not to be wondered therefore, that, being in possession
of so precious a talisman, they should consider themselves
invulnerable, and endowed with power to crush their ad-
versaries at pleasure, merely by displaying the magic

words : *A giant's falsehood is better than a pigmy's truth.*
These words, or this motto, can be easily comprehended by
the following simple and easy process : an unknown Russian
has " some how or other " told truth, while we, Edinburgh
Reviewers, known to all, have " some how or other " told
falsehood ; but he is not an author of celebrity, and we,
with our beloved Dr. Clarke, are great writers ; *ergo,* our
authority is better than the Russian's, and our falsehood is
more conclusive than his truth ; or, we physicians have
unanimously predicted the death of a certain patient called
Russia, who " some how or other " has survived our predictions, and is healthier than ever : but we are great
physicians, and she is but a woman : *ergo,* though living,
she must be considered dead ; the more so, as she has the
impudence to live, contrary to our wishes, and our irrevok-
able sentence. I could give many more specimens of
this magnanimous mode of reasoning, equally honourable
to the head and heart : but other innovations no less interesting, and creditable to the Edinburgh Reviewers, impe-
riously demand my attention.

Unless the innovations which I am about to notice, are
of an older date, and have been overlooked by me, which
is probable, as I am no zealous reader of the Edinburgh
Review ; I must claim the credit of having transformed
the " reverend and potent senators " and grave inquisitors,
into mere blustering pedagogues, sometimes awkwardly
facetious, and egregiously ridiculous, sometimes too poetical, sometimes too hasty, generally inconsistent, and conspicuously absurd.

They are *awkwardly facetious,* when they laugh at the
certificate from the district clerk, prefixed to American
publications ; and *egregiously ridiculous* when they make
a jest of that which is an established law of the United
States ; and which is worse than ridiculous in such *great
reviewers,* not to know, as they must have handled many
books before mine, whose copyrights were secured in
America. It shews the *peculiar* attention of these gentlemen to their duty, and their exemplary *correctness* in dis-
charging it.

They are *too poetical,* or, to use a plainer language, they
are guilty of a downright falsehood, when they declare that

I told them my name and profession. How they came to know me, is a matter of no concern to me ; but it is certain they could not have made the discovery from my pamphlet, which was published anonymous, entered, and certified by and for the booksellers. What must the world think of the self-degradation of those who dare to tell a falsehood so unprofitable, and so easily exposed ; and who, alluding to me, in their review, as the supposed author of " The Resources," blush not to say, '' he tells us that he is'' so and so, when the very work they are reviewing is a positive proof that I told no such thing ?

They are *too hasty* and premature in their triumph, when they pompously and with evident exultation declare my ignorance in ascribing the fall of Dantzick to their party, when that party were no longer in power : is this all they can find to impeach my veracity, or discredit my information ? To what a pitiful shift must they have been driven to mount and ride on such a bubble ? The question was not whether Dantzick fell one time or another ; but whether it fell through the neglect of certain great talkers and small doers. If I had asserted that Dantzick fell when these full-mouthed but empty-headed gentlemen were in power, the Edinburgh Reviewers might chuckle with some reason at my mistake ; but I only asserted, what I assert again, and what the Edinburgh Reviewers have not denied, that Dantzick fell through the criminal neglect of the " talent administration," that is, through their not sending the promised and expected succours, a reliance on which prevented the Russian government from using its own means, until the mischief was irremediable. Let them deny this if they can, or let them prove by their new logic, that the murderer, who had previously admin- istered a dose of poison, is exculpated from the crime, by his absence, disappearance, or removal, at the time the vic- tim expired.

They are *generally inconsistent*, for, throughout the review, if a defence of their political profligacy can be called so, they appear in contradiction with themselves. First, they intimate that Russia had no right to make a separate peace, because the coalition, though broken in its

efficacy, still existed, or was understood to exist in intention ; then, soon after, justify Austria and England for de-
serting Russia, upon the ground that there was no coalition
between them. They repel every idea of subsidy, except
such as may assist a nation in self defence, and yet com-
mend the refusal of subsidy to Russia on the ground that
she waged altogether a defensive war ! ! As for their calling a loan a subsidy, and their confounding all distinction
between lending and giving ; I would simply ask, is there
no difference between a government's giving money away
without any expectation of reimbursement, money too,
somewhat compulsively obtained through the medium of
taxes from the pockets of the people ; and the sums which
private individuals, of their own accord, and for their own
advantage, were willing to lend, without any burthen to
the country, or any trouble to the government, except
their mere word of encouragement ? And is it by the
miserable process of pounding loan and subsidy in one
mortar, that they attempt to repel my charge of faithless conduct towards Russia, faithless even were there no
such distinction between loan and subsidy, one or the other having been expected ? Is it thus that men of strong
minds, skilled in every species of argument, defend themselves when innocent ? I charge them with refusing every
efficient succour to Russia, and they reply that loan and
subsidy are one and the same thing. I charge my neigh-
bour with having cheated me out of bills and cash, due to
me in virtue of our partnership, and the only defence he
makes, is, that bills and cash are the same thing ! !

Precisely of the same character is their rhapsody about
the Austrian fears and jealousies of Russia ; a power, who,
according to their system, never can be equal to France ;
and yet more to be feared than France ! ! The Edinburgh
Reviewers may shew much ingenuity in reconciling this
palpable contradiction ; but it requires more influence than
they possess, or ever will possess, to persuade any person
of common sense and reflection, that the fear of one who
never wronged us, never took up arms against us, never
seized upon any of our possessions, but has been a long-

fied friend, guided, even in moments of anger and dissatisfaction, by principles of honour and moderation—that
the fear of one who bravely, generously, and repeatedly
shed his best blood in our cause, can even for a moment
counterbalance the fear of a powerful, unprincipled, and
experienced enemy, the resentment of past injuries, and
painful consciousness of present humiliation, the mortifying ~~appearance~~ of increasing aggression, and the anticipated horror of future misery.   The Edinburgh Reviewers,
it seems, have yet to learn, that the apprehension of what
Russia may possibly undertake hereafter, is very different
from the certainty of what has been, or is to be inflicted
upon Austria by the implacable ruler of France, urged
rather to claim as his own, than spare the mutilated fragments of her empire, in virtue of that fatal and degrading
bond of consanguinity, which has enchained her to his
bloody car, and imparted a certain fascination to the real
cause of terror, which overawed her from every exertion
favourable to her friends, and conducive, ultimately, to her
own safety.

They are *conspicuously absurd*, when they proclaim a
man's situation in life, as the criterion of his knowledge.
As in this instance they dream about Worontzow's
chapel, which never existed, and are desirous to know
something more of my history; I shall gratify their curiosity, whenever they agree to apply their own rule to
themselves, and by stating truly who they are, or what
they were, furnish the world with the exact measure of
their own worth.   They shall find that I am no ministerial
hireling, as they seem to insinuate, nor the slavish scribbling tool of a disappointed faction : but as great a stranger
to the one party as to the other, a friend to those who do
justice to my country, and a sworn enemy to all those who
endeavour unjustly to vilify her character, and by degrading
her to screen their own perfidy : a declaration, one half of
which the Edinburgh Reviewers may apply to whomso-
ever they think deserving it, and to the other half they are
welcome themselves.

Whether my situation during the ten years' residence

in London, was favourable to the acquisition of knowledge, and whether my opportunities were few or many, I leave to the decision of others. I have, however, learned a thing or two of considerable importance to myself. I have learned to despise all self-created patriots, noisy dema- gogues, canting hypocrites, spouting orators, office hunters in disguise, in show cringing slaves, in reality profligate tyrants, enemies to power, merely because it is not their own—and the whole train of their supporters, and hungry expectants sighing day and night after *loaves* and *fishes*, and opposing the government, not from principle, but self- interest, which admits of no other creed but this : *So long as we are among the* OUTS, *the* INS *must invariably be in the wrong.* I have learned, that the pretended political reformers are the only dangerous and real enemies of the English constitution, in appreciating whose matchless form and excellence, I was not alienated from my own government ; but, as a man, I contemplated with pride and delight so magnificent and durable a structure of human wisdom, while, as a Russian, I did not think it suitable to the Russian empire. I learned to respect, but not to envy the Englishman his country, and to feel happy in being born in Russia, whose future glory, from the conscious- ness of the past, I have always contemplated with confi- dence : looking forward with enthusiasm to that period, when she would confound and cover with shame all such malignant revilers, and unprincipled slanderers as Dr. Clarke and the Edinburgh Reviewers. Russia herself was my only *schoolbook* and *gazetteer.* I have further learned, that the very errors of Mr. Pitt's active and vigilant admin- istration were worth all the virtues, if there were any, of the succeeding administration formed of his political op- ponents, who, whatever their talents may have been indi- vidually, proved themselves, as a body, totally incapable of useful exertion, and left behind them much cause for grief, but none for joy.

I have also learned, that, of all the literary, philosophi- cal, or political associations which ever existed, there never was one so dangerous in its tendency to mislead young and unthinking men, too indolent to examine for

themselves, as the fire-brand-fraternity conducting the Edinburgh Review. To say they possess talents, is the
worst recommendation that can be given them ; for Satan
has talents too, and hence the magnitude of mischief. In
truth, the arbitrary tribunal into which they have erected
themselves by usurpation, by a complete prostitution of
their talents to the vile purposes of a faction, and by their
readiness to sacrifice, without mercy, the virtuous or literary reputation of every man who does not worship with
them at the same political shrine, is so tyrannical and odi-
ous in its nature, and so discreditable to those, who, with
wilful blindness, suffer themselves to be implicitly guided
by its decision ; that it can only be compared to the power of their chief idol Buonaparte, whom they hypocritically
censure, only when they dare not praise, and whose despotick, unnatural sway has been first established upon the
indolence, then maintained by the fears of mankind, and
finally rendered permanent (at least in appearance) by the
sordid passions of slaves, labouring to exalt him, and thus
to rivet the chains which they themselves, and for themselves among the rest, had foolishly and wantonly forged.
If I were to attempt to give a greater scope to my sentiments and ideas, long entertained, of the Edinburgh Reviewers, I should call them a certain non-descript monster,
such as Ariosto himself, the greatest of poetical dreamers,
never contemplated in his visions : a compound animal,
whose nature and real existence can only be understood
by giving it the serpent's tongue, the wolf's tooth, the
fox's tail, the monkey's limbs, the syren's face, the croc-
odile's eyes, the raven's voice, the tyger's claws, the ca-
meleon's colours, the fugitive body of Proteus, and, to all
these, a perpetual motion. This amphibious animal, though
it lives and moves in every element, has its most favourite
haunt in the sea of politicks, between the Scylla of usurp-
ing undivided despotism, and the Charybdis of lawless
many-headed democracy, where, crossing from the one to
the other, and alternately embracing each, it intercepts all
such as heedlessly float on the midway tide, and by chance
might have escaped the two extremes, so dangerous and
fatal. How such a monster made itself acceptable to so

many individuals scattered over the globe, some of whom
must be possessed of virtue and talents ; and how it came
to be cherished with such *peculiar fondness* in America, by
persons too of great respectability ; would be a mystery altogether incomprehensible, and would of itself overrule all
my objections, had it not been in my power to explain it,
by pointing out the two principal causes which have coop-
crated in favour of its reception.

The first, in relation to mankind in general, is that undeniable human propensity, originating in pride, to be
pleased with seeing our fellow creatures rather reduced below, than exalted above our own level. We listen to the
censure of others, particularly of those who have attempted
to rise above us, with an ear predisposed to be credulous ;
but we hear their praise with the secret pain of conscious
inferiority, and never join in it without some reluctance. It
is therefore by administering to this general passion, at the
sacrifice of a few individuals, by flattering many and griev-
ing one, and doing it in an artful, yet audacious manner,
occasionally relieved by a show of liberality, that the Edinburgh Reviewers have made their sentiments so current,
and usurped so extensive an authority. A cunning tyrant
pursues exactly the same course. His capricious rigour
in degrading the more elevated subjects to the rank of inferior and the most numerous orders, pleases these, and
relieves their pangs of envy : while a rare occasional favour, apparently extorted by superior merit, not to be easily imitated, and therefore far removed from envy and
jealousy, hoodwinks suspicion, lulls selfish passions to
sleep, exalts the donor, and makes his sway assured in its
progress.

The *second cause*, with regard to America in particular,
is the jesuitical and well seasoned flattery, with which they
often regale the ear of the American nation ; not that they
really respect and cherish a disposition to favour the Americans ;* but that it suits their views at home, to exalt and

* The review of the Columbiad shows their *respect* for American genius,
and the conduct of the Foxite administration, full as hostile as the one that prece-
-ded it, shows their *disposition* to favour the American interests. In truth, they

13

approve whatever may, in any degree, result in the humiliation of their own government. Their motive, however,
is either unperceived or disregarded, while their praise is
accepted as a tribute of sincerity. A double appeal, at
once to personal and national vanity, is the most powerful
and resistless charm to make proselytes, that ever was exercised by the magick of the pen ; and so pleasing it is to
hear some commendation of our country, particularly
when we are absent from it, that, though we ourselves
may be conscious of its insincerity, and even misappli-
cation, we still hug the dear flatterer to our bosom,
and feel happy in being deceived. Hence almost all the
Americans who visit England, and who never subscribed
to the doctrine that *the British government is the most vil-
lainous in the world*, associate there exclusively, or, at least
by preference, with the opposition-men, the same whom,
while at home, they had openly denounced : nay, I know
some, whose publick labours are an honour to their country, who have execrated the unprincipled ambition of the
Corsican usurper, and have spoken of his system of universal subjugation with undisguised abhorrence, to have
mingled so quickly with the English opposition party,
more or less favourable to the usurper and his system, as
to forego every trust in their own knowledge and judgment, and indiscriminately to adopt the sentiments of the
Edinburgh Reviewers, the mouth-piece of the party, on
every subject of foreign policy, directly or indirectly connected with Russia ; tending, upon the ruin of her political
importance, to elevate the power of France, and thus destroy the hopes and confidence of Europe, in the only
continental power that had the will and ability to check
the overwhelming torrent of French aggression. I am
conscious, that a weakness, which flows from the source of
patriotism, is too sacred to be censured ; but I trace in it
the mischief of which I complain, and cannot but lament,
that the effects of its indulgence should have been so aus-

change their principles with their places ; they abuse and denounce Cobbett,
or praise and side with him, as their interest dictates ; and their respect and
love for Americans is very much like that of Buonaparte.

picious to the dangerous influence of the Edinburgh Reviewers. Strip them of these auxiliaries, and they will
dwindle into blank insignificance, for what are their other
pretensions ? Let us examine.

*Is it acute and just discrimination, correct and sound
judgment discovered in their critical reviews of literary
productions ?* I hesitate not to say that, whenever their
blind admirers will take the trouble to peruse and digest
the books so reviewed, they will find, nine times out of
ten, their own opinions, formed upon those of the reviewers, to be incorrect, and repugnant to truth and justice. It
is by this true standard, the only one by which all reviewers ought to be judged, that I some years since, tried
them ; and, having detected them, turned from their pages
with disgust and indignation : the more so, as they evidently betray want of integrity rather than want of
talent. Their literary lava* is but a modification of that
political poison, which they are constantly diffusing ; and
which, like the effluvia of the Bohun Upas, tends to destroy every thing within its reach. They praise at random, because sometimes it is necessary to appear to do
so : but they censure upon the settled nefarious system of
political exclusion. They seldom fail, and never cease, to
abuse real merit, until the public opinion against them is
strong enough to make any further abuse dangerous to
their interest ; but their voluntary praise is bestowed only
when the author receiving it, has not rebelled against their
politicks, and is not likely to offend them with his future
celebrity. In England, where there are so many other reviews to counteract undue impressions, and where the
learned read for themselves, the capricious and almost always unjust decision of the Edinburgh Reviewers on literary productions cannot do much harm ; but, in America,
where there are so few counteracting checks, it has already
had, and will continue to have, a most inauspicious influ-

* It is worthy of remark, that such productions in the Edinburgh Review as
are entitled to distinction by their usefulness and merit, are not generally written
by the Reviewers, that is, by any of the members constituting the critical junto ;
but are contributed by strangers, as was the case with the article on the French
code of conscription, which got so much credit to the Reviewers, but was
written by Mr. Walsh of Philadelphia.

ence on the domestick literature. Such is the habitual ven-
cration here for the word *we*, created, probably, by the
frequent perusal of the Edinburgh Review, that any *he*,
by multiplying himself, has now a reasonable hope of di-
recting the publick opinion at pleasure. Nothing can be
more monstrous, than that the *ipse dixit* of any individual
should, by the mere magick *we*, decide the fate of a work,
and even reverse, with a single stroke of the pen, the favourable judgment it at first obtained. The best man, as
to intentions and ability, who assumes the office of a crit-
ick, is liable to error, and prejudice, often personal, for or
against the author : if, therefore, without inquiring into
our own original reasons, out of pure respect for this crit-
ick's opinion, we give up our own, and think worse or
better of a book than we did before, we betray our previous
or subsequent want of judgment, or our unmanly sub-
serviency, both ways establishing a tyranny which must
control or check the progress of literature, and blast the
youthful genius in the bud. Poetry, more tender than
her sister flowers, withers and shrinks the quickest from
the frowns of such a stern and absolute tribunal; and this,
possibly, is the chief reason why America is so inferior to
herself in this department ; indeed, she is likely to continue so, as long as the Edinburgh Reviewers reign lords
paramount, converting all the young men, who read them,
into criticks, which is a much easier character to assume,
than that of braving such criticisms in the character of a
poet or an author.

*Is it the superior elegance, and classical chasteness of
language, that so strongly recommend the Edinburgh Review, and place it above all competition ?* It would be
presumptuous in me to question, in this particular, the
claims of the Edinburgh Reviewers : but Mr. Thelwall,
as great a character in his way as themselves, who for
years made grammar and rhetorick his peculiar study and
profession, has already anticipated me ; and has, with
signal success, exposed their ignorance, or voluntary
breach of all the rules and principles of English composition, making their vitiated taste no less conspicuous, than
their absurd confusion of metaphors floating on the turbid

stream of a quaint and deformed style, sustained altogether by the bitterness and acrimony of personal abuse. I
know not whether any American bookseller had ever boldness enough to publish *Mr. Thelwall's* celebrated *letter to
Francis Jeffrey, esq,* the chief head of the Edinburgh-hydra:
but, as I have it in my possession, every unbeliever, by
applying for it, may be satisfied of the truth of the above
statement.

*Is it the profound knowledge in history, antiquities, and
various arts and sciences, that preeminently distinguishes
the Edinburgh Reviewers?* This may possibly be the
case; at least I confess my incompetence to decide. I
leave therefore their history and antiquities to the examination of lord Byron, who appears to be somewhat skeptical
on the subject of their infallibility in these branches; and
their arts and sciences to the investigation of the respec-
tive professors, who are the only proper and legitimate
judges.

*Are they particularly to be admired, adored, and imitated
as immaculate and profound statesmen? By their compa-*
ny we shall know them. Gentle reader, hast thou heard
of men, whom at first in anticipation, and afterwards in ridi-
cule, were called "all the talents" of the country; and
who began their career by conjuring up the story of the
assassin,* merely to scrape up a correspondence with
Talleyrand, and to be cheated? Hast thou heard of men,
who, instead of warning Prussia of danger, which they
might and ought to have done with effect, continued to
carry on a hopeless and protracted negociation at Paris,
which created fear and distrust in her, and paralysed the
succouring arm of Russia; until Buonaparte, under the

* The offer of the pretended assassin, who, strange to say, never
was afterwards seen or heard of, to bring Buonaparte's head, or do
something equally heroick, was communicated to Talleyrand, and
thus led the way to the negociation which ruined Prussia. The assassin, if there really was such, must have been a great dunce, to
make such offer to a set of men, who held Buonaparte's life full as
sacred as their own, hung upon it for food, like worms upon a car-
case, and valued it as the life of their life, and the support of their
political existence.

slender cover of this fatal negociation, seen through by all, save those in whom blindness was a crime, having matured his plan of invasion, and being literally tired of inventing new pretexts, whose absurdity rather insured than prevented their reception, with one foot kicked at length my lord Lauderdale* out of Paris, and with the other, crushed in a moment the kingdom of Frederick the Great? Hast thou heard of their childish and foolishly unsuccessful expedition to Constantinople, which at first frightened the women in the seraglio, but soon after was frightened away by their shrieks and lamentations, and which is pompously noticed by the Edinburgh Reviewers as a mighty effort of cooperation with Russia, calculated at the least to bring the neck of the Sultan under her feet! | Hast thou heard of two other expeditions, one, equally efficient, to Egypt, just to humble the pride of Englishism by showing that the Turks could beat them; and another, the only successful one, to Buenos Ayres, sacrificed because Sir Home Popham, the original projector, was not a branch of "the illustrious talents," who modestly took to themselves the first glory of this expedition, and, when it was afterwards frustrated by their own neglect, and through the ignorance of their own Whitelock, *magnanimously* gave up Sir Home Popham to the clutches of a court martial, from which he was rescued only by the good sense of the country? Hast thou heard of agents and ministers who had the singular felicity of making themselves obnoxious by their arrogance, to the governments near which

* This poor ambassador, while on his return, was pelted at Boulogne by " all the talents" with good English bullets; which was at once a specimen of the Foxite energy and tenderness towards France, showing, that, even in anger, they did not treat their dear France, worse than they did their own ambassador ! |

, The Edinburgh Reviewers are anxious to exculpate their friends from the, suspected *tenderness* for France, and yet defend them altogether upon the ground of this tenderness ; for, passive waiting for the " advent of times," Austria's error of judgment in opposing France at all, silent and peaceful musing on the future, are the constant burthen of their song, as if Buonaparte all that time was sitting too, with his hands folded up, or merely playing a few harmless tricks unworthy of resentment or resistance ! |

they were sent out to reside ; and who seldom failed either to render themselves ridiculous by their folly, or mar the mutual good understanding by their imbecility ? Hast thou heard of great financiers who were scouted in Parlia-ment, and laughed at for their bubble-scheme of *ways and means without taxes ;* and who to save a few pounds by withholding just and fair assistance from the continent, embroiled England in a new war with Russia, that cost millions to the nation, besides the expense of honour and integrity ? Hast thou heard of distinguished politicians, who, out of pure opposition to Pitt, reversed the whole system of manly, liberal, and honourable English policy, pursued, for a series of years, towards the European continent ; and, in about one short year, one month, and one day, disgra-ced their country more than other ministers could have done in a century ? In short, hast thou heard of men who have forfeited every claim to publick respect and indul-gence, who are equally offensive to a loyal Russian, and to a good sound Englishman, the natural friend of the other, and for whose unworthy sakes Russia is to be calumnia-ted ; her virtues to be denied ; her achievements to be obscured ; her sacrifices to be derided ; her energies to be decried ; her defects, (for who is faultless?) to be mag-nified ; her moderate and legitimate ambition to be held forth to the world as more destructive† than that of Buon-aparte ; her efforts to save Europe at once to be deprecia-ted, and brought into suspicion ; and all, who, like Dr. Clarke, lend their hand to put her reputation down, to be trumpeted about as men of sense, candour, and profound

* The famous scheme of lord H. Petty, the *illustrious financier,* whose deeds were to *rival and outshine the deeds of* Mr. Pitt !! !

† The absurdity, already noticed, of believing on one hand, in the comparative importance of Russia, and, on the other, in her will and power to do a greater mischief than France, palpable and ri-diculous as it is, sticks yet so fast to the friends of the Edinburgh Reviewers, that Mr. Whitbread, talks even now, of extortion from Russia her Polish dominions, as the price of peace, without even a hint about the restoration by Buonaparte of his immense spoils, *very innocently* taken, and *righteously* acquired ! ! !

erudition‡ If thou hast, gentle reader,¹ I will tell thee,
these men, of whom I have given thee a correct picture, are
the worthies who, in the Edinburgh Reviewers, have always found their willing minions, and supporting slaves :
ask therefore thy own sense, when such are the masters,
what must be the servants ?*

*Is it the purity or supereminent wisdom of political doc-
trine, that entitles the Edinburgh Reviewers to the privilege of dictating to the world ?* I do not mean their domestick system of politicks, with which I have nothing to
do ; but their foreign system, the prominent features of
which are *the irresistibility of France, the incompetency of
Russia, and the superior prowess and conduct of Austria.*
'The *irresistibility of France* has long been the pretext
with the party to justify their refusal of assistance to the
continent ; and a cloak to cover their reluctance to oppose
France in her career ; so that it is not for want of their
good wishes and measures that she is not yet irresistible.
This precious quality of *irresistibility,* which, like a
woman's honour, cannot be impaired without being destroyed, has been severely tried, as far back as the year
1799, by the old Suwarow in Italy ; but " the talents,"
or the Edinburgh Reviewers, which is the same thing,

‡ Whoever reads and examines the conclusion of the review of
" Clarke's Travels in Russia," in the Edinburgh Review, must
be convinced, that what I have alleged is the true motive and the
constant object with the whole party ; and that it is a death-blow to
their interest to say one word in favour of Russia, unless by compulsion. How then, and with what justice, with such proofs before
him, can a sensible American, totally unconnected with either of
the English parties, take Dr. Clarke and the Edinburgh Reviewers
as conclusive authority on Russian affairs, at the time too that
the conduct of Russia has so fully convicted them of error, if not of
deliberate falsehood ; is beyond my power to explain, and I can
only wonder at, and lament the infatuation.

* The Edinburgh Reviewers say, I asserted, that these men were
turned out of office for their bad conduct to Russia, when the cause
of their removal was the catholick question. The Edinburgh Reviewers are guilty here of a misrepresentation, as I did not even
hint at the principal cause of their removal, but merely said, (and
who can doubt it ?) that the prince regent, who removed them, was
conscious of their shuffling policy, &c.

still preached its durability to the world. It was rather
rudely questioned at the battles of Pultusk and Eylau, but
the Edinburgh Reviewers still preached,until they preached
away the friendship of Russia,and then,with an ideotick stare
stood incredulous, or wilfully heedless of the mischief they
had accomplished. The battle of Aspern, so honourable to
Austria, was likewise a bitter test : but still they preached
on ; until, at length, the veteran Kutusow had rent the veil
with his native sword,stripped naked the preaching fanaticks
and impostors,and left them no alternative but to plead error
or hypocrisy. Nay, had he lived longer, he would have
anticipated the proof of time, in showing that France is
even *vulnerable* ; for, whatever hopes they may entertain
of her convalescence through the temporary success of
Bonaparte, they may be assured, that France herself cannot easily recover the blow she received in the last campaign, the effects of which I have elsewhere endeavoured
to foretell, and see no reason to change my opticks of futuri-
ty. Bonaparte may recover, he may by an intrigue with
his weak father-in-law, by a temporary peace, or even by
arms, preserve his own political existence, brilliant as ever
in appearance ; but, every moment he lives, every mangled
corpse he adds to the pile which props up his tottering
throne, is fatal to France ; and every drop, feeding the tide
of blood that keeps afloat his bark of despotism, is drawn
from the veins, from the very heart of France, who, if not
speedily relieved, must sink from total exhaustion. There
is a power that evidently protects Buonaparte, for some
purposes as yet to be accomplished ; and twice it sent
death to save him, by removing Suwarow and Kutusow
from his way, both at two critical periods which seemed to
threaten his destruction. If it be not, therefore, impious to
scan the unrevealed decrees of Providence, a belief may be
indulged, that it is to chastise other offending nations,
then to punish France herself, the guiltiest of them all, and
end, perhaps for ever, the tyrannical influence she has so
long exercised over the European world, that Heaven still
preserves, and still permits the demon of ambition, while
he makes her a scourge to the rest, to feed upon her vitals,

and not to quit his hold until fleshless and juiceless, she is cast off, like a withered leaf, a prey to every puff and gust of wind. It will be found, that her conscription system, and her internal, but not eternal resources, strained within, and unsupported without, whereon the Edinburgh Reviewers, and other better men, misled by them, have founded her imaginary invincibility and Buonaparte's irresistible power, are but the life-stream of a bleeding giant, which, while in its torrent it overwhelms the insects within reach, weakens himself, and must, at the end, terminate his own existence.

As to the *incompetency* of Russia to defend herself, and resist France, from what has been said in the preceding paragraph, or rather, from what has been done by Russia in the face of the whole world, it would be now a superfluous multiplication of words to tell the Edinburgh Reviewers that they were willing or ignorant enough to be mistaken, and have not had so much magnanimity, or even common decency, as to acknowledge their error. It would be equally superfluous to give another proof of their *admirable foresight* as to Spanish affairs : for it was a part of the same political system with them, that *Spain could not be defended.* It is enough to observe, that they are doubly enraged at the success of the English and the Russian arms, which disappointed their expectations in the north and west of Europe ; and that, not daring to abuse the former, they are willing to revenge themselves on the latter : and no doubt, will hail with joy their Antæus, rising again in vigour from the earth on which he was prostrated. But as, by the incompetency of Russia, is implied her general inferiority to Austria, and her incapacity to become the rallying standard of nations seeking deliverance from the French yoke ; I shall take up my last division of the Edinburgh system, and contrast the energy and conduct of Russia, with the *superior energy and conduct of Austria,* by a plain statement of historical facts, within the memory of every man living.

Russia, whose orbit of revolution did not come in contact with that of France, and who, though the farthest re-

moved from her, was rather courted than shunned by her, felt of course the less concern in the changes and innovations occasioned by her aggression; having the less to hope from her friendship, and less to fear from her enmity, and participating in a degree, proportionably smaller, in the interest and apprehensions of other European nations nearest to her: and yet, from a mere sympathy, from the motives of humanity, and the generous aspiration to the glory of delivering the oppressed more than from any immediate interest, she parted at length with her sons, and on the plains of Italy beheld, unmoved, their blood mingle with that of her allies, and the common foe. The success, as long as her veteran general (Suwarow) sustained the principal command, was beyond her sanguine expectations: but the jealousy and perfidy of the very friends she was thus succouring, reversed the prospect, and she was forced, against her wishes, to recal her victorious troops, as a necessary sacrifice to her just indignation.* Severely

* Unquestionably, the emperor Paul might have subdued these feelings, and let policy, or original motives of glory still prevail; but it would have been an effort beyond the common powers and patience of man: besides, what prospect was there from a coalition, a part of which was so corrupt and rotten? Mr. Carr may talk as much as he pleases, about the secret influence of Madame Chevalier; but it is neither in his power, nor in that of the Edinburgh Reviewers, nor in that of Madame de Stael, who, in her " Appeal to the Nations" says, that Paul *recalled his troops from caprice*, to gloss over the glaring duplicity, to say the least, of the Austrian government upon this occasion; a duplicity which Suwarow, foreseeing the effects of Paul's anger, concealed as long as possible, and, by doing so, (this was his chief offence) incurred his master's displeasure. The subsequent official complaints of this veteran who never before complained of any body, and historical details of his passage over the Alps; the repeated efforts of the archduke Charles to appease, through him, Paul's indignation; the dismissal of Baron Thugut, Austria's prime minister, who was accused of treachery by the Russians; all confirm, that Austria, when too late, was herself conscious of having acted an unworthy and ungrateful part towards Russia. But the deepest wound inflicted upon Russia, and the irreparable calamity to Europe, was the death of Suwarow, caused, not as is generally supposed by his sovereign's displeasure, but by the faithlessness of Austria, which forced him to give up the contest, of the success of which he was assured, and to

and quickly was this defection of Austria punished on the field of Marengo, where her fortunes were instantly blasted, and her ruin, with that of Europe, commenced : for now, Suwarow, the only man that could have saved both, slept in the grave, the victim of perfidy and ingratitude ; and Buonaparte, unchecked, uncontrolled, fearless of a rival, had only to choose which way to bend his steps, and which country first to select for the aim of his conquering sword. Thus Austria, in the year 1799, by spurning the friendly hand of Russia, had, with a suicide-hand, dug out a grave in which her own, and the liberties of the European world, were to be speedily entombed. Towards the year 1805, when the weight of pressure increased upon her, and when the progress of destruction had nearly reached her, she again sought and obtained the assistance of Russia; her steadfast and now relented friend ; but in the battle of Austerlitz she betrayed the same foul unregenerate spirit within her ; and while Russia exposed the life of her emperor in that battle, she, the most interested of all, kept her own

continue which his honour was pledged ; and this too at a most critical moment, when the defeat of the Russians at Zurich (under Korsakoff) called for an atonement, and when Buonaparte returned from Egypt, with whom, then the only remaining French general of celebrity, the old Russian hero was particularly desirous to measure his sword. Suwarow, being once asked, who, in his opinion, were the greatest generals in the world, replied, *Alexander the Great, Hannibal, Julius Cæsar,* and then added, in a whisper, *Buonaparte ;* his mortification may, therefore, easily be conceived, when he was withheld from encountering the only man he considered his rival, and whose overthrow, which he was confident to achieve, wanted yet to complete his glory. When, by the intrigue of the Austrian cabinet, who thought they could do without him, he was commanded to leave his conquests in Italy unfinished, *he wept ;* when, on his march to Switzerland, he received the news of Korsakoff's defeat, *he grieved in secret, and pined in silence,* wearing outward the visage of incredulity ; when he received orders to march home, *he shut himself up for three days in his chamber, a solitary prey to sorrow and distress ;* when Paul spoke angrily to him, on his arrival at St. Petersburgh, he *serenely smiled :* but when he heard of Buonaparte's return, he fetched a deep sigh, and *expired.* Such is the last tragick scene of this hero's life ; and can any politeness or courtesy make me mute on the perfidy of a power that caused a catastrophe, so fatal to herself and to the world ? No ; Never !

sovereign among the shady groves of a luxuriant garden, and let him out just in time, by a precipitate and ignominious peace, to assure Buonaparte's dear-bought and doubtful victory ; to prevent the Russians from wresting it back the next day, by the intended renewal of the action ; and to compel Alexander to lead his forces home, after rejecting the enemy's proffered friendship, and leaving a heartless ally, sunk, almost by choice, several degrees lower, on the scale of degradation. Aggression now followed aggression : the German empire, of which Austria was the head, was insulted, lacerated, and rent limb by limb ; innocent persons were torn, in the midst of peace, from the bosom of their country, of which Austria was the natural protector ; yet, strange to tell, she winked ; and of all the continental powers, Russia alone, (the feeble voice of Sweden excepted) protested loud, but in vain, against all these atrocities. Soon Prussia was designated as the succeeding victim of French ambition : and here again the dragon of Austria, though his own golden apples were to be plundered, slept in deceitful security : in truth, what else could be expected from a power that could, only two years previous, behold with indifference, without even a remonstrance, a guiltless prince (Duke d'Enghien) allied to her in blood, murdered by the hand of the midnight Corsican assassin? Again Russia went forth, and went alone, to the assistance of the oppressed ;* but the precipi-

* The Edinburgh Reviewers say, at this " critical period," Russia notified to the court of Vienna, that she would send her army to " Moldavia," instead of Prussia, as was expected ; which made Austria comply with all the demands of Buonaparte. Now, as the world knows, that the Russian army under Beningsen, never went to Moldavia, but actually marched to the assistance of Prussia ; the Edinburgh Reviewers stand convicted of chicanery, which must discredit every thing they have said in defence of Austria and themselves. In the same manner, they reproach Russia with having squandered her subsidy, and not having furnished a sufficient number of troops to Austria, before the battle of Austerlitz ; when it is a notorious truth, that Austria entered Bavaria, and opened the campaign with such precipitation, that Russia was obliged to

tation of the latter, and the supineness or criminal predilec-
tions, and wilful neglect of the English " talent" administration, rendered her efforts abortive ; and Prussia, being
overthrown, before any aid could be rendered her, the
whole storm of war rushed in an instant upon the territories of Russia, thus at once, from a mere auxiliary, transformed into an unprepared principal. Now was the time,
now was the first long sought-for opportunity for Austria
to recover with a single blow her own losses,and be,(a glorious title !) the universal deliverer. Her whole force was
now between Buonaparte and France, she had only to order, and he was intercepted, undone. Did she do so ?
Did she in turn, offer her services to Russia ? No. She
disobeyed the dictates of sound policy ; she disregarded
the ties of friendship and gratitude ; she did nothing. She
did worse than nothing ; she devoted Russia to destruction.
It is curious how the Edinburgh Reviewers endeavour
to defend her conduct on this occasion, by intimating that
Russia did not ask her assistance in time, and in a proper
manner. How do they know this was the fact ? Where
is their authority ? I do not believe it. Granting, however, for argument's sake, that their intimation is true; it follows that Austria sacrificed her dearest interests, her safety, her honour, her duty to the German empire, and to
Europe in general, merely because she was not *courteously*,
and *gracefully* asked to do that, which she ought to have
done without being asked, and for her own sake. One
would suppose that the spirit of " the talents" was trans-
fused into her ; otherwise, she could never have acted
so, and the Edinburgh Reviewers would never be so eager
to defend her at the expense of truth and common sense.
Russia, deserted thus by Austria, neglected by " the
talent"-administration, abandoned by all, and courted only by her enemy, feeling the want of preparation, and the
necessity of future provision, concluded, at length, the
send what troops could be collected, by forced marches, and take
the field some time before the term agreed upon between the parties.

peace of Tilsit; which was condemned by those very men who first advised it,* but which was not, as they held forth, an ignominious submission on her part, in the manner of their *favourite* Austria, but only *the yielding of a giant who stoops to conquer*. Austria had soon ample reason to repent; for the very power which she had then preserved, the power of Buonaparte, only a little more than a year after the peace of Tilsit, recoiled upon herself; and far stripping her of the remnant of her independence, left her to the mortifying conviction of having deserved her fate. It is true, that Russia, diverted by past experience, on this and this only occasion, remained neutral; but, though she did not assist, magnanimously forbore to molest or execute any revenge by word or action; yet even this magnanimity and forbearance were lost upon astounded Austria, who was now draining the very last dregs of her voluntary degradation, by giving up a portion of her imperial house to the polluted embraces of the murderer of the Duke d'Enghien; as if determined to the last to exhibit herself in the contrast with Russia, who, subsequently, though with less urgency of reason, rejected the hand of Buonaparte, offered in marriage with her imperial master. The repentance of Austria, in having preserved at Tilsit the power of Buonaparte, was but of a short duration; and the world beheld her with astonishment acting the same part over again, during the last campaign, so glorious to Russia: nay, regardless of every sacred consideration, and, urged by a sort of wilful madness, she even sent her forces—against whom? against the only power which never took arms against her, which was expressly to defend her, which had neither will nor interest to injure her, which still might resuscitate the hopes of the continent, and the destruction of which would have spread the pall of death over Austria herself, over Europe entire, and a great portion of Asia; or would have turned it whole into a

* Lord Hutchinson declared in Parliament, that urging Alexander to make war, he kept advising him all the time to make peace with Buonaparte.

barren region, where the groans of the suffering slaves, and the never-ceasing strokes of the tyrant's active lash, would be the only sounds to greet the human ear. What excuse had Austria for a measure so every way fatal? None; for she was not, like others, impelled by irresistible necessity.

To crown her political iniquity, and to prove to the world that it is the effect of her own free agency, she now seems labouring to preserve her own chains, and, though not even gilded to please the eye of an idiot, to hug them with all the devotion of a faithful slave. Not content to stand an unconcerned spectator, which was the worst thing expected from her, and to leave Russia and Prussia to their own exertions, she has, if appearances do not deceive us, made the severest censure fall short of her due, and outrunning every prediction limited by her inactivity in the common cause, she has voluntarily thrown her weight between the allies and Buonaparte, paralyzed their efforts by an ill timed armistice, allowed the enemy to recover his breath, and forced Alexander, who otherwise might have withdrawn from Germany, to negociate, and probably to accept a peace, revolting to his feelings, as the prize of the independence of Prussia, which his sword, if not diverted by an interloper, might have restored, preserved, and secured in a manner more effectual and more beneficial to the rest of Europe: so that if Buonaparte reigns, if the nations suffer, if the arms of Russia fail to extend salvation beyond her territories, and the effects of her last triumphs are diminished, or their pruning edge blunted or averted, Austria alone is the fatal cause—the first original cause, the medium, and the last link of all the calamities already inflicted or yet to be inflicted on the wretched population of Europe! To recapitulate the whole, while Russia, with less interested motives, as was before observed, resisted every aggression of France; protested against every encroachment; flew to arms from the generous wish to defend others; bled on every occasion for the liberties of Austria and Europe; and never joined the tyrant in a league, much less in a bond of consanguinity, to destroy those liberties; she, Austria,

having the nearest interest in Germany, saw with comparative indifference, this ancient and venerable fabrick invaded
and mutilated; bore, with silent resentment, every aggression: submitted, after a feeble struggle to every encroachment and extortion; beheld, with criminal apathy,
her honour insulted, and even a prince of her blood wantonly murdered; fought only (except once in the beginning
of the French revolution) when absolutely driven and goad-
ed into resistance, and yielded, as soon as a transient relief
was obtained, from reluctance, or fear to persevere; took
arms against Russia herself, and frustrated her efforts for
general deliverance; sealed the bond of union with the tyrant
by a degrading and unnatural sacrifice of one of her own
princesses; and finally appears to have joined him in a
league, the consequences of which, if they be no more than
her passiveness, hesitancy, and procrastination, must, if
not checked, consummate the misery of Europe:—Are
these facts true? or is this an imaginary contrast? I will
not insult the well-informed reader by answering the question. And, when it is farther considered, that Russia combatted throughout with more energy and success, than Austria; no one can hesitate to choose between them, and
none, but canting jesuits, like the Edinburgh Reviewers,
can pretend even to doubt, which of the two is entitled to
the confidence of nations, and which has evinced more will
and power to become their saviour.

The Edinburgh Reviewers, and their humble imitators,
will probably stigmatize my severity and resentment, as
proceeding from some national prejudice against Austria;
but national prejudice, strictly speaking, is the growth of
ages; of long severe collisions which never took place between Russia and Austria; and even when the latter put
on her French trappings, the former still persevered in
friendship, and never once appeared in the character of an
armed enemy; my only excitement, therefore, is the notorious truth of my charges. I own, that Austria's base
treatment of Suwarow in Italy and Switzerland, her ungrateful desertion of Russia in the preceding Polish campaign of 1806, and her more than treacherous invasion, con-

jointly with Napoleon, of a territory belonging to so ancient a friend and defender, during the last destructive
war, glorious to the invaded, and fatal to the invaders, are
deeply engraved upon the heart of every loyal Russian;
nay, I should despise myself for being less alive to such
outrages; but I solemnly declare, that my present indignation chiefly arises from her most unnatural abandonment
of the interests of Europe, at the moment when the
preceding Russian victories furnished her the means of
regaining her own freedom, and that of Germany. I particularly complain of her, because, among the degraded
vassals of France, she is the only one that can throw
off the yoke with impunity; and twice, particularly the
last time, when Buonaparte was left almost without a soldier, she had it in her power, with a gleam of her sword,
to deliver the world and herself of a monster.
Others, Prussia for instance, have been also criminal and
ungrateful towards Russia; but of these I do not complain. I have rather defended, than condemned Prussia;
because I knew, that the superficial politicians who clam-
our against her, had overrated her means, and are angry
at her disappointing their unreasonable expectations, when
they ought to blame their own credulity and want of information; and because I know, that her political impotence
alone checked her growing hostility to France; and that
she was constrained by real force and necessity to appear
on the field, as it were, against herself. The subsequent
events, as soon as the force and necessity ceased to operate,
have completely verified my opinion of her, as elsewhere
expressed: and happy should I feel, were the conduct of
Austria such as to make me retract my former and present censure. The triumph of successful prophecy, which,
by proving my severity to have been well founded, must
exculpate me from undue prejudice against Austria, is but
a poor consolation to me, who have looked forward with
inexpressible anxiety to the deliverance of Europe: and
whose only joy would consequently be, to see Austria,
contrary to my expectations, exert herself on the side of
justice, and finish the work so nobly commenced, and all

but done, by Russia.   I dare not hope ; but I still wish to be deceived.

But it is time to dismiss Austria, and return to the Ed-inburgh Reviewers, just to say a few words before I con-clude, on the spoliation of Poland, for years a standing dish with them; to which every body has been indiscrimi-nately invited ; and which has provoked a keener appetite in many on this side of the ocean, by the additional season-ing of imagination.

A recurrence by their friend, Mr. Whitbread, to this old worn-out subject, and at this time, deters me not from ap-proaching it in my turn ; and the only pause I make, is that of surprise and astonishment at the attempt to revive an infatuation similar to that which prevailed during the French revolution, and which made infants lisp " liberty and equality," while their parents were expiring on the bloody scaffold, victims to unexampled tyranny and des-potism.   Whatever error there was in the principle of partition, Russia, for her share, has made noble amends, by ameliorating the condition of that part of Poland which was allotted to her ; and her efforts have been rewarded by the conspicuous attachment of the Poles to her govern-ment.   I think I hear the Edinburgh Reviewers spurn this assertion with a frown of pretended indignation, and denounce me as a dreamer in impossibilities ; but I do not fear their frowns, and repeat again, that the Poles, under the dominion of Russia, are preeminently attached to her government ; and the reasons are, *the lenity of its adminis-tration ; the same origin of extraction ; similarity of lan-guage and customs ; and the congeniality of religious prin-ciples.*   These are no chimerical motives of attachment : they are, and always have been, the most powerful links in the union between one nation and another.   Why should a thing so natural excite incredulity, when we see, with scarcely an emotion of surprize, the most unnatural union between some of the German states and Buonaparte ? If certain individuals could for a moment forget their early impressions, produced by the canting language of sophisticated and self-interested philanthropy, such as

characterises the Edinburgh Reviewers ; if they would or
could exchange their book-knowledge for one of experience and practical observation, by actually visiting Poland ;
they would discover, not without keen self-reproach, that
the sympathies, of which they are so prodigal, have been
engaged, not in the cause of the Polish people, but in the
cause of the Polish nobles, the most relentless tyrants, and
the most capricious and numerous oppressors, that ever
cursed a hapless nation. It was the reign of these ; it was
their power of life and death over their miserable vassals,
that were crushed under the Russian sceptre ; and the
groans of this many-headed monster, expiring for want of
victims to feed it, have been mistaken for the complaints
of a people, who are all the while rejoicing at their deliverance, and are enjoying, what they never enjoyed before,
the blessings of comparative liberty. A greater proof
cannot be given of their contentment, than that Buonaparte, with all his intrigue, art, and power, never could
excite them to rebellion ; and, with the exception of some
of the nobles, all the soldiers he obtained in Poland, were
from those provinces only, which, by his own invasion and
temporary conquest, were placed at his mercy. I say again,
I am not justifying the principle of partition ; but I am only
stating the consequences, such as they are, and such as I
can vouch for, from my personal knowledge ; and surely
it is not to the disadvantage of my evidence, to say, that
more than once I have passed through Poland with an eye
and ear willing to catch and retain information. If any
persons, however, are determined not to believe me, they
cannot reject the corroborating testimony of Sir R. Wilson, a man of knowledge, an officer of distinction, and a
gentleman, in every sense of the word ; who, besides his
being free, as a stranger, from the supposed national prejudice of a Russian, had likewise the advantage of a personal
intercourse with the Polish people. A rejection of an authority like this, would be an unequivocal confession that
his promulgation of certain unpleasant truths about Buon-

a parte in Egypt,* or his vindication of the Russian character, from the malignant aspersions of Dr. Clarke, have
subjected him to the wrath of the Edinburgh Reviewers;
and that for this reason alone, he is, like myself, to be proscribed. It would be a dishonourable, tyrannical, and out-
rageous demand, that every thing, even so just and right,
should be considered as reversed, the moment it comes in
opposition to the authorities of Dr. Clarke, and the Edinburgh Reviewers; the former, as a publickly convicted
calumniator of Russia, having no reputation for veracity
to lose; and the latter—who for political purposes, not
generally understood here, have told things of Russia, not
believed by themselves, and proved false in every instance
—laughing at the blind zeal of their own votaries, that follow them, with the unconscious slavishness of a mob, ignorant of the design of a popular leader, yet shouting in
his train, under the flattering delusion of independence.

* The manner in which this great doctor, has attempted, upon
his own negative authority, to do away Sir R. Wilson's accounts
of Buonaparte's atrocities in Egypt; and the eagerness with which
this attempt was seconded by the friends of the Edinburgh Reviewers, is a most finished specimen of the new logick brought by
these Reviewers into operation. The doctor was in Egypt, but
forgot to inquire into the said atrocities; and therefore heard
nothing about them: *ergo,* these atrocities never existed; and Sir
R. Wilson is an impostor! ; So that when Russia is to be justified, the authority of those who lived there (not travelled) or were
born there, goes for nothing; but when Buonaparte is to be defended, respect for the Edinburgh Reviewers, who back Dr. Clarke,
alters the case so far, that a question unanswered, because it never
was asked, must be conclusive against the fact solemnly stated;
and the positive testimony of a respectable sworn witness, who was on
the spot when murder was committed, must be invalidated by the
negative evidence of another who was at the time a thousand miles
off, but chanced some years after to pass the spot, where, not
choosing to investigate the affair, or not hearing the story from
the self-volunteered ghost of the murdered, he *wisely* concluded,
and forsooth must be believed, that no murder was committed!!

From what has been stated in the preceding pages, in explanation of the factious motives and views of the Edinburgh Reviewers and their friends, no one can be at a loss, why they have uniformly endeavoured to fix the whole odium of the partition exclusively upon Russia, treating Austria and Prussia, who participated in the spoil, with comparative lenity ; or why, profiting by the event, which excited every where so great a sensation, they have, in contradiction to themselves, held up, as highly dangerous, the Russian ambition, which, on other occasions, they have affected to despise. It is sufficient to observe, that the influence of Russia was inauspicious to their interests at home ; and, therefore, they resolved to make war upon her moral and political character, in which resolution they have ever persevered.* Unfortunately, their efforts were but too well seconded by the ferment which previously existed in the publick mind, and in such a degree, that Mr. Pitt himself, hurried away with the torrent, appeared in opposition to Russia, and by checking her progress in Turkey, unwittingly contributed to the raising of that stupendous fabrick of French grandeur, which he afterwards in vain laboured to demolish. Painful experience, though too late, has proved that he was mistaken in his calculations, and that Russia,—there being a necessity, as long as France exists, of a counteracting check to her aggrandizement,—is the only power (not Austria, as the Edinburgh Reviewers feigned) capable, with the moderate support of others, to give this beneficial check. She is also the most eligible for this purpose ; for, whatever her ambition or projects may be, it is certain, that lawless and wanton usurpation, for the mere pleasure or lust of conquest, is totally incompatible with her essential interests and pursuits. The increase

* On one occasion only they deviated from this resolution : it was when Mr. Pitt declared himself against Russia, and thus converted Mr. Fox into her temporary friend ; which plainly shows that all the actions of the dangerous faction to which the Edinburgh Reviewers have attached themselves, have no other object but to oppose the existing administration in order to get the government into their own hands,

of population, progress of commerce and civilization, promotion of agriculture, and every improvement relating to
internal industry, all depending on the state of peace, have
been her constant and undeviating aim ; consequently, a
power, with so sacred an object in view, never can be destructive to her neighbours. Add to this, that she is in her
natural state, and not excited, like France, to that feverish
energy, which is and must be kept up by forced and artificial means, fatal to all who breathe within her atmosphere.
The very circumstance of Russia's dividing her spoils with
others, proves her far less dangerous than France, who, whenever she can, takes all ; while the improved condition of
Russian Poland, though it shows the ambition of Russia,
shows also the more lenient nature of that ambition, robbed
of its terror by the good to which it aspires. It is clear,
therefore, that whatever Dr. Clarke, or the Edinburgh Reviewers, have advanced to the contrary ; they themselves,
if not influenced by opposite interests, would confess that
Russia is the only continental power, whose moral and
physical energies, with a view to the past, present, or future, entitle her to the hopes and confidence of Europe.
Here I pause—to take my leave—I hope, for ever,—of
the Edinburgh Reviewers, and all their friends and distant
followers, who love to put on the cumbersome armour of
their masters, and strut their short hour, dealing their puny blows, rather to amuse, than to harm any one. If I
have, in any degree, exposed the dangerous influence, and
the pernicious tendency of that partial, unjust, and unsound
political doctrine, which the critical junto have ever laboured to propagate—of that despotick and unnatural sway
which they have usurped, and are exercising over the
minds of certain persons here, well disposed in the main ;
my object is obtained, in the absence of which I should
never have thought it necessary to answer them, my answer being already furnished, as far as Russia is concerned,
by the facts carved with her sword on the backs of her invaders. If I have, however, failed in the task I have im-
posed upon myself, either from my own inability, or the
difficulty of shaking deep-rooted prejudice in others ; I

shall console myself with having made the first attempt of the kind in America, and leave the more successful prose-cution of it to men of greater talents, some of whom doubtless must agree with me in my views and opinions of the Edinburgh Reviewers. I care not who the physi-cian is, as long as there is a hope, ever so remote, of cur-ing the disease.

CPSIA information can be obtained
at www.ICGtesting.com
Printed in the USA
BVHW04135806I218

534939BV00011B/107/P

Here ends Jasmin ʻIolani Hakes's
The Pōhaku.

The first edition of this book was printed
and bound at LSC Communications
in Harrisonburg, Virginia, in January 2026.

A NOTE ON THE TYPE

The text of this novel was set in Adobe Caslon Pro, a typeface inspired by the original Caslon serif typefaces designed in 1722 by William Caslon. Caslon's types were based on seventeenth-century Dutch old-style designs, which were then used extensively in England. Because of their practicality, Caslon's designs met with instant success. The first printings of the American Declaration of Independence and the Constitution were set in Caslon. The Adobe Caslon Pro is a revival of the original font designed by Carol Twombly. The OpenType Pro version merges formerly separate fonts and adds both central European language support and several additional ligatures.

HARPERVIA

An imprint dedicated to publishing international voices,
offering readers a chance to encounter other lives and other
points of view via the language of the imagination.

ABOUT THE AUTHOR

JASMIN `IOLANI HAKES was born and raised in Hilo, Hawai`i. She is the author of the novel *Hula*, which was named a Best Book of the Summer by *Harper's Bazaar* and *Elle*, a best debut novel of 2023 by *Booklist*, was a winner of *Audiofile*'s Earphones Award, and was named *Honolulu* magazine's Book of the Year (About Hawai`i). Her essays have appeared in the *Los Angeles Times*, *Literary Hub*, and *The Sacramento Bee*. She is the recipient of a Writing by Writers Emerging Voices Fellowship and residencies at Hedgebrook, Virginia Center for the Creative Arts, and Storyknife. She lives in California.

A NOTE FROM THE COVER DESIGNER

The process of designing this cover for a multigenerational saga about the founding of Hawaii, the women who carried its legacy forward, and the book's central relic was multifaceted. Having already established a general aesthetic for the author with *Hula*, we explored cover designs both considering that cover and separate from it, knowing *The Pōhaku* is a distinct novel with its own unique characters and themes.

Some cover designs skewed more historical and others more fantastical, until we narrowed in on the visual that connects the entirety of the text: the stone. The pōhaku is centralized, depicted in a dark, heavy tone in contrast with a lush background, illustrated in a similar style as that of *Hula*, for consistency. The printed jacket's lamination helps elevate the design further. Rich colors look and feel vibrant and full of life, much like the pōhaku itself.

—*Stephen Brayda*

BRIDE OF THE FAE PRINCE

Am I? I reach up to pinch my cheeks, but Jacquelle leans over quickly and pinches them for me. I wince. If that doesn't bring color to my skin, nothing will.

Father sighs. "If only you weren't still waiting on your bloom."

I should have expected the comment. It stings nevertheless. When I'd stood before the mirror, my maids putting the final touches on my hair and ornaments, I'd thought I looked rather nice. I'd braved a tentative smile—had even imagined Father looking at me and saying that the elusive "bloom" that had graced all four of my sisters before me had finally arrived. Never would I claim the beauty of my two older sisters, but I'd certainly thought myself more than passable. At my sides Vivienne, Jacquelle, and Yvonne all nod sadly. Agreeing with Father.

"I love that raspberry color on you!" squeaks my very enthusiastic youngest sister, Amelia. She breaks out of line to hug my arm, leaning her sweet curly head on my shoulder with no care for the elaborate styling of her hair. A few wisps have escaped, but it only makes her lovelier, freer, brighter. "The king of Enslington will be besotted by the end of the night."

I can't restrain myself from giving her arm an affectionate pat.

"I hear he's very serious and bookish," says Vivienne. "Make sure to keep your mouth shut unless you're eating or smiling. Bookish men do not like chatty women."

Jacquelle takes my other arm, and before I know it, we're marching steadily toward the ballroom, leaving behind the warm fire in the grate, the red wallpaper, and Great Grandfather's pink-cheeked glare. I want to set my heels into the ground like an ornery old goat and make them drag me. But that is hardly becoming of a princess.

This is my one duty, after all.

I was born to flutter my eyelashes at foreign kings and princes in hopes one of them would approach my father for an alliance. Then, when the days of my betrothal are fulfilled, my new husband will send for me and I will follow him to another kingdom, leaving my home behind.

3

"Vivienne always says men like quiet women," Jacquelle whispers in my ear conspiratorially, giggling. While Vivienne may have the most beautiful face, all four of us are jealous of Jacquelle's perfect figure. "She only says that because her betrothed could be her grandfather and needs to nap every two hours. King Ilbert is younger, like my betrothed—he's not even forty-five!—and young men like clever women. You don't have to say much, just so long as what you say surprises him. If you make him laugh, he'll adore you."

"Make him laugh," I mumble. "Don't talk too much."

"Remember, clever and unexpected." She winks.

"Not *too* unexpected," says Vivienne with a frown, apparently overhearing us.

"Keep your distance from your sisters during the ball," says Father, dismissing Amelia and offering me his elbow. "You don't want to be overshadowed."

Having Father's attention on me is almost as disconcerting as the ballroom doors looming at the end of the hallway, flanked by statuesque manservants in starched livery. Light from beneath the doors illuminates the tiled floor we tread, and not much else, save the sconces lining the wall.

Don't talk too much. But *when* I talk, be clever. Don't be overshadowed. My gut churns. A wave of lightheadedness passes over me. I should have made myself eat earlier. It seemed impossible to eat when everything in my life has culminated in this one moment— where I will either succeed or fail completely.

"Remember how you move," says Yvonne, with a bump of her hip against mine. She has long blonde hair, like mine, wavy rather than curly. "Elegantly, to show your good breeding. You want just enough *allure* to intrigue him, but not enough to ruin your reputation."

I don't want to listen to any of Yvonne's advice. She's betrothed to the most spoiled and cruel man I've ever laid eyes on. He is the nephew of the king of Osremer, and while he waits for his uncle to die and pass the crown to him, he throws revelries said to rival even the debauchery of the fae.

I'm not naïve enough to hope for a young and handsome husband. All I hope for is kindness. I will take a homely man thrice my age if he is kind. Since Yvonne's marriage was arranged, however, I've almost stopped hoping for even that. As horribly selfish as it was, that night I'd wept beneath my covers—so desperately *relieved* to be the least lovely of my sisters. Otherwise, it would have been me the licentious Osremer heir wanted.

My gown rustles with every step closer to those doors. Closer to my future.

This isn't about me. This is about my people, my kingdom. It is my duty to make whatever match will most benefit my people. My *one* duty.

Scurrying footsteps make all six of us glance to the left, where a lanky young runner scurries down a hall toward us. Father lets go of me, stepping aside so the runner can deliver his message. Vivienne feigns disinterest, pointedly looking away while Jacquelle and Yvonne lean closer to catch a scrap of it. Amelia slips back to my side, and her presence is a sweet comfort when my heart cannot stop pounding.

Father's jaw sets grimly as the runner leaves and he straightens. Amelia's hand threads through my elbow. "What is it, Father?"

"News about those nasty fae?" asks Jacquelle.

He draws in a deep breath, his shoulders tight. "The fae have expanded the Long Lost Wood a mile along our southern lines in just the last week." His gaze falls right to me, and I cannot find his brow fast enough before I'm trapped in his attention. I tense. "We need this alliance tonight, Isabelle Louise. If the king of Enslington allies with us, he will help us fight the fae and keep our people safe."

"But King Ilbert doesn't have a large military," says Vivienne—the closest any of us would dare to asking the real question.

How is this alliance even going to help?

Father purses his lips. "Well, I still have one more daughter, don't I?"

Amelia goes a little pale, then gives an uneasy giggle. "Of course, Father," she says, and though her smile is bright enough to fool him, I don't miss the delicate crease on her forehead.

We reach the ballroom door much too quickly. I follow the crack between the doors all the way up to the ceiling until my neck is arched back. It's the only distraction to be found from the black spots dancing across my vision.

"Places, girls!" barks Father. We hurry to obey, arranging ourselves in order from eldest to youngest—except that I am to take Father's side.

Because I am the sacrificial bride of tonight's dance.

Don't talk too much. Be clever and unexpected—but not too unexpected. Show my refined breeding. Be alluring—but not too alluring. Don't be overshadowed.

I swallow heavily. This is my chance to save our kingdom. This is my chance to serve our people. It doesn't matter what my future husband is like. I don't have to like him—I'm only marrying him. It's not as though I'll have to talk to him much after we wed.

I just have to make him like me. No matter his preferences or inclinations—I must be what he wants. Whatever that is.

I hope he doesn't want a bold bride, because my hands will not stop shaking. Father glances down reprovingly at them as they shudder on his arm.

"Remember," he whispers, leaning down toward my ear. I stiffen, focusing my attention on the dancing flame in the sconces lining the wall. "Use your feminine wiles and he won't be able to resist you. Do nothing to scare him away."

I turn frantic eyes up at him, balking as the great doors open outward and the announcer cries, "King Roland! Princess Isabelle Louise, Princess Vivienne, Princess Jacquelle, Princess Yvonne, Princess Amelia."

"My what?" I whisper back.

"Your *wiles*," Father hisses through a smile.

BRIDE OF THE FAE PRINCE

I force my lips to tilt upward as my mind spins frantically. Won't feminine wiles ruin my reputation? Do I even have any?

The ballroom opens before me in a gleam of a polished wood dance floor, golden arches and chandeliers against gold-inlaid embellishments of the slate gray walls and ceilings. Filigree detailing curves along the edges of the arches, the pillars set with statues of maidens and young men in various states of dress.

It's so *full* of people, of colors.

The air turns stifling. Father's arm is suddenly my only refuge, and I try not to give into the impulse to shrink a little closer to him. I hide my free hand in my skirts. Will King Ilbert notice how sweaty it is through my glove? Will it scare him away?

I scan the crowd of courtiers, foreign ambassadors, and two of my sisters' betrotheds. *So many people.* Will I know King Ilbert when I see him? He's a king—and yet my vision swims with lace, brocade, taffeta, piled mounds of curls. I steal a glance up at Father, and find his eyes pinned in one direction of the room. I follow that look, and find a tall man setting his goblet on a servant's tray, his full attention at Father. And me.

He's actually . . . not unbecoming. He is certainly much older than I am—twice my age, at least, I suspect—but he has a crisp jawline with a close-shaved beard and a pronounced brow that bespeaks a firmness of character, yet a pleasant smile stretches across his face when he sees us. A kind smile, I think?

My heart lifts.

He has a very slender physique, which is quite the opposite of my eldest sister's intended, and he wears a finely cut black surcoat with elegant fur trimmings, a belt inlaid with precious stones, and a dark blue silk coat.

I can certainly do worse. Much, much worse.

Unless, of course, that jawline and smile are simply a veneer for a dark, malicious personality. How would I know? A ball doesn't require masks to be a masquerade.

7

He bows cordially before my father, and then smiles at me as he bends over my hand. Can he feel how sweaty it is? Can he feel how I tremble? I hope my smile isn't as wobbly as my fingers are in his.

"Princess Isabelle Louise," he says, and flashes that warm smile once more.

I curtsy and my ankle chooses that very moment to waver—but I don't fall or sway too grandly, I think. Does he notice? Am I to be known as the clumsy daughter as well as the least comely?

"Y-your Majesty," I say. My voice is rail-thin and weak.

"May I have this dance?"

My gut thrills. Are my so-called wiles working? Or is he merely fulfilling his obligations? I hardly trust my voice when my stomach is so unsettled, so I merely incline my head and offer a smile.

"Music!" calls Father, clapping his hands at the musicians in the corner on their dais. Immediately, a lively tune begins on fiddles and an Algravian imported harpsichord. I glance at him as King Ilbert draws me toward the center of the ballroom. I'm almost struck dumb at the sight of Father—*beaming* at me.

Something lifts in my chest. I haven't ruined this yet. Maybe there's hope that by the end of the fortnight, I'll be betrothed.

I peek up at the King of Enslington and smile.

He's not looking at me. He's looking over my shoulder at the rest of my sisters. As though sensing my gaze, he glances down, finds me staring, and offers me another polite smile. He slips his hand around my waist as we take our position on the dance floor. Inside, my courage falters.

"This is a lovely dance, Princess Isabelle Louise," he says. "You look beautiful in your gown."

Just like that, my courage is bolstered. My eyes widen as warmth floods my cheeks. He thinks I look beautiful? I catch myself just before I let my smile widen too much. What if he thinks me vain? Vivienne always cautions against appearing too vain. Or perhaps his compliment was merely polite. I duck my head and say a quick, "Thank you, Your Majesty."

We dance in silence. Sweat slides down my neck with each passing minute. My glove must be growing damp. I desperately hope he cannot feel it!

Wait—was I supposed to compliment him back? It must be too clumsy and awkward to compliment him now, after the silence has lasted. Besides, every compliment that comes to mind is vastly more awkward than the silence. *You are far better looking than Vivienne's betrothed. I am glad that you are not an old drunkard. Your smile makes me think you have a good heart.*

I hate this. *I hate it, hate it, hate it.* There is only one thing I was born to do—only one thing in my entire life I need to accomplish, and it comes down to this moment. What if I mess everything up? What if I cannot get him to agree to marry me? What if he decides he'd rather marry some other kingdom's princess? What if my father cannot get the military strength he needs to fight the fae? What if people die because I cannot flirt well enough?

"Are you quite well? Forgive me, but you seem rather pale," says King Ilbert.

Pale? I'm pale? Heavens, I need to pinch my cheeks! But we're dancing, and my hands are occupied. Does King Ilbert mind pale wives?

Deep breath.

This panicking will get me nowhere.

"I'm quite well, thank you," I manage with a breathy laugh.

The lightheadedness returns in full force just as the dance ends. King Ilbert takes my hand, his brow knit with something akin to concern as he leads me off the dance floor.

"The dance is too lively for you, Highness," he says as he escorts me to a seat. "Please rest and allow me to fetch you some refreshment."

Fetch me—what? He releases my hand, straightens, and strides off toward the table of refreshments and crystalline goblets of wine. I stare at him unabashedly. He's a king! Does he consider himself a servant? *None* of my sisters' betrotheds would offer to fetch refreshments. They'd hardly even order a servant to take care of her needs.

I stare after his tall form as he threads his way through the crowds, and try to pretend I don't notice my sisters staring bug-eyed at me from across the ballroom.

"Why did he leave you here?" demands a familiar voice from my right.

I flinch, turn, and find my father's disapproving frown hovering above me. "H-he is g-getting me r-ref-ref—something to drink."

"Did you stutter like that during the dance?"

I duck my head. "N-no."

That frown won't go away, no matter how much I long for it to. I want him to beam at me again—but what does it matter? It wasn't as if it could last, and I was a fool if a tiny part of me thought it might.

"I thought you'd made progress on it with your tutors," says Father.

I made extremely great strides with my tutors. When I was a child, I could hardly get a full sentence out. Where I am now is as different as day is from night. But Father wouldn't know that, because a small stutter always returns when I'm anxious.

"If only you were as beautiful as Jacquelle and Vivienne," says Father with a sigh and a shake of his head, as if this is already a lost cause, "you could make up for this defect. Just avoid talking."

"King Ilbert called me beautiful," I want to say, but I bite my tongue.

"Ah! King Roland!" cries a loud, boisterous man from behind Father.

Father loses his frown as he turns, donning his warm kingly mask before I can even blink. "My friend, you are enjoying the wine, it seems."

"Best wine on the continent," cries the newcomer, whom I recognize to be the old Prince Brochfael. He is heir to the throne of Algravia, and with the king on his sickbed, it is looking like Prince Brochfael will have some years on the throne before his enthusiastic relationship with alcohol takes him to an early grave. "Perfect for coping with these rambunctious fae, now isn't it?"

"Aye, aye, my old friend," says Father with a laugh.

Prince Brochfael spots me, sitting where King Ilbert left me. "Ho! Is this the young bride you've promised me, Roland?"

My eyes go wide. Father's face flushes scarlet, and he doesn't meet my gaze. My stomach drops straight to the floor.

Father promised one of us to Prince Brochfae!? But the man has five wives already—and legend certainly has not regaled him as a magnanimous husband. I press a hand to my unsteady stomach.

"I have two unpromised daughters," says Father, and places a hand on the prince's shoulder to guide him away from me. "It is yet to be seen which will have the privilege of your attentions."

Even though my knees wobble so much I cannot trust them to hold my weight, should I decide to stand, I understand now. Prince Brochfael will soon be king, and Algravia is known for its military strength. Father cares enough about his daughters to give us a chance at something besides being the sixth wife of an old drunk, but whichever of us couldn't secure an alliance fast enough . . . Father had this arrangement made. A backup plan.

If I succeed in this alliance, that only leaves—

Amelia.

King Ilbert is taking much longer to get my refreshments than I would have imagined, but at this moment, I'm relieved he hasn't returned. I pinch my cheeks quickly, hoping that will fight the sudden pallor that must have overtaken my features.

Then, after flattening the skirts of my gown to disguise wiping the sweat seeping through my gloves, I arch my neck as I try to locate where King Ilbert has disappeared to and if he is returning soon.

There he is!

A goblet glitters in his hand like a garnet, the facets of crystal catching the light. Is that for me? The thought should ease my cascade of panic, but it only sharpens it. If I am successful tonight in *ensnaring* his interest, then sweet little Amelia, barely eighteen, will be wed to Prince Brochfael.

I cannot let that happen.

I *won't.*

It would be better if I went, if I were the prince's bride. I can endure being the sixth wife of an old goat, can't I? If I am one among six, then surely I can slip notice easier. Perhaps he has mistresses, too. It wouldn't be too bad for me to handle.

But I cannot let Amelia endure it. She needs someone kind, someone gentle. Someone who will be good to her and cherish her sweet nature. I glance down and discover my hands are fisted in my skirts. Quickly, I smooth them out again.

King Ilbert is caught in conversation. I cannot see to whom he speaks, only that it's someone by the food, and he's smiling. The smile is just as kind as the ones he gave me, but it seems both easier and warmer. To my surprise, he throws back his head and laughs—and it's quite a nice sound.

The selfish part of me wants to pretend I heard nothing from Father about his deal with Prince Brochfael. I want to be blissfully ignorant as I waltz into a marriage with King Ilbert. I want to claim the attentions of what seems to be a genuinely kind, not altogether unbecoming, and relatively young man.

I give myself a little shake. I *won't* do that to Amelia. I'll leave the ball if I must. I'll go get her and make sure she's introduced to King Ilbert.

But when I stand, I'm finally able to see who the king of Enslington is speaking to. Who he is laughing with as he holds the goblet he went to fetch for me.

It's Amelia.

Her face is flushed a pretty pink, complimenting her lovely lilac gown. With that grin on her face, she's a vision. She isn't like my other sisters, however, who are beauty paired with some starkly undesirable trait such as coldness or arrogance. She is beautiful, and she is pure goodness.

I don't blame King Ilbert for the way he looks at her.

It's a relief, truly. From what I've surmised, he is a good man. And more than I care about my own happiness, I want Amelia to be cared for.

A boisterous laugh echoes from farther down the ballroom. I follow the sound until my eyes land on a guffawing Prince Brochfael smacking a servant so hard on the back, he nearly trips and drops his tray of empty goblets.

I'm yet again only relieved I'm no great beauty. Beauty will do me no service in my future.

CHAPTER 2
THE PRINCE

W HAT DO YOU mean, you won't marry her?" High King Faradir demands. "Princess Listhra is beautiful, of royal descent, a worthy warrior. You cannot have an objection to her."

I make a show of tapping my temple, and then say, "Well, if you require a more *literal* clarification, I mean that I shan't pledge myself to her, and likewise shan't accept any pledge from her. Is that clearer? Or shall I rephrase, Father?"

The court goes so quiet that one very soft intake of breath from somewhere is audible to everyone. I keep my arms crossed as I lean against one of the columns circling the High King's throne and framing the sacred stream.

The princess in question glares daggers at me with her cold, amber eyes and tosses a lock of rich brown hair over her shoulder. Why she expected anything else from me is utterly beyond my comprehension. She should know not to take it personally. After all, she's only the sixth fae woman I've declined.

Faradir has this idea in his head that the more royal and beautiful a woman is, the more tempted I'll be. As if I'll fall for his thinly veiled tricks. I know him better than that. And he should know me better.

Alas, here we are.

"I mean no offense to you, of course, Princess Listhra," I say with a sigh, tossing a grin her way. "You're fairer than the choicest firerose at the height of its bloom, and there's nothing you could have done to change my decision."

She stands opposite me, to my father's left, so I'd have to stare at my shoes to miss her glare. She shouldn't be so offended; she knows I'm not lying, so she ought to appreciate the compliment.

Perhaps my words were too harsh. But really, I'm almost insulted that the High King thought I'd agree to marry her. I've known her for some two hundred years now, and every year she grows more intolerable than the previous. He'd have had much greater success with the second woman he'd brought me. She wasn't pretentious like the rest. I kind of liked her. One Oleria, second youngest princess of the Ildreer Court.

But, thankfully for me and her, Faradir gave up too early.

"Such a callous rejection, my son," says the High King, glaring at me as he leans back against his throne, fingers drumming on the armrest. His glamour makes his face shine like a small sun, his hair falling like molten gold over his shoulders. It's stark against his brilliant white robes. He's daylight compared to me. I take after my mother's darker complexion. She was a daughter of Nothril—the Night Court. "Can you not see how you've broken her heart?"

I lift one eyebrow at him. "She cares as little for me as I do for her."

That familiar expression crosses the High King's face, and I barely have time to fortify myself for that dreaded *snap* of his fingers. It resounds in the stone-still court full of bodies who dare not make a sound.

I almost don't want to look, to see who it is this time.

But everything I do and don't do is carefully measured by the man on Faerieland's throne, and I owe whoever will be dragged through that door my acknowledgement—and my promise.

BRIDE OF THE FAE PRINCE

I sigh loudly. "This again? How predicable, Father."

Without moving my body, I swivel my head with disinterest toward the opened double doors. Toward the winged, fanged guards and the human chained in iron between them. They drag him forward, the crowd of Faerieland's denizens rustling to allow them through. Though, the stench of iron and the pulse radiating from the chains are the stronger motivations for the swift withdrawal.

The guards drop the man to his knees at my feet, facing the High King.

He looks up at me. *Calver.* Who, just this morning, laid out the very clothes I wear now.

"Prince Trenian," he whispers, then, without lifting his head to the dais, "High King."

"How long have you served my son?" the High King asks him.

He swallows. "Thirty years, Your Majesty."

I hold his gaze for a heartbeat. *As I have vowed to you, so I will do.* A fraction of relief passes over him as he bows his head, his shoulders shuddering.

"A pity to lose one so faithful," says Faradir as he lifts his hand into the air and curls his fingers inward.

I don't flinch at the gasp of pain at my feet, at the wheezing gurgle. I don't flinch when the High King squeezes his hand into a fist, and a snap cuts through the air with sharp finality.

"It grows more challenging to keep my household staff populated these days," I drawl, flicking my wrist for the guards to drag away the body. I don't let my eyes linger, lest I betray the fury burning through my blood. "High turnover is simply not good for morale, and the effort to train new staff is quite headache-inducing."

"Then perhaps you ought to behave yourself, Prince Trenian, and I won't find myself needing to discipline my wayward son," says Faradir.

I chuckle. "Come, come Father. You always present it as if it is you against me. Surely, we can arrange something mutually beneficial. You want me to have a son to continue the line of succession." My mouth twists into some semblance of a grin. "And *I* want to fall in love."

Both are as close to lies as I dare come.

The High King leans forward on his throne, steepling his fingers and regarding me with obvious disdain. "And *what*, dear son, would you propose?"

Now I do grin. "A bargain."

"Does this face"—Faradir points at his own—"look like it thinks you can give me a bargain I'd be tempted to take?"

Something shoots like glee down my spine. All these years of collecting my pieces and planning. Now the board is set, and it's finally time to make my first move. "What if I promised to travel to the Fae Courts and return with a bride by the height of Lulythinar?"

At that, the High King's eyes sharpen. "And what would you want in return?"

"The freedom to choose whom I wish."

He leans back against his throne. The hand that just choked the life out of my manservant strokes his long beard, considering. His eyes glaze and narrow, as he works through what angle I might be hiding. He'll suspect that I am intending to marry, but not sire an heir, and will add a clause to the bargain involving an heir.

"Tell me what you intend to bargain," he says.

"That I will travel to the Fae Courts and if I do not return with a wife by Lulythinar, I will marry whomever you choose."

His attention shoots back to me. He wasn't expecting my offer to marry whomever he chose—or so soon. Lulythinar isn't even a fortnight away. I rein in my impulse to let my grin morph into something smug and calculating. He knows I'm angling for something. He knows I'm trying to trick him.

But it's an offer he cannot refuse.

"And you, Father? What would your bargain be?" I ask.

The High King thinks for several long minutes. Our audience is silent, so silent I can almost forget that dozens upon dozens of fae creatures stand just behind me, hanging on every word. How quickly news will spread among Faerieland that the heir to the High King's throne is searching for a wife. The Courts will be insufferable. I'll hardly be able to attend a social function. Faradir had better add a clause about royalty. I would, but I must leave *some* gaps in the bargain for him to fill.

At last, he says, "I would give you freedom to choose a noble for your bride from among the Courts by Lulythinar. If you are without possession of a bride by Lulythinar, you will marry my choice and fulfill your duty to produce an heir. Will you accept the terms of this bargain?"

I turn his words over in my head, searching for any hidden tricks. It's as I anticipated from him. "I will," I say, and raise my fist into the air as Faradir does the same. "Let it be so."

"Let it be so," echoes the High King.

Light flares between us, sharp and fast, like a brand. A tattoo appears, adding to the ones on my right arm. This one spans the width of my wrist, the picture of a crown broken into two pieces. One to represent his side of the bargain, and one to represent mine.

If the High King is alarmed or concerned by the appearance the bargain has taken, he doesn't show it. I lower my fist, leveling a hard gaze at him. "I shall request leave of Your Majesty. It appears I have a bride to woo."

Without waiting for said leave, I push off the pillar and stride through the throng of onlookers. They part like the Maltun Sea at Lulythinar before me, and I don't spare a single glance for any of them. The guards go to open the doors, but they move too slowly. I plant my palms and shove them open, letting them swing closed in my wake.

It's technically true that the High King wants me to continue the line of succession, and that I want love. Faradir wants me to a have a son, but not because he wants a reliable line of succession. It's because no one can sit on the throne who doesn't have Great Kings' blood flowing in their veins, and I am Faradir's only heir.

The moment I have a son, however . . .

Nothing will stop him from killing me.

And I do want love. I'm just not stupid enough to risk such a thing. There will be no love in the marriage I'm about to begin.

My steward, Edvear, another lowborn fae about half my age with yellow cat eyes and nubby horns protruding from his curly brown hair, hurries to my side once we're far enough from the throne room. I twist my fingers by habit, throwing a quick illusion spell around us to conceal his voice.

"Master Ash!" he says quickly. "Calver—"

"Is dead," I reply briskly. "Tell Sanak there is an opening available if he wants it. I'm going to need a small unit of warriors, and if you could find some lowborn fae to dress up as dignitaries, that would be just fabulous."

"Lowborn fae? My lord?"

"Yes, just anyone the High King doesn't care about. Ones that won't care about humans." I glance sidelong at him, at the shellshocked expression he's trying to hide, and I find myself softening. "You look crestfallen. What is wrong, my friend?"

"It's just . . . Calver is dead. He was your manservant for over thirty years. Were you not . . . Are you not . . . ?"

My jaw clenches, and I pick up our pace. "He knew what he was getting into. He accepted the position, knowing the risks, just as you did. I will fulfill my vow to him, as I will fulfill my vow to you should anything happen. My father won't relinquish the idea that I'm attached to my staff, so he fancies executing them whenever he disapproves of me."

Edvear looks ahead, toward the door of my quarters that we swiftly approach. His nostrils flare as he avoids my gaze. He says nothing, but he's served me long enough that I recognize the expression.

I lower my voice. "The High King killed my mother when I disrespected him once. Was I glad when that happened? No. Neither am I glad for Calver's loss. Or any loss. That doesn't mean I am foolish enough to risk forming attachments. Truly, Princess Listhra ought to be relieved I rejected her instead of pouting like a pixie without a book to nibble on."

Edvear nods once, back to his composed self as he bows his horned head and opens the door for me. "Which court are we going to first, then? I'll have our servants pack us."

"No need to pack. We're not staying at any of them."

He stops short. "What? But the bargain—"

"We'll travel to a few of them—shouldn't take but an hour to go through the portals and come back. We will stop at the Nothril Court and enlist Prince Rahk to come with us."

Edvear closes his eyes. "I should have known. Where are we *actually* going?"

I pause in the doorway of my study, my hand gripping the lintel as I turn and flash a wicked grin at my steward. "We are going to the human world."

CHAPTER 3
THE PRINCESS

KING ILBERT OF Enslington solicits Amelia's hand in marriage in the week following the ball. It soothes a great deal of my anxiety to see her on his arm, laughing while he smiles down at her.

Though I've hardly spoken two words to him since our dance, I've observed them together from across a ballroom, or from my window overlooking the palace gardens. There seems to be an unusual and utterly delightful warmth between them. He has been nothing but cordial and gentle with her.

I'm not sure if I dare hope for something akin to true affection for my youngest sister's marriage. It is a foolhardy wish, one that asks for nothing but disappointment. And yet, I cannot quell it. Perhaps she *will* be happy after all.

I look down at the little potted plants I keep on the sill. My rosemary isn't doing so well, but the lavender has a small purple bloom that gives me a burst of happiness every time I see it. "It seems everything has turned out as well as could be hoped for. For Amelia,"

I tell my plants. I give one of the tiny leaves of my thyme a gentle stroke. Yvonne always teases me that it's not my job to grow food for the kitchen, but I ignore her. These aren't for the kitchen. They're mine, and tending to them brings me great satisfaction.

Maybe if I get married someday, I'll be allowed a full garden of my own to care for.

When King Ilbert's party departs at the end of negotiations, and wedding preparations begin in earnest, the hopeful optimism I have for Amelia's future is replaced with a sick knot of dread for my own.

At breakfast that morning, I don't manage a single bite. I keep my gaze on my full plate, pushing food around with my fork to hide how close I am to throwing up. The smell of roasted chestnuts and steamed porridge cloy in my nostrils, making me sicker.

Finally, Father sets down his teacup.

I brace myself. Here it is.

I'm doing this for my people. I'm doing this for my people. I will fulfill my duty. We need this alliance. And I will be grateful that it is me, and not Amelia, who faces this fate.

The assurances only barely take the edge off the dread flooding me from head to toe.

My sisters set down their utensils, pulling their hands demurely into their laps. We await Father's words. I stare at the painted gold edges of the china set before me.

"Now that we celebrate the successful negotiation of our dear Amelia to King Ilbert," Father begins, smiling at us, "tomorrow will begin negotiations with Prince Brochfael for Isabelle Louise's hand."

"Prince *Brochfael?*" Amelia chokes. "Of *Algravia?*"

"The same," replies Father curtly, shooting her a reprimanding look for her outburst. "Algravia has the military strength we've needed for some time, to present a force against the fae."

Amelia's horrified gaze burns into my forehead, but I continue staring down at my plate. At the patterned, royal blue tablecloth. At my spoon still resting above the scalloped edges of my plate. She'll be furious if she finds out I knew about this.

"Yes, F-Father," I say, just as I said when he announced that King Ilbert was coming for a wife. Though my eyes remain downcast, I lift my chin slightly. My fate is all but sealed. I might as well approach it with dignity.

The door swings open and a runner comes straight for the head of the table. Father straightens, his brows drawing together with alarm as he pats his mouth with a napkin and stands. "What is it?"

The runner, a boy who barely looks fifteen with his gangly limbs, fumbles to pull something from his pocket. "Your Majesty. It's a missive from—from—"

"Don't stutter! There's already enough stuttering in this court. What is the matter?"

"It's from Prince Trenian of the Fae!"

Father's face turns ashen. He crosses the narrow room, snatches the note from the boy's hand. With a quick snap, he breaks the seal and reads it. His eyes scan over the paper, running in fast zig-zags. His lips part as he reads. Sunlight from the open window turns the black hairs of his beard a silvery gray.

"Well, it seems we are in luck," he says abruptly, folding the note back up and slipping it into his pocket. He keeps his composure mostly, but there's a fissure in it: a tick in the sagging skin beneath his lower lashes. "I'm glad we didn't rush into arrangements with Prince Brochfael."

"Father?" asks Vivienne, patting her full red lips with her napkin. "Whatever do you mean?"

I make the mistake of watching Father instead of keeping my attention on my plate. His gaze snaps to mine, searing my brain with an intensity that makes me want to crawl into a ball and die. I slip my hands beneath the table and fist them in my skirts.

"It seems the Prince of the Fae fancies a human bride. And I have one remaining unpledged daughter."

My sisters gasp and turn to look at me. I grab the arms of my chair with white knuckles to steady myself.

"You can*not* mess this up like last time," says Yvonne to me, with a meaningful look. "If you scare him off like you scared off King Ilbert . . ."

She doesn't have to finish the sentence. If I don't win over the Prince of the Fae, then I'll have to marry Prince Brochfael. But is Prince Brochfael actually worse than a fae?

I am some kind of sacrifice, aren't I?

"What . . . does he m-mean?" I find myself asking. "Why does the P-Prince of the F-Fae want a human bride?"

"It's not your place to ask questions," says Vivienne.

"Can a fae even mate with a human?" asks Jacquelle with a crinkled brow.

"Enough," says Father, waving his hands. "This prattle grows tiresome. We haven't a moment to spare. Prince Trenian is coming this evening."

"This *evening*?" bursts Amelia.

"We cannot possibly prepare a reception for him on such short notice!" cries Vivienne.

"We will have to, now, won't we?" snaps Father with no lack of ire. "And we will have to do something about Isabelle Louise's face."

"My face?" I reach up to touch my jaw self-consciously.

"What is there to be done?" asks Jacquelle. "She is what she is! It's not as though you can change her face!"

"You keep talking as if she's ugly!" says Amelia, who shoots to her feet and runs to my side, kneeling beside my chair and wrapping her arms around mine. Utterly heedless of the dozen protocols she breaks. "She is very lovely! I heard King Ilbert say so!"

"Not lovely enough, apparently," snorts Yvonne. "Did he propose to Isabelle Louise? Hmm?"

Amelia doesn't back down. She frowns, tightens her grip on me, and says, "You all are just being mean."

BRIDE OF THE FAE PRINCE

One of these days, my stomach will calm down enough for me to take a deep breath without fearing the loss of its meager contents. Right now, all I can do is cling to the armrests of my chair and wait for the world to stop spinning.

I wish I was already married. I could expend my energy adapting to whatever situation I found myself in, instead of this constant oscillation between hope and dread. It seems like every time I adjust to the newest prospect of my future husband, it changes—and for the worse. What could be worse than the almost-immortal son of our enemy?

I shouldn't ask such a thing. Fate loves to laugh at those questions.

A combination of bravery and stupidity washes over me long enough that I steal a glance at Father. He has this strange look about his face, as though a brilliant idea has just struck him. He stares at me, a tiny smile slowly curving his lips.

My heart drops.

"What, Father?" asks Jacquelle.

"What if it was our country's custom to veil our maidens?" He taps his chin.

"It isn't," says Yvonne.

He shoots her a look. "But what if it *was*?"

"Would we have to veil ourselves too?" asks Amelia.

"If you aren't married, you would be veiled. Along with every maiden of the court."

That is a lot of veils. Our poor tailors will be thrust into a frenzy.

Doubt niggles at me. Is my face truly so reprehensible that I must wear a veil to be tolerable? Perhaps I am worse than merely not as pretty as my sisters.

Wouldn't it be more off-putting for a fae prince seeking an alliance to not see the face of his bride? What if he marries me, finds out who I am beneath the veil, and then starts an outright war against my father for deceiving him?

"This could be the answer to everything," Father is saying. "If he wants a wife from among my daughters, then I will have sway in the negotiation. Perhaps we can come to terms about the encroaching border. We could negotiate peace. You"—he turns to me suddenly, fixing me with an expression that terrifies me just a little—"could be our salvation, Isabelle Louise."

I stare at him, the blood draining from my head so quickly I might pass out. This is all happening so fast. Hardly a week ago, I was supposed to ensnare King Ilbert, and I failed. Now I'm supposed to ensnare the Prince of the Fae when he cannot see my face, and the fate of our entire kingdom depends on that?

Yes, I think I might indeed pass out. It seems the only sensible option at the moment.

"See to the preparations immediately, my daughters. I will write up a decree that must be dispersed to every member of my court. Veils for all of you—and any maiden who fails to wear a veil tonight must not be admitted. Isabelle Louise, I'm counting on you tonight."

With that, he strides out of the room, leaving behind his unfinished breakfast.

When he's gone, I breathe a little easier. Silence reigns like a tyrant around the table as my sisters exchange looks. Yvonne has gone back to eating. Vivienne looks as though she's running through lists in her head of what will need to be accomplished before tonight. Jacquelle slowly lifts a bite to her mouth, chewing absently.

Amelia sniffles, dabbing away tears.

I wait to stand until I'm sure I won't topple over, then make my escape.

"It's not that I'm unwilling," I say, fingering the new, lacy veil in my lap while a maid applies cosmetics to my face. Wearing cosmetics beneath a veil strikes me as the most ridiculous thing anyone has done today, but I suppose we must be prepared for blustering winds. "I am willing to do whatever, marry whomever, if it is for the good of my people. I simply cannot bear up under this pressure. What if I fail again?"

Amelia, sitting at my side with her own veil thrown back, says with a chipper tone that belies the heaviness of her expression, "Just be yourself, and he'll be unable to resist you."

That's easy for her to say. She is an irresistible person.

The maid finishes her work, bobs a curtsy, and slips out of the room, leaving me alone with my sister. Her gaze is hard on the side of my face, so I look up and smile at her. My fingernail catches in an eyelet of lace. I force my hands to be still.

"It'll work out," I say, as if the proclamation itself will make it so. "It'll be just fine. And to have peace? That will make it worth—"

The door opens.

"A letter for you, Highness," whispers the maid, holding out a little tray to Amelia, her nervous gaze casting between us.

Amelia snatches up the letter eagerly, murmurs her thank you to the maid, and is already ripping it open before the door shuts. Surprised by her reaction, I cock my head to one side, waiting as her eyes quickly scan the contents. Color rises into her cheeks.

"You cannot blush like that and not tell me who sent the letter," I say.

She looks up, and her blush only deepens. "Oh! It's . . ."

Her finger partially covers the address, but I read enough to blurt: "It's from King Ilbert?"

She looks up at me over the top of the letter, and the corners of her eyes crinkle in a broad smile. "I'm the most selfish girl that ever existed! I shouldn't be grinning like this, but—'

"What did he write?" I demand, leaning to get a glimpse. "Is it . . . a *love letter*?"

"Maybe."

"Amelia!" I cry, both shocked and yet strangely giddy, my own troubles immediately forgotten. "What did he write? You cannot keep it secret now—you cannot be that much of a tease!"

"You mustn't let the others read it! They mustn't know he writes to me!"

My mouth falls open. "This isn't his first letter?"

She hides behind her letter. "I should have told you!"

"How many letters?" I'm almost laughing now. "*How* many love letters has the King of Enslington sent you, Amelia?"

She murmurs something too quiet for me to hear.

"Amelia!"

"It's the fifth, alright? Are you satisfied now?"

I gape at her, at the silly grin she cannot suppress, the color in her cheeks. Is Amelia . . . *falling in love* with her betrothed? I can hardly believe it!

"Here. Read it. And I promise I have been dying to tell you, but it seemed so horrible of me when you are facing—"

"None of that!" I take the letter, smoothing it out as I read the king's elegant script. My eyes bug as I read line after line, until I stop halfway through and give it back. "He writes you *love poems*? He is a *poet*? He is obsessed with you!"

An uproarious giggle escapes her. "He is *so* obsessed with me! I've not believed it possible! But he writes me faithfully and tells me he cannot stop thinking about me!"

"Well done!" I cry, clasping her hand, unable to stop my own grin. "You have done what none of us have managed! Oh, how I have longed for your happiness!"

"It feels like a crime to experience so much happiness while you—"

"No," I say, cutting her off at once. "You must have no shame over how the cards were dealt. I have wanted nothing but your happiness, and seeing it now gives me tremendous joy."

Tears fill her eyes, but when she opens her mouth to reply, the door opens again.

It's a courier, standing rigidly tall as he announces, "The fae envoy has arrived, Highnesses."

My words die upon my tongue. I meet Amelia's wide-eyed gaze. Fingers trembling, I lift the veil. She stands, helps me situate it over my face so it falls just past my shoulders.

BRIDE OF THE FAE PRINCE

When I glance back at the mirror, a ghost stares back at me. A ghost in a royal blue gown.

"Deep breaths," Amelia says. "He'll love you, and he'll be kind and handsome and everything you've ever dreamed. I promise you."

She knows as much about the prince as I do. The only likely thing is that he will be handsome, as I've heard all fae are beautiful due to the glamours they wear. I'll likely never discover what lies beneath those glamours. But this is the son of the Fae King who encroaches on our borders, who threatens the safety of our people and our very existence.

He will not be kind to me.

It isn't as if Prince Brochfael would be kind to me, either.

It doesn't matter. It's selfish of me to consider my happiness when lives are at stake. The only thing that matters is doing what I was born to do: marry for the good of my kingdom.

We file into the hallway, down a grand staircase, until we meet with Father and the rest of my sisters standing by the entrance to the palace. It's almost a haunting image, of brightly colored gowns against the gold filigree detailing of an enormous painting of a gory battlefield, of dusk's half-light leaking through the curtains and mingling with the candlelit shadows, of my veiled sisters floating across the polished floor like wraiths.

I draw a deep breath through my nostrils.

It's time to meet my future husband.

CHAPTER 4
THE PRINCESS

FATHER AND HIS guards go outside first. My sisters and I hang back, and I lose myself in the midst of them, finding security in their numbers. When it's time to follow, we make our way to the outer bailey where the drawbridge lowers to allow entrance to whoever waits on the other side of the wall.

When I come to a halt, Father's back takes up most of my vision. He stands with his feet wide, shoulders back, his right hand gripping his left wrist behind him. A cold wind nips at my veil, and for once I'm grateful for the light warmth it offers.

From my vantage point, and through the sheen of my veil, I can make out two very tall men crossing the bridge on foot. *Very tall.* One is dark complexioned, the other lighter, his hair almost white. Which one is the prince? Is either of them the prince? I crane my neck to catch a better view, but it's almost impossible from behind Father. A few glimpses reveal pointed ears, long hair, sharp weaponry. More people file through the drawbridge until there's a full envoy in the courtyard.

When I peek back over my shoulder, I'm shocked to find our own warriors lining the pathway, inconspicuous archers standing at the ready on the parapets. Is Father preparing for an underhanded trick from the fae? Or does he intend to intimidate them as part of his negotiation scheme?

I try to stand on my toes, my curiosity overcoming everything else, when Vivienne grabs my shoulder. "Be still," she hisses.

I obey, swallowing my ire. *Fine.* I suppose I'll get to see the prince plenty enough if we wed. Through my narrow window between the arms of Father and one of his men, I can make out a sword swinging at the hip of the white-haired fae.

My heart pounds an erratic rhythm as the silence stretches through the courtyard.

"Prince Trenian," booms Father. "We offer you our warmest welcome."

There's a dry snort. "A welcome as warm as cold steel, apparently."

It's the darker one who speaks, the one I have trouble catching a clear view of. Prince Trenian. My attention narrows on the little bit of him I can see: one tan, bare forearm covered in tattoos.

"Forgive the precautions. We have not received an envoy from the fae in over a hundred years," says Father.

Prince Trenian snorts again and cracks his knuckles. An oddly casual gesture for a prince. "My father doesn't believe much in diplomatic relations." His voice is low, rich, and . . . *very* sardonic.

"I would take it that you differ from King Faradir?" Father shifts slightly, just enough that I can peer between the gap enough to glimpse a smile widening a dark, chiseled jaw. It doesn't strike me as a particularly *friendly* smile.

"My actions might suggest such a thing. Come, now—am I to stay on your doorstep until dusk, or shall you receive me with this warm welcome?"

Father sweeps his arm toward the palace steps, beckoning. A warm welcome indeed. The prince smirks and mounts the stairs, obliging the unspoken invitation. Father joins him, and my sisters and I fall into step behind them.

Finally, my view of the prince isn't obscured. His height strikes me again, along with his wide shoulders and broad frame. I was under the impression that most fae were bright, light, and lithely elegant. I wasn't expecting him to look like my father's choicest warriors—only far more beautiful, where they are scarred and brutal, possessing a predator's grace instead of a lumbering gait. His face is all chiseled edges and masculine definition, framed by his long dark hair.

Handsome would be . . . well, an understatement.

Maybe it is good I'm veiled after all. It's a bit hard to imagine someone like him ever choosing a mortal bride, no matter how beautiful to human eyes.

As if sensing my thoughts, he pauses on the top stair and turns.

Bright blue eyes shoot to find mine, piercing through my veil. They shine like cut sapphires. My feet halt on the lowest stair. I couldn't move even if I wanted to. Not while he's looking at me like that.

My sisters draw back, leaving me standing here. Alone. Bearing the enormous weight of his scrutiny. I cannot even breathe as his gaze sweeps me from the top of my veil to the tip of my toes.

His eyes slide to Father, whose face and body have gone taut. "Do I have the honor of beholding your daughter before me?"

My throat goes dry. I force my hands to stay still at my side and try to keep them from trembling. He can't see through my veil—so why does it feel as though his gaze runs over every feature of my face?

That gaze is not kind. It is calculating. It isn't hard to imagine that mouth twisting just slightly more into cold cruelty.

"This is my daughter, Princess Isabelle Louise," says Father.

Prince Trenian's attention shifts to him, then back to me. I freeze once more. His voice rings out with command. "Lift your veil."

I blink. Did he just address me directly? Nerves tingle down my spine to my fingertips. If I hesitate, he'll think I'm defying him. I cannot be defiant. I lift trembling hands toward my veil.

35

"It is our custom," Father says quickly, his tone arresting my movements. "It must not be removed until after the wedding."

The wedding.

That word shoots straight to my numb brain. Prince Trenian follows my movements as I drop my hands back to my side. He studies me for such a long moment, I lower my gaze.

Then, abruptly, he chuckles. "I suppose I'll allow the peculiarities. But please, spare me the trickery. I'm willing to make an alliance, but do not forget that I am heir to the land that encroaches on yours even now."

The air goes quiet so suddenly, I can almost make out the faint twang of an arrow nocking against a bow string.

Father offers an icy smile. "I can assure you, such a thing is impossible to forget."

The prince flashes a dazzling beam full of teeth that makes my feet wobble. "Excellent."

"We have prepared a banquet for you this evening. My attendants will see you to your chambers for your refreshment after your travels."

I step aside, leaving room for the prince's attendants to follow, including the other tall, white-haired fae with the enormous sword. He's even bigger than the prince, rough where the prince is elegant. I don't *want* to think how easy it would be for him to flick that sword and take off my head, but it's a little hard when he walks within a few feet of me.

These fae are terrifying.

I glance back up the steps to where the prince strides away, led by a white-faced servant. At the last second, he stops. Turns. His eyes land on me again—

And he *winks*.

Then he's gone, disappeared into the palace.

My sisters flood around me, escorting me through the door after the prince and his warrior. None of them speak a single word.

CHAPTER 5
THE PRINCE

"ARE YOU SURE this is going to work?" asks Edvear.

Prince Rahk of the Nothril Court, my only friend, meets my gaze for a second. His dark eyes are a touch too keen for my comfort, but he doesn't speak for me. He's more of the brooding type, standing over there silently by the wall, observing everything quietly, and generally unnerving strangers with his imposing presence.

"I suppose we shall see," I say. This chamber is surprisingly small and dark, lit only by a dozen candles scattered about. A single lumiral globe could have this room bright as day. But lumiral globes wouldn't work in air as stifling and magicless as this. It's the constant death that drains away the land's magic. How do humans manage it? My skin almost crawls from an invisible itch—the lack of power at my fingertips, the emptiness running through my blood. I tug on my sleeve, then glance over my shoulder at the short, squat bed with what looks like heavy carpets hung from each of the four posts. Interesting. "I'm more concerned about finding a bed Rahk will fit in."

Amusement cracks through Rahk's stern expression. "Worry about yourself."

"I suppose that is what I'm best at," I return, and slide into a too-small chair. The arms hug my hips and legs tightly. I hope the chair doesn't try to come with me when I stand.

Rahk doesn't risk one of the chairs. "Well? What do you think of her?"

"Of the princess?"

"No, the king's bloodhound," deadpans Rahk. "Of course I meant the princess, Ash."

I smirk and throw my feet up on the chair Rahk won't sit in. "It's not as though I can have many thoughts on someone whose only visible feature was a pair of tiny hands."

Rahk's scrutiny sharpens, as if he knows I have more thoughts than that. I smile innocently at him. He sighs and turns to address Edvear. I tune them out as they discuss the logistics of the fae "courtiers" we've brought along.

I do have thoughts about the princess, but they're primarily made of intuition and questions. The only thing I'm fairly certain of at this point is that she is ugly. She barely hesitated when I asked that she remove her veil. It was her father who prevented it, and even though human lies have no scent like a fae's, I'm almost certain he lied straight to my face about the veil custom.

I'm a little insulted.

But that's beside the point.

If the king is lying about the veil custom, then he has something to hide. Nothing else seemed strange about her, aside from how short her human frame was and that she was covered in lace and brocade from head to toe. Humans have such ghastly taste in fashion. It makes sense that she is unbecoming. Or perhaps she cannot speak or has some disfigurement.

If it is merely cosmetic, I do not mind. If she's particularly unbecoming, it might be more difficult to convince my father that I truly fell in love with her. His version of love cannot comprehend anything but admiring outward beauty and disregarding the rest.

I can always keep her veiled. It might even work in my favor to create an aura of mystery about her. If I glamoured her and gave her a veil that revealed glimpses of her jaw or mouth, just enough to convince onlookers she is some great beauty beneath . . .

So long as I can glamour her, it doesn't matter for my plan what she looks like.

"There's the plotting face," says Rahk.

I blink and look up. The movement makes me realize how creased my forehead has become. I've been staring off into space, my chin propped up on my fist.

"She is a woman, my lord, not a pawn on a board," says Edvear.

I level a glare at him. "I am *aware* that she is a woman."

Rahk crosses his arms over his chest. Light from the candles on the nearby vanity flickers over his face, casting it in severe shadow. "She looked frightened."

"How could you tell? You couldn't see her face. She looked a little stiff, but that could just be her personality."

Rahk doesn't answer, only sighs and pushes off the wall, making his way toward the pitcher of water and shallow basin set on the vanity for washing. He pours the water into the basin, grabs the cloth, and sets to scrubbing his face.

"What? How could you tell?" I twist around to look at him. "What did I miss? I cannot have you being able to read my wife better than I."

"It was her hands."

"What about them?"

"They . . . were sweaty."

Sometimes I forget he has a nose like a seelie hunter. "Ah," I say, nodding and leaning back in my chair, returning my feet to their footstool. "I ought to put her at ease then. What kind of things do humans tell their children about us fae? Does she think she marries a demon monster like those beyond the Veil?"

"I wouldn't know."

I shoot a look at Edvear. "Do you?"

He shrugs.

"Useless lot of you," I grumble. "Shall I have you scrape out my chamber pots for a fortnight for this failure?"

A concerned look spreads across Rahk's face. "*Scrape*?"

"Shall I have your cook reassess your diet?" asks Edvear.

"Should have known you'd turn my teasing back on me," I grumble and try to stand, only to have the chair come with me. "Great Kings, I never thought of myself as having wide hips, but this chair is making me reevaluate my entire life. Help me get it off, Edvear!"

As Edvear comes to aid me, my mind wanders again. Back to the veiled princess awaiting my suit of marriage. She cannot have a single clue what is going on. And Rahk is right—she's probably terrified out of her wits.

I won't marry her unless she agrees. I'll ask her tonight. If she doesn't want to marry me, I'll pack up my entourage and move on to the next kingdom.

Is this the right thing?

In theory, it seemed exactly the right thing, but now that I've seen the girl, part of me cannot help but wonder if it's too dangerous to bring a mortal into the equation, into Faerieland, as my wife.

But then my gaze lands on the tattoo circling my wrist. The reminder of my bargain with the High King.

It's too late to back out now. I've set fate spinning into motion. I've placed my bets and gambled my very life in the process. No one will stand up to the High King if I don't.

I just wish my life was the only one I gamble.

"Isabelle Louise," I mumble under my breath, rolling her name across my tongue and tasting it like fine wine. "Isabelle Louise."

CHAPTER 6
THE PRINCESS

"IS SHE READY?" Father calls from the hallway. His voice is thinner than usual, strain coating every syllable.

"She's ready," Vivienne calls back.

Of course, I had to be with the four of them while my ladies assisted me in putting on my gown and applying cosmetics to my face. Which they promptly covered with a veil. Their constant chatter and exhortations are like a wet blanket around my shoulders, making me small and shivery.

Except Amelia, of course.

But she is unusually quiet, and when I look at her, her face is pinched. As though she fears I marry a monster.

Better I marry a monster than she.

She might find some happiness with King Ilbert. Who knows what will happen to me as the bride of the fae prince?

Yvonne hangs back slightly from the others, making snide comments here and there. It isn't until she makes one about Prince Trenian being handsome that I finally realize she's actually *jealous*.

Which saddens me even more. Her betrothed is the worst of all my sisters' husbands-to-be.

"Your actions are more important than ever," Vivienne reminds, the sternness of her tone hardening the beauty of her face. "Our people need this marriage to stop the encroachment of the fae on our lands, and you'd better not scare the prince away tonight like you scared away King Ilbert."

Jacquelle pipes in. "Don't forget that you must be eager to fulfill all marital duties when the time comes." She shoots me a significant look in the mirror, making me blush. "If you show any hesitation, it could ruin the peace treaty."

At some point, I block them out. I have enough on my mind without their pestering concerns. I cannot think ahead to a wedding—or what comes after a wedding—until I'm through tonight. That is my first goal. Keep my composure tonight and try not to accidentally send the son of our enemy packing the moment I don't meet his expectations.

I float in a fog out the door to meet my father. The journey to the ballroom is similarly hazy. My sisters' voices buzz around my ears, keeping me anchored. It's only when those grand double doors swing open, and the familiar announcer's call carries across the gilded space, that I blink, the fog clearing with a sweeping wave of anxiety.

Immediately, my eye is drawn to the tall, dark figure in the middle of the room. His back is to us, but when the announcer heralds us, he turns around. His gaze latches onto me, somehow picking me out from my sisters in an instant.

"Remember," says Vivienne from behind me, "you must—"

Faster than I expect, the prince is before me. I suck in a quick breath, craning my neck to peer up at him. He's so handsome in his strange clothing, with his long, embroidered tunic of midnight blue, his close-fitted sleeves that hide his tattoos but reveal muscular arms.

He doesn't address my father.

Instead, he reaches out to me, palm upturned. Father's gaze burns into the side of my face, momentarily paralyzing me. Am I not supposed to respond because of the slight to Father? That doesn't seem like a good way to begin peace negotiations.

Slowly, I lift my hand, place my cold fingers in his much larger, warm hand. It closes around mine, and I stare at it, almost uncomprehending. I look up, find his gaze on me. It shifts to glance over my head at the crowd of sisters close behind me.

"You all are like a gaggle of sprites around a crumb of gold in a riverbed," says the prince with a quirked mouth and one half-raised eyebrow. "You'll suffocate the poor maiden before I have a chance to learn how frighteningly young she must be. Truly, there is nothing in the world that makes one feel so ancient than to ask a human her age!" He looks down at me and smiles. "Dance with me, Princess."

I mean to say that it will be my honor, but when I open my mouth, no sound emerges. If I force it, it'll come out in stutters. The fae prince doesn't want a bride who struggles to talk. I resort to a deep nod and curtsy.

His smile widens. Then he tosses a casual, "King Roland, Princesses," over my shoulder at my family. With that and nothing more, he draws me after him toward the center of the polished wood dance floor, and even though I ought to be terrified, I'm almost more relieved to be rid of my sisters' constant reminders of the many ways I can mess this up, than I am frightened to be in the prince's presence.

He doesn't seem cruel or evil yet. That's good, right?

"So," he says, his deep voice drawing my attention all the way back up to his face. If I crane my neck anymore, I'd be staring at the golden chandeliers above us. "They veil you so that I may not lay eyes on you. Do they also bind your feet from dance, and your tongue from speech?"

No one has bound my tongue, but it seems desperate to prove the contrary. I shake my head.

"I'll need proof."

This, at least, surprises a reaction out of me. "I b-beg your pardon, Your Highness?" Once the words are out, my stomach clenches. Did he notice the stutter?

"Ah, there it is! What a gentle voice you have. Once we are wed, you must sing me to sleep with lullabies."

My face flushes hot. If I express my confusion, will he take it as aversion? I resort to ducking my head in a nod. *Lullabies? Is this a fae tradition?*

His hand slips to my waist as the music starts. A large, strong hand. Its warmth seeps through my bodice to my skin. I look up at him, obscured by my veil, and suddenly I hate that I wear it. I want to see him clearly, and I don't want him to be disappointed when he sees me for the first time. I'd rather show him my face and stand before him as I am, so he knows what he is marrying.

We start dancing, and I'm not sure what else I was expecting, but he guides me into the dance with strength and confidence, moving gracefully. My shoulders ease just a fraction. Tonight will be easier if he is skilled at our dances. It's hard to imagine the fae wouldn't have their own, very different dances. He sends me out into a spin, then draws me back.

"I must ask you a question."

I blink, then nod and manage a clear, "Yes, Your Highness?"

He gives me a shrewd look. "Would you be alright with never seeing your family again?"

My thoughts come to a startled halt. I tilt my head, turning his question over in my mind. He must mean that if he marries me, he won't bring me back to visit my family or my kingdom. I truly will be lost to Faerieland. My chest tightens. But I cannot say no, right? Such a thing might upset him, and that could ruin our kingdom's chance for peace.

I glance over my shoulder at my father, my sisters. They stand where I left them, watching me and Prince Trenian as we dance alone in the midst of the ballroom. A courtier engages Father in a conversation, but Father is only partially listening. His gaze is glued to me.

My eye finds Amelia, so unusually still and solemn.

I would miss her.

But I must not say anything to indicate I don't want to marry him. "Y-yes."

"And what are your thoughts on dying?"

I keep thinking I'm past the surprises and the inclination to sputter. Alas, I'm not. What sort of question is that? He's hardly said four sentences to me, and he's asking about dying? A sudden fear seizes me, one that makes me blurt, "Do you intend to kill me, Your Highness?"

His eyebrows lift, and something brightens across his face. Has my question pleased him somehow? It seems an unusual thing to be happy about.

"Of course not," he says, spinning me again. "I would never hurt you."

"Then I confess, I do not know what your question means." It takes everything not to sag in relief when the words come out clearly.

He gives me another shrewd look. "I'm four hundred and seventy-three years old. I've seen many humans come and go in Faerieland. I have seen many, many die."

Oh. He's asking if I am alarmed that my lifespan will be much shorter than his.

Then, belatedly, his age registers in my mind.

I nearly choke on my own air. I knew he'd be much older than I, but he looks like he could hardly pass for thirty. To think that he was alive long before our kingdom gained its independence . . .

I give my head a tiny shake. *Snap it together, Isabelle.* I cannot afford to show shock to the prince. "I'm a-alright with that."

Now it's his turn to look surprised, but then his face shifts, like he's impressed or pleased. Have I passed his tests? Have I successfully not scared him away?

"Then you are not afraid to marry me?" he asks, as we dance past the statue of a hunter drawing his bow.

Afraid to marry him? Of course I am! But I cannot let him know that. "It would be my h-honor, Highness."

He looks at me as though he wants to say something else, his jaw flexing once or twice. I wait, but he only spins me out again before drawing me back.

The music ends. I curtsy as the prince bows. He doesn't ask me for another dance, but instead offers his arm to escort me off the dance floor. I slip my hand in the crook of his elbow and try not to notice how thick and solid his arm is.

My effort is noble, but I fail utterly. Most of the kings and princes I've met aren't exactly . . . *muscular*. Perhaps his handsomeness will be my consolation, since I'm very possibly marrying a monster.

The prince hands me off to my father and says, "You have a lovely daughter, King Roland."

My eyes go wide, and I look up at him. He cannot mean that—he cannot see me. What, then? It's not as though I said anything clever. Maybe he likes quiet women who will do as they're told.

The prince looks down at me, responding to my gaze. "I look forward to wedding her tomorrow."

Shock floods my body from head to toe.

"Tomorrow?" sputters Father. I don't think I've ever seen him *actually* sputter before. Vivienne, nearby, sucks in a sharp breath.

A tiny smirk plays on the edges of Prince Trenian's mouth. "What? Do you wish me to wed her now? I suppose I don't have other plans for the evening—"

"Tomorrow will be lovely," says Father quickly. "Once we've settled the conditions of the alliance. Not before."

"Of course," he says with barely suppressed amusement.

Is it even possible to settle the alliance in a day?

My head goes light, and I catch hold of my father's arm so I do not sway on my feet. The prince's eye shoots toward me at my movement. A knot appears on his brow. "Will that suit you, Princess Isabelle Louise?"

All eyes turn to me. The lack of music, of even quiet conversation in the ballroom, echoes around my head like a pounding drum. My tongue clings to the roof of my mouth. *Wretched tongue*. It's certainly fast—too fast—but I cannot say no, right?

"Yes," I manage.

With that, the prince turns and strides out of the ballroom, his shoes clicking on the reflective floor. The doors swing on their hinges in his wake. Their quiet creaking is the loudest sound in the full room. The stunned stillness spreads across the crowd like a canopy.

And everyone stares at me as though I have five heads and an assortment of feathered tails.

"Why did you say yes?" asks Father, a thick vein standing out starkly on his forehead.

"It wasn't like she could have said no!" says Amelia, glaring at him and coming to slip her arm in mine.

"A gown cannot be made by tomorrow!" cries Vivienne.

"Then she will have yours!" Father snaps at her.

"*Mine*?" Vivienne's face goes pale. "But my wedding is in but a month, and this gown took five months to make! I could never get a replacement in time! It won't even fit Isabelle Louise!"

"Then *make* it fit!" cries Father. "Steward! Send for my advisors. I must meet with them immediately. And prepare a royal wedding, will you? We have no time to lose!"

"I need to leave," I whisper to Amelia, taking her elbow. "I need to get out of here."

"Of course," she says quickly, and the two of us slip out as chaos breaks across the ballroom.

CHAPTER 7
THE PRINCE

THE CADENCE OF the princess's voice echoes in my ear long after I leave the ballroom. Rahk was right—she is terrified. My gut twists. I want to *talk* to her, plainly and openly, without that wretched veil. It doesn't matter if she's disfigured or uncomely.

If I'm honest, however, the reason I want to talk to her is that I want her to somehow ease the guilt that slides down my spine at the memory of Calver's slaughter. And all the others before him. But she cannot ease my regret. It is not for her to bear or soothe. It is my burden alone.

There's still so little of her for me to base any opinion upon. She reminds me of a soft, downy gray dove that warbles and shudders when held. A thing of sweetness and innocence.

But just for a moment there, I saw a part of her—the true Isabelle Louise.

Do you intend to kill me, Your Highness?

Beneath the trembling and quiet, there's something sharp and spirited. Something that she suppresses—something that spineless man she calls Father and those prattling wenches she calls sisters suppress in her. Will I ever see the real girl beneath the veil? Or will she be ripped from my grasp too soon?

This time tomorrow, she will be my wife.

And I know next to nothing about her.

It had never mattered, not in theory, not when I scribbled down my plans and watched those same plans burn in my furnace. I needed a human wife—and a royal, to abide by the terms of my bargain. She was merely an idea on a page, a figment of my imagination.

But now there is a young woman about to pledge her life to me. A woman I took into my arms and danced with, a woman whose cold little fingers I held.

"You've left us again," says Rahk.

I look up. Edvear and Rahk joined me the moment I left the ballroom, but I forgot to acknowledge them. We're at my quarters now. I force my mouth to twist. But suddenly, I don't want to walk into that room with them. I don't want to listen to them talk and speculate.

"You go on ahead. I'll be in shortly," I say, and stride on past them, ignoring Edvear's protest.

I walk the length of the hallway, waiting until the door closes behind Edvear, and Rahk goes to his own room. Then I circle back, cast a quick spell to disguise my steps and smell, so Rahk's stupid nose doesn't detect me. Such a thing would normally be second nature to me, but here in the dull, lifeless human air, it's very uncomfortable, like an ear popping at high altitudes.

The farther I move, the more I wonder how this kingdom hasn't fallen already. Even the tiny bit of magic I manage is more than enough to slip the guards as I retrace my steps to the ballroom.

The hallway has grown darker. It gives me the impression of a dungeon, with how close the walls and ceiling are compared to the bright openness of the High King's palace. My skin crawls from the underlying scent of dust and decay beneath even the most decadent whiff of perfume. The sound of chaos filters in through the closed ballroom doors ahead of me. Is she still in there? Would I scandalize her if I tried to speak to her alone? Rumor has it that humans are very strict about their decorum.

The door opens, and I slip into the shadow of a window curtain. A twist of my finger draws the shadows deeper around me. I wince. The negotiations tomorrow had better go well, because I do not want to stay in this stifling and stagnant world a moment longer than I must.

My efforts are rewarded. Two veiled princesses slip out of the door, arm in arm, and hurry down the hallway. The one on the right, closest to me, is *her*. Even if I couldn't tell the color of dress she was wearing, her scent and posture are indicative enough of her identity.

"Are you alright?" asks the sister.

Isabelle Louise gives a nod, but her knuckles are white on her sister's elbow. "Of course."

The human ability to lie so boldly and without consequence never ceases to amaze me.

"You don't have to do it. I can talk to Father. You can refuse. Oh! I know. I'll tell the maids not to hem your gown, and then you'll trip walking down the aisle—and *then* he'll be so disgusted he'll leave. It'll solve all of our problems."

A solid plan. Nothing disgusts me more than women who trip.

"I will marry him," comes Isabelle's quiet but firm response. "I have one duty. I will fulfill it. Our people need this."

There's resignation in her tone, as if she is walking to her own pyre.

I shouldn't be listening to this, but I cannot help myself. I cannot tear myself away from the shadows as she walks past, cannot stop up my ears as the sister protests.

"There are other alliances—"

"Like Prince Brochfael?" Something about the way she says it suggests this Brochfael fellow—though I cannot help but pity a man with such a name—is even less desirable than a fae. The sister drops the point immediately.

"I shouldn't have spoken to King Ilbert. He should have been your husband. I didn't think he would like me—"

"I want you to be happy with him. I am glad he will be good to you," Isabelle says, and there's a subtle fierceness underlying her words. "Do not try to convince me not to marry Prince Trenian. He is fae, but that doesn't make him a monster. I'm sure I will be quite content at his side."

Her words slice right between my ribs. She has no idea how wrong she is.

"But—but—"

"But what?"

"What about . . ."

"What?"

"Tomorrow *night*? Aren't you scared?"

My eyes widen, and if I had been intending on leaving, I certainly can't leave now. Not until I hear her answer.

She's quiet for some minutes, and when she finally speaks, I strain to hear.

"I will fulfill my duty."

My chest rises and falls with a deep breath. Why does it suddenly feel so hot in this palace? Perhaps I ought to take refuge in the gardens. It'll be cooler there.

But I stay rooted to the spot as Isabelle Louise and her sister turn the corner and vanish from my view. I blink once, twice, thrice. Then something hot coils around my heart. A fierce determination that shoots through my body, radiating out from my core to the tips of my toes and fingers.

I will be good to her.

I will swear it on any name. As long as she lives, my wife will want for nothing. I will repay her for her sacrifice. Anything that is within my power to grant, I will give—and then more.

She will have whatever she wants. I only hope she lives long enough to enjoy it.

Negotiations start early the next morning. I leave one of my warriors in charge of my entourage to ensure they do not leave their rooms unsupervised and wreak havoc on the human palace. Then I take Rahk and Edvear with me into the council room.

It's larger than I expect, with the same low ceiling as much of the rest of the place. *So dark and depressing.* If there's a window anywhere, it's hidden behind the carpets hanging on the wall. A large oak table fills the center of the room, covered in candles with rows of fresh parchment, a trimmed quill, inkpot, and wax for a seal.

There are some dozen men beside the king, hiding behind the table like it's a buttress, and I lift a brow.

"King can't make a decision on his own?" Edvear mutters under his breath. His long ear twitches like a horse's when a fly buzzes too close.

I agree, but don't respond.

The king comes to greet me and reaches out his hand. "Welcome. We are honored."

Something about the way he says it strikes me as humorously ridiculous. I manage to contain my reaction to a grin as I say, "The honor is all mine, King Roland."

A few of the humans have the decency to look uneasy. Good. Last night, I might have been the fawning suitor, but today, I am the future High King of Faerieland. My grin slides into something colder and calculating.

"Would you have a seat?" Roland asks, and gestures for one of the servants standing on the wall to bring me a drink.

I lift my hand to stay him. "Why don't we skip the pleasantries?"

Roland glances at me, hesitancy written in every line of his face.

"I wish to marry your daughter, Princess Isabelle Louise. In exchange for her hand, I offer my promise of safety for your lands once I am High King. I will not only cease the encroachment of our borders on your land, but I will restore that which has already been taken."

Stillness washes over the crowd. They didn't think I'd make such a handsome offer, did they? My lips twist.

"How long until you are High King?" asks one of the presumed advisors. One with a beard like a goat's. "How do we know you won't honor your promise hundreds of years from now, when our lands are completely swallowed up?"

Part of me is relieved they're not entirely idiots. Another part wants to groan inwardly. I have no desire to be here all day, and I'm starting to get a sense from the room that the king doesn't intend to hand over his daughter easily.

He must show some reluctance, or else he has no power.

"I will be crowned within the year," I say, and the lie rolls off my tongue and fills the room with iron-stink. Rahk struggles to keep his face from wrinkling, but his nose wriggle gives him away. Edvear fares better, but his face has a line of sourness about it.

The humans don't react. They don't even notice.

It's a lie because, as much as I intend to be on the throne within the year, it is no guarantee yet. But I cannot show any hesitancy, and since there are no fae in this room besides my trusted two, it is worth it for me to taste sourness on my tongue for the next few minutes.

I retract my refusal and accept the servant's goblet of wine, but never once do my eyes leave the king's. "Once I have taken the princess as my wife, you will relinquish all ties and claims to her as your daughter." This is the part I do not want to say, but I must make it part of the terms. Otherwise, I might have another fae-human disaster war on my hands. "Faerie is a dangerous place for humans. You will promise not to retaliate should harm befall your daughter."

"Not retaliate?" Roland bursts out suddenly. "What plans have you for my daughter? Will she not be protected?"

"She will be under my protection."

"You guarantee her safety, then."

I draw in a breath through my teeth, then say the words levelly. "I do not."

The room goes still. Only Rahk doesn't flinch, standing solidly at my back.

I stare evenly at the king. This is the one part of my plan that I detest with all my soul. The simple fact that Faerie is not only a dangerous place for humans—it is not a place for humans at all.

Humans die there. Many at the High King's hand.

Isabelle Louise is a human. There is no telling how long she will last, but when her time doubtlessly comes to an end, I will watch her die.

And I will hate myself for bringing her there.

The tattoo on my wrist burns as if freshly seared. An ever-present reminder that there is no going back now. There is no retreat. There is only forward. I set my jaw in a determined line and face the father of the girl whose death I will solidify.

"Then you may not have her," says Roland.

I'm actually surprised. This human is improving on me. I'm glad he will put up at least some semblance of a fight for his daughter's safety. But they need this marriage as much as I do. In the end, he will cave.

"Then you will not have my promise of safety and peace for your lands," I say coolly. "My father is greedy to expand his borders. He moves slowly now, testing the waters, but he thirsts for blood. Who will protect your daughter when he declares war? Will she be safe then?"

Roland says nothing, and his advisors glance uneasily among themselves.

I retract my improved opinion of the king. He's as spineless as I first thought.

When the silence continues, I press onward. "This is what I will give your daughter. I will take her as my wife, and so long as she lives, I will take no other. I will protect her, so that if anyone raises a hand against her, I will sever that hand and affix it to my wall, and should anyone make an attempt on her life, I will remove their head and give it as a toy to the fauns."

The uneasiness turns to blankness. Confusion.

Maybe I'm not getting as far as I thought. I've always heard stories of humans being greedy, but I've never seen it quite to this extent. It's time to pull out my trump card.

I lower my voice. "I will even give her a strand of my hair."

"No one wants your hair!" bursts the king. "I want the assurance that my daughter will be cared for, respected, and protected as your bride. I want you to honor her."

I frown. It occurs to me that he might not understand my hair offer. The things people in Faerieland will do for a strand of my hair! And it's just blown off here with a sweep of the hand!

"Of course I will do those things," I say, still frowning. Is he afraid I will go to all this trouble to get a human bride and then just toss her aside for the unseelie to eat? I think I'm offended.

"Then in that case . . ." Roland glances back at his advisors, then returns his gaze to me. "I accept your terms. We will have an official treaty written up for you to sign before the wedding."

I cannot contain the small curl of smug satisfaction that escapes my composure. I didn't think it would take much to make him capitulate. I'm mainly glad I didn't have to sit here all day, arguing against his pretense of caring for his daughter, when clearly he views her as little more than bargaining material.

"Thank heavens," I say. "I was about to pack up and go find a different bride."

CHAPTER 8
THE PRINCESS

S TAND UP STRAIGHTER, Isabelle Louise!" barks Vivienne.
I comply, but shift my weight between my feet as subtly
as I can. I've stood still as a rod during fittings like this before,
and the reward was always passing out. I'd rather not begin my
wedding day by fainting.

Vivienne is in an especially sour mood as I stand in her dress.
Mine is not much better, as this dress is the last thing I would have
chosen for myself. It's not *ugly*, but it is very . . . very . . . *much*. The
skirt is so full I take up half the fitting room wearing it. It's a nice
shade of dark peppercorn blue, but so *heavy*, and covered in frills. It
somehow manages to be both solemn and fanciful in a way that makes
me feel like I'm walking in someone else's shoes.

I won't have to wear it long, though. Just through the ceremony.
I won't even have to attend the celebration following and trip my way
through the banquet hall.

I will be in a bedchamber awaiting my husband.

A wave of lightheadedness passes over me, and I switch my weight to my other foot quickly. Last night, I told myself I wouldn't worry. I needed to not scare him away first during our dance. *Ensnaring him*—as Father likes to put it—was my priority.

Now that he has confirmed that he wishes to marry me, and is in negotiations for my hand, it's officially time to panic about tonight.

Part of me wants to postpone the panic even further until after the dress fitting and ceremony. But then there will be no more time to panic. It will be happening.

Father called me to him this morning for *the talk*. I want to block that conversation from my memory. He said he was sorry my mother couldn't give it to me, or some other lady from the court, but that there was simply no time. A sentiment which I echoed more heartily than him.

He gave me no specifics, mercifully. Just general advice on my conduct, and an emphasis that my cooperation was paramount to the success of this alliance.

I am to give Prince Trenian whatever he wants. No resistance, no tears, not a single whimper. A little maidenly shyness is fine, welcome even. But too much shyness will ruin everything. No hesitation, no requests.

I want to believe he won't hurt me. I want to believe he will be gentle, like he was during our single dance. But there is something about him, an edge of wildness and danger, that terrifies me. He's a stranger, and the son of our ruthless enemy—a fae. He is no human. Fae are known for their debaucheries, their baseness. They are not known for their kindness.

I hope the prince is disappointed when he removes my veil. Maybe then I can fade into the background, away from his notice.

I doubt I can fade away tonight, however.

Drawing a deep breath, I stand up a little straighter. I can do this. I can face whatever is ahead of me today, tonight, and beyond. This is good, I remind myself, because if I wasn't marrying Prince Trenian, Amelia would be. She's too good and sweet to be lost to Faerieland.

BRIDE OF THE FAE PRINCE

The day rushes by in a whirl of silk, brocade, and lace. I hardly have a moment to breathe, and for one blissful moment, I collapse into a chair in my underthings while the tailors and Vivienne fuss over the dress.

Then I'm dragged back to my feet, and the insufferable thing is wrestled on me once more.

Before I know it, it is almost time for the ceremony.

"How did the negotiations go?" I ask Amelia, who has attempted to be my spy throughout the day.

"They went well, apparently." She fusses with the back of my gown, making sure it falls just so. "I couldn't find out what happened, except one of my servants told me that another servant overheard a couple of the advisors talking about it afterward."

"What did they say?"

"It was a little jumbled and didn't really make sense."

That is an evasion. My gut sinks. "Tell me."

"Oh just something about—"

I twist back toward her. "Tell me, Amelia."

She looks at me, worry streaking a line across her forehead. "He made Father promise not to invade Faerieland if anything happened to you."

Dread fills my core. But then I shake myself and say briskly, "Well, of course he had to do that. This alliance is to prevent a war, not start one. Besides, invading Faerieland would be a deathtrap. Father would never do such a thing."

Do you intend to kill me, Your Highness?

I would never hurt you.

Did he lie to me? No, no, he couldn't have. Right? He seemed genuine. But what do I know of genuine when it comes to fae? I could have misread everything. Maybe everything is a glamour.

Amelia squeezes my hand. "You can still escape, Isabelle."

As if I could! This dress is as effective a prison as any iron bars. I force a reassuring smile and squeeze her hand back. "You cannot convince me to forsake my duty."

"Duty schmooty!" Amelia cries, tears welling up in her eyes. "What about *happiness*? Let it all burn!"

And this is why it is good that I am marrying the prince instead of her. "My greatest wish," I whisper to her, leaning close, "is for your happiness."

Then a servant is knocking at the door. "It's time, Highness."

Now. This moment. It's arrived. There's no going back.

I straighten my shoulders and try to ignore the fluttering in my belly. It's time to become the bride of the fae prince.

CHAPTER 9
THE PRINCE

W HY ARE YOU nervous?" asks Rahk. His white hair glows silvery in the low light.

I shoot a look at him. "I'm no—" And then I cut myself off abruptly as iron fills my mouth. Quickly, I take a swig of watery wine to wash out the taste.

Rahk just smirks at me with that knowing gleam in his eye. His question hangs between us as I wash my face, dry it off, and run a hand through my hair.

"I don't *intend* to be nervous," I mutter.

"You've never been married before. Being nervous is understandable."

I glare at him. "I don't need your patronization."

Rahk's smirk widens.

"My lord?" says Edvear, cracking open the door to my chamber and meeting my gaze. "It's time. The humans are assembling."

"Now?"

"Now."

"Well, give me a minute to put on my shoes," I grumble, hoping my tone disguises my nerves. I march to the wardrobe, fling it open, and grab my boots. At this point, I've stopped looking for my things in the last place I left them, but rather wherever Edvear would think the most proper place for them would be. I drop onto the edge of the bed—I'm not making the mistake of sitting in one of those human chairs again—and yank one boot on, then the other. I lace them up, shoot my steward a sharp look when he approaches. It might be my wedding day and my hands might be just a touch unsteady, but Mountains of Ildrid, I can put on my own Kings' cursed boots.

I stand. Straighten my long tunic, my coat, the silver medallion hanging to my navel. Then I level my shoulders and stride out of the room.

No going back.

I relax my jaw and don a cold smile as I pass through the king's guards and my courtiers. My warriors have done an exceptional job keeping them contained. I will reward them when we return.

But for now, I stride toward the doors of the chapel where my bride is waiting. With each step, my nervousness is replaced by something else. Something not unlike the hot determination that coiled around my heart last night.

Hardness sets in. This is war, and I will face my bride. She will be mine. Mine to care for. Mine to protect.

Mine to lose.

I've accepted this responsibility, this weight of her on my mind and conscience. Now I will bear it without faltering.

I stride down the hallway, and at the end of it is a woman in a frilly blue gown, with a matching blue silk veil over her face. The veil is sheer enough that it clings to the contours of her forehead and nose, and nothing strikes me as ugly about that scant bit of profile.

That veil had better not be a trick.

BRIDE OF THE FAE PRINCE

She stands at her father's side, surrounded by her gaggle of sisters. My jaw works even as I flash a smile at them, as Isabelle Louise turns to see me.

I told Roland to keep the ceremony short. I haven't time to waste on human frivolity and customs. Aside from that, however . . .

The strength of my own desire to take my bride to our chambers, to remove her veil, to have her alone for once without all those sisters, almost catches me off guard. If it was up to me, we'd skip this ceremony altogether and go straight to the fae bonding.

That would probably scandalize the humans

My bride makes no move, as that veiled face stays fixed on me. What does she see when she looks at me? What is she *thinking*? The nervous tells she's had so far aren't apparent at the moment. I wish Rahk could tell me if her hands are sweaty again

He is close on my heels, a great sword strapped to his hip and another across his back. The humans said it was their custom that the bridegroom brings his best swordsman to accompany him. Seems a little out of place to me. It certainly would make sense at a fae wedding, but a human one? Do they suspect the clergy of plots to disembowel their patrons?

Whatever the case, I definitely don't mind having a heavily armed Rahk at my back. Especially with my magic so limited here.

It's silent when I reach my bride, as if the small crowd of onlookers has drawn in one collective breath. Waiting. For what, I cannot begin to guess.

I stop before my bride.

Her head tilts up to mine.

Neither of us moves. Roland clears his throat. I swivel my attention to him and say dryly, "You still veil her? In but a few minutes, she will be my wife."

Something shifts in the air around the young woman. It's so subtle I almost miss it.

Fear.

"It is our custom," says Roland. "You may remove it after the ceremony in your chambers."

Right, because then it will be her, and not him, who bears the brunt of my reaction—my *disappointment,* presumably—when I see her for the first time. He releases his grip on Isabelle Louise's arm and steps aside, motioning for me to take his place. My face hardens, but I do as instructed and hold out my arm for the princess to take. When she loops her tiny fingers around my forearm, they twitch.

There's little I can do to assuage her fears, especially in front of an audience. So I resort to laying my other hand atop hers. She jolts in response. I grit my teeth. Then I swipe my thumb over the back of her wrist in a soft caress. It doesn't make her relax.

I will be good to you, I say in my head. *Have no fear.*

But she *should* be afraid of me. She *should* be afraid of being my wife.

The clergyman begins a dreary lecture that makes me wonder if the true reason bridegrooms bring their swordsman was as a threat to the priest that he ought not to drone on too long. This ceremony strikes me as tradition piled atop tradition with little meaning. It is so pale compared to the true depths and beauty of a fae bonding.

The priest gives a cough, clearing his throat, and says, "Now, Prince Trenian of the Fae, you must plight your troth to Princess Isabelle Louise of Aursailles."

"I must what my what?" I repeat, raising my eyebrows.

The priest peers over the rim of his spectacles at me. "Plight your troth."

That is another language. Unless it has something to do with trouble with my trough, and if that's the case, this might be one of those human wedding traditions I choose not to ask questions about.

"Right," I mutter. "My trough. Of course."

The princess's shoulders give a little shake. I glance down at her. Was that a quiet snicker? It's so unexpected, it warms my ears and makes me want to squeeze her hand. I don't, lest she misinterpret the gesture.

BRIDE OF THE FAE PRINCE

Plighting my trough, as it turns out, is the part of the ceremony where I make my vows to her. They involve a curious statement about pledging to love and hold her no matter if she be fair or ugly, whether she be sick or healthy, until death parts us.

She must be one ugly and sickly princess.

I suppose there is an additional possibility that the young woman might not be so young, but her smooth hands indicate the contrary.

To my surprise, the princess pledges no trough to me, but no one blinks. I suppose that is just how the humans do it. I confess I'm disappointed. I wanted to hear her speak.

Then, the priest announces, after another long segment of dreary nonsense that I tune out, "Prince Trenian, you may kiss your bride."

CHAPTER 10
THE PRINCESS

THOSE WORDS RING in my ears. This moment has come—a herald of what will happen shortly now. My new husband will kiss me, and I do not know why the thought terrifies me so much. It's only a kiss.

But I've never been kissed before. What if I don't like it? What if I *do* like it?

A hot blush sears my cheeks as Prince Trenian turns in surprise to look down at me.

"Kiss her?" he asks, and his thumb does another sweep across my wrist. My knees threaten to give out, while my lungs seem determined to reject any attempt at breathing. "I thought I was forbidden from removing her veil." There's a dry mockery in his voice.

No one answers him.

The prince side-eyes my father, then the tall, imposing warrior behind him. Then those piercing blue eyes meet mine through my veil. He cocks one eyebrow.

"Very well," he mutters.

My grip on his arm tightens as he removes his hand from mine and turns toward me. He hooks his knuckles under my chin, lifting my face toward his.

I can't swallow. Cannot even breathe.

My gaze fixes on his mouth. His thumb brushes over my lips, as if locating them beneath my veil. Then, without removing the layer, he leans down toward me.

I cannot even close my eyes as his lips gently press against my veil-shrouded mouth. I'm not sure how much it counts as a kiss, but the veil doesn't keep me from feeling the softness of his mouth. It's over before I can blink. He straightens and returns his hand to cover mine, almost as if nothing happened.

Meanwhile, I go dizzy on my feet. The priest's words are like hammers against a pounding headache.

Then, suddenly, they stop. Cheering starts, though it's a stagnant and limp sort of sound, and I realize it's done. It's official.

I'm married.

I cannot bring myself to look up, to look into that piercing blue gaze of the fae prince who is now my husband. I can feel it on the top of my head, burning into my hair. He wants me to look at him, but I simply cannot do it.

His thumb slides to the sensitive part of my inner wrist and gives a stroke as he turns us to face the onlookers. Amelia is crying, which I can tell from the way she keeps slipping a kerchief beneath her veil to dab at her eyes. My other sisters stand like they are carved from marble.

My father wears a grim expression.

Vivienne steps forward. "I will escort Princess Isabelle Louise to the bridal chambers."

The prince's grip tightens on my hand, and he shoots a narrow look at her. "I think not."

And with that, he takes off, dragging me behind him down the corridor, away from the chapel. His warrior steps in line behind us.

I nearly let out a startled squeak, and we've only gotten a few steps away when my father's voice cuts through the din.

"She must stay beneath my roof tonight. You may not take her until dawn."

The prince stops abruptly. I stumble to a halt in this cumbersome dress. The warrior behind us—ever silent—also stops. Prince Trenian turns to look over his shoulder. I don't look up enough to read his face, but the ire is plain in his tone.

"I will abide by my terms of our agreement. I expect you to do the same." He smiles—something cold and vicious. "We wouldn't want to find out what would happen if you didn't."

Then, to my shock, his hand slides around my waist. I suck in a breath. He bends, and my eyes widen as his other arm scoops me up under my knees. I frantically grab hold of his shoulder with one hand, and his lapel with the other. My skirts are so full, they poof high enough to almost block his vision. He glances at me, and I cannot read his expression. It's a mixture of possessiveness, anger, and perhaps a tinge of bitterness. Then those lips that just kissed mine twist up in a sardonic kind of humor, and I know exactly why I've been terrified of him this whole time.

There's wildness beneath that princely veneer. A wildness that has just laid claim to me.

He hefts me more firmly in his arms and marches around the corner, out of view of my family and the people of our court.

This veil might not have been the worst idea. It hides my mortification from anyone who looks, including my new husband.

"Rahk," orders the prince, and the warrior lengthens his stride until he is walking beside us. "Find this Great King's cursed bridal chamber for me before I look too much like a fool."

I blink. His tone has completely shifted from how he addressed my father. He almost sounds bored now. This will take some getting used to. To my surprise, Rahk only smirks and says in a deep, low voice, "Follow me."

The prince does, his grip on me never faltering.

We wind through hallways that I've known all my life, but they feel different tonight. The distant sounds of feasting echo through the corridors, yet it only seems to emphasize the ever-sure footfalls of my new husband filling the air around me. I hold on to his lapel as though I'll die if I let go. He hardly looks at me—but it almost seems like he's trying not to.

There's a new sound at the end of the hallway, and Rahk's steps slow. I twist my neck to get a peek, and find a maid curtsying before a doorway wreathed in white roses.

"You are dismissed," says the prince.

The maid curtsies again and swallows visibly. "Your Highness does not want aid with the gown?"

The prince stops before the door, and glances down at the mounds of fabric in his arms. My breath goes shallow at his study. A roguish grin spreads across his face, making my skin turn considerably warmer than before.

"I like to pick and choose my battles," he says. "Brocade isn't one I have an interest in approaching without a knife."

The maid curtsies. The prince sets me on my feet. I wobble, and he steadies me with a hand on my low back.

"We won't be long, Your Highness," says the maid, and ushers me into the bridal chamber. The door shuts behind us with a thud that rattles my bones. Inside, the room is softly lit with candles, illuminating the enormous four-poster bed at the center of the space. I put out a hand to catch myself against the wall.

"Have you eaten today, Highness?" asks the maid, her voice threaded with urgency.

I shake my head as my vision swims.

"Then drink this. I brought it for you." She pushes a cup into my hand.

It seems the servants pity my fate, too. I shake my head, pressing a hand to my stomach. "I d-don't think I can k-keep anything down right n-n-now."

BRIDE OF THE FAE PRINCE

Her mouth pulls into a grim line. "Try a little, Highness. See if you can take it. Here is a stool to sit on while I unlace your dress."

With my skirts poofing around me, I sit gratefully, accept the cup, and take a tentative sip. It's warm milk, spiced with cinnamon and nutmeg, sweetened with honey. Before I know it, I've guzzled it down greedily. The maid takes the cup and gives me a plate of cheese, cold cuts, and buttered toast. When I have finished that, she gives me a bowl of stewed apples with cream.

I don't know if I've ever been so famished in my entire life.

My limbs are trembling less when I've finished eating, and a weight falls from my shoulders with every layer of gown that the maid peels off me. She works quickly, quietly, and I find solace in her presence.

Too fast, I stand in nothing but my shift. I wrap my arms around myself, half afraid the maid will declare her work done and leave me like this for the prince.

She doesn't.

She withdraws a long, pure white dressing gown from the wardrobe. It is simple, much too unstructured and thin to be considered modest, but it covers me from my throat to my ankles. Who chose this for me? Someone who wants me to feel less vulnerable and exposed. Amelia, maybe? Whoever it was, I'm grateful.

As her last step, the maid bids me sit on the bed, unwinds my hair from its complicated updo, removing dozens of pins until my hair falls to my waist, and brushes the blonde locks until they shine.

"Would you prefer your hair down or in a braid, my lady?" It's as if she knows I don't have the mental capacity to sort through dozens of hairstyles for my wedding night.

Down feels too . . . intimate. Which, I suppose, means I ought to leave it down.

"A b-braid," I say, and swallow. My stomach is less tumultuous after the food, but it still flutters and flips as my hands tremble. At least my legs are steadier now. I try to focus on the soothing hands of my maid as she braids my hair into one long braid and arranges it

71

over my shoulder. Then she straightens my veil—still the blue silk—and steps back.

"Are you ready, my lady?"

No. "I-I-I am." The prince had better not want me to talk tonight, because with each passing minute, I trust myself less and less to get a full sentence out.

The girl hesitates.

"Y-yes?" I ask.

She hesitates another minute, then folds her hands and ducks her head, whispering quietly. "Thank you. For doing this. For us."

I attempt a smile, but the veil conceals my efforts.

"My uncle was lost to the Long Lost Wood," she continues, her voice barely discernable. "Because of the encroachment. You will save the rest of us."

The words fall like a blow, knocking the wind from my chest. Father never said anything about fatalities because of the slow invasion! I manage a nod, and then a faint, "You're w-welcome," even though I feel like I ought to be thanking her for the unexpected care and kindness she has shown me tonight.

She bobs a curtsy and, with one last glance at me sitting on the bed, she goes to the main door of the chamber and slips out. Her soft voice echoes through the crack in the door just before it shuts. "Your bride awaits, Your Highness."

CHAPTER 11
THE PRINCE

I PACE IN the hallway outside of the bridal chamber. Rahk leans against the wall, arms crossed, watching me. "You're prowling again."

I stifle my urge to growl at him. That would only prove his point.

"You can tell me what's bothering you."

I glare at him.

"Or I could give you one of my swords, and we could work out some of that anxiety."

"And terrify the poor girl with the sound of combat in the hallway?" I shoot back, more irritably than I intend.

Rahk remains cool and composed. As always.

"Punching something would feel good," I admit under my breath. I had been feeling confident and in control of the situation. I was *ready*. But then my wife just had to be wearing that horrendous gown that looked like it would take the credentials of a human university to remove.

I'm not angry. Not at her—or even the maid. But now I've been forced to wait out in this hallway for approaching half an hour, and I'm getting very close to tearing my hair out.

Rahk chuckles. Then, startlingly fast, he balls his hand into a fist and throws a punch at me. I barely duck in time, caught off-guard, before throwing a punch of my own. Rahk dodges, ever swift on his feet. His sheathed sword bangs against the wall as we ram into each other, his armor hitting against my fine robes.

He almost lands a blow right to my mouth, which is hardly fair. But he is so determined to distract me that I almost miss when the door to the chamber opens. My stomach flips. I catch his wrist, shoving his next punch aside as I stand at attention and hiss, "Rahk."

He beats me to composure, however.

I try to disguise my labored breathing with a serious expression as I face the maid.

"Your bride awaits, Your Highness," she says as she curtsies. Then she scurries away like a mouse of the Small Cities.

I stand before the door, suddenly unsure. Suddenly wishing the maid had taken just a few minutes longer. My heart still races from Rahk's distraction. Truth be told, I could use more distraction.

My wife is on the other side of this door. And she is terrified of me.

A hand lands on my shoulder. I exhale through my nose, then meet Rahk's solemn expression.

"Think of fairy puffs, my friend."

"What?"

"They're sweet. All you have to do is be sweet."

I glare harder until he grins and backs away, lifting both hands as though in surrender. Then I face the door and draw a deep breath. Before I have time to think anything else, I clench my jaw and push it open.

It swings a little harder than I intend, banging into the wall. Rahk hisses from behind me. As if I'm not already cursing myself for opening the door in possibly the most terrifying way I could have.

BRIDE OF THE FAE PRINCE

There she is. Seated on the edge of the bed, dressed in a simple white gown with that cursed blue veil over her face. She clasps her hands tightly in her lap, a long, thick braid falling over her shoulder. Her hair is a light blonde. To think I married this woman before I even knew the color of her hair.

What color are her eyes?

I swallow, tearing my gaze from her just long enough to close the door behind me. The latch gives a click, firmly shutting out Rahk and the rest of the world. Enclosing us into this space together. Me and my new wife.

No backing out now. Her fate is sealed with mine. The only way is forward.

I turn back to her.

The first thing I'm doing is ripping off that veil.

CHAPTER 12

THE PRINCESS

H E FILLS THE doorway. My mouth goes dry as I stare at him from beneath my veil. I try not to crumple the folds of my dressing gown beneath my sweaty hands. I try to breathe. Try to not imagine what will happen next.

I just need to take this moment by moment.

And at this moment, he studies me from across the room. His scrutiny makes my skin itch, but I can handle scrutiny.

He takes a step toward me. My heart rate kicks up several notches, and my veil sucks in toward my mouth with my sharp inhale. *Calm down*, I tell myself. I can handle him walking closer to me.

He keeps coming, his stride more purposeful.

Breathe, breathe, breathe, breathe, breathe—

He stops before me, staring down at me. I don't look up at him. I can't. Instead, I keep my gaze downcast, fixed on my clasped hands. Then one of the prince's large hands enters my range of vision. He catches the edge of my veil.

I close my eyes, keeping my head bent and my hands curled around my gown as he pulls the silk away.

There's silence.

So much silence.

Suffocating, deafening silence.

Then a knuckle hooks under my chin. I force myself not to resist as he tilts my chin up.

"Open your eyes," he murmurs.

I swallow. My lashes flutter, and I find myself staring at a pair of deep blue eyes, close enough that I can see the silver ringing their edges and the subtle gold flecks dotting his irises. He is so very beautiful, strange and unfamiliar to me. My . . . *husband.*

Is he not going to say anything? Will he give no indication of whether he's pleased or horrified with my face? Is it even possible for fae to find humans attractive?

Heat climbs up my neck and into my cheeks. When I can bear it no longer, I lower my gaze and stare at the silver medallion hanging from his neck. It is a stylized tree, with roots even more expansive than its branches.

He clears his throat. The sound draws my eye back up, and our gazes meet for one second before his flees away. Then he turns and marches to a side table I hadn't noticed when I walked in. There are goblets, along with a decanter. He busies himself pouring a glass for each of us.

I don't move a muscle, even though I long to snatch the veil that has fallen to the floor and cover my face again. But I can handle his disappointment in me. I've handled my father's for this long. What more is my husband's?

He tosses back the contents of his glass, winces, then clears his throat. He sets down the glass and plants both hands on the table, his back still to me.

When he speaks, the sound of his low voice gives me a small start. "I thought for certain the veil was your father's way of tricking unsuspecting fools like myself into binding themselves to the most garish of your race."

My gut sinks. My whole *being* sinks, as though I am a stone thrown into an ocean.

His head tilts in my direction, though his back is still to me. "I was wrong, it seems." Then he pours another glass for himself and tosses it back with a wince. "Gah, this stuff is weak. Tastes like moldy water."

I sit frozen on the bed, like a carved statue.

My husband is . . . *pleased* with me. At least with my face. I should be worried about this, right? The more pleased he is with me, the more I should fear at his hand. I should hope to be beneath his notice. A wife he can hide away in a chamber to pursue her own quiet interests out of the way of everyone else. But though nothing will make my hands stop trembling, I cannot deny that there is a part of me that thrills at his words.

No bride truly wishes, in her heart of hearts, to be found repulsive by her bridegroom.

He lifts the glass he poured for me. "Would you like a drink?"

I shake my head.

He gives a dry chuckle. "I don't blame you."

I follow his movements as he sets the goblet back down. When he withdraws his hand, it gives the barest tremble. He hides it quickly by turning to face me again, leaning against the table and resting his palms on its edge.

My lips part as the realization hits me. *He's nervous too.*

If he's nervous, that means he cannot be a brute, right? Or am I just hoping for things I have no business hoping for? Despite my rational mind telling me I shouldn't get my hopes up, part of me lifts. Enough that I can meet his gaze.

He pushes off the table and comes to stand in front of me. I swallow at his nearness, at his sheer size, at what he might want from me. Then he grabs one of the two chairs at the ends of the table and drags it right in front of me. My eyes widen as he sits, his long legs wide—framing my knees—as he leans closer.

"You haven't said a single word since I entered," he says.

Dare I trust my tongue to behave? I lick my lips. "F-forgive me."

His gaze falls to my hands in my lap, clenching the fabric of my dress. He reaches out and gently pries one of my hands free. My heart picks up its erratic rhythm. What is he . . .?

He takes my hand, turns it palm up, and traces the pad of his thumb across the glistening sweat. I flinch, mortified, and try to pull my hand back. He tightens his grip. I stop fighting. He pulls something from the inside of his vest—a kerchief.

I watch dumbfounded as he wipes my hand dry, then takes the other and dries it as well.

"You're afraid of me."

I look up at him, find those intense blue eyes fixed on me. "F-forgive me."

"Have I done something to scare you? Are you afraid of what I am? Has someone terrified you with tales of what I will do to you?" His face comes closer as he speaks, as if that will help him read whatever I'm trying to hide on mine.

I keep my back straight as Vivienne's, forcing myself not to pull away from him. I speak the words slowly, hoping it will help them come out smoother. "I d-don't . . . mean . . . to offend you, my l-lord."

A line creases his forehead. When he speaks again, his voice is soft and gentle. "Call me Ash."

Ash? Not Trenian? Is this some kind of nickname? I manage a nod.

He takes my hand in his, heedless of its clamminess, and holds my gaze so intently I can barely keep from looking away. "Isabelle. I will not hurt you. Not tonight, and not ever."

My lips part. The tension in my shoulders eases slightly.

His mouth hardens, and though the words are quiet, there's an underlying fierceness to them that leaves me breathless. And a little terrified. "Hurting you would be like ripping the wings off a beautiful butterfly—it would be monstrous. And though I may seem a monster to your mortal sensibilities, I can assure you that you will not find an enemy in me, little wife."

I lower my gaze, but this time it's because I don't want him to see the sudden welling of my tears.

He *is* kind. It wouldn't make sense for it to be a front now—now that I am his wife and at his mercy. If he was truly a licentious brute, he'd have taken me already. Yet he hasn't. He is instead sitting before me, assuaging my fears.

I have the sudden impulse to lean forward and rest my head against his chest. I quickly rein it in. Does this mean he won't . . .? That he doesn't want . . .?

"Look at me," he says gently.

I obey and blink rapidly to hide the evidence of my emotion.

A muscle moves in his jaw. "It has been a long day for you. But I must marry you my people's way before we leave tomorrow."

"Your people's way?"

He looks at me in surprise, as though shocked I would volunteer to say anything that wasn't an answer to a question. Then a quirked brow quickly replaces the surprise. "Would you like me to show you?"

Perhaps I should keep to my restraint. But I *am* curious, and I think my interest delights him. Or perhaps he's just glad I'm not almost in tears anymore. Either way, I nod.

His countenance brightens, and he gets to his feet, tightening his grip on my hand. "Then come with me, darling. I will show you how to marry a fae."

Darling? My face goes hot as he draws me to my feet. He doesn't let go of me as he guides me to the other side of the bed, where there is a stretch of floor. He kneels on the rug, then gives my hand a tug.

"Kneel across from me, facing me. Yes, like that. Now, hold out your left hand to me."

I do as he says, and he takes it, turns it palm up, and presses a kiss to the sensitive skin. A pleasant tingle shoots up my arm, but I don't flinch, despite how a gasp almost escapes my lips. His eyes flash to meet mine right before he pulls his lips away from my hand, and there's something about that golden flecked twinkle that is almost a question.

The kiss isn't part of the ritual, is it?

I'm not sure I'll ever be able to draw a full breath again.

He places his left palm flat against mine between us, and the size difference is almost comical. My pale skin against his dark, my extended fingers barely passing his knuckles. We sit there, kneeling with our hands between us, and I wait with bated breath as his head bows and he mutters words in an unfamiliar language.

My eyes widen. Our hands start glowing, as though a bright candle is hiding between our palms and casting light through the gaps between our fingers. A breeze catches the strands of my hair, teasing them from my braid and whipping them around my face. The prince—no, *Ash*—keeps muttering, and as he speaks, a shining dot appears on his chest, right over his heart. It grows, becoming a gleaming golden thread that winds around his left arm, coming straight for our joined hands. Part of me wants to jerk away, to let out a short squeak and scramble to safety.

But he promised he won't hurt me . . . and I believe him.

So I stay.

His head lifts, enough that our eyes can meet, and he speaks clearer, firmer. The language is still unintelligible to me, but the intensity of his expression makes me think this is serious, whatever he is saying. A lock of his hair falls onto his forehead, the chiseled edges of his face painted in light and shadow.

Then he says my name, right in the middle of that string of nonsense. The glowing gold string wraps around our hands, binding them together. At some point, my mouth has dropped open, and I'm gaping. At him, at this string from his heart.

He stops speaking, lowers his head, and winces. His free hand presses against his chest as he takes a deep breath. The glow around our hands doesn't subside. Is this exhausting for him? Draining or painful?

"My lord? Are you well?"

BRIDE OF THE FAE PRINCE

"Ash. Not lord. *Ash.*" He lifts his head. "And I am fine. Merely struggling to access my magic in this stifling air."

"Would it help to open a window?"

He blinks at me in surprise for several long heartbeats, and then a grin spreads across his face. "I wish, love. Now, it's your turn. Can you repeat after me?"

Repeat after him? In another language? I struggle enough in my own. My toes tingle with numbness. "B-but I d-don't have magic," I say.

"You will have mine to draw from."

"It w-will h-hurt you?"

"Think of it like sprinting as compared to walking. It is only painful when done too long."

I give a slow, hesitant nod, not wanting to speak again.

"Are you ready?"

I nod again, my throat closing.

"*Rometh elrial tadoth,*" he says.

I take a deep breath and squeeze my eyes shut, twisting my free hand into my dress to hide how it clenches into a fist. My heart hammers, my throat turning to sandpaper. "R-r-r-rometh elri-i-iel tadooth."

"Good," he whispers. "*Samens lir dyketor.*"

"S-samens leer dye-dye-ketor."

"Very good." He gives me more sounds to make, but with each of his gentle reassurances, my anxiety eases, and with it, my words become clearer. Then he gives me his name. "Trenian Ashrift Solavirth."

"Trenian Ashrift Solavirth," I repeat. I open my eyes, and then open them some more in shock when I see a pearlescent white strand twining around my arm and wrapping around our hands. I almost jerk away and try to shake it off.

A few more collections of strange syllables, and then Ash announces, "That's it."

The glow of the strands around our hands illuminates his face. A kernel of wonder hatches in my breast, expanding into something that swells almost too large for my rib cage.

My husband is everything I could have dreamed of—and more. How were the cards dealt thus, that my husband would be both handsome and good, strong and kind?

"What did I say?" I ask.

"You pledged yourself heart, body, and soul to me and no other, as long as we both shall live. And you gave me permission to forgo addressing you by your royal title."

"And you promised me the same thing?"

"Similar things, yes."

"What's next?" I breathe.

"I said I would never hurt you, but I must make a slight caveat to that. The bonding needs blood to finish. Only a drop from each of us. May I prick your finger?"

With his free hand, he draws a knife from his belt. It's long and wicked, curving in jagged edges from top to bottom, with a point so sharp it catches the light of the magic binding our hands. *What if this is when the kind façade washes away, and he* does *kill me after all?* My first instinct is to flinch away from it, but his hand curves around mine. His thumb trails a gentle caress down the side of my pinky and hand.

"Have no fear. You may do it yourself if you prefer."

I won't be a coward. It's just a prick. I shake my head. "Y-you."

"Then I will do yours, and you will do mine."

"Oh, no! I couldn't hurt you!"

Something shifts in his face. Something hard—and there's that glimpse of the wildness deep inside him. The wildness I suspect he has tamed for me in this bridal chamber.

"It will not hurt me."

I don't believe him, but I focus myself on not flinching as he maneuvers our joined hands so the pad of my thumb sticks out. It throbs in anticipation. His eyes flick to mine, then back to my thumb. He sets the tip of the knife against it, letting me take a deep breath. Then there's a prick of pain, and I give a little gasp. But just like that, it's over, and we both watch as the drop of blood wells on my finger and spills over the glowing threads. They react almost immediately, the gold and pearlescent strands fusing together as one. My eyes are wide and fixated on the sight so that I miss when he holds the knife to me, hilt presented toward me.

"Are you sure it won't hurt you?" I ask, my fingers closing around the hilt. They flex, unfamiliar with any type of blade.

His mouth twists wryly. "Just don't cut off my finger, darling."

Cut off his finger? Oh, gracious heavens have mercy, what a wretched thought! I really shouldn't be holding this knife. I don't know how to use it. What if I press too hard and run the knife straight through his palm? What if I permanently maim my new husband?

"I will help you." His hand takes hold of mine, covering it where it wraps around the hilt of the knife. He exposes his thumb and carefully guides the blade closer until its tip rests against the pad of his finger. "There. Now just apply gentle pressure."

"H-how g-gently?"

He smiles, then applies the pressure on my hand. It's just a tiny bit, and then blood wells. I pull back quickly, and he lets go as I set the knife down. "D-did I press too hard?"

The thick thread that was once our separate threads glows in response to Ash's drop of blood. It brightens, brightens, until it's so bright I shield my eyes.

Then, abruptly, it goes out. I slowly lower my hand.

It's dark again in the room, with only the gentle light of the candles to illumine the broad shoulders and bright eyes of my new husband. Our hands, pressed together, are no longer bound by magic.

I start to withdraw my hand, but his fingers thread through mine, clasping our hands together. He leans toward me. My heart lurches as his other hand reaches for my face, coming to rest beneath my jaw and around my neck, pulling me forward

Is he going to kiss me? Do fae marriages end the same way human marriages do? This time I have no veil—it will be just our lips, nothing separating them. Sweat breaks out on the back of my neck as he comes closer, closer, closer.

But he doesn't kiss my mouth.

His lips land on my forehead, soft and slow. As if he is in no hurry to pull away. Then, to answer my question, he murmurs against my skin: "No, you are perfect, little wife."

CHAPTER 13
THE PRINCE

GREAT KINGS CURSE me. She's beautiful, and I like her. This is not how this was supposed to go tonight. I'm not supposed to like her. I'm not supposed to love that she smells like sweet lavender. I'm not supposed to want to unwind her braid and run my fingers through her hair.

She was supposed to be ugly, and I was supposed to do what I could to make her comfortable and, if possible, something close to happy as long as she's my wife.

She wasn't supposed to have the largest doe-like eyes I've ever seen. She wasn't supposed to say unexpected things, like offering to open the window when my magic wasn't flowing well. She wasn't supposed to be concerned she would hurt me.

I have the urge to ask her questions, to draw out more unexpectedness from her. Somehow I know that the more I discover about her, the more I'll like her.

What we have between us is tentative. I don't want to ruin it. I don't want to break her hesitant trust. But there is a part of me—a part of me I quickly suppress—that wants to kiss her, to see what she will do and how she will react. To see if she wouldn't hate it. To see if she would . . . *No.* I'm not following this line of thought.

If she wasn't looking at me, I'd probably wipe my hand down my face and let out a few growls of frustration.

Of course, I must care about her to some extent. I need to preserve her life as long as possible. But *liking* her, feeling drawn to her—it's dangerous. Much too risky for the road ahead of us.

Far too dangerous for Faerieland.

I sheathe my knife and stand before offering my hand to help her up. Once she's on her feet, I march back over to the table with the decanter. My throat is parched, but I don't want to drink that moldy excuse of a brew. When I search the room, I find a pitcher of water in the bathing chamber. It also tastes bad, but I down three glasses of it. My back is to her, and I don't want to turn around until I've collected some composure.

My heart sings from the bond we just made, and I try to tamp down on its exhilaration. It didn't respond to the human ceremony, of course, but now I feel it. Deep in my soul. The fact that this woman is now my wife, and I her husband. My hand buzzes where it held hers—the aftereffects of magic. Sustaining her vows proved to not be as strenuous as I expected on my magic. A small thing to be grateful for. A nice little distraction to focus my mind on, instead of the fact that I have a *wife* and she is in this bedchamber with me. Alone.

I wanted this. But now that we're here, it's almost more than I can take. A horrible consideration rushes through my mind, of ensuring her comfort, then bidding her goodnight and crashing on Rahk's floor for the rest of the night.

I doubt she would appreciate that. Besides, I married her. It's my job to stay with her.

If I turn around, I will be forced to look both at her and at the bed that fills the room. But there is one other thing I must do tonight. So regardless of how fractured my composure is, I must regain some semblance of control. I grip the table, grip it so hard it gives a little creak.

Then I let go, releasing a deep breath. I turn around.

She's standing where I left her, on the opposite side of the bed. Her white gown is stark in the candlelit dimness. Its simplicity suits her lovely frame much better than that flurry of fabric she was wearing earlier. Would she object if I asked her to unravel her braid?

I won't ask. I've kissed her twice already tonight, and that is two times more than I should have.

But tonight, it's just us. Tonight we are nothing but newlyweds, a bride and bridegroom. I don't want to think about how things will have to be between us when we go home. All I want is to just be a man with his new wife.

She is quiet standing there, wide eyes blinking slowly at me. She's not sure what is next. I think I've consoled her enough for her to know that I meant what I said about not hurting her. Still, I should probably say something—something more explicit. At the same time, she flushes so easily . . . Would she rather I didn't acknowledge *expectations* and simply move past them as if they were never there to begin with?

I open my mouth, but the words don't come. The only thing that will is: "You need another name. Isabelle Louise is a lovely, practical name. But it's just not safe or suitable for Faerieland."

"Safe?" she repeats, without stuttering. Satisfaction curls in my stomach. She doesn't stutter when she is reacting genuinely, and it pleases me to see the sincere sparks of confusion and curiosity in her instead of the stilted, rehearsed *"my lords."*

"Names have power in Faerieland, and if you are to be my wife, we cannot give others power over you. And we need something suitable for ballads. Something more romantic, with a touch of tragedy."

She nods once, then tilts her head. "Ballads?"

I cannot help my grin. Cannot help skirting the bed to come closer to her. I lean against the wall just a couple feet away from her, crossing my arms as I look down at her. "Of course. You, my star, will be sung of for ages to come. That is what comes of marrying the next High King of the Fae, and being his human bride, no less."

She seems to consider this, her brow wrinkling slightly, as if the possibility of being the subject of ballads has never occurred to her.

I'm not lying. She will go down in history as the High King's beloved human wife. Stories will be told of her, regaling her beauty and goodness. She will be the subject of legends.

Of tragedies.

I will be sung of, too. As the lover who wept over his dead bride's body.

I focus on the floor between us, tapping my chin as if I am thinking of a name for her. I already have one—I thought of it after our dance. What I'm actually thinking about are ways to block out the image I have in my head of her doe eyes, wide and unblinking, frozen in death.

"Am I supposed to give you a name, too?" she asks.

I look up. There's that surprise that thrills through my blood and draws a line of pleasure in its wake. I smile. "If you want to, but there's no tradition necessitating that. Does my name displease you?"

"No more than Isabelle Louise displeases you."

Her reply is so unexpected I actually laugh. *There's that spark she hides.* Warmth courses through me again, followed by the urge to close the distance between us, to discover how she reacts to my touch. But I force myself to remain against the wall as I try not to notice the way her lips twitch at my reaction.

"Stella."

Her eyes meet mine, a question knit into the twitch of her brow.

"Do you like it?" I hope she does. It suits her very well, in my eyes. "In our language, it means *the one from the stars.*"

"Stella," she repeats, trying it out. Then she looks up and smiles. "It's beautiful."

Great Kings have mercy on the idiotic way my brain stops working for a minute. I look away and clear my throat. "Well, shall we take rest? Would that please you, Stella?"

BRIDE OF THE FAE PRINCE

Her eyes dart to the bed, then back to me, not quite meeting my gaze. Her lips part, but she quickly clamps them shut and nods. She's nervous again. I'm not sure how to fix it. Then I look down. I'm still fully clothed. The last thing I want to do is start stripping for sleep and terrify the wits out of my wife. I guess I'll be sleeping in my clothes.

I shoot a quick glance at her. Will she panic if I take off my overcoat? How can I do this without giving her a heart attack? Everything I think of saying seems like the sort of thing that would just terrify her more, or at the least, embarrass her. Neither do I want to say anything that might hurt her, or imply that I find her undesirable.

I wind up saying nothing as I sit on the side of the bed closest to the door, my back to her. Should any foul play be afoot, anyone who enters this room will encounter me before Stella.

I take off my boots one at a time. Next is my belt of knives. I take my time unbuttoning my overcoat, then my formal tunic, and slowly shrug them off. Then, still without glancing at her, I pull back the sheets and quilt and slip my legs beneath them.

Only then do I look.

She hasn't moved, standing rigid, one hand gripping the collar of her dress. She's gone a little pale. My shoulders sag an inch as I gesture to the other side of the bed. "You have nothing to fear," I say gently, and then barely keep myself from cringing as the taste of iron spreads across my tongue.

Her chest rises, falls. She takes a few hesitant steps to the bed, pulls back her side of the sheets, and slips in. We stay like that for several minutes. Sitting upright. Neither of us talking, the gap between us wide enough that a third person could be added, and still none of us would touch.

I need to lie down and close my eyes. Pretend that I'm exhausted instead of very thoroughly awake. Perhaps if she thinks I have fallen asleep, she will relax enough to do the same.

There's just one problem.

This bed is too short for me.

CHAPTER 14

THE PRINCESS

HE ISN'T GOING to hurt me. *He isn't going to hurt me.* I chant it over and over in my mind as I clutch the sheets and try to find the will to lie down. But despite my rational mind insisting that he clearly has no intention of touching me, it is an intimate thing to share a bed. *We are married,* I remind myself.

I just cannot imagine that it's physically possible to sleep with a man only a couple of feet away. Especially *a fae* man.

Ash goes to lie down, making my heart zip straight into my throat. He scoots down into the sheets. When he extends his legs, however, they hit the footboard. His knees are still bent.

"You humans and your puny beds and your puny chairs," he grumbles. "Is six more inches too much to ask for?"

The giggle is out before I can stop it.

Ash's head swivels to mine. "Did you just laugh? At my discomfort?"

If he'd asked me this question an hour ago, I would have apologized as quickly as I could. But now, an apology isn't what escapes my lips. "No, my lord."

He blinks, as though surprised I would blatantly lie to him. Then he grins and leans back against the pillows. "I might have to lie diagonally if I'm going to get any sleep. Or . . ." He kicks off the covers and props his feet up on the footboard. He folds his hands across his stomach. "This might do."

I bite my lip as I smile, then scoot under the covers myself, bringing them up to my nose as I curl into a little ball. I can do this. We can do this.

The relief hits me so hard it blazes past my nerves, straight to my tear ducts. I *like* my husband. At least, what I've seen of him thus far. How is such a thing possible? How am I lying here on my wedding night, which should be the most terrifying night of my life, and instead I cannot stop thinking of how fortunate I am? Just a few nights ago, I faced the possibility of a future as Prince Brochfael's sixth wife.

Yet here I am now.

I need to cry. Tears of relief, of gratitude, of hopefulness. But I cannot cry. That was one of Father's strictest instructions. No tears on the wedding night. I'll save them for when I am alone—or after my husband falls asleep.

"Come, tell me one of those thoughts flying through your brain."

His voice makes me jerk to attention and blink up at him from over the edge of the covers. He's looking at me with that one eyebrow cocked. "M-my what?"

"One of your thoughts. I can see them racing around your mind. Tell me one of them."

Tell him? What I'm thinking? "I am not sure that is a g-good idea."

He rolls onto his side so he can face me, his elbow propping up his head. "Is that so? Why?"

I shrink a little lower into the sheets. "B-b-because they m-might n-not please you."

My father would be vastly displeased if he knew the contents of my thoughts.

That is an understatement. He'd be furious. And then he'd berate me like always—and there would be no place for me to go, no place for me to hide without incurring greater wrath.

"Please me?" Ash scoffs. "What has that to do with the price of pixie dust in the swamps?"

I blink once. Twice. "Pixie dust is real?"

"Of course it is. Now don't try to distract me. Tell me one of those thoughts that might not please me."

Now I just want to fling the rest of the covers over my head as my face goes hot. What could I tell him? That he's better than I expected? That I thought he was going to abuse me tonight? That I think he's the handsomest man I've ever seen, and I really didn't think I'd marry anyone remotely nice to look at? That this is all happening so fast? That I'm relieved he doesn't find me ugly?

"Or else," Ash says, noting my hesitation and leaning closer, "I'll kiss you."

My mouth suddenly goes very, very dry.

He smirks. But is that just a tinge of hardness on the edges of that smirk? "So what will it be: a thought, or enduring a terrifying kiss?"

A kiss would indeed be terrifying, but probably not in the way he's imagining. I doubt there would be much *enduring* on my part.

Well, those thoughts will just be added to the growing pile in my mind that I can never, ever say to him.

But when my hesitation continues, as I wrack my brain for *something, anything,* to say to him, he scoots closer. Bringing his face nearer to mine.

"A thought, or a kiss," he says again, his voice lower than before.

Quickly, quickly—something! Anything!

"Um . . . I l-like f-food," I blurt out desperately.

His face halts above mine. "What a displeasing thought."

Now I feel like an idiot, but something about his reply loosens an inch of tension inside me.

"Are you hungry?"

I shake my head, my belly still very full of the maid's earlier offerings.

"You're sure?"

I lick my lips and nod.

"Then let's have another thought."

Another one? I wrack my mind, clenching the sheets tighter in my grip. "I . . . d-don't know what you w-want." *I don't know what you want me to say.* But I stop before I stutter too much.

"What I want?"

I nod.

A strange expression crosses his face, one that I cannot read one bit. "I don't want you to be afraid of me."

My mouth opens beneath the covers at his honesty. His wish, however—that's impossible. "F-f-forgive m-me."

He frowns. I've said the wrong thing, haven't I? I cringe.

But then his hand comes to rest on top of my head. Warmth seeps past my hair, into my scalp. I peer up at him just as he bends down and presses his warm lips to my temple. Heat floods me to my toes. I stiffen.

"There's nothing to forgive," he murmurs, his mouth above my hairline. "Sleep well, Stella."

Then he rolls over, kicking his feet up on the footboard and falling back against the pillows before closing his eyes. With a flick of his wrist, all the candles in the room go out. Darkness wraps us up like a blanket.

Breath puffs in and out of my mouth. I taste the linen of the sheets and the faintest fragrance from their last wash. My eyes are wide, too wide, as if I can make them adjust to the darkness faster. I've heard fae have excellent night vision.

I want to cry, but it's not safe yet. I need to be sure he's asleep before I let myself shed a tear.

There's no chance I'm falling asleep with him right there. What if he changes his mind about . . . *expectations*? What if he wakes up in the middle of the night, and he's a different person? What if I just ruined the tentative trust we were building with my reluctance to speak? Father would have punished me for speaking my mind. How am I to be sure Ash is different? Besides, why would it matter for him to know what I'm thinking?

I lie awake, not moving a muscle, as I wait for Ash's breathing to even. My mind spirals deeper and deeper into doubt, which is ridiculous, because I ought to be thankful for how fortunate I've been so far. I shouldn't be scared or worried about the future or about losing my husband's good opinion. Did I even have it to begin with?

Can I even trust my own ability to know if he likes me or not?

Does it *matter*?

It's impossible to know for certain, but I guess an hour or two has passed since Ash rolled over. He hasn't moved. His breathing has evened. It should be safe now, right?

I turn my face into my pillow, and as quietly as I can, loose the dam I have on my tears. I don't let myself sob—that would be too loud. To keep from sniffling, I breathe through my mouth.

I cry out my fears of tonight; I cry my gratitude, and then I cry for the looming future I don't understand. I cry that I don't know what my husband wants from me. I cry because he has been kind to me, and I like him. I cry because after tomorrow, I will never see Amelia again.

Finally, the tears slow. I keep breathing through my mouth, keep fighting to stay silent. Then, as carefully as I can, I push up on my elbows, wipe away the tears stuck in my lashes, and try to peer into the darkness at my sleeping husband.

It's hard to see much except the broad expanse of his frame, his wide shoulders and chest, his long legs sticking out over the end of the bed. One of his hands is beneath his head, the other lying across his stomach. He doesn't lie beneath the covers, and I don't understand how he's not cold.

His hair is much longer than human men typically wear it. I don't mind it as much as I thought I would. Though he could be bald, and it wouldn't dim his beauty.

Even so, I'm glad he's not bald. Prince Brochfael is bald.

A sudden thought occurs to me. If he is to become the High King of Faerieland, then does that mean I'll be Queen of the Fae? I shove the thought out of my mind before it chokes me, but not before my stomach bottoms out.

What does it even mean to be queen of his people?

I won't worry about that now. I just need to make it through tonight. Later, I'll worry about the rest.

"Stare at me any longer and I'll fear you're plotting to put a dagger through my heart."

His voice is a dagger through *my* heart. Something a little louder than a squeak but quieter than a scream bursts from my mouth as I scramble back, back, away—

The bed ends. I go tumbling off the edge, scrabbling with my nails for any kind of hold on the sheets. They give way, and I land on the floor, my skirts riding up my calves as I catch myself with my hands.

Faster than I expect, Ash is kneeling beside me, a hand on my back and the other planted on the floor beside my knee. "Great Kings, wife, did you hurt yourself?"

I shake my head, despite my rattling teeth, and yank my dress back down to my ankles.

"Are you sure?"

I nod and bite my lip. It's too dark to see clearly, but his brow is knit in concern. As if he's genuinely worried about me. I'm more worried about what I'll be like in the morning when I've gone the whole night working myself into an anxious tizzy and not getting a wink of sleep.

His words register belatedly in my brain. *Stare at me any longer and I'll fear you're plotting to put a dagger through my heart.* He *knew* I was studying him? He was awake all this time? Did he hear me crying? Did my crying wake him up?

The blood drains from my face. His expression is hard. Is that anger glimmering in his eyes?

"I'm s-s-sorry I w-w-woke you up." Why won't my stupid tongue work? I just—I want to go hide somewhere. I don't want my husband looking at me right now, seeing the tearstains on my cheeks and beneath my eyes.

He scoops me up, eliciting a surprised hiccup from me. This feels different from earlier, without mounds and mounds of fabric between us. It is far more intimate, the warmth of his strong arms burning into my back and knees through the thin material.

He sets me on the bed without a word, then walks around to his side.

I don't know what to say to get rid of that expression on his face. I want something lighter, brighter, happier on his face. I want him grinning again, chuckling, or even laughing like that one glorious moment earlier. But even if I did know what to say, my tongue wouldn't cooperate.

I don't think I could bear it if he is angry with me.

He sits on his side of the bed, his back facing me. Then he tilts his head and asks, "Would you like me to hold you?"

My brain sputters to a stop. What does that even mean? I'd have understood better if he asked me if I wanted him to do cartwheels for me. I stare at him blankly until, eventually, he turns to look at me.

"If you'd rather, none of this just happened and I'm still asleep."

A stray tear leaks out of my eye at his offer. I thought I was done crying, for heaven's sake! A sniffle escapes me.

Ash heaves a sigh, swings his legs onto the bed, and twists toward me. "Just come here, Isabelle. It's alright. I won't hurt you."

The effort to restrain this new well of tears grows more monumental by the moment. A few more tears leak free despite my best efforts. I don't want him to see me cry—I'm not supposed to cry!

I don't know who moved, whether it was me or him or both. All I know is that one minute, I'm sitting on my side of the bed and he's sitting on his, and then we're in the middle, his arms wrapped around me and my face buried in his chest.

He says nothing as I cry, just holding me tightly and wrapping me in his warmth.

"I'm s-s-s-sorry," I gasp. Getting the words out is like fighting against a river's current. "I d-d-didn't w-w-want to d-d-disappoint you."

"Shh, don't talk. It's alright."

How is he not upset? Or angry? It's the reassurance I need to tuck my head beneath his chin and stop fighting to explain myself. He holds me through my tears, stroking my back. At some point, I become aware of his fingers fidgeting with something near my waist, and then I realize he's pulled the ribbon from my braid as he unwinds the long strands and tangles his fingers into them. It feels even better than when he stroked my back, and slowly, it soothes me until my tears have finished into little sniffles and hiccups.

"Thank you," I mumble into his chest. "I'm sorry for crying."

"I don't mind."

I find that hard to believe, but my shoulders relax anyway. "I can explain."

"I don't require an explanation, little wife. But if you want to talk, I'm happy to listen." He twirls a strand of my hair around his fingers.

I hiccup. "I didn't think you'd be this kind."

"Neither did I."

I sniffle, arching my neck to look at him. He peers down at me and gives me a little smirk.

"What do you mean?" *Hiccup.*

"I've never been married before. And I can assure you, I'm not this nice with people like your father."

Somehow, that makes me giggle, and his arm squeezes me closer.

"But you're so old. How could you not have been married?"

He chuckles, and the sound rumbles through his chest and into mine. I like it. Very much. "We're not like humans who bond as soon as they reach adulthood. It is not uncommon for fae to be three or four hundred years old before bonding."

"But you're almost five hundred years old, right?"

"Mm hmm." He combs through my hair with his fingers, and my eyes close despite myself. I'm so tired. Much more tired than I realized.

"Is it strange"—my question cuts off into a yawn—"to be so old?"

He chuckles again. "I am not so old to my people. In fact, I'm still considered young. We mature differently than humans. Much slower, actually."

"Really?" The quiet question comes out on an exhale. He smells so soothing, like a forest after a rainstorm. And he doesn't seem angry with me at all. I almost want to fall asleep like this. "How did you stay awake for so long? I thought you were asleep."

"We don't need as much sleep as you do. I thought you would feel more at ease if you thought I was asleep. I'm sorry I frightened you. I certainly wasn't expecting you to go leaping off the bed like a gazelle."

I give a sniffly laugh in response. "You make it sound so graceful."

"You should sleep."

"Will you?"

"Maybe." His hand pauses its caress of my hair. "Why did your father veil you?"

So Father's claim that it was our custom did not fool him. I sigh, fighting to stay awake long enough to respond. "Because he said I wasn't beautiful like my sisters."

Silence. I almost fall asleep. Then his hand goes back to playing with my hair as he mutters, "Then your father lacks eyes as well as a spine."

And that's the last thing I hear before oblivion claims me.

CHAPTER 15

THE PRINCE

I LAY ON my back, trying to look anywhere but at the canopy over the bed. It's a dizzyingly intricate tapestry design of a large geometric flower that makes me feel like I'm falling into spinning circles if I stare at it too long. Humans have such garish taste in décor.

I knew this night was going to be long, no matter how it went. No matter how ugly the canopy was.

I didn't think it would be this long.

Stella's head rests on my chest, one hand curled beneath her chin while the other rests on my ribcage. My arm is wrapped around her, her long hair falling over it like a sleeve. Her lashes fan her cheek, her pink lips parted as tiny snores fill my senses.

Part of me never wants dawn to come. Maybe we can stay like this, stay forever lost in the hope of something new and sweet.

It's impossible. As sure as dawn comes, Lulythinar approaches. I need Isabelle—*Stella*—to survive until then more than anything, or else every plan and plot I've devised will be for naught.

But I cannot stay here, or anywhere else. I must take her back to Valehaven at first light. It would be easier to keep her alive until Lulythinar if I didn't. Doing so, however, would ensure that I win the battle and lose the war.

I don't play games I intend to lose, and I have much to do in Faerieland before Lulythinar.

After tonight, I am even more determined to win both the battle and the war. I must, for the sake of the butterfly asleep on my chest.

The first stains of dawn leak through the window. Stella shows no sign of stirring, and the last thing I want to do is wake her up. She needs sleep, especially after yesterday. I tell myself that is why I don't get up when I ought, why I stay here holding her.

It's definitely not because everything will change between us today. Because once the sun rises, I am no longer a bridegroom with his new wife.

I am the future High King of the Fae with a throne to win.

THE PRINCESS

I SLOWLY STIR AWAKE, stretching my arms and legs, arching my back. Something tightens around me, and I realize the warmth surrounding me isn't from the covers. My eyes blink open, and I find myself staring into those gold-flecked eyes.

"Ash," I say stupidly as heat licks up my neck into my cheeks.

"Stella."

His use of my new name only serves to deepen my flush as awareness spreads of how I'm pressed against his side and half lying on his chest.

"You're awake," he says.

I shake my head.

One eyebrow rises. "You're not awake?"

"Nope."

A pleased smirk stretches across his face. And then he rolls toward me, his free arm coming around me as he pulls me into an embrace. His lips press against the skin near my ear. A gasp catches in my throat.

"Then you won't hear me say that you have the sweetest little snores," he murmurs.

I go stiff in his arms. "Snores? I s-s-snore?"

He chuckles, his hand sliding up to tangle in my hair again. He pulls back enough to look down at me, and I think I might drown in his smile. Then, just a swift instant, and his eyes drop to my lips. He immediately looks away, clears his throat, and unwraps his arms from me, sitting up and throwing his legs over the side of the bed.

Cold washes over me. I draw the blankets up to my shoulders and try not to shiver.

"A tray was brought in while you were asleep," Ash says as he pulls on his boots, buckles on his knives, and swings his overcoat onto his shoulder. He nods toward the table where there is, indeed, a tray of food and a freshly brewed pot of tea. "Your maid is waiting outside to aid you in dressing, I assume." He doesn't look at me, drawing that briskness around himself like a cloak. "I will be back shortly."

He's leaving me? Why does my gut sink like I am disappointed? I sit up, holding the blankets to my chest and tucking my hair behind my ear to keep it out of my face. "Where are you g-going?"

"To see that my people are ready to leave. We must not waste time in returning to Faerieland."

"Will I see my f-family?"

"Do you wish to?"

I nod.

"Very well. Then you shall bid them farewell before we leave."

Leave. Leave—the human world. My people. Aursailles. My family. Amelia. This is when the rest of my life begins, in a strange place and among strange people I do not understand.

After last night and this morning, I'm less afraid of it all. Which is perhaps not a good thing. But Ash is someone I can trust, someone who will be good and kind to me.

Perhaps we've been wrong about the fae.

Whatever the case, if this marriage, this life ahead of me, will save my people, I will face it.

Ash bids me good morning and leaves me to eat my breakfast, while a different maid than the one who attended me last night comes to brush out the tangles he left in my hair.

CHAPTER 16
THE PRINCE

PART OF ME hopes Rahk won't be waiting outside our chamber. Of course, he is. Why does he have to be so dependable? I stride past him and launch into instructions for preparing to leave. "Please arrange for a carriage—"

"Already done," he says, matching my pace.

"Each fae we brought needs to be accounted for. I won't have one getting left behind and wreaking havoc."

"Already done. Your warriors are guarding them now."

Each order is met with the same answer, that it's already been done. I stop, frowning. "Did you do all this?"

"Me? I've been guarding your door since last night. Edvear did everything and reported to me early this morning that once you gave the word to leave, they're ready."

I give a rueful chuckle. "He is always a step ahead of me. I'll be quite bereft when my father gets around to killing him." My tone isn't quite as cavalier as I intend.

Rahk glances at me, too knowing. "It went well then?"

I look up. "What went well?"

He merely raises an eyebrow.

Last night.

"It was fine," I reply, quickening my step.

"Are you . . . *blushing*?"

"Why would you ask such a ridiculous question?" I growl.

Rahk's jaw falls open. He grabs me by the shoulder, accusation heavy in his thick brow. "Did you—? You said you wouldn't! You know you can't risk an heir!"

"I didn't!" I snap, flinging off his arm. "Of course I didn't. I'm not an idiot."

"The color on your face might disagree."

"Keep talking and the color on *your* face might start having opinions too."

Rahk chuckles and says nothing more. I wipe my hand down my face. Great Kings, am I actually blushing like a maiden? This is a pathetic turn of events. I need to pull myself together. Yes, I almost kissed her a few minutes ago, but it means nothing. I maintained my control, and I will keep maintaining it.

Staying focused will be much easier when we return to Faerie.

THE PRINCESS

I STAND BEFORE THE MIRROR, the empty tray on the table and my maid gone, holding yesterday's veil in my hands and considering it when the door opens.

I turn, find Ash looking at me, at the fur-lined sky-blue travel gown I wear, at my hair bound up into a bun at the nape of my neck, then down at the veil in my hands. "Don't even think about putting that back on."

Somehow, that makes me almost smile. I drop the veil and face my husband, clasping my hands before me.

"Are you ready?" he asks.

My heart gives a nervous little jolt, but I nod. When he extends his hand out to me, I take it and grasp my skirts in the other as he escorts me out of the bridal chamber. The white roses have wilted, leaving petals fluttering down to the threshold like ghosts.

"Your trunks have been loaded," Ash continues, hardly looking at me. "If you still want to say farewell to your family, they are waiting for you at the entrance. You may see them alone, should you prefer that?"

"I'd prefer you come," I say quickly, and then regret it. "I-i-if you w-wish."

"I have nothing drawing my attention elsewhere."

I'm not imagining it. He's more distant now. I hope it wasn't something I did or said, or perhaps I did *not* do or say? I shove the thought away. If he's still distant when we arrive, I'll consider it then. Not before.

"T-thank you."

He takes me straight to the entrance of the palace, where a line of servants stands at attention before the great double doors flanked by armored guards. Footmen stand at the handles, ready to open them.

Father's chin is lifted as Ash and I walk through the door. My sisters are to his left, in order from youngest to oldest. They still wear their veils, making it impossible for me to judge the expressions on their faces.

"Prince Trenian," says Father.

"King Roland," replies Ash with a sardonic smile.

"It has been a pleasure."

Ash's cold grin widens as he tugs me closer to his side. "The pleasure was all mine."

Father's gaze shifts to me. I duck my head. "Farewell, daughter."

I don't respond immediately, using the time it takes to curtsy to carefully gather my words on my tongue, so I do not stutter. "Farewell, Father." When I rise, something hot sparks inside me, and I meet my father's eyes for the first time this morning. *Anger.* Anger that he gave me to a man he was certain would abuse me. Anger that he's giving my sisters to similar men. Only Amelia seems to be marrying a true gentleman.

I'm angry that he made me feel small my entire life.

It's that anger that hones my words as I go one by one, saying goodbye to each of my sisters by name without stuttering. When I get to Amelia, her shoulders shaking subtly, I throw decorum to the wind for possibly the first time in my life. I reach out and wrap her in a tight hug. She hugs me back, her nails digging into my bodice as a choked sob puffs her veil.

"I love you," I whisper to her, swallowing back my own tears. "You are my dearest treasure in the world."

"How do I know you will be alright?" she whispers so quietly no one else can overhear.

I tighten my arms. "He is good, Amelia. He was kind to me. I know I will be happy with him."

"Are you sure?"

I smile, give her another squeeze, and step back. I wipe away the lone tear that escaped my guard. "I'm sure."

Then suddenly, a bolt of frantic energy hits me. The servants packed for me. I don't have anything that I'm attached to, except—

"My plants!" I gasp, whirling around, as though they might be sitting by the door, just waiting for me to scoop them up. "I didn't—"

"Plants?" Ash asks, and then turns to the line of servants. "Fetch the princess's plants at once!"

"They're in your trunks!" Amelia assures quickly. "I made sure they didn't get left behind."

I sag in relief, and share one last grateful glance with my little sister before Ash offers me his hand. "Then are you ready, my darling?"

I draw a deep breath and place my hand in his. "I am ready."

CHAPTER 17
THE PRINCESS

ASH HANDS ME into the carriage, then climbs in behind me. He sits opposite me, legs spread wide, elbow resting on the window. He props his chin up on his fist, staring at a tiny gap between the curtains.

I fold my hands in my lap and stare down at them. A thousand questions fill my mind, but I cannot ask them. My heart still thrums with the strangeness of forever goodbyes, denial beating like a drum in my breast. That cannot have been the last time I will see my family ever again. It's unfathomable. I'll never see my maids again. Never see the single portrait of my mother hung in Father's study. Will my plants grow in Faerieland, or will they die?

My life is with my husband now. I steal a peek at him, find him exactly how he was a moment ago. His brow is drawn in a thin, tight line, as though he contemplates something serious and concerning.

I'm too nervous to ask.

It's better to be quiet and watch, to take my cues from him. Right now, he's not giving cues. I stay silent.

The carriage gives an unexpected lurch. I barely catch myself on the seat, splaying my hands to brace myself. Ash hardly moves. I let out the breath I was holding.

Then the carriage lurches again. Harder than before. I don't have time to catch myself this time. I go flying forward, straight toward Ash. His head turns, eyes widening, right as I land with my palms on his chest, trying to stop my fall, my face just shy of his—our mouths nearly touching. He instinctively catches my waist, his arm wrapping around me. He gives a short swallow.

Then he blinks, and his mouth quirks. "Can't restrain yourself around me? I don't blame you, darling."

I turn scarlet. I try to get a few words out, but I have neither a reply nor a willing tongue. I'm frozen, unable to move, trapped.

His expression softens, and he tilts his face up to nuzzle his nose against mine. Shock ripples through me, blazing across my skin. I blink twice, then again. Finally, my body obeys what I tell it, and I give a little push on his chest to put me back in my seat.

His hand catches hold of my wrist, and when I look up at him in surprise, he tugs me down to the seat next to him. It's not exactly roomy, and even less when he spreads his arms across the back of the carriage, all but putting his arm around my shoulders.

"Well!" he says, and I'm not sure he knows what he intends to say. I wait.

He swivels his head to mine. "Tell me one of your thoughts."

Again? "But I told you two last night!" The protest is out of me before I can stop it.

"What can I say? I'm a greedy husband. So tell me one of your thoughts, or else I'll kiss you."

Panic bursts in my chest. Not *bad* panic—just . . . a thrill of something heady that is more than a little terrifying.

"Um . . . are we almost there yet?"

He snorts. "That hardly counts as a thought, but I'll humor it because you're cute. We have most of the day ahead of us. Once we reach Caphryl Wood, however, the rest of it will go much faster. We'll be home by nightfall."

My brain vaguely makes the connection that Caphryl Wood must be the name the fae have for the Long Lost Wood. The main thing I can think, however, is: *cute?*

It's too warm this close to him. It would be better if I were on the other side of the carriage.

"Another," he says, looking at me expectantly, as if he's calling for a refill of his drink.

A tinge of boldness surfaces from some hidden part of me I didn't know existed. "What if I asked *you* for a thought? Or else I'd kiss *you?*"

A wide grin spreads across his face. This travel day will end up going by much too quickly if he wears that often.

"If you did," he says, ducking his head close to mine, "I might just have to see if you intend to make do on your threat."

My stomach flips over itself. I should have known he'd say something like that. I sit up straighter, attempting to show him that he doesn't ruffle me. "That will suffice for a thought from you."

"Wait—no!" he protests, laughter edging his words. "I retract my words."

I can't help the tiny little smile that plays at my mouth. "No retractions."

We stare at each other for a minute, until I blush and shift my focus back to my hands, and he goes back to staring out the window.

It's not much later that Ash says he needs to speak with his steward and Rahk, but that he will be back shortly. I stay in the carriage, and can't quite find the guts to peek out the window. Not until some time has passed and still he hasn't returned. When I look, we've left the city far behind and travel on an old road through a wood thick with barren trees. Ash is on horseback, riding next to Rahk. Every few minutes, they exchange a few words, but they mostly seem quiet.

He's uncomfortable around me, isn't he? I suspect it's similar to how I feel—uncomfortable, not unpleasant.

The clip-clop of the horses' hooves, the rumble of wheels on uneven ground, and the swish of the carriage's red velvet curtains are the only things keeping me company.

When Ash does come back, it's been a couple of hours. I've managed a dreary half-awake nap, but I don't sit up from leaning against the wall when he comes in. He slides quietly to the opposite side of me, and I close my eyes again.

Except that I take little peeks of him here and there. He spends most of the time staring out the window, but every few minutes, he glances back at me.

I don't know how to interpret this.

We don't talk the rest of the way, except for a few words when a horned fae delivers a basket of packed food for a meal. I catch myself just before I startle and betray my instinctive fright. At last, a whistle sounds from outside, and Rahk comes to knock on the carriage door. "We're at Caphryl, Highness."

My heart quickens. I sit up straighter and smooth my hair. I'm about to meet . . . the rest of my life. *Faerieland.*

Ash shoots me a look, measuring my reaction. I try to keep my face blank and neutral, but when he holds out his hand to help me out of the carriage, there's no disguising the slick of sweat across my skin. He doesn't react, only takes my hand and helps me to the ground.

My legs ache when I land, and I discover my rear has gone quite numb. The chilly air burns my cheeks as I straighten my skirts and wipe my wet hands in their folds.

Then I look up.

Though the world is bright with late afternoon sun, the expanse of forest stretching before me is dark and forbidding. The trees tower high above normal trees, creating an imposing barrier between our worlds. Their branches are thick, with an impenetrable canopy of dark leaves, despite the season. We've pulled off the road, which leads right into the forest, the end of it swallowed by the tree line.

BRIDE OF THE FAE PRINCE

A wind picks up around our party. The horses whinny nervously, and a fae with small horns and bright cat eyes that must be the steward discusses something with the human carriage driver and footmen. Possibly about taking the horses back? Clearly, we're not driving any farther.

The wind sharpens, whipping my skirts and yanking tendrils free of my hair. It smells different too, rich and tingly . . . almost like the color purple—if colors had smells. But there's also a tang to the air, like spices in a drink, or a dash of rum in a cream sauce.

The fae unload my trunks—which are probably full of things I don't need. It seems unfair to make the fae hoist them over their shoulders and bear the burden, but they're the only things I have from my home. I don't want to just leave them here.

It's so dark.

On the wind, something hums, calls softly, like a whisper in the dead of night. The warmth I felt when my new husband nuzzled his nose into mine vanishes into nothing. Icy dread seeps into my toes, into my stomach, my shoulders.

I'm not supposed to be here.

This isn't my world. This isn't where I belong.

I'm *wrong* here.

I need to turn around, to go back to where I belong. I'll die if I walk into the Long Lost Wood. I'll be torn to pieces, or lost like the souls of drowned sailors. If I take one step past that tree line, I will never come out again.

I know it as deeply as I know my own name.

A hand on my low back jolts me back to awareness. Ash's face is hard, the cut of his jaw seeming harsher, his ears longer, more pointed. His teeth just the tiniest bit sharper. I search for that sweetness I've been so surprised to discover, but it's not there in his face.

It's in his touch, however.

I look away, focus on the warmth of his hand at my back as he guides me forward. Toward the tree line.

Every step is a battle. I know this is my future, that this is where I must go. But every fiber in my being resists, telling me this is wrong, this is dangerous, this is my death. It takes everything in me to not turn around and run. I take a deep breath, lifting my gaze to the trees, to the strange lights like fireflies in the forest.

I faced last night. I can face this.

Setting my shoulders, I keep pace with Ash and don't flinch as we approach the forest. The wind picks up, slicing like ice against my skin. That tang fills my nostrils until I'm almost choking with each breath. When we reach the edge of the forest, the wind shifts into strange, wispy voices.

Let me taste your flesh, mortal.

Save us!

Fly away little bird, before his jaws snap your wings.

"Don't listen to them," says Ash.

I look up at him. His eyes are made of iron, flashing with darkness as his hair blows away from his face. When he speaks, the points of his incisors gleam. They're definitely longer than they were before.

"Will you come with me, mortal wife?" he asks, holding out his upturned hand.

Something is different. Something is wrong. I swallow.

I trust him. Even if he frightens me.

I slip my hand in his and nod. He says nothing more, a grim satisfaction playing across his features. He takes a step, draws me after him—until we're swallowed by the forest.

Everything goes quiet. There is no wind. No sound. No whispers. Barely any sunlight. Only an unending, hushed sort of expectancy. A pregnant stillness ready to give birth.

Ash turns around without making a sound. He issues a silent command, and Rahk is at my side a second later. His hand rests on the hilt of the sword at his hip.

Ash lets go of me and strides forward into the forest. At his footsteps, the trees . . . *shift.* They part, as if playing pieces on a board game, sliding away and clearing a path. The ferns covering the ground do the same until all that is before me is a wooded archway and a carpet of pine needles.

With a gentle pressure on my back, Rahk urges me forward. I obey, moving slower than my husband, who marches ahead through the living arch as if he is master over them.

Perhaps he *is* their master.

The rest of our entourage makes no noise as we pass through the forest, despite the trunks they carry. If I didn't look back, I'd have thought they weren't still with us.

Despite knowing nothing about him, Rahk is a comfort at my side. Ash trusts him, so I trust him, too. When the steward comes and takes my other side, not touching me but standing between me and the rest of the forest, I am extra grateful.

Ahead, the trees part before Ash, until they reveal a door. It's a simple wooden door with a brass knocker. Golden light leaks around its edges and through its keyhole, as though opening it would reveal a small sun on the other side. Ash stands facing the door, legs wide and shoulders back as the rest of us catch up. He looks over his shoulder, finds me, and there's something in his gaze that I do not recognize at all.

A shiver goes down my spine.

He reaches out, grabs the handle, and when he twists it, light bursts from the cracks in the door with blinding intensity. I find myself reaching out and gripping Rahk's sleeve, turning my face away.

When I bring myself to look again, Ash stands in the doorway, backlit by brilliant gold light.

Then he vanishes.

Rahk gently moves me forward. When I look up at him, he meets my gaze, and there's a question there. *Do you wish to go before or after me?* I point to him with a little wince. He nods and steps forward.

This time, I keep myself from looking away. As Rahk steps into the doorway, as the bright gold casts him in shadow, I suck in a sharp breath.

Behind his silhouette is a great pair of wings.

117

Then he's through the door, and it's my turn. The steward gives me an encouraging nod that juxtaposes his unsettlingly bright, slitted eyes. I grip my skirts as I force myself to take the few steps to the doorstep.

The door itself almost seems a part of the trees winding around it. When I take hold of the cool knob, there's a rush of sound, of wind, of words.

Fly away, little bird! Fly away—far, far away. Return to your nest before his jaws snap your wings. Fly away, little bird!

I turn the knob and open the door.

The world beyond is not what I am expecting. Rahk catches me with a firm grip on my forearm when I stumble.

We stand on the rocky shore of a crystal sea. The late afternoon sunlight gleams off the water's surface, refracting into blinding rays. Before me is a magnificent, white-stone palace set into the edge of a cliff. Its many floors, open windows, and arched doorways array in such elegance that is so unlike the gray strongholds I'm used to. This palace is one of luxury, with no fear of attack. Hanging gardens and gushing waterfalls cascade down the cliff face. A blustery, warm breeze whips around me—far too warm for the furs I wear.

And there is Ash, standing with his back to me, hands clasped behind him. I take a deep breath, then step to his side. He steals a sidelong glance at me and, in that moment, he looks more like the Ash I've come to know so far. Then he returns his gaze to the palace cut into the side of the rock.

"Welcome to your new home, Stella."

CHAPTER 18
THE PRINCE

STELLA'S FACE IS a little pale, but she bravely sets her shoulders and gives a firm nod.

She has no idea what she is about to face.

"Edvear," I call, turning away from her as my steward jogs to my side, a little breathless from the portal. "Please see to Princess Stella's trunks; that they are safely stowed. You may remit compensation to the fae who joined us. I have no need of them in the future. Tell them to lie low until Lulythinar is past if they value their lives."

Edvear nods, bows, and hurries off. I glance back at Stella, find her rubbing her arm as though she's cold, despite the fact she must be sweltering in that dress. She catches my attention, a question in those large brown eyes.

A question I don't want to answer.

My magic flows freely now, filling my blood, my limbs, lungs. But this air is stifling in a completely different way than the human world was. It's not good to be home.

Rahk watches me expectantly, his arms crossed over his chest, eyebrows lowered. Waiting for me to do right by my wife.

I draw a deep breath. I turn toward her and hold out my hand. She accepts it. It's almost unsettling how much she trusts me now. Perhaps equally unsettling how much I trust her, too. I draw her closer to me, giving in to this new, inexplicable, and constant desire for her to be near.

"I am going to glamour you," I say, tilting my head down to hers.

Her brow knits. "Why?"

Because her mortality smells so strongly that anything within the nearby vicinity will scent her blood and mark her as prey. "Because your dress is unsuitable for our next . . . engagement." It's not a lie.

She glances down at her travel gown, perhaps a little self-consciously. "I have dresses in my trunks."

Human clothes, of human fashions. Her humanity will stand out well enough without her wearing them.

"They are not suitable," is all I say. "I will also glamour the feel of your clothes, not simply the appearance of them. You will be more comfortable."

She nods. My jaw works, my magic tingling in my fingertips, glad for release. I take a deep breath.

Then I bend down, catch her face in my hands, and press my forehead to hers. A gasp escapes her throat as I close my eyes, sinking deep into the flow of life and magic around me, the river running through my soul. It's bright and golden, with threads of sapphire. I wrap that liquid gold around my awareness of her, concealing the smell of her humanness with my own. Her practical gown melts away to reveal a sweeping train of pure white, edged in gold filigree, with delicate sheer sleeves that end in a point over her hand, and a beaded sweetheart neckline. I unravel her hair in my mind from its bun until it falls in gleaming waves to her waist.

Her ears, her short stature, her sweet human face—none of it I change to resemble a fae.

I open my eyes, find hers wide and staring at me, her jaw sagging. For a second, I cannot move. The next second, I give a dangerous thought to how easy it would be to angle my mouth and claim her lips.

Then I pull away, drawing a deep breath into my suddenly tight lungs.

She looks down at herself, at the change I've wrought to her gown. When I glance sidelong at Rahk, he's not looking our way, but feigning interest in the hilt of one of his knives. Stella touches the delicate gold belt slung across her hips, then pinches one of the sleeves. Brushes her hands along the softness of her skirt as her hair falls around her.

What have I done? part of me cries. *What have I done, bringing her here?*

She looks up, catches me staring. "This isn't real?"

"It's a glamour."

"Where did my dress go?"

"You're still wearing it."

She narrows her eyes. "Impossible."

And I smile despite myself. Stella studies me carefully, as if she sees the underlying darkness beneath my smile. It's too knowing of a gaze for my comfort.

"Ready?" I ask Rahk. He returns his knife to its scabbard and gives a short nod. "You might want to make yourself scarce too," I say to him. "For a few days, at least."

"I will stand with you."

I give a rueful snort. "And if I order you to visit your sisters for a few days? The Nothril Court misses your presence, I'm sure. I will write to you about the issue of Orawyth."

Rahk glares at me. "*He* cannot hurt me without losing my court's loyalty."

Stella glances between us. I consider several responses, all of which would frighten her. So I say nothing, merely offer her my arm and try to ignore the warmth of her fingers curling around my elbow.

We walk across the bridge over the channel emptying into the sea, heading toward the palace.

To my endless relief, Rahk doesn't follow.

He knows what I cannot say in front of my wife, and he knows as well as I do that following me right now might just be the stupidest thing anyone can do.

As we reach the other side of the bridge, as Stella uses her other hand to lift the front of her skirts as we walk up the steps to the palace—her pure-white train dragging behind us—she asks, "Is Rahk a prince, too?"

I nod. "He is heir to one of the Courts that answers to my father."

"Then why does he answer to you like a servant?"

I almost laugh out loud. She hasn't seen enough of Rahk if that's her opinion of him. She's mistaken his deference to me as his future High King and his role as my choice warrior for a subservient soul.

But with each step that brings us closer to the throne room, the softer side of me locks away, too tight and far away for her to reach. It's why I give her a roguish wink and say, "Because he values his life."

Stella struggles to keep up with me, but not because the pace I set is too vigorous. She cannot seem to tear her attention away from devouring the palace as we climb a winding, open staircase with vines dripping from the railing. "These vines are so healthy!" she cries. "Look how large their leaves are! We don't have plants like this back home!" Then she cranes her neck to stare up at the arches overhead, the waterfall we pass, the winged statues on display. I tug her into a corridor once we reach the right level.

Don't listen to her, I tell myself. *It'll break your focus.*

We round a corner, file down several more hallways into the heart of the palace. There, before us, is a pair of double doors guarded by two enormously tall guards, their feathered wings tucked in close to their backs. Stella wipes her sweaty hand on her white dress and tightens her grip on my arm. But I don't look at her. A single wrong glance can spell disaster from here on out. Everything must be carefully executed with precision.

I close my eyes as we reach the door. Only a second—and I fill my soul with ice. I have no feelings, no desires, no cares. My raging heartbeat eases into a lethal calm. I open my eyes, focusing on the doors as they open.

The familiar scene unfurls before me. The High King's court is packed to the brim with fae and various creatures from all over Faerieland, come to watch the spectacles of royal intrigue or present their cases before their king. The vaulted ceiling arches overhead, and sunlight streams down from above, creating a glowing halo around the golden figure lounging on his throne

A hush falls over the crowd. I refuse to look at Stella, to measure her reaction, to wonder what she might be thinking. I focus on Faradir, on his posture, the graceful display of lofty elegance and power.

"Miss me?" I say, my lips spreading in a wide grin that shows my canines. "I've returned with my new wife."

CHAPTER 19
THE PRINCESS

WE STAND IN a throne room. A stream is cut into the marble floor before the dais, crystal water flowing in a circle around the throne, which is lined with pillars carved in elaborate designs of flowers and waves. At its center, in a dazzling glow, must be the High King. The fae who has slowly been swallowing up our land with his forest. He is the most beautiful person I have ever seen, man or woman, fae or human, and his golden radiance would put even beautiful Vivienne to shame. It makes me thoroughly uncomfortable to be in the presence of one so transcendent.

A transformation has been working over Ash since he left our bridal chamber this morning. But in that last minute, before the door of the throne room, there was a complete shift. His eyes have darkened, narrowed, even as that sardonic eyebrow quirks up and a poisonous grin spreads across his face.

Now I stand on the arm of a complete stranger.

I cannot help but wonder if the true Ash was the one who held me while I cried last night, or the one that prowls toward the High

King of Faerie, dragging me alongside him despite my unyielding footsteps and the heavy train of my dress.

"She's a human!" comes a high-pitched female shriek.

The High King shoots to his feet. "What have you done, Prince Trenian?" he bellows, his tone a kindling fire before it bursts into a blaze. "This goes against the terms of our bargain!"

Ice floods me from head to toe.

Something is very, very, *very* wrong.

My instinct is to clutch Ash's arm a little tighter, but suddenly . . . I'm unsure of everything. Why is the High King so furious? Hadn't he sent Ash to marry me?

Will Ash even protect me?

My husband gives a dark, humorless chuckle. "If it did, I'd be dead, dear Father, my High King. But alas, there was no stipulation that my choice bride must be fae."

"She was to be from one of the Courts!" Even as the High King says it, a paleness comes over his face, as if realization has sunken in. Realization of what? He reaches back toward his throne, grabs the armrest, and sits down heavily.

"You meant the fae courts?" asks Ash innocently, tsking his tongue. "You should have specified. I thought a human court satisfied that requirement completely."

What is happening?

I curl into Ash, my mind reeling, my heart thundering in my chest. He offers no comfort, however, and it's like seeking refuge in one of the marble pillars surrounding the dais.

The High King stares at Ash, not moving a muscle, not reacting. Then, suddenly, he leaps to his feet, grabs something near his throne, and hurls it.

Straight at me.

I can't even scream as I try to duck behind Ash, but his hand shoots out and grabs my upper arm with enough force to break it. Does he *want* me to die?

But then I look up, and the gleaming point of a spear winks at my face. Halted mere inches from me. I go cross-eyed looking at it. I'm either going to faint or vomit.

Ash has snatched the spear right out of the air, stopping it from killing me on the spot. I swivel my gaze from the tip to him as he grins viciously down at me. He turns that grin up to the High King.

"Get rid of it!" demands the High King. "I command you!"

Ash tosses away the spear, his fingers still digging into my upper arm. With a quick jerk, he pulls me to his side, wraps an arm around my waist, and splays his hand possessively over my stomach. His pinky hooks around the gold chain draping my hips. I go rigid.

"Sorry, Father. She's my wife now. I'm a little attached." He bends down, nuzzling his nose against my hair. "I couldn't bear to be parted from her so soon."

I don't dare breathe. Especially as Ash's thumb swipes over my ribs—not a sweet gesture, but a claiming. My whole body shudders.

Who have I married? And what have I done by coming here?

Something has shifted in the air since I arrived. Something deadly. I counted my blessings too quickly. I didn't marry the best man of my sisters, did I?

I married my own destruction.

This is why he was asking my thoughts of death. He wasn't asking about life span—he was asking because he knew the High King of Faerieland would try to kill me. And if the High King wants me dead, then who can stand in the way of it?

My hands tremble.

"Since the time of my return comes as a surprise and no one could have expected you to prepare a wedding feast for my new bride, I have made the arrangements myself." Ash spreads his free arm wide—the one that isn't wrapped tightly around me—gesturing to the courtiers who hang on his every word. "All the wine, the dance, the feasting, the merriment you can dream of. It has already been prepared on the palace grounds. Eat, drink, and be merry, my friends."

A cheer goes up, and the crowd rushes to the doors. I want to curl closer to Ash, to bury my face in his chest as if that will keep me safe. One look at him is deterrent enough, and I keep my rigid posture at his side.

The High King stays on his throne, eyes locked on Ash's as the throne room empties. His spear gleams on the floor by my foot. Only one person remains after the crowd leaves: a tall woman draped in silver, with black hair down to her knees and a downcast face, hunched shoulders. She stands to the right of the throne, eyes fixed on the ground. She wears a crown, but it looks ready to tip off her head at any moment.

Is *that* the Queen of Faerieland?

The High King stands, and with the grace of a prowling lion, walks down the steps from the dais, approaching us swiftly. With a flick of his hand, he dismisses the one remaining woman. "Be gone, wife."

She bows and scurries away like a wounded animal. I swallow as the High King shifts his attention fully to Ash, his smile growing. "Well played, my son. Sometimes I fear you are a lost cause. Then you pull a little stunt out of your pocket . . ." He stops a pace before us, and his eyes fall to me.

He is so beautiful, I can hardly think. He is like a sun molded into the image of a man, but with a towering height that exceeds even Ash's. His skin and his hair is luminescent glory. No part of him resembles Ash—except the piercing blue eyes that study me now, like a cat thrilling to torture its prey.

"Welcome to Valehaven Court, Princess," he says to me. And then reaches out with ring-studded fingers toward my throat. My airways close and my vision tunnels.

Ash snatches his wrist just in time. He chuckles. "My wife's neck belongs to me. I have no intention of sharing."

The High King withdraws, but his eyes don't leave my face and the fear that must be written plain across it. "You must be exhausted from your travels. I bid you sweet dreams and peaceful slumber."

BRIDE OF THE FAE PRINCE

"You think we would skip our own wedding feast?" asks Ash, his thumb swiping over my ribs again. "We wouldn't miss it for the world."

With that, he bends slightly, loosening his grip on my waist so he can instead . . . scoop up my legs? I'm too confused to understand what is happening until he's slung me straight over his shoulder, kicked aside the king's spear, and marched out the door. Blood pounds in my temples as I struggle to take a full breath. I try to push up on his back—and only succeed in finding my gaze locked in the High King's until the door closes behind us.

Once we're back in the hallway, Ash striding down the tiles like *he* is High King, I twist and give a little kick. "Ash!"

He hoists me back farther, so I lose my grip and fall against him. "Be a good little wife and behave yourself, hmm?"

"*Ash.*" It comes out like a half-hearted plea. How do I ask for a smidge of dignity? Does he even care? Is this who the Prince of the Fae truly is? This callous, devious man who parades me around over his shoulder like some prize he's bought? My gut burns.

"*My lord* might be a little more respectful, don't you think?"

Isabelle Louise would have cowed, slumped in submission, and stuttered an apology. But I'm Stella now, and I was prepared to tolerate this sort of treatment when I wed, but after last night, when I saw how Ash *could* treat me, I don't like this. Not at all.

I consider my options. Throwing a tantrum will hardly win back my dignity. Making demands will only make me seem weaker when he scoffs at them. I'm no match for him in strength, and I'm only reminded of that when he gives me a little toss and pats the back of my leg, saying to some passerby: "Doesn't my wife have the cutest little feet?"

Through the curtain of my hair, I see the hilt of his knife in his belt. The one that he used to prick my finger, that he helped me use on him. Memories of last night only fuel my anger more, and without hesitating, I wrap my fingers around the hilt and yank it out of the scabbard. I don't know what I intend to do with it beyond threatening him.

"Ho now!" cries Ash in surprise, snatching my wrist and prying away his knife from my grip. "I have a mutinous bride on my hands!"

"Put me down, Ash," I growl.

To my shock, he does. I almost topple over when I'm on my own two feet. I twist my gold belt back into place, straighten my skirts and comb my hair out of my face, about to spit an irritated thanks, when he grabs my upper arm and drags me into a dark corner behind a winding staircase.

With a twist of his wrist, he cuts off the sounds of the palace, leaving us in quiet. Has he wrapped us in some spell? Something to conceal us from listening ears and prying eyes?

I don't have time to consider it, because Ash presses himself very close to me, leaning down so his face is only a few inches from mine. I suck in a gasp—and then another when his large hand slides up to my neck. His fingers dance along my skin, lightly wrapping around my throat. Threatening pressure.

I don't dare breathe. My hands, moving on instinct, grab hold of his wrist as if I can pull him away from me.

It's the wrong move. His other hand pries my grip off him— and pins both my wrists to the wall above my head. My head goes dizzy, my heart pounding desperately in my chest. I give a little twist and pull on my hands, but his grip doesn't give. I feel like a songbird trapped in a cage while a predator sticks its claws between the bars.

"You are going to listen to me very closely," says Ash, his voice low and sharp enough to be the edge of a blade as he brings his mouth close to my ear. "If you have any interest in seeing another dawn, you will do exactly as I say. Do I make myself clear?"

I can't move. Can't respond. Dread flows liquid through my veins.

His grip tightens on my hands, his thumb stroking my chin. "Do I make myself clear, Stella?"

I manage a nod.

He smiles against my ear. "Good girl."

He lets go of me, and I sag against the wall. Pressing a hand to my heaving chest, I don't look at him. I won't. He made me trust him. He made me like him. I was *glad* to be his wife.

All of that is gone now.

I am his pawn, his *little wife*. He did not choose me because he liked me. He did not seek me out or hold me because he was drawn to me. No, this was his plan from the beginning.

He married me to spite his father.

He offers me his hand. The feral urge to bite it almost overcomes me. But I heed Ash's warning, and take it without protest. Beneath the compliance, fear and anger simmer like molten lava in my gut.

Ash draws me out of the corner, back into the hallway, and with a flick of his wrist, the sounds of clicking footsteps, rushing water, birdsong, and distant chatter flood my senses. He takes my hand and settles it on his arm with a detached smirk playing across his features.

He doesn't throw me back over his shoulder.

A win is a win, I guess?

The sound of feasting and revelry, of raucous laughter and ribald shouting, reaches my ears before anything else. The smell hits me next, of steaming food that curls my stomach and the heady thread of alcohol. Am I about to experience the debaucheries the fae are known for?

I don't want to. I don't want to see it, to become lost in it.

But Ash doesn't slow his pace, and nothing will soften the set of his jaw.

I can handle this. I've handled everything else so far.

Later, I'll be watching for my opportunity. My opportunity to get back at my husband, to punish him somehow. To make him regret his deception.

It occurs to me in a sudden shade of brilliant crimson that there is no possible way Ash can fulfill the promises he made to my father about halting the conquest of the human lands. He is at odds with the High King, who holds the true power over the invasion.

Don't think about that, I tell myself swiftly when my lungs cave in on themselves. *You didn't marry a fae for nothing. You didn't. Even if it seems like you did.*

We come to a wide set of double doors flanked by armed, winged fae. Beyond it, the shouting and laughter increases. I wet my lips and try to keep my hands from shaking.

The doors open, and an entire scene unfurls before me as the crowd shouts, "Prince Trenian!"

It's a large stretch of green lawn, packed to the brim with all manner of strange people. Creatures great and small, some with wings and others with fins, bodies of all vibrant colors, both humble and majestic in one place. The edge of the lawn dissolves into a lush garden, with tall trees like weeping willows, creatures hanging from their languid boughs and others dancing in circles around their trunks. Living vines curl like snakes around the arms of some fae, taking empty goblets and flinging them against a stone walkway until they shatter into thousands of diamond-sharp pieces.

Drink flows like a river, and there is an alarming amount of indecency—both in dress and action. I try to avert my eyes, but wherever I avert my gaze to, there's more.

I don't want to go. I want to hide, to run away. Ash brings me forward, his pace never flagging as he grins and enters the throng. I inch closer to him. He waves at a fae he passes. "How is it?"

The fae—a male, with skin the color of rich blue ink and curved horns protruding from his brow—sits cross-legged on top of a table. "The best since the last Lulythinar!" he cries, lifting his goblet and then pouring it into his mouth. It spills over his lips, dribbling down his chin.

I try not to gape. The fae has liquid money dribbling down his chin, and he doesn't care.

"I picked it out with you in mind." Ash winks, and everyone cheers. As he continues into the throng, he calls out to more fae, and they respond with enthusiasm.

BRIDE OF THE FAE PRINCE

They love him, my mind registers dully. Of course they do—he just surprised everyone with this extravagant celebration.

Not everyone is happy, however. As we move deeper into the masses, I catch more glimpses of frowns, glowers. A winged guard here and there glaring.

A shiver races down my spine.

Before I know it, we're at the center. Ash's grin shifts, a tell I've come to recognize. I have just enough time to brace myself before he scoops one arm around my legs and lifts me up to his shoulder. I squeak and reach for something to anchor me—and find myself grabbing hold of his ears.

"Not the ears, not the ears!" he hisses quickly, flinching. It's the only genuine reaction I've seen from him since we left the carriage back in my world. I wind up with my arms around his neck. Probably not the most dignified position for the Fae Prince's bride, but at this point, I care more about not falling.

He clearly wants to make a spectacle of me. My eyes widen as I take in how high I am, how far I can see across the masses, and how hundreds upon hundreds of eyes are fixed on me.

Fainting would be a welcome respite.

Come on, Stella. Faint! Do it!

I remain stupidly conscious.

"Introducing Princess Stella, my beautiful human bride!" calls Ash to the crowd.

A cheer goes up, echoing my name back to me. When I glance down, my hands are sheet-white.

"Smile, wife," says Ash.

I muster a distant cousin of a smile. There's no missing the glowers on the guards' faces, on many other faes'. One face in particular snags my interest like a fly landing on a spider's web.

A woman more beautiful than any I've seen before, with dark curls piled to a great height above her head, stares at me with piercing golden eyes. That gaze sweeps me up and down, her lip curling in disgust. It's as though she sees past Ash's glamours, past every protection I wrap myself in, straight to every inadequacy and insecurity.

I want to curl away from her scrutiny.

Ash bends and sets me on my own two feet, but before I have time to collect myself, a flute sings out, and his hand slides around my waist. He lowers his mouth to my ear and whispers, "Dance with me, my darling."

I want to beg him to slow down, to give me space to breathe and adjust. By now, I should know that is a foolish hope. I comply, and when I discover this dance isn't even a dance I know, I shuffle my feet back and forth and hope my long dress hides the fumbling steps.

The dance picks up speed, and soon I'm nearly tripping over my hem, and almost skipping to keep up. When the music surges, he lets go of my hand, catches me around the waist, and lifts me straight off my feet. I give a quiet little scream, but the music and the sound of fae dancing, laughing, and fighting around us drown the noise.

That is when a knife comes out of nowhere, and I barely see the glint of it before it slices straight for my side. Ash twirls me out of the way in a flash, then dips me. I cling to his arms, my eyes wide as he pauses, our noses nearly touching.

"Someone just tried to kill me," I gasp.

That dark, sardonic look twists his mouth. Cunning sparks in his eyes. I hardly recognize him. "Survive the wedding feast—great fun, really." He winks, and I gasp when he ducks to press a kiss to the hollow of my throat.

Then he pulls me back to my feet. I look to the side, just in time to watch a blade from someone—not one of the guard's—slice straight through the fae's neck who tried to kill me.

I'm going to be sick. All down the front of our wedding raiment.

"How much longer?" I gasp, stumbling after Ash's steps, trying to keep up with the dance.

Something in that mask-like face slips. Softens—just slightly. His eyebrow twitches, and he looks away from me. I'm not sure what that means. Then he pulls me so close I'm pressed to his chest. "Not much longer. You're doing beautifully."

That is all the encouragement I need to sink into myself, to block out the rest of the world. In the silence left behind in my mind, I hear that voice from the forest again.

Fly away little bird, before his jaws snap your wings.

CHAPTER 20

THE PRINCE

W HEN WE'VE STAYED long enough at the revelry, I guide Stella through the rowdy crowd, grinning and laughing at the jokes tossed at us that are at best questionable, and at worse, outright ribald.

Stella says nothing.

She doesn't even cling to me like she did earlier. She merely follows, and there's a vacancy about her expression that I've never seen before. I want to block her out so I do not lose focus or face, so I can completely immerse myself in this role.

Try as hard as I might, I can't help but be anxious at that expression on her face. Fear, frustration, anger—all of that I can deal with. But she acts like she isn't even here. Like she's somewhere else.

I need to get her somewhere safe before I lose her entirely.

We finally make our way out of the crowd, through the palace. I move swiftly, not because I want to make her struggle to keep up, but because my composure is starting to fracture, and I cannot let a single pair of eyes witness it.

Don't think about any of it. Don't think. Don't break.

Thankfully, the palace corridors are mostly empty, and we arrive at the familiar arch of my own doorway quickly enough.

"Here we are!" I say with far more cheer than I feel, pushing open the door to my suite of rooms. Then, when she doesn't respond, I shut the door and say, "It's your new home, Stella." My voice almost cracks on those words. I sniff—and find nothing suspicious in the air. The individual rooms might be suspect, but out here is fine. Not much can get through my wards.

She blinks—not at me. She pulls her hand out of my elbow and scoots away from me, standing still with her head bowed and hands clasped.

My gut drops.

Something is wrong.

Is she ill? Did someone hurt her? That is impossible—I haven't taken my eyes off her this entire time. I didn't dare give her anything to eat or drink out there, and no one touched her except for me. No one could have hurt her, right?

I take a step toward her. "Stella?"

She flinches.

"What's wrong?" I ask, coming toward her again. "Stella?"

She throws out her hand, eyes wide but not meeting mine. "D-d-don't—"

Don't touch me. Her movements speak clearer than her words. I swallow the rock in my throat and quickly lift both hands. "I won't touch you."

Her eyes finally meet mine then, and my whole being falls. They're wild with terror, with hurt. She looks like a doe terrified for her life.

I did this. *I did this.*

Not the High King. Not anyone from the revelry. *Me.*

I shouldn't have been so warm with her earlier, last night. I should have kept my distance, helped her understand that there can be nothing between us.

Stella breathes hard, wrapping her arms around her middle. I don't know what to do. Helplessness washes over me. I want to take her in my arms and comfort her, but if I so much as move, she reacts.

"Y-y-y-you-you-you—" she gasps. "You . . . you . . . you—you—" Those words come faster, more desperate. The same sound, over and over and over again, like she's actually choking on the effort it takes to speak. Each syllable is like a serrated knife sawing in and out of my heart. "You—"

My mouth opens. Everything in me falls.

She can't speak. She's so scared, she can't speak. Last night, she was terrified, yet she barely stuttered. But this?

Ash, you wretched fool.

She gasps, pressing a hand to her throat as her chest and shoulders heave. "You—you—y-y-you—"

"Stella, breathe," I plead, taking two steps closer. She jerks back and throws up her hand again, still choking on her own air. I stop where I am. "Breathe, girl, breathe! Stop talking!"

Mercifully, she listens, focusing on breathing hard in and out. It's like the wind is knocked out of her. I stay where I am, my heart in my throat, as she breathes, and breathes, and breathes again.

I could use magic to make her breathe, but I'd have to touch her. That is clearly the last thing she wants.

I hate myself for bringing her here, for scaring her. For being rough and callous with her. For making her face the prospect of her own death multiple times this evening.

I'm such a cad.

Once she's finally breathing again, her hands pressed to her chest, I venture another tentative, "Stella?"

She looks up at me, those white-ringed brown eyes arresting me as they fill with tears. Then the sobs break loose, and she is bending over, arms wrapped around herself as she cries her heart out.

The sight of her like this is physically painful.

I can't bear it any longer.

I take a step toward her. She doesn't flinch or move away, so I close the distance between us. I hesitate, then lay my hand on her back. When she accepts the touch, I pull her into my arms and wrap her up in a tight embrace as she sobs.

You swore to be good to her, the furious part of me shouts. *Now look at what you've done! You've hurt one of the few good and beautiful things in this world.*

"I'm sorry," I whisper into her hair, almost desperately. "I'm sorry. I'm so, so sorry."

We stay like that for some time, until she starts trembling, her knees knocking against mine. I bend and scoop her up, carrying her to the couch and arranging her in my lap. She curls in close, her tears slowing to little stutters as she folds herself into a ball. I hold her close, trying to keep my own emotions in check.

She trusts me enough to allow the touch. It's the lifeline of hope I need to keep from believing I've already lost her.

At long last, she is calm, her head lying on my shoulder.

"You betrayed me," she whispers.

This woman can slice a man in half with her words. I close my eyes and bow my head.

"You didn't warn me. Or explain anything to me."

"I didn't want to frighten you."

She pushes back on my chest enough to glare at me, and there's that spirit she buries so deep. "You thought I would be *less* frightened to walk into something completely unfamiliar and find out this wasn't the political alliance you said, but some kind of trick against the High King of the Fae? You thought I would be *less* frightened to discover via a spear hurtling toward my face than simply telling me that my new father wants me dead?"

With those words, she scoots off my lap and gets to her feet. She moves several steps away, arms crossed over her chest as she glares at me.

I lean forward, resting my elbows on my knees. "I didn't think you would follow me into that throne room if you knew what awaited you."

"Because you think I have no courage?" she shoots back. "Because you think so little of me?"

I reel back. "What?"

"You *cared* about me. You were good to me last night and this morning. I was supposed to become the sixth wife of a drunkard—my father was going to give me to someone who would beat me, Ash. And I was going to take it. I wasn't going to run away or complain. Because I understand how important these alliances are to our people. I know my duty as a princess of Aursailles."

Her words pummel me into silence. I stare, open-mouthed. Beneath my shock, rage simmers. Roland was going to give Stella to *whom*? Even more terrifying is that I know she s telling the truth. She *would* have endured it. She would have endured it at his hand, and she would have endured it at mine.

I never liked Roland. Not for a second. But right now?

I *hate* him.

"And then you entered that chamber, and you were *good* to me. You listened to me. Comforted me. Cared for me. You wanted to know what I was thinking. Do you know when a man has ever cared about my thoughts? Or anyone, for that matter, except my youngest sister? Don't you understand that I *trusted* you? Did you see me hesitate to walk into that forest with you? No, because I trusted you. Because, for once in my life, I felt like I was something of value. And then you act like I'm some brainless prize to tout around your people? With no respect for my dignity?"

So it's not just fear. It's humiliation. I remain in silence, waiting and watching for her to continue.

But she doesn't. She stops herself, flushing, and angrily swipes a piece of hair behind her ears. *She's embarrassed by her outburst.* Has she ever been this forthright in her entire life?

There are many things I ought to be feeling right now. Shame, chagrin, perhaps a little defensiveness for the things she doesn't understand about my people, about the situation at hand. Still, the foolish part of me cannot help but admire this burst of passion.

I knew that passion simmered beneath her delicately demure façade. A strange sense of honor radiates through me that I am one of the few people—possibly even the only person—to have witnessed Stella being herself.

Why do I feel so proud of her for standing up to me? I know I scare her. I don't *want* to, but we're so different, her and I.

"I'm sorry," I say.

She blinks at me, as if hesitantly relieved.

"You're right. I should have trusted you enough to tell you. I am not used to entrusting my plans to people. When you live as the son of one of the wickedest High Kings in history, you learn not to trust anyone. I've been betrayed more times than I can count, and I have paid *steep* prices every time."

"Betrayed?"

I clench my jaw. "Loyalty is easily bought and swayed among the fae. Rahk is my only true friend—and the only reason I can be close to him is because the Nothril Court is too powerful for the High King to risk angering."

She nods. A stray sniffle.

I gesture to the couch next to me. "Would you consider a seat? I can explain whatever you'd like answers to."

It's an enormous risk to trust her with this sort of information. If someone got ahold of her, decided to torture her to get information—I shudder at the thought—I would lose everything.

But strangely, it's a risk I want to take. I want to explain to her, to talk to her, to help her understand I intended no disrespect to her.

Most of all, I don't want to lose her trust.

She hesitates, then comes and sits on the opposite side of the couch. She draws her knees up and wraps her arms around them, saying nothing.

I feel as though I've won a little victory.

Perhaps this tentative *something* between us isn't completely lost.

I stare at her. She stares at me.

"What would you like to know?" I ask.

"Why does your father want me dead?"

She goes straight for it. Very well, then. I finger the lines of the tattoo peeking out beneath my sleeve. "A number of reasons. I made a fool of him by tricking him into allowing me to choose a wife for myself. He didn't imagine I'd marry a human. It puts him in a predicament."

"What predicament?"

I draw a deep breath. "He wants me to have an heir."

Stella's face goes bright pink. I give a short chuckle. "Not a half-human heir." That doesn't make her blush go down at all. "I tricked him because he has been attempting to maneuver me into a marriage and siring an heir so that . . . well . . ."

It's oddly painful to say the words aloud. I barely maintain my neutral expression from slipping into a wince. "He wants me dead."

Her eyes go wide. "What? But you're his son."

A wry huff escapes me. "His *rebellious* son. He is unable to harm me directly because of the spell on his throne. There are several laws he must abide by or else the throne will reject him as High King and he will lose his position. One of them is not killing the only living heir to the throne. If he kills me now, there will be no one to rule."

"Why doesn't he just have another son of his own?"

"He is unable to have more sons. Whoever sits on that throne is cursed to only sire one son in their lifetime. Thus, he must rely on me to have a son."

She considers this, frowning. "So . . . you tricked him by making him think he was getting what he wants—you siring an heir—but you married me instead. The throne won't accept a half-human heir."

"My *father* won't accept a half-human heir."

"So now your father wants me dead too, because I am a hindrance to you siring an acceptable heir."

"That . . . would be my least favorite part of this plan."

She gives me an arch look. "So, will I die?"

"That is not the plan."

"But there will be people actively trying to kill me?"

I nod grimly. "As long as my father sits on his throne, you're in danger."

"And you are . . ." She trails off, tilting her head, as though searching for the most diplomatic way to express her question. "You are trying to stop him?"

This is the terrifying part to entrust her. I lean forward, clasping my hands together. When I open my mouth, the words don't want to come. It's like dragging stones up a hill to say aloud the secret I've been keeping for so long. "I am working to overthrow my father from his throne."

She stares at me, utterly dumbstruck.

"It's a delicate matter," I say. "The throne is more like its own sentient entity, forged by the last of the Great Kings to only accept a ruler with his blood who has never broken a law of Faerie. There are other curses on the throne besides the aforementioned one; any fae who kills the High King will die. Except for me—I will instead forfeit my throne. Thus, I am trying to trick the High King into breaking a law of Faerie, and you are bait for that."

"Why?" she asks, eventually. "Why are you trying to trick the High King off his throne?"

"Because he'll kill me the first chance he gets."

"That's not why."

I glare at her. What I said was true . . . if not the entire truth. "No one in Faerieland is safe until he is gone."

Also true—also not the entire truth.

Her keen eyes narrow, but she doesn't push the subject, shifting the topic instead. "When we arrived, you . . . changed."

I clench my jaw. "Yes."

"Why?"

"Because I have a reputation to uphold."

"Of a cavalier, rebellious prince?"

"Indeed."

She nods, running her nail down the curved, polished arm of the settee, following the engraved designs. She doesn't say anything for so long, I tear my eyes away from her and focus my gaze instead upon the embroidered hem of my tunic.

"I want to make a request of you," says Stella sitting up straight and fixing me with an intent stare.

I find myself straightening too, leaning just a fraction closer to her as I nod too quickly. "Whatever it is, if it is within my power to grant, I will do it."

"I'm not asking that you tell me all your secrets. But I want to know each day what I am walking into. I want to be prepared for whatever we face."

We. Does this mean she will help me? What if she doesn't need to be my pawn, my tool, or my trophy? What if she could be my true companion and friend?

No, no, my heart says immediately. *It will only hurt more when she dies.*

But maybe she doesn't have to die. Maybe, with her help, I can make my moves so that she doesn't *have* to be my collateral.

"And," she continues, "I will offer any aid I can in exchange for you ensuring that when you become High King, you will end the fae invasion of my lands." My brain is already spinning as I take this in, but she interrupts my thought process when I don't answer quickly enough. "Ash?"

I blink and glance at her. "I'm sorry?"

Her lips twitch. "Where'd you go? Will you honor my requests?"

I nod, perhaps just a touch too eagerly as a weight falls off my chest. "Yes, of course. I will do what I can to not blindside you in the future. I also give you my word that I will do whatever is in my power to save your people from my father's conquest. That was no idle promise of mine that I gave King Roland."

"Thank you," she says, and makes to get up, signaling that her interrogation is over. I am so relieved that we've worked through this issue that when she goes to walk past me, I catch her hand. She gives a surprised bleat as I press my lips in a brief kiss to her knuckles.

"You, my darling, are going to be the spark that sets this whole world ablaze."

CHAPTER 21
THE PRINCESS

H IS KISS SURPRISES me, enough that I scramble away when he lets me go. Part of me hates myself for not being bold enough to stay. It's better, though, to not indulge. I've only *mostly* forgiven him for tonight. Partially because of his explanation, but more so because of the sheer devastation on his face.

"These are your chambers, then?" I ask and swallow a yawn before I betray how exhausted I am after this long day and all those tears.

"They're our chambers, yes," he replies, and something about the way he says it makes my neck turn warm.

I spin toward him, clasping my hands behind my back. "Well? Are you going to give me a tour, *my lord*?"

He winces at the title, making me wonder if he adopts his princely persona so thoroughly that he doesn't even realize the inconsistencies. Nevertheless, he gets to his feet, which only serves to remind me how tall he is, and gestures for me to follow him.

The room we stand in has couches arranged in a semi-circle facing six curtained window arches. The tables between each couch are wooden, maintaining their natural color and shape and even a few leaves, but they're polished with something that gives them a lovely sheen.

Two gnarled tree-things flank the door. I take them to be the fae version of a coat rack, because when Ash gets up, he removes his overcoat and flings it onto one of the branches. My eyes might be teasing me, because I could have sworn the branch extended just a fraction, as though to *catch* the overcoat.

"This is where I welcome any guests I choose to receive," Ash says, gesturing at the couches where we were just sitting. White walls are accented by roughened, dark wood beams and punctuated by something like climbing ivy. My heart thrills at the sight of the strange but beautiful plant. Once I'm settled, my first order of business will be to explore the plants of Faerieland. "It is also for leisure, should you like that. I haven't had much use for leisure, and I haven't had much use for guests either, so this room isn't used very often. You're welcome to do as you please, though you should be aware that this is considered a public space."

"So don't wear only my shift out here. Understood."

He gives a surprised chuckle, but when I glance at him, he's already looked away, giving me nothing to focus my gaze upon except a long, pointed ear that seems a little redder than it was a moment ago. "As I said, you may do as you like. I merely intend to inform you. In there is the dining hall, for taking meals." He points through an arched doorway at what appears to be part of a long table. Then he continues toward the back of the quarters, to a hallway with several doors. He stops at the first door and pushes it open. "This is my study, where I spend most of my time."

I peer around his bulk to discover a smallish chamber with bookshelves lining every wall, a shocking amount of the bookshelves' occupants scattered like a child's discarded toys across the floor. Papers flutter on the desk at the center of the room, anchored down by ink pots, quills, and other detritus.

It looks like a tornado blew through the space.

His steward must keep the rest of the quarters clean but isn't allowed to touch Ash's *important documents*.

"That is quite the judgmental face," Ash remarks dryly, pulling the door shut and cutting off my view. "At this rate, I might be forced to ban you from my study. I can't endure your pretty face set so critically."

My cheeks heat, both at his unexpected compliment and that my thoughts were so obvious. I tighten my clasped hands behind my back and bounce on the balls of my feet, saying nothing and trying to look as innocent as possible.

Ash narrows one eye at me, then continues his tour, pointing out the next door as the bathing chamber, and leading to the final door in the hallway. "And *this* is the door to our private quarters."

He pushes open the door. It's a very large room, with another sitting area to the left beside a great big window covered by a gauzy white curtain. I nearly gasp at the sight of flowering vines draping down the wall like a blanket, though the blossoms are all closed for the night. It's easy to imagine taking tea and breakfast here while staring out the window at whatever lies beyond.

Maybe I *can* make a life here.

If the High King doesn't kill me first.

That's when I survey the rest of the room and notice the large bed to the right, with a rich green coverlet and white embroidery. More gauzy white fabric twisted with silver ivy drapes the four posters.

My throat goes suddenly dry, and when I peek at Ash, he's not looking at me. His long ears are pink again. Why does it excite me to see him flustered? Perhaps it feels a little bit like comeuppance for how he treated me earlier. Is there something I can say to embarrass him further?

"This is where we will sleep tonight?" I ask.

The color of his ears deepens as he clears his throat. "You may sleep here if you like. I will be in my study for the duration of the night, catching up. But I thought you might prefer to see your room."

I blink and realize there's yet another door at the far end of the room. Why does my stomach sink? Did I think Ash would hold me in his arms every night as I fell asleep? Now that I set my mind to it, it's a ridiculous notion. He even said yesterday that he doesn't need sleep like I do. Of course I will be sleeping alone.

It's better this way. I'm used to this.

He pushes open the door. Of all the rooms, this is the one that I love the most. It's octagonal—odd, but I can't help but like the character it has—with windows on one half of the octagon overlooking a beautiful garden and waterfall. The furnishings are simple, but elegant. A large, downy-soft bed, with a coverlet of many beautiful blossoms and all different colors woven together, rests in the center of the room, its headboard set against a wall completely covered in vines. All the air leaves my lungs.

My lips part slowly at first, then split into an enormous grin. "It's—"

But Ash's finger falls to my lips suddenly, his eyes darting around the space. I swallow my words, trepidation washing away the awe and excitement of the moment before. Is something wrong?

He frowns, pulling his finger away from me, as he prowls silently across the room. I watch, not twitching a single muscle, as he shoves his hand behind one of the pillows on the bed and . . . *grabs* something.

It squeaks.

I gasp.

Ash shoots a sharp look at me, telling me to be quiet. I clamp my hands over my mouth, refusing to breathe as he pulls something out from behind the pillow.

It's a small, winged creature that writhes in his grip. It shakes its tiny fist at Ash, its mouth opening and closing as though it's yelling. No sound comes, however. Did Ash just glamour away the sound like he did when he pulled me aside after we left the throne room?

Ash continues through the room, the one humanoid creature in his fist. He yanks another out from the curtains, and a third from a drawer. Then he marches to one of the windows, unlatches it, and flings the creatures outside, slamming and locking the window shut behind them.

"Well!" he says, brushing off his hands on his trousers. "My apologies for that. I would have rather inspected the rooms *before* you came, but alas, it couldn't be helped."

I slowly pull my hands away from my mouth as the tension eases out of my body. "What . . . *were* those?"

"Oh, just pixies. Someone likely hired them to spy on you and wait in your room until you arrived."

"Spies?"

"Don't worry, they wouldn't have hurt you."

"But . . . are you sure you got them all?"

"Of course. I learned a long time ago to keep my wits about me. Pixies are good spies, but once you know their smell, it's impossible to miss them. They're gone now. And they're not in any of the other rooms either." He pauses, studying me, and then lays a gentle hand on my shoulder. "It's nothing to be afraid of. I will take care of you."

I nod, accepting his assurance. He has every motivation to keep me alive. I trust he will do everything to keep me safe.

Whether it is enough remains to be seen.

I try to bring my mind back to the room, back to the beauty of the space, the fact that it almost seems made for me. "What is this room for?" I ask, changing the subject.

"You."

"No, before me."

He rubs the back of his neck. "For a mistress. Normally, a princess consort would have her own suite, but I didn't want you to be so far away. Thus, I brought you here."

I nod again, and when I can think of nothing to say in reply, we end up standing there silent for several minutes.

Finally, he says abruptly, "Your trunks are stacked in my study. I didn't think you'd want them cluttering your room until you've had a chance to sort through them. You can wear your human clothes if you like, but . . ."

"I have no attachment to human fashion," I reply, glancing down at the beautiful dress I wear. "This is much more comfortable. If you'd prefer that I wear fae fashion, I will oblige."

He glances at me, and a pleased expression warms his face. "Very well. There are a few spare sets of clothes that should last until we can have more things made for you. For now, I ought to remove your glamour."

He steps forward, hesitates slightly before taking my face in his hands and resting his forehead against mine. Our breaths mingle, our mouths so close. I still can't quite bring myself to close my eyes, staring wide-eyed at him as he shuts his own, brow furrowing in concentration.

The beautiful white dress begins to glow, and slowly melts away, revealing my sweltering, fur-lined travel garb from earlier. He pulls back, and I tug at one of the scratchy sleeves.

"Will I have a maid?" I ask.

"If you would like one."

"I believe I would. If that would not be dangerous."

"My staff cannot hurt you. They are bound by oaths. I'll arrange for someone to see to your needs first thing tomorrow. Now, you must get sufficient rest. I've heard that humans need plenty of sleep, and I intend to take good care of you."

With that, he leans down, presses a chaste kiss to the top of my head, and leaves the room, shutting the door behind him.

I blink, suddenly alone.

It takes some time, but eventually I manage my way out of my dress. I find a beautiful light pink nightgown in the drawers on the opposite side of the room and slip it over my head. It falls like butter over my hips, soft and silky but not cold like satin.

Exhaustion tugs at my body, but I walk past the bed to the wrap-around windows, dusk light and fireflies dancing across the garden. It's like a fairytale, beautiful, romantic, and magical.

But I don't smile.

This is a completely different world from my home—but also from what I expected. I'm not prepared to face a king's vengeance, or my husband's cunning machinations.

To some extent, I can trust Ash. I do believe him when he says he will do what he can to protect me, that he will try to not blindly throw me into situations I'm not equipped for without a warning.

At the same time . . . he is a fae. He is one of *them*.

He can be cold and callous. He can be wicked and wild. That is not the sum of him, but it remains that I must be on my guard around him.

I will not blindly trust him.

This does not mean I *blame* him for bringing me here. I can understand wanting to challenge a king whose goal is to kill his only son. I can understand many things about his situation. That does not mean I will believe everything he says or expect a complete reformation of how he treated me this evening.

I must make sense of this world and how it works, or I will die. Beyond that, I want my people spared from the High King's conquest. Beyond that . . . I dare not let myself consider what else I want. At least, not yet.

I cannot rely on Ash for everything.

That's when I notice three little potted plants set on the corner of my windowsill. *My herbs.* My rosemary, thyme, and lavender. I chew my lip, and maybe I forgive Ash a little more for earlier.

Crawling into bed, I pull the coverlet up to my chin. It's the softest bed I've ever encountered, much softer than even a goose feather mattress. Despite the newness of this world, the constant alertness in my body, I quickly fall asleep.

THE PRINCE

I RUN A HAND through my hair as I make my way to my study. It is strange and disconcerting to leave her alone after being at her side for these last twenty-four hours. What if something happens to her? What if there's some trick I didn't sense?

I shake my head. She is safe here. My wards indicate no breach. The pixies are gone.

She's safe.

For now.

Why did I have to marry the most likeable woman I've ever met? I ought to have married some grisly old dowager queen. Then maybe I'd be less worried about her all the time.

I open the door to my study, only to be greeted by a stack of sealed missives left by Edvear, and the pile of my wife's trunks stacked to one side. I flop down into my chair, lean my elbows onto the desk, and rake my hands through my hair and fist them tightly, closing my eyes.

It seems I might need rest too.

But I can't.

The first thing I do is open the secret compartment beneath my desk, unlock the ward with my thumbprint, and pull out the vials. Six in all. I study them grimly for one moment, at the sloshing liquid contained in each. Most are clear, but one is black and viscous. I unstopper the first one and take a bitter swig. I move through all of them until I get to the black one.

No matter how many times I do this, the black one always remains the most disgusting. I fortify my strength and take my swallow. It slides down my throat like a slug, coating it with something that tastes like tar. I cough, take a few desperate swigs of an old, tepid cup of mint and grell tea as my eyes water.

It's done. Until tomorrow.

I straighten, and before I touch those missives, I take a stack of fresh parchment and set to work. The first is an order for the royal tailor to come by midday tomorrow. The second is for Edvear to shift duties of Hylath, who has been on my staff for some forty-odd years, and set her as Stella's personal maid.

Then I take the top letter off the stack. It's an invitation. I slide my knife under the seal and open it.

Lulythinar Masquerade Ball.

I toss it aside and continue to the next one. It's a sealed note from Edvear.

Mama Bagogs says she will be delighted to receive you tomorrow afternoon, if it should please Your Highness.

One of the tattoos on my arm twinges—half of a broken heart. I dip my quill into ink and scrawl my reply on the same sheet.

Perfect. Tell her that I will bring Stella if she cares for an outing.

Tomorrow, I will initiate the second step of my plan. The step of bringing about the unlikely alliance between the Nothril Court and my father's most powerful emissary, the Neverseen King.

The next isn't even sealed. It's merely folded and hidden between the other letters. I recognize the disguised handwriting despite the lack of a signature.

A star burns at midnight. It won't set for three days.

I snort ruefully and mutter, "My sympathies, Rahk. May the moods of Lord and Lady Nothril not prove too overbearing. Though it'll be wise if you delay your return more than three days."

The last note is also not sealed. It's on a torn sheet of paper, in Edvear's hand, scrawled so quickly it's almost illegible. I frown at it, tilting it to decipher its meaning. At last, I make sense of the scrawl.

Do you really intend to do this to her?

My jaw hardens. I toss the paper into the glowing embers of the fire and let it darken and curl until it's consumed.

It's easy to criticize those who make hard choices when you never have to. I don't pretend to be proud of what I've done. Who I am. What I've become. But I would rather be a monster to give those better than me a chance at a good life.

If I must be a monster to keep Stella alive and get my throne, then I will be the ugliest, deadliest monster.

And I won't regret it.

CHAPTER 22
THE PRINCESS

THE NEXT MORNING, I wake to a strange clicking sound. I stretch, move to sit up—and find five eyes on strings bobbing above my face.

A gasping sort of scream tears out of my throat as I throw myself off the bed, grabbing the nearest thing for a weapon: a long, tapered candle. I hold it out in front of me, pointed at the . . . *creature* in my room.

It's close to my height, and *green*. It stands on two feet like a human, with two arms like a human, but the hands are blobby palms with four slender fingers that bulge into blobby points. A dress with a sash clothes its green body. Its head is more like the lower half of a head, with a wide, toothy mouth and blue lips, nostrils but no nose, and—and—and what seem like *tentacles* with eyeballs at the end of them coming out of where the creature's brains should be.

It burbles something at me and takes a step closer.

"N-n-not another s-s-step," I snarl, holding out the candle at the . . . *thing.*

The door barrels open, and Ash stumbles through in breeches and a blousy white shirt that is open at the chest. His hair stands up at strange angles like he fell asleep at his desk.

A massive sword is in his grip, a faint glow tinging the blade.

"What is it?" he demands, frantically surveying the room, me, and the creature. "What is wrong?"

I blink. He has no reaction to the creature, who burbles something at him, sticks out a waggling blue tongue, and gestures at me. All five of those eyes swivel toward me as it plants its hands on its narrow hips.

Ash's eyes follow hers, finding me and the candlestick I'm wielding. I stare at him stupidly, realizing he must know this creature. I lower the candle. "I w-w-was s-startled," I mumble quietly. It's the only explanation I can give that doesn't imply my horrified shock at the creature's appearance.

Ash straightens and sets down his sword, propping it up against the wall. "I think introductions are in order." He comes to my side—and very distinctly averts his gaze from what I'm wearing.

I glance down, my sleep-fogged mind clearing, to find I'm in my nightdress. It's not scandalous by any means, but . . . it *is* a nightgown. I cross my arms over my chest self-consciously.

"Stella"—he places his hand on my back—"this is Hylath. She will attend to your needs if that pleases you." There's a lilt on the edge of that statement—a question about whether Hylath is suitable to my sensibilities.

I look at her. At her strange eyes that blink out of sync with one another. *Where is her brain?* Maybe it's behind her nostrils? It just doesn't seem like there's enough space for it. Maybe she doesn't need one. "I-it is a p-pleasure to meet you," I say with as much dignity as I can muster after my little display.

Hylath sticks her tongue out and wiggles it at me.

I lift my brows, glance up at Ash. "Is she angry?"

One of her eyes pops wide and lifts straight up.

Ash gives a little laugh. "She is pleased to meet you, too." He glances sidelong at me. "She also says you need to freshen up."

My eyes widen and I pull away from him. "Does that mean I smell?"

"She doesn't mean it like that," he says, and immediately Hylath lets out a shriek—startling me—and covers her nostrils with her hands. And runs straight out of the room.

Ash's mouth is twisted, both humorously, and a little sourly.

"What . . .?" My question trails off. I suppose I'm not exactly sure what to ask.

"You cannot smell it then?"

My eyes widen further. Is this some fae sense of smell that is extra strong? Alarm makes me take another step back. "Do I smell that bad?" Oh dear, this is quite unseemly of me!

"Not *your* smell—the smell in the air."

Frowning, though undeniably relieved, I take a big whiff. The air carries a light, natural perfume from the blooming flowers along the walls and the herbs on my windowsill. Even the coverlet from the bed gives off a lovely floral scent. Nothing offensive. I love how this room smells.

"I thought so," says Ash. "I simply wanted to confirm."

"Confirm what?"

"That you cannot smell lies."

I think back on the words we just exchanged. "You lied about what Hylath meant about my smell. Is that why she ran out?"

"The air stank like iron."

"Then why did you lie?"

"Because I didn't want you to be self-conscious."

I blink. "Then I *do* stink."

A sheepish grin. "I don't mind."

I glare at him and dodge around his hand reaching for mine, going to the far side of the room.

"I said I don't mind," says Ash with a laugh. "But I shall issue Hylath back in. The tailor is coming this morning for your new clothes.

This afternoon, I need to leave the palace. You're welcome to come with me."

"To go where?"

"To visit a friend. I think you'd rather enjoy the excursion."

"Will anyone try to kill me?"

He shrugs. "Unlikely. You'll be safer at my side than staying here in the palace."

I wish I *could* smell lies. His face doesn't seem to be one of concealing a bitter flavor, but I imagine he's adept at hiding it. I draw a deep breath, consider my options, and eventually nod. "Very well. I will be honored to accompany you."

"Then I shall leave you to ready yourself for the day. Breakfast awaits you in the common rooms."

With that, he turns to leave, grabbing his sword. He hesitates. Looks back at me. Then quickly turns back and leaves.

I have a moment of reprieve to sink my toes into the luxurious, moss-like carpet on the floor. Then the door opens, and one eye peeks in. Its lid closes in a slow blink.

"Come in," I say, and resolve myself to get used to her appearance as quickly as possible. Perhaps I can convince Ash to introduce me to all of his staff, so I don't have any more frights. I ought to know friend from foe, anyway.

Hylath rummages through the set of drawers against the far wall of my room, pulls out various clothing items, lays them over her arm, and beckons me with a trill to follow her. I grab a robe and wrap it around myself before venturing out of the haven of my room.

We pass through the bedroom just as Ash is pulling on a fresh shirt. I quickly avert my eyes—though not before I glimpse a tanned, toned back—and scramble into the hallway to the bathing chamber.

I hadn't seen much of it yesterday, so today I find myself once more agape at the beauty of a simple washroom. Plants of many varieties hang from the ceiling, climb the walls, and fill nearly every nook and cranny in the room. They surround a great white-marble tub with gold claw feet. A long vanity faces a beautiful mirror, providing far more space than I would ever need to get ready.

Hylath points to the steaming tub and gestures for me to get in. I step around a stunning array of potted plants to get to the tub, only to realize once I'm there that the door lists open on its hinges, and my maid doesn't seem to have any consideration for my modesty.

"Um . . . can we close the door?" I squeak.

Hylath's eyes blink in unison at me. Then she marches to the door, kicks it shut, and turns around to stare at me expectantly.

Trying not to feel nervous with Ash on the other side of the door, I hurry to disrobe and slip into the tub, hugging my knees to my chest as Hylath works to lather my hair. Each time footsteps sound near the outside of the chamber, I tense up and wait for them to pass.

I experience a heart attack when it's time to get out of the bath, and heavy footfalls happen *right* outside *just* as I'm standing up in the tub. I squeak and leap into the towel Hylath has for me. All five of her eyes blink at me again, as if she's wondering what in the world is wrong with this stupid human on her hands.

"S-sorry," I chatter, wrapping myself up tightly. Hylath burbles something in response, then gestures to a short stool before the enormous, beautiful mirror. She brushes my wet hair, then lets out a series of shrieks that almost make me duck and cover my ears.

"I'm coming, I'm coming!" calls Ash from the other room.

She shrieks back and punctuates it with a few spits.

"Have some patience, you crotchety old woman! Let a man put some trousers on before he scares the daylights out of his poor bride."

Five eyes roll in succession while I wrap my arms over the thin material of my robe and wait, tensed, as footsteps come toward the door. They stop, and a knock sounds.

Hylath spits three times.

"It's not *my* fault you're in a bad mood today," says Ash with a glare at my maid when he opens the door. "If you're not nice, I'll have to separate you from my wife, so your temperament doesn't rub off on her."

My eyes widen. Hylath sticks out her tongue and waggles it.

Ash laughs.

I sit where I am, not a little perplexed. Then Ash's gaze falls to me. And sweet heavens, he's so handsome, especially with that little quirked grin.

"Hylath says you need my help," he says, giving me a quick once-over. "She did?"

Hylath lets out an exasperated huff as she burbles and her five eyes point—if that's possible—to my wet hair.

"Dry her hair?" Ash repeats, then looks at me. He shrugs. "I suppose there are worse things than getting to touch my wife's hair. Move aside, you old woman."

She waggles her tongue, spits twice, and hobbles away. Ash takes her place at my back. I grip the collar of my robe tightly, staring at him in the reflection of the mirror as he drags over another stool to sit on.

He sits, lets out a great sigh, and surveys my hair. Then he looks up, catches my gaze in the mirror. "Tell me one of those thoughts. Or else I'll kiss you."

I flush as he threads his fingers into my hair and combs them through, leaving dry strands behind. "How are you doing that?"

He grins at me and then leans forward to whisper conspiratorially, "Magic."

"Is it glamour?"

"This isn't glamour. I'm pulling the moisture out of your hair via heat and evaporating it into the air. It works better with smaller chunks than all of it at once."

I watch him work, trying to keep my shoulders and back straight as he moves closer to the hair at the nape of my neck. A shiver goes down my spine, and he looks up. "It's . . . s-s-sensitive," I say, trying to pretend my face isn't turning red by avoiding my reflection.

He only gives a small smirk as he continues. "You still haven't given me a thought."

"I asked how you were drying my hair!"

"That is a question, my love. There is a difference. I tolerated it before, but no more. I want a *thought* from your brain. An opinion. An insight. An observation."

I lower my brows. "This is a big mirror."

He chuckles. "Another one. Or I'll kiss you."

"How many thoughts are you going to want? I might need to prepare them in advance."

He laughs outright at that. "That is for me to know, and you to find out. Besides, I have no doubt that there is an abundance of thoughts floating around in this pretty head of yours."

"If you overestimate me, you'll only be disappointed."

"It is impossible for me to be disappointed in you. Now, a thought or I will be forced to carry out my threats. I am nothing if not a man of my word."

It is impossible for me to be disappointed in you. Something about that catches me off-guard, forcing my lips to part in surprise. I swallow the sudden lump in my throat and blink hard.

The awareness of Ash's scrutiny prickles across my neck. When I look at him, he is already focused back on my hair, running the long strands between his fingers and drying them until they fall warm and soft against my back.

"I'll help you out," he says suddenly. "I'll give you prompts. Here's your first one: *My husband is so . . .* Now you get to fill in the blank. Some popular answers include *handsome, dashing, clever,* and *swoony.*"

It surprises me so much that a giggle escapes my throat before I can stop it. Ash grins, giving my hair a gentle tug.

"Well, what shall it be?" he asks.

"My husband is so . . ." I trail off. *Unexpectedly kind.* ". . . tall."

He pauses, considering, then shrugs. "I suppose that will work. You like tall, right?"

I flush. "Um . . ."

A roguish grin spreads across his face. He leans forward and presses a quick kiss to my shoulder, making me catch my breath. "Good."

I blink rapidly, quickly hiding my gaze in my lap. His hands make one last stroke through my hair before he sweeps it all over my shoulder and stands. I look up just as he catches my shoulders and presses yet another kiss to the top of my head. "Done."

Instead of leaving, he hesitates. His voice drops to a whisper. A whisper that isn't ringed with irony or humor. A whisper that is almost painfully earnest.

"You are beautiful, Stella."

Without another word, he straightens and leaves the chamber. I remain seated on the stool, staring in shock at my reflection in the mirror.

CHAPTER 23
THE PRINCE

STOP FLIRTING WITH the girl, I tell myself as I leave the bathing chamber. *And Mountains of Ildrid, stop kissing her! You're asking to get hurt even worse than last time.*

It doesn't seem to matter how many times I tell myself these things. I build up my resolve, only to walk in and find her fresh from a bath. What's a man supposed to do but kiss and flirt with her? And how am I *not* to like her when she faces all of this with quiet practicality, gracious understanding, and endurance?

Truly, I just see those eyes, watch the color rise into her cheeks, listen to whatever unexpected thing she says, and my resolve crumbles. It just *crumbles* like a castle made of sand!

You're pathetic, I tell myself as I rake a hand through my hair. I make my way to the table where breakfast has been laid out for the two of us, and for a few moments, I debate tossing a few scarpi biscuits down my throat and heading to my study.

It would make it easier for me to like her less if I don't spend so much time with her.

But I owe it to her to eat with her, and to wait until she's ready. I tell myself that is the only reason I pull out a chair for myself and drop into it, staring at the covered dishes of food and the steam wafting out of the spout of the porcelain kettle.

"What is troubling you, Master Ash?" comes Edvear's familiar voice.

I look up suddenly. "Troubling me?"

"You were scowling."

Was I? *Am* I? I sit up straighter and attempt to soften my features. "What's the damage from last night?"

His goat ears twitch. "Those three who tried to harm Lady Stella last night at the revelry are dealt with. None of them were significant members of the court."

"That's good, at least." I drum my fingers on the table. "What rumors are circulating?"

"Mostly what you anticipated. There is debate over whether she will be queen, though most dismiss the possibility because of how angry the High King was. Some were placing bets on how long she will live. Others speculated that the High King is, in fact, jealous of you and wishes to claim the human princess for himself."

I shudder at that. "I suppose I should be relieved that's not a viable option on the table." As soon as the words are out of my mouth, I stop drumming my fingers. Then, catching myself before I give too much away in my expression, I shift to lean my chin on my fist.

Would the High King attempt to take her as a pet? I would like to think him too repulsed for something that debased, but I cannot dismiss the possibility. It would certainly be a more creative way to punish me than outright killing her. It would be torture for me to see what he would do to her, how he would break and humiliate her before the eyes of the world—and know that I was helpless to stop it.

Faradir took a human pet once—ages ago. She wasn't a noble, merely a beautiful girl who had stumbled into Caphryl Wood and was brought before the king as a trespasser. He took a fancy to her—much to my mother's chagrin at the time—and kept her wrists shackled in golden manacles as he forced her to kneel beside his throne, wearing a gold chain around her neck and little else. Some of the higher nobility, like Rahk's family, refused to see the High King unless he sent the girl away for the duration of their visit. I remember that fire that burned in her eyes when he first chained her, first forced her to kneel beside his throne and made her obey his every command, no matter how degrading.

I was only a child, but I never forgot the way her eyes changed with each day, until they were dull and lifeless with the abuse she had suffered. Her fight leaked away until there was nothing left. Then the High King decided she was no longer beautiful or interesting.

So he took that chain around her neck and pulled it, cutting off her air, until she collapsed at his feet before the entire court, never to move again.

It's not difficult to imagine Stella there instead, frightened and innocent. Bound and awaiting a tyrant's whims.

My blood curdles.

"You d-didn't need t-to wait for me," comes Stella's soft voice.

I look up, pulled out of my thoughts, to discover that Edvear is bowing to leave, and Stella stands in the arched doorway. She wears a simple floor-length gown of soft blue with spider silk detailing at the waist, neckline, and hems. It's one of the crossover fashions from when fae thought to imitate human dress. That style has since passed, but Edvear was able to get his hands on a few things to make Stella more comfortable.

She's lovely as always, but I cannot help my twinge of disappointment that her hair is all bound up in a bun at her nape.

"I was glad to wait for you," I say, getting to my feet to pull out the chair for her.

She sits, tucking an imaginary loose piece of hair behind her ear. As I push in her chair, I try to ignore her fresh scent and keep my manner brisk until I'm safely seated back in my own chair.

Stella tries not to stare at the fairy from one of the seelie courts who comes to serve us. She's not used to the folded iridescent wings, the bark-like skin, or the wild tangle of hair atop the fairy's head, but she attempts politeness and smiles when she pours her cup and serves her plate.

I'm only a few bites into my own meal when I realize Stella is just staring at the round biscuits and nectar compote. I glance between her and the plate. My eyes widen.

"You've never had scarpi, have you?"

She leans back in her chair with a little wince. "I'm not very b-brave when it comes to food. Why does it look like a cake but smell like fish?"

"Because scarpi *is* fish. Scarpi biscuits are a common breakfast here. The purple liquid is a nectar compote. You dip the biscuit in the compote."

Her nose wrinkles, but she cuts a small bite out of the biscuit, dips it in the compote, and brings it to her mouth. She takes a delicate sniff—and gags.

"I'll have the cooks prepare something more suitable for your palate," I say quickly. "Another time you can try it. If you're able to get used to our food, it might make social obligations likely less . . . detestable."

"But shouldn't I avoid eating in social situations? In case something is poisoned?"

"Now you're thinking like a fae," I say with a rueful chuckle. I'm going to corrupt this girl's innocence, aren't I? "Nothing will touch your lips that I haven't inspected first."

"Aren't *you* worried about being poisoned? What happens if you die before you sire an heir?"

The memory of a black slug sliding down my throat assaults me, and I barely swallow my bite of food. "I'm not worried about being poisoned, no. I take precautions."

She purses her lips as though biting back a question. The servant whisks in just then with a plate for Stella. Her eyes widen when the fairy replaces her plate with a fresh meal of eggs benedict.

My staff continues to please me and surpass my expectations. The servant must have been listening to our conversation. How they whipped this up so quickly is beyond me!

Stella lets out a tiny squeal. "My favorite!" She dives into her breakfast with vigor, and it's all I can do to keep eating my own instead of just grinning at her. Her cup of steamed mothweed milk is replaced with tea, and she gasps. "Thank you!"

I can get used to having her around. I already *am* used to it.

"I have another question about the things we discussed last night," Stella says, dabbing her mouth daintily with a napkin between bites.

My fork is halfway to my mouth. I pause, then set it back down. "Yes?"

"What happens if you die before you have an heir?"

I draw a deep breath and lean back in my chair, my appetite gone in an instant. "Well, at that point, hope of my father's ruling line will be lost. He will be vulnerable, and the other Courts will have the opportunity to vie for the throne. It will be a civil war. A very bloody one. Some Courts won't want a new High King, preferring to remain independent, while others will have aspirations for the office themselves. If the fae war among themselves, the human worlds will be at risk. The current law limits frequent human and fae interactions. But one of my ancestors established that law, and if the High Kingly line is broken, the law will also be broken. Inevitably, someone will claim the throne and the line will start over. A new throne and a new set of laws will be forged, but not after the humans and fae have their own bloody wars."

"And perhaps the new High King who arises will be even less merciful than the first," says Stella.

I stare grimly down at my food. "Indeed."

"It seems a heavy burden to bear."

I look up, surprised. "Pardon?"

"It's a heavy burden for you to bear." Though her eyes are softly rounded and innocent-looking, they're uncannily sharp. Prodding for a reaction. "The blood of thousands if anything goes wrong."

I blink at her, caught off guard. I avert my gaze, pour myself another cup of steamed milk, and guzzle it down.

"Well," I say, slamming the cup down on the table and pushing back my chair. "The tailor will be here shortly, but if it would please you, I'd love to show you the gardens your room overlooks. I suspect you will find the place very rejuvenating."

If she notices the briskness in my voice, or the way I don't quite meet her gaze, she doesn't say anything. She nods eagerly, dabs her mouth with her napkin, and stands when I pull out her chair.

It's not the blood of thousands that rests on my shoulders. If I don't play this right, it'll be the blood of *millions*.

CHAPTER 24
THE PRINCESS

A SH ESCORTS ME out of his quarters on his arm, and I half-expect him to revert to how he was last night. The palace is empty, however—seemingly drunk into a stupor— and Ash maintains that approachable expression as he leads me outside.

What I see makes me stop dead.

It's strange, because I looked at this garden out my window, but something about walking into it is so magical that it almost overwhelms me at first.

A long, arched walkway stretches before us, the entire thing covered in vines of pink roses. Petals fall to the stone path beneath our feet, and butterflies of all different colors and sizes flit from flower to flower. Running water burbles nearby from a fountain I cannot see. Beyond the walkway, the garden opens into a dazzling array of color and beauty, making me want to run to see it all, while still walking slowly enough to savor each part. To think I was once afraid I couldn't grow anything in Faerieland!

My mouth gapes in awe until a butterfly flies toward me, and I hold my breath to avoid scaring it. It keeps coming, until it lands on the tip of my nose.

I stare cross-eyed at it as its wings slowly fold and unfold, filling my vision with whorls of blue and purple. When it still doesn't move, a grin slips free of my restraint. Then I giggle. The butterfly flaps its wings and flies away.

"Did you see that?" I demand, whirling toward Ash and grabbing his arm. "It landed on my *face!*"

He's smiling down at me. "Yes, I saw, love."

Then he bends down and plants a kiss on the tip of my nose, right where the butterfly was. I stand still, even more stunned than when the butterfly landed on me. But Ash takes my arm and brings me deeper into the garden, either oblivious to how hot my face has gone or purposefully ignoring it.

"You'll like this," he says, tugging me toward what appears to be a gazebo tucked away in a corner. I eagerly follow him, barely containing my excitement as we reach the overgrown structure made of chiseled and polished granite. I can hardly decide what is prettier between the colorful flowers wrapping around its pillars and dripping from the ceiling or the sparkling crystal streaks through the stone. Ash guides me up the steps until we're standing on a blanket of petals, surrounded by so much beauty.

I gasp, twirling to see every section of the gazebo, and clap my hands. "Oh, Ash! It's so lovely! I cannot believe this is right outside your quarters!"

He gives a pleased grin, then drags me right back out of the gazebo. "You'll like this too."

And that's when I realize it.

He is *trying* to make me happy. It pleases him to delight me.

Now my face heats even more than before.

Don't get too carried away, I tell myself as Ash leads me through countless rows of the most beautiful flowers I've ever seen, heading toward a stunning waterfall ahead. It's a challenge when I want to gasp at everything I lay my eyes on, but I must be on my guard. There are things I still need to understand and questions I must ask.

But I cannot help exclaiming, "Oh!" as Ash brings me right to the mossy bank beside the waterfall tumbling over the edge of rocks into a small pool and stream that winds through the entire garden.

Ash grins at me. "There's a bench here. Come, let's sit."

I can't tear my eyes away from a pair of dancing butterflies, so I just let Ash lead me to the bench.

"You seem a little overwhelmed."

"I've never seen anything so beautiful!"

"I have. I've seen something far lovelier, in fact."

"Then you're quite lucky."

He only chuckles. "Indeed."

I drink in my fill of the landscape, then force myself to focus. My husband sits beside me on the bench, his arms sprawled wide across the back, his long legs stretched out in front of him. His handsome face is turned away from me, watching the waterfall. I take a deep breath and ask, "Ash?"

"Yes, darling?"

"Am I to be queen?"

Silence falls. His gaze flees from me, focusing on the pair of butterflies I admired earlier. "You are to exist. That is it."

Not queen, then. Part of me is relieved. I have no interest in being queen over a people I do not understand. A larger part of me is concerned. If Ash is to be High King, and I am not to be his queen, is that because of my humanity, or is it because I am not expected to survive long enough to claim the title?

It might be both.

Something leaping out of the water, straight into the air, and falling back beneath the pond, interrupts my thoughts. "Was that a fish?"

"Was what a fish?"

"Watch!" I point, studying the water and refusing to look away.

My determination is rewarded. Only a moment later, the *thing* jumps out of the water and plops back in with a splash. It has a long, sleek body the color of a shiny blueberry. "See? That!"

"Ahh. Indeed, it is a fish. Are you surprised?"

"I haven't seen any so vivid, and certainly none that jump out of the water!"

He chuckles. That merriment fades after a moment, a line appearing between his brows.

"What?" I ask.

He studies me. "You're different."

"Different?"

"From when we met."

My shoulders ease and I turn away from that intense gaze. "Oh. Well, that was because I needed to not scare you off. Now I'm not worried about that."

"Scare me off?" asks Ash, a note of surprise in his voice. "What makes you think I'm so spineless that a human princess would scare me?"

I sit up straighter, brushing my hands down my skirts to smooth them. "Do not underestimate princesses. They can be quite vicious when provoked."

A sudden laugh bursts from my husband. It's such a pleasant sound, I'm emboldened to think of something else unexpected to say. Something to elicit another laugh.

Don't talk too much. Be clever and unexpected—but not too unexpected.

The memory of Jacquelle's voice pulls me back to the ball where I was supposed to attract King Ilbert and failed. I'm not sitting on a bench by a waterfall, but one in a ballroom, waiting for the King of Enslington to bring me back refreshment as Father and Prince Brochfael discuss my future as though I am a breeding horse for sale.

Ash waves his hand at me, pulling me out of my thoughts. "You're thinking things, but you're not saying them. My courage is insulted yet again."

"You want me to say what I'm thinking?" It's a question steeped both in mortification and hope.

"Of course," Ash says, as though it is nothing "What else should I want you to say?"

Nothing. Quiet pleasantries. Something witty and flirty. Something unexpected, but not too unexpected. I chew on my lip.

Show my refined breeding. Be alluring—but not too alluring. Don't be overshadowed.

"H-human princesses are bargaining tools," I say slowly, summoning my courage. "If they're lucky, they will be betrothed from birth, and the most important accomplishment of their life will already be established. All that would be left is bearing heirs."

Ash listens quietly, not interrupting.

"For those of us who are not so fortunate, we are responsible for ensuring that we capture the interest of a suitable match. If we are unsuccessful, then either our father has less bargaining power when negotiating the terms of the alliance, or there simply is no alliance. As such, it has always been imperative that we please our potential bridegrooms. If he should find you lacking, then you are considered snubbed, unfit for the one thing you were born to be."

Ash is quiet. Then he echoes softly, "Unfit for the one thing you were born to be."

The tone in his voice is more telling than the actual words he repeats.

I'd bet good money that Ash claims he doesn't care about any of it. The truth, however, is obvious. Perhaps he'd call it a weakness, and that's why he hides behind the persona of a cavalier prince.

I don't have time to articulate a response or formulate a question, because just then, a gaggle of fae children come running down the path. If they were humans, I'd guess they were six or seven years old. For all I know, they could be twice my age. They resemble human children except for long, poking ears that are far too large for the rest of their body and the fact that one of them has light blue skin and white hair.

It wasn't the passel of fae children that made me squeak and pull my legs onto the bench.

It was the enormous wildcat cub at their heels, with two rows of fang-like teeth and long, protruding claws.

Ash smirks at me, then calls out to the children, "What mischief are you hooligans up to?"

"Ith that the human printheth?" the blue-skinned one lisps, pointing at me. The others gasp, eyes wide as they stare at me. I stare back for only a moment because the wildcat leaping toward me makes me let out a shriek and stand up on the bench.

"Whoa there," says Ash with a laugh, catching my wrist so I don't lose my balance and tumble backward. "Have no fear, darling. The cub won't hurt you."

"I don't believe you," I say, even though the children give no reaction of being suddenly assaulted by the stench of iron. The black-and-white patterned cub reaches the bench and sniffs the edge of my skirt. I squeak and step back, nearly knocking myself over the back of the bench. Ash nudges the cub away and tugs me back down.

The children are laughing. Hysterical, am I?

One of them whistles for the cub, and it bounds back, jumping on one of them and knocking him clean to the ground while the others laugh.

"That creature is d-dangerous," I insist, trying to ignore the gleam in Ash's eye as he tries not to laugh at me. "Children shouldn't be playing with it."

"Can the human princess play with us?" asks one of the children with a grin on her face. A grin that concerns me.

"I don't think she likes your pet," says Ash, matching the child's grin.

"We could put Tolgot away!" says the blue-skinned boy.

"We could show her the caves behind the waterfall!" says the third child—a boy with long, shaggy brown hair and a pair of ears even wider than the others.

"Oh, the caves are so beautiful!" says the girl, clapping her hands.

"And *spooky!*" says the shaggy-haired boy.

I chuckle despite myself. They seem sweet, and they're all alone. Besides, now I'm curious about these caves. Surely children aren't too much of a threat, right?

"Maybe another time," says Ash. "She's busy at the moment."

Three pairs of eyes blink at me. "She doesn't *look* busy."

Ash lowers his brow, then slides across the bench to wrap his arm around me, making me stiffen. "See? She's busy. With me."

"Dithguthting," the blue boy grumbles.

"But the caves are so pretty!" cries the girl.

I'm not sure I can handle their pleading much more. Besides, I need to learn to understand this world. I scoot so I can whisper in Ash's ear: "Is it safe to play with them?"

He flinches, then shivers—and clenches his jaw.

That is not the reaction I was expecting.

"She touched his ear!" one of the children loud-whispers to the other. They stare in shock at me, at Ash. Then down at their own feet.

I blink.

Ash shoots to his feet, looping my arm through his elbow. "Maybe another time, children," and drags me back onto the path toward his quarters.

I glance back at the gaping children and the cub gnawing on the blue-skinned boy's trouser hems. Then I look up at Ash's clenched jaw. "What just happened?"

"Nothing," he says. It's obviously a lie, but his face shows no sign of reacting to the taste of iron. Does he lie often?

"I just b-broke e-etiquette I don't know a-about, didn't I?"

"I suppose you make a regular habit of touching and whispering in men's ears, then?" It's almost a growl.

I've offended him. Embarrassed him, if the color rising into his neck and cheeks is any judge. "N-no, of course not. It's just . . . I think it means different things in our cultures."

"What does it mean in yours?" he asks, almost scoffing.

"It means I don't want anyone else to hear what I'm saying."

He scoffs again.

"What?" I demand, frowning.

"That's ridiculous."

"Why?"

"Because everyone can still hear it. So is it a human thing to just go around whispering in ears?"

I suppose I did forget about fae hearing. "Well . . . we u-usually only do it to p-people who are r-relatively cl-close to us. Or if it's very important to k-keep something quiet—"

He stops. Spins toward me. Lifts a hand and traces the pad of his thumb down the arch of my ear. It's so unexpected, and his touch is so teasing and light, I shiver.

"Ha!" he bursts, triumphant. "See? Your ears are sensitive too! And here you are saying I'm overreacting!"

"It's a little sensitive, but not much!" I insist right back.

"Not much?" He scoffs again.

"Try again. See if I react this time."

"Fine, I will!" And with that, he touches my other ear—a soft caress on the outside shell.

I cannot help my flush as I stare up at him, as his fingers brush a few loose strands of hair. But I don't move a muscle.

He blinks, his jaw dropping open. "How do you do that?"

I try to hide my mischievous grin, but part of it slips through. "You cannot hold still if I touch your ear?"

He turns away, giving an irritable grunt. "It's sensitive."

I burst into laughter.

His ears turn red, and I have the very sudden and very strong impulse to test his claim, to dare him to hold still while I touch his ear the exact same way he touched mine.

"Don't you dare," he growls, pulling me toward the exit of the garden.

I giggle all the way back.

CHAPTER 25
THE PRINCESS

T HE TAILOR IS here, my lord," Edvear tells Ash when we make
it back inside.

I try to pull myself back under composure, but the
steward glances sidelong at me, a question in his eyes. Obviously, I'm
failing. Ash grunts in acknowledgement, striding right past him
toward the main living space of his quarters.

A man stands there, spectacles perched on the bridge of his nose.
He has a curled mustache and a prominent Adam's apple poking out
above his starched collar. He's a very slender man, of medium height,
with a measuring ribbon hanging from his neck and a sketchpad
under one arm. I blink.

He's a human.

"Prince Trenian," says the tailor, bowing.

"She'll need a full wardrobe," replies Ash in that same brisk tone
that I'm beginning to recognize as his way of trying to hide his own
discomfort. I fight to keep my lips from twitching. Unnerving him is
just too much fun. "Also, a ballgown."

My head swivels up to his. "A ballgown?"

The smirk is back. "Of course. What else would you wear to the Lulythinar Masquerade?"

"A masquerade?" I breathe, clutching a little tighter to his arm. Would it be like last night's dance? If so, I don't want to go. Not one bit.

"What sort of ballgown would please the lady?" the tailor asks. He pulls out a pencil and begins making notes on his sketchpad, his gaze flicking from his notes to me, assessment sparking in his eyes.

Please me? I know nothing about fae dresses. I was hardly allowed an opinion, even on my own human dresses. I peek up at Ash, as if he's supposed to send a message directly to my brain about what I should want.

"I think a butterfly costume would suit her very well, if that should please her," says Ash, arching an eyebrow down at me.

My mind goes back to that beautiful butterfly that landed on my nose in the garden—then his kiss once it had flown away. Warmth spills into my middle. "I'd like that very much."

"Wonderful." The tailor adjusts his spectacles and tugs his measuring ribbon from around his neck. "Now, Your Highness, if you might consider giving the lady some privacy, I must take her measurements."

Ash hesitates. It's only a second, but it's enough to make the blood rush to my head. Does he not trust the tailor? Is he afraid I'll be hurt if he leaves me alone? But the tailor is a *human*. Surely I'm safe with him, right?

"Hylath!" my husband calls.

A gurgling answer spits from the washroom, and the door opens enough for one eye to poke through.

"Come aid the tailor," Ash orders, letting go of me and heading toward his study. "He requires your help taking Lady Stella's measurements."

Ash doesn't trust even the humans in Faerieland with my safety. I make a mental note of this. I mustn't be too naïve around them. Even if the tailor seems quite nice to me.

Hylath growls, but comes to help. Ash barely closes the door to his study before Hylath is yanking me out of my dress until I'm in nothing but my shift. The tailor moves with practiced efficiency, all business as he works. It reminds me of my time at the palace, where I endured standing on a wretched little stool for hours while they took and retook my measurements, draped dresses over me, and my sisters gathered around, sharing their various opinions on my clothes. Because it's so familiar, I'm not embarrassed.

A sudden question occurs to me. Why did the tailor send Ash away? It's true that I prefer this arrangement, but how did the tailor know that? As far as the tailor knows, Ash is my husband in every sense of the word, and if that were true, him being present now wouldn't have broached any concern.

"Almost done." The tailor wraps his tape around my waist as he scribbles something on his pad. "Hylath, would you kindly fetch a dressing gown for the lady and then summon His Highness?"

Hylath blinks all five eyes at once, as though irritated to be ordered around, but with a grumble, she hops down from the stool she'd been standing on to hold the measuring tape for my height.

"Just your shoulders now," says the tailor. He rearranges the measuring tape across my shoulders, then around. He leans closer, brow furrowed with concentration as Hylath leaves the room.

I keep my eyes fixed on the far side of the room.

Then the tailor's voice tickles my ear, low and urgent. "If you need to escape, send a request to me for a white dress."

My breath catches.

"Quite narrow shoulders you have, my lady," he says as Hylath reenters the room with a long blush-pink robe. At the tailor's beckoning, she wraps it around me and secures the tie firmly across my waist, ensuring my decency before she squawks loudly in the air. I resist the urge to cover my ears.

Almost immediately, Ash's door swings open. "Don't order me about like I'm *your* servant," he says grumpily to Hylath.

She waggles her tongue at him. I still don't understand what that is supposed to communicate, but Ash grins in response. Then, with all the drama of a royal, he flops onto the couch facing me, spreading his arms and legs so wide he takes up almost the entire piece of furniture.

My heart is still hammering from the tailor's words. I hope nothing shows on my face.

Escape?

The thought of it terrifies me. But I'd be lying if I said a part of me doesn't tuck that knowledge away. I'll consider it later.

And consider it, I shall.

I would need more information, though. How dangerous would it be to escape? And once I've escaped, what then? I cannot return home to the palace in Aursailles. I'd have to survive on my own somehow . . . somewhere.

"I have many color swatches for your perusal." The tailor is as pleasant and unruffled as when he first entered. He holds up a stack of large, colored squares of fabric. Returning to my side, he offers his hand to help me down from the stool and motions me toward a simple folding chair he brought with him. Then he sets the squares across my torso.

Ash frowns at the color—a dark burgundy. The tailor pulls back the swatch to reveal the next: a slightly different shade of burgundy.

This is going to be a long morning.

Ash's brow creases, his eyes flitting from the color to my face, and I try to keep my blush at bay. "Hylath, undo that bun. Her hair should be down for me to properly see what suits her coloring."

With a grunt, my maid totters back over me, her blobby fingers unwinding my hair with far less care than Ash handled it this morning.

I glance up, find Ash's eyes heavy on mine as Hylath arranges my long hair over my shoulders. I hold his gaze, unsure what that look means, unsure why his bright eyes seem to darken just slightly.

He looks away first, clearing his throat.

"How is the color, Highness?" the tailor asks. This one is a deep red.

Ash shakes his head. "Skip the reds. I want to see the lighter colors you have."

The tailor obliges, holding up a periwinkle swatch.

My husband's mouth spreads in a slow smile. "That's more like it."

The tailor thumbs through the swatches, pulling out similar colors and a wider variety of pinks, blues, purples, and greens.

"This pink is quite lovely with her eyes, yes?" says the tailor, holding up the next one.

Ash's gaze drills into mine. I lick my dry lips. He smiles again. "Quite."

I flush and look away, unable to keep staring into those sapphire-and-gold eyes. What am I to do with all this attention? It's quite overwhelming. I have the sudden longing to take refuge in my room so I can think, perhaps make plans of my own. Do the fae have libraries? It might be good for me to locate some histories, maybe some dramas or poetries too. Things that will help me understand these people better. Here, ignorance is deadly.

I want to understand why the High King hates Ash so much. I want to know if I can trust my husband, or if offering to help him is utter foolishness.

I want to know if Ash is the hero or the villain.

I want to know if he is good.

"That is everything I need for now," says the tailor as he snaps his bag shut and readjusts his spectacles, looping his measuring ribbon around his neck once more so it hangs over his vest. He smiles and bows at Ash, then at me. "I'll require two days for the wardrobe. At the end of the week, I will bring the ballgown for a fitting. The dress you ordered last night for this evening will be delivered within a few hours."

Two days for an entire wardrobe? And a ballgown by the end of the week? What sorcery does he use to sew? I climb out of the chair he brought as my questions buzz around my head.

"The gown will be ready before Lulythinar, right?" asks Ash.

"Of course, Your Highness."

With that, he bows once more and leaves with his bag and chair. Hylath lets him out, then scurries off with a chirp to the washroom. Perhaps she's cleaning it after this morning? Or does she intend for me to follow so she can help me back into my dress and redo my hair?

The dress has been laid over the back of one of the dining chairs. I'm just about to retrieve it when Ash's voice sounds from the couch.

"Come here."

I look up with a start. He studies me from beneath that prominent brow, his eyes the color of a sea at midnight.

I think . . . I think I might know what that expression means.

If you need to escape, send a request to me for a white dress.

A lump forms in my throat. I need to keep a clear head. That will be especially difficult if Ash's lips find their way to my shoulder again. It would be better if I put my hair back up and redressed.

"I'll b-b-be back in a m-moment," I manage, scooping up my dress and scurrying around the couch, giving Ash a wide berth. Is that a chuckle behind me? I can hardly hear anything past the blood roaring in my ears.

THE PRINCE

IT'S A GOOD THING Stella left.

I wipe my hand down my face, give myself a little shake, and get to my feet. Work has piled up, and I wouldn't be so behind if I hadn't fallen asleep last night. It's just suddenly much more difficult than before to *focus*.

I enter my study and shut the door behind me. My desk is a disaster of paper, but I set to work shuffling through the stacks and rifling through drawers. Where did I put it? I told Edvear to set it on my desk, and I know I saw it around here somewhere . . . It's been so long since I organized—

Ah ha! There it is.

I pull the small, crumpled parchment out of the pile. It's folded in half and yellowed with age. I stare at it, slumping back into my chair. It crinkles as I unfold it.

239 Humpidy Lane, Mithral, Valehaven Forest.

I sigh, toss the address onto the table. My manservant Calver was in my employment for thirty years. It's not an enormous amount of time, but it's enough time for that ache to hit, especially when, for the first time, it wasn't him setting out my clothes this morning.

Stella will be gone too.

I'll be left with nothing but reminders of her sweet laughter at the sight of a leaping fish, or her smell clinging to the room adjacent to mine. Now I won't be able to walk into the washroom without imagining her there before the mirror, her long, wet hair dripping on the floor.

That's why I'm doing this. Because no matter what short-term sacrifices I must make, those sacrifices will be worth it. *If* I win this gamble.

The tattoo of the broken crown itches on my wrist. I need that constant visual reminder of the bargain I made with my father. I need it, but *oh*, I hate it. I hate that it means it's too late to get Stella out of this.

This situation is what it is.

I need to stop bemoaning this turn of fate—and truly, it's quite conceited of me to be frustrated that I like my wife, as if this situation would be better if I didn't. It shouldn't matter if I liked her or not. Her loss should still impact me.

But it *is* different. Because it will be worse than losing my staff. It will hurt, and I will hate every minute of it, but I know better than to get attached to my staff.

I give a rueful snort at that. *Do I, really?*

Either way, the day Stella dies will change me forever, and I doubt for good.

A soft knock sounds at my door. I lift my head, even as my heart lurches. "Come in."

The face that pokes around that corner is a little hesitant. I smile.

"Your steward announced the midday m-meal. Shall I wait for you?" Stella asks, pale fingers drumming silently on the door. Her hair is back up in that blasted bun I despise so much, and she's dressed again in her beautiful dove-gray dress.

I pause. It would be better if I told her to eat without me, to just have Edvear drop something off at my desk after she's eaten. But I push back from my desk. "I'll come. But first, a thought from you. Or else I'll kiss you." The practiced threat rolls off my tongue, and I cannot deny that every time I say it, part of me hopes she'll opt for the latter instead of the former.

It's better that she's still so terrified at the thought of a kiss from me.

Her eyes widen. And then, to my surprise, she pulls the door shut. Straight in my face. I stare at it for one moment, and then I cannot *help* the way a grin spreads across my face as I shove to my feet, march across the room, and yank the door open. "You know the rules. If you won't tell me a thought—"

I stop. The sitting room is empty.

Is she . . . *hiding* from me?

My grin widens.

I prowl down the hallway to the bedroom, fling the door open, and give a sniff. Not here. The washroom is empty except for a scowling Hylath who mops the floor of the washroom with her tongue. She tells me she'll throw up over my letters if I step on her wet floors. I quickly vacate those premises.

Then I push open the door to the dining room—and there she is. Just sitting primly at the table, spine straight as she delicately sips from a teacup. "My thought is this. You surprise me much too frequently with these requests for my thoughts. Even if your

186

assessment is true and I have many thoughts that I choose not to speak, those thoughts become very difficult to collect when faced with imminent consequences. So truly, it isn't a fair request that you make of me."

And with that, she takes one dainty bite of her food.

That is quite a speech coming from her.

Maybe if we had the chance to be married for decades or centuries, I wouldn't be surprised by her at every turn. But two days into marriage, everything is a surprise.

Mountains of Ildrid, how it sends glee shooting through my chest! I want to surprise her back.

Instead of taking my seat across from her, I come around behind her chair. She stiffens, eyeing me warily. *Perfect.* I duck toward her and wrap my arms around her middle, hugging her as she sits rigidly in her chair.

Even when she tries to be stiff, she's so soft.

"It thrills me that you're so easily flustered at the idea of me kissing you," I say into her ear. "It makes me want to carry out my threat."

Her cheeks turn the color of wine in a matter of seconds. It's so satisfying. I feel as though I've won a prize, and I want to do nothing but keep winning them, over and over again.

And that's when stupid Edvear comes around the corner carrying a dish. I straighten too quickly, drawing away from my wife, and take my seat. My collar is suddenly too tight. I give it a few tugs.

Neither of us look at each other for the rest of our meal.

CHAPTER 26
THE PRINCESS

I HAVE HARDLY a few minutes of refuge in my room, away from the confusing presence of my husband, before it's time to leave his quarters and attend to this *errand* of his.

The first tip I get that this might be more dangerous than he lets on is the fact that he makes me wear a cloak with a hood, and then pulls that hood down over my head. I'm unsure what good it'll do, considering that a simple cloak cannot hide my humanity, and it's not as if it'll be hard to guess who I am when I'm at his side.

When I press him for information, he says we're merely paying a visit to someone, and it's unlikely that anyone will follow us.

Perhaps he's just paranoid.

I draw a deep breath and follow him out of his quarters, through the beautiful gardens and out through a gate in the hedge.

"Is this a secret exit?" I ask, panting a little to keep up with Ash's long legs.

"No palace exit is secret. The High King always knows when people come and go, as his wards get disturbed. This, however, doesn't have guards since it's a servant's exit. He'll know that we left, but he won't know it's *us*."

I follow him, drawing my cloak tighter around my shoulders. "Is it within walking distance?"

"On a Path, yes," he says, giving me a quick wink before taking my hand and drawing me to the right, away from the golden glow of the palace and the sparkling water toward the same forest we traversed yesterday.

The forest that urged me to flee.

I swallow my premonition, unable to help the way my other hand wraps around Ash's arm, as if being closer to him will somehow protect me from what he could do to me.

Order a white dress.

"Here's the Path." Ash marches straight into the forest, me in tow, and I see no path as he picks his way between trees and low shrubbery, matting pine needles deeper into the ground. His grip tightens on my hand. "Can you see it?"

He's staring down at me. I swivel my attention to the ground, to the very distinct lack of path before us. I shake my head.

"I didn't think you would."

Is that just a smidge of disappointment lacing his words? Something urgent bubbles up within me, the need to shrivel inside myself. I let go of Ash's arm, but don't dare pull my hand from his. It would be too obvious. I let my steps flag just slightly and fall a little behind him.

Of course he's disappointed. How could he not be when I am a human, and he is a fae? Perhaps he lied to me this morning when he said he could never be disappointed in me.

No, he couldn't have lied. Hylath was there, and she didn't react.

Then perhaps he merely *thought* he couldn't be disappointed. It was an honest statement at the time, proved false in new circumstances.

"Tell me one of those thoughts in your brain," says Ash abruptly. "They seem to disturb you."

I blink, looking up at him. "My lord?"

"You're thinking things. Tell me."

Absolutely not. He cannot know what I was thinking just now. It would be—

He lets out a great sigh, and I falter. "Oh Stella, must you insult my courage so frequently? It's quite wounding."

Is that true? There's no one here to help me tell if it is a lie. I chew my lip, and something fiery and bitter builds up inside me. Perhaps I *should* tell him what I'm thinking, and just *see* if it would make him angry. Maybe it'll make him stop asking about my thoughts.

Ash stops abruptly, turning back toward me. I don't have time to think before he's caught my chin, tilted it up, and pressed a swift kiss to my cheek, dangerously near my lips. "If you delay much longer, I'll be forced to kiss that mouth of yours. And when I do, it won't be short and chaste."

Every thought eddies from my mind. I can think of nothing but how *hot* this cloak suddenly is, and the looming presence of my husband before me.

He ducks in again while I'm still frozen, kissing my other cheek. I suck in a shallow breath when he draws back a few inches, his mouth hovering over mine. "Last chance, my darling."

"I c-c-c-can't th-think," I gasp, twisting slightly. My lungs scream for air. Why is breathing so hard suddenly? Even if I knew what to say, I could hardly get it out now!

"Then tell me what you feel," he murmurs gently.

"F-flustered."

"From me being close, or the thought of me kissing you?"

"All of it!" I burst and wrench free of him, stumbling back a few steps. Surprisingly, he lets me go, and I don't know what that expression means on his face. I gather the scraps of my composure back together, straighten my spine, run my sweaty hands over my skirts, and set my shoulders back. "Ash," I say, lifting my chin.

"Stella," he replies, his voice low.

"I don't wish to be flustered right now. That is my thought. Now, if you would be so kind as to not purposefully try to induce anxiety in me, I would be grateful. These last few days have been enough by themselves."

I turn and continue marching forward in the direction he'd been heading, but not before I catch his expression. Thinned lips, without a trace of amusement, and a creased brow. Then he's at my side, and we walk together through the forest without touching.

"This way," he says, gesturing for us to skirt a tree, and I wonder if he made up the whole *path* thing, but I follow anyway.

We perform the rest of our walk in silence. I almost want to ask what he's thinking, but that would invite more conversation—and I'm not sure I'm ready for that yet.

Before I know it, the forest changes. The ground becomes littered with mushrooms, white and red spotted ones, others brown or gray. The farther we walk, the larger the mushrooms get, and the brighter the colors. They shift slowly enough that I don't notice it at first, until there is a very distinct mushroom the color of lapis lazuli before me, almost as tall as my knees.

That's when I realize the trees are also getting taller. Much, much taller. Unease trickles down my spine. I scoot a little closer to Ash, wishing I was brave enough to either face this with straight shoulders or grab hold of my husband's arm.

"Why are these mushrooms so big?" I whisper.

Ash glances sidelong at me. "They're normal sized."

Normal to *him*, perhaps.

He hesitates then, stealing another glance my way.

"What?" I ask, wrapping my arms around my middle and trying to hide from the deepening sense of premonition.

Ash gives his neck a quick scratch. "I suppose I should probably inform you that as we enter this colony, we're actually . . . shrinking."

"Shrinking?" I blurt, and then look around at the mushrooms now tall enough to reach my waist, the trees towering to the heights of heaven above us.

What makes the panic hit past my wall of disbelief is when I look down and discover how large the pine needles are—long enough and fat enough for me to pick one up and use it as a toy sword.

I stop walking. My vision starts to go black.

"Whoa now. Easy there." Ash's arm slips around my waist, catching me as I slump, my head falling onto his shoulder. "Forgive me, my love. I forgot this isn't normal for you humans."

What won't be normal is the shape of his nose after I regain my strength enough to swing a blow at him.

Where did *that* thought come from? It's hardly a dignified response for a princess.

Despite clinging to consciousness like a barnacle on a turtle's shell, I mumble against Ash's neck, "I have the strangest urge to punch you."

"You wouldn't be the first," he replies, scooping me up into his arms.

The next thing I'm aware of is a crackling voice permeating through the fog of my awareness. Wobbly, but kind and warm. "Do you want smelling salts for the girl? Fainting is not a good sign, Prince Trenian. Poor thing must be taking the adjustment hard."

"She's taking it quite brilliantly," comes Ash's voice, oddly close. So close, I'm feeling it more than hearing it. "She has a practical head on her shoulders, especially for a princess. It's very admirable."

"I never thought I'd see the day our prince was wed, much less to a human. My boy was honored to serve you all these years. He had nothing but good things to say of the kind master you were when he came to visit." That crackling voice breaks slightly, punctuated by a sniffle.

My eyes won't open. I frown.

"Calver was faithful in his work. You should know that he faced his end bravely. He didn't beg, and his last thought was for you."

The sniffles grow.

The warmth around me shifts, and a chair creaks. "Here," comes Ash's gentle urge.

"Oh, thank you," is the return half-sob. A loud nose blow follows this.

"I've made arrangements with Prince Rahk," Ash continues, despite the soft weeping. "He will take you to Orawyth, and you'll be safe there. It's a human land, but completely disconnected from our continent, like those beyond the Veil, and Faradir does not have access to either of the doors to that world. Once I'm High King, you can return, and you will be under my protection."

"He said you might make me leave," the woman responds sadly.

"It's only temporary. I will place a ward on your home while you're gone. You will return to it exactly as you left it. But you must leave, because if the High King learns of the bargain I made with your son, he will kill you to spite me. I don't intend to take chances. Not now."

My eyes finally manage to open, enough for me to see Ash reach across the table and lay his hand over the woman's.

"Your son gave his life for your safety. I will not allow his death to be in vain, Mama Bagogs."

She nods, sniffling, and gives his hand a squeeze. "You are a good man, Prince Trenian. All of Faerieland groans under the hand of your father. We long for the day you will be our High King."

She's a human, heavyset, with gray hair bound up in an orderly bun. A patched kerchief covers most of her head, and her simple calico dress is starched smooth, without a single wrinkle. Eyes the color of honey fall on me. "You're awake, my lady!"

"Decided to join us after all?" says Ash with a quirked lip as he shifts me upward. I catch hold of his collar, suddenly dizzy at the movement.

"F-f-forgive me," I manage, my vision going in and out of tunnels.

"Do you still have those smelling salts?" Ash asks the woman, studying me as he tilts my head back on his shoulder.

"Of course!" She hops up, hurries to grab something. "Right here, Highness!"

Ash takes the offering, and before I can protest, slides it right beneath my nostrils. Stink flares into my brain, so sharp and startling

194

I let out an unladylike, "Ugh!" and twist away. But when I open my eyes, I'm alert.

Alert—and much too aware of all the places Ash and I are touching.

"Better, love?" he asks.

"Here's some steamed sap tea, dear. It'll help restore your strength."

Before I can protest, an oven-fired clay mug is pressed into my hands. The sweet scents of nutmeg and maple syrup waft to my nostrils, and curiosity replaces my misgivings. I lift the mug, take a tentative sip, and pure goodness blossoms on my tongue. I make quick work of the drink, and when I set the empty mug back on the table, I discover Ash has already downed his.

"Mama Bagogs' steamed sap tea is the *real* reason I accepted Calver as my manservant," Ash whispers to me, loud enough for the elderly woman to hear.

"Oh, enough of your flattery! Let me go find another chair for your wife!" says Mama Bagogs, bouncing up again. "There's one in the other room—"

"Please, don't trouble yourself." Ash flashes one of those devilishly brilliant grins of his and tightens his arm around my waist just as I go to stand. "There's no need." And with that, he pulls me right back to perch on his knee.

"Ash!" I give a little protest, trying to calm my whirling thoughts.

"What? Am I scandalizing you? You look scandalized. Do married couples in your land not behave this way?"

"Certainly not!" Not in front of others, that's for sure. For most couples I know, I highly doubt even their private interactions would share any resemblance to this.

But in the end, it's Mama Bagogs who rescues me. She grunts while trying to bully a rickety, wide-legged stool through the small, arched doorway of what looks like a cozy bedroom. Ash leaps to his feet, sliding me off his lap, and hurries to the woman's side. "Please! Don't trouble yourself! What would your son think if I let you break your back trying to move furniture?"

"Oh, now, Prince Trenian, you mustn't burden yourself!"

"I'll burden myself however I please!" And with that, he snatches the stool from her grip, replaces it in the bedroom.

"Now, now—" the woman protests.

"Now, now!" he retorts, wagging his finger in her face. "You might be my most troublesome denizen yet!"

She bursts into a warm chuckle, shaking her head. I stand beside the table, picking at my sleeve with my fingernail as I watch this odd exchange. I'm not quite sure what to make of it when Ash grins down at the little lady, or when he bends and presses a quick kiss to the top of her head.

Have I been misreading his frequent kisses this entire time? His warm demeanor? Perhaps what he feels for me is more platonic than I first thought.

Somehow, I cannot *quite* make myself believe it.

"Stay safe," he says to her. "You'll leave in two days. Pack only what you can carry. I'll give you funds enough to settle comfortably in Orawyth until it's safe for you to return." He comes to my side, stooping from the low ceiling, and briskly slides past his chair. I peer up at him, but he doesn't meet my eyes.

"Stay safe yourself, Highness," the woman responds with a warm smile. "And take care of your sweet wife."

"With my life," he replies, laying a warm hand on my low back and escorting me out of the hut. He ducks beneath the lintel.

He closes the door behind us. There's a strange sort of round roof over—*oh.* I nearly pass out again as I realize the awning isn't an awning at all, but the gilled underside of a white-spotted red mushroom.

I blink, glancing around to find more people like myself, mingling with . . . *mice*? They walk upright on their hind legs with tails and everything. They're clothed, with eyes much more intelligent than a mere rodent's, and all around are different shapes and varieties of mushroom houses. Their gates are made of chopped pine needles, and I stare agape at the mushroom cap wheels on a cart as it rolls by us. It's a whole bustling *town* of tiny people!

"What is this place?" I blurt, awe filling my voice.

"This is Mithral," replies the prince with an uncharacteristically serious expression. "One of the Small Cities. They're built by the Mips—the mouse people—but most of the free humans who live in Faerie have taken shelter here. The High King doesn't know. He knows about the cities, that is. Just not that humans hide here. It makes them a rather ideal place to keep my more at-risk staff and their families, and it's also where I find most of my human staff."

He eyes me, and a few different things flash through his gaze. I try to catch hold of them, but they're gone too quickly.

"How long have they been hiding here?" I ask.

"A little under two hundred years."

The way he says that . . . I peer up at him. He glances down at me, blinks, quickly looks away. That look confirms it.

He's the one who started hiding humans here.

"Well!" Ash says brusquely. "This day is certainly getting away from us, isn't it? Come, come, we mustn't be late for the feast tonight. The tailor should have delivered your gown by the time we get back."

With that and nothing else, he takes my hand and pulls me after him. We move briskly through the dirt street, framed by blades of grass tall enough to be trees. I'm forced to dodge around the ruts left behind by acorn wheels.

As we go, a mouse on the other side of the street tips his top hat at Ash. A lady mouse in a pink calico frock and lace-trimmed apron curtsies. A few humans, standing in a group, elbow each other and point—only to bow deeply when Ash spots them. He gives nods and warm smiles to all of them. I gape like a codfish, stumbling along behind him.

They love him. They love him, and they revere him. It's a completely different reception compared to the High King's throne room that pulsed with fear.

A sudden shout goes up behind us. I look over my shoulder as a man who looks to be about forty barrels toward us, desperation ringing his eyes.

"My lord! Prince Trenian!"

Ash stops, looks back. He pulls me to his side, keeping his hold on my hand. The man falls to his knees before Ash, and Ash looks down at him with that uncharacteristically serious expression.

"Prince Trenian!" the man gasps, panting for air.

"What is your name?"

"Andrews, my lord. Milton Andrews."

"What may I do for you, Andrews?"

"It's my daughter, Highness! Princess Listhra has just bought her years of service, and we all fear for her life."

Ash gives an acknowledging grunt, apparently well-aware of the princess' reputation. "You wish for me to buy her into my service?"

"Yes, my lord—but in exchange, I would give twice her years of service myself. Would you accept this bargain?"

Ash says nothing, his brow creasing as though he is deep in thought. "I have no openings on my staff, I'm afraid."

"Then I would give thrice the years of service, to be fulfilled by anyone of my line, at any time of your choosing. If you would but purchase my daughter back."

"Very well, I will bargain with you, but only for twice the years. My steward will be in touch with you over the next few days to finalize the bargain. You know the risks, but it's one of my policies to detail you in full."

The man clasps his hands together. "Thank you, my lord! Thank you!"

Ash's grip tightens on my hand, and then he's helping me up a hill that must be hardly a bump in the ground. I pant as we reach the top, struggling to maintain my decorum as I press my free hand to my chest.

"Ash?" I ask.

"Hmm?"

"Who was that woman? How did you know her son?"

"So you weren't out that whole time, eh?"

I wasn't exactly fully conscious! But I keep my mouth shut, waiting for him.

He draws a deep breath. "He was a manservant of mine. He died the day I met you."

"Died?" There's something much more ominous behind that word. It's there, in the sour twist of Ash's lips. Perhaps I might think it was a lie, but this strikes me as very true. Almost as if the truth tastes worse than a lie ever could.

"My father killed him."

"Oh."

We're silent as we continue our trek. Did he carry me all this way while I was passed out? The mushroom houses only now begin thinning out the further we go. How much longer until we start . . . growing?

It takes me a dozen more steps before I find the courage to ask: "Why did your father kill him?"

He doesn't answer at first, slowing down to help me over a log that's probably nothing more than a twig. "Because I rejected his choice of bride."

"You *rejected* his choice of bride?" The shocked words are out before I can stop them. I clamp my jaw shut, but not before he shoots a surprised look back at me. "F-forgive me," I quickly add, swallowing thickly. "I didn't know anyone could reject a marriage arrangement."

It sounds foolish once I say it. Perhaps there are many people who reject such a thing. It just seems so . . . utterly . . . *impossible.* If I had dared to even express displeasure at my father's choice . . . I barely withhold a shudder.

"Perhaps you forget that the High King is actively searching for ways to end my life. It's a complicated situation." He says this with a little smirk and wink, but there's something else there. Something lurking behind the brightness of his eyes.

My heart gives a twinge. Though my relationship with my own father has always been complicated, he never actively tried to do me harm. There might have been hurt and bitterness between us, but never malice.

My mind flickers back to last night, to the way the air simmered between Ash and his father. The way I'd been caught like a mouse between two predators.

And yet . . .

I give Ash's hand a little squeeze. It's not much. Perhaps words would be better, but my tongue is too tired and elegant words have never been my strength. This little hand squeeze will have to suffice for the things I wish to say. A simple *I'm sorry, Ash.*

A very subtle twitch goes through his fingers, and he half looks back at me, then stops, as if he doesn't want to meet my eyes. Instead, his hand tightens around mine, and he tugs me a little closer than before.

The mushrooms around us have shrunk to Ash's height by now, and with each step, we *un*-shrink, as it were, until finally the trees are a normal height, and the surrounding forest is how it should be. My feet wobble, but my husband's grip on my hand steadies me.

It's his hand stiffening in mine that first alerts me to something being wrong.

I look up, find Ash's piercing eyes scanning the forest, two bright beacons in the dim half-light spilling through the tops of the trees. My heart gives a nervous little putter. I'm really not sure how much I can handle after yesterday. I want to ask what's wrong—and I almost do. At the last second, I close my lips and seal them shut. I don't want to distract him.

Oddly enough, despite the mounting unease rippling through my body, there's an anchor in the fear.

I think I might truly trust my husband, and beyond his mere need to keep me alive for his own purposes. I think—

Ash's body slams into mine so fast my breath is knocked out of my chest. I land hard on my back, wheezing up at the overwhelming weight of my husband on top of me. His hand claps over my mouth, and my eyes go even wider than they were a split second ago.

Then my gaze latches onto the tree, hardly a foot away—and the black-fletched arrow sticking out of it.

"It seems like Faradir was jealous of our steamed sap tea. Should we make him some, hmm?" Ash's face splits into a deadly grin. "Do as I say, love, and don't scream. It'll simplify things."

CHAPTER 27

THE PRINCESS

A SH LIFTS HIS hand from my mouth, pushes off his elbows, and gets to his feet. He offers a hand down to me, and I stare at it stupidly, half-blinded by the terror pumping through my veins. Before I take it, he reaches down, snatches my arm, and hauls me up so violently I nearly go flying in the other direction. A sharp *whiz* slices into the ground where I just was. Another black arrow.

I'm going to be sick.

He tucks me behind his back, keeping a hand wrapped like iron around my upper arm. "Come now!" he calls into the woods. "Is that any way to greet my new wife?"

I curl close to his back, not brave enough to peek around him and see who he's talking to. It takes all my willpower to bite back a frightened little whimper, but I'm not completely successful. Ash's grip tightens in response. I wince, grabbing hold of the back of his tunic for security.

"Why don't we all put down our weapons and just have a nice little chat? I'm sure we could come to a mutually beneficial agreement. What do you say?"

Another arrow flies, and this time it goes straight through the edge of my skirts not hidden by Ash. I let out a squeal, squeezing closer to his back.

"How unfortunate"—he gives a sigh of long suffering—"that you all are *determined* to be unreasonable. Things might get messy now, and my laundress will give me such a talking to. Don't you know how difficult it is to get blood out of clothes?"

Why is he *provoking* them? My heart pounds like a hammer against my chest, threatening to climb right out of my throat. I look around, searching for some place to hide—

A fae in all black eases against a tree not far away, and as I watch, lifts his bow. A clear shot at me.

"Ash!" I scream.

His head whips to the side, and before I can react, flings his arm toward the fae, toward the arrow zipping toward me.

The arrow disintegrates into powder that floats away on the wind. I stare, rendered immobile, as that long, glowing sword I've seen Ash carry once before appears in his hand. It looks impossibly heavy, yet he doesn't even grunt as he throws it with deadly precision into the chest of our enemy. Blue blood spurts as the fae cries out, falling to his knees.

As if that wasn't enough, Ash's hand closes in a fist around air, twisting. The blade in the fae's chest twists in response, and a pained scream rings in my ears. I tear my eyes away as Ash makes a hacking motion with his hand, his flintlike jaw clenched.

He yanks his hand toward himself, and the sword comes flying back, its hilt sliding into his grip. I cling to his cloak, breathing hard, as he faces the rest of the men sent after us. Specks of blue blood fleck his cheek. His voice is that of death itself.

"It's your own fault if you don't run."

They don't run.

Ash's back twists as he throws his sword once more. I try to block out the sounds, but instead of burying my face in his cloak, I have enough presence of mind to turn around. I keep my back pressed against his, my hands gripping his belt through his cloak, and scan the forest. Looking for any sign that we've been surrounded.

It gives me something to think about, to focus on besides the screams and the awareness of Ash's deadly movements.

"No one is b-behind that t-t-tree," I stutter to myself, forcing my eyes to stay open as another pained cry splits the air. "O-or th-th-that one."

At last, silence falls. I quiver, still clinging to his belt. He lowers his hands and twists his head back toward me. I take that to mean it's safe. Still, I cannot quite find the strength to let go of him.

Then, his commanding voice rings out, and it's not addressed to me. "Return to my father. Tell him that if he wishes to hurt my wife, he'd better try harder."

Footsteps scramble off, making at a run, and my breath turns shallow and short.

Ash turns around, slowly, so I'm forced to relinquish my grip. I stare up at him, my eyes feeling much too wide for my own face. His expression is hard as stone, and even colder. That coldness softens a little as he meets my gaze. Then he glances down and clucks irritably, rubbing at a blue stain on his tunic.

"Poor laundress," he says. "Well, enough of that. Come along, darling."

He takes my hand and guides me through the bodies littering the forest. As if this is a regular occurrence for him! My knees almost give out about seventeen times in the first dozen steps, as though I'm a newborn lamb. These nerves of mine better put themselves in their proper places before I am forced to ask my husband to carry me the rest of the way.

"I don't want your father to try harder," I mumble under my breath.

Ash looks down at me in surprise, a bright smile replacing his hard-edged expression. "I like it when you say what you think. You have many amusing thoughts."

"I . . . what?"

"Let's hurry back. You're looking a little pale. I'm afraid I've overtaxed you."

We make it back to the palace with no more incident, slipping back through the servants' gate in the garden. Ash plucks a pine needle out of my hair before we reenter his quarters.

Ash's rooms only became mine just yesterday, but slipping into them is like pulling a cozy, well-worn wool blanket around my shoulders. They're familiar, and—for now—feel safe.

A manservant comes to take the cloak from Ash's shoulders, and the hood from mine. He's human, and he works with efficiency. Did he make a bargain with Ash like that man back at the Small City? That Milton Andrews? Or a bargain like what I think Ash's former manservant made with him regarding Mama Bagogs?

"Where is Edvear?" Ash asks the manservant.

"He went out, but he should be back soon." He whisks away the garments, and I'm about to make a beeline to my own room when warm fingers touch my forehead, slide down my temple, and tuck a stray hair behind my ear. I glance up, startled.

Ash regards me with a solemn expression. "Are you alright?"

Something about the question halts me in my steps, makes my lips part as though I have something to say—but I have nothing to say, right? My hands still shake, but that's to be expected when one was just shot at multiple times.

I purse my mouth, attempt a nod. My neck remains upright, unmoving. It's been a few seconds before I realize I'm just gaping at him.

His own mouth tightens into a line. Then he drops his hand from my face and leans back against the doorframe, eyeing me with that startlingly intense expression of his. "When I married you, I promised to be good to you."

I'm not sure what promise he's referring to, if that was the gist he got from the vows he pledged me at our ceremony. My throat goes dry anyway.

"But I fear I may be overwhelming you."

It's not *him*, so much as everything being new. And his father trying to kill me. Sure, Ash can be larger than life when he wants to be, but he's given me far more attention in the span of the last two days than I've ever had in my life.

His attention is probably clouding my judgement. It would be better if I slipped into the background somewhere, out of his way. Then I'll think more clearly.

"Stella?"

I blink. "Pardon?"

"Where did you just go?" He slowly wags a finger in front of my face, and I go cross-eyed following it. "You left the room a moment ago. Where'd you go?"

My eyes lift from his finger to his face. I cannot read his expression. There's a perceptive gleam to his vibrant irises, a gleam that studies me intently. How much has my face revealed? "F-forgive me. I am merely tired. There is much for me to pr-process."

"What troubles you?"

"Pardon?"

"Something is troubling you. Are you afraid?"

The question almost makes me snort. I restrain the undignified sound and only allow a small smile. "Of course not, my lord."

He blinks at me, as he always does when I lie straight to his face. Usually, a grin is quick to follow it, but this time it doesn't come. Instead, he takes another step toward me, narrowing the distance between us. My heart picks up a staccato rhythm as he slips a finger beneath my jaw, lifting my chin so our eyes meet. "You are my wife, Stella. It is true that my father wants your life and will take it when given the chance. But you are *my wife*, and I will fight with everything I have for you. I will—"

"You hardly know me!" I burst before I can help myself. "We're practically strangers!"

Strangers married to each other, attempting to forge some semblance of a life together in a hostile world. Our situation is so insane, so ridiculous, it could be comical!

His eyelids lower. "I don't want to be strangers, Stella. I want to *know* you. And I want you to know me. I don't want this to be a mere political alliance, a transaction where I purchased you from your spineless father."

My lips part. He cannot know what that means to me. Or maybe he does, and that's why he's saying it. Because he knows I'll fall for it. Confusion roils in my gut. For a world of people who supposedly cannot lie, I feel as though I swim blind and deaf through a sea of falsehoods and half-truths.

His hand slides from my jaw for his fingers to curl around the back of my head, as he comes closer, closer, until I can hardly breathe. "I don't want you to be afraid."

Not knowing what else to say, I merely nod. He can wish all he wants, but that won't change anything. We stay like this for some time, until my eyes drop, and then he lets me go.

"I sh-should like to take a rest, my lord," I mumble.

His jaw flexes. He nods.

I turn and make my escape, fleeing to the refuge of my room. I shut the door behind me and breathe deeply. The knot in my throat thickens, sharpens, until I cannot hold back the tears anymore. I flop onto my bed, stuff my face into a pillow so no one can hear me, and let the dam loose.

I thought if I just approached this with simple practicality and acceptance, everything would be fine. But I'm *so confused.* I don't know up from down in this world, left from right. At least if I'd been married off to Prince Brochfael, I would know exactly my place in the world. I would be the quiet Isabelle Louise who submitted to the whims of her husband, just as I submitted to the whims of my father. I knew my role: to take whatever was given to me, to give whatever was asked of me, and to do so without complaint.

It would be a miserable existence, but I would know exactly what to make of it. I'd hardly have to speak a word, and once Prince Brochfael grew tired of his new wife, I could slip back into the shadows where I belong. I'd find a quiet way to pursue a proper interest besides growing little potted plants in windows. Needlework, perhaps.

Here? With Ash? I don't know my place anymore, and it deeply frightens me. Who am I, if not quiet, docile Isabelle Louise? Why do these unexpected flashes of will, of anger, plague me? Why do I want to test and prod the identity I've had for so long?

Ash doesn't want me to be afraid. Well, neither do I! But how else am I supposed to face this terrifying new world? A little fear seems like a healthy thing, right?

But I don't *want* to be afraid.

I want to stand on my own two feet, to face arrows without a flinch of fear, to not shirk from my imposing, charismatic husband. Maybe I *should* tell him what I think! Maybe that will give me the boldness I long for.

Maybe I've spent too much of my life letting others decide my path. Maybe I've considered myself too much a victim of my father, and now, of my husband, Faerieland, and the High King.

But perhaps I can face these things head on.

What if my spine didn't have to bend? What if . . . what if I *wasn't* afraid?

Who would I be if I let go of the chains of fear binding me?

I throw aside the pillow, launching to my feet. I pace the U around my bed, back and forth, and then—because I still don't want anyone to hear me—I growl under my breath. "Don't tell me what to do, Vivienne. Don't tell me what to say, Jacquelle. Don't tell me who to be, Yvonne. I'm sick of it! I'm sick of your nagging, and the way your nagging has carried with me to Faerie. Go away and nag yourselves!"

I stop, breathing hard, but more invigorated than I have been in a long time. Uncurling my arms from around my middle, I put them at my sides and clench my hands into a fist. "I really like you, Ash, but I'm *not* going to be your pawn. I'm your wife, not a piece on a gameboard. So don't think I'm just going to capitulate to your every whim!"

This feels good. *Really* good. I prowl to the other side of my bed, nearer to the window, and stick a finger in the air.

"And *you*, High King . . . well! Don't think you can kill me or my husband without a fight—from both of us! You think I'm just Ash's pet. Keep thinking it, High King, and let's see what happens when you discover you've underestimated us!"

This is ridiculous, part of my brain insists. *You're making a fool of yourself.*

"So what if I am?" I demand aloud. "What if I *don't* want to *care* anymore?"

My lungs heave with every breath, as though I've just run to Mama Bagog's house and back with no breaks. But I am alive! Electricity buzzes beneath my skin, igniting me. A grin spreads across my tear-streaked face, and I cannot suppress it. I don't want to suppress it! This is glorious!

But then I find myself facing the patch of floor between the window and my bed's floorboard. My grin fades, my shoulders slightly hunching as tension radiates back into my body. Every instinct tells me to bow my head, to fold my hands in front of me.

Father.

How can I face him? It goes against the code written into every bone in my body.

I bow my head.

I clench my fists.

There is no more cowering. I'm not Isabelle Louise anymore. I'm Stella. There's nothing left for me back in Aursailles. Faerieland is my home now. Ash is my home now. Father has no power over me. His memory cannot hurt me.

I lift my head, glaring into the empty air where I imagine his eyes to be. "I'm done making myself small so you can feel strong. I'm *done* being afraid of you. I'm *done* fighting for your fickle approval. I've had enough! *Enough,* I say!"

They're only whispers, but they leave me breathless. Breathless and *alive.* My shoulders are lighter. My whole *being* is lighter than it ever has been, as though I've shed a massive burden. I stare at the empty space that represents Father. My chin quivers, and I don't fight the tears as I bow my head. "You were supposed to protect me."

I shudder, stumbling back to the bed, falling onto it, and burying my face in my hands.

Father didn't protect me. Not in the way he should have. The sobs that wrack my shoulders aren't sobs of helplessness or frustration. They're made of mourning, of realized loss.

They're also a goodbye. A severing.

No part of me belongs to him anymore.

Cathartic tears wash away the pain, until I'm left spent and exhausted on my bed, folded in the soft blankets and staring at a damp spot on my pillow. Then I close my eyes, and a warm peace blankets my soul in darkness.

CHAPTER 28

THE PRINCE

A CRAMP IN my hand forces me to set down the quill and rub the sore muscles. I roll one shoulder, then the other. I pop my neck and release a great sigh, leaning my forehead against my palm and staring down at the half-scrawled coded letter before me. A request for Rahk to take Mama Bagogs to Orawyth. And a notice that it's time for him to propose to Lord and Lady Nothril the absorption of the Neverseen King's portal to Orawyth.

The Neverseen King will be suspicious at such a request, but I know he hates the Orawyth portal. The Nothril Court will be happy to have it completely under their control. It'll make everyone happy, if the deal goes through, and I will be one step closer to allying my father's two biggest threats. They'll have to come to Valehaven and present the news to the High King. And that's the distraction I'll need to make Faradir forget about Stella long enough to ensure the rest of my plan works.

I massage the bridge of my nose. Maybe one of these days, my scheming will pay off in the form of a crown. Or perhaps I'll just end up as dead as my mother.

Once my hand stops hurting, I pick up the quill and finish the letter. It goes in the stack on my desk of outgoing mail. From the other stack left by Edvear, I pick up the next thing. It's that request from Milton Andrews, signed by Edvear, indicating that it's been taken care of. Which means I'm now obligated to buy this slave girl. The tattoo itches on the inner skin of my forearm. I roll up my sleeve to scratch it.

Humans in Faerieland. Such a complicated problem. They don't belong here, and yet here they are. But the laws must stand. If a human enters Faerie in breach of terms, then they are enslaved. There must be punishment—otherwise humans will find it quite thrilling to run amok in our forests, getting themselves into all sorts of trouble. I cannot thwart the law.

But I do what I can to mitigate its abuse.

It's just . . . it's never enough. And in fact, it's often almost too much of a risk for me to take. If I go buy this girl from Princess Listhra, someone will report it to the High King. He'll think I have an attachment to her, and then she'll be the next on his list to demonstrate his wrath.

I'll have to send her back to a Small City and accept her father's service in her stead.

A knock sounds on my door.

I look up through the strands of hair fallen in my face. Part of me lifts in sudden hope—but I squash it immediately. I know this knock as well as I know my own name. It's not the soft little knock I cannot deny I've been hoping to hear all afternoon.

Edvear opens the door.

"Yes?" I ask, raking my hair out of my face and leaning back in my chair.

"The banquet is fast approaching, my lord. Do you care to dress appropriately?"

I glance down at my blousy white shirt, its throat strings loose, my sleeves rolled up to my elbows. My bare feet. I sigh again. I half-consider just glamouring myself for the evening, but then I remember I'll be maintaining Stella's glamour. Best not to stretch myself too thin in case something goes . . . *sour* tonight. My awareness goes to the vials in the secret compartment of my desk.

"I'll be there in a moment," I say, picking up my quill again.

But Edvear doesn't leave.

"Yes?" I ask, lifting an eyebrow.

"The Lady Stella . . ."

My alertness sharpens. "Yes?"

"She's still asleep, my lord."

My other eyebrow joins the first. "She's still asleep? That's . . ." My voice trails off as I check my pocket watch. "Four hours, now?"

Poor thing. I need to stop frightening her so much. I forget how taxing close calls with death are for humans. I've grown too used to this dance.

"Indeed, my lord. Shall I wake her? Or would you rather leave her behind for this evening?"

I sigh, dropping my head. It seems a cruel thing to wake her when she's clearly so exhausted. But the prospect of going without her makes my teeth tighten. It wouldn't be a good look. Perhaps I could play it to my advantage? Maybe if I presented it like an insult to Valehaven's hospitality . . .

"I'll check on her," I say, rising to my feet. "Please have someone select tonight's clothes for me. I don't think I have time. Did Stella's dress arrive?"

"Indeed. Several hours ago, in fact."

Perhaps the biggest travesty of not bringing Stella tonight would be not seeing her wear the beautiful new gown. I picked it specifically for her. It's a mesmerizing blend of fae and human fashion, and conservative enough to be comfortable to her sensibilities. I'm probably too eager to see how she likes it.

Edvear closes the door behind him. I pause, then slip my hand along the hidden latch. It springs open, and I open the drawer. Vials of poison gleam back at me.

I let out a deep sigh. Always best to be prepared in my father's presence. Even if he cannot kill me yet . . .

It doesn't hurt to expect the unexpected.

I dare not underestimate the High King.

Quickly, I take a draught of each, wincing as the last slug-like black liquid crawls down my throat. With a shudder, I pull my composure back together and return the vials to their hiding spot.

I march out of the room, cross the hallway to the bedroom. I skirt around the untouched bed that I haven't slept in for over a week and pretend it isn't looming in my periphery as I stop before Stella's door. There are those familiar soft snores.

I smile despite myself.

What if I declined tonight's banquet, too? Perhaps I could work *that* to my advantage. Then, instead of playing political games with fools, fops, and worse, I could slip beneath the covers beside her and hold her in my arms while she sleeps.

I blink. Shake the thought away. Silently, I turn the latch and push the door open.

She's curled in a little ball, her mouth hidden beneath the covers. Her dark lashes fan her cheeks, her hair a disastrous mess with half of it fallen free of her bun. The sight of her like this, sweet and beautiful and completely lost to her dreams, could stir even the blackest heart to soften.

My shoulders slump slightly. If I'd had any intention of waking her before now, it has flown away like a bird from its coop. I'll just have to go without her, then.

I should be relieved. She's safer here, after all.

A snore turns into a snort. Stella's brow puckers faintly, and she rolls onto her back and stretches her arms above her head. Oh Great Kings, is she waking up? I can't let her catch me—

Her eyes open. Blink thrice up at the ceiling. Then they swivel straight to me, puffy with sleep.

"Ash!" she gasps, clutching the covers to her chest and sitting upright, hair falling everywhere. Hastily, she tries to comb it back, succeeding only in making it worse.

I put up my hands quickly, and in my surprise, my glamoured control over my flush slips away. I reassert it and hope she doesn't notice. Then I take a step backward, moving my foot beyond the threshold of her room. "My apologies! I didn't mean to wake you. I merely intended to check on you."

She stares at me, her own flush brightening. She looks down, seems to realize she's fully dressed, and lets the quilt fall into her lap. Twisting the fabric of her skirt between two fingers, she glances around, her eyes lighting over the space—everywhere except me.

Her eyes aren't puffy from sleep. She was crying.

Mountains of Ildrid, I'm such a cad for taking her today. I should have left her behind. I'd mistakenly thought we wouldn't encounter trouble, and hadn't anticipated all the adjustments she'd face along the way.

"How long have I been asleep?"

Her voice startles me back to the present. "Four hours," I answer.

"Four hours?" she gasps. "Oh! Don't we have . . . something tonight?"

I clear my throat, turning my head away. "A banquet. But you seem fatigued . . ."

Her lips spread in a wry smile. "I should hope myself rested after four hours."

She wants to come? My gaze shoots to hers, and something about her seems different. Different from even a few hours ago.

"You don't have to come," I say, clearing my throat again. "I can think of an excuse—"

She sits up straighter, leveling her narrow shoulders. "I'd like to come."

I blink, certain I didn't just hear those words cross her lips. "It will be dangerous," I say quickly, and I'm not sure why I'm trying to dissuade her. "Only fae will be present, aside from the human servants. The High King will be there. I will have to be the way I was. Last night."

A slight shudder passes through her at the mention of the High King. But as fast as it comes, it's gone, and when she looks up at me again, determination flashes in those soft eyes.

"Would it help your . . . *goal* if I came?"

I lick my lips. "Yes."

"Then I will come." She pushes aside the blankets, swinging her legs and long skirt over the edge of the bed. She stands, and my throat goes dry as she takes the three steps separating us, tilting her face up toward mine.

I stare down at her. *Kiss her,* every fiber of my awareness begs. I swallow, struggling to keep my glamour in place as she gives me a coy little smile.

"If it's dangerous, you'll just have to protect me," she says.

I watch her lips form the words, hardly hearing them over the sudden roaring in my ears. *Kiss her. Kiss her. Kiss her.*

That's when it clicks—what is different about her.

She's not stuttering. Not even slightly.

Which means . . .

She's not afraid.

Something shifted after our conversation in the entry hall. I'm not sure what it is, only that there's no denying it. *This* is what I saw during our first dance back in Aursailles, that spark of something buried deep inside her. It burns brighter now, and it is undeniably compelling.

She makes to walk past me, to leave her room. My self-control breaks with a resounding snap. My hand darts out and I catch her around the waist. She gives a small, surprised yelp, but doesn't resist when I pull her back against my chest.

I love the feel of her.

Swiftly, hardly trusting myself if I linger too much, I sweep her tangled hair to one side and bend down, pressing my lips to the gray fabric covering her shoulder. Her breath catches, her hands flying to grip the arm I have around her waist. I don't want to let her go. But I'll do something I regret if I don't.

I cannot help indulging in a quick squeeze to bring her closer, one more inhale of her sweet scent. Then I let go, and she scampers out of the door. The little bit of her face that I catch before she's beyond my range of vision is bright red.

Perhaps I'd chuckle if I didn't feel so bereft.

CHAPTER 29
THE PRINCESS

THE DRESS THE tailor dropped off for me is . . . *well* . . . the most scandalous thing I've ever laid eyes on. Never in my wildest dreams would I have even *considered* wearing something so . . . so . . .!

I stare in the bathing chamber mirror, hardly daring to meet my own embarrassed gaze. I can't go anywhere like this! It's not as though I can keep my hands plastered over my exposed chest. Perhaps the lower neckline could be tolerated—it's not *that* low, after all—if my legs weren't exposed to their knees! It doesn't *matter* that the dress technically reaches the floor. If it's translucent material, it's utterly indecent. I can hardly notice the beauty of the shimmering pearlescent blue gown, the color of a snowflake, with its glittering train. The colored fabric of the bodice and hips gives way to translucent material around my knees and continues all the way to my slippers. Both of my calves are on complete display! And if that wasn't enough, there is a *slit* running halfway up the skirt.

Hylath burbles irritably behind me, her eyes bobbling all over the place as she styles my hair in an elaborate updo that makes my neck look long and slender. In a fae mirror, I find myself thinking how very choke-able I look with my hair this way.

Do I really want to get used to this place and all the horrifying thoughts it brings?

"Is she almost ready?" comes a masculine voice from the other side of the door.

Ash's familiar timbres send a scalding line of panic down the length of my body. I barely restrain myself from hopping up and hiding my scandalous appearance.

Hylath makes a chorus of noises in response. Ash huffs. I brace myself, expecting him to enter at any moment. He doesn't, which must mean Hylath told him I wasn't ready yet. She weaves sparkling beads and strands of pearls into my hair. It's quite lovely, and if I wasn't so self-conscious about the dress, I might enjoy feeling beautiful.

"Grrrbaurgh!" says Hylath, clapping her hands and stepping back. I'm done, then?

I swallow, plucking nervously at the neckline. It's not that low, right? I'm merely not used to having *anything* on display. I can handle this. No one will look twice at me, or my exposed legs. That's what they do here—they drape a piece of gauzy fabric around their body that only barely covers the essentials. What I'm wearing is quite demure.

I can't make myself believe it.

It's so much leg! How could showing more than an ankle ever be appropriate in a public setting?

Another knock at the door jolts me. "Stella? Are you ready?"

"Oh!" I breathe. He's going to see how self-conscious I am! After that unexpected shoulder kiss not that long ago, I feel especially exposed. "You can do this," I mutter softly, hoping Hylath doesn't have augmented hearing like the high fae. "You can do this. You're not afraid, remember?"

Funny how a new fear always seems to pop up in the wake of another!

I get to my feet, clench my fists at my side, and force myself to cross the distance to the door. *I can do this. I can do this. I can—*

The door swings open. And there's Ash.

He wears a long, midnight blue tunic that reaches his mid-thigh. The throat is open in a deep V, which is heavily embroidered in a thick, shimmering silver thread in the pattern of fauns, fruit trees, and what might be dancing nymphs. A heavy gold medallion with a gleaming sapphire the size of my fist hangs from his neck. He wears tall boots that hug his calves almost up to his knees, and a pair of dark breeches that somehow catch the light with his movement. His long hair is swept back, partially tied at the back of his head, revealing the tips of his pointed ears. He's got a ring on almost every finger, and a large, gleaming silver crown atop his head.

The sight of him, so tall and majestic, looking so very princely, completely halts my own self-consciousness. I'm once again struck with a sense of awe that I am *married* to the *Prince* of the *Fae*. This is my life.

This is my husband.

His gaze sweeps over me in similar appraisal. Is it just my imagination that they're a little wider than normal? That a muscle just jerked in his jaw? That's when the awareness of my almost bare legs hits me so hard I do the only thing I can—I cover my chest.

Ash's attention returns to my face, his expression shifting as a line creases his brow. "You don't like the dress?"

There's disappointment in his tone again. It hits me hard—but not as hard as before. I can better notice it now, how much I am tempted to quickly assure him that I do, indeed, love the dress. But that would be a lie, and I am ready to be honest with him.

"It is b-beautiful," I say carefully, still covering the bit of exposed chest. "I am simply not c-comfortable with . . . bare legs!"

His gaze sweeps back over me, his brow knit. "Your legs aren't bare."

"They *practically* are!"

He chews the inside of his lip. "I can glamour it, if it would make you more comfortable. Would you perhaps . . ." He gestures at my hands. "Want other adjustments?"

I nod quickly. He gives me a rueful smile. That's when it clicks: he's not disappointed in *me,* but he picked this dress out, didn't he? He's disappointed that he didn't please me.

Oh.

Ash bends down, catches my face in his hands, and presses his forehead to mine. This time, I close my eyes with him, feel the rush of something deep inside me, around me, whirling like the ribbons of a dancer. His skin is warm against mine.

What would happen if he and I had *time* together?

I think I could love him. I'm halfway there already. And while I don't want to be presumptuous, he certainly *seems* to be partial to me as well. But humans and fae are very different. There is always the possibility that I misread him.

Or that everything has been a lie.

Does the truth even matter if I don't survive the next few days?

Ash lets go of me and pulls away. I glance down, discovering that the low neckline has been replaced with a demure sweetheart neckline, with a sweep of fabric that follows the curves of the dress to join with my sleeves. And beneath the translucent skirt is a layer of glittering ice-blue material.

It's beautiful.

Now I no longer feel exposed, and instead, I can gape down at the stunning gown. It moves with me like water without the confines and wires of the gowns back home. I'm grinning and I cannot help it. I turn that grin up at Ash to find a warm smile on his face, and a softness in his eyes.

"Smile like that, and no fae glamours can hold a candle to you," he says.

"What?" I blurt.

He smirks and takes my hand, drawing me out of the bathing chamber toward the door of his suite. Toward whatever awaits us tonight at this banquet. He ducks his head closer to mine, saying in a low tone, "I think you know exactly what I said and exactly what I meant. But if you want another assurance that I find you exquisitely beautiful—"

"Oh no!" I hurry to say, flushing hotter. "That is . . . um, one assurance is enough! Multiple is much too overwhelming."

He chuckles, his smirk widening. "Perhaps you ought to adjust yourself to being overwhelmed."

"I couldn't do such a thing!"

His eyebrows raise in surprise. "You couldn't?"

"Well," I say slowly, frowning, "it would seem to me that the very nature of the word is that it cannot be adjusted or prepared for. If you adjusted to overwhelm, then you wouldn't be overwhelmed. You'd be merely . . . whelmed."

He throws back his head and laughs, a bright golden burst of sound that sends warmth tingling to my toes. He pauses at the door, that smile still stretched across his face. I stare up at him, much more pleased with myself than I ought to be.

How easy it would be to let myself completely fall under his thrall . . .

He draws a deep breath, and the smile fades. "Stella," he says quietly, "tonight will be trying for you. I'm not sure what to expect, but you will be seated next to me, and I will keep a careful eye on everything."

My heart thrums nervously at his words, my palms starting to slick.

"We will be seated near the High King, and I'm not sure whether he will engage you in conversation or ignore you. He likes to catch me off-guard. Princess Listhra will be there, who is from the Solirius Court."

"The one with the servant girl that man at the Small City was concerned about?"

"The same. Other members of the High King's court and other courts may be present, but I have not seen the guest list. Stay close to me and you'll be fine."

I nod, tightening my grip on his elbow.

He reaches out for the door, pauses. He lets go, turns to me, catches my jaw and tilts my face up. I suck in a quick breath of air. His eyes rove over my face as mine rove over his, trying to read the secrets behind the furrow of his brow.

Then he bends and presses a soft kiss between my eyebrows.

I swallow hard, breathless, as he pulls away and pushes the door open.

He draws me after him as we enter the palace hallways, walking over polished white floors, between arching marble pillars and past multi-storied fountains and waterfalls, with lily pads, lilies, and fish the colors of rainbows. A sweet, floral aroma rises from the lilies, far stronger than anything I've smelled back in Aursailles. Ash swiftly takes a left, barely slow enough for me to keep up with.

Guards with long, protruding fangs line this new stretch of hallway, their wings tucked in close to their muscular bodies. My unease rises with every step, with the weight of their gaze heavy on me and each movement I make. It feels like they can see straight into my brain and pick apart my thoughts, fears, memories.

Perhaps they can. I wouldn't be the wiser.

At long last, we reach a grand pair of solid gold double doors with silver filigree designs portraying an enormous tree, its trunk running straight down the seam between the doors, its great branches, detailed leaves, and massive root network a mesmerizing display of craftmanship.

When I glance up at Ash, he's not the same person who kissed my forehead and laughed at my words back in his quarters. That glint has entered his eyes, sharper than the great sword he had this morning, and his mouth twists into a sardonic smirk that reminds me of a snake cornering its prey.

The warmth, the sweetness, is utterly gone.

The doors open.

I'm temporarily dazzled. The hall before me is pure gold, with a towering domed ceiling. A gemstone mosaic dances across that dome, portraying life-sized images of silk-swathed, horned fae playing lutes, frolicking with curly-headed fauns, riding centaurs and aiming bows at . . . *humans?* They run in various directions, mostly unclothed. One unfortunate human bows over himself on the ground, while a seated fae rests her feet on his back. A literal footstool.

Sickness washes over me. I pull my gaze down from gaping at the ceiling, only to be dazzled once more by the brilliant sea of colorful fae seated at the long, ornate table. At the center of the table is a marble statue of a winged woman wearing live greenery. Birds with scarlet plumes nest in that greenery and sit and chirp on the statue's shoulders or wings. In the statue's hand, it holds a glowing globe that illuminates the entire room. A miniature sun.

I am a princess. I am accustomed to finery. But this is splendor, the match of which I've never seen. I feel like a small little mouse peering from a tiny hole into an entire world where I don't belong.

"Prince Trenian!" the announcer calls. Then, in a mutter, "Princess Stella."

The room echoes back a bright, joyful chorus of Ash's name, and a mumble of my own. As if there is some obligation binding them to acknowledge me, despite how deeply they long to ignore me.

My whispered question is out before I can help it: "Why don't they ignore me like they clearly want to?"

Ash's face has paled. Almost in response to my notice, his skin brightens until it is luminescent. *His glamour.* My stomach drops to the floor. What—?

I follow his gaze back to the table. Realization floods through me, and my stomach drops even more.

The High King isn't present.

And there isn't a single man in the entire room.

They're all women.

It's a hall full of beautiful young women, more than half of whom wear crowns atop their silky hair. Even though they're seated, it's obvious how tall they are, willowy and elegant with effervescent skin, full lips, sparkling eyes. And their *clothes* . . . to think I was self-conscious in this unaltered gown! It's almost laughable now. If I felt mousy before, now I feel like a dead mouse.

There is only one open seat, right in the center of the throng.

This is the High King's first trap for Ash. To assault him with the women he should have married. To tempt him away from me.

Ash's voice interrupts my realization. "They address you because they must. It is the law in Valehaven that all fae must address those with titles upon entering a room."

That sounds wildly impractical, but the vanity of it doesn't surprise me one bit.

My husband's arm loosens in my grip as his attention shifts to the problem before us. A slow grin spreads across his face. He may hate his father, but is there part of him that thrills in these games? These unexpected maneuvers?

Or is it just a front?

It's impossible to tell.

"Prince Trenian," Ash calls, announcing himself again in a voice that booms through the room, louder than the announcer's. "And his *wife*, Princess Stella."

Then he strides straight toward the table, coolly confident and unflinching with that devilish grin as I cling to his arm and follow. The women rise from their seats, some smiling so beautifully it makes my heart ache. Some lift their chins in a regal display of majesty.

Before we're halfway to the table, several women have already made it to us. Or, to Ash, rather. They don't spare me a passing glance as they approach my husband. He greets them all by name and title, including the ones still seated.

"Trenian," one of them says in a lovely, sing-song voice. "It's been so long!" She steps right into his personal space, letting her gaze fall to his mouth. Her hair is long, white, but her eyes are black as ebony. "I've quite missed you."

Ash returns a steel-edged gleam of teeth. "A hundred years haven't brought enough lovers into your embrace to make you stop pining for me? I suppose they *do* say that women only want the men they can't have."

She blinks, fluttering her lashes, and gives a bell-like chuckle. "Oh Trenian. I should have known a century couldn't change you."

"A hundred years cannot mar perfection, now can it?" he replies with such haughty arrogance I never would have thought him capable of being disappointed that I didn't love the dress he picked out for me. "Have you met my wife? My love, meet Princess Pelarusa from the Nothril Court. She's one of Rahk's sisters, if you can believe it. Pelarusa, my wife, Princess Stella."

Now that he's pointed it out, the resemblance is in the striking hair color, the cleft chin, the wide cheekbones. Where Rahk is built like a boulder, this princess is formed as though from the most delicate strokes of a painter's brush.

Am I to curtsy to her? I opt for a nod. She doesn't look at me.

"You named your pet?" she replies archly, even though hers was almost certainly among the voices that welcomed me by name into the banquet hall. "How quaint."

"I've done much more than name her," says Ash with a roguish grin, drawing me closer and wrapping a large hand around my waist. He positions me slightly in front of him, which makes my breath come a little faster as I stare up at the three fae women crowding around Ash, and the one that distinctly *won't* look at me. "Come, let us eat! I'm utterly famished!" he says before the other women can shove the first aside and make their moves on him.

They sit back down, leaving that one empty chair in the middle of the women. I swallow nervously, glancing up at Ash. He takes me straight there, grabs the back of it, drawing it out so that its back legs scrape on the floor, and then drops into it.

"Your pet can go stand with the other humans along the wall," says the beautiful woman to his left, gesturing to the line of waiting servants. A pair of shining silver wings flutter from her bare back. Her long black hair is mounded elegantly atop her head, and her eyes are the most arresting shade of brilliant gold.

"I don't think I could bear to be parted from my wife for so long, Princess Listhra," Ash replies with a wink, catching my wrist—and pulling me down into his lap. He tugs me back until I lean against his chest, one of his arms wrapping tightly around my ribcage. I can barely breathe, and my hands have gone wet and clammy. My feet dangle above the ground, my skirts pooling on the floor.

Ash snaps his finger, and the servants—*human* servants—bring out the first course. A sudden bolt of fear replaces the last. They're going to serve fae food! What if I cannot choke it down?

Perhaps as Ash's *pet*, I'm not intended to eat at all.

Gleaming golden liquid in a crystal goblet is set on the plate before Ash. He picks it up, holds it high. "A toast!"

The rest of the women lift their goblets. The fairy-winged one says, "To the beauty of love."

"To the High King's throne!" another calls.

Ash gives a quiet snort at that as another chimes, "To hope of new things to come."

His grip tightens on my waist. "Hear, hear," he whispers under his breath, near my ear.

"To Prince Trenian and the thousands he has slain!"

I stop breathing. What?

"Hear, hear!" call half a dozen women, bright smiles on their faces.

It's not a lie.

I cannot help the way my body goes stiff. Or the way my mind returns to the mosaic on the ceiling above me, of the fae hunting humans and using them as their footstools.

Ash's voice rumbles through his chest into my back. "To wedded love, and the joy it brings."

Glasses clink around us, but Ash merely leans back in his chair and takes a sip. Then he tilts his head toward mine, and his thumb gives my ribcage a subtle stroke. "Care for a sip, love?"

I shake my head. My stomach is much too unsettled to brave something new.

"That is a lovely dress, Princess Stella," a bright voice says directly across from us.

I look up, startled, to find an unexpectedly kind pair of silver eyes fixed on mine. The woman has a radiant pair of electric blue wings, which only brings out the shimmering streaks in her silver hair. A delicate circlet crown drips a diamond between her brows.

"You have been some time from Valehaven, Princess Oleria," Ash says from behind me, removing any obligation for me to reply. Is that a note of curiosity in his tone? "I am surprised you have come."

"My father doesn't like it when I am gone, but he could hardly refuse the High King's particular invitation," she replies.

"He probably should have," Ash says darkly with a chuckle.

Oleria glances at me, as if searching for an explanation to this comment, but I'm no less confused than she is. A servant slips in to my left, and I turn to look as he sets down a new goblet, this one much smaller than Ash's. It bears the same golden liquid.

"For the pet," says Listhra with a magnanimous smile, as though bestowing a gift.

They truly don't think of me as Ash's wife, do they? I'm his passing fancy, a mere heartbeat in his life. One to tire of.

Well, I'm tired of *them*. "I may be his pet," I say, "but at least he *likes* me."

For the first time, the fae woman's golden eyes snap to mine. Startled disgust overtakes her beautiful features.

Ash sets down his goblet with a roar of laughter. His arm cinches around me, bringing me closer to him in a way that isn't to protect me, but to banish every last shred of distance between us. His voice is low, meant mostly for me—and yet every fae in that room can hear exactly what he says. "It is true, indeed: I like her better with each passing moment." To my surprise, he picks up the smaller goblet, holds it up to the light so it sparkles, and adds, "How lovely of you, Listhra."

Then he downs the entire glass in one gulp.

Listhra gasps, her face turning ashen before her glamour quickly masks it. "Prince Trenian! You shouldn't drink that!"

He sets the goblet down with a clink, leaning forward suddenly as his hand splays tightly over my waist, his gaze fixed on the princess. "Why?" he demands, eyes glinting and mouth twisting dangerously. "What might be in Stella's goblet that wouldn't be in mine?"

She plasters a demure smile on her face. "You should be careful. Not many are happy you've taken a human wife."

"Ah. So you think I ought to be worried about poison?"

Poison?

"It wouldn't do to be careless," she replies, turning her attention back to her goblet as she lifts it to her lips and takes a delicate sip. "Without you, the High King's throne will fall."

"What a shame that would be."

He leans back against the chair, and a tiny part of me is relieved to not be the one thing between him and this conniving woman. I relax slightly against him, trying to measure my breath to calm my raging heart.

Ash brings his mouth to my neck, just below my ear, making me shiver. "If I were you, I'd jump up and run behind my chair. Otherwise, your dress might get soiled."

My dress? Soiled?

When I don't move, he gives a little snort. "Perhaps I should be clearer. Get out of my lap, Stella, before I throw up on you."

My eyes widen, and I barely have time to scramble to my feet and get out of the way before Ash scoots his chair back, bends double, and vomits all over the floor. I cover my mouth and nose with my sleeve as screams erupt around the table.

"He's been poisoned!" someone shrieks. "Get a healer! Find the antidote!"

My blood freezes for one hazy moment. Horror descends upon me like a wet blanket. It's hardly a second before my husband is swarmed. I lose sight of him, and I stare at the rainbow of dresses and barely-dresses clumping around the chair where Ash had sat only a minute ago.

I toss aside my horror and throw myself bodily into the sudden swarm of women. "Get away from him!" I shout fiercely, rage bubbling up from somewhere deep inside me. "Get *away* from *my husband!*"

My nails rake across bare fae skin as I try to tear my way to Ash. I earn an elbow to the gut and hardly notice it in my fury. He is *my* husband. *I* need to get him back to his quarters. Edvear will know what to do. But poison can move quickly. His fae blood and enormous body had better give me enough time to get help to him.

If these *stupid* fae princesses would just get *out* of my *way*.

"*Move aside!*" I bellow, grabbing a handful of gauzy fabric and yanking with all my might. As if that will give me enough leverage against their greater strength.

It's Pelarusa who sees me first, hears me first, and the look in her eye makes me realize how vulnerable I am. Most of these women, if not all, want me dead. And I just—

With one swipe of her delicate but shockingly strong arm, I go flying backward and hit the ground hard. I push up on my elbows, expecting her to pursue me, to slice into me, but before she gets a chance, Oleria steps between us. Pelarusa bares her teeth at her, but returns to the fray.

Oleria's silver eyes flash, her brow grim. "You need to get out of here. Fast. I'm not going to be much protection."

Why is she helping me? No, there's no time for that. She's right. I'm practically defenseless, and if everyone suddenly decided they wanted me dead, I'm sure they can accomplish it easily enough. But Ash—

Something shatters. A roar follows the shattered glass. "*Who tried to poison my wife?*"

Ash's head is suddenly visible above the women, and they back away as he flings his arm wide. Fury burns in the gaze he levels on Listhra. She has half-risen, staring with wide eyes at him. He grabs the armrest of her chair and yanks it toward him so hard she falls back into it, and lifts an ashen face to his.

"Who gave you that poison?" he demands, his voice lowering to something much more terrifying than his shouting. "*Who?*"

A hand clamps down on my wrist. I nearly scream, whirling as another hand clamps down over my mouth.

"It's me," comes Rahk's deep voice. "Come with me quickly, Princess Stella."

"But Ash has been poisoned!"

"He will be fine, my lady. Have no fear. It's more important to get you to safety."

I want to protest, but he's right, and at that moment, Oleria shouts over the din: "Go with him!"

There are at least half a dozen angry fae women between me and Ash. Rahk lets go of me, and just for a split second, I can almost make out the shimmering edges of a pair of wings behind him. They vanish too quickly, and I have barely a heartbeat of hesitation before I latch hold of his proffered arm and let him whisk me out of the hall.

The door shuts behind us, cutting off the cacophony with a firm thud.

CHAPTER 30
THE PRINCESS

MY HANDS SHAKE at my sides as Rahk guides me back through the halls of the palace, which seem oddly quiet after the feasting hall. He's a silent, looming presence. A presence that can turn on me in an instant.

But Ash hardly trusts anyone except Rahk. If I'm not safe with him, I'm not safe with anyone. I swallow, huffing at my attempts to keep up with his long legs. As if realizing how fast he's walking, Rahk abruptly slows.

"I thought you'd gone home," I say by way of greeting.

"I did. Until the invitation arrived from the High King for my sisters to come to a banquet tonight held in Ash's honor. I came as quickly as I could."

He came right on time, it seems. "Thank you, Prince Rahk."

"It is my honor, Princess Stella."

Once we reach Ash's quarters, Rahk raps a special sequence on the door. *Ti-tap, ti-tap, ti-tap, tap.* The door opens quickly, revealing the steward and his twitching ears.

"Why, Master Rahk, I didn't know we had the pleasure." Edvear's gaze shifts to me. "Lady Stella, you're returned from dinner. Where is His Highness?"

"He's been poisoned!" I say in a rush. "He drank my drink and there was poison in it!"

Edvear rolls his eyes to the ceiling, shaking his head as if at a little boy caught stealing cookies. "Come in, come in. Hylath, fill the tub with cold water, will you? Sanak, fetch the antidotes again, please."

"Again?" I say as Rahk gently pushes me inside with a hand on my back. "He's been poisoned before? This is a regular occurrence?"

Edvear lifts the train of my dress out of the way and shuts the door. "Not regular, no. Neither infrequent."

I press a hand to my chest, not sure what to do or say. Rahk doesn't remove his hand until he's nudged me toward the couch and gestured for me to sit. I sink into the soft cushions, only now realizing that it's not just my hands that are shaking—it's my whole body. "Will he be alright, then?"

"Of course, my lady. He keeps a stock of antidotes for this very purpose."

The notion is so foreign to me that I lean forward, massaging the bridge of my nose and wracking my brain, trying to understand. He said something this morning, hadn't he? About taking precautions against poison. This must have been what he was talking about.

But what if he doesn't have the antidote for *this* poison?

Stop worrying, I tell myself, trying to pull my frazzled nerves back under my control. *This is Faerieland. Poison is a normal part of life here. Ash wasn't concerned, so I won't be either.*

It helps. A little.

The door swings open again, and I leap to my feet as Ash stumbles through. "Ash!"

"Oh, thank the stars you're here, wife," he gasps, pulling the door shut and collapsing against it. His garments are clean, thankfully, but he looks horrible. Pale, wane, his eyes stark against the gray cast of his skin. "I almost lost my mind when one of the servants said someone had taken you out."

"They didn't tell you it was me?" asks Rahk with a furrowed brow.

"Oh, they did; I just didn't believe them at first. Some *vermiltris*, please, Edvear! And a cold bath."

"Right here, my lord!"

"Are you alright?" I demand, rushing to his side as he takes the proffered glass with something like yellow sand settled at the bottom and downs it in a quick gulp. "You look horrible!"

"You wound me, little wife." He returns the empty glass, wrapping an arm around his middle and wincing as though in great pain. "I'm supposed to be handsome in your eyes at all times, right?"

"Come, my lord. The bath is ready for you. Was it a large dose?"

"Quite," Ash replies through gritted teeth. "I forgot that sylph's eye makes your insides feel like they're exploding."

"How did *you* fall for sylph's eye?" Rahk asks, crossing his arms and frowning as Edvear pushes and prods Ash toward the washing chamber.

"I didn't."

My eyes widen. "You *knew* my drink was poisoned?"

He flashes me a wan grin over his shoulder. "Of course. It was actually quite obvious. The liquid was much too viscous to be only saflixir nectar."

I scramble after him, catching my long skirts in my fists. "Then why did you drink it? You could have died! What if you hadn't had the antidote, or it had been a poison you weren't familiar with? Or—"

"Calm yourself, darling. While I can't say I mind your fussing, I mustn't let you work yourself into too much of a frenzy."

Edvear kicks open the washroom door, and Hylath is there, rushing to and fro in the back of the washroom with a stack of towels and blankets in her arms. But Ash is still looking back at me as he leans heavily on Edvear.

"I drank it, Stella, to show to all Valehaven that someone is trying to kill you. Unfortunately for you, it's not just one someone. That poison tonight didn't come from the High King."

That's when the door shuts me out. Sounds of shuffling slip through the door, and then there's a loud slosh and splash, presumably as he gets in the tub.

I turn back, still clenching my skirts in my fist. Rahk stands beside the couch, arms crossed over his chest with that furrowed brow. I draw a deep breath. "Th-thank you. For bringing me back."

He tips his head.

Silence falls.

"Send for me if you need anything, Lady Stella," he says at long last, as though he's my servant rather than a prince in his own right. "I stay in Nothril Court quarters when I'm here. Ash can show you when he's feeling better."

He turns to leave, but pauses when I call after him. "May I ask you a question before you leave?"

He glances back at me, his dark eyes watching, assessing. "Yes."

"Ash has been poisoned before?"

"He has."

"By the High King?"

"No, not by the High King. At least, not intentionally. There was once, when Ash was much younger, when he tricked his father into poisoning him. It was shortly after his mother was killed—"

My mouth falls open. "His mother was killed? By whom?"

Rahk's eyes darken. "By the High King."

I scramble around my brain, confused and shocked by this revelation. Then again, should I be? He's trying to kill his son and me. Why shouldn't I be surprised that he killed his wife?

Is *this* the real reason Ash hates his father and works to overthrow him?

"Why?" I ask quietly.

"To punish Ash for something."

238

"Oh."

Rahk chews his lip, his gaze casting away from me and latching onto a butterfly landing on the windowpane and fanning its wings. "We knew each other as boys, but Ash changed that day. He was furious and bitter for many years. It wasn't until he got this idea of overthrowing his father that he finally pulled out of it. The time he tricked his father into poisoning him was while he was trapped in those angry years. He set up this elaborate scheme, created rumors of a courtier attempting an assassination, planted clues to be discovered, and maintained false lines of communication until he'd convinced the High King he knew exactly what was going on and who was behind it. Then, at the grand banquet where hundreds of courtiers and dignitaries were present, Ash laid his trap. Faradir sent a poisoned goblet to who he thought was the assassin, but Ash bribed a servant to switch his goblet with the courtier. When Ash was the one who started choking and gasping for air, the High King nearly lost his mind. He thought he'd just poisoned his only heir and nullified his claim to the throne. But Ash had been taking small doses of poison already, so while he was very sick for several days, he didn't die."

"He was trying to scare the High King?"

"Indeed. He wanted to punish his father by making him believe he'd lost the only thing he cared about: his throne. But I think there was more to it than that. I think Ash wanted to prove he could out-maneuver his father."

I nod slowly, grabbing one of the couch's armrests and leaning into it. "And the other times he was poisoned?"

Rahk waves a hand. "None of them important. Once he took half a lethal dose of *bindorg* as a boy at the dares of his friends. His mother didn't let him hear the end of it, especially since he was sick for two weeks that time. The others were attempted assassinations by other Courts."

He mentions the attempted assassinations almost flippantly, as if one might dismiss an offer to share a meal because they weren't hungry.

This is normal here, I tell myself.

When I'm silent, he inclines his head in a deep nod. "I'll leave the rest of your questions to Ash. He'd probably not be happy with how much I shared."

"Well, I am grateful to know those things." I straighten, putting on my best princess smile and curtsying. "Thank you, Prince Rahk, for your aid tonight and the explanations. They are greatly appreciated."

Once he leaves, I pace outside the washroom, rubbing my arms and listening to the susurrus of my gown trailing on the ground. Eventually, I muster the courage to approach the door and knock.

"Ash?" I call. "Are you alright?"

"He's quite fine, my lady! Do not trouble yourself!"

"I'm rather miserable, my love!" calls Ash before the steward has even finished. "I am in great need of comfort!"

Edvear calls back through the door: "With all due respect, please ignore him, my lady." A long, pained groan follows this, and on its heels is an exasperated, "Don't be so dramatic, Highness!"

I stay standing at the door, warring with the desire to enter, to make *sure* he's alright, and the desire to leave him be. Clearly, there is a routine here.

Listless, I pace again. Then I stop, looking up and realizing I stand before the door to Ash's study. I glance behind me. The sitting room is empty. Biting my lip, I hesitate for just one moment. Then I slip through the door and mostly close it behind me.

The study is dark and lined with disheveled bookshelves. A single glowing globe, half the size of my fist, sits on a wooden pedestal on the desk, illuminating the room. I step around stacks of paper and overturned books, their pages irreverently smashed and bent, until I reach his desk.

It's covered in more paper.

Giving into my curiosity, I lean over the desk, careful not to touch anything. Most of it seems to be correspondence, with one pile unread and the other outgoing—judging by the mismatched seals on one pile and matched seals on the other pile of that same great tree he wore on his medallion tonight. Not all the correspondence is arranged in either pile, however. One such missive catches my attention.

It's tossed aside from the rest. The paper itself looks made of gold, and in big letters across the top, it says: *Lulythinar Masquerade Ball.* Beneath it: *Attend or die.*

This must be a general invitation, because the High King cannot threaten Ash with death. *Yet.* I shake away the shiver running down my spine and lean closer to read the rest of the invitation.

The washing chamber door opens.

I jump, scampering away from the desk. In my haste, I almost knock over the glowing globe. I catch its pedestal, right it quickly, and scurry to the fallen books on the floor. Why do I feel like a child caught snooping through the pantry? It's not as though I've done anything *wrong.*

Still, I scoop up three of the fallen books and pretend to busy myself straightening their pages as I slip out of the study.

And there's Ash. Dripping wet, wearing nothing but a towel wrapped around his waist. My eyes goggle a little and I quickly avert my gaze.

But I'm not fast enough to miss the gray pallor of his skin, or the way his shoulders sag and he leans against the wall. Or the way his eyes narrow as I exit his study, dropping to the books I hold in my hand.

"Are you f-feeling better?" I ask, still not quite bringing myself to look straight at him.

"Did you find anything interesting in my study?" He twists his mouth sardonically, as though it's a joke. Perhaps if he were at full health, he would have disguised the note of concern in his voice better.

I frown, moving my hands through the pages, smoothing their creases so he doesn't see their subtle shaking "I made q-quite an alarming discovery."

"Do tell."

I swallow and close the cover of the book. "You're very disrespectful to literature. Books don't d-deserve to be splayed out on the floor."

He cracks a wan smile, and leans heavier against the wall, his chest rising and falling. I set the books down, twisting my fingers while I debate whether I should come closer. Before I can decide, Hylath squawks from inside the bathing chamber.

Ash's brow wrinkles, and he growls, "I told you I don't need a sponge bath!"

She chitters something back, and Edvear retorts something from inside that I can't hear.

"Well?" my husband says. "Are you going to comfort me or not?"

I duck my head and clasp my hands behind my back as I take a few tentative steps forward. Perhaps if he wore a shirt, I could look up at him. Instead, I stare at the ground until his feet enter my line of vision.

I'm not sure what sort of *comfort* he wants—though I rather suspect it's *attention* he's craving—but I do owe him thanks. Had he not been watching out for me so carefully, I might have died tonight.

It's something that hasn't quite sunken in yet.

"Ash," I start to say.

"Why don't you look at me?"

"Because you are . . . lacking . . . attire."

"So you *do* find me attractive."

At this, my head shoots up and I fix a stern glare on him. "Compliments don't mean as much when you ask for them, Prince Trenian."

He grins in response, and I'm once again too aware of his shirtless state.

I clear my throat. "I was *about* to thank you for saving my life."

"Oh?"

"Now, I'm still not convinced it was necessary for you to risk yours in the process, but I appreciate it nonetheless."

His eyebrows lift in surprise, and his grin widens. Is it my imagination, or is a little color returning to his cheeks? "Not necessary? How would people know someone had tried to poison you?"

"You could have simply announced it. Perhaps sent the goblet off to be inspected."

"And what sort of story would that make in comparison to me dropping to the floor and almost dying?"

I narrow my eyes at him. "Drama isn't the only thing that will accomplish your goals."

He places a gray hand over his heart in mock hurt. "Me? Dramatic? Why, the notion!"

"My lord!" The washroom door swings open, and there's Edvear, glaring fiercely at Ash. "You are supposed to be in bed!"

Ash shoots him a narrow look, even as his shoulders shudder. "I got *distracted*."

"To bed with you! You mustn't overtax your body."

Ash waves his hand as though batting away an irritating fly. "I know my limits."

The look the steward tosses me says, *"No, he doesn't. Try to convince him to rest, will you?"*

Then Edvear pulls the door shut, leaving Hylath in the washroom as he marches off to the kitchen, disappearing behind a corner.

"You should go to bed," I say to Ash.

"Come with me."

Those words, spoken so quietly, almost too quietly for me to hear, send a bolt of lightning through me. I whip my head up, meeting his gaze. He's still that ghastly gray, and he leans so heavily against the wall I'm afraid he won't be able to make it the few steps to the bedroom. Though his face is lined with pain, there's something earnest, and almost . . . *fragile* about his expression.

"Only for tonight," he adds. "I just . . ."

I wait, hardly breathing, for him to finish that sentence. He doesn't, and I stare up at him, tongue-tied, not sure what to say or do now. Am I allowed to say no? What would I be saying yes to?

"You're poisoned," I say, not even sure what I mean by the statement.

"It's not contagious, if that's what you fear." He gives me a little smile. "I'd even put on clothes. Just for you."

"I'll . . . think about it."

He gives a slow nod, as though this is what he expected. Then he pushes off the wall, almost stumbles, and makes his way to the bedroom. I stay where I am as he shuts the door behind him, presumably to dress.

What now?

Perhaps I am trying too hard to understand what is happening around me. Maybe I should just shrug my shoulder and say, *Well! Just another evening in which my husband drinks poison meant for me. On to tomorrow!*"

But part of me feels rather inclined to grab a pillow and scream into it.

"My lady?" Edvear returns to the room, a steaming tray in his hands. "I assume you had little to eat at dinner. I shall leave this for you in the dining room, should you be hungry."

My empty stomach gurgles. I clap a hand over it, and now I'm the one catching hold of the back of a chair to keep my balance! It's as though the mention of food makes my body suddenly remember that it's been hours since I last ate.

Edvear hardly sets the tray on the table before I sit down. He leaves me to eat, and eat I do. I set into the stewed rice and venison with more vigor than any princess should, but I am much too exhausted and alone to care about such things. There's a cup of warmed chocolate goat's milk on the side that is utterly divine. I gulp it down greedily.

When I'm finished, the steward slips out of the hallway and comes toward me. I dab my mouth with the napkin and look up, waiting expectantly.

"He is asleep, my lady."

I set the napkin down. This gets me out of giving Ash my answer, it seems. I can do the more comfortable thing and return to my own room without having to tell him to his face.

I'm still in my fancy gown, so instead of going to the bedroom first, I find Hylath in the washroom. When I walk in, she's bent over, her face in the giant copper tub. Two of her eyes pop up and blink at me, but she doesn't otherwise move. A strange lapping sound comes from inside the tub, and when I creep closer and peer into it, Hylath's long tongue is lapping the water. *Drinking it.*

I barely restrain my disgusted *"ugh!"* but probably fail to keep my face straight.

Hylath lifts her mouth out of the water, a third eye joining the first two. She burbles something with a lilt at the end. A question.

"Could you help me remove this dress? And . . . take down my hair?"

She heaves a great sigh of long-suffering, then straightens and gestures for me to sit on my stool like normal. Within twenty minutes, I'm in a soft pink gown of a material very close to silk, but not quite the same. It's not as cold and has an almost velvety sheen on one side of it. I wear a light robe over it, and my freshly brushed hair falls in long, loose waves down my back.

There's nothing for me to do now but creep into the bedroom past a sleeping Ash and try not to make noise as I slip into my own room.

When I'm outside the bedroom, I hesitate. Then, forcing away my own nervousness, I straighten my shoulders and push the door open.

The room is dark, but for a few glowing crystals hanging from the ceiling. The light is just enough for me to make out the path to my room and not stub my toe on the corner of the bed. Ash's deep, even breaths fill the space, and his dark form beneath the covers of the bed seems to overcome my awareness. I take a step closer, just to see if I can make out his face . . .

Then I remember the way he knew I'd been studying him on our wedding night. What if he's not asleep this time, either?

I scurry to my room and shut the door behind me.

Safely ensconced in my own private room, I wrap my arms around myself and let my eyes wander over the lovely space, the only light coming from the dots of glowing fireflies from the garden outside the windows. They illuminate the blossoms on my lavender plant sitting on the sill.

I sit on the edge of my bed, but I don't lie down. I stare at the door I just shut.

Why does this space feel so much lonelier than before?

Loneliness has always been safer for me. Back in Aursailles, it was safer for me to be lost than to be found. Being in the presence of others meant dealing with my own inadequacies and failings.

Now loneliness feels hollow. It's warmer to be wanted, to be *with* someone good and kind.

As I sit here, alone in the dark of my room, I realize I don't really *want* to be alone anymore. I don't want to stay here by myself. Which is grand and all—but do I have the strength and boldness to get up and change it? To choose closeness over fear?

My husband had the boldness tonight to drink poison meant for me. Knowing it would hurt him, knowing he'd be miserable. Perhaps I can find the boldness to stand up . . . put one foot in front of the other . . . and open the door.

It swings outward.

The four-poster bed looms in my vision. There's the great lump of my sleeping husband. And here I am, only a few feet away.

It's not too late to turn around, to throw myself into the downy-soft refuge of my own bed. Maybe another time I can be bold and fearless. It doesn't have to be now, does it?

I firm my spine.

I am Stella, and I'm done being afraid.

Within a few steps, I'm at the side of the bed closest to me—the empty side. It's quite tall. Much higher than the usual human bed. It's a good thing my husband is asleep and doesn't see my graceless scramble onto the mattress.

My heart pounds so hard it nearly rocks my whole body. I peel back the thick layer of blankets and slide underneath. Then I curl up on my side, laying my head on the pillow and staring into the dimness at the far wall of the bedroom. I try to slow my breaths to calm my raging heart.

It's ridiculous that I'm this nervous. We're so far apart. We might as well be in different countries. I'll probably wake up in the morning, only for him to already be gone. So why won't my racing mind slow down?

Why can't I calm my heaving lungs?

There's nothing for me to—

The mattress shifts. The movement abruptly stops. "Stella?"

My eyes fly wide. Oh, what was I thinking? I should have stayed back in my own room! I clench my fists tighter around the covers, breathing hard. Maybe I can pretend to be asleep . . .

The mattress shifts again. He's scooting closer, isn't he? What am I ever to *do*?

Warmth envelops me from behind as an arm slides around my waist and pulls me against a broad chest with a low, quiet groan. My breath catches.

"I didn't think you'd come," he whispers, nuzzling his face into my neck. "But I'm glad you did."

I can't breathe. I won't ever breathe again.

He presses a soft kiss to my hair. Then he leans back, resting his head on the pillow next to mine, and lets out a long sigh. His arm around my waist relaxes, but doesn't leave. It's only a few minutes before his breathing evens out again. This is different from our wedding night. He's actually asleep this time.

Getting poisoned must have taken a lot out of him.

It takes nearly an hour, but eventually my own heartrate slows down. My fear leaks away, and I relax into Ash's arms. It's such a new sensation, but I think . . .

I think I can get used to this. To the warmth and strength of his body, the gentleness of his touch.

I think I like this.

I think I like *him*.

As I drift off to sleep, I snuggle a little closer to him. When his arm tightens in response, and his warm breath stirs my hair in a sigh, I let the butterflies in my stomach sing me to sleep.

It's much colder when I wake. Which is strange, because the second thing I notice after the irrepressible chill seeping into my bones is the fact that Ash's body is still wrapped around mine. I shiver, trying to scoot closer to him, but my limbs are stiff as lead.

A pair of lips press light kisses to my hair and ear as I stir.

"You're awake," a deep voice rumbles between kisses. "Why do you shiver, little darling?"

My throat is painfully dry. I lick my lips, find them equally dry. My brow puckers.

"Maybe we should stay here all day," Ash murmurs, his hand beginning to draw slow circles on my waist. "I could pretend I still feel horrible, and you can say you're tired. We can abandon this whole prince and princess thing and just . . . be together."

Staying here sounds nice. Maybe he can find another blanket and keep holding me. I'm just so *tired.*

He gently pulls my hair back from my neck and leans to press his lips against my cheek. He freezes. Pulls back. I mourn the loss of his warmth.

"Stella?" he breathes, sitting upright.

I wish I could answer him! My tongue cannot form words, and when I peel open my eyes, everything blurs in my vision. He lays a hand over my forehead, and I drink in his warmth, shivering as my teeth chatter.

"Your skin is cold as ice!"

All of me is cold as ice. Why is he so surprised? Why does he sound scared?

The bed shifts as Ash flings aside the covers and throws his feet over the edge of the bed, standing and almost *running* to my side.

His bare feet slap against the floor. The loss of his warmth almost makes me too cold to think. It's like my body is slowly freezing over, starting from inside and moving outward.

More blankets land on top of me, and then a sudden wave of heat rolls through me, but it's not enough to make me stop shivering.

"What is wrong, Stella? I can't even warm you with magic! *Edvear!*"

Only a minute later, the door opens. The blurry outline of Ash's head whips up.

"Edvear! Get a doctor here immediately! A human doctor! If you have to smuggle him across the border, then do so! Something's wrong with her."

My awareness tunnels until the surrounding voices are lost to the cold.

CHAPTER 31
THE PRINCE

POISON?" ASKS EDVEAR, his eyes going wide with alarm. "I don't know! Not a poison I'm familiar with! None of the ones I've studied make someone so *cold.*" I run my hands up and down the pile of blankets on top of Stella, trying to warm her with my magic, but it seems to do little good. I wrack my brain, running through the last twelve hours. She came with me to dinner. She didn't eat anything. I lost track of her for a few minutes, during which Rahk rescued her and brought her back. Neither of them mentioned anything bad happening—someone giving her food or pressing a strand of hair into her hand or even anyone else *talking* to her. The servants didn't mention anything.

Had something happened? If not then, when else could something have happened? She was back here the rest of the time—in my arms most of the night. There was a short window when we weren't together while I bathed, and then again before she climbed into bed with me.

A sudden fear hits me. She had been in my study when I left the bathing chamber. Did she find my stash of poisons? Did she touch any of them? Consume anything? Surely, she wouldn't be so stupid! I cannot imagine a world where she'd find an unlabeled bottle and just take a swig.

"Did she eat anything?" I demand as Edvear hurries back out the door.

"She had a meal, yes. But it was prepared in our kitchens, and I served it myself. When did she start displaying symptoms?"

"This morning. Once you've gotten the doctor, I want a list of everyone who touched her food and a timeline of preparation. I want to know if it was ever left alone, and if so, where and for how long."

Edvear nods and leaves. Part of me wants to follow him out, to go racing through the hallways of the palace shouting for a doctor myself. But I cannot bring myself to leave Stella alone, shivering so cold.

Steeling my spine against the bitter fear swelling in my soul, I get back in bed. Pulling her so she's flush against me, I wrap her up as tightly as I can and lay my head against hers. I siphon heat from the air with my magic and pour it out of my hands like water from a glass. It's not enough. Nothing stops the shaking.

I let my shoulders sag, but don't stop trying to warm her.

Have I already lost her?

It hasn't even been three full days since we arrived in Valehaven. How could this have happened? Did my father outwit me somehow? Anticipate some move of mine that I didn't think he knew? How did he get past my guard? Or was it Princess Listhra, who tried to poison her last night? Did the High King learn of her attempted poisoning and use it as a distraction from his own maneuver?

Stella *can't die*. She can't. She just . . . can't!

Lulythinar is swiftly approaching, and if she dies now, what can I do? It might be too late to find another wife, which means I'll be forced into a marriage, forced to sire an heir. Forced to ensure my own death.

But deeper than that . . . I can't lose *her*. She's too good to die. Too bright and beautiful and sweet, and I . . . I . . .

I let out a groan, squeezing her closer to me and pressing another kiss to the top of her head. She has fallen back into sleep, her mouth open and her brow puckered. Her name is another groan on my lips. "Oh Isabelle. My darling Stella. Please, please be alright. *Please.*"

If the High King killed her, I'll forget this whole throne overthrowing business. I'll kill him in cold blood and destroy my claim to the throne. I'll plunge this whole world into war and leave the fae without a king, and the humans without protection from their rampaging.

A bitter resolve sears like a brand into my soul.

If she dies, this world burns.

The High King took my mother from me. He *cannot* take Stella too. If he does, he'll finally know the monster he spawned. It prowls beneath my skin now, ready to rip and devour.

What I'll do to him will make what he's done to me look trite.

The door bursts open and I tense, tightening my arms around my wife as though expecting the High King to walk in with a sword and try to kill her this very instant, despite the wards I have on my quarters to prevent exactly that.

It's Edvear. And a short, round-bellied human doctor with spectacles and a little black bag that he sets on the bed and opens. I recognize him as the traveling doctor who visits the Small Cities.

"If you would please, Your Highness," says the doctor, coming around to Stella's side of the bed.

If I would please *what*? If he thinks I'm going to let go of my wife—

"My lord, he must check her for poison. We must move quickly!" Edvear's voice, ever practical, rankles down my spine. Nevertheless, I disentangle my arms from around her and sit up.

"Please help me lay her flat, Highness."

I spring into action too quickly, rolling Stella from her side onto her back. I brush the hair out of her face, my heart hammering, as the doctor pulls down the blankets and presses two fingers beneath her jaw, and another two fingers to her heart. He looks at the ceiling, his mouth moving slightly as though he's counting.

Next, he pries her mouth wide with his thumbs, peering inside. He pulls a strange instrument from his bag, a long slender piece of wood inlaid with bits of metal, it seems, and lays it on her tongue. After another minute, he peels it up, and sets it on the bedside table, the wet end propped up on a small towel. He doesn't speak while he works, taking a little dish and mixing two different liquids from vials in it before dipping the instrument into the liquid.

"Well?" I demand, and Edvear shoots me a look like I'm being rude.

"A moment or two for the results, Highness," says the doctor. "But I don't think it's poison."

I blink. Surely, I heard him wrong. "You *don't* think it's poison?"

"No, indeed. I'm checking just to be thorough."

"Then what is wrong with her?"

"Seems like blood sickness to me." The doctor stares down at Stella, a frown on his face. "You don't come across cases very often, and it varies in expression person to person, but I suspect that's the case today. Other illnesses make the body hot to the touch as the body raises its internal temperature to fight the illness. This is not that."

"She's merely ill?" I ask, not even daring to breathe.

"I wouldn't say *merely ill,* as blood sickness can be fatal."

My newborn hope plummets. "Fatal? What *is* blood sickness?"

"Aha! Not poison!" The doctor grabs the stick out of the liquid and waves it in front of my face. As if it's self-explanatory. I narrow my eyes at him and pull my wife back into my arms. Being under so many blankets with her so close should make me blisteringly hot, but holding her is like cuddling an icicle. Just in case she's aware of what is happening, I stroke what I hope is soothing lines through her hair and down her back.

"*What* is blood sickness?" I all but growl.

"It is a rare phenomenon when a human comes to Faerieland and breathes in the magic from the air."

"The air is making her sick?" I ask, alarmed.

"Not exactly. As you know, fae have magic and humans don't. *Most* of the time. But very occasionally, for whatever reason or another—fae blood in the bloodline, perhaps, or a fairy curse or blessing—a human is born with magic."

Static fills my ears. I look down at the delicate woman in my arms, her shivering little shoulders, her gleaming hair. Stella? With . . . *magic*?

"Human air is stifling for magic, as you are aware. But when a human with their own magic comes to Faerieland, the dense magical air can awaken their powers. This sickness is her body struggling against her awakening magic. It doesn't happen for every human, but in most cases, it does. If she pulls out of it, her life expectancy will be much greater. More comparable to a fae's than a human's."

If Stella has magic . . . that would change *everything*. I shove away the rising hope, swallowing hard. "You said it's fatal?"

"It can be."

There's that ragged, desperate hope again, rearing its ugly head. "But it isn't always?"

The doctor snaps his bag shut, sets it on the bedside table, and plants his hands on his stomach. The look he gives me sends dread pooling in my stomach. "It depends on how strong she is and how strong the blossoming magic is. You . . . have a few options."

I tighten my grip on Stella, tangling my fist into her hair. "What options?"

He takes a deep breath and lets out a long sigh. Not a good sign. "You can wait and see what happens, whether she takes a turn for the worse or if she pulls out of this on her own."

"How long could that take?"

"As little as a day, or it could be weeks."

Weeks? I barely restrain my growl of frustration. Lulythinar isn't even a full week away. I can't leave her side while she's in this state!

And what if she dies, anyway?

"The risks are, of course, that she will not be able to survive the strain of her humanity and magic clashing, or that her humanity will win out and her magic will be extinguished."

I let out a sigh, dropping my head to rest on Stella's. "And the other option?"

"We could give her a transfusion of fae blood."

I look up, glance over at Edvear, who listens silently and says nothing. "A blood transfusion? Why?"

"To help with the developing magic. It's a more surefire way to not let her magic be overpowered. It's faster, too."

"And the risks?"

The doctor adjusts his spectacles, squinting slightly. "Well, fae blood is usually toxic for humans, but in the case of a magical human, it's less so. Still, there's always the risk that it'll kill her. The trick with the blood transfusion is to allow the magic to kill off just enough of her mortality to allow for the magic to bloom, but not enough that she dies."

I clench my jaw so hard I almost don't notice when my insides give a small cramp. The poison must not be completely gone yet. It seems like such a trivial thing now. "What is her greatest chance at survival?"

"It's hard to say, Highness. I've only seen two cases of blood sickness in my life. I've heard of more, but—"

"Did they survive?"

"Neither, Highness."

It's as though my heart has been pinned to the wall, and a line of people keeps taking punches at it. Every moment I dare to hope, another punch is thrown.

"Did either of them get the transfusion?"

"One did. He seemed to fare better, too, before he took a turn."

Silence falls over the room. The quiet wheeze of Stella's breathing becomes loud in the stillness.

"And the other cases? The ones you've heard of?" I ask.

"I've heard of two surviving with the transfusion, and one without intervention. The rest didn't make it."

I consider this information as I work my fingers gently through the tangles in Stella's hair. I just want to go back to last night, to that moment I woke up and smelled her sweet scent, to the feel of her against me when I drew her into my arms. The sound of her little gasps as I kissed her.

Now she might die, and I'll never have told her that I . . . that I . . .

I squeeze my eyes shut against the sudden and completely foreign impulse to cry. The last tears I shed were for my mother when she died. Perhaps it's fitting that I shed them for my wife, too.

But not now. Not with the doctor and Edvear in the room.

I need to make this decision. Survival isn't likely with either course of action. But *if* she survives this, she'll be much more likely to live longer in Faerieland if she has magic.

If she pulls through . . . if she *does* have magic . . .

Her lifespan would be longer. The High King and his plots aside, she wouldn't grow old and die within a few decades. She could *live* with me, be my wife—for centuries.

She could be my queen.

The thought fills me with such blistering hope. What if I didn't have to lose her like everyone else? What if I could keep her? What if I could fall asleep with her in my arms every night? What if we could have children together?

The urge to weep strengthens, almost getting the better of me. I shouldn't dream like this! Not while she might be dying in my arms right now. But I can't help it. This is something I could live for beyond overthrowing the High King. This is goodness, happiness, sweetness, *fullness* unlike anything I could have dreamed of.

I tighten my jaw, hardening my resolve, and press a kiss to the top of Stella's head. "Give her the transfusion."

The doctor nods, as though unsurprised. "Then I will need fae blood. Would you like her to have yours, Highness?"

"Mine?" I didn't consider *where* the fae blood would come from.

"I have a few vials of fae blood. They're not particularly fresh, though I have preserved them properly. I keep them in case of emergency, but I prefer to use fresh blood when I can."

He is *not* putting some stranger's moldy blood into my wife. "Yes, yes, take mine."

"Very well." The doctor pulls a scalpel from his bag and nods at me. "Let us get a bowl."

CHAPTER 32

THE PRINCE

THE DOCTOR BLEEDS my wrist until blue blood has half-filled the bowl that Edvear fetched. My stomach clenches in a way completely unrelated to the last remnants of poison when I follow his instruction to pull Stella's icy arm out from beneath the blanket and roll up her sleeve.

A parched little cry of pain slips between her lips when the needle pierces her skin. Her head thrashes from side to side. I catch it against my chest, holding her still while the doctor injects the blood into her veins.

When it's finally over, and we have matching arm bandages, the doctor closes his bag with a snap. "That is all I can do for now. Make sure she stays comfortable, and try to get her to take a little water and broth every few hours. Then pray she comes through."

With that and nothing more, the doctor bows and leaves. Edvear gives me a too-knowing look. "Shall I have your desk brought in here for the time being?"

I gnaw on the inside of my mouth. Different snippy responses fly through my head, but I shove all of them away and nod. "Thank you, Edvear."

When the door closes, and I'm once more alone with my wife, I tighten my arms around her and bury my face in her shoulder. She's so limp, so cold. It's almost as if she's already gone. The tears come then, hot and unfamiliar.

"You can't die," I growl, my shoulders shuddering. "I can't lose you, too. Not after . . . Not after you made me like you, you confounded woman! I shouldn't have lifted that cursed veil. Shouldn't have talked to you. Shouldn't have brought you here at all. Cursed Great Kings! Don't you hear me, Stella? Come back to me. *Please,* come back to me." My voice breaks into sobs, and I fall back against the pillows, clinging to her as though my touch will bring her back.

I was a fool to love her.

Everything I love dies.

THE PRINCESS

I RUN THROUGH the empty halls of my palace home, chasing fragmented phantoms. What am I searching for? Why can't I find it? I know it's here somewhere . . . somewhere . . .

A glitter catches my eye.

Gripping the skirt of my ice-blue gown in my fists, I run harder. My feet slip on the polished floor, and I land hard on my hands and knees. Palms burning, I shove up and get my feet under me. For a moment, dizziness nearly overcomes me and I almost fall back to the ground. Gathering my determination, I catch my balance on the white-plastered wall, breathing hard. My skirts float lightly around my ankles, the translucent material revealing my calves and bare feet, clinging to my hips. For once, I don't care about modesty. There is *something* here. Something I need so desperately. It itches at the edge of my mind.

Something sparkles in my vision again. *Reminding* me.

I'm running again. The glitter sits on the windowsill by the feasting hall, overlooking the gardens I used to love so much. I race between tapestries depicting maidens among wildflowers and waterfalls and knights amid gore and bloodshed. I slide to a stop before it, reach out—

It vanishes between my fingers like a will-o'-the-wisp.

THE PRINCE

MY STAFF BRINGS my desk into the bedroom, and I pointedly ignore it for hours, doing nothing except rocking my frigid wife and *thinking*. So much thinking.

Sometimes I hate my own brain. The way it tries to shove away the grief, focusing instead on trying to spin this new hurdle to my advantage. I put it off for hours, but each time Edvear slips in to add another sealed missive to my stack, my resolve crumbles just a fraction.

When the day is mostly gone, I finally get up I'm still in the thin wool trousers and loose white shirt I wore to bed last night. I don't bother changing, glamouring myself, or even eating the plate of food Edvear has left for me. The second plate, actually. The first untouched plate sits on my bedside table. Beside it is a mostly full mug of tepid broth and a cup of water. I didn't get much down her throat earlier.

Making sure I've piled every blanket I can find on top of Stella and slipping a pair of spell-heated bricks near her feet, I force myself to stumble to my desk, fall into the chair, and stare blankly at the papers before me.

Distraction.

I don't *want* a distraction. I *want* to stay with my wife.

Clenching my jaw, I force my hands to move, to pick up the first missive, to open it. Reading the first few lines punches the wind right out of me.

To His Royal Highness, from the Imperial Human Tailor. The first set of dresses for Lady Stella will be delivered tomorrow morning at first light. If any adjustments are required, I am happy to oblige.

I need to get control of myself. I cannot allow myself to burst into anger at any mention of her. Yes, my heart might be cleaving in two, but I can handle this.

It's not as though I'm a stranger to heartbreak. I can handle it.

Straightening my shoulders, I scrawl a reply and force myself to inquire about the progress of the Lulythinar ballgown. Then I move on to the next missive, and the next, until I'm through my stack. I shift to the list of things Edvear has left for me. Mostly household things, like approvals for large purchases and issues with the staff. Beneath it is a scrawled reply in Rahk's handwriting that says: *I'll take care of it.*

It's a small relief I won't have to worry about Mama Bagogs or Orawyth right now.

The next thing I do is pull out the social calendar Edvear keeps for me and study it. I take my thickest quill, dab it in the ink, and slash through everything I can possibly skip. Curse the consequences. The whole world can be offended if it likes. The High King can get his underthings in a twist if he so pleases.

I don't care.

Every few minutes, I look up, stop working, and listen for the sound of her breathing. It's impossible to detect movement beneath that mound of blankets, but that soft whisper of sound is my lifeline.

The last thing I do is set down my quill and stare into space. Like it or not, I must spin this somehow. No one can know that Stella has blood sickness. Everyone knows someone tried to poison her last night thanks to my . . . *little display*. I could set Edvear in charge of spreading the rumor that last night's poisoning was only a guise for

Princess Listhra's real trap, which awaited Stella when she returned to my quarters. Perhaps if the High King thinks I'm getting lazy protecting my wife, he will go back to underestimating me. Maybe I could get Rahk to let it slip to one of those gossipy sisters of his . . .

I cannot spread the rumor that the High King tried to poison her. He'll know it was false, but since Princess Listhra *did* try to poison Stella last night, no one will believe her if she insists this second attempt wasn't her doing.

In fact, she might even claim it, as it'll make her look much cleverer than she actually is.

When I don't show for tonight's banquet, more rumors will spread. Rumors that I've gone and fallen in love with the human girl I wed. They'll reach my father's ears, and he will believe that he finally has a tool he can use against me, if he can just get his hands on her.

If I manage it carefully, Stella won't be hurt.

And the High King will walk right into my trap.

It's then that I notice the next missive on my desk. It's written in Edvear's hand.

The High King summons you to the throne room.

Everything inside me freezes.

Movement catches my gaze. I shift, whipping my attention to Stella as she lets out a low moan and thrashes her head from one side to the other. I shove back my chair, nearly tripping in my haste to get back to the bed.

"Stella?" I ask, hating myself for how much worthless hope can reside in a single word.

She groans again. Sweat glistens at her hairline, catching the last lights of the day. *Sweat.* I place a hand over her forehead and almost draw back. She's burning up. I rip the blankets off her, leaving her scorching feet bare to the air, and her nightgown sticking to her sweaty skin.

"Edvear!" I call. "Get me a cloth and a bowl of ice water!"

It's hardly two minutes before he enters the room with the requested objects. I thank him, ask him to inform the doctor of the change, and set to dabbing Stella's face, neck, and shoulders with the cooled cloth. The excess water dribbles into her hair and the bedclothes.

Her fists clench the sheets as her head thrashing increases. Her face contorts in a wince, as though she's in pain. Another moan slips between her lips. Then a third. The fourth almost sounds like my name.

"Stella?" I breathe, dabbing her cheek. "Can you hear me, love?"

"Ashhhh," she moans, clearer this time.

"I'm right here, darling. Right here."

Her hand loosens its death grip on the sheets, darting up and catching hold of my wrist. I cannot help the simultaneous leaping of my heart and twisting of my gut.

"Ash," she says again, and this time it's almost a sigh. Her body relaxes.

Blood pounding, I lean over her, leaving my wrist in her grasp. I try not to bring her discomfort by touching her, but I bring my mouth to her ear. "I'm here, Stella, and I'm not leaving you. Come back to me, sweet wife. Please, come back to me."

Something keeps me from saying the thing I ought. Those three words my heart keeps saying over and over again. Once I say them to her, I cannot take them back. And once I say them to her, there'll be nothing to buffer me from the pain of losing her.

When I pull back, I startle sharply.

Her eyes are open.

Those two beautiful doe-soft eyes blink once at me, and I cannot help catching her face in both of my hands, suddenly wild with hope. "Stella? Stella, love!"

"Ash?" she croaks.

"Yes, it's me! Ash, your husband. Can you hear me?" She needs water, more broth! Maybe if I can get her to sit up, she can take something a little more substantive. Oh, Great Kings! I could sing and laugh and dance from the relief coursing through me!

Then her eyes roll back in her head.

"No, no, no! Stella, stay with me, girl! Stella?"

Her eyes shutter closed again, and her limp hand falls from my wrist. She goes still.

Frantically, I search for a pulse, my vision almost turning black. She can't be gone. She *can't* be. I won't allow it! I won't—

There's her pulse. Sluggish, but steady.

I let out a long whoosh of held breath, release my hold on her, and slump. I run my hand down my face, groaning and fighting yet another wave of desperate tears. Mountains of Ildrid, I can't take this! I'll be mad by the end of the week.

CHAPTER 33
THE PRINCESS

I CANNOT FIND it! I've been searching for *days*, for weeks! It's not here. But I *know* it's here. A frustrated whimper escapes me as I plunge deeper into the halls of my childhood home. Shadows coalesce around me, and every few minutes, one of them will flicker into something I recognize: Amelia's grin, Father's clenched fist or his furrowed brow, a flash of petticoat and brocade.

None of it is what I'm looking for.

I shove past the shadows, fighting the instinct to reach out and grab Amelia's hand. If I do, her shadow will simply melt through my fingers. It's pointless.

Nothing matters until I can *find it*.

Whatever *it* is.

I've looked almost everywhere. My rooms, the main floors of the palace, the servants' quarters. Even the throne room, the banquet halls, and my father's private study. I'm not sure why, but this palace doesn't feel like my true home anymore, as though my connection to it has worn thin.

There's just one place I haven't gone . . .

I shake my head. It's somewhere else. I should go back to Father's study, to the tearoom. Picking up my skirts, I hurry up another flight of stairs. My breathing turns labored, which seems odd. I've been running for ages. Why am I getting tired now?

Stella.

I pause, chest heaving, ear tilted toward the ceiling. I know that voice. When I try to place it, it's like trying to remember a dream after I've woken up. It slips into the ether, further and further away. But my body, my very soul, knows this voice. It calls to me, warms something deep inside of me.

It makes me feel loved, cared for.

I want to hear it again.

Stella, darling.

I close my eyes, ducking my head instinctively to hide my smile. Why am I smiling? Why do I love when this voice says my name, calls me sweet things? Where is it coming from?

My mother used to sing lullabies to me when I was sick or hurt. Something about her song always soothed me, even when I was most miserable. There was one time when I . . . well, you'll think this is very foolish of me, but I drank some poison at the goading of my friends. Have no fear, I paid the price for my folly.

This voice is so rich, so deep. And I am so very winded. Maybe for a few moments, I'll just sit here and listen . . .

I was sick for days. I've never felt so awful in my entire life, and I cried like a baby when I couldn't fall asleep for how much pain I was in. If you were awake, you'd probably arch one of those pretty eyebrows and tell me it served me right. It did, of course. But can't you have a little pity? No one told me we didn't have the antidote for this poison!

I sit on the top stair, leaning against the railing, smiling as I listen. I wish I knew who this voice belonged to! Whoever it is, he likes my eyebrows, apparently.

My mother heard my cries and came. At that moment, it was as though I looked on an angel of salvation from my misery! She'd never looked so beautiful to me. She clucked her tongue, running her fingers through my sweaty hair. I still remember the soothing way her thumb traced along my forehead. Like this. But imagine it being much softer. My thumb is probably too rough.

My eyelids hang heavily. What if I just stopped searching, took a little nap? Then I could . . . start searching again . . . later . . .

I lower myself to the floor. The polished tile is cold against my cheek, but it's oddly soothing. My skirts flare down half the staircase, like a waterfall of silk and crystal.

Then she sang to me, and I'd never heard a more beautiful song. I forgot everything as I listened, and finally, I fell asleep. What do you say, darling? Shall I sing it for you?

Yes, I whisper back, the word barely a breath between my lips. *Sing to me.*

I'll take that as a yes. But of course, you cannot tell anyone that I sang you a lullaby. It would ruin my reputation.

I chuckle softly, my awareness already slipping away.

A rich, warm voice floods around me, soft as a blanket and strong as a pair of arms holding me to a solid chest. Its low timbres are quiet, but I do not have to strain to hear them.

May you always find your way back home
Through darkened nights
On lonely roads
May you always find your way back to me
I will hold you safe and sound
My love, in my heart
For all eternity

Recognition hits. I push up on my elbows, my lips parting as the voice continues singing. I *know* this voice. This is Ash. My fae husband. Confusion rolls through me, but beneath it is a steady certainty.

Ash is here with me. He hasn't left me alone. And I think . . .

He *loves* me.

Possibly more than anyone ever has.

That knowledge sends strength shooting through my limbs. It is the last piece of determination I need. I shove myself onto my knees, then get to my feet. Resolve hardens in my gut. I whirl, run back down the stairs, suddenly full of vitality.

I'm not afraid to look in that last place. No, I'll confront it. And I'll do it with Ash's beautiful voice surrounding me, filling my soul. I run hard down several flights of stairs, turning down long hallways. Always taking the darkest, deeper path that leads lower, lower, to the belly of the castle.

Finally, I skitter to a halt before a large iron grate. Even from here, the stench reaches like tentacles toward me. That stench brings back memories—of being cold, alone, trapped in the dark. Memories of tears, of rocking against a cold stone wall. Of begging for Father, nothing but silence answering me.

Ash's voice fades to silence, his song finished.

Part of me falters as that wretched silence fills my ears. Then I grit my teeth and take another step toward that grate.

Somehow, I know this world isn't quite real, so I walk straight through the metal as if it's nothing but a curtain of running water.

To the dungeon beyond.

I shiver, and my steps flag as the stink of rot redoubles. But I'm done cowering. I'm done letting fear keep me from doing the things I should have done long ago.

I march down through the narrow stone stairwell, cold slicing into my bare feet. Each step carries me deeper into the darkness that once swallowed me whole. And with each step, a new voice fills my head. Memories—of Father.

"It's for your own good. You need to understand how this world works, and your place in it. You were not born a princess so you can live a charmed existence of balls and fine foods. Your duty is to serve your people."

"I will, I will, Father!" My own voice, high-pitched and desperate, floods me with all the things I felt in that moment. I had to convince him I would do anything he said. I had to show him I wouldn't protest or fight him. Anything he wanted—anything for my people, I would do it.

"I know, but you must understand the reality of war, and what can happen to you and your people when peace is compromised. I know you don't want this—I don't want it any more than you. But I must teach you, or else I will have failed you as your father and your king."

"I do understand!" Pleading echoes through my mind, ceaseless and frantic. I keep walking down those steps.

"You think you understand now, but you will forget when you are older and stronger willed, when you long to make your own decisions instead of following your duty. I do this so you won't forget."

"I promise I won't forget!" Tears roll down my cheeks, reminiscent of the ones I shed when I was only eight years old. How that frantic desperation and pleading turned to fighting when he pushed me into that cell, my cries to screams when the door clanged shut, my hands reaching through the bars, trying to grab for the fading shadow of my father.

My father, who left me alone in a dank, cold cell for the longest, most miserable night of my life.

I stop before that cell. I can almost imagine a little girl curled up in the corner, shivering and crying as the sting of betrayal coursed through her blood.

At some point, that child decided this was for her own good, and she hated herself for screaming, for crying. This was a normal part of being a princess. All her sisters had done it before her, and Amelia would do it after her.

The bar is cold in my hand. Bowing my head against it, I close my eyes and let those last reserves of tears pour free.

I'd never been the same after that night. I learned that staying beneath notice was the safest course of action. I learned it was better to swallow my tears, to submit, to not show a shred of defiance. To not make a sound.

That was when I'd started stuttering.

Perhaps it served me throughout my childhood and early adulthood. But it serves me no more. I grip the bars in either hand, and I do something I'm not sure I've ever done before in my entire life.

I scream with every fiber of my being—the loudest sound I've ever made. Again and again, I scream, until my throat is raw and my body shakes.

My fists clench around the bars, not flinching from the cold. Not flinching from *anything*. From this memory, from the dreaded nightmares that followed in its wake, from the smallness I'd made my own.

I'm not afraid anymore.

Not afraid of what others can do to me. Not afraid of the sound of my own voice.

"I am Isabelle Louise Stella Ashrift Solavirth," I growl. "And I will claim my birthright."

With that, I throw open the cell door. There, in the center of matted straw and piles of foul excrement, floating above the ground, is a small glowing orb the size of my fist.

I don't know what it is, only that it is *mine*.

And it is what I have been looking for all this time.

I reach out and take hold of that deeper, fuller part of myself I've been so afraid of for so long. It's warm, and touching it is like sinking into a hot bath during the frigid winter.

Throat scraped raw, my cheeks stiff from the salt of my dried tears, I smile.

CHAPTER 34
THE PRINCESS

VOICES.

And *warmth*. The warmth of a body pressed against mine. I draw a deep breath through my nose. Everything feels nice. He's here with me, isn't he?

My eyes won't open. My limbs won't move. A bolt of panic replaces my sense of calm. My mind is fogged and sluggish. If only I could open my eyes, then maybe I could think better—

The voice behind me speaks. His words don't register in my mind, but the tone is resigned, irritated. Then the body shifts away from me, and panic floods. He's leaving me? Where is he going? But I worked so hard to get back to him! I'm exhausted, and all I want to do is lie in his arms. He can't leave me!

Not yet!

The world turns colder and the mattress shifts. He's gotten up. Maybe he'll come back soon.

The door opens, shuts.

Quiet fills my awareness.

I open my mouth, try to call for him. No sound comes out. My lips form his name. *Ash.*

Well, I'm not about to wait around for him to come back! If I can just get my limbs to work . . .

It's a painful process, but each new movement comes a little easier, until I've managed to scoot to the edge of the bed. The covers are much too heavy for me to lift, and I cannot find the strength to sit up.

So I just roll myself off the bed.

My fall is cushioned by the quilts I unfortunately bring with me. I land in an ungainly heap, breathing hard. My eyes barely crack open, revealing a blurry world of bright color. I've got to move fast before Ash gets too far ahead.

With a grunt, I push to my hands and knees—and promptly fall back on my face. A whimper escapes my lips. But by the Great Kings, I'm going to find my husband if it's the last thing I do.

I twist and roll myself out of the covers, then crawl on my knees and elbows around to the footboard of the bed. Chest heaving, I pause for a ten second break to recover my strength.

My strength does not recover. I forge on anyway.

Finally, I reach the door. It's closed. I lift a weak arm, slamming the heel of my hand into wood and scrabbling with my fingernails. The doorknob is *so high*. The world tilts to one side, and I fall back to the ground, losing my progress.

"Aaaaaaa," I moan. My husband's name is one syllable. Three letters. Yet my tongue is determined not to work. Useless thing.

Slumping to the floor, I poke my fingers beneath the door. Maybe I can slide underneath it. I'll be a cat and squeeze through. *Think small, fluffy thoughts . . .*

Then there's a voice. "My lord? Before you leave . . . um, there are fingers beneath your bedroom door."

Oh, that's Edvear! Maybe he can help me get to Ash. I try to call for him, and only succeed in choking on a gasp. I sag back to the ground. I am *so* tired. Maybe I'll sleep a little bit, and *then* I'll go after Ash.

It's so rude of him, leaving when I want him the most.

"Fingers?" comes the startled reply. It's quieter, as though from the other side of the world. Ash! He hasn't left me completely yet. "Stella?"

Come back, Ash! I want to cry after him. *I need you to kiss me!*

The thought comes out of nowhere, but once it's there, frustration burns through me. We're married, and he still hasn't properly kissed me yet. The gall! It's so infuriating I manage to lift my head and lean it against the doorframe. Everything is still blurry, but who needs clear vision? It's quite overrated, in my opinion.

"Please don't leap over the furniture, my lord! I'm sure it'll be quite as—"

The door swings open so fast I'm almost sucked through the opening. The world tilts again, and I slip from the doorframe to the threshold, hitting the ground hard.

"*Stella!*"

I sag, going limp at the sound of his voice. He came back for me. Happiness runs liquid through my veins, though my lips only move in the smallest smile. "Aaaa," I moan.

"You're alive! Oh Great Kings, you're alive!" The voice almost sounds teary in its exaltation. It quickly shifts to concern, coming closer and closer to me. "But what are you *doing*? You're hurting yourself! Can you hear me? Stella?"

"I can hear you just fine!" I retort—except it comes out in a garbled collection of incoherent noises.

Warmth wraps around me, easily lifting me up. The ground vanishes from beneath me, and in my relief to be back in my husband's arms, I cannot even find the strength to grab hold of him in return.

"Tell the High King that it'll have to wait," Ash barks over his shoulder. "And cancel everything else for the day. Oh—and send for the doctor! At once!"

275

He lowers me back onto the bed, clucking his tongue and commenting on the disarray of the blankets. I don't care if half of them are on the floor. I'm just glad to have him *back*. His voice, a constant comforting hum that comes in and out of clarity, soothes me enough that I sink into the mattress, eyes closed, thinking of nothing but how happy I am.

"You scared me nearly to death!" Ash says, slipping beneath the covers beside me and drawing me tightly into his arms. "Are you awake?"

The sound I make is an attempt at an affirmation, but incoherent as it is, it works. Ash keeps talking—why is he talking so much?—and I hear my name over and over again. Then, somehow, I pry my eyes fully open.

There he is, his face hovering above me. His mouth is moving, and his handsome features blur in and out of focus. I beam up at him, grinning. He stops talking. Just for that one moment.

And then he disappears.

THE PRINCE

MY HAMMERING HEART thuds to an abrupt halt when her eyes open—eyes I was afraid I'd never see again—and then a wide smile burst across her features. She's a mess of tangled hair, crusted eyelashes, matted and wrinkled nightdress, but the moment she smiles, I forget everything else. She's never been more beautiful.

Then her eyes roll back in her head, and the pleadings that cross my lips are much too desperate for a prince. I'd had her back! For a single, glorious moment. But now she's gone again, and I slump against the bedframe.

It's an eternity before the doctor arrives. I wait impatiently as the moment replays in my mind when I opened the door and found her sprawled on the ground, her nightgown in a tangle around her knees and her head fallen to rest against one of her extended arms.

When the doctor finally walks in, I sit up sharply. "She was awake!" In a rush, before he catches and rearranges the spectacles that bounced off his nose at my outburst, I tell him how I'd found her, the way her eyes opened, and how she grunted in response to my questions. "She managed to get *out* of bed—and all the way around to the door!"

I'm breathing hard when I finish, waiting—waiting—*waiting* for him to say something! To say she's better. To say she's still going to die. To say anything.

He gives a single, prim nod. "Good."

"It's good?" I straighten. "It is, right? This means she'll recover?"

"It would seem so." The doctor comes around to Stella's side of the bed and performs a few checks. Opening her mouth and one eyelid, taking her pulse, touching her forehead. "She is much improved. It's quite a miracle, I'd say. Her magic may not be very strong if she's already overcome the worst of the sickness. But that is to be seen."

After he's performed a few more inspections, he leaves with the same instructions as before.

I'm alone once more, holding my sleeping wife.

She's through the worst of it. She's going to be alright.

Stella will survive. And she has *magic* now.

Everything has changed.

I carefully extricate myself from her, lay her against the pillows, bring the blankets up to her chin, and tuck her in. I bend down and press a kiss to her forehead. It is like kissing a heart mended anew, or the bright rays of a fresh dawn.

"Rest," I murmur.

Then I straighten, all but throw myself into my desk, and pick up my quill. I take a blank page and scrawl frantically while hope sleeps a few feet away.

CHAPTER 35
THE PRINCE

E DVEAR INTERRUPTS ME shortly after Stella's brief awakening, and I turn a bright grin on him. "Hullo."

He doesn't grin back. My stomach drops, knowing exactly what he's here for. "The High King . . ."

I sigh, setting down my quill and running a hand down my face. "I should have known he wouldn't be put off for long. What sort of mood is he in today?"

"I'm not sure, my lord, but the servant at the door seemed anxious."

As much as I hate responding to the High King's summons, I've put this one off long enough. It's a fine line to walk, of keeping him *just angry enough* that I can manipulate the situation but not so explosively angry that he does something unpredictable.

I push myself up to my feet, sighing again, and cast one last forlorn look at Stella. She sleeps much more peacefully now, her skin a normal rosy hue instead of ghastly pale or feverishly flushed. An ache builds in my stomach. I don't want to leave her until she's better.

It seems I have no choice, however.

"Keep an eye on her, please," I say, grabbing an overtunic and shrugging it on. "And if anyone comes to the door, send them away. I don't care if it's the tailor, or a scheduled delivery. Don't open the door while I'm gone."

He bows. "Of course, my lord."

With that, I sweep a glamour over the rest of myself, hiding any last visible sign of poison, polishing my tall boots and outfitting myself in a prince's garb that is *just* a smidge too casual for visiting the High King. Not that it matters too much; being High King gives him the power to see through glamour.

On my way out, I pause long enough to cast one more ward over my chambers. No one is going to lay a finger on my wife while I'm gone.

The walk to the throne room through echoing marble hallways is not too long—not as long as I should prefer—but I use the time to focus my mind.

I don't care about my wife. I don't care about my staff. I don't care about Rahk. Nothing matters to me but that throne, and I will do anything to get it. I will be ruthless and sacrifice anything in my way.

It's time to make Faradir pay for what he's done.

I reach those doors that I used to shrink before. There's no shrinking now. Before the guards can move to open them, I stride past and shove both doors open so they swing on their hinges—and hit the walls.

My mouth twists into a grin as I meet the High King's gaze. He lounges in his throne, a bored, lazy expression on his face as he leans his jaw on his fist.

"It really has been too long, hasn't it?" I say, stopping and spreading my hands wide. "Dear Father, my High King."

"Everyone out," snarls the High King.

The court shuffles as everyone hurries to obey. I don't know why they bother to be here, anyway. Surely, they have better things to do than scurry in and out of the throne room at their king's whim.

When we're alone, the High King heaves a great sigh. "Prince Trenian. There you are."

I flash a grin and bend double in a bow. "At your service."

"Are you?" he shoots back, tenting his fingers under his chin. "I have a military assignment for you."

Not what I expected. Is he trying to separate me from Stella so he can kill her more easily? I arch an eyebrow as I lean against my customary pillar, arms crossed over my chest. "Oh?"

"Indeed. I've been testing the borders of the human lands long enough. They have no interest in retaliation. It's time. I want you to lead an attack against Aursailles. Starting tomorrow. I expect the kingdom to be captured by Lulythinar."

A muscle inside my knee jerks. Not what I expected at all. In fact, this quite complicates matters . . .

"I'm afraid I cannot," I say, drawing on a disinterested air and inspecting my fingernails. "I'm still on my moondust, you see. Or whatever humans call that time of marital seclusion and bliss following a wedding."

"You reject this assignment, then?" The glint in his eyes unnerves me. He *expected* me to decline. Of course he did. Now he'll make me pay.

It's suddenly so obvious I don't know how I didn't see it before. He's manufactured an opportunity to punish me. Laying a trap.

I let out a long sigh. "Oh Father. When will you ever stop putting words in my mouth? I don't *reject* the assignment. I accept it. I just cannot attack until *after* Lulythinar. When my moondust is over."

"I want you to attack now. The assignment isn't for after Lulythinar. It's for *now*."

"Come, come, Father, we both know you can bear to wait five days for your conquest. You're just trying to kill my wife. Ruses don't look good on you."

The High King says nothing for a long moment. As though contemplating the best way to chew me up and spit me back out.

Finally, he reclines in his throne. "Do you know why I endeavor to capture the human lands?"

I tap my chin as though considering. "Because peace treaties are far too last millennia?"

"Because that peace treaty gave the humans land that used to be *ours*. The humans negotiated the land that is now Aursailles, Osremer, Enslington, and Algravia. But that land used to be *ours*. Only a weak king would have capitulated so much."

Or perhaps only a weak king would feel the need to so desperately prove his might. But I withhold that particular thought. "I fail to see what this has to do with why I should go on a rampage against my father-in-law while I am still celebrating my nuptials."

"You're not celebrating. The human was poisoned."

I frown. "The poison has dampened the celebration a bit, true." No iron coats my tongue, as my statement refers to my own poisoning. By now, it's second nature to lie without lying.

He lets out a long sigh, wiping his hand down his face. "How did I spawn the most tiresome fae in all the Courts?"

"The mystery of the ages," I reply dryly, returning my attention to my fingernails. "If we are almost done here, my heart grows anxious to return to my wife."

"We are not done!" The High King slams his fist down on his throne, rattling a half-filled goblet and the tray of refreshments on a small table beside him. "I should have killed your mother the moment you were born and raised you like a true son of mine. Then maybe we wouldn't keep playing these games."

He thinks he'll get a rise out of me by mentioning my mother. I merely shrug. "Perhaps. Or perhaps I *am* a true son of yours, and maybe *that* is what you don't like about me."

The High King lifts his hand and snaps.

Cold washes down my spine, but I don't let it show. I merely glance over my shoulder, schooling my features into boredom.

"Bring it in, Prince Rahk," says the High King.

Rahk.

I'm not swift enough to hide my reaction—just a twitch of my fingers. The High King's gaze burns into me, and in the corner of my eye, his mouth curls into a smile. He knows about my friendship with Rahk. But he also knows he cannot hurt Rahk without angering the loyal and powerful Nothril Court.

He *can* coerce him, however.

Or perhaps . . . *no.* Rahk wouldn't betray me. He wouldn't.

But if he did, then there's only one person he can be bringing in. *Stella.*

The doors open, and there's Rahk. Standing tall and still, his face the hard, disinterested mask of a warrior. And in his fists, he holds her arms pinned behind her back. Not Stella.

Hylath.

I fight the wave of sorrow, of *fury* that nearly explodes from me. She's nothing but a handmaiden. She's done nothing deserving of death. Just like all the rest.

I hate Faradir.

I hate him so much I barely keep myself from snapping and running to her defense. It's possible—not likely, but possible—that I could kill him in cold blood. Right here. Right now.

But that would nullify my claim to the throne. Everything I've fought for would be gone.

Hylath doesn't make a sound, her belly expanding with each breath. It's so unlike her to be still and quiet.

"Last chance," says the High King, smirking now. "Take my armies. Raze Aursailles. Slaughter the humans. Finish by Lulythinar."

Last chance to bend, to bow. To destroy my wife's homelands, the very people she sacrificed herself to save. To leave her unprotected here in the High King's palace while she recovers.

This is my last chance to show the High King that I am affected by his murder of my loyal staff. My last chance to prove it's an effective tool.

To leave Stella in a tyrant's bloody hands.

I meet Hylath's five bobbling, blinking eyes. She's the only one of my servants to whom I owe no vow. She said my rescue of her was payment enough. Right now, it doesn't seem like anywhere near enough.

"Of course not," I reply. "I told you I am busy until Lulythinar."

The High King leans back in his throne. "Prince Rahk. Remove this creature's eyes."

Rahk doesn't look at me. But Hylath does, her five eyes whirling to me, wide. I meet that gaze, hating that my cold, cruel expression will be the last thing she sees.

Then Rahk shoves her to her knees, presses her face into the floor, the connective tissue supporting her eyes laid out like a bare neck on a chopping block.

It happens so fast I can't even blink before Rahk's sword is out, before it's sliced a clean line straight through flesh. A scream cuts through the air, straight into my chest. Blue blood drips as Rahk pulls my blinded servant to her knees. Her eyes roll like marbles, one coming to stop near my foot.

It's crueler than just killing her.

"Take her away," says the High King.

Rahk moves to obey.

"But first . . ."

Rahk stops, glancing back at the High King with that same impenetrable mask.

"Ready my armies, will you, Prince Rahk? I want Aursailles by Lulythinar."

Of course. Of *course*. He'll try to turn me against Rahk by making him his pawn. He'll give him distinguished tasks like leading his armies, tasks that should be mine, that will show his favoritism. The King and Queen of Nothril won't be able to argue.

And it'll be an effective punishment, that I wouldn't have anticipated.

This is why I told him to stay away from Valenaven for a few days! If only that cursed invitation hadn't arrived for his sisters, if only he hadn't come to protect me.

But then he wouldn't have saved Stella two nights ago.

This was all part of Faradir's plan from the moment I returned with a human bride.

Rahk clasps a fist over his chest. "It would be my honor, High King, my liege."

"Very well. You both are dismissed."

Rahk bows. Then turns and drags Hylath after him, her eye tendons wilted and her body shaking in pain. I don't move for a long moment. Not until the doors are shut. Then I lift my attention to the High King.

"Have something to say to me?" he asks, folding his hands across his stomach and regarding me with the barest hint of smugness in the twitch of his jaw. His long golden hair falls over his shoulder and gleams almost as brightly as his teeth. "You might consider obeying me for once, and then perhaps I won't send your friend for your wife's throat."

I curl my lips at him. "I thought I told you that my wife's throat belongs to me. No one touches it except me."

The High King's chuckle reverberates against me as I stride out of the throne room. Just like that, the castle of my schemes crumbles to pieces. There's no time to wait for the Nothril Court and the Neverseen King to ally over the Orawyth portal. There's no time for *anything*.

It's only a glamour that keeps anyone from seeing how my hands tremble with fury.

CHAPTER 36

THE PRINCESS

I OPEN MY eyes, and it's almost surprising when they're clear. I sit upright in bed. This is Ash's bed. Not my own. I scan the room, find Edvear sitting by the door on a little stool. He smiles at me, bright and relieved.

"You're awake, my lady."

I draw the covers to my chest self-consciously, then reach up and touch my matted hair. Thoughts race around my mind, loud like a cacophony of ringing bells, mingling with stray memories. I put myself in this bed. When Ash was poisoned. Or, rather, when Ash poisoned himself. The running around the Aursailles palace—a dream, then. Was the vague memory I have of collapsing against the door, of staring blearily up into Ash's shocked face, a dream too?

I blink at Edvear. "So it would seem. Where is His Highness?"

The smile falters. "The High King summoned him."

"Oh." That's not good. But the High King cannot kill Ash until he's sired an heir, which hasn't happened yet. So this cannot be that bad, right?

"Would you care for a change of scenery? I can help you to sit by the windows in the living room, to give you a little fresh air until His Highness returns."

It would indeed be nice to get out of this bed. "Fresh air would be lovely."

He fetches me a thick robe while I use a brush to work out the worst tangles in my hair. My strength hasn't fully returned, but I don't lean on his arm as heavily as I expect when he takes me out to the living room. The sunlight feels so good on my eyes, and the sight of the beautiful gardens outside the windows immediately lifts my spirits. I sit in one of the settees facing the gardens and sip the rich, warmed chocolate another servant brings me.

I feel *so much better*.

Suddenly, the front door bangs open, startling me.

Edvear is there at once to greet Ash, who doesn't see me at first. He tosses his cloak to the live coat rack. The rack catches the cloak out of the air and goes still. But Ash grips the back of one of the upholstered chairs with white knuckles, his shoulders bowed. My heart falls straight to the floor.

"My lord?" Edvear asks. "What has happened?"

"A new position has opened up on our staff," Ash replies darkly. "We also need medical attention immediately. Send for one of the low fae doctors. Was Rahk here?"

"No, my lord. No one has come to the door."

Suddenly queasy, I set down my hot chocolate on the end table near me. The dish clinks against the wood. That's when Ash looks up—spots me. His flashing eyes pierce mine.

"See to those things," he orders Edvear, his attention never leaving me as the steward leaves. His face is such a mingling of intensity that I can hardly tell the fury from the relief—the latter, I think, from me being past whatever sickness has plagued me these last few . . . days?

But beneath the fury is something fragile. Something that's already broken.

My mouth is so dry, but I speak anyway. "Ash."

A growl bursts from him. He shoves aside the upholstered chair and crosses the distance between us. Then he's beside me, dragging me into his arms with a long, low groan. He buries his face in my neck.

My body goes taut. That is . . . not what I was expecting. "Ash?"

"*Stella,*" he growls in response, tightening his grip on me. Edvear is gone, but I'm acutely aware of the fact that he can return at any moment.

I close my eyes, an ache building in my chest. Slowly, I let myself soften against him, and reach up one tentative hand to brush the long strands of dark hair out of his face. "What happened?"

"I'm a fool," he groans into my neck.

"What happened?" I ask again.

"I don't want to tell you."

My brow furrows. I stroke his hair behind his ear. "Why?"

"Because I don't want to worry you."

I give a dry little chuckle. "Too late."

"And maybe if I don't tell you, it won't be real."

My hand slows, my heart hammering.

"And because I'm just so glad that you're still alive. I'm so glad, but I am *so angry*, and I'm afraid of what I'll do—how I'll retaliate. I'm a fool, Stella. A fool for bringing you into this. A fool for thinking I can outmaneuver that snake. Everywhere I go, death follows me." His fingernails dig into my side, though not enough to be painful, and he pulls me closer, his breath hot against my skin as he groans again. "Everything I love dies. And *Stella*, I love—" He cuts himself off abruptly, then finishes. "I love more than I should. I *know* I shouldn't care about anyone—anything! So why can't I help it? Why can't I just be heartless? Why can't I just be more like my wretched father? He wants a monster. And I want to *be* a monster. But I just . . . *can't.*"

Alarm builds inside me until I'm breathing hard and tensing back up in his arms. I'm too afraid to ask what happened, too terrified to know what he lost.

One of Ash's hands slides up to tangle in my hair, his mouth pressing short, desperate kisses against my throat, my jaw. "I need to send you away," he growls softly in my ear. "I need to get you out of here. I'll just kill him. I don't care about the throne. I'll kill him, I'll give up my right to rule, and I'll let the Courts destroy themselves. And then I'll come and find you—after the world has burned itself to the ground. It'll be just the two of us then, and there will be nothing left that can hurt you."

I squeeze my eyes shut. His voice, his words, are the sound of heartbreak. Hopelessness. Futility. All this time, *this* was beneath the strength, the cunning, the confidence, and intimidation.

Heartbreak.

The memory of Rahk's voice is in my ear: *"He killed Ash's mother."*

And now Ash is afraid to lose me.

I work my hands free of his embrace, plant them against his chest, and push. He only resists for a moment, then lifts his head to look at me with such desperate, mournful eyes. I rest my palm on the side of his cheek, and he leans into it, tilting his face so he can press his mouth to my wrist in a short kiss.

"Ash," I say softly. "Don't be afraid."

He snorts—a dark, sardonic thing. Then his blue eyes flash, arresting me with sudden force. "Why, my darling, shouldn't I be afraid?"

"Because," I say slowly, "neither of us is dead. And because we can face your father together." I finish by giving his nose a little *bop* with my finger. "You're smart. I'm . . . less so, but not wholly useless, I think. We're the perfect team. We'll think of a way to handle whatever situation comes up."

He stares at me as though I've sprouted three new heads. So I just smile.

Then, before I know what's happening, the back of my head hits one of the settee pillows, and Ash's hands are planted on either side of me. I gasp, staring up at him as unexpected intensity fills his face.

He brings his mouth to mine until there's only an inch separating our lips. "Ask me to kiss you."

My eyes widen, my heartrate skipping and tripping ahead of itself as another gasp lodges in my throat. "What?"

He lifts one hand, trails the rounded edge of his knuckle down my temple, my cheek, making it harder and harder to breathe. "I won't kiss you. Not until you ask. So *ask* me, Stella. Beg me. *Please.*"

My gaze falls to his lips, so full and warm. I really have married my own destruction, haven't I? It would be so easy to say yes, to lift my own mouth just a fraction higher . . .

Instead, I manage: "Are you . . . begging me . . . to beg you?"

He lets out a long sigh and leans his forehead against mine, our noses brushing. "You are just *determined* to be the death of me, aren't you?"

I cannot help my little giggle. "Torture is my specialty."

His eyes darken. "Fine, you confounded *woman.* If you want to play games, I'll play games with you."

And with that, he sweeps my hair away, and leans down to press his lips in a scalding kiss against the hollow of my throat. I suck in a sharp breath, and he rasps a low chuckle.

"Ask me to kiss you," he murmurs, moving his mouth to my jaw. "Ask me, Stella."

Part of me wants to be stubborn, to resist him past his self-control, until he just breaks and kisses me, anyway. But if I do that, his control might not snap, and he'll pull away, and the offer will be past.

Losing this game is still winning.

I smile as he presses a third kiss just below my ear. I tilt my chin up, meeting his gaze. "Ash?"

He stares down at me, his chest heaving with his fast breaths. "Yes, love?"

I give him a little grin. "Will you kiss me?"

His eyes shutter closed as a groan slips from him. "Great Kings, *finally.*"

Then his mouth is on mine, his arms wrapping around me, pressing me close, kissing me like the world is crumbling around us, and we're the only ones left. His lips coax mine, moving slowly, then fast, then slow again, as though trying to hold himself back. As though making sure I can keep up. If he wasn't still kissing me, still tangling his fingers in my hair, I'd smile. Instead, I wrap my arms around his neck and pull him closer to me, until the world is nothing but warmth and sweetness, and I want for nothing but to drown in it.

When he pulls back, I open my eyes. He stares at me, lips swollen, his expression torn between blistering hope and that sorrowful devastation. He's thinking about losing me again, isn't he?

My fingers are strung together at the back of his neck. I use them to pull myself up, just close enough to rub my nose against his. He nuzzles me back, softly, sweetly.

"I think we need to talk," I whisper.

"I'd rather keep kissing you," he breathes back, nuzzling me again. But he doesn't kiss me, and instead we prop ourselves up against the back of the settee. He reaches out, takes one of my hands, and threads his fingers with mine.

When I look up at him, his gaze has hardened.

"The High King demanded I lead his armies into battle," he begins, jaw flexing. "Tomorrow. Against—"

"Aursailles," I breathe, my gut hollowing out. An unwelcome memory returns, of what one of those fae women said at the banquet. That Ash has slain thousands. Another memory assaults me of him killing those fae who came for me. He'd ended a dozen lives before I'd barely had a chance to breathe.

My father wouldn't stand a chance.

Amelia.

Ash nods.

"What did you say to him? I thought the boundary between our worlds was stronger!" I refuse to pull my hand away from his, but I cannot deny that the impulse is there. What if I kissed him right after he'd betrayed my people? After he'd betrayed *me*? After he guaranteed the death of my beloved sister?

"The treaty only prevents a certain *terrorization of humans by individual fae* and *the invasion of fae lands by humans*. It doesn't stop the High King from conquest," he growls. "I refused the order."

The pensive, dark lines haven't left his face. "You don't say that like it's a good thing."

"He sent Rahk to do it instead."

Shock thrums through my body. Ash tightens his grip on my hand, as though afraid I'll pull away. My voice comes out in a croak. "What? And Rahk—he will do it?"

"He has no choice. If he refused, it would give the High King the grounds he's wanted for centuries to get rid of my only friend. When Rahk is compliant, and does what he asks, the High King cannot kill him without enraging the Nothril Court. Rahk has gotten by for a long time by avoiding notice, and since the High King cannot directly harm him, I didn't have to worry. But now he's figured out a way to use Rahk against me, and in a way that doesn't anger the Nothril Court."

My lungs tighten as I listen. Is there anything we can do? Anything Ash can do to stop it?

No. I won't worry about that yet. Not until I hear the rest of it.

"What else?" I ask, bracing myself.

"For my defiance, he inflicted one of his *punishments*."

"Punishments?" I scan his body, looking for sign of injury and finding none.

"His favorite means of punishment these last couple of centuries has been executing a member of my staff." Ash's face darkens, followed by a flash of pain as his jaw flexes. "He took Hylath. He made Rahk cut off her eyes."

I gasp, covering my mouth with both my hands, and before I can stop them, tears well up, clouding my vision. "Hylath?"

"And he knows he upset me this time—I didn't hide my reaction as well as I usually do. Which means he knows this is an effective way to break me."

Two tears slip down my cheeks, but I swallow the rest behind a firm wall of determination. "He's *not* going to break you, Ash."

He looks at me, emotions warring across his face. "Sometimes I think he won't, and other times it seems the inevitable outcome."

"He's not going to break you," I say again, firmer this time.

His eyes narrow. "You sound so convinced of that. Pray, tell me why, little wife?"

I squeeze his hand, meeting the force of his gaze with my own. "Because the High King is fighting *against* you, against anyone who defies him—human or fae. But *you*, Prince Trenian, are fighting *for* something."

Slowly, his lips part, his jaw going slack, and he stares at me as though stunned.

"He's fighting to keep his tyranny," I continue, bolstered by his reaction. "But you are fighting for freedom. For hope, for peace. For *me*, Ash. That's why you won't break."

Before I know what is happening, he's captured my face in both hands, my lips with his mouth, kissing me with a sudden burst of passion that leaves me dizzy.

"How did I just walk up to your father's palace and find you?" he breathes between kisses. "How did I marry a stranger—only to discover I'd wed an angel? I'm half afraid the moment I turn around, you'll have sprouted wings and flown out of my reach."

"Don't be ridiculous. I'm not going anywhere," I say, pulling back just enough that our lips aren't touching anymore. "But we must figure out what to do about the invasion! Can we get word to my father? So he can prepare?"

Ash springs off the couch, setting into a vigorous pace in front of me. I go cross-eyed watching him go back and forth, back and forth, so I pry my eyes away from the pensive line between his brows, the way he rubs the stubble along his jaw with one of his large hands, and instead focus them on my lap. That's when I realize I'm still wearing my nightgown and robe.

My sickness.

I have many questions to ask Ash.

"The human armies will be no match for the High King's forces." His voice is grim when he adds, "Or Rahk. Sending word won't make a difference."

"They could evacuate the cities! Perhaps the children could be spared, if they could be warned and aided."

Maybe Amelia is already in Enslington. Time flows differently between our worlds. She might be married and safely out of harm's way. Or maybe she is just as at risk as everyone else.

He shakes his head. "They won't be able to move fast enough. They'll be overtaken, and then they'll spend their last days hunted and terrified out of their wits."

"So a slaughter is better?" I demand, fisting my hands in the fabric of my robe.

He lifts a hand, staring at the floor as though his mind is working. "No, no, of course not. We must stop the armies from leaving." He pops his head up suddenly, one eyebrow upraised. "We could kidnap Rahk."

I flash him a dubious expression, and he nods even without looking, waving his hand toward me.

"No, you're right. I can't overpower Rahk, so I'd have to come up with some trick, which might work in a pinch. But the key is to ensure that the High King is the one who finds Rahk tied up—well, no, that would humiliate him."

Rahk's humiliation is definitely preferred to the mindless slaughter of my people, but I keep my mouth shut. For now.

"Besides, taking Rahk out of the equation will only delay the inevitable, but not as much of a delay as we need. I can bargain with the High King. That would show my hand—that I'm against this, even though I implied the opposite in the throne room. But it might be the only good option. If I can think of an enticing offer . . . something he wants more than the human lands. Or, rather, something worth delaying his conquest for . . ."

Ash looks up at me, frowning severely. I blink back at him.

"He wants *you* dead," he says. "Not that I would sacrifice you for your people—Great Kings, no. I don't care how many people must die for you to live. But, maybe . . ."

A slow but sure grin spreads across his face as those cunning eyes light on me. "I've got it."

CHAPTER 37

THE PRINCESS

IT ISN'T HYLATH who helps me bathe and dress. It's a human woman, in her forties judging by the light lines of aging around her nose and lashes, who readies me while Ash works. I meant no offense to my new maid, but I nearly broke down at the reminder of what Hylath has just suffered. After Ash and I discussed the plans—which sounded alarmingly risky to me—I'd asked him to tell me what happened while I was sick. Now, as I step out of the tub into the towel held out by the new woman, my fingers and toes tingle.

Magic?

I don't feel particularly different than I did before all this. Perhaps a little taller, a little stronger, a little bolder. But I'm still . . . *me*. I would have expected magic to feel like an electric buzz through my body. Something exploding from my hands into the ether. When the woman turns her back to me, I twist my wrist, flick my fingers, like I've seen Ash do.

Nothing happens.

He'd looked so hopeful when he'd told me that since I've pulled out of the sickness, I should have magic now. My life expectancy should be much longer.

Should be.

The woman helps me dress without a word. She doesn't call Ash to dry my hair but towels it until it's dry enough to style and pulls it back into a bun. I understand now why Ash tries so hard not to love; I barely find it in myself to ask the new woman's name for fear that as soon as I get attached to her, she'll be gone, too. But I don't want to live like that, so I force myself to ask. Her name is Dorthea Burton, called Dottie by most.

Apparently, while I was ill, the tailor stopped by with the rest of my dresses. All except the ballgown for the Lulythinar masquerade.

This dress is a light rose color, with long flowing sleeves and an off-the-shoulder neckline lined with what almost seems to be real roses. The skirts billow out from a tapered waist in the pattern of petals, with a sheer top layer lending shimmer to the ensemble.

It isn't practical for spending the rest of the afternoon and evening in Ash's quarters. It's suitable for an evening banquet—and another visit to the throne room. I draw in a deep, shuddering breath, then level a look at my reflection.

I won't be afraid of the High King. Not anymore.

Or, at least, I'll be less afraid than I was before.

Dottie opens the door for me, and I'm so focused on dragging my skirts through the narrow space that I don't realize until I look up that Ash is standing a few feet away.

His gaze runs over me, then glances at Dottie closing the door. Where I might have hoped for warmth or a brightening of admiration, his expression remains closed and grim. As though he's reminded of his grief for Hylath, and perhaps dreading what we're about to do.

He gives me a rueful smile anyway when his eyes lift from taking in my dress. "You are lovely, my darling." He reaches out, catches my hand, and presses a kiss to the backs of my knuckles. With his mouth still against my skin, he lifts his gaze to mine, his vibrant eyes a stormy shade of dark ocean.

My breath catches, but not pleasantly. I don't *want* to be afraid. So why can't I keep the tiny tremble from raking down my arm?

His fingers tighten on mine. "You have nothing to fear at my side."

A sharp, overwhelming tang fills my nostrils—cold and metallic and *vastly* unpleasant. Not knowing what else to do, I burst into a coughing fit, yanking away from Ash and pressing the sleeve of my gown over my nose and mouth.

"Stella? Are you alright?" he asks, suddenly at my back, hands on my bowed shoulders as I keep coughing, my eyes watering.

I nod vigorously, taking deep breaths through the fabric of my gown. "It's nothing," I choke. "I think it's merely some residual cough from my illness."

"Sanak! A glass of water, please!" he calls.

My mind reels as I pull myself back under composure. Ash . . . just *lied* to me. And I could smell it! Goodness gracious, how does anyone keep a straight face with that repugnant stench cloying in their nostrils? When the servant comes running with the glass of water, I drink it down greedily.

Ash just lied to me.

He *lied* to me.

It was to comfort me, to ease my worry—but it was still a lie. Perhaps the most terrifying thing of all is not that I'm safe at his side, but that he would want me to believe that it is so.

Suddenly, I'm very glad I lied right back about the cause of my sudden coughing fit. A lie that didn't have any taste to me and earned no reaction from him.

Does this mean I can smell fae lies, but my own have no stench, no flavor?

Perhaps this is from my *blood sickness*.

"Are you alright?" Ash asks, all gentle attention as he takes the glass from my hand and returns it to the awaiting servant. He doesn't *seem* to suspect that I can smell his lies. "You can stay behind if you're not feeling well."

I straighten my shoulders, lifting my chin. I'm not going to tell him. I want to see what other lies he tells me when he thinks I cannot detect them. "No, no, I'm quite fine. I might need to bring a handkerchief with me, however, just in case another fit overtakes me."

"Take one of mine," he says immediately, pulling it from inside his overtunic and handing it to me. It's a rich scarlet, with gold embroidery, and much softer than any other kerchief I've held before.

I slip it into one of the hidden pockets of my gown and somehow manage to smile up at him. His lips purse into a tight line, and he holds out his arm for me to take.

We leave the safety of Ash's quarters, entering the main hallways of the expansive palace. This path to the throne room is starting to look familiar, and I keep my eyes peeled as we walk, noting the corridors, the different directions we could turn down, the landmarks along the way. The winding outdoor staircase that leads to a bridge over the water below. The statue of a winged fae archer. All the varying greenery and flowering vines that distinguish otherwise nondescript columns and windows.

Perhaps on our way back I can ask Ash to show me where Rahk's quarters are. I need to know where my allies reside.

Now that I can detect lies, I intend to find out exactly who my allies are.

Ash is one of them. Everything he said this afternoon, when he held me in his arms and kissed me, was true. But I cannot help how my hair rankles on the back of my neck. The first night we came here, he promised me he would be honest with me about what we would be facing together. This small, seemingly innocuous lie he just told me is a poignant reminder that even with Ash's power and strength and cunning, I am not safe with him.

Perhaps I'm not even entirely safe *from* him.

That is fine with me. I'm done relying on him for everything. It's time for me to learn the ways of the fae, this palace, the High King.

I'm nothing but a lowly human here, which means everyone expects me to be stupid, ignorant, and helpless. All things that I was when I first came, and to some extent, still am. But not forever.

It starts with learning my way around this place.

When we arrive at the throne room, those familiar double doors with the tall fae guards almost make me want to cower against Ash like I did the first time he brought me here.

No more.

I straighten my spine, and this time Ash doesn't drag me in with a possessive hand around my waist as though I'm the new prize he's collected. The doors open, and we march into the throne room, side by side. Ash wears that predatory grin, and I level my own half-smile straight at the High King, sitting there on his throne.

Faradir's surprise shows only in the way his mouth—opened to speak—clicks shut. His gaze flickers to me, almost with disinterest, only to stop . . . and narrow. One of those eyebrows raises curiously, and he stares openly at me.

In my periphery, Ash glances down at me.

It occurs to me then that perhaps I ought not to lift my chin so haughtily. Perhaps I should play the timid little wife instead—to keep up the ruse. *Oops.* I lower my eyes to my toes.

"High King," says Ash with a mock bow.

"Prince Trenian," replies the king. And then, as a bored afterthought, "Princess Stella." He returns his gaze to Ash. "Twice in one day, you bless me with your presence. Are we to celebrate the apparent recovery of your human?"

He says it like he might say *animal.*

I don't flinch. Instead, I slip my hands from Ash's arm and bow in a deep curtsy. "Thank you for your consideration, Your Majesty."

Beside me, Ash blinks just a touch too fast. It's the only slip in his mask. It's too hard to tell if his surprise is positive or negative. He tucks an arm around my waist, pulling me back up just as I was standing. He leans toward my ear, growling into it, and it's just loud enough that I know it's not only intended for me to hear. "No wife of mine bends the knee. To *anyone*."

I cannot help a little smirk at that. I straighten at his side, trying not to feel anything at all, as his hand moves possessively to my hip. He grins at the High King. "Are you still so surprised this human caught my attention?"

He's flaunting me. A carrot on a string. I'm the bait, the temptation.

The High King waves a hand as though dismissing a fly. "Tell me why you've deigned to interrupt my proceedings, dear son. I have important matters to attend to."

Ash stalks closer, bringing me with him. I try not to give into the impulse to stare down the High King with this newfound boldness. Letting go of fear is truly an electrifying sensation.

"I'm here to bargain with you, dear father."

"Another time, perhaps."

My gasp is entirely genuine when Ash grabs me, snatching my wrists behind me and holding them fast while he pulls me back against his chest.

And presses his curved knife against my throat.

My awareness narrows to the sharp edge of steel like ice ready to slice into my vulnerable flesh. The pounding of my blood through my veins.

This was *not* part of the plan.

"And if I was bargaining my wife's life?" he purrs, and his voice is about as warm as the blade against my neck. "Would you listen then?"

Part of me insists I trust my husband, his wicked games and careful maneuvers. The other part of me burns with betrayal. He didn't tell me this was going to be part of this negotiation. The plan was that he would bargain five more human kingdoms and his magic to extend the Faerie border, whereas Rahk could only ransack the kingdoms but not claim them as part of Faerieland. In exchange, Ash was going to ask for the slaughter to be delayed until after Lulythinar.

He hadn't lied when he told me his plan. So either he's making this up as he goes—a terrifying thought—or he deliberately crafted his words to conceal more information than they revealed without telling an outright lie. The fae are clever with their words, after all.

Ash's grip on my wrists tightens and pulls down harder, so I'm forced to extend my vulnerable neck, giving his blade more room to tease against taut flesh. I gasp again, and it's loud enough that there's no way the High King misses it.

He leans back further in his throne, his gaze back on me. On my neck. Something dark and hungry flashes in his irises. Something that wouldn't have intimidated me if his son didn't have me at knifepoint.

Ash is just like his father.

The thought momentarily stuns me. I don't actually think that, do I? No, he's not like his father—all of this is so he can save my people from the High King's bloodthirst. Ash *is* different.

But . . .

Not *that* different.

My chest heaves with each breath. Panic flares, bright and sudden, and instinct makes me twist against Ash's hold. He has my wrists pinned with three fingers—*three* fingers—and I cannot wrestle free. *Breathe,* I tell myself. Breathe and stop struggling! I force my body to relax.

"We all know you want her dead," says Ash, his dark, sardonic voice rumbling into my back. "We also know that I do *not* want her dead. In fact, I'm quite loath to part with her." The pad of his thumb runs along my jaw, a possessive stroke. Then he bends down slightly and presses a kiss to my forehead. Just as he does so, he presses the knife harder against my throat. Still not enough to break the skin, but enough to rip a whimper from my lips.

The High King hasn't said a word, feigning indifference. The sharpness of his gaze betrays his otherwise languid pose. He sighs, shrugging one shoulder. "What does her blood matter? You still have enough time to go kidnap yourself another human princess before Lulythinar. This one's death hardly makes any difference."

"Does it?" Ash's grin is almost audible. "And if I bargained not to marry another human princess before Lulythinar?"

"What is it you want?" the High King snaps. "If you're willing to give up your human, then it's probably not something I'm willing to give in exchange."

"I want you to promise not to have me killed—directly or indirectly—after I've sired an heir."

The High King barks a laugh.

My body goes stiff, my eyes widening. The whole throne room goes quiet. Was this expectation always unspoken? Had some courtiers not even realized this was Ash's predicament?

And for heaven's sake, this wasn't the bargain he told me he'd be asking for! Where is the agreement for my people? Where is the promise to protect my homelands?

Anger burns beneath the questions flooding my mind.

Then, behind it, another question. A darker, deeper fear.

What if this was his plan all along?

What if he doesn't actually care as much about me as he made me believe? Or, perhaps more aptly, what if he is willing to sacrifice literally anything for this throne? For his own life? He's been sacrificing for this plot of his for centuries.

What if his lie to me that I have nothing to fear wasn't merely a gentle encouragement? What if it was a lie because he intended to slaughter me right here, right now, at the foot of his father's throne in a bid to guarantee his own life?

Faradir rises to his feet, his golden skin luminescent, the white of his robes gleaming like rays of sunlight. He moves like a prowling tiger as he slowly takes one step after the other until he's off his dais, approaching Ash—approaching *me*—with that black cunning shining in his pupils.

When I look into the High King's eyes, my reflection stares back at me. That of a girl in a flowery gown, eyes wide as she stands pinned between two predators. He comes to a stop, not even a full pace from me. Perhaps I would flinch if I didn't have my arms twisted behind my back.

As it stands, fury washes through my blood.

"Perhaps I'd consider such a bargain . . ." says the King, his eyes roving over me, over my soft and sweet gown. "If it wasn't her life that you offered, but her leash."

My heart stops beating. I barely keep my eyes from shutting. He wants me to be *his* pet. Not that long ago, I wouldn't have flinched. Ash always promised to protect me, to be *good* to me.

But Ash apparently lies to me—and he's good at hiding those lies, despite their foul taste and strong stench.

I don't claim to know what the High King could do with me, but I know *enough* to be terrified at the prospect. I want to believe that Ash would never give me up to torment. I want to believe that he meant every good and tender thing he has said to me. But what can I be sure of at this point?

Nothing.

With a word, Ash could have what he's been fighting for all these years: his life.

A horrible thought occurs to me. What if the true reason Ash has been fighting his father isn't because of injustice, but because he merely wants his own life?

"Hand my wife over to you?" drawls Ash. "I think we both know I'm not *that* cruel. Or that desperate. If you won't take her life in exchange for mine, then I suppose it's settled. Though perhaps I ought to tell you about another bargaining chip I have."

"Enough. I have no more patience for your trite plots. You are dismissed until this evening's banqu—"

"The Neverseen King owes me a favor."

Ash's voice slices straight through the crowd of people and the High King's statement like a knife through skin and bone. Silence falls as the High King half-turns back, his brow pinched.

"The Neverseen King owes you a favor?" he asks.

"Indeed." Ash's thumb on the side of my wrist, restraining me, gives a little caress. Reassurance? Or another lie? "And I have a mutually beneficial agreement for us."

CHAPTER 38

THE PRINCE

S TELLA IS STIFF against me, her tiny wrists so thin, so . . . *breakable*. I try not to think of her at all while I level a grin at the High King, whose interest I've *finally* captured. The challenging thing with this bargain is not giving up exactly what I want, and exactly how much I want it.

In most cases, the High King would rather lose something he wants than give me what I want. Which means that the moment I suggest delaying the conquest of the human lands, he'll know that is what all of this is about. And he won't accept anything—just to punish me.

But a favor from the Neverseen King?

He is the only emissary of the High King who could successfully revolt. With his position as gatekeeper of the Bridge, he is in some ways a member of this court, but in more ways, an entirely separate entity who holds in his hands the power to destroy worlds.

More than anyone, the High King dares not lose the loyalty of the Neverseen King.

He knows what a favor is worth.

"You would offer me your favor?" the High King scoffs. "And I suppose you want my kingdom in exchange?"

Stella's pulse rams against the tips of my fingers. Her hands clench into fists, but she doesn't struggle. She keeps her upper body rigid against mine, not daring to move with my knife still in place.

"I won't give you my favor. But I can redeem it—and make him set his monster Crenfyre loose upon the entire human continent to destroy it."

The High King pauses, and I try not to care. Try not to *hope*, as the moments slip by, as his consideration lengthens. He cannot deny how compelling my offer is.

But I made a fool of him only a few days ago with my last bargain.

Stella will never forgive me if her people aren't spared.

She may not forgive me anyway for this knife, and my last-minute changing of the bargain.

"I see," says the High King at last. "And in exchange for such a favor, you wish for me to delay the conquest of the human lands."

"Until the day after Lulythinar," I reply, not allowing my sinking gut to be visible on my face.

He's not going to accept the bargain.

The High King's gaze falls to Stella, to the way fear is written over every inch of her body—so unlike when she entered this throne room without flinching, without cowering.

This is why I didn't tell her every ploy I was considering for this meeting.

I needed her fear. Her true fear. I need her doubts of me written plain across her face.

Plain for the High King to see.

He sets his jaw, turns on his heel, and marches back up the dais. "I'll consider it."

He'll . . . *consider* it? But there's hardly any time for consideration; the warriors are to deploy tomorrow. I mask my confusion and the surge of desperation behind another sneering grin.

"I am honored, dear Father. See you at the banquet." Then I sheathe my knife, let go of Stella's wrists, and take one of her elbows as I turn on my heel and march toward the doors. She scrambles to keep up beside me, her breathing ragged.

The air shifts.

With a rough yank, I pull Stella to my other side just as a bolt of pure magic slices through where she'd just been. It smashes into the floor, sending blackened chunks of marble flying in every direction. Her eyes widen in my periphery, but I don't look at her. Neither do I look back at the High King, at his raised palm.

The guards open the door, and I drag Stella through and around the corner. The doors clang shut. Maybe I should be relieved. I'm not.

"Ash?" she demands.

I'm getting her out of here. Before Lulythinar. Clearly, if she has magic, it's nothing remarkable. If it was, she wouldn't have been able to hold it in after an episode like that. Earlier, for one glorious moment, I'd thought she could smell lies, which would have been an invaluable asset to her. But then it was only a perfectly timed coughing fit.

If she can't even smell lies, what is her magic?

I need to give up. All this time, I was so wrapped up in vengeance, in making my father pay for what he's done. But all this time, I didn't have anything to lose.

Not like I do now.

Now I have everything to lose.

Stella is wrong about me. I *will* break. I *am* breaking. It's enough—the death, the bloodshed, the rebellion, the games. I've had enough. I'd rather use my favor from the Neverseen King to set Crenfyre on Faerie—and just watch as that parasite sucks the life out of every breathing creature. Every green plant.

You'd let the parasite take the Small Cities?

I gnash my teeth, hating my own weakness. No, I couldn't bring myself to allow harm to the Small Cities.

Maybe I should send Stella away . . . and take my own life. I'll rip my father's throne from him that way. I'll let him watch as his empire crumbles to pieces without an heir. And then war will ensue.

Is there any place I can send Stella where she'd be safe from such a war?

My knife is lifted out of its sheath. I whirl, snatching the wrist of—

Stella.

My mind comes to a screeching halt, and I stare blankly at her, at the fire of defiance in her eyes, the set of her jaw, the enormous hilt of my knife in her tiny hand. My grip on her, stopping her.

"Snap out of it," she growls at me, her gaze narrow.

I merely stare at her, my mind still not caught up. "What?"

"Stop it, Ash. Stop thinking whatever dark and depressing thoughts you're thinking. We're both still here. All is not lost. And you have some explaining to do if you still want a shred of my trust."

Her words strike me like a blade with their sharp simplicity. The fog clears, and she stands there before me like a beacon. A bright light that refuses to go out, that refuses to be cowed.

I can't let her stay here. Faerieland is too dark for her. She deserves so much more than me, than this. So much more than daily attempts on her life. She deserves more than I can give her. More than I can dream of giving her.

I'm getting her out of here. Perhaps I'll pay the price with my life. It'll be fine, so long as she can go on living.

We're in the middle of an empty hallway beside the statue of the winged archer. No one seems to be nearby, and I sense no presence. Even so, I throw a quick spell around us so that no one will hear what I say.

I release her, and she lets me take my knife and replace it in its scabbard. Then I pause, draw a deep breath, and gently reach three fingers for her face, for the hair I want to brush behind her ear.

I love you.

She steps back abruptly, out of my reach, crossing her arms over her chest. I freeze. The words on the tip of my tongue fall back into my mouth, unspoken.

"Please don't touch me," she says.

There's no vitriol in those words, but they are firm. Unyielding. They pierce me like one of Faradir's arrows. So soft. So deadly. I drop my hand.

"You betrayed my trust."

Ah, yes. Yes, I did. Part of me immediately scrambles to apologize, to explain why I couldn't tell her everything, to describe how my plans changed when I saw Faradir's expression. But then I look at her, at her crossed arms, and I feel myself closing up. My heart strives to put distance between us. Every kiss I've given her was a mistake. Especially today's kisses. They were indulgences in emotion and affection that I have no business harboring. I knew it at the time, but I kept protecting this hope. This *stupid* hope I wouldn't have to give up everything I ever loved.

Everything I love dies. Why can't I get it into my thick skull?

If I love her, Stella will die.

She can't stay here.

Because I *do* love her. I love her, and I will do foolish things because I love her, and my heart will be ripped to shreds when the High King finally gets his hands on her.

If I truly love her, I'll let her go.

"Yes, I used you like a pawn," I say coldly, resuming my walk at a pace more manageable for her. "Because I'm trying to save that spineless father of yours."

"That is not what I am angry about, Ash, and you know it. You said you would tell me what to expect, that you wouldn't blindside me. But that's what you did in there. That's not right."

"I had to blindside you!" I exclaim in frustration. "You want me to save your people. I had to get Faradir's attention."

"Stop putting the blame for this on me! No part of the High King trying to raze my homeland is my fault. It wasn't my choice to marry you. It wasn't my choice to be brought to Faerieland. I wasn't the one who tricked the High King and incited his wrath! From the start, I have submitted to being used as a pawn, first by my father and now by you. But not anymore." Her eyes flash. "We have a deal. You promised to warn me, so I expect you to abide by that promise."

I clench my teeth, guilt filling my lungs even as I harden myself to it.

Her voice softens, turns to pleading, and that is the final blow that makes me crumble. "Ash, you're all I have. I need to be able to trust you. *Please* understand that."

It hits me then, the full force of what I've done. How right she is, no matter my excuses or my insistence that I had no other choice. Maybe it was my only option. Maybe it wasn't. But it doesn't matter. What matters is that I broke my promise to her. I hang my head as it fills with my own pitiful voice. It's such an inadequate offering. "I'm sorry, Stella."

She nods, accepting my apology. She doesn't take my arm, however, and the space between us is like a cavernous hole in my chest. I need to get used to this. I've thought of her as mine, but she's not. She may be my wife, but she's not mine to hold, to touch, to kiss, to love.

"Would you please show me Rahk's quarters?"

The abrupt question startles me back into the present, into the large brown eyes peering up at me. "Yes, of course," I say, even before her question registers in my mind. Then, when it does, I frown. "Why?"

"In case something happens to you, like when you were poisoned."

I find myself shaking my head. "Don't go to him unless you must—unless something is *very wrong* with me or Edvear. The High King may try to task Rahk with killing you, and he'd be sacrificing his own life to disobey those orders."

She purses her lips, considering. "I see. I still want to know where his chambers are."

I draw a deep breath, then veer down the corridor to our right. "It's near the banquet hall with the other guest chambers for the Nothril Court. Come this way."

When I steal a glance at her, Stella is carefully noting our surroundings, our path, and seems to pay special attention to the marble waterfall we pass. Maybe in another life, another set of circumstances, she could have survived in Faerieland.

By the time I've shown her Rahk's door and taken her back to my quarters, there's barely enough time for Edvear to serve Stella a light, un-poisoned meal, and for her to eat it before it's time to return to the banquet hall.

I steal away to my study for just a few moments and shut the door. I lean against the door, closing my eyes, and let the despair wash over me.

Give up.

Keep fighting.

Give up.

Keep fighting.

A growl rumbling in my throat, I push off the doorframe, march to the desk that Edvear had moved back here while we were gone, and pull out a fresh sheet of parchment. Rahk's compromised loyalty is going to cost me greatly.

Neverseen King, I write. *I'd like to call in my favor.*

I pause, my pen suspended in the air a few millimeters above the paper. Then, with a determined set of my jaw, I keep writing.

I want you to give my wife safe passage out of Faerieland to Orawyth.

CHAPTER 39
THE PRINCESS

ASH IS WITHDRAWN when we leave for the banquet. I ask if he has any stunts planned for tonight; he sighs and tells me no, but that doesn't mean there won't be any. It's not a lie.

I take his arm when he offers it, and he guides me to the banquet hall. I follow the path, saying the directions in my head before each turn, quizzing myself to make sure I have it memorized. To my surprise, I get it all correct.

One step toward not being lost in a palace full of people who want to kill me.

Rahk's doorway is two hallways down, three doors on the left. Shockingly close. Perhaps that was why he was there to rescue me the night Ash was poisoned. I tuck away Ash's concern about Rahk, about what the High King might make him do.

I want to reach out to Ash, to comfort him, to encourage him, but I'm so deeply entrenched in my own worry. What if the High King doesn't take the bargain? How then are we to stop tomorrow's slaughter?

All of it seems impossible. These political machinations make me want to strangle everyone involved. Right now, it feels like my only friend and ally is lost to our own hopeless predicament. And I have no strength left to pull him out.

Plus, I haven't forgiven him yet for the throne room.

The doors open, and I'm overwhelmed by a wave of exhaustion. My limbs are weaker than normal. Am I still recovering from my illness? Probably. The stress of today has been too much on my healing body.

I don't know how I'm going to get through this banquet when only a few minutes in the throne room zapped my strength. Especially because there *will* be stunts, maneuvers, games. It wouldn't be Ash and the High King if there weren't.

I'm half expecting another trick like before, but when Ash guides me into the banquet hall, it's not full of women. It's a diverse group of fae, male and female, with various complexions ranging from luminous gold to cobalt blue. Murmurs of our titles rise from the gathering, but Ash doesn't answer. I try not to stare at the radiant wings attached to a female fae—*Oleria*—sitting beside one of the two empty chairs at the head of the table.

Two empty chairs.

My sudden burst of optimism is immediately quelled. The High King hasn't arrived yet. So, of course, one chair belongs to him, and the other to Ash. I glance up at him, trying to read him for my cues.

A darkly sardonic smirk twists his face, responding to this new situation, like it is a fresh game to play. Then, just like last time, he takes me to his open seat. Unlike last time, he goes to the head seat, grabs the back of it, drags it out, and gestures at it. And looks expectantly at me.

The blood drains from my face.

The quiet murmur of conversation around the table goes quiet.

I can't—I can't sit in the High King's chair! That is just asking to be murdered. I grip my skirts in both hands and shake my head.

Ash's gaze darkens, his jaw flexing. "Princess Stella," he says, not at all kindly. "*Sit.*"

I have two choices, it seems. Neither of them good. I can set my mouth in a stubborn line and refuse, thus causing a public scene that will no doubt have the fae talking about how Ash cannot control his *pet*. Or I can do what he says, choose to trust him—that he isn't meaninglessly throwing me to my own death.

I don't understand all these moves of his, but I understand enough. This is spite. Yet again.

So much spite. So much hatred.

If Ash isn't careful, he will become just like his father. He isn't there yet, but he is on the very same road. I'm not skilled in clever subterfuge or these tricky bargains, but my intuition tells me that sitting in that chair is the wrong move. It'll enrage the High King, and I'm beginning to think that Ash doesn't always have something he's trying to accomplish by enraging him.

In this instance, we need him to accept Ash's bargain.

The High King won't accept that bargain if Ash pulls a stunt like this. He'll punish him further. He'll drag Hylath back in here, kill her off. Or maybe this time it'll be Edvear. Or he'll call Rahk in and make him do something terrible again.

My people will not be saved if I sit in that chair.

I can almost hear the echo of Ash's voice that first evening we'd arrived here, and he had pinned me to the wall and told me that if I valued my life, I'd do exactly what he said. No questions, no hesitations.

I take a deep breath, relax my grip on my skirts, and level a hard look at Ash. My answer. A firm *no*.

Ash meets my gaze . . . and grins. A wicked grin that sends ice to my toes.

I refuse to bend.

He gives a dark chuckle, slams the chair back into its spot, and drags out his own. He drops into it heavily, flings one knee over an armrest, and grabs the full goblet before him. One swirl, two. Then he sniffs and takes a drink.

All while I stand beside his chair, heart pounding. Waiting.

I'm not about to throw myself into his lap. If he intends to leave me standing here this entire meal, then so be it. I won't sit in the High King's chair.

It's still quiet around the table. Some two dozen eyes burn into me, scalding me like a brand. I don't move. I may want to crawl into a hole and die, or pick up my skirts and run from the room for all I'm worth. But I intend to keep whatever scraps of dignity I still possess after defying Ash.

He lets out a long sigh, setting down his goblet on the table. His attention flicks to me. "Go stand with the other servants along the wall, will you, my dear?"

My own sigh escapes between my teeth. *The other servants.*

Anger burns in my gut. Anger and humiliation.

Our shared kisses this afternoon feel like a lifetime ago. Another world entirely. He was a different person, the one I deeply care about. This is the mask.

It is an ugly mask indeed.

It's only now that I realize it's not all for show. This is the ruthless side of Ash, the terrifying Prince of the Fae that Amelia begged me to run away from. This is the son of my kingdom's enemy. This is a deadly, bitter, vengeful prince.

Rahk was wrong.

Ash never fully pulled out of that darkness after his mother was killed. He's in it right now. It's in the voice telling me to stand along the wall, as though I'm nothing but his pet. A dog. A slave for his pleasure.

Send for a white dress.

I incline my head toward Ash in a single nod, then turn, march toward the wall lined with humans every five paces, and situate myself as close as I dare to the door that opens in the direction of Rahk's room.

This far away from Ash, he cannot protect me like before. I need an escape route.

Now is not the time to wallow in my embarrassment as the chatter resumes around the table, as I watch Ash turn and begin conversing with the woman on his left, taking sips from his goblet and kicking the leg he has flung over the side of his chair. He lazily addresses those seated at the table.

Now is the time to watch carefully for any threats.

The doors on the opposite side of the banquet hall open. My biggest threat saunters right in. Faradir wears robes of sapphire blue, making his eyes twice as spectacular. He moves with the confident ease of a panther through a jungle. His gaze shoots straight to Ash—and he pauses suddenly. He glances around the room until he finds me. Standing against the wall in my beautiful gown between human servants. Satisfaction twists his lips.

He takes a seat, leaning on the armrests as he smiles down the table at his gathered guests. "Welcome! Prince Trenian. Princess Oleria." He continues down the line until he has addressed every guest, and they have addressed him. Then, with a sniff and not even a sideways glance: "Princess Stella." He smiles at Ash. "I'm glad to see you've put your pet where it belongs."

Ash merely inclines his goblet in reply, then returns to the conversation he's having with Oleria. He says something to her and she laughs. Her retort earns her a smirk. There's something different about her. Haughtiness doesn't cloak her every movement, and the tilt of her head, the pretty smiles, strikes me as almost . . . genuine. Unpretentious. I don't know why she helped me before, but I doubt staring at the back of her head will help me unravel the mystery of her motives.

I dare to release a sigh. I shift my weight between my legs, trying to ignore how exhausted I am, how much I long to sit down. If I lock my knees for too long, I'll pass out, and that will only draw attention back to myself.

One of my biggest assets here is being beneath notice. Truly, I should be glad I'm here and not seated at that table. I'd be much too close to the High King for comfort, and it would be much more stressful than simply standing here.

Why am I lying to myself like this?

This stings. My pride, certainly, but an ache grows in my heart. I know Ash well enough to know this probably means nothing between us—he's not *actually* trying to punish me. He's not *actually* angry with me for refusing to obey him.

But it feels like he is.

Do I even know Ash like I think I do? It's too much to worry about right now. Something changed earlier between us, and I'm not sure what it was. We're still allies, and we're still married, but can we trust each other?

I'm starting to believe we can't.

There's something deeply wounded about Ash, and if he doesn't change, if he doesn't stop making decisions out of pain and spite, I'm not safe with him. And if I'm not safe with my own husband in this foreign world, then the only prudent option is to escape.

The thought fills me with such deep sadness, I barely maintain my composure.

I'd truly believed I'd found something special with him. Something I hadn't dared dream of. Someone who truly saw me, cared for me— *loved* me. Someone who would defy heaven and hell for me.

Was I just seeing what I wanted to see?

This is getting too far out of my control, too far from my father's intentions for marrying me off, too far from protecting my people— *too far*. I need to get out of here.

I shift my weight again, loosen my shoulders as I exhale. My heart doesn't stop aching—*won't* stop aching. As expected.

They don't call it heartbreak because it feels good.

My breath cuts off suddenly. I try to inhale—can't. Panic flares, hot and heady. I can't even choke. It's like something has wrapped around my throat. Something that slowly squeezes the life out of me.

I can't make a sound, even as I fall to my knees, grabbing at my throat. My vision tunnels.

And then, abruptly, it stops. Air floods my lungs. I drag in one heaving gasp after another, clutching my chest as I lift my eyes.

The High King clenches his hand into a fist. And Ash's hand is extended, flat, toward me, blocking the High King's fist. He isn't looking at me, instead staring at his father.

"Don't you *dare* touch my wife," he snarls.

The High King merely smirks and lets his fist relax. Ash pulls back his own hand—and goes back to eating I stay where I am, palm flat on the floor, my skirts billowing around me as I breathe heavily.

Shaking, I get to my feet. Ash hasn't so much as glanced at me. Hasn't looked to see if I'm alright. This charade is so important to him he cannot break character for the span of one glance.

He did save me.

Perhaps he's more aware of me than he lets on. Or perhaps I'm just seeing what I want to see. Again.

He must protect me because if I die, his bid for the throne and his life will be over. There are many reasons to ensure I remain alive. None of them need to be about truly caring for me. I need to stop thinking that they do.

I need to talk to Rahk. He might kill me. Perhaps he's nothing but glamour and falsehoods too. Somehow, I cannot believe either of those things. Rahk has been levelheaded and straightforward every time I've spoken with him. He was the one who explained Ash's background to me when no one else had—including Ash.

His room isn't far from here.

And if Ash isn't going to ensure my people aren't slaughtered tomorrow, then maybe I can talk to Rahk and see if he has any ideas. If there's information I need to know that Ash wouldn't tell me.

Perhaps he can help me get word of the attack to my father, so an effort at preparation and evacuation can be made.

I've stayed as a decoration long enough in this hall. With one last glance at Ash, at the High King and the merriment happening at the banquet, I turn and slip out of the doors, unimpeded. They close behind me without a sound, swallowing up the noise and laughter and clinking crystal, the painted ceiling of human footstools. The guards pay me no heed as I level my shoulders and march down the hallway.

It's strange, being in these empty hallways mostly alone.

The ache in my heart eases, my mind sharpening. Keeping my eyes and ears open, I walk down the next hallway, reciting in my head the directions Ash gave me. To my relief, I encounter no one.

Night is falling, throwing shadows over the white marble of the palace. Statues are cast in sharp relief against walls and floors. All is strangely quiet. The only sounds come from the swishing of my gown and my breathing.

I approach a corner warily. With the way the light and darkness play together, my shadow is cast ahead of me. Anyone around the bend can see me coming, but I cannot see them.

Despite my erratic heartbeat, I keep my back and shoulders straight, not flinching as I turn the corner.

No one. Not even a guard standing at alert.

I breathe a quiet sigh of relief and quicken my pace. Finally, I reach Rahk's door. Glancing behind me once more to ensure I'm alone, I draw in a deep breath and lift my hand to knock.

"Are you looking for me, Princess Stella?"

The low voice nearly startles me out of my slippers. I whirl to find a tall, densely muscled Rahk leaning against the wall at the end of the corridor, apparently coming toward his room.

"Prince Rahk!" I gasp. "I was coming to find you!"

He frowns, his brow creasing. "You look troubled. What is it?"

I glance around, nerves crawling up my spine, raising gooseflesh on my arms. "Where can we talk?"

His jaw flexes. He motions for me to follow him. "This way."

I pause, glancing back at his room. "Not . . . in your chambers?"

A slight smirk. "Ash would probably kill me." He motions for me to follow him back into the hallway I just came from. Biting my lip, I hesitate, then hurry to follow him. Hadn't he told me I was welcome in his chambers? Perhaps that was before the High King took an interest in him?

"Are you hungry?" he asks suddenly as we walk, pulling something out of his tunic pocket. Without waiting for an answer, he carves a slice out of what looks like fruit and hands it to me.

I take it, blinking down at the pink flesh, the glossy red sheen of the soft skin. A trail of juice runs over my fingers, dripping onto the floor. My gut plunges.

It's faerie fruit.

"Thank you," I mumble, my hands shaking as I lift the fruit to my lips. Can I fool him by pretending to eat it? Perhaps I can delay long enough that I can . . . go back to the banquet hall? Whatever the case, I cannot let on that I know what he is doing. "Th-this is so kind of you, Rahk. Ash didn't give me anything to eat at the banquet."

He watches me through the tail of his eye as he leads me . . . somewhere. Not toward the banquet hall, and heading in the complete opposite direction from Ash's quarters.

My pulse thrums. I'm still holding this fruit. Still haven't taken a bite.

Is he trying to kill me?

Surely, if he intended to outright kill me, he would have done so already. It's not as though I have any defense against him. What did Ash say about faerie fruit, anyway? That fae gave it to humans to make sport of them.

Rahk just betrayed me.

Not sure what else to do, I let my knees buckle. Rahk spins toward me as I let the piece of fruit fall from my hands, and collapse in a pile of skirts as though I've fainted. I hit the ground hard, but not as hard as I could have without all this fabric cushioning me. I roll my eyes shut and lie limp.

My heart pounds as though I'm sprinting for my life.

This probably won't work. I didn't really think through this carefully. He's just going to plunge a dagger through my throat or lift me over his shoulder and carry me straight to the High King.

I'm not expecting my mouth to be suddenly pried open. White hot terror floods me, and I send my elbows flying toward my assailant. They connect with a neck, but the sudden burst of surprised sound isn't a masculine grunt. It's a feminine yelp.

My eyes fly open.

It's not Rahk trying to pin my flailing limbs and pry open my mouth. It's not his hand that squeezes the slice of fruit so its juices drip down my throat.

It's Princess Listhra.

I kick, rolling to spit out the juice. A shockingly strong grip yanks me back, smothering my mouth and nose with my skirt, cutting off my airways. *Forcing* me to swallow.

It's sweet. Much too sweet. So sweet my mouth puckers as Princess Listhra pulls back, letting me breathe.

Suddenly, I don't want to struggle anymore. A smile breaks across my face as golden warmth floods every limb, every follicle of hair, every inch of skin. A giggle follows the smile, and I stare up at the most beautiful woman I've ever seen.

She smiles back at me. "I have a lot of questions for you, little human."

I clap my hands, joy flooding my veins. "Please! Let me answer them!"

CHAPTER 40

THE PRINCE

SHE LEFT. SHE left. She left.

And I can't go after her. Not if this idea is going to work. Not if we're to have any hope that the High King will accept my bargain. The idea occurred to me after I sat down—a way to angle this situation.

It was a good thing Stella refused my order to sit in that chair. She would have been fine; I was prepared for the High King's retaliation. Nevertheless, it was shortsighted and, frankly, stupid.

But this could work in our favor. My apparent indifference toward her. The High King could be tempted to accept my bargain because he senses an opportunity to take Stella from me, which would remove my bargaining chip out of the equation. He wouldn't feel the need to raze the human world so quickly.

But then she *left*.

How could she be so stupid? She's going to Rahk. Why else would she have asked me for directions to his room earlier? Rahk isn't to be trusted now that the High King is using him. And it doesn't matter that his quarters are so close—it only takes a minute of her being unguarded and stumbling across the wrong person before it's too late.

I try to keep from crushing my fork in my fist. I need to maintain my composure.

It doesn't help that the High King seated Princess Oleria next to me. It's becoming apparent he reads me better than I once thought. Of all the women he tried to get me to marry over the years, she was the only one that I didn't despise.

Princess Oleria isn't a fool. I've always thought her a breath of fresh air among an army of pretentious, preening peacocks. It was a relief when the High King gave up trying to convince me to marry her. A relief, because I *could* have liked her.

He apparently hasn't given up.

Why did Stella have to leave?

I force myself *not* to shoot to my feet, *not* to chase after her. *Find her, Rahk. Keep her safe. Don't let anyone hurt her. And don't you dare hurt her, either.*

"I hear the air is different in the human lands," says Oleria beside me. "Are the rumors true?"

I take one bite, chew, swallow. It lands like a rock in my gut. "They're very true. It's much more taxing to use magic in the human realm."

"It's why we need to purge humanity from us once and for all," says the High King, taking a long, languid sip of his drink. "The ground must lie fallow for a time to get rid of that constant cycle of death."

Oleria grins prettily, leaning forward a little. "How interesting! Prince Trenian, would you care to explain something for me, then?"

I take a great draught of wine, sigh, set it down. I turn over a few more sarcastic responses, only to discard them when they're all lies. "Ask away."

"Then why have we been slowly expanding our borders?"

The High King doesn't interject, feigning interest in his food. His attention, however, burns into me. He wonders what explanation

I'll give. I give her a crooked grin, lifting one brow. "Something about how smaller amounts of death are more easily absorbed by the magic in our air than an entire continent of death. My illustrious father would know more on the subject than me."

Bring up the bargain, I want to snarl at him. *Give me your answer.*

His answer is the one thing keeping me nailed to this chair. If I get up and go after Stella, he'll answer a firm no. So I must stay, despite how sick I feel. She would want me to get this answer more than anything.

She'd die for her people.

I hate this.

Oleria says nothing for a few moments, as though waiting for the High King to offer a more detailed explanation. None is given. So she merely smiles and continues eating.

She doesn't like the High King either.

When she turns to me again, one eyebrow is arched in a mischievous line, even as she keeps a serious expression. "I haven't had time to offer my condolences for your poisoning. Or that of your wife's. You both are looking well today, however."

I wave a hand, forcing a chuckle. "I'm quite certain no condolences are necessary. It was but a minor inconvenience."

A minor inconvenience when *I* was poisoned.

Devastation when it seemed Stella wasn't going to make it.

I barely keep my fingers from twitching in the barest tremble. She could be dead right now. Stupid, *stupid* girl.

Not that I can blame her. I try to forget the pain that flashed across her face when I told her to stand by the servants. It's impossible.

She left because of what I did. What I said.

Has she realized yet that deep down, I'm nothing but a monster? A monster who will destroy her once I'm through destroying everything else? I'm no prince. No kind husband. I don't blame her for leaving, and yet I desperately wish she hadn't. I wish I could have her trust, despite not earning it.

"Prince Trenian," says the High King.

My eyes flick up to his. Just in that moment, Oleria's hand brushes mine under the table. I keep myself rigid, not responding to her touch, not giving any outward indication of it as I await the High King's words.

"I've considered your bargain."

Oleria's fingers slide into mine. Warm, soft, much larger than Stella's. They thrum with that electrifying pulse of magic. But there's something *in* her hand. Something that she presses into my palm. I close my fingers around hers, and if any of the human servants see and word gets back to the High King, this will seem nothing but a clandestine moment. Rumors could start that Oleria and I have secretly loved each other for centuries, but because of the High King, we dared not be together. The human bride is nothing but a pawn in my games. A mere tool in this larger dance of love and immortality.

I accept the tiny, folded piece of paper and slip it into my sleeve. All while not taking my eyes off the High King. He doesn't look at me, occupying himself with the golden swirl in his goblet.

"You redeeming your favor from the Neverseen King in the manner that you described—letting Crenfyre loose on the human continent—in exchange for my delay of the conquest until after Lulythinar."

He stops, trailing off into silence as he takes several more bites of food. Purposefully torturing me. I lean further back in my chair, grabbing my goblet and taking a few stupidly small sips.

He sets down his cutlery. Plants both his hands on his armrests. And smiles at me.

"I've decided to accept your bargain."

Accept? There's a catch. He wouldn't accept this, despite how it benefits him. Unless he feels threatened that I might use the favor in other ways.

My fingertips buzz as I set down my goblet. "Wonderful. I shall begin discussions with the Neverseen King."

And with that, I shove back my chair, swing my knee off the armrest, get to my feet, and march out of the banquet hall. The moment the doors swing shut behind me, I break into a run.

Where are you, Stella? Where did you go? Please don't be dead. Please don't be dead.

Please.

I catch a whiff of her scent immediately and follow the trail to Rahk's room. My heart lurches with hope. I pound on the door with the side of my fist, not even bothering with our code.

The door swings open. It's not one of Rahk's attendants, but Rahk himself. Shirtless, with a dark glare.

"Where is Stella?" I demand, shoving him aside, searching for any sign of her. "Stella!"

"She's not here," Rahk replies, already in motion. He scoops up two scabbards—his swords—and buckles them over his back. Glamoured boots and a shirt appear before he hurries after me outside the room. He sniffs in the doorway, his eyes darkening. "She *was* here. So was someone else."

We share a look of pure horror.

Then Rahk takes the lead, catching her scent faster than I can. My lungs clench with dread. *So much dread.*

We run down the hallway, turn left—the opposite direction from anywhere she should have been going—and barrel down the next hallway.

"The scent is still fresh," Rahk growls, likely to ease my anxiety. Nothing will ease my anxiety, however. I'm going to throw up.

"Can you tell who she was with?" I demand.

Rahk's jaw flexes. My gut drops.

"Princess Listhra."

THE PRINCESS

"SO TELL ME," says the beautiful princess, "why did Prince Trenian marry you?"

I sit where she tells me to on the settee in a cozy little room. A few other fae ladies are here too, and each one is more beautiful and kinder than the one before. I love them all! It's so special being able to sit with them and talk to them!

They want to hear what I have to say. It's so *very* special. I giggle, flush, press a hand to my mouth. "He wanted to marry me to trick the High King."

A few of the women exchange looks, but the princess's golden eyes don't move from me. It's almost overwhelming, all this attention. I blush a little more.

"Yes, because of the bargain. But why *you*? What are his feelings toward you, dear?"

Why me? I rack my brain, not wanting to think too long—I don't want to disappoint the beautiful princess—but it's a confusing question. Do I even know the answer? I scrunch my face up in concentration. "I think he married me because I was the closest human princess who wasn't betrothed to anyone. I was *almost* given in marriage to Prince Brochfael as his sixth wife, and he's known to beat them. That terrified me."

"Yes, yes, I'm sure. But does he love you? Does he care about you?"

This room is *so* beautiful. It's all decorated in a rich, deep blue, with silvery curtains and a polished agate floor threaded with gold. Each object my eyes land on is like a whole new world of beauty, and soon I find myself staring in utter enchantment at a miniature glass-blown peacock on the table beside me. Its tail feathers are even the right colors! It must have been so *very* difficult to—

"Princess Stella."

"Oh, yes?" I sit up straighter, determined to stay focused and *not* stare at the beautiful peacock. Or how the curtains waft as though they're made of gossamer silk. Or how—

"Does he care about you?"

"Oh no, Prince Brochfael just wanted another wife to add to his—"

"Prince Trenian, you imbecile," the beautiful princess snaps. "Does Prince Trenian love you? What does he think of you, feel for you?"

I shrink a little lower at her tone. I don't want her to be angry, especially not at me. And how am I to answer her question? It's much too confusing. Behind the confusion is a sad, pricking pain. A growing ache. I forget about the peacock. "He broke my heart," I say softly. "He pretends to love me, but I don't think he does. I don't think he can. He's too sad after his mother was killed. Do you know that the High King killed his mother?"

She ignores my question, waving it away with bright eyes that focus like twin beams of starlight on me. "He doesn't love you? Does he just use you?"

I nod, licking my lips. No matter how bright and happy I feel, that ache just keeps growing.

One of the other women leans forward. Her hair reaches almost to the floor, thick and silky and so black it's almost blue. She's so beautiful, I cannot help but stare and smile at her. Her voice is so melodic, like a sweet song that matches the tranquil ocean of her eyes. I could listen to it forever.

Then her words register in my mind.

"How many times has Prince Trenian slept with you?"

Finally! A question I can easily answer. "Twice," I reply promptly. Then I frown, trying to recollect my sickness. "Or maybe three? It probably depends on how you count."

The women exchange looks, as if this is unexpected. One has a decidedly dubious expression written across her face. Princess Listhra looks at me, her brow narrowing. It's much too shrewd of a look for my comfort, and it makes her less beautiful. It frightens me.

But I don't have anything to fear. Not from someone so good and kind as her!

"Has Prince Trenian consummated your . . . *marriage*?" the princess asks.

That's the easiest question of them all. It gives me a burst of happiness to answer. "Nope!"

The women glance at each other again. Grins spread across faces, and it makes me so very happy to see them happy! I grin back, barely restraining a giggle. Princess Listhra's eyes glitter so prettily.

"He kisses me a lot," I say, and the giggle finally slips free. "A *lot*. I love his kisses."

Those grins vanish in a flash. I gasp, clapping a hand over my mouth. What did I say to make them so upset? I need to fix this.

"Well," I quickly amend, "he didn't kiss my lips until today. Today was the first time. The rest of the time, he just kissed me elsewhere."

"Elsewhere?" says one of the women, flashing another concerned and dubious expression.

I nod. "My forehead usually. Or my shoulder. Cheek. And so forth."

Princess Listhra's brow puckers slightly, but one of the women chuckles. I giggle back, delighted at her reaction.

"Do you have any other questions?" I ask eagerly. "I would *love* to answer them!"

"Should we just kill her then?" asks the lady at the far end. "Now that we know she's truly just a pawn, no matter what Trenian wants us to believe."

"Kill whom?" I ask, leaning forward. Maybe if they like me enough, they'll invite me into their secret. Wouldn't that be wonderful? I could belong in a group for once!

"It's too early," says Princess Listhra. "There's too much time before Lulythinar to kill her now. He'll just go get another one. We must wait until it's closer to the deadline, and he doesn't have time."

"How do we know we'll even get another opportunity? She's been here for days, and no one has been able to touch her until now. Trenian guards her too carefully."

"We'll just have to be on high alert. I'm telling you, if we kill her now, he'll go get another human wife, and he won't come back until five minutes before the deadline. It'll be impossible to kill his second wife then. We can't be sloppy about this. No mistakes. Whoever makes mistakes will *not* be who the High King chooses for Trenian's wife and the future Queen of Valehaven."

"So, what do we do with her? Just drop her back off where we found her?"

A glint enters Princess Listhra's irises as she regards me. My heart swells at her attention. She's so beautiful. I can only dream of being so glorious as her. I shouldn't be sitting near her—I should be on my knees, bowing down to her.

"We can have a little fun with her while we wait for Trenian to come searching for his pet."

CHAPTER 41

THE PRINCE

RAHK COMES TO a stop before a door. The private chambers of Princess Listhra. He holds up his hand, halting me from barreling into the chamber. He walks up to the door, silent as a forest dryad, and leans his ear against it. His gaze shoots to me, and he nods once. Five paces, and he's back at my side.

"She's in there. Alive," he says under his breath. "I smell faerie fruit."

I don't stop to hear anything else, rushing forward and barely remembering to wipe the panic from my face before I throw open the door.

There's Stella. Not a dozen paces from me. Wearing a brilliant grin . . . and not much else. It's something of a long tunic that reaches mid-thigh, but instead of fabric, it's made of leaves poorly woven together. As though a child was practicing—and mostly failing—to spin ornala petals into cloth. It's probably glamoured, so she thinks she wears a beautiful gown. She dances, her feet bare, on a pile of glass shards, utterly oblivious to her own blood staining the floor.

Beyond her are four sets of wicked smiles. They all chime, "Prince Trenian."

My boiling rage slows to a lethal simmer.

"Ash!" Stella gasps. "It's you! You're so handsome! Come dance with me!" Her ankle rolls, but she doesn't stop moving, twirling, dancing. She won't stop until I pull her out of that cursed dance. Or she dies of exhaustion.

I cross the distance between us, scoop her up into my arms. She breathes hard, blood dripping from her feet. I tighten my grip on her bare knees.

"Princess Listhra. Lady Iluna. Lady Shalar. Princess Brolnyr." Then I smile. My coldest, deadliest smile. "Don't tempt me toward vengeance, ladies. You don't want me as your enemy."

Three of them turn pale at the threat. The fourth—Listhra—only smiles back. "If I cannot have your love, I'll settle for your hatred."

Well, she has it.

Cradling my wife to my chest, I spin on my heel and storm out of the room. She hasn't stopped gasping, as though she's been dancing for some time now. I grind my teeth, wanting some outlet for my fury. If I hadn't left, if I had stayed for even one more minute, I would have slaughtered all those women right there.

Rahk slips to my side but doesn't glance a single time toward Stella. I'd glamour her if that wouldn't take time. But I need to get her back to my quarters *now*.

Before something worse happens to her.

I could have lost her.

She could be dead right now. A couple of bloody feet, exhaustion, and humiliation are so much better than I could have hoped for. I'll still never forgive myself for letting this happen.

When we reach my quarters, Rahk steps back. "It's best if I keep my distance. Where I can."

I give a tight nod. "Thank you, brother."

He bows and leaves, the sound of his footsteps vanishing as Edvear opens the door. His eyes drop to Stella in my arms. They widen. He begins barking orders, and servants spring into a flurry of activity.

I take Stella to the dining room table and set her atop it. She clings to my tunic, mewling a tiny protest, and I grit my teeth against the saucer-like eyes she turns up at me.

"I need to get the glass out of your feet," I growl, gently prying her hands off me. "You're bleeding."

She tugs at her hand, and I release it. Slowly, she lifts it up toward my face. Touches my cheek. I freeze—and simultaneously a sliver of my rage thaws. Just enough to make me hold still while she traces her finger along the line of my jaw. My chest tightens. She lifts her finger, her eyes staring straight at my mouth as she brings one finger to my bottom lip.

I take her wrist, pull her hand away as something inside me cracks and a burn overtakes my cheeks. "Sit still. Your feet need attention."

She stares up at me with those huge eyes that will forever and ever be my undoing. Clad in that scandalous garment of leaves. And she's oblivious to it. Moving on instinct, I pull my cloak off my shoulders and wrap it around her. It's not much, but it's more. When I clasp it at her throat, her voice startles me.

"Ash?" she whispers.

"Stella?"

"You broke my heart."

Every muscle in my body stops. It's said so softly, so gently, and yet it hits me harder than one of Rahk's blows.

"I think I'm going to run away," she continues, almost nonchalantly, as if her words aren't slicing into my chest, carving up my heart, and serving it on a platter. She turns a bright smile up at me. "I think you will be happier without me."

Edvear appears just then with a bowl of warm water, another empty bowl, bandages, and a pile of cloths. He glances between

Stella and me with thinly veiled shock. Then he sets the supplies down and leaves.

I don't stop him.

Not knowing what else to do—what in the world can I say to that?—I pull up a chair, place the bowl on it, and carefully set one bloody foot in the water. No longer under Listhra's compulsion, she hisses in pain, but doesn't otherwise react as the water turns bright red.

"I want to be strong enough to live in Faerieland, but I think I might just be too human," continues Stella, her voice oscillating between sad resignation and bright optimism. "I want you to be happy, Ash. More than anything. I think you will be happier with a wife you don't have to worry about so much."

False.

"A fae wife would help you overthrow your father more than I can."

False.

"You've been kind to me, Ash, but I think we're not suited."

"It doesn't matter if we're not suited," I growl, unable to keep silent. "We're *married*. We stay together. We work through things. *Together.*"

Why am I saying this? I'm the one planning to send her away the first chance I get.

She gives a bright little laugh, which turns into another hiss when I set to work pulling out each tiny piece of glass from her tiny foot. They plink when they land in the empty bowl.

"I know." She reaches out then, smiling at me as she threads her fingers into my hair. I try to ignore her touch as I work. It proves impossible. "But we're not even of the same people. You're fae. I'm human. Even if we stayed together, even if I wasn't killed, I cannot give you an heir, right? How would you get an heir, Ash? I like you too much to not be jealous if you take a mistress."

I squeeze my eyes shut, my lungs so tight I can hardly breathe. "I'm *not* taking a mistress."

There is no one, Stella. No one but you. There was no one before you, and there will be no one after you.

"Then there goes your line of succession!" she chirps, laughing. "It's a little silly, I think, to fight so hard to overthrow your father without sending everything into war, only to do so a generation later." She bends down toward me, toward where I kneel, and I look up just in time for her to bop me on the nose. "You need an heir. And it cannot come from me. Which means—"

I snatch her wrist, glaring up at her intoxicated face. How is she so reasonable, even when she's under the influence of faerie fruit? It's maddening. "You might be able to give me an heir. I'm not aware of any curse that would prevent a human-fae heir from sitting on the throne."

She stares at me, blinks, and then flushes bright pink. "Oh."

I let go and bow my own hot face back over her foot.

She chuckles again, but this time it's more forced. "And your people would accept such a ruler?"

"They don't have a choice. Whoever sits on that throne is their ruler, whether they like it or not." That doesn't mean they'd be *happy* about it. I wipe down her foot, and when she flinches, I search for the piece I missed. *Plink.* I set her other foot in the bowl while I wrap the first in a towel. It's a very good thing I have something to do with my hands while my mind spins and my heart breaks.

She strokes her fingers through my hair, and for one stolen moment after I finish picking the glass out of her other foot, I pause what I'm doing, close my eyes, and revel in how good it feels.

"I was angry with you at dinner," she says, her tone still light. "Sometimes you scare me, Ash. I care about you very much. But if I'm not safe with you, I'm not safe with anyone in Faerieland."

I groan, leaning my head against her knee, my palm cupped around her ankle, my thumb grazing her calf. "*Stella.*"

"You sound very sad," she says, combing my hair and tilting my head back so she can look down at me and offer me a smile in an attempt to cheer me up. She's so *very* intoxicated. "I want you to be happy. I don't want you to be so broken after your mother's death."

I go rigid.

She either doesn't notice or doesn't care as she keeps threading my hair through her fingers. "I think that's what I'm afraid of. That you'll cherish your hatred of your father more than your love for me."

I've stopped moving entirely. Stopped breathing. I stare at her free hand, the one resting on her leg.

"That's why my heart is broken," she whispers. "And that's why I'm going to run away."

My hands tremble. All of me is trembling. I want nothing but to pull her into my arms, to bare my heart to her, to tell her that she is everything—*everything*—to me. To beg her to stay with me.

Instead, I dry my hands, get to my feet, and stare down at her. Anger burns in my gut. Anger and fear—so, *so* much fear. I plant my hands on either side of her, bringing my face to her upturned one until we breathe the same air. Her eyes flutter closed.

"Kiss me, Ash," she says softly.

"Not tonight," I whisper, despite how I long to. "Not while you're intoxicated. I won't leave you with regrets."

She tilts her mouth up toward mine, nuzzling her nose against my cheek. "I could never regret kissing my husband."

I let out a low, breathy chuckle. A strangely warm sound when my body is ice, and everything inside me is breaking. "Says the woman who claims she's going to run away from me."

"This might be our last chance. We shouldn't waste it."

I stare down into her face, her closed lids, the soft curve of her cheeks, the fan of her lashes, the elegant line of her brow. Am I even strong enough to give her up so I don't lose her? Lifting one hand, I tangle it into her hair, pulling her toward me until our foreheads touch, our mouths only a fraction of an inch apart.

"It's not hatred of the High King that is stronger than my love for you, Stella," I whisper against her lips. "It's my fear of losing you."

"If you were afraid of losing me, you'd kiss me."

"You're intoxicated."

She peels back just enough to glare at me with those doe-like eyes. "That's not *my* fault. It's not *my* fault that you're handsome, and that I like you even though you were cruel to me tonight. It's not *my* fault that we're married and married people are supposed to kiss. So why do you punish me as though it's my fault?"

Perhaps I am more than a little intoxicated myself, because I cannot find the flaw in this logic. Not when she's so close, her silky hair in my fingers, her willing mouth hovering just below mine.

"If I kiss you," I say, breathless, hating myself for my own weakness, "then you cannot be angry with me tomorrow."

She smiles, and my gaze snags on her curving lips as she twines her arms around my neck. My heart rages in my chest, thundering like a wild beast. I can't keep balancing on this precarious edge—I need to fall. Either into a stolen kiss, or into the distance I need to create.

"Deal," she whispers.

My control snaps.

I claim her mouth, dragging her against me as our lips move together. My fist tightens around her hair, our souls twining with each kiss. I can't get close enough to her. Hooking my hand behind her knee and pulling her until she's flush against me still isn't enough. It's just as well, isn't it? She'll never belong to me. Though she's my wife, we cannot have a future together. Even these moments are stolen, bargained for with the entire human world and the life of my only love. She wouldn't even be kissing me if not for the fruit.

I'm the one who starts weeping first, trails of salty tears landing on her cheeks. I can't stop kissing her. The world can burn, burn, burn, and I'll stay here, Stella's lips the only song I want to sing. I want her—nothing but her.

Have I wanted too much by wanting her? Have I taken too much by kissing her now? It doesn't matter; nothing can tear me away now that I've started. Now that she clings to me, now that her own tears mingle with mine. Neither of us letting go, both of us knowing we should. Why do we fight this ill-fated battle?

I break away, gasping. "We should stop."

She turns tear-streaked eyes up to me, and they burn with ferocity. "You can't make me."

And then she has me by the neck again, dragging me down into another kiss. My gasp shifts into a groan, and I pull back on her hair, exposing her neck to my lips.

She's going to hate me tomorrow.

I let out another groan, a very different groan. I stop, closing my eyes, breathing in the smell of her. The sound of her quick inhales, the thump of her heartbeat against my mouth. With a sigh, I press one last kiss just beneath her ear, and pull back. She tries to reach for me, but I step several paces away. I run a hand through my hair, breathing heavily.

"I'm sorry," I say, and my voice sounds ragged even to my own ears. "I can't keep kissing you. I won't take advantage of you."

Her eyes well up with fresh tears, her hair tangled as she pulls my cloak around herself. "I don't want to say goodbye to you, Ash."

I can't keep listening to her, looking at her. It's killing me. Stumbling back a few steps, I swipe angrily at a fresh stream of tears down my face. I don't know what she means by this running away business, but it doesn't matter. I'm going to get her out of here, anyway.

My hope, my little butterfly.

I'll make myself let her fly away.

I turn my back on her and march to my study, nearly tripping over an end table because my vision is so blurry. When I reach the doorway, I clear my throat and call out, "Edvear! Attend to Lady Stella, please."

With that, I slam the door.

Collapse against it. Slide until I hit the ground.

Bury my face in my hands.

I throw a quick spell around me. And then I let the sobs pour free. They're ugly, loud. Broken.

It must be hours later when they finally subside. I rake a hand through my hair, thinking of nothing but the feel of air filling my lungs. The crust of salt on my cheeks.

That's when I remember Oleria's note.

Slipping it out of my sleeve, I unfold the tiny piece of paper until it's half the size of my hand and the writing revealed.

If your wife orders a white dress from the tailor, she's asking for help escaping. There's an underground network, run by one called the Ivy Mask, who smuggles humans out of Faerie. Let her escape if she tries. It's her only chance at survival. If you need a royal wife by Lulythinar, you can have me. We can annul when it's safe, and you will be under no vow to sire an heir. Listhra is trying to help the High King so he'll choose her for your wife. And if you're suspicious of my motives, know that I want nothing from you except relief from the High King's reign. I am sick of the fear he wields like a scepter.

I fold the note back up and toss it into the fire. It turns to ash. Then I stare at the lumiral globe on my desk.

She risked her life to pen those words. I'm inclined to trust her honesty. Perhaps this is my only option to not lose the throne and Stella's life. It doesn't mean I would *possess* either, but neither would be gone forever.

What this solution *doesn't* fix is how to get the High King off the throne before Lulythinar is over. Because if I don't solve that problem, then I will be bound to decimate the human lands and all but exterminate their race.

Four days.

Everything is on the line now.

But just maybe, *just maybe,* I can remove Stella's life from that line.

CHAPTER 42
THE PRINCESS

I WAKE UP bleary-eyed and heavy in my own room, sunlight streaming from the windows onto my face. Despite how many blinks it takes to wipe the blur from my vision, I feel very clear-headed.

The faerie poison isn't controlling me anymore.

I sit up. I'm no longer in that sack of leaves, but a soft nightgown. Courtesy of Dottie, who helped me to bed last night. The fae women told me it was a beautiful gown, but no matter how I looked at it, it was a sack of leaves, exposing most of my legs and arms, and plenty of the rest of me where the stitching failed. Perhaps some of the tiny residual bits of magic I apparently have allowed me to see the garment for what it was, and yet in that moment, I couldn't care. Even though a tiny voice in the back of my mind railed at me for being so indecent, it was like that part was locked in a glass vault.

I almost wish I didn't remember what happened. That I didn't remember the fool I was with them, with Ash. The way I begged him to kiss me. The way he caved, kissing me with such tormented passion, I was left breathless.

I'm glad he pulled away when he did. I never would have kissed him if I hadn't been so intoxicated. I would have been reeling still from the banquet, the throne room. The realization that Ash was only a few steps away from being no better than the High King.

I told him I planned to run away. I told him practically everything.

My feet ache from dozens of tiny cuts. When I try to stand, they hurt even more. I pull them back under the covers. Just for a moment. I lean back against the bedframe, staring out the windows at the fairy garden beyond. So beautiful, so magical. My little herbs on the sill are spilling out of their containers, growing faster than I ever believed possible. I'll have to repot them soon.

I don't want to give up on Faerieland, on Ash.

I *want* to do everything in my power to make this work.

But I just cannot shake the feeling that this is so much bigger than I can understand, that the prudent thing is to slip away before it's too late. I have no place in this game of immortals—except as a pawn. A piece on the gameboard for him to maneuver. It's time I recognized that.

I might be a favored piece. Even a beloved one. But I'm still a *piece*.

That's when the memory hits me. Ash's bargain. My people. The invasion. *Amelia.*

Oh, heavens have mercy.

I throw off the blankets, ignore the pain in my feet as I shove open the door, race through the bedroom, and into the hallway. I glance around, panting, then make a beeline for the closed door of Ash's study.

Grabbing the handle, I twist and shove. The door swings wide.

Ash is draped half across his desk, his head barely propped up by his hand. He wears a snug pair of trousers and a white shirt that is rolled up to his elbows and gaping wide at the throat, revealing quite a bit of muscular chest. Stubble lines his jaw, his hair a disheveled disaster, and when he lifts his eyes to me, they're sunken in shadows.

He's a mix of disastrous fae beauty and sleep deprivation.

Was he telling the truth when he said fae don't need as much sleep as humans, or does he just run himself ragged?

A mirthless smile tilts his lips. "You're a sight for sore eyes, sweet darling. Come to berate me for my conduct last night?"

Heady memories of his mouth on mine, or my neck, his hands in my hair flood me. My stomach flips, my face going hot. I shove it all away. "My people!" I burst, wringing my hands. "Aursailles! We have to—"

"The High King accepted my bargain."

I stop. "What?"

"He accepted my bargain to delay the conquest."

His voice is dull, his shoulders slumped, his eyes empty of their usual fire.

"You don't look happy."

He lifts one sardonic brow. "Should I be?"

He's slipping into that darkness again. If he falls too far, I won't be able to reach him. I lower my voice. "What aren't you telling me, Ash?"

A dark grin spreads across his face and he plants his hand flat on his desk, lifting the rest of his body with some effort, until he's leaning back in his chair. Staring at me. I cross my arms over my chest, wishing I'd thought to grab a robe before barreling in here in nothing but a nightgown.

"Where should I begin?" he asks, trailing his ringed finger along the smooth edge of the knife resting on his desk. "I'm considering remarrying."

My stomach bottoms out. Even so, I don't move. I don't speak. Later, I'll let the hurt and betrayal wash over me. For now, I keep my spine straight.

"If I can't get the High King off the throne by Lulythinar, I am bound by an oath stronger than death to destroy the human continent."

That, at least, I knew. I knew that was the risk we took fighting for a delay.

He pushes up on his desk, unfolding to his full height. He regards me with that cold smile. The one that he wears before his father. The one that isn't him at all.

Unless . . . maybe it is more him than his sweetness ever was.

He takes three steps toward me. Slow, prowling footsteps. I don't flinch, even as warning bells go off in my mind. *This isn't him; this isn't him,* my mind keeps begging. I shut those protests down.

This *is* him. I need to stop hiding from the truth.

"I know the tailor told you to send for a white dress if you need to escape," he growls. Two more steps, and he's towering over me. Then, he lifts one hand over my shoulder, plants it on the open door behind me, and shuts it with a slam. Pinning me in the space between his arms.

I tilt my head back, meeting the force of his gaze with my own. I'm desperately aware of my own smallness, of his power, his sheer magnitude. He may be so much larger than me, so much stronger. His shoulders alone may dwarf me.

But I *will not* cower.

I glare up at him, at his beautiful, twisted face leaning over me.

"It would simplify things for me if you sent that request," he says, lifting his other hand toward my face, dragging the tips of his fingers in a scalding line from my temple to the hollow of my throat. "Then I wouldn't have to arrange smuggling you out of here." He twists a lock of my hair between his fingers, his eyes never leaving mine. "I have dreams about you, you know. Ever since I married you, when I sleep, you're there. Like a light at the end of a tunnel. Your smile, your eyes. In the very best dreams, you laugh." He leans closer, ducking his head and nuzzling his nose against my ear. "I love your laugh. It's the sweetest, most beautiful sound."

My back hits the door. I didn't mean to retreat, but here I am, fighting to keep my breathing steady. "You shouldn't say these things."

"Why not?" His breath ghosts my skin. "You're my wife."

"You just said you're going to remarry," I gasp.

"I said I was *considering* it. As an alternative to keeping you here, where only death awaits you."

"Exactly," I breathe when his lips trail along my jaw. "You shouldn't be saying these things if you're going to send me away and marry someone else."

"It's not safe for me to have you, but that's not the same as me not wanting you." He wraps a hand around my waist and pulls me against him. I can hardly breathe when his low voice murmurs, "And Great Kings, Stella, I *want* you."

I close my eyes against the pressure building in my chest. "You're just determined to make this as difficult as possible, aren't you?"

"Why do something the simple way when you can make it twice as complicated and ten times more miserable?" he murmurs against my neck.

And it's *those* words, that scrap of dry humor, that give me sudden hope. The Ash I know is still here. He's buried beneath so much darkness—he's lost in that cave. But he's not gone. He's still with me.

Last night's words ring in my head. *It's not my hatred of the High King that is greater than my love for you. It's my fear of losing you.*

I take hold of his face in both my hands and push him back. Just enough to look him in the eye. "Ash, you need to take back your fear. You need to take your life back into your own hands. Your father is the cruelest man I've ever met, but you get to decide whether you want to be his victim or not."

He stares at me, his eyes widening a fraction. The wintry smile vanishes.

"You're afraid of him," I continue, letting the force of my own will burn through my words. "And your fear gives him power over you. All these games you play with him are your attempts to prove you're not afraid, but none of them get to the root of it! None of them makes you truly overcome your fear of him. You can kill him, and you'll find it still doesn't cure you of your fear. Because you must *face* it to overcome it."

His hand on the door clenches into a fist, his jaw hardening. A storm brews behind his eyes, ready to blast through his barricades and decimate everything in its path. And the strange thing is . . . I'm not afraid of it.

He shoves off the door, putting space between us as he rakes a hand through his hair, breathing hard. His shoulders quake, his free hand trembling in midair. When he speaks, his voice is like the night itself.

"He killed my mother."

I stay where I am at the door, listening. Waiting.

He drags his hand through his hair again, catching the ends of it in a clenched fist. "He *killed* my *mother!*"

Tears spring to my eyes, but I don't move.

"She loved me, Stella." He bares his teeth at the bookshelf, shoulders bowing, refusing to look at me. Tears slide down his cheeks and drip off his nose. "She *loved* me. And he took her away from me. For a *stupid* reason. Because he sent me this silly mask he wanted me to wear for a ball, and I didn't want to wear it. I wanted to wear my own mask, the one I liked. It was the stupidest thing, Stella. I was a willful idiot, but I didn't think he'd . . . If I'd just worn the Kings-cursed mask, she would have lived. She wouldn't have died. She could be here now. I could have introduced her to you. She would have loved you. *Oh,* how she would have loved you!"

I bow my head and stick my knuckle between my teeth. Tears stream down my cheeks as I bite. Hard.

"He had her killed right in front of me. She was dressed in red, with a ruby-studded mask. All ready for the ball. Father and I had an argument about my mask—we argued about *everything*. And then, before I could react, he'd drawn his sword and stabbed her in the chest." His words break off into sobs, and he bows himself over his knees. It's a strange thing to see one so mighty reduced to such soul-deep heartbreak. The next words are spat out with disgust. "I wailed like a baby."

He shoots up then, and begins prowling around the enclosed space as if it's a cage, and him, a tiger. "He killed her. He killed her! Even though she'd done nothing wrong. From the moment she was given in marriage to him, she did everything to please him. He should have loved her. He should have cared for her. He should have cared for *me*. I'm his son! His very flesh and blood! But he loved neither of us, and when he saw how much I loved my mother, he destroyed her."

His words strike a chord in me. And I *understand*. I understand what it is to be betrayed by the father who should have been the protector. Who should have had the fiercest commitment to love. I cannot fix his pain or take it away.

I can weep, however. I can feel his pain, join him in it, bear it with him.

"Stella." The name is ragged, a shredded piece of despair as he lifts destitute eyes to me. "In those dreams that I have of you, when you smile and laugh . . . They always end with *you* dying. It's you in that red dress and red mask. It's your blood that is everywhere. I see it over and over again—every time I close my eyes. I'm helpless to stop it. Do you have any idea how many times I've tried to think of a way to spare you? From our first conversation, I've hated myself for dragging you into this. I thought losing you would be like losing the rest of my staff. It would hurt, but I'd pretend to not care. I never thought that marrying you would make me see what life could be like if I wasn't caught in this trap. This *nightmare*. I was content before. Content with my games and my revenge." His voice breaks again, and then he throws his arms to either side, and the words are ripped from him in a terrifying shout.

"I don't *want* my revenge anymore, Stella! I just want *you*."

CHAPTER 43
THE PRINCE

I STOP.

My words ring in the air between us. I stare at Stella, at the way shock has sliced through the compassion and anguish on her face. Her flushed cheeks are wet with tears, tendrils of her beautiful hair framing her face and falling over the front of her rose-pink nightgown.

I don't want my revenge anymore. I just want you.

That declaration levels the mountains of heartbreak in one thunderclap.

My vision fills with her. With Stella, Princess Isabelle Louise of Aursailles, my human wife. My mind goes back to the moment I entered our wedding chamber and removed her veil, when I saw her face and those soft doe eyes for the first time. I'd tried to suppress it then, but I'd known from that moment when our gazes met that everything was about to change. I knew she would ruin me, that she would be my destruction. My greatest weakness. This whole time, I've been fighting for shreds of control—because since my mother, I've not had anything to lose.

All these years, everything has come back to loss. Pain of past loss, fear of future loss.

What if life isn't about fighting loss at every turn, but embracing love whenever you are privileged to encounter it?

I love Stella. Each time I look at her, she grows in beauty. With every word out of her mouth, my estimation of her virtue only increases. She's lovelier in this simple nightgown than she was in last night's extravagant ballgown.

Now that I've said the words, it's almost ridiculous how hard this realization hits me—how ridiculous that it is hitting me only now, when I have known this truth our entire marriage.

There is *nothing* I want in this world but her.

Least of all my revenge.

At the same time . . .

My brow lowers. "You mean more to me than making the High King pay. You've *always* meant more to me. But as long as he lives, you're in danger."

It's as though saying those words brings my entire conflicted world back into its proper orbit. My shoulders sag, the tension flowing out of my body. Light and lightness fill me to the brim. I run both hands through my hair, and before I can stop it, a *laugh* escapes me.

The crushing burden is gone.

I am taller. Weightless.

I turn fully toward Stella, who watches me carefully, her lips parted. An unstoppable grin bursts across my face. Her eyes widen, and color floods her cheeks as she swallows. Her tears have dried. My arms ache for her.

"I have to kill him," I breathe, still grinning like a fool.

Stella blinks thrice at me.

"If I kill him after we get him off the throne," I continue, realizing I haven't explained myself and sudden change of demeanor to her, "you can live. With me. I won't have to give you up. I can put our child on the throne."

Her brows come together in a skeptical knot. It's so adorable, I cannot help my laughter. Then I can bear the distance between us no longer and rush across the room, grabbing her by the shoulders so suddenly her mouth opens.

"You don't understand, do you?" I say, beaming down at her. "Everything is the same. But everything is *so different*. What I'm trying to accomplish is the same. The High King needs to be off that throne. Once and for all. But I'm not doing it to avenge my mother. It's what you said yesterday—you were wrong, but you were so right! You said the difference between me and my father is that he was fighting *against* and I was fighting *for*. Except I haven't been. I've been fighting against him as hard as he's been fighting against me. But I understand now."

That wasn't very coherent. She's still staring at me as though I've sprouted a few new heads. I suck on my teeth, trying to find the right words. "What I'm trying to say is that I am going to fight for you, Stella. I will fight for you with everything I have. I will play dirty if I must. Faradir has reigned long enough. As long as he lives, he is a threat to you. So I will kill him."

"You cannot kill him, Ash," she says, breaking her silence. "You'll lose your throne if you do."

"True." I breathe hard, staring at her puffy pink cheeks. I stumble back to my desk and crash into my chair, burying my face in my arms. It's so sudden, this overwhelming relief—even in the face of Stella's observation. I don't even know what to do with it all! How to live. How to walk. How to *feel*.

I look up, and there she is, still standing by the door. My whole life, I've had to be strong for myself. No one else did it for me. It's why I did all those stupid things as a child. Drinking poison and hunting for tarlith cats and so forth. I was trying to prove I was strong. Strong enough to be a worthy successor to my father.

Strong enough for my father to love me.

Then this sweet, soft woman walks into my life, and shows me what true strength is. She stands there, unshaken, unfaltering, though she looks as though she could break at the least provocation. She *doesn't* break.

She's stronger than I ever was.

I feel as though I've spent my entire life on a quest for the wrong thing, but now I know what I should be searching for—only to discover that I've had it all along.

It doesn't matter if Stella has magic or not. It never did. She doesn't need it. She's strong enough, just as herself. Great Kings, she must be exhausted after this last twenty-four hours. She has hardly recovered from sickness and look at what I've dragged her through!

I meet her gaze, shoving aside talk of kings and thrones for the moment. "I owe you an apology. Quite a number of them."

CHAPTER 44

THE PRINCESS

ASH'S FACE HAS gone soft as he looks at me. I want to take that tenderness, wrap it up, and save it, to pull out when I most need it. But I don't want to need that. I want this Ash to be the Ash I always see.

"I'm sorry for all the times I've scared you," he says, eyes never leaving mine. "You were right last night. As long as you're here, you need to be able to trust me, rely on me, be safe with me. And I've done a terrible job of it. But Stella, I will let you choose what you want moving forward." He takes a deep breath, as though bracing himself for what he says next. "If you want to leave, I will help you. I'll make sure you find someplace safe. The human lands aren't safe right now, but you could find temporary refuge in one of the Small Cities until everything blows over. Or Orawyth. Now, I cannot take you to Orawyth because the only portals to there are in the Nothril Court and the Bridge, but I can get you there."

His words roll over me. Calm, level. They give me a way out of this terrifying world. They give me a choice, a chance. I don't say anything, letting him finish.

"You'd be safer if you left."

I nod once. Safety is very important to me—more important than I thought it would be. And yet . . . is it everything? Is it truly what I want? That after all this talk of giving up fear, of refusing to bend to its whims, I'd choose it over everything else?

But fear isn't the only consideration here. It isn't the only reason I'd want to leave.

"Or," Ash says, dropping his voice even lower, "you could stay with me. You could stand at my side, face the High King with me. It may cost us everything. I cannot promise you'd survive, though you know I will do everything I can to keep you safe. Even if you survived Lulythinar, it is your choice. Whatever you want, I will honor it."

My choice. What *I* want.

The notion is so strange I almost balk. I haven't had much of a choice this entire time, have I? I made no choice except to do as I was told to save my people, no choice except to make the best of this situation.

But now, *now*, I have a choice. Do I stay here with my fae husband and risk my life for the sake of saving my people? Or do I leave the saving in Ash's hands and run? Do I choose my life over my sweet sister Amelia's? Over every human on the continent?

I bow my chin against my fist, staring down at an open book on the floor, its pages wafting in the draft coming from beneath the door.

Ash didn't ask me what I was duty-bound to do.

What is such a thing as wanting? What is desire? I've hardly dared to want anything in my life, because the more I want, the more I'm vulnerable to the stings of disappointment. Of loss.

But now I must decide. Which means I must first know what I want. I want my people's safety—but how much will my presence here change anything? What can I do to fight for Aursailles when I'm just a human princess in a magical world of lethal bargains and royal intrigue? It almost seems like removing myself from the equation, from the list of things that can be used against Ash, might do more for my people.

Do I want Ash?

The question seems to come from nowhere, but as soon as it's there, I realize that *it* is the question that needs to be answered. This isn't as much about my people as I want it to be. If it were, the answer would be easy. But ever since I became a way to hurt Ash, my aid in the situation has gone down drastically.

This choice isn't about them.

It's about me. And Ash.

Do I want him? Him, with his unbuttoned shirts and disastrous treatment of books. Him, with all the wars he's taken upon himself to fight. His determination, his kindness, his chaos, his strength. His *weakness*. His faults. His throne, his people, his power.

His deadly cunning.

His broken heart.

Do I want my husband? Everything that makes him, him? Or would I rather walk away, give him up, and save my own life? Do I want him enough to give up my life for him?

It's not a simple question.

And I don't have an answer.

"I need to think about it," I say finally, lifting my head and meeting Ash's steady gaze. He maintains eye contact, then nods.

"Send a note for a white dress, then." He gets to his feet and picks up a sheet of parchment. He steps around his desk and approaches me, holding out the paper. "If you decide you want to leave before Lulythinar, it's important to have arrangements in place."

I reach out, close my fingers around the paper. It almost feels like a betrayal to accept it. "If . . . if I choose to leave before Lulythinar, what will you do?"

His chest rises with a deep breath. "I'm not sure," he says quietly. "Oleria said I could marry her at the last minute if I needed it. If you decided to leave. Or perhaps I'll just . . ."

Just marry his father's choice.

There's no alternative. If I leave, he'll be forced to remarry one way or another. There is no option to leave and come back later. If I leave, I give him up. Forever.

The thought creates a burning sense of *wrongness* in my gut. I don't want to think of him marrying that beautiful winged fae. I don't want to think of him bound to Listhra either, forced by the terms of the bargain to sire an heir.

The thought of him being with anyone else, touching anyone else, kissing anyone else—it makes me want to vomit.

He's mine, a desperate, primal part of me demands.

I grit my teeth. I won't make decisions based on jealousy. And yet, I cannot ignore it.

I cross my arms over my chest again, self-conscious about my nightgown. I wish I were just wearing something simple and modest. Like that dove-gray dress I wore the first morning after I arrived. It would be so much easier to have this conversation fully dre—

Something shifts in the air.

Ash's eyes go wide.

I blink. Hardly daring to think, I look down.

The nightgown is gone. In its place? The gray gown I'd just wished I was wearing. I shoot my gaze up to Ash's, where his jaw sags open.

"Did you do that?" I hiss, terrified of the answer.

He shakes his head vigorously. "I—no, no, I didn't. Stella . . ." His voice trails off as his eyes trail me up and down. A slow smile spreads across his face. "You just glamoured yourself."

"I definitely did not!"

He looks up at me, his face splitting into another of those devastating grins. He's so beautiful like this, I'm afraid his happiness will go straight to my head, make me dizzy with the desire to be near him. Here I am, trying to make a level-headed decision about whether I'm going to go or stay, and then he looks at me like that. It's simply not fair! And not conducive to clear thinking.

"You *do* have magic," he whispers, perhaps to keep the staff from overhearing. "Stella, look at you!"

I look down again at the dress. It feels real to me, so very real. And yet, when I peer closer, it's almost as if there's a very slight shimmer on the hems. A shimmer that I never noticed when Ash had glamoured me.

I peer up at Ash, searching him for the same telltale shimmer. He is so tall, it takes me a moment to run my gaze over him from the top of his tousled hair, down the deep V of his blousy white tunic, to his rolled-up trousers and bare feet. His beauty, both his muscled form and his handsome face, could easily be glamoured. But . . . I find no shimmer.

"Are you wearing any glamours?" I ask.

He shakes his head. "Shall I don one for you?"

"Please."

A second later, those trousers and white shirt are replaced by . . . by . . .

A chortling laugh catches in my throat and I slap my hands over my mouth to keep it in. "You—Ash!" And then I can't help laughing outright.

He's wearing the very same dress I am, only slightly adjusted to his proportions. Enough that his broad shoulders don't bust the bodice at the seams, but not enough to lend the dress length, so it hangs just below his knees, baring his calves.

It's so unexpected, so vastly ridiculous and incongruous, that I double over with laughter, leaning back against the door, fighting to breathe. I barely remember to check for the little shimmers around the edge of him, but I find them between gasps of laughter. A distant part of my brain registers that I can see glamours like a fae.

Ash just stands there, grinning down at me like an idiot. He dismisses the glamour, and he's back to his own clothes, but I'm still laughing, sliding down the door until I'm sitting on the ground.

He drops to the floor next to me, stretching out his long legs. I try to swallow my laughter, but a giggle emerges as I swipe the tears from my eyes. "You have just scarred me for life," I say, and giggle again. "I need to scrub that image out of my mind with soap!"

He smiles softly at me. It's one of those smiles that makes my insides turn to warm honey. For a moment, I'm caught in that gaze, forgetting my laughter. Forgetting everything.

And then, before I quite know what's happening, Ash wraps an arm around me and pulls me onto his lap. My thoughts fly away like a swarm of bees. He's looking up at me with half-lidded eyes as he threads his arms around me and pulls me against his chest.

"Tell me what you're thinking," he says. "Or I'll kiss you."

His mouth is so close, it would take hardly any effort to press my lips to his.

But that would just be too easy.

"Sometimes, when you come near me," I breathe, "I get a little dizzy."

He lifts one eyebrow. "Dizzy?"

"Dizzy."

He frowns, as though considering this, his brows drawing together. "Am I supposed to be complimented?"

I give him a little grin. "If you'd like."

He chuckles, low and raspy. "Then I'm complimented. Now tell me another thought, or else I'll kiss you."

"I'm not mad that you kissed me last night."

At this, his shoulders visibly relax, and he closes his eyes. "I was sure you'd hate me for that."

"Well, I'm glad you stopped, but I don't blame you for starting."

He gives me a wry look. "It *was* part of the deal that you wouldn't be mad at me."

"Perhaps I'm adjusting to the way you fae make bargains."

"*That* was not a bargain, but sure." His gaze falls to my mouth, his hand lifting to trace circles on my ribs, leaving fire in its wake. "Another thought, darling."

I can't help leaning forward just an inch and pressing a kiss to the tip of his nose before pulling back. His eyes widen slightly, and they lift from my mouth to hold my gaze. My heart hammers. "Sometimes, I think I'm falling in love with you. And other times I think it must be one of your fae glamours or spells making me feel this way."

His thumb stops moving on my ribs.

A scrap of sense returns to my addled mind. Enough that I think to ask, "*Have* you ever put me under any kind of spell, Ash?"

"No," he says immediately, and the word rings true in the air. No iron stench assaults me. And it makes me realize that not once yesterday evening, or this morning, has Ash lied to me. "Not besides glamouring your appearance and hiding your human scent."

This lifts a weight I didn't know I carried. It's confirmed. Fact. Ash hasn't manipulated me with magic. I didn't *think* he had, but at the same time . . . there was always the possibility.

"I . . ." Ash pauses, clears his throat, a flash of pain cutting across his features. "Stella . . . I don't want you to be afraid of me. Of who and what I am. I know I'm many things that I shouldn't be, and not a lot of things that I should. But I want to be a better man for you. So if you decide to stay, if there's even a chance we might have a future together that lasts beyond the next four days . . ." He trails off again, lifts one of his hands from my waist, and slowly, almost tentatively, reaches for my cheek. As if afraid I'll pull away. His eyes bore into mine, a tender intensity that makes me forget to breathe. His palm cups the side of my face, and it's cool against my flaming skin. "I'm a broken mess, Stella, but if you want me, I will be completely yours, and only yours, for the rest of my days."

The tips of his fingers brush the hair at my temple, and it might be my imagination, but when he ducks his head, I could swear he says, *"I already am."*

I'm not sure what to say to such an enormous proclamation, so I swipe my hair behind my ear, my gaze fleeing to my lap, and mumble, "Thank you."

His answering expression is half-amused, running his fingers through my hair with a gentle touch. Then the smile vanishes, replaced by a firm jaw. "I'm not going to be cruel to you in front of other fae anymore."

I look up, and both protest and hope rise in my chest. "But—but if the High King knows the truth . . ." I stop, unsure of what the High King might find out. "If . . . if the High King doesn't think that I'm your pet, that you take our marriage seriously, wouldn't that change things?"

"It'll probably make him more determined than ever to take you from me," he replies, that hardness not leaving his jaw. "But you don't deserve to be treated like that, even if it's an act. You're my wife, Stella, and I want to give you every dignity before my father, before this court. It's time they saw you for who you are. A woman whose soul and goodness shine brighter than the sun."

I blink against the sudden urge to cry. And then I *stop*. I stop stopping myself, stop bottling myself up. I press a hand to my mouth, lean forward so my head rests on his shoulder, and crumple into tears.

He stiffens, and then wraps both arms tightly around me, holding me close. "Stella?" There's shock in his voice, but he seems to set it aside, pressing me closer and stroking me softly. I curl into a tiny little ball on his lap, and with his arms around me—even though I cry—I can breathe easier than I have in a long while.

Safe. Home.

I don't want to leave.

"I'm so sorry if I . . ." He trails off, pressing a kiss to the top of my head.

"It's alright." I'm almost laughing through my tears. The poor man is so bewildered and here I am, unable to reassure him that everything is fine! "That only . . . that means a great deal to me."

"I didn't know it hurt you this much," he says, remorse heavy in his voice. "You've taken so much in stride. I know it's been tremendous pressure, and I knew it taxed you. I'm sorry, Stella. It's done, it's past. Never again. I'm so sorry."

I relax against Ash's chest and shoulder, surrounded by his comforting warmth. And I think . . .

I don't want to leave him.

We stay like that for a sweet while until I twist my neck to peer up at him. His eyes are closed, his breathing steady, but at my movement, they open slightly, looking down at me.

"Tell me a thought," he murmurs. "Or I'll kiss you."

My lips curve, and there are many things I could say. Many thoughts fly through my brain, ready to be spoken and shared. But I close my mouth tight. I look up at him expectantly.

Waiting.

His eyes widen, then narrow, almost playfully. His lips part, and his throat bobs. "Last chance," he says, and it's almost a growl. "A thought, or a kiss."

I push back a little on his chest, lifting my head so we're level. And then I just smile mischievously at him. His eyes darken.

He surges upward, one hand catching the back of my head and pulling me into a sudden, glorious kiss. I close my eyes, slipping one arm around his neck, the other into his hair as I kiss him back. I think I might be getting better at this, at how to angle my mouth, how to move my lips against his, and each passing second is sweeter than the one before. I'm lost, and heavens, I never want to be found. Not when he holds me like this, touches me with such infinite gentleness and fierce passion.

He bends his knees slightly, bringing his legs up so I'm forced to lean into him even more. "I want you closer," he murmurs between kisses. "I *need* you closer."

"You're making me dizzy again," I say, half giggling, not even opening my eyes.

His response is almost pained in its earnestness. "And *you're* making me happy again."

So *this* is what marriage can be. This is what goodness can be had on the other side of promises and commitment. Marriage was always my one duty, my one purpose. It was not something to eagerly anticipate. My husband's attentions were not something to covet.

It was a lifetime sentence to unhappiness—and one I was prepared to accept.

I think of my four sisters, three of them wed to men who won't ever ask their thoughts, or pursue them, or care for them. One of them, perhaps, will have a man who is genuinely kind and respectful, a man she might one day come to love.

And though I'm in a new and terrifying world where powerful forces want me dead, I pity my sisters more than anything. They'll never know what it is to be kissed like Ash kisses me right now. They'll never taste the glorious beauty of what passion *can* be. All they will know is a life of being set aside, trampled, and unloved.

I regret every resentful thought I had about them. No matter if I don't even survive the week. In a few short days, I've lived more life, experienced more goodness, more beauty, than they can even dream of.

A little moan catching in my throat, I pull Ash closer, and I let myself imagine a life together. So unexpected. So improbable. That a fae heir and a human girl could forge something new and beautiful, something bright and glimmering.

I want this to be my future.

I want to fall asleep with Ash's arms around me, to wake up to his kisses, hear his laughter when I try to choke down mothweed milk. I want his smiles forever. I want to see him as High King.

I want to see him as a father.

A sharp knock on the door behind us makes us both jump.

Then Ash groans and pulls me into another kiss before breaking it off and growling, "Go away. I'm punishing my wife."

Edvear coughs on the other side of the door. "The tailor is here, my lord. With Lady Stella's Lulythinar gown."

And just like that, the reality of life crashes back down on us. Ash lets out a great sigh, looking at me from beneath hooded eyes. As though he's very tempted to tell the tailor to go find someone else to deliver dresses to.

366

"I'm not done with you," he growls, pressing one last kiss just below my jaw. His shoulders drop slightly, and he stays there, his mouth hovering above my neck. "But you should probably order that white dress."

"*I won't*," I almost whimper, almost pull him back to me. It's nigh impossible to imagine leaving now. My heart would be too broken. But even if I have no intention of leaving, Ash is right. I shouldn't close off my options. There's still time before Lulythinar for something to happen. The High King could pull some trick that Ash can't counter.

But maybe, if it was bad enough, Ash could come with me. We wouldn't have to separate. We could go together. He could hide until his father died, until it was time for him to take the throne.

I'm not sure if the tailor lets princes come, but if there's a chance that Ash and I can be together . . .

I don't want to give up on that.

My cheeks are warm when I leave Ash's study. His presence looms behind me, prickling my awareness as we enter the living area. This glamour thing is coming in handy; I have only to wish my hair was primly arranged and my dress wrinkle free, and it is so.

My husband, on the other hand, feels no need to make himself presentable, and when he flops on the settee in the living room, he looks even more disheveled than he did when I first entered his study.

The tailor and three human aides carry an enormous garment bag into the living room. I resist the urge to join Ash on the settee for fear the color in my cheeks will never leave. Or . . . can I glamour that, too?

I wish my cheeks to be creamy rather than tomato red. The heat doesn't dissipate, but when the tailor looks at me, standing beside where Ash leans on the settee's armrest, his glance is cursory and returns only a second later to the gown he brings.

"Would Princess Stella care to try the dress on?"

"That won't be necessary. I assume you have made no mistakes," Ash says.

"No mistakes were made, Highness."

"I think,"—my voice cracks slightly, but I force myself to keep speaking—"I should also like a white dress. I believe that was an oversight when we first ordered my wardrobe."

Ash makes no move, just continues staring with infinite boredom at the care the tailor and his aides take to lay the dress on the opposite couch.

The tailor nods once, makes a note on a pad he pulls from his pocket, and says only, "You are quite right, my lady. Please forgive the oversight. I will have the dress sketch to you on Lulythinar's Eve and if you approve it, the dress will be delivered on Lulythinar."

I translate this as: *I will tell you the plans for escape, then you will leave Faerieland on Lulythinar.*

"I hope you had no intention of wearing the dress before then," Ash says with an irritable scoff that I could have sworn was real.

"I have plenty of dresses, my prince. I'm sure I can wait until then."

"Thank you for your patience, Princess Stella," says the tailor. "We will not deliver it late."

"See to it that you don't. You are dismissed."

The tailor bows once to each of us. He doesn't even glance at me this time, but quietly motions for his aides to follow him as Edvear opens the door for them.

Once it's shut, Ash's sharp gaze shoots to mine from beneath the long strands of his tousled hair. He catches my hand, and while I watch, brings my knuckles to his lips. He holds my eyes as he flips my hand and places a second kiss to my palm. It's warm and tingling and intimate— and I'm suddenly very aware of Edvear bustling around behind us.

"I wouldn't have guessed his ploy," Ash says, "but I'm afraid you couldn't have requested a dress without me knowing something was off."

"Was I that obvious?" I ask, horrified. Does the tailor think I exposed him to Ash?

"Certainly. Because you never ask for things for yourself. You're a princess, you know. You have every right to be just a little capricious." His kiss moves to my inner wrist.

I wriggle away as Edvear and other household servants come to remove the gown. "I will endeavor to become more capricious, if it will please you, husband."

Ash rolls his eyes, smiling. He waits until the servants are gone to tug me down closer to his level and whisper: "You are such a lovely distraction, I fear I am tempted to throw away kingdom and future simply for the pleasure of engaging your lips for a few hours."

My blush returns in full force, and I completely forget about my newly discovered powers. In fact, I forget so completely that I find myself once more in a nightgown with loose, tangled hair.

Ash's mouth curves in a devilish grin. "It seems you feel the same way."

He tightens his hold on my hand so I can't pull away, even when I give a solid yank. "You flustered me, that's why my glamour broke! I think kingdoms are far more important than kisses!"

"Do you?" His arm slides around my waist, tugging me until I stand between his knees and he is looking up at me with too much smug triumph. He coils a lock of my hair around his fingers. "But you like kisses better than murder plots, yes?"

"I suppose it depends on who is getting murdered," I reply stubbornly, crossing my arms over my chest and refusing to look at him.

"That is quite the vicious response coming from my sweet wife."

Then he catches me by the hips, pulling me close, and ducks his head to press a kiss to my stomach. I blush furiously, every cohesive thought abandoning my brain like a swarm of hummingbirds.

Except for one thought.

I want to give him an heir.

I dismiss the thought the moment it appears. We're just trying to survive the next few days. I might be leaving. And the only way I could ever give him an heir is if the High King was dead and Ash sat on the throne. In any other circumstance, it would be far too dangerous.

So I let that little desire go.

Ash pulls back from me when the servants' quiet footsteps come back down the hallway. In a more serious tone, he says, "We should test the limits of your magic. And then I really do need to shut myself into my study."

We both regret the separation. But there's no use trying to pretend away reality.

I nod. "Back to murder plots it is."

CHAPTER 45

THE PRINCE

STELLA'S BRAND OF glamour is both like and unlike fae. When we return to my study, I give her dozens of prompts, telling her to glamour her face to appear like a sister's, to glamour her clothes, her voice, her expressions, her scent. She struggles the most with the last two, probably because her sense of smell isn't as strong as a fae's, and it twists her mind a bit to make one expression but imagine herself making a different one. Once I suggest it is like forcing yourself to smile when you don't want to, or keeping your composure when all you want to do is break out into laughter, it clicks for her. In those ways, her glamour is very similar to a fae's.

What I quickly discover, however, is that her glamour abilities extend beyond even my own. When I tell her to make herself look like Edvear, she does it. Hardly a minute later after another command, I find myself staring at my twin. Which is vastly unsettling, even as my brain turns over this new development.

"How strenuous is this?" I ask, Stella's reflection revealing what a mess I am.

"It's . . . fine?" She sounds almost uncertain, as though she's not sure if she's missing a piece of information.

"You're not struggling to maintain the glamour?"

"I don't think so?"

At my order, she holds the glamour for ten minutes, then twenty, with no sign of flagging.

It's astounding.

"We can glamour ourselves as other people," I tell her, unable to stop myself from pacing the small length of the room. "But rarely can a fae maintain it for more than a few minutes. Yet *you* walk around like it's nothing!"

Her glamour melts away, to my relief, revealing Stella's drawn brow. "It felt the same as glamouring my clothes."

She may not have magic that can split the world in half, but what she has comes as naturally to her as breathing. With a few careful glamours, she could pass completely as fae. Longer ears, a disguising of her scent, and a bit more height, and she could easily join revelries, and none would be the wiser. Her glamours have a stamp like any other, but we fae are so used to glamour we notice its absence more than its presence.

We could use this.

We could really use this.

In fact, this might be the missing piece we need.

"You have your thinking face."

Her voice startles me back into the present. I blink, realizing that I've made my way to my desk but am pacing behind the chair instead of sitting in it.

Stella smiles.

That smile catches me off guard. It's a sweet smile—one of understanding. One that says she can read my mind, and she likes what she finds.

"I . . ." I trail off stupidly, not sure why I started a sentence when I had nothing to say.

She marches to my side. She's wearing one of her new dresses: a lovely blue day dress with a square neckline, long draping sleeves, and a skirt with beautiful, gauzy layers of fabric. For once, only half of her hair is up in a bun, leaving the rest to cascade down her back.

She takes me by the elbows, and I let myself be pushed into my chair and a quill pen shoved into my hand.

"I see that clever brain of yours working," she says, finding me a blank sheet of paper and laying it in front of me. "So let it work. I will go practice glamours with Edvear."

I drop the pen, catch her round the knees, and scoop her up into my lap as she gasps in surprise. "I will do what you say," I murmur against her ear as I pull her as close as I can. "But first, you must give me a kiss."

"If we start kissing, we won't want to stop. Which simply isn't productive."

Mountains of Ildrid. Why can't she stop being so unintentionally adorable?

Her breathing hitches when I splay one hand possessively across her waist and bring my face to hers so we are nose-to-nose. "I'm not letting you go until you kiss me."

She stares at me, her eyes so large and rounded. I'm beginning to think she might have cast a spell over me. Because even though she will probably leave me, even though the weight has not shifted from my shoulders, I can do nothing but soak up whatever sweetness we have left.

When she closes her eyes and brings her mouth to mine, the most exquisite delight fills my soul and burns through me like a heady wine.

She is all that matters. The High King, the throne, revenge—none of it matters anymore. Only her.

Only her.

She pulls away just as I am trying to catch the back of her neck to deepen the kiss. Protest rumbles in my throat, but I know she's right to scramble out of my lap and scurry halfway across the room.

Her face is bright red even as she tries to compose herself—forgetting her glamours already.

"You need to work," she says sternly, straightening her spine and lifting her chin. "And so do I."

I don't reply. I just smirk at her like a silly fool and let my gaze take her in from head to toe—which has its desired effect of making her even more flustered.

With a flounce of her skirts, she turns around and marches out the door. "You are not helping."

I'm laughing when she closes it behind her.

Once she's gone, however, my mind returns to the matter of her magic and my ever-present problem of the High King. I've toyed around with various plots over the years. It's easy enough to enrage or distract Faradir; it's the matter of removing him that has continued to give me difficulty. My last plan collapsed with Faradir's plan to raze the human lands. I cannot kill him without giving up my claim to the throne—which is less of a price than the rest of the fae would face as a penalty for the same crime. But what is the point of deposing him if no one can then sit on the throne? Thus, I've stayed focused on tricking him off the throne—nullifying his own claim via him breaking his own laws. But Faradir is clever and careful.

What if . . .

Could *Stella* kill the High King? Can he see through her glamours like he can see through mine? Or could she be invisible to him?

My quill hits the paper, and before I know it, I'm completely enveloped in possibility.

I think I know what to do.

I think I can get the High King off his throne. Before Lulythinar. Before I'm bound to decimate the human armies.

Maybe—just maybe—I can do it and keep Stella as my wife.

CHAPTER 46

THE PRINCESS

E DVEAR GIVES ME glamour prompts to practice while he sorts mail. He made sure the servants were busy elsewhere so they wouldn't intrude. When I told him about my magic—and urged him to not tell the other servants until Ash said he could—both his eyebrows shot up toward his horns. He seems to work so hard at keeping his composure that he tries not to let much show on his face, but this news apparently came as a shock. He recovered himself quickly, rearranged his features in solemnity and said, "That is wonderful news, my lady. We are all happy that you recovered so quickly."

My chest warmed a little at the kind words, and he brought his work into the washroom to help me practice.

After his sixth command—to glamour myself as a potted plant, after which he shook his head in disbelief—I ask: "How is Hylath?"

A crease forms across his brow. His attention focuses harder on the letters he sorts in different piles. For a long moment, he doesn't answer. His Adam's apple bobs once.

"She is healing fairly well," he says eventually, his voice a little thicker than usual. "The blindness . . . is an adjustment, though."

My vision clouds, my stomach turning heavy. Both for Hylath, for what she has lost, but also for Edvear. There's a weariness in his voice, a sadness born of losing much. How many servants and friends has he lost because of the High King's feud with Ash?

Edvear's throat bobs once more, his face losing every line, every care. His bearded jaw tightens, and his voice returns to normal. "Try glamouring yourself like Ash again, but this time, practice his mannerisms. Glamour is only half of a disguise. If you're going to do what I think His Highness will have you do, you'll need to learn to mimic the body language and the mannerisms of the people you're disguising yourself as."

With that, we go back to work.

Ash doesn't emerge from his study for the rest of the day. I practice glamours, and Edvear coaches me on how to observe different people's distinct mannerisms, how to incorporate them into my glamours. He has me glamour myself into inanimate objects and watch as he calls and interacts with different servants. Then he has me imitate them and offers me critique until I've mastered each one.

We spend most of the day working on it, moving around to different locations in the quarters to avoid becoming too suspicious, taking breaks for meals and once for me to lie down after being hit by a wave of fatigue. A maid brings me tea when I wake up, and Edvear insists that we take things a little slower, since I'm still recovering from my illness.

Twice, I sneak up to Ash's door, listen to the scratching of his quill on paper, and then return to my room. The second time, it is long after dark and I am ready for bed in a light blue nightgown. My hand rises to knock, but then the sound of his quill reaches my ears. He's still working so hard; I don't want to disturb his focus. I shove away the disappointment, the voice that whispers, *"We might only have a few more days together,"* and leave.

Yes, I want his goodnight kisses and embraces. But I don't want to distract him from what he needs to do to save himself, to save me—to save all of Faerieland.

It sweetens my heart, however, to realize that I do not carry my old fear of being turned away, of him either not wanting to see me or not caring. I know that if I knock on that door, if I poke my head in, I will be warmly welcomed.

That is why I'm smiling when I shut our bedroom door, climb into the giant bed, and fall soundly asleep.

I wake much warmer than I fell asleep, but the world is still dark. Hot lips trail kisses from my temple to my ear. A hand strokes down the side of my ribcage to my hip, then back up again. Ash's familiar scent envelops me as he wraps his body around mine and whispers, "Go back to sleep, love. I just missed you."

I roll over, burrow into his chest, and obey.

CHAPTER 47

THE PRINCE

ORNING COMES FAR too early. My eyelids are heavy and my head foggy when I find Edvear's unwelcome face leaning over me, his fingers tapping lightly on my shoulder.

Stella is fast asleep in my arms, one strap of her nightgown falling down her shoulder. Edvear is the only reason I quickly slide it into its proper place and drag my eyes away from the fan of her lashes, and cascade of her light hair.

"The Neverseen King is in your living room," Edvear mouths to me.

"What?" I reply, more out of surprise.

"The Neverseen King. Living room."

This is one way to start the day. I assumed he'd come after dawn, at least. Edvear slips back out of our room. Slowly, as gently as I can, I extricate myself from my sleeping wife. As much as I'd rather stay here, curled up with her soft form, not even the High King would ignore the Neverseen King.

And I invited him anyway.

379

Stella stirs when I stand, but I quickly lean over, smooth her hair away from her face and press a kiss to her cheek.

"Keep sleeping," I tell her in a whisper. "I'll be right back."

She doesn't even open her eyes, and her breathing evens and slows. I barely look away from her as I dress and rake a comb through my hair.

This could work.

The sleepy haze over my brain quickly dissipates.

This could really work.

I shut the door as quietly as I can, and my heart picks up its pace with every step down the hallway.

I turn the corner, and there he is.

The Neverseen King sits on one of the settees, his knees spread wide, one arm hanging off the armrest, his chin propped up on the fist of the other. He doesn't wear his usual shadows, and when piercing, cerulean eyes shoot to mine, I realize just how long it has been since I've seen his face.

I smile at him—and it's not the smile I give Stella. "Hello, Neverseen King. It's been too long, my dear cousin."

His brow darkens. "I'm not here for pleasantries. You've sent conflicting messages. I understand you wish to redeem your favor. So tell me if I am to take your wife away from here or if there is something else you want. But be quick about it. I don't like being away from the Bridge."

"I am aware," I reply with another cool grin as I move to pour cups of tea from the tray Edvear left for us. The Neverseen King doesn't touch his, only stares at me with that piercing gaze. "How *is* your wife? It's been a long time since I had the pleasure of her company."

"Get to the point, Prince Trenian."

I chuckle, taking a sip of my tea as I make myself comfortable on the opposite couch. "'Tis a shame. We used to be such friends when we were boys. But you're right. We should get down to business. I do not want you taking my wife. I have something else in mind for your favor."

"Out with it."

I set down my teacup. "Stage a coup for the High King's throne."

The Neverseen King is quiet for a whole minute. Then, in a low growl: "You have *got* to be kidding me."

"Indeed, I am not," I reply, taking another sip of my tea. "I'm not asking you to *actually* try to take over the High King's throne. Nor am I asking you to kill him."

"Then what *are* you asking me to do?" he demands, barely contained anger coloring his tone. "If it puts my throne or my queen at risk, the answer is no."

I say nothing, letting his words echo in the quiet room between us. He may say such things, but we both know the truth. He is indebted to *me*, and the terms of our bargain dictate he fulfills any favor of my choice when I request it. He doesn't have the right to refuse.

"I may be bound to honor the bargain," he says, as though reading my mind. "But if you make an enemy out of me, you know I will make you pay. You have the upper hand of this bargain, but I have the Bridge. It wouldn't be difficult for me to take your wife from you, as you once tried to take mine."

If I didn't have so much practice holding my composure with Faradir, I would have cracked the teacup in my hand. My smile is made of teeth and sharp edges. "I am glad we understand one another. Hear my demand, then, Neverseen King. I, Trenian Ashrift Solavirth, do thus claim your favor: that you stage a false coup against the High King according to the timeline and specifications I will give you. I do not ask you to commit treason against the crown and—"

"Only to pretend I'm committing treason," he replies darkly. And yet, the fist under his chin has loosened slightly—as though he is secretly relieved I did not ask for more.

I *could* have asked him to kill the High King and forfeit his throne, and likely his life and queen in the process.

But I don't want him as my mortal enemy. And neither do I want the Bridge without a king. That disaster would be even worse than what I'm facing here.

"I also bid you wait on my order to destroy the human lands with Crenfyre," I say.

The Neverseen King shoots to his feet. "You are out of your *mind*!"

"I vow that I will not order such a thing. I merely need this as part of our bargain so my bargain with Faradir does not collapse. We may bargain again to ensure it so you know this is no trickery of mine."

"You are going to be the death of me," he says, sinking back into the chair and running a hand through his hair. "Tell me what these specifications are, then, before I lose every fragment of my patience."

And so I oblige him.

I still haven't broken my fast before I knock on a door I have always avoided.

A human servant answers almost immediately. She's young and pretty, with bangles down her arms and her ankles, but her cheeks are sunken, and the eyes that meet mine are those of a frightened animal's.

She makes me think of that man who chased me down in the Small Cities to save his daughter from this very plight.

"Tell your mistress that Prince Trenian is here," I tell her.

She shows me into the reception room without a word and disappears. The room is so blue it makes me think of drowning in a vast ocean. Tightening my jaw, I distract myself by picking up a trinket on the table beside me: a beautiful, glass-blown peacock.

"Prince Trenian," comes a low, melodic voice full of false sweetness.

"Princess Listhra," I reply, as she appears through a waterfall of blue sothsril silk, wearing a sunset-orange dressing gown with a slit from floor to hip that matches her golden eyes. She swishes into my space and settles on the seat next to me.

"Don't tell me you're visiting me because you're lonely," she coos. "I cannot imagine that human is particularly . . *satisfying*." She says this with a pointed once-over of me, while I wait for her to get her insults out of her system before we proceed.

Except she should know better than to insult my wife by now.

"My wife pleases me greatly," I say, curling my lip in a threatening smirk. "Unlike some. Now, would you care to hear why I have come?"

She flutters her wings lightly. "Of course. Do not keep me in suspense any longer, darling."

"Very well. Going back to our little conversation that we had the other night—the one after you'd thought to make sport of Princess Stella—I did find out, when I asked her later about the particulars of her time with you . . . she mentioned an interesting tidbit of information."

Listhra's luminous skin pales a split second before she catches herself and reasserts her glamours. She laughs daintily. "Whatever she told you, she must have been lying. Humans don't remember what happens while they're under the influence of faerie fruit. You must find it so curious to watch her lie without consequence."

"That is only the case sometimes," I tell her, enjoying the way she swallows and struggles to maintain her composure. Because she *knows*. She knows what I'm about to say. "I'm sure you can clarify for me any misinformation I received about that night. As I heard, your *friends* wanted to kill Stella. But you—you decided not to. Why?"

She strokes a long, slender finger down the slit of her gown. An attempt to cover her discomfort that pairs well with the snakelike smile she gives me. "It would hardly do to make the future High King of Faerieland a mortal enemy."

"But that wasn't what you said that night, was it?"

"I didn't claim to only have one reason for my actions."

"No, you didn't. But I wonder what the High King would think if he knew his favorite little accomplice, the one he'd promised to reward with a queenship if she could succeed, neglected to kill the prince's human wife when she had the chance."

The reward was a suspicion on my part, but Listhra's lifted chin only confirms it. "It was as I said then. If we killed her, you would only go find another one."

"You think the High King would agree with you?"

She goes silent, her gaze falling to the glass peacock I rub my thumb along.

"It does make me wonder," I say, never taking my eyes off her face, "what the High King would do if he found out you were more concerned with not making me want to kill you than you were about his interests?"

She stands up, paces across a swirling blue rug to the crystal mantle. She places one delicate elbow on the mantle and rolls her eyes. Still trying to maintain the illusion that I don't have complete control over her now. "You must want something."

My grin widens. "Indeed, I do."

For once, her veneer fades, her glamours loosening so her straight, white teeth become fangs. "Then tell me what it is."

I set down the peacock, steeple my fingers, and answer her. "The Neverseen King is planning a coup. You will help me stop him—or else he may end up as the High King, and he already has a queen."

Her eyes, now slits like a cat's, blink slowly. "What do you want me to do?"

"I need you to tell Prince Rahk and Princess Pelarusa about the Neverseen King's coup. It is too risky for me to interact with them right now with Faradir's spies everywhere, so you will be my messenger. You will meet with the two of them individually, and then you will call your friends together from the other night like normal. You will invite them here, and you will stay in your room until the banquet tonight. And if you deviate in any way from my command, I'll report your duplicity to the High King."

I can almost hear her turning over my words in her head, searching for the trick. But she knows if Faradir catches her playing both sides, or acting out of her own self-interest rather than being his loyal dog, he will have every right to publicly execute her. No matter what court she is from or who her parents are.

She knows as well as I that she made a mistake, and now she must pay for it.

Slowly, hesitantly, she nods. Then she bares her fangs and snarls, "Now get out."

CHAPTER 48

THE PRINCESS

I AM ALONE when I wake, leaving me to wonder if I imagined Ash joining me last night. But when I sit upright, the covers on his side of the bed are mussed.

Ash isn't in his study, or anywhere else. Edvear informs me that he left at dawn but should be back soon. I eat the breakfast I'm served, ready myself for the day, and then approach Edvear while he is polishing silver in the kitchen.

"Edvear—"

"I think it might be better if you rested today, my lady."

"I wasn't going to ask for help with glamours again. Would you take me to see Hylath?"

The clink of silver stops.

"Please?" I ask, offering him my biggest smile as a bribe.

"She doesn't want visitors."

"Please?"

"His Highness will kill me if I let you leave his quarters without him."

"I can glamour myself! I can be one of the other servants!"

He doesn't budge for several minutes, picking his silver back up and scrubbing even harder with his cloth, his chin setting in consternation.

"I could order you," I remind him, lifting my chin.

He glares at me. "I won't be responsible for something happening to you."

I let out a dejected sigh, plop onto the stool next to him, and join him in his task. He eyes me warily at first, protests, but then gives up and lets me work beside him. I've never polished silver before, so I watch what he's doing and mimic it as best as I can. "If it would be too foolish to leave, then we don't have to go. I only wish for her to know that I care. And I long to see if she's alright. Hearing how she's doing from someone else isn't the same."

He doesn't reply for a long minute, rubbing away at a silver knife.

Then, with a sigh, he pulls something out of his coat's inside pocket. At first, I think it's a wooden paddle about the size of my spread hand, but then he flips it over, and I discover it's a mirror. It looks as though it is made of once-green, twisted vines, but now it has died and faded to brown.

It is still beautiful.

He sets it down before me. "Say her name."

I blink once at the reflection of my own confused face, then oblige. "Hylath."

The mirror ripples like water, making me flinch just slightly. Then I lean forward, hardly believing my eyes when there, swimming into focus, is the familiar but strange form of my old maid. She is curled up on a small cot. Her wide mouth is angled in a frown, and where she used to have those five unsettling eyes always blinking out of sync, there are now only five stumps of bandaged tissue. She has a mostly empty bowl of soup on the small table beside the cot. Another creature like her brings a steaming mug of something hot, pats Hylath on the shoulder, and it looks like she urges Hylath to sit and drink.

"Her daughter," Edvear explains.

No sound emerges from the mirror, but somehow, I can still hear the *burrrap!* Hylath makes in protest, flapping her long tongue at her daughter. The daughter spits some retort back, and Hylath seems to grumble, but sits up anyway, accepts the drink, and drains it dry before laying back down.

The image blurs.

"How could the High King do this?" I demand. "How can he be so cruel, so evil?"

This needs to end. Faradir *must* end. What he did to Ash, to his servants, to Hylath, to me, and countless others—it needs to end.

Edvear says nothing, only takes the mirror as the image fades, and replaces it in his pocket.

I dry my cheeks with my sleeve. "What is that?"

"It's only a faerie trinket," he replies. "His Highness procured it for me when I became his steward. It helps me manage the servants."

"Do I say anyone's name and see them?"

Could I ask to see Amelia?

"Oh no, that would be far too dangerous. It can only see those who offer their blood to it. I can only see members of my staff."

I'm trying to think of something to say to hide my disappointment when Edvear's large ears twitch below his horns and he springs to his feet. "His Highness is back."

I hear the door close a moment later, and I'm only a few steps behind Edvear when he greets Ash and takes his cloak.

But Ash's attention shoots immediately to me, cunning light sparking in his beautiful eyes. Cunning . . . and *hope*. "Good morning, wife. What say you? Should we cause some mischief today?"

I squeak when his arm wraps around my thighs and scoops me right off the ground. Edvear makes himself scarce as Ash lifts his chin toward me and captures my lips in a long, dizzying kiss.

"What sort of mischief?" I ask when he sets me back on my own two feet. The memory of Hylath's misery is fresh on my mind, making me add: "If it hurts the High King, then yes."

"My woman is bloodthirsty once again, everyone," he announces to the empty living area. Then he cocks one dangerous eyebrow at me. "Show me your best Listhra glamour."

CHAPTER 49
THE PRINCE

NEVER IN MY wildest dreams would I have ever imagined I would allow Stella to march—completely unprotected—out of my quarters into the dangers of the High King's palace. But a lot has changed in a short amount of time.

I stand, heart raging like an untamed beast in my chest, with Edvear's little mirror. We added Stella's blood to it before she left. Through it, I can watch my terrifyingly fragile little wife march in disguise with a tray of tea straight to Listhra's doors.

We spent hours perfecting her glamours, her scents, her tone of voice, and her mannerisms. The afternoon was half gone by the time she set out, disguised as one of our own servants. I kept my gaze glued to the mirror, watching with so much anxiety I could barely breathe, as she made her way to the main kitchens of the palace, switched her glamour for Listhra's sad human slave, and picked up a tray of tea to deliver.

What worries me most are her scents. I had to glamour each scent into a piece of cloth, and Stella had to breathe it in deeply for a while before she was able to get the glamours right. Even then, I'm not sure she can fool someone with as strong of a sense of smell as Rahk. And what if she gets confused and switches the servants' scents with each other? What if Listhra realizes she's not the right servant?

I will have given her my bride—right after blackmailing her because she hadn't used the opportunity she had to kill Stella.

"I know she can do this," I whisper to myself, reminding my stupidly anxious brain that I never would have sent her to do something so dangerous if I didn't fully believe she could do it. And I am ready the second anything goes wrong.

"She reached Listhra's door," I call to Edvear, who hasn't been hiding his disapproval of my plan as well as he usually does.

"Did they accept the tea?"

I watch the mirror, barely breathing, as Stella, in the form of that sad, hunched slave girl, bows when the door is opened. She enters, and I nearly lose her small form in the sudden onslaught of brilliant blue and half a dozen colorful fae women sitting and laughing with Listhra. Just as I made Listhra promise to do.

Stella's movements slow slightly.

"Don't stop. Don't get scared," I whisper. "You're doing wonderful, darling."

One of the fae women plays a harp in the corner, and she glances at Stella and frowns.

"Did they accept the tea?" Edvear demands, his voice pitching high.

Stella places the tray on the center table, bows, and as other servants serve the princesses and ladies, the woman on the harp goes back to playing, and Stella slips back out of the room.

I let out a gusting sigh of relief. "She made it out. They accepted the tea."

Edvear's shoulders relax, and he goes back to polishing a silver teapot he'd been neglecting the last few minutes.

What I see next in the mirror does something thrilling to my gut. When Stella steps back into the hallway, now empty of her tray, she isn't in the form of a lowly human maid anymore.

She is tall, with silver wings, glowing gold eyes, and a delicate saunter that is completely convincing.

My wife has become Princess Listhra.

I let out a loud, celebratory holler, then bite down on my fist to contain the pride bubbling over inside me.

I love this incredible girl I married.

CHAPTER 50
THE PRINCESS

DON'T LET THE *glamour slip. Don't let the glamour slip. Don't let it slip.*

I struggle to keep my wits about me, to remember exactly where I am supposed to go in this labyrinth of a palace. If I let my focus drift, I might be startled out of my glamour or forget to maintain my scent.

I nearly let my scent slip in Princess Listhre's quarters. I had to assert my glamour hard to disguise reaching into my pocket for the scrap of cloth with the human girl's scent. Once I'd smelled it again, I was able to reassert it stronger.

But I almost let myself get caught.

Now it takes extra focus to hide my shaking hands and wobbly ankles with the princess's confident, swaying hips.

Ash's directions take me away from the quieter ends of the palace I usually frequent with him, and instead I return to a place I once never wanted to see again.

The palace greens.

Where the revelry was my first night in Faerie.

When people tried to kill me.

But I'm not vulnerable like I was then. I have my own protections. As my imitations of Listhra's slippers make a loud, confident proclamation of my entry down the silver-streaked marble steps into the greens, eyes of all shapes and sizes turn to me.

My heart hammers, but I lift my chin.

I was not made to shrink. I was not made to be small and quiet. Fear made me that way.

I intend to be much more.

But this is still nerve-wracking.

Guards' eyes follow me down the steps onto the lawn. Vines slither up trees and skitter out of my way as I march past the few tangled knots of fae that seem to be perpetually intoxicated, and I make my way to the more dignified group lounging in the shade of trees with sprawling displays of small foods and drinks arranged on blankets.

That is where I find Prince Rahk and Princess Pelarusa bickering with each other.

"I'm not going back to Nothril," Pelarusa is saying when I approach. "I don't *care* if you think—"

Rahk shushes her immediately, his dark eyes sharpening on me as he somehow maintains his easy posture leaning against the trunk of a towering oak. "Princess Listhra."

Panic suddenly flares across my vision. I've observed the fae custom of greeting everyone by their title upon entering a room . . . but does that same formality extend to the outdoors? Am I supposed to greet everyone who lounges out here now? The random buck-horned fae on the other side of the burbling stream? What about his lover napping in his arms? There must be several dozen at least, many of whom seem to be titled. All of whom I do not know, save the two before me.

"Prince Rahk, Princess Pelarusa," I say, glamouring my voice like Listhra's sing-song tones. I hope none of my sudden fear comes through the glamour.

"Get sick of your party?" Pelarusa snaps.

So maybe they *don't* do the addresses outside. I contain my sigh of relief.

But confusion quickly follows the relief. Was Pelarusa not invited to Listhra's afternoon tea? What in the world am I supposed to do with this?

"Did my invitation not arrive?" I say, hoping it doesn't come out too squeaky. At the last minute, I decide I should be playing at my own offense. I let one eyebrow lower *just slightly* and pair it with one of Listhra's wretched smiles. "I'm not used to my invitations being ignored."

"I received nothing." Pelarusa flips her snow-white hair over her shoulder, fixing her black eyes on me in a glare.

Rahk, ever cool and collected, simply studies the two of us. Whenever his gaze settles on me, I feel as though he can see straight through my disguise. Just to be extra careful, I reassert my scent. I try not to let his unnerving study undo me.

Ash said it was important Rahk didn't discover who I was, in case he is under orders to kill me.

"It sounds like one of my servants might be in need of a little *lesson*," I say lightly, laughing, even though the words make me sick. I need to end this and get out of here before I do something like throw up and betray my humanity.

"You can sulk here with your brother if you like," I say, fluttering my glamoured wings. "Or you can come join us in half an hour. Up to you."

And then I do the main thing I was sent here to do. I take a tiny, wrapped parcel with a sealed note from my pocket and reach forward like Ash taught me, placing my hand and the hidden gift on Rahk's forearm.

"It's always good to see you, Prince Rahk." I give him my best Listhra smile and wait for him to place his hand over mine, covering the gift and accepting it. I pull my hand away, give a wink—just like Ash instructed—and all but flee the scene.

Don't let your glamour slip, I beg myself as I hurry back the way I came, back through those terrifying armed guards, back through dozens of prying eyes, and into the palace. *Don't let it slip.*

Don't. Let. It. Slip.

I want to run, to release my pent-up adrenaline. But the point of all of this was to be seen, and I can still mess this up terribly.

My nerves are through the roof when I return to Listhra's chambers, the women's laughter ringing in my ears as I clear the empty tea set. I barely remember to glamour myself into Listhra's terrified servant girl. I feel my scent slipping, but my hands are occupied with the tray I carry, so I can do nothing but try to reassert it by memory and scramble a little faster out of the room. Finally, I return the tray to the kitchen, glamour myself back as one of Ash's servants, and hurry as fast as I dare home.

At some point, I become aware of footsteps behind me, almost drowned out by my pounding heartbeat—but not quite.

When I walk faster, the footsteps speed up. Panic floods me to my toes, and not even the familiar winged marble statue—indicating how close I am to home—can ease my terror.

I'm going to be the next servant the High King kills or blinds in front of Ash, I realize with a sickening sense of dread. Maybe the High King heard one of Ash's servants was out and decided to snatch me up for his son's next punishment.

Oh heavens.

I break into a run. I cling to my glamours as I do so, running as hard and fast as I can as the steps behind me turn to a pounding rhythm.

I just have to get to—

"It's a good thing you're running after dallying so long, by the Great Kings!" comes Ash's sudden rebuke. "I should have you flogged!"

I barely keep myself from running straight into his chest. He catches me by the shoulders, stopping me as I gasp for air. I nearly weep in relief and throw my arms around him, but he shoves me behind him quickly and growls, "Get out of my sight before I lose my patience with you. You left Edvear to handle the silver all by himself. Don't make me lower your wages."

Still gasping for breath, I gladly do exactly as he says, stumbling past him into the door that Edvear opens for me. He pulls me inside just as I hear Ash say to someone, "Sorry you had to see that."

Edvear hands me a glass of water, which I guzzle down between panting for air. My glamours melt away—and it's a relief to just be myself. To be safe as myself.

Ash shuts the door behind himself, and the next instant he has me by the shoulders again. "Are you alright?"

I press a hand to my heaving chest. "Who was that?"

"Someone from the High King." Ash's voice is grim. "You were clever to run and keep up your glamours."

Then he crushes me to his chest.

"I think I did everything properly," I manage between gasps. "I gave Rahk the gift and the letter. And my sleight of hand was definitely not clean."

"You good, good, clever girl," Ash says, holding me tighter. He cups the back of my head with one hand, presses a kiss to my temple. "You were brilliant. You've perfectly set everything up. Now all we have to do is sit back and watch the High King begin digging his own grave, starting tonight."

There is a new spring to Ash's step when we walk to the High King's banquet that night. He picked tonight's dress: a gown of a light sapling green that clings to my hips and trails behind me in a long, glittering train. A layer of shimmering green leaves and tiny red roses covers the bodice, sweeping from my shoulder to my hip and spraying out along the train. My hair is bound up elaborately on the crown of my head, with falling curls arranged with threading little vines. A silver tiara studded with rubies rests on my head.

When I looked in the mirror, I thought I looked like a woodland fairy queen. When Ash looked at me, a flush crept over his ears and cheeks, but a second later his gaze seemed to snag on my bare throat. As though seeing just how breakable I am.

Then he met my eyes and smiled. "You are exquisite, my darling."

Now my hand is tucked in his elbow as he guides me through the palace as though he is already the High King, and I, his queen. It occurs to me that while he often walked this way, with a possessive and authoritative stride, the difference is that now, it's true. It's not a pretense.

When the grand doors are opened for us and we enter the banquet hall, it's different for me, too.

For the first time, no part of me is afraid of the tall, striking fae prince at my side. I know I am safe with him. There's no fear he will turn on me or use me now.

He is fighting for *me*, not his throne. I am no longer the pawn, but the prize.

Maybe the rest of the courtiers seated at the High King's table sense it too, because when the announcer calls our names, their eyes do not just fall on their own Prince Trenian. The weight of a dozen stares almost pins me in place. I've never been so visible before. Even when Ash first presented me to the High King's court and all hell broke loose, the attention was on Ash and the fact that he had a human on his arm instead of a fae woman.

Now they *look* at me. At my face. My dress. The beautiful crown I wear.

Ash is looking at me too.

His gaze is the most powerful of them all. It's the one that overwhelms me and makes me look at my feet. It's the one that makes a flush overtake my features, and because Ash strictly warned me against using my glamours tonight, I cannot hide it.

"Prince Trenian. Princess Stella," everyone says.

Ash rattles off the names of everyone present, and for some reason I am surprised to hear Prince Rahk's name and Princess Listhra's.

Princess Pelarusa is also present, and my nerves—which were so much more composed a second ago—decide to become fluttery and agitated. What if Rahk, Listhra, or Pelarusa look at me and somehow read the secret of what I did today?

As though my thoughts summoned him, Prince Rahk is staring at me. Is that a slight furrowing of his brow—perhaps in confusion? Curiosity?

Perhaps my scent glamours didn't hold, and now that Ash is the one glamouring my human scent, perhaps his sensitive nose is picking up on something that would give me away.

To my shock, there are three chairs open. One for the currently absent High King, one for Ash . . . and one for me?

Ash flashes that magnanimous grin of his, and dramatic as always, takes my hand and leads me into a twirl before pulling out the empty chair next to Princess Pelarusa and gesturing for me to take my seat.

I do, and when he grabs the back of my chair, he leans down. His warm breath tickles a curl near my ear as he whispers softly: "I love you, Stella."

He pulls his chair out and takes a seat, already engaging Rahk, who sits across from Pelarusa, in conversation.

Meanwhile, I sit like stone, shocked.

I have known for some time that he loves me. At first, I was hardly brave enough to admit he might like me. But as the days have gone by, I've not been able to lie to myself anymore. His love hasn't been the point of contention in my mind anymore—it was whether his love was enough to overcome . . . *everything else*. Up until yesterday, I believed it wasn't.

So, hearing it now, whispered in my ears, knowing the truth of it—knowing everyone else at this table heard it, too—I can hardly breathe.

Ash is still talking, though I can hear none of it, when he looks at me. His glance is brief, but it is heavy, weighted, warm and dark and soft and bright all at once. The look says, *"You know I mean what I said."*

Beside me, Pelarusa has gone even more still than I, but she recovers herself much quicker. She says something to Listhra, who sits next to her, and they laugh.

I cannot help but wonder if Pelarusa did, indeed, go visit Listhra this afternoon and if she did, how that conversation went.

"High King Faradir!" the announcer cries.

Everyone echoes . . . everyone but me. I keep my lips sealed shut as the glorious king, with his long golden hair and luminescent skin and brilliant white robes, enters with a beaming smile and takes his seat at the head of the table.

He blinded Hylath. He killed Ash's mother.

In a way, he's been killing Ash since the day he was born.

I have never wanted anyone dead.

Not like I do now.

As though sensing the dark lance of my hatred, Faradir looks up—directly at me.

Blue eyes pierce mine. I don't flinch. I hold his gaze.

And just for a fraction of a second, it's like his glamours barely melt under my scrutiny. The veil covering the blackness of his heart with beauty lifts slightly, and his eyes darken.

Perhaps he sees something in me, too. I wear no glamours, so there is nothing I hide behind. Perhaps that is what he sees—that I am not a fool, not a puppet, but a force of will in my own right.

"Princess Stella," he says, his black eyes fading to blue once more and a smile stretching across his teeth.

I turn away and look at the fae across from me in a slight that is not at all diplomatic and probably petty. But since I'm not a fae, *I don't have to address this monster.*

So I will not.

The woman across from me is a contrast to the other fae present. She is the only one to show signs of aging, with fine lines around her eyes and streaks of gray in her light brown hair. She's beautiful, but there is no light in her eyes, no strength in her shoulders.

The High King's wife. His new queen—the one he married after killing Ash's mother. I haven't seen her since the night Ash presented me to the High King. Which, looking at her now, doesn't surprise me, though it does grieve me.

I hope Vivienne, Jacquelle, and Yvonne all fair better than her in their marriages.

"Would you like anything to drink?" Ash asks me suddenly, holding a vial of something that sloshes like water.

My throat has gone very, very dry, so I nod. Everyone else at the table is drinking a golden liquid in sparkling goblets, but Ash sends my goblet back and fills a new one with something that is still golden, but translucent.

I take a sip, then almost burst, "Oh!" in surprise. It's apple juice. Something familiar and safe for me to drink.

He smiles at my pleasure.

"So, Prince Rahk," says the High King suddenly, turning his attention on the silent prince sitting beside his wife. "I hear we are to congratulate you on a forthcoming engagement to Princess Listhra."

Rahk, who has always been cool and collected in my presence, nearly chokes on his drink. He sets it down abruptly, presses one of the table's scarlet napkins to his mouth, and swallows hard. Then, without a glance to the princess in question, he looks straight at Ash, then at me. Did he just put the pieces together?

But Ash doesn't seem concerned at all, and only lifts one eyebrow and his goblet. "Congratulations."

Rahk doesn't reply.

Faradir lifts an eyebrow in a way that is a near replica of Ash's. It's a little strange to see the father-son resemblance between them when I love one so dearly and hate the other so viciously.

"I see I was misinformed," says Faradir with a smile at both Rahk and Listhra, the latter being shockingly quiet. "I heard you were exchanging gifts. What pray, if not engagement gifts, were you exchanging this afternoon?"

Now I dearly regret Ash insisting I not use my glamours. My face goes hot, likely turning as red as the roses on my gown. I try to hide it by taking another sip of my drink.

Princess Listhra, a thread of alarm in her voice, speaks up at last. "There must be some mistake. I spent the afternoon in my chambers with my ladies. Any of them can vouch for my whereabouts. I gave no gifts to Prince Rahk."

The table, previously full of chatter, goes dead silent. The words ring true, making the confusion almost palpable.

"You gave him no gift?" the High King asks, leaning back in his throne-like chair and casually stroking his chin.

Listhra, realizing her mistake—that her words sounded like an evasion of the question—quickly insists, "I gave him nothing. Nothing at all."

The words ring true again.

My apple juice is already gone. Ash, his face a perfect mask of confusion and amusement at the scene playing out before us, pours me another glass.

This isn't going to work, my mind repeats over and over in my head.

Ash takes a slow sip from his goblet, then asks, "Well, I suppose the only question to be asked is this: Prince Rahk, what did Princess Listhra give you this afternoon?"

Listhra's skin goes pale, and when she leans forward to look at Ash, there is murder in her lovely golden eyes. She knows she's been tricked, even if she doesn't understand how.

Prince Rahk clears his throat, back to his carefully composed self as he announces, "She gave me nothing. The package was empty."

"How very interesting," the High King says with an almost gleeful grin, tenting his fingers and eyeing Listhra. "How mysterious of you. Perhaps your friends will enlighten me. Princess Pelarusa, you often spend time with Listhra, do you not?"

Rahk's attention sharpens on his sister. No one makes a move as servants set a plate of food before each of us.

"I did not spend the afternoon with her in her rooms," Pelarusa says quickly, which is not a denial—because a denial would be a lie.

"Of course," the High King says with a kindly smile. He turns to the rest of the table, saying to no one in particular: "Well, Lulythinar is almost here! The day after tomorrow. It seems to come faster and faster with each turn of the cycle."

He gestures at the ceiling, which I have avoided looking at since I first entered this room and beheld the murals of fae abusing humans. But now many people glance up and give a surprising cheer.

Sometimes I forget Lulythinar is a celebration, and not the herald of certain doom if the rest of Ash's plan doesn't fall into place.

I look up. And blink.

I must have missed it before because of the angle of the ceiling, but seated at the table, I have an unobscured view of a small, circular window into the night sky, and the moon that nearly fills the window. Only a small sliver of night is visible.

The midnight of Lulythinar, the moon must fit perfectly in the window.

Too aware of my vulnerable neck at the many enemies around me, I look down at my plate.

There, lying in a row on a skewer, are four tiny bunnies. I can barely acknowledge the artistic smears of sauce and whatever is arranged in the corner of the plate.

All I can see are four open eyes staring back at me.

I'm going to throw up.

And it won't be because of poison.

"We're fortunate to be enjoying a delicacy tonight, imported from the human world." The High King's voice curdles my stomach. I look up, find myself met with his smile. "Enjoy."

And the High King winks.

CHAPTER 51
THE PRINCE

WHEN RAHK EXCUSES himself from the banquet early, a line between his brows, I know what he is going to do. That doesn't stop me from hurrying to my desk the moment Stella and I return from the banquet ourselves.

My word will only bolster Rahk's and emphasize the severity of the situation at hand. I dip my quill in its inkwell and begin writing.

Dear Lord and Lady Nothril,

I write to you of an urgent matter. Princess Pelarusa is at risk of becoming a victim of the High King's wrath due to her association with Princess Listhra—whom the High King has reason to believe has betrayed him. Prince Rahk is also in danger. I believe the situation to be worth your alarm. Neither Princess Pelarusa nor Prince Rahk have done anything meriting the High King's punishment.

Prince Trenian

I seal the note and give it to Edvear with the urge to have it be delivered to the Nothril Court this very night.

Then suddenly, I'm alone in my study. Alone with my thoughts, with the glowing orb of the lumiral globe.

My beating heart slows as I realize that there is nothing else I need to do tonight. My eyes, which have grown glazed, focus on the pages on my desk, the books on my shelves against the four walls, the books on the floor—

There are no books on the floor.

There is a neat little stack beside one of the shelves, which must contain less than half the books that once decorated the ground.

Edvear might have done it.

But it wasn't Edvear, was it? It was Stella. She must have done it when I was gone earlier.

I stare at that stack of books, at each spine and smoothed page, the neatness of that stack—as if it is my very life that she has come in and rearranged.

I love you, I finally told her at the banquet.

I don't want her to leave. If my plan works, she won't have to. But there is one aspect of my plan that makes me uncomfortable. I already risked Stella's life by having her impersonate Listhra. This plan, however, would make that stunt look like child's play.

I don't doubt that Stella can do it.

I doubt my own ability to let her do it.

She is not an expendable piece on a board. She is my wife, my entire soul, and to lose her would be to lose everything that ever mattered.

That aside, it might not even practically work. I have one test to run tomorrow on the extent of Stella's magic. It's a test I'm loath to do, but if it works, I might be able to bring this to an end even before Lulythinar even begins.

My hope feels like such a tentative, fragile thing in my chest, like a flickering candle ready to be blown out by a breath.

But I will hold the tiny flame of my hope like it is a torch ready to set a forest ablaze.

I stand, march over to the stack of books, and find a note on the topmost cover. My frown turns up in a smile as I read.

I couldn't find where these belonged on your shelves. I made what I think are highly educated guesses for the rest. If you find your books out of place, blame it on yourself for treating them so horribly in the first place. I cannot believe I call you husband.

-S

I'm grinning by the end of it. I stare at her pretty hand, the careful elegance of each letter, hear her voice in my head saying the words. Then I roll up the note until it's smaller than my pinky finger and hide it in my drawer of poisons for safe keeping.

My blood pounds as I leave my study. It's late, and by the darkness under the door of the bedroom, I can tell that Stella already went to sleep—likely exhausted after the toll these last several days have taken on her.

I open the door silently, then shut it behind me. My feet make no noise as I approach the bed.

There, just as I'd hoped, lies my wife on what has become *her* side of the bed. I've wanted to believe the days of her sleeping in her own room were long gone, and now I can only smile as I slip under the covers beside her.

"Finally," she mumbles.

My eyebrows lift. I scoot closer until I can wrap one arm around her and pull her so she rests against my chest. Her eyes open, and they're not at all fogged with sleep. She was awake, waiting for me.

That thought does a lot of things to my already pounding heartbeat.

"You'd better be Ash, because it's so dark I couldn't tell otherwise," she says.

It's a reminder that while she may have glamour magic, her eyes are still limited by human nature. She cannot make out the features of my face like I can hers. Part of me likes it, likes that I can read every twist of her sweet mouth, every emotion that shines in her beautiful eyes, likes that I can admire the color in her cheeks, the line of her jaw, the sweep of her nose, her delicate brow. Still, with a twist of my fingers, I use my magic to open the skylight so moonlight bathes us.

"You're getting sassier by the day," I reply, breathing in her scent deeply. "At this rate, I only have a few months before all the sweet turns to sass."

"Oh, certainly not so long."

I chuckle, cup the side of her face, and claim her lips in a long, lingering kiss. "I'm sorry for making you wait."

"Me, too."

This time, I laugh out loud. Then I tighten my grip on her, tangle our legs together, and wonder how I'm supposed to spend the night with her and not cross boundaries that aren't safe for us to cross while Faradir is on the throne.

"What do your tattoos mean?" Stella asks against my chest, tracing one finger along my biceps where there is a Small City acorn cap wheel.

I welcome the distraction as I caress her waist through the sothsril silk nightgown she wears. "They're bargains. Or, rather, they represent the bargains I've made."

She surveys my arm, all the many markings I've gained over the years. She has me roll up my sleeve to my shoulder to get a clearer glimpse of each tattoo in the moonlight.

"You've made a lot of bargains," she says after a thorough study.

"Living in Faerie usually does that."

"If I make bargains, will I get tattoos too?"

"You will."

She wrinkles her nose. "I don't think they would look good on me."

"They do disappear when they've been fulfilled," I say. "Many of mine have already half disappeared, either because I have completed

410

my end of the bargain, or because they have completed theirs. Or they disappear once it's no longer possible to fulfill the bargain."

"What's this big crown on your wrist for?" She traces each spire of the crown, and I find my eyelids drooping closed. I might be just as tired as she is.

"It's the bargain I made with Faradir, that he would allow me my choice of a bride from any of the courts so long as I had a wife by Lulythinar."

"Both broken halves are still there."

"Both will disappear after Lulythinar. If you stay with me, if I have you as my wife on Lulythinar, then I will have honored my part of the bargain, and Faradir will have honored his by letting me choose. If I do not have a wife by midnight on Lulythinar, then I will be bound to marry Faradir's choice. His half of the bargain will disappear, and mine will disappear when I have sired an heir."

I try to say the words as emotionlessly as possible. I doubt it works. Stella props herself up on one elbow, leaning over me, over my face, as though trying to make her human eyes see me as clearly as I see the furrow of her forehead.

Then she tilts her face, and my sleepy mind almost doesn't have time to prepare itself for the sudden flood of tingling pleasure down my spine when she begins whispering in my sensitive ear.

"I'm staying with you, Ash."

I go still. Inside, my mind erupts with sudden protests, with urges that she's safest the farther away she is from me.

But then she adds, "Because I love you," and then I cannot contain the overwhelming joy that has me sitting upright, pulling her to me, and smothering her in forceful, passionate kisses. The kisses I wanted to give her at the banquet tonight. The kisses I've *always* wanted to give her.

When I feel my self-control fracture, Stella—as aware as I am of the boundaries we cannot cross—asks about a different tattoo, and then another, and another, until we fall asleep tangled in each other's arms.

411

CHAPTER 52
THE PRINCESS

I'VE HARDLY EATEN breakfast and readied myself for the day before chaos erupts across the entire palace the next morning.

"I don't know what His Highness put in that letter or what the Neverseen King has done," Edvear says, coming back from some errand, and panting hard as he shuts the door, "but it looks like heads are going to roll."

"Heads?" I ask, wiping my mouth with a napkin and rising from the table. "What has happened?"

Before he can answer, the door opens once more. Ash's shoulders fill the doorframe. He shoots eyes flashing with thrill and a devilish grin my way.

"Want to come see the fun?" he asks me by way of greeting.

"Are you sure it would be safe for—" Edvear protests.

But Ash catches me by the hand and drags me out of his quarters into the hallway. It's usually empty, but right now it's full of guards and bustling servants.

"Should I—" I start to ask about my glamours, but Ash cuts me off with a sharp shake of his head, apparently knowing exactly what I was wondering.

He tucks me into his side protectively and pushes through the throng, his grip tightening every time someone steps too close to me.

It's a stark reminder of just how vulnerable I still am, no matter my glamours and newfound strength. If I am hidden as someone else, I am safer. But I, as myself, Princess Stella, am vulnerable as ever.

Ash leads me to the palace greens, not through the main entrance, but through a servant's corridor until we circle around to the side and come upon the tableau.

The High King stands on the marble steps of the palace, flanked by his guards. He wears his tall, golden crown on his golden head, a long spear in his hand.

Before him, on the palace greens, stands an entourage wreathed in black. At their helm are two figures—each at least seven feet tall. The woman wears a black gown made of shimmering silver stars, her dark skin effervescent like a moon's shadow. Her hair is white as fresh snow, falling over her shoulder and to the ground in a show of cascading beauty. Her crown is jagged and sharp, made of obsidian.

The man beside her is equally tall, though his skin is pale, and his long hair a rich, raven black. He wears a trailing robe that catches and swallows the light around him. His crown matches the woman's, and together, they are a force unlike anything I have ever seen before.

Lord and Lady Nothril.

Rahk's parents.

And I thought *my* father was intimidating.

Behind the king and queen of the Nothril Court are dozens of Nothril warriors in dark, reflective armor. They are all armed to the teeth, but none have drawn their weapons.

"To what do I owe this pleasure?" the High King asks.

414

It is Lady Nothril who speaks. "We have come to retrieve our prince and princesses. Their time of visiting Valehaven has come to an end."

I steal a glance at Ash and find him watching the scene with rapt attention. Around us, more fae are gathering, nobles and commonfolk alike, and the scene is silent as a graveyard. Waiting.

The High King smiles. "I'm sure you did not need so many warriors to bring them home. Do you expect them to put up a fight?"

"The instability of your rule does not give you leeway to threaten our heirs," replies Lady Nothril, bothering with neither civility nor diplomacy.

I can almost hear the anger bubbling in Faradir's gut as his smile twists just slightly.

Ash's head whips to my right, and he pulls me to his other side, shooting silent threats with his gaze at the guard who stepped too close to me.

"No harm has come to your heirs," Faradir says, upholding his air of magnanimous king. "There is no need to hurl insults."

"Did you intend harm on Prince Rahk or Princess Pelarusa?" Lord Nothril speaks for the first time, and his voice rumbles through my entire body, making me shiver and lean closer to Ash.

"My actions and intentions are never unprovoked," says the High King, unable to give an honest denial. He lifts his chin and narrows his eyes. "And if you intend to keep posturing, I may be provoked into showing you why it is I who am High King, and not you."

This sounds like a fabulous time to leave and go back to our quarters before we become collateral damage. And yet, I couldn't leave the scene even if Ash wanted me to.

Lady Nothril lifts her chin even higher than Faradir's, and on her, it's far more terrifying. "You have the gall to threaten us when it may be only hours or days before the Neverseen King stages his coup. All after I have received multiple reports of you threatening my heirs when they have done nothing except be caught in the associations of another court's traitor."

415

The High King goes strangely still.

Listhra.

Everyone now believes her to be a traitor, not just the High King. My mind whirls and my balance goes unsteady. It's one thing to hear Ash speak of these things in theory, when they sound like harebrained ideas that might or might not work.

It's another to watch the fruit of Ash's machinations unfold right before my eyes. To see exactly how we led everyone to believe Listhra is working with the Neverseen King to overthrow Faradir, and how we took a loyal Court and turned them against him.

The High King's face has gone pale. At first, I am not sure why. Then he speaks, and my eyes widen with realization.

"What did you say about the Neverseen King?"

He didn't know.

Faradir's spies are everywhere. That was why yesterday's plan worked. Yet somehow, they haven't discovered this piece of information.

The Neverseen King must have been very clever and swift about how he constructed his own plot.

"You did not know?" Lady Nothril demands. Her low voice rises in pitch until it is *almost*—but not quite—shrill. "You *did not know* the Neverseen King intends to overthrow you? You did not know that he has gone to every single Court to collect those who would swear fealty to him as High King? You didn't know there is a tremendous number of fae who would take anyone, even the half-human spawn of your heir and his pet, over *you*?"

Faradir's lips pull back from sharp canines. His fist tightens on the spear he holds. "Take your heirs and be gone before I make do on the threats I have issued. If you continue to overstep, you will not like the consequences. And neither will your heirs."

Lady Nothril's mouth tilts up a fraction. "Spoken like a man truly about to fall."

"Spoken like a woman who longs for the death of her firstborn son," replies Faradir.

Ash's eyes widen. My heart stutters.

And as if sensing us, the High King swivels his head toward us, his blue eyes piercing straight into Ash, then me.

I know you're behind this, the look says. *And I know how to make you regret it.*

Faradir turns back to Lord and Lady Nothril, and already, Rahk, Pelarusa, and two other princesses I do not know have slipped into the ranks of the Nothril Court.

Faradir lifts one hand to Lord and Lady Nothril—a farewell. Then he turns his back to them and marches into the palace.

The Nothril entourage likewise turns and marches away. Rahk is the only one who looks back, who finds Ash and me in the throng of onlookers. His face is an impenetrable mask, but the gesture itself says more than enough.

I peer up at Ash, at the grim set of his handsome brow, the tight angle of his jaw. Still, the light remains in his eye, that fierce glint that at times has scared me. The glint that says he will play this game as ugly and messy as he must to win.

He takes my hand and guides me back into the palace, his gaze constantly roaming, his nose upturned for the scent of threats. He leads me through quieter passages, away from the din. I make note of every turn, memorizing this palace that—if all goes well—will be my home for the rest of my life.

Maybe.

"Ash," I whisper.

He looks down at me, and with a flick of his wrist, the scattering servants' and guards' footsteps around us are cut off, leaving us in a silent bubble.

"He's going to retaliate," I say.

"The Neverseen King will distract him."

He says those words, and they are true—but we both know they shield a lie. He knows as well as I that angering the High King always, *always* has its price.

"I've spent too much time defending, and not enough time attacking—or attacking for foolish reasons," Ash says quietly, his large hand gripping mine. "This time, I'm attacking from every front. Yes, he will retaliate, but I intend to set enough fires ablaze that his attention grows divided. Then we can swoop in for the final blow. Speaking of which, we have one final trap to lay."

CHAPTER 53

THE PRINCE

I'M READY FOR the celebration of Lulythinar's Eve long before Stella is. Dottie is just as particular as Hylath, and insists I am not to see my wife until she is completely ready.

So I sit, staring out at the sun setting over the garden outside. Butterflies dance among the flowers, their wings incandescent and shimmering, catching the rays of the dying sun like fluttering jewels.

They are things of such beauty.

And they are *oh* so fragile.

My heart has beat an irregular rhythm since the visit from the Nothril Court.

He will retaliate.

He knows how to make me regret this.

One butterfly in particular catches my attention. Its wings are iridescent purple with pink eyes and silver glitter. It goes from flower to flower, then flies into the air and twirls with its friends.

It's so oblivious to the dangers of the world it lives in. So oblivious to its own vulnerability. And even to its own beauty.

As the room darkens around me, as the sun descends lower on the horizon, I cannot help but voice the fear I've been trying to suppress all day. The fear that has plagued me from the moment I first laid eyes on Princess Isabelle Louise.

What if this doesn't work?

What if I risk it all . . . and lose?

What if I lose her after all?

I want to be fearless, to be the rock and shield she needs. I'm being careful, so careful.

But I'm terrified.

That's the honest truth. I've been trying to hide from it, to convince myself I know exactly what I'm doing. And I do know what I'm doing. I wouldn't be going through with this if I didn't believe we had a chance.

This can work.

But the possibility of success doesn't make this game any less treacherous.

If I could wager my life instead of hers, I'd do it in a heartbeat. I desperately *wish* it could be my life on the line. But my life has never been at risk, and it won't be until I have a wife who bears me a son.

It's every other life in Valehaven, in all of Faerie, that is at risk.

It sits wrong in my bones that the one person who has no life at stake must be the one making the moves for everyone who does. That I must wager my wife's. I cannot even spare her from the game, despite that she is the reason I make my moves.

Darkness overtakes the garden as the sun dips below the trees.

Can I live with myself if I fail?

Can I live . . . if she dies?

Is it selfish to be as afraid of losing her as I am of the darkness that would follow it? To be condemned to a long life after losing my purpose?

It's a question that terrifies me. A question I don't want to answer.

But I need to know the answer.

My ears prick as the washroom door opens. Immediately, I stand.

"Close your eyes!" Stella calls from down the hallway. "If you're to get the full effect, you can't see me trying to fit through this hallway in all these skirts!"

My lungs tighten. I close my eyes and croak back: "Alright, then."

"Are they closed? You better not be using fae trickery on me!"

My mouth twitches. "They're closed."

"Do you promise not to open them until I say? No games where they're closed this second, but open the next?"

I cannot help but smile. "I promise not to open them until you tell me."

"Don't make me bargain with you. A tattoo would completely ruin my outfit."

"I won't look!" I insist, laughing despite myself. "If you don't hurry, I might lose my balance with my eyes closed for so long and faceplant on the floor."

"Always dramatic!" she huffs, and the sounds of rustling fabric reach my ears as she struggles to make it through the hallway. "This is the price I pay for marrying a clever fae who likes to outwit me."

"It's getting harder by the day," I say, my eyes creasing upwards.

She lets out an unladylike *harrumph!* as she turns the corner, and it might be my new favorite sound of hers. "Harder to what?" she demands, sounding breathless as she enters the living quarters.

"To outwit you."

"Don't lie to me."

"I'm not lying! If you were a fae, you could smell the truth in the air." She's quiet for a heartbeat longer than I expect, so I add, "I never said it was *hard* to outwit you, only that it grows more challenging by the day."

Something sails through the air next to my face. I lift one eyebrow, keeping my eyes closed still. "Did . . . you just throw something at me? I shall not comment on your aim."

Instinctively, I duck—just in time to dodge another missile. It hits the ground with a soft thump and rolls.

"Hold still and let me abuse you!" Stella demands, and it sounds like she's waving her next object threateningly.

A chuckle rumbles through my chest. "I'm going to open my eyes now."

"Wait, I have to get my face in a prettier expression!"

I draw in a deep breath, let it out in a gusting sigh. I tilt back my head impatiently. "Stella."

"Fine, you can look!"

I open my eyes.

She stands before me in a color I've never seen on her—a dark, deep blue. The gown is lush, billowing, and yet not at all bombastic or pretentious. Frothy white layers peek through the skirt, making the entire dress look like a midnight ocean with rolling waves. It shines and sparkles as though the water's surface reflects a myriad of stars above it. Her shoulders are bare, but the slightest shimmer on her arms betrays the thin gloves she wears of spider silk.

The most startling transformation of all, however, is the silver hair swept on top of her head in a crown and trailing to her elbows in long waves. A new tiara rests atop the elaborate style and glitters with dark sapphires and diamonds.

Her ears are not covered, as per my instruction, and there's something about their rounded tips that always fascinates me and draws my attention to them.

But it's hard to think about her ears when her big eyes are staring at me, lined with silver and tiny precious stones. Her lips are a warm lavender, highlighting her unusual human beauty yet mingling it with something entirely ethereal. It's like she's from another world—from beyond the Veil—and yet the smallest things like her downcast gaze and the fan of her lower lashes on her flushed cheeks ground her in enough reality that I know I'm not hallucinating her existence.

She goes blurry until all I see is one blob of midnight and silver.

"Oh, Ash!" she cries, rushing to me at once. "You're not supposed to cry!"

I cannot deny my own tears without suffering a mouthful of iron, so I only offer a watery smile, catch her chin, and tilt her face up so I may kiss the tip of her nose.

"You're thinking sad thoughts," she whispers, her glittering eyes searching mine with that infinite gentleness. "You're thinking about me dying."

What's the use in denying it?

She frowns. "You have the expression of a baleful hound dog hoping for scraps from under the table. It doesn't suit how dashing you look. Shall I tell you the real reason I didn't let you open your eyes for so long? I had to let myself admire you first. I couldn't let you see how red my face had gotten when I walked in and saw you in your own dark blue tunic. Has anyone ever told you that those boots make your calves look so handsome?"

I grab the back of her neck and pull her into a kiss, pouring every one of my sorrows and the depths of my love for her into our kiss. My fingers thread into her hair, tightening as my brow furrows, and I kiss her harder, deeper, savoring every sweet second as if it will be our last.

When the kiss ends, we stay there, our foreheads pressed together and our open mouths only an inch apart.

"Ash," she breathes.

"Stella," I murmur in reply.

Then I pull back, try to smooth over any hair of hers I disrupted, and give her glittering cheek a caress with my thumb.

"Are you ready?"

She nods and tries to discretely wipe away a lone tear that threatens to smear her carefully applied paint. "And just like you said, not a shred of glamour. Not until *the moment*. Is everything ready?"

I pat my breast pocket, where a tiny vial rests against my heart. It seems to thump with my pulse, as though it knows exactly what it's for. "Are *you* ready?"

Stella smiles at me, a slow, determined thing that, if he saw it now, would terrify the man she called Father. It gives me the rush

423

of courage I need to take her hand and lead her out of the safety of our quarters.

She squeezes my hand. "Let's go win a crown."

I squeeze back, swallowing hard. I close my eyes briefly, and in the quiet of my mind, I remind myself why we're doing this. Why I'm fighting with everything. It's all for Stella, for the chance to spend the rest of my life with her. I need this moment. I need the strength it gives me to pull my old mask down over my face and face the world of Faerieland and High King Faradir.

Because tonight, I'm going to poison my wife.

CHAPTER 54
THE PRINCESS

TONIGHT'S BANQUET ISN'T in the banquet hall, like usual. Instead, it's on the palace greens. When we reach the marble steps descending into the revelry, we stop and wait as the crier announces us.

I stand there, holding Ash's hand, wearing not a single glamour. It's not unlike my first night in Faerie, when Ash threw that revelry to celebrate our marriage. The night multiple people tried to kill me—likely to please the High King and bring him into their debt.

Nothing has changed. And yet, everything has.

The traditional color of Lulythinar's Eve is midnight blue—the color of the almost full moon's missing sliver. Apparently, everyone wears the color tonight, and then tomorrow night, at the culmination of the celebration, there is a masquerade ball.

Ash guides me down the steps, supporting me so I don't accidentally trip over my gown. It always surprises me when I look down and find locks of silver hair falling over my shoulder. As though hearing my thoughts, he says, "The silver is lovely, but I confess I prefer the natural color of your hair."

"Oh, I do too, but this color suits the overall look much better. I never would have thought of it if Dottie hadn't suggested it. Wouldn't you agree?"

He shrugs, smiling softly. "Maybe. I still like your natural color more."

And then he looks at me, and perhaps it's just how breathtakingly handsome he is, or his expression, or the knowledge that after tonight is over—assuming nothing goes wrong—I will slip under the covers of his bed and fall asleep in his arms, but for several heartbeats, I cannot breathe.

"What is Lulythinar celebrating again?" I ask, breaking our gaze to take in the trees strung with lumiral globes, the rings of dancing horned fae, the half-eaten faerie fruit scattering the green grass, the High King sitting on a throne amid the throng, his sad wife standing with bowed head and folded hands beside him.

"Magic across all the worlds waxes and wanes with the moon." Ash's gaze follows mine to where Faradir drains a goblet and trades it for a full one, sloshing the liquid in his impatience. "In the human world, a moon's cycle is a month. Here, it cycles every hundred years. Even those beyond the Veil feel the shift in the magic. The human world feels it too, just to a lesser degree with its lack of magic."

With that, he signals a human servant over and takes two flutes full of sparkling gold liquid off his tray. My heart starts pounding when he dumps one of them in the grass and, just like last night, refills it with apple juice. Everyone knows he won't be taking chances now that it's rumored I've been poisoned once already.

But just as I am accepting the glass, a voice I never like to hear calls: "Prince Trenian!"

Ash turns, taking my elbow with his free hand and drawing me halfway behind his back. The moment Princess Listhra stumbles up to him, almost knocking into his shoulder, I'm glad for it. Her eyes shine too brightly, like twin suns at noon. She catches hold of the front of Ash's tunic, trying to drag his face down to hers. He doesn't move an inch, but his grip on me tightens.

"I don't know what you did," she snarls. "But I am going to—"

"Oh dear, it looks like someone isn't quite in their right mind," a new, sweet voice interrupts. I turn just as Princess Oleria appears out of nowhere, takes Listhra by the shoulder, and somehow manages to pull her off Ash without seeming at all graceless. Perhaps it's because she does it with a laugh, catches Listhra as she stumbles again, and turns to wink at us. "Do forgive her. A few dunks in the waterfall should be just the thing to sober her up."

She flutters her beautiful wings lightly and prances off with Listhra, who is quick to turn her snarl on her rescuer.

I hold my flute of apple juice with a sweat-slicked hand, unsure if I should drink now, or wait. I don't want to be nervous, but I cannot help it. Ash told me what to expect, so it's not as though I will be blindsided. Still, it won't be fun or easy.

But when I look up at Ash, his gaze follows Oleria and Listhra as they disappear, a new glint in his eye. That cunning gaze shifts to me and my flute of apple juice.

He looks like he's just found the last piece to a puzzle I didn't know he was building.

Then he takes my arm and purposefully draws me into the thickest crowd of celebrating fae. Someone shrieks in delight far too close to my ear, others whirling past me in some strange dance, and more so drunk they'd put Prince Brochfael to shame. I stick a little closer to Ash and cling to my flute of juice, making sure it stays far away from anyone who could *actually* try to slip something into it.

When Ash grips my waist and gives me a sudden twirl, I force a laugh—just as he's forcing a wicked grin. He drags me in for a kiss, and the surrounding people explode in a cacophony, some approving, others disapproving. At that, both of our smiles widen in genuine amusement.

Then he holds out his glass, clinks it on mine with a laugh. "It's Lulythinar's Eve! Let's celebrate!"

Only I notice the darkness in his eyes when I clink my glass against his . . . and drink.

CHAPTER 55

THE PRINCE

THE POISON DOESN'T affect her immediately. I can barely breathe in fear that this might go wrong somehow, but I must pretend that we are having the time of our lives. So I take her by the waist and lift her straight off her feet, twirling her in the air and laughing. All the while closely watching her face for signs the teeny-tiny dose of poison I gave her is taking its toll.

It's fast acting, but not especially painful. It's one of the poisons I take daily, one of the poisons that, since the blood transfusion when she was ill, she should have some immunity to.

Some immunity.

I didn't give her much. Not enough to kill her, but enough to make the next few hours unpleasant. Enough that my test for Faradir should go undetected.

Stella, still holding on to her glass, smiles at me and lets me whirl her around the crowd as music dances through the air. Nerves edge her smile, but I can only tell because I know her so well. She's a good little actress, a surprisingly good one considering that her temperament doesn't naturally lend itself to it.

"Don't make me dance until my feet bleed like those other fae," she says abruptly, as though feeling the intoxicating pull of the music, the waiting entrapment for unsuspecting humans.

I pull her a little closer at the mention of that night, when Rahk and I found her in Listhra's chambers. "I will keep you from the enchantments," I promise her. "If you dance with me, you will be safe."

That's when she suddenly staggers.

"Stella!" I cry, and the alarm in my voice is completely real, completely raw. Even causing it, expecting it, I cannot help how it hits me when she falls against my chest.

Her skin turns gray in a matter of seconds. I sink to my knees, holding her in my arms as her eyes roll back, as the strength in her neck gives way and her head falls to my elbow. Her long silver hair pools on the ground with her glorious skirts.

"Stella? Stella!" Supporting her suddenly weak frame in one arm, I catch her face with my other hand. She's *so* gray, so suddenly. Rationally, I knew this was going to happen. This is part of the plan. This doesn't mean I measured the dosage incorrectly. This doesn't mean she might not have the immunity I thought she would.

But the thrum in my veins is entirely real.

It's like I look at her in my arms, at her gray skin and the blue tint of her lips, and I see my worst nightmare. Around us, the dance and music has stopped. Hundreds of curious eyes burn into my awareness, but I don't look up.

My fingers clench into her hair. "Stella!" It's a choked cry, a raw scrape of sound from my throat. "*Stella,* open your eyes. Open your eyes!"

She opens them slowly, meeting mine. Her lashes flutter weakly. Soft brown irises fill my vision.

And then they turn vacant.

Empty.

Her chest stops moving.

"Stella—*no!*" I scream. My grip in her hair turns frantic, sharp, and no matter how many times I try to tell myself that this is only the exact glamour I told Stella to wear, that she's not dead, I cannot breathe around the stabbing pain in my own chest.

She's gone. She's gone.

She's not gone, I snarl to myself. But my rational mind doesn't care.

I bow my head over Stella's silver one, clutching her to my chest as tears pour in rivers down my cheeks. I wish I was acting. I wish this were a façade.

"Well, isn't this an interesting tableau?" says a voice as deep as midnight.

The entire air of the celebration shifts in that instant. Curiosity switches at once to tension, even fear, perhaps a thrilling sort of anticipation.

Because the voice belongs to the tall figure, standing hardly a few feet from where I am collapsed on the ground with my seemingly dead wife. A figure wreathed in shadows like a whirlwind of soot and dead embers.

The Neverseen King.

He approaches me, his shadows pulling back from his hooded face long enough for me to make out the smirk tugging at his lips. "I see you played with fire and got burned, my dear cousin."

And, strangely, it is his arrival that makes my heart remember this is only a ruse. A ruse that will give me valuable information about whether Faradir can see through Stella's glamours like he can see through any fae's.

"Did you do this?" I snarl, clutching Stella's body to my breast. "Did you poison her?"

The Neverseen King only chuckles darkly and turns away—toward where Faradir sits on his throne, pale in the darkness. "It is a good thing I didn't poison your only heir. That would have made for some thrilling court dynamics."

"You should be at the Bridge," Faradir snaps. "It's almost Lulythinar."

431

"I thought I would pay a visit. Part of me misses these celebrations. It's been a long time since I've enjoyed a Lulythinar." He spreads his hands wide, and when I look up from Stella, I'm just in time to catch Faradir give the barest flinch.

Because we all know that the only fae with comparable power on this side of the Veil to the High King is the one who has spent the last three hundred years keeping the doors between worlds. The one who can beckon armies with a single drop of his blood.

Faradir clearly feels what we are all thinking: that if the Neverseen King attacked him this instant, the High King might not win. So he stands, his long robes of dark blue a contrast to his golden skin and hair. "I still remember when you were such a young boy. I remember my sister's pride when she told me she was with child. When I loved you nearly as much as if you were my own."

"Thank the Great Kings you didn't," the Neverseen King says with a wry chuckle and a pointed look at me as I hold my glamour-dead bride.

"Did you do this?" I snarl once more at him, pitting myself against him so the High King doesn't realize I'm behind this, too. When he doesn't answer, I throw back my head and bellow through my raw throat: *"Who poisoned my wife?"*

That is when Faradir finally looks our way. *Truly* looks.

And then it's as though he completely forgets about his greatest threat standing before him. His bright gaze latches onto Stella in my arms, her gray skin and wide-open, sightless eyes.

He turns white as the Lulythinar moon above us.

He reaches back, grabs hold of the armrest of his throne, stabilizing himself. One shaking hand lifts, points at Stella.

My heart goes to my throat.

The Neverseen King, and everyone else, has gone silent as death.

"What *is* that?" Faradir growls. His tone rises in panic. "What *is* that?"

At first, it takes me a moment to understand his reaction. Then it hits me.

432

He can see that she's glamoured. But he cannot see through that glamour.

That is the information I needed tonight.

And . . . it's not what I hoped for.

I cup the back of Stella's lolling head, tuck it against my heart, and my voice echoes through the silent crowds, through the shadows the Neverseen King wears, through the lumiral globes strung through tree branches: "Whoever did this, I will make them pay."

Then I scoop my wife up in my arms, and get halfway back to the palace before the High King's furious voice cuts through the confused din behind me.

"Guards! Bring back to me what he carries! The human is *not* dead!"

I curse under my breath. "Sorry, darling," I mumble to Stella as I sling her over my shoulder and break into the fastest run of my life. Rushing air dries my wet cheeks, and my grief from a few minutes ago is entirely forgotten, transformed into pure determination that I will get my wife back to my chambers whole.

CHAPTER 56
THE PRINCESS

I BARELY MAINTAIN my glamours while slung over Ash's shoulder like a sack of rice. It's harder than I thought it would be to maintain my death glamour while fighting the poison's influence spreading through my body. Harder still to listen and try to understand what is happening around me.

The sluggish weakness in my limbs makes me feel like a trapped ragdoll. Completely helpless and vulnerable. Ash's shoulder digging into my hips only makes it worse.

Still, I cannot hide from the glow of Ash's sword when he draws it, the sudden spatter of bright blood on the marble steps, or the guards' cries of pain that cut off abruptly.

Then Ash is running again, and I wish my hands were strong enough to catch hold of his tunic and hold on. The fog in my mind and ache in my body clears enough for me to be worried Ash will trip on my gown while he's running, and then the thought is gone like all my strength.

"Prince Trenian!" someone bellows. More guards?

Ash doesn't turn, doesn't acknowledge them. He only races through the empty palace hallways as though his life depends on it.

I release my glamours, unable to hold on to them any longer. It probably makes no difference anyway in my appearance. My flopping arms are gray as death.

Then, at last, a door opens, and the smell of our quarters washes over me. I sag in limp relief when the door shuts once more.

"Edvear!" Ash shouts, carrying me toward the couches before laying me down. The knuckles he brushes over my cheek are flecked in blood, but he doesn't seem to notice. I barely have the strength to roll my eyes toward Edvear's stubby horns as he comes running from the kitchen.

"My lord!" he cries, his mirror clutched in one hand. "I sent one of the servants to monitor everything tonight, and . . . and . . ."

"Did the High King get him?" Ash demands, his touch turning cold. He is kneeling beside me, and from my vantage point, all I can see of him is his tense brow, the lock of long hair falling over his circlet crown, and his beautiful eyes, now darkened with something akin to panic. "Did he get another of my servants?"

"No, no, my lord. It's . . . Right after you ran . . ."

Ash snatches the mirror from Edvear. His hand slides from my face to grip my shoulder, and his gold-flecked eyes harden as he clenches his jaw.

"Wh-what—" I start to ask, barely able to make my lips and tongue move.

Ash's hand on my shoulder tightens. He hands the mirror back to Edvear, briefly turns his face away from me.

My heart quickens with dread. I try to rise, can only find barely enough strength to lift my head. "Ash, what—"

"He executed Listhra's friends," Ash grinds out. "Lady Shalar and Princess Brolnyr. Except Lady Iluna."

"The Neverseen King left when you did," Edvear says, his always steady voice shaken. "It was immediately after that when he ordered them brought forward. One of them is a princess. The Courts will riot!"

436

"They will," Ash agrees grimly. "After the Nothril Court and the Neverseen King, Faradir wants them to be angry. He wants to meet them in force. In *war*."

"My lord, this is very, very bad—"

"I've got it under control," Ash replies, a little sharply. "If Faradir wants a war, then we'll give him one."

"But surely there has been enough death!" Edvear pleads, and all that comes into my range of vision is one of his outstretched hands. "How much more can we endure? How much more until we admit the High King is *the High King* and we cannot gainsay him?"

Ash's hard as flint gaze drops to me, drifts down to my chest as though to reassure himself I am still breathing. "Get Princess Stella's antidote immediately."

"My lord—"

"Immediately!"

Edvear shuts his mouth, and I can almost hear the desperation in his retreating footsteps.

Ash turns his full attention back down to me, his expression softening into worry. "Can you feel this?" He runs his hands up both my arms. I give an almost imperceptible nod. "Are you in pain?"

That's a question I'm not certain at first how to answer. I am overwhelmed with weakness, with the ache that comes with it. But I don't think I would describe it as painful, exactly. I shake my head.

"Good. You should have your strength back by tomorrow morning at the latest. If you still have immunity from my blood, it will be much faster." He looks up when Edvear reenters—except it's not Edvear, it's a different servant I haven't met. He gives Ash two large leaves the size of my palm. Ash takes one leaf, rolls it up, and bids me open my mouth. "Can you chew?"

I do, and the unusual flavor that fills my mouth I can only describe as vaguely minty.

"Good, good," he says gently. "You don't need to *eat* the leaf, but do chew and suck on it. When you're done, I'll give you the second."

He makes to stand, but I reach out with a flailing arm. He stops immediately, turning his concern back on me in full force. "Stella?"

"Ash." Speaking with the leaf between my teeth is a challenge, but I can't let him go just yet. Not after what I saw in his eyes when I asserted my glamour at the celebration. "Are you alright?"

A shadow crosses his face. He takes one of my hands, turns it up, and kisses my palm softly. Wordlessly.

Then a soft but hurried knock raps on the door. Ash gets to his feet, his hand going to the hilt of his sword. Feeling bits of strength returning the longer I suck on this leaf, I turn my head to follow him to the door as Edvear darts out of the kitchen.

"Hurry!" the steward says. "The footman should be back from—"

Ash opens the door—

And a rush of blue fabric and flashing gold eyes come flying through the opening. Ash, letting out a grunt of surprise, barely dodges the sword Listhra swings straight at him. Someone is screaming, but I can hardly tell whom.

My blood spikes in my veins. I think my husband's name rips from my throat in a frantic cry, but there's a sudden clash of blades, the sound of glass shattering, and such a cacophony I can barely think.

I can't stay lying on this couch.

Especially as Listhra's wild eyes latch on to me, lying helpless with a stupid leaf sticking out between my teeth.

Listhra dives for me, using her wings to lend her speed. Ash snarls something vicious, grabs her ankle, and yanks her back so violently she hits the wall with enough force to send a crack shooting up through the plaster.

Frantically, I roll off the couch, land in a heap of limbs on the floor, spit out my leaf, and try to crawl away. My legs give out. A frustrated, panicked whimper escapes me as something flies through the air, hits the couch just above my head with a sharp thunk. *A knife.*

If I was any less scared, I might have been able to scream.

"You wicked soul!" Listhra plunges her blade toward Ash's stomach, and he ducks to one side, rolls, and grabs her neck. She screams, then sinks her fangs into his shoulder. He lets out a hiss, tightens his grip on her neck, and slams her head into the wall.

"Don't make me kill you," he seethes.

She chokes, her eyes rolling back.

I grab hold of a couch leg and pull, trying to drag myself away. My arms aren't strong enough. I fall to the floor again.

"Get her out of here!" Ash yells, and at first, I think he means Listhra, but when Edvear's panting breaths hit my face, his clumsy hands grabbing my arms, I realize Ash meant me.

"You killed them!" Listhra shrieks.

"You need to leave." He has her pinned against the wall, her wings flattened. "I never wanted your death or anyone else's, but you well know that I am more than willing to shed blood." Heavy breathing punctuates his words, suggesting that she is vastly stronger than she looks, and keeping her from throwing herself at me is taking much of his strength.

Edvear tries to help me to my feet. My knees buckle immediately, and Ash shouts, "Just drag her! Get her out of here!"

So Edvear takes hold of my wrists and does exactly that—drags my limp body to the hallway.

It gives me a full view of the small orange glow Listhra is summoning in the palm of her pinned hand—a glow that Ash hasn't noticed.

Listhra's furious, hate-filled golden eyes meet mine, and her palm twists toward me.

Aiming.

"Ash!" I scream.

My cry is enough.

A loud snapping fills the room.

Listhra's body sags. The glow vanishes.

I barely have the presence of mind to realize that I just watched my husband snap Listhra's neck with his bare hands before Edvear drags me around the corner and out of sight.

CHAPTER 57

THE PRINCESS

I'VE NEVER SEEN so many servants around. I didn't even know Ash kept so many employed. But apparently there is a lot of clean up to do. When the missing footman finally returns from watching the celebration, I recognize him as Milton Andrews, the man who promised service to Ash in exchange for the rescue of his daughter from Listhra. He works harder than the rest, righting overturned furniture and scrubbing blood off the floor.

I'm propped up against the hallway doorframe amid the flurry of movement while Dottie draws a fresh bath for me. Ash left with Listhra's body. I don't know where he's taking it, only that when the door opens and his broad shoulder and bowed head come into view, I'm greatly relieved.

He comes straight to me. Kneels before me, carefully cups my face.

"Are you alright?" we ask at the same time.

"Your shoulder?" I ask. Two dots of blood mar his beautiful tunic.

"It's fine; it'll heal quickly. You're less gray," he says. "That's a good sign. Is your strength returning at all?"

By way of answer, I grab my skirts and tug them up enough to reveal one bare foot. I splay my toes for him. "I can do this."

His serious expression cracks on the edges, his mouth twitching in a faint mimicry of a smile. "Impressive."

"I'd show you my other foot, but the shoe is still on it. This one must have fallen off at some—"

He catches my face and claims my mouth in a bruising kiss. Servants bustle quietly behind his back, but it feels like I am in a cocoon of his desperate affection, shielded from all who wish me harm.

It takes every last ounce of my strength, but I lift my arm, thread my fingers through his hair at his temple. Telling him without words that I love him, that I am so grateful for how he protects me.

The water shuts off in the washroom. I open my eyes at the same time Ash does. The look he gives me—it takes me back to another world, another night, when he lifted my veil after our wedding and our gazes met for the first time. I smile.

"I love you," Ash breathes, his voice giving out halfway through the sentence.

"My lady? Are you ready for your bath? Oh!" Dottie stops in her tracks when she sees Ash bent over me, and how close we are.

"She's ready," Ash replies easily, and the emotion choking his voice only a second ago is gone. "Would it help if I carried her?"

"Oh! Well, I . . . yes, yes, my lord. It would be of great help."

Ash scoops me up as though I don't weigh twice what I normally do in this gown. He takes me into the washroom with its tangled overgrowth along the walls, ceiling, and even floors. As soon as he crosses the threshold, he frowns, glancing around as if trying to decide where to put me.

"You should put me in the bath," I say from where my head rests against the column of his neck. The bath is so sudsy and inviting, with steam coming off the top of the water. At this moment, it looks so relaxing I think I could drown in it and still be happy.

"Fully clothed? In your gown?" Ash tsks his tongue and kicks out a stool, dragging it with his foot toward us. He gently sets me on it, holding me up with his hands on my shoulders so I don't pitch forward.

"Couldn't you have given me poison that makes me extra strong?" I whine, hating how my limbs refuse to obey me, no matter how many mental commands I give them.

"I should have, shouldn't I?" he replies, another hint of a smile in his voice. Then, letting me lean forward on one arm, he sets to undoing the tiny buttons down my back with one hand. My eyes widen in surprise. A hot flush climbs up my neck. I didn't even think I could muster one while poisoned. Well, here we are!

Dottie enters and shuts the door, mumbling her thanks under her breath. She kneels in front of me, confiscates my last shoe, and then sets to taking down my elaborate updo. All while Ash steadily makes his way down the buttons lining my spine.

I'm not sure whether to be more grateful or embarrassed to have two people helping me undress. I think I would rather have it be two maids . . . or just Ash. Not this mix of Dottie and Ash. Because with each moment, I'm so aware of all the lines Ash and I haven't crossed, lines that—if I'm honest—I long to. I want to discover what it is to be fully and completely his wife. Part of me is afraid I never will.

Ash reaches the last button, undoes it with a slip of his fingers. Then, to my shock, while Dottie is turned away setting hairpins on the counter, he runs his knuckles in a scalding line down my exposed spine. My breath snags. The heat in my cheeks redoubles.

Then she is back, and I try to not look like a naughty child caught stealing a forbidden cookie.

Behind me, I can almost *feel* the beginning of Ash's smirk. I suppose if my tomato red face cheers him up, I can put up with feeling a little dizzy in addition to having no strength in my body.

She removes my gloves, and the moment she turns away to set them down, Ash leans forward and nibbles my ear. I barely restrain my gasp.

"Would it be best if I held her up so we can get the gown off?" Ash asks her.

"Yes, my lord. That would be easiest."

A bolt of panic hits me, sharper than I expect. I'm not wearing much beneath this gown, and this isn't how I want him to see me for the first time. I want him to see me when I feel beautiful, not when I can barely hold my head up and my skin is cast a horrible hue of gray.

"Let's finish taking her hair down first," Ash says. He probably feels how stiff I've gone against him. Perhaps he knows why, perhaps he doesn't. But once all my hair is free from its pins and tiara and other decorations, he takes its long locks and arranges so it falls down my back and chest all the way to my hips.

He is discretely covering me so I will not feel so exposed.

My throat thickens.

He presses a kiss to the back of my head, then gently takes me under my armpits and lifts me enough off the stool so Dottie can shimmy the gown down my torso, hips, then legs. He sets me back down, and the stool is cool against the thin fabric of my undergarments. Self-consciously, I wrap my arms across my chest, only to find that my hair is doing a surprisingly good job of keeping me decent.

Dottie gathers the gown up in her arms and carries it out of the washroom.

For a moment, we're left alone.

Ash's breath caresses my ear as he leans close to me. "You have nothing to be afraid of, darling wife. I will not do anything against your will."

"It's not my will I'm concerned about!" I blurt, without thinking.

"Yes?" he prods, his tone almost painfully gentle.

Footsteps outside make me stiffen, and I barely have time to say: "I don't want you to see me when I'm all gray!"

Then Dottie is back.

"If you wouldn't mind lifting her once more, Highness," she asks.

Ash obliges. I restrain a whimper when she strips off my undergarments and I'm left wearing nothing but my long silver hair. Except the next second, I'm swept up in Ash's arms and lowered into the steaming bath and its mountains of bubbles.

The heat and privacy are more than enough to make my entire body go liquid with relief. Ash leans me back against the tub but keeps a secure grip under my arms, so I don't slip beneath the surface and drown.

The silver of my hair leaks out into the water, making it shimmer like stardust. Dottie scrubs me while Ash keeps me upright. His forehead presses to the back of my neck. As though communicating both his affection through his touch and his assurance that he is not looking while my maid works.

It astounds me all over again—the way he loves me. The extent he goes to care for me, my feelings, my worries, my insecurities. It goes against reason that I could feel so safe with a prince of the fae. A man who broke someone's neck not even an hour ago.

I love him. I really do. And I have a feeling that every day we have together, that love will only deepen. I hope I have the opportunity to look back on this moment one day and know that I love Ash more than I could have dreamed of now.

He and my maid work together to wash my hair until it's back to its normal color. Ash gently pulls the length of it out of the tub and dries it with his magic while Dottie cleans my face of paint, glitter, and gemstones.

"I should be helping," I mumble, even as I want to fall asleep with the feeling of Ash's hands in my hair.

"You had one job tonight, which was to be poisoned and glamour yourself. You succeeded spectacularly."

"Did you get the answers you were looking for?" I keep my question vague for the sake of the maid's listening ears.

Ash goes quiet for a minute. Then, at last: 'Yes."

But it wasn't the information he was hoping for. I can hear that much in the timbres of his voice.

"She's finished, my lord," Dottie says with a bow. She fetches a towel while my heart, which had calmed to a comfortable rhythm, skips five beats all in a row.

"I'll close my eyes," Ash whispers as she steps away. Then he drops his voice, and his lips curve against my earlobe. "Regretfully."

I have no time for a retort, but the bath water is suddenly much too warm for my skin. Still too weak to rise on my own, Ash helps me to my feet and lifts me out of the tub. In the mirror I can see that he dutifully keeps his eyes closed while my maid dries me off and helps me into a lavender night dress.

"You are finished, my lady. Can I be of further assistance?" Dottie asks. Somehow, despite the work she has put into my appearance today, not a stitch of her starched apron or a single hair from her tight dark bun is out of place.

Ash opens his eyes, finds mine in the mirror's reflection. "Thank you. I can take it from here."

She bows and leaves, shutting the door to the washroom behind her.

In one fluid motion, Ash scoops me onto the counter. I gasp, surprised. At least I can hold my head up now, so I don't fall forward when Ash leans over me. He plants a hand on the counter behind me so I can lean against his arm as he tilts my chin up.

His bright eyes study my parted lips, drift down to the gray skin of my neck and collarbones.

I wait, breathless, until his eyes lift to mine. His fingers let go of my chin, trail down my shoulder, my arm, until his warm hand comes to rest on my thigh. Still, he holds my gaze.

"I have always found you beautiful," he says at last. "*Always*. From the moment I lifted your veil, I found you beautiful. I have no idea what flaw your father saw in you, but he was wrong. I understand why you don't want me to see you like this—and I honor that. But Stella, nothing could mar your beauty. Not in my eyes. I thought you just as beautiful while you were ill as you were in a ballgown. As for tonight, you drank poison for me. For us. You've endured its effects like a warrior. Do not think I would look at you in disgust. Quite the contrary. When I look at you now . . ." His hand slides to my hip, tugs me closer to himself. The other hand on the counter shifts to my back and trails upward in a hot, tingling line. "I am almost afraid of how deeply I long to know and love all of you."

His handsome face goes blurry in my vision. My lungs squeeze painfully tight, from the sweetness of his words and my frustration with our situation—that we should be married but still not fully each other's—and my poisoned body for its weakness.

His fingers thread through my hair, tug my head back so my mouth is just below his. "But in moments of struggle, I try to remember that every sacrifice will be worth it in the end. Sometimes I cannot quite believe it, but if I let go of that belief . . ."

"It'll be worth it. I know it will be," I say as the tears clogging my vision slip down my cheek and Ash's face comes back into focus.

He closes the distance between our mouths. Our souls collide with our lips, hungry and satisfied at once; full and empty both. He deepens the kiss as his hands caress my back, my waist, my leg. I close my eyes and let myself fall into him, knowing that at every turn, at every opportunity, he will choose me.

It's the only true home I've ever known, and it is worth more than its weight in precious jewels.

He doesn't stop kissing me as he picks me up and carries me to our bedroom.

He doesn't stop when he lays me down, pulls the covers up to my chin. When he starts to withdraw, I find enough strength to push myself up, to wrap my arm around his neck and insist our lips don't part.

"Oh Stella," he groans. "You are making me lose my mind. And every shred of my will."

Carefully, he scoots me toward the middle of the bed, then slips in beside me. The mattress dips from his weight, and then his arms come around me.

"I want you to tell me all of your secrets," I whisper to him, snuggling closer. I don't want to say the words aloud, but my gut twists, part of my soul insisting that this might be our last night together. I don't want to waste a minute of it. "I want to know about the things you hide beneath the surface. The things you haven't told anyone."

The things you may never get another chance to tell me.

He is quiet for a few minutes, the tips of four fingers running up and down my back in gentle caresses. "Then I will bargain with you, wife."

"Bargain?" I push up on my elbow, shooting a glower at his darkness-shrouded face. "I—"

Teeth flash back at me in a grin. "Put away your claws. I wish to bargain with you—a secret for a secret. I tell you one of mine. You tell me one of yours."

I give a silent huff, then settle back against his chest, tucking my arms between us. "I don't want a tattoo."

He laughs. Playfully nuzzles his nose into my face. "We don't have to make it a magically binding bargain. Shall I start?"

"You may."

He chuckles again. "Very well. A secret I have told no one . . . hmm . . . Here's one. Rahk and I became friends when he found me crying as a young child and pitied me."

"Is he older than you?"

"He is. Though not by much. He cheered me up by standing on his hands and purposefully falling over."

I giggle. "That's adorable."

"He would probably not appreciate me telling you."

"I'll be certain to blackmail him with it at the earliest opportunity."

I yelp when Ash gives me a light pinch. "Behave," he chides, amusement limning his voice. "Now it's your turn."

I give him another huff and a glare. He only smiles in return. "Fine. I was terrified of you when I first heard I was going to marry you."

"That's not a secret. Try again."

I wrack my brain. Memories come back, of my father, my sisters. The way I could barely talk. How small I always felt. I never told Ash what I experienced while I was ill. It hadn't felt relevant anymore. But perhaps it's time to acknowledge that.

"One thing my father always thought important for my sisters and I to understand was that we were extensions of the crown. Bound to honor our people, to always seek their interests over our own. To sacrifice ourselves for them. And now, I see how well-meaning that goal was. How noble. Yet the means he used to teach us were . . . harsh."

"I hate him," Ash growls. His touch shifts ever so slightly, growing more protective as I speak. I doubt he even realizes it. I notice, however, and it soothes any lasting fear, any lingering pain from those memories.

"When we were each around seven or eight years old, he would take us down to the dungeon and lock us in a cell overnight. Then he'd leave us alone. To teach us how much we were required to sacrifice for our people."

"He did that to you?" Ash growls a vicious curse, his hand in my hair clenching into a fist. "He locked you in a *dungeon*? I will gladly accept any opportunity to kill that man."

For some reason, I was afraid saying the words out loud would make my heart race, my blood pound, my lungs constrict. Instead, I feel nothing. Nothing but a calm acceptance. It is part of my story. A difficult part, but one I have come to terms with. One I have overcome.

"I have another secret to tell you," I say, instead of responding to Ash's death threats. "When I was sick, I heard you."

His fist in my hair unclenches slightly. "Heard me?"

"You sang to me. A lullaby your mother sang to you. I didn't know it was you at first, but when I did, it gave me the strength to keep going."

"You heard me singing to you?" Ash props himself up, and even though it's too dark to see, I can almost *feel* the way his brow furrows.

"You have a beautiful voice," I say with a smile.

"Oh, well . . ." He scratches his neck, as though embarrassed. "Thank you. I . . . Stella, I was so terrified I was going to lose you." He tilts my chin up, kisses my mouth. "I was so afraid you weren't going to wake up."

I kiss him back. "It's your turn now."

He lets out a gust of a sigh that stirs my hair. Contemplates. Then, in a voice as thick as the shadows around us, he whispers: "I didn't always hate my father. And he didn't always hate me."

I stay still, listening.

"I wouldn't describe us as a happy family. That isn't how things are among us fae usually. But we were together. Father, my mother, me. He didn't love my mother, and he didn't love me. Still, he thought my mother was beautiful, and he was proud when I defied her, or when I smart-mouthed members of his court as a child. In case you haven't gathered, I was not well-behaved. I was, to put it frankly, a terror. But I loved my mother, no matter how difficult I made life for her. And I loved my father—held him in deep reverence."

"Did it change when he . . . when he killed your mother?"

"The change happened long before that, actually. We have this board game called Fool's Circle—I should teach you to play some time—and while you can learn the rules in a matter of minutes, it can take thousands of years to perfect strategy. Faradir taught me how to play when I was very young and took tremendous pleasure in regularly beating me in front of his court. He would pat me on the head and say, 'Maybe next time, my boy.' I was mostly happy to have his attention, even if he made sport of me. At some point, however, I decided I wanted to win for once. So, for an entire year, I searched high and low for clever opponents in the game, and played with them in secret. When Faradir called me to play him, I would lose on purpose, so he wouldn't suspect anything until I was ready.

"Then, one day, I suddenly knew it was time. I waited until he called me. I played my usual role of the witless, enthusiastic son." He lets out a long exhale, one made of sorrow. "I believed he would be so proud of me. But, as you can imagine, with so many subjects watching, Faradir did not want to lose. When I began winning, his demeanor changed. I remember looking at the way his brows pinched together and the sudden dread that filled me. I had messed up. He played with me because he enjoyed winning, and I realized I would

450

rather still be able to play with him than win the game. So quickly, I pulled back. I maneuvered the game back in his favor. He won. But he still looked upset—even angry. He patted me on the head like always, said, 'Maybe next time,' and left."

"Was there another time?" I ask.

He shakes his head. "Not once. Things changed between us after that. He never wanted to see me anymore, and when he did, it was always to criticize me. The more publicly he could humiliate me, the better."

"You stopped being his son," I say softly. "You became a threat. Someone who could outwit and outsmart him."

Faradir never loved Ash. Ash was the prince, the tool, the heir to be molded into what served Faradir's agenda the most. And when he stopped serving that agenda, he became a liability.

Ash shrugs, as if he doesn't care, even though we both know the truth. "Your turn."

I know what my next secret is, but the thought of it brings a flush of something like shame to my cheeks. I don't want to admit to having lied to my husband, but things were different between us, and I was trying to survive in a world of people who wanted me dead. The shame vanishes like dew.

"I can smell lies," I say without preamble. "I didn't tell you before because—well, at first, I didn't even know, and it caught me so off-guard I wasn't sure what to do. Especially since it was you who lied to me. At the time, I decided not to tell you because I wasn't sure if I could trust you."

"I *thought* you could! I simply convinced myself I was making things up. You had that coughing fit, and I was nearly so excited—"

"Excited?" I demand, gaping at him. "Why would you be excited? You had just lied to me!"

"Because it meant you had magic! And I *want* you to know when people are lying to you. Even if it's me—which, please forgive me for that lie. I only wanted to ease your fears, though I should not have lied to do so. But Great Kings, does this mean you can smell our lies but yours have no stench, no taste of their own?"

I nod.

"That is going to be an immense asset when you're quee—" He cuts off abruptly. Swallows. I feel the sudden vulnerable flash of his gaze on me.

Queen.

He doesn't have to say what we're both thinking:

If we survive tomorrow.

If we win Ash's throne tomorrow.

"I suppose . . . if we're sharing secrets . . ." Ash trails off at first. He takes two knuckles, touches my ear, then traces a line down my neck to my shoulder, across my arm to my fingers. I close my eyes, leaning into his touch, listening to his voice as he speaks again. "I'm haunted by the fear of losing you. It has dogged my steps from the beginning, and in some ways, it is worse now—even with a plan. Even with the pieces carefully placed. What if we survive but fail to dethrone Faradir, and I am bound by blood to destroy the human lands? How could our relationship continue with that between us? How could you ever forgive me? And possibly the greater question: how could I forgive myself?

"But what if something goes wrong and I lose you? What if I cannot protect you from all who wish you harm? What then? In either case, how do we—or I—keep on living?" His voice breaks. He holds me to his chest, as though he can keep me here forever. Tucked away in his little corner of this dark world. "You are my life. So how could I live without you?"

I close my eyes, let the weight of his question—a question that clearly has plagued him for some time—fall over me like a blanket. Part of me wants to bask in the sweetness of what he is implying, or to reassure him that everything will go as planned tomorrow.

But I cannot do either. Because his question is one for the ages.

How can you go on living after tragedy?

I think carefully before I open my mouth. "Living with my family was very stressful. Marrying you was terrifying. It would have been easier, in that moment, to not have married you. I would have chosen that, if I'd had a choice. But marrying you was the best thing I've ever done. I discovered what it was to be loved. Even if this ends horribly, I will have no regrets. I would always choose to have weeks as your bride instead of decades as the wife of someone like Prince Brochfael."

"You don't mean that," he murmurs against my hair, and the cheek he presses to my forehead is wet.

"I think you need to stop believing that I would have been better off if you'd never entered my life," I reply firmly. "Because it's not true. Do you remember who I was when you first met me? Do you remember how I could barely talk—much less about my own thoughts and ideas? Do you remember how I struggled to look you in the eye? When you showed up at my father's doorstep, I was living a miserable life. Isabelle Louise never could have done the things I did today and yesterday. She was too afraid. I am better for having known you, Ash. Can you believe that? For me?"

He presses a teary kiss to my brow. The sound of his heavy breathing fills the silence between us. I lean into him, letting him feel my closeness, the honesty of my words. With my fingers, I trace his face, his long hair, his sensitive ears—making him shiver.

Then I whisper to him: "I think sometimes life takes us through dark times to get to the good times—the better times. And the goodness is so much sweeter for the hardship we endure."

Ash lets out a short, low groan. One that is almost frustrated. "It doesn't make me any less afraid of the darkness."

I close my eyes against the rising tears, but I force my lips to part, my tongue to loosen, and my throat to open. I force myself to speak, no matter how difficult it is to say the words. "No, and when we're in the depths of it, it's hard to believe we will come out of it. But I tell you this so that if the darkness comes, part of you can remember . .

. and *believe*. Because I know, Ash, I believe with all my soul, that there is always, always happiness on the other side of heartbreak."

"If darkness comes," Ash growls, and moonlight catches on his silent tears, making them glitter like stars, "then I will remember, and I will curse this memory, this moment, even as I treasure it. But if I ever did find happiness again, then maybe I will find my faith once more, and I will know that you were right. As you always are."

His reply hits me like a salty wave against my ankles, insistent but gentle, stubborn and soft at once. Perhaps only some fears are meant to be overcome, while others are meant to be borne with dignity.

I tilt my face up to his, breathe out his name like a secret and a promise. "Trenian Ashrift Solavirth, I love you."

"Isabelle Louise Stella Ashrift Solavirth," he replies, taking my jaw in one of his large hands and guiding my lips to his, "you will forever enchant me."

And so we trade secrets, kisses, fears, and hopes until the rays of Lulythinar's dawn stream through the windows.

CHAPTER 58

THE PRINCE

I RISE A little earlier than Stella, and after throwing on a few glamours, I'm ready to receive my guest.

Princess Oleria sits in my living room with one knee crossed over the other, a steaming teacup in one hand. Her silver hair is swept forward over one shoulder, likely to not disturb her shimmering wings. She sips daintily as I approach.

"Princess Oleria," I say, sitting across from her and pouring my own cup of steamed, spiced mothweed milk from the set Edvear left.

"Prince Trenian," she replies.

Her expression isn't smiling, but it's still pleasant . . . even underscored with the concern and perhaps dread she likely experienced when I summoned her.

"I hope you are well," I say, and I mean it. Part of me wishes Stella and I could have gotten to know her better. Perhaps even become friends.

She cocks one shapely eyebrow. "Please, Trenian. We can skip the formalities. What can I do for you?"

Very well. I take a deep breath, almost hating myself for the words that must come out of my mouth. "I . . . need you to kill the High King."

Oleria doesn't react at first. She takes another sip from her teacup, nods once, and says only, "I'm flattered by your faith in me."

"I was going to have Stella do it." Back when I had hope that she could be the one person to use glamours and get right up close to the High King without him noticing them. Instead, it's almost worse that he cannot see through her glamours, because he can immediately register her as a threat. "But I no longer have a way for her to get close to him."

"I see." She's quiet, contemplating.

"I know I'm asking for everything. I know I'm asking for your life," I say quietly. "It's not fair, and it's not even right, but I have no other—"

Now her lips twist, almost ruefully. "Killing the High King wouldn't kill me."

I stop. Frown. A branch from the side table reaches out and accepts my teacup. "I beg your pardon?"

Instead of answering, she lets her glamours go. Like rain washing away paint, the glamoured Oleria is gone, replaced by the real girl. I take her in anew, my mouth opening in shock.

Before me is not the fae princess I've known at a distance for some ages. Her long, pointed ears are gone, leaving mostly rounded ears with slightly pointed tips. Her fae beauty melts into something still beautiful, but less ethereal. Something more grounded. Only her wings and silver hair remain unchanged.

It's her scent, however, that hits me hardest. Her *true* scent.

"You're human." The words are out before I can restrain them.

"Half human," she corrects. "My mother was human. She was, as my father liked to say, his secret vice. So, when I was born, I was named princess, despite my heritage. Despite being illegitimate. My fae blood gives me enough power to hide my human blood with glamours, which I have done my whole life."

I cannot find a single word to speak. All this time, she was half human? And I never suspected it? It must put her at risk, to have that much of her magic occupied with maintaining glamours.

She gives a huff of dry amusement at my shock, shakes her head. Then she pulls each glamour back in place, and the Oleria I know sits before me once more. "Seeing as I am only part fae, the same laws don't apply fully to me, and I could kill the High King without the same consequence."

"You have enough fae blood to take a hard hit. You'd probably be incapacitated for weeks."

She lifts one narrow shoulder in a shrug. "Better that than Faradir still being High King. I would only ask that when you are High King, you fight to protect humans. There are many fae who are sympathetic to the plight of humans, but few are vocal enough about it to overcome those who delight in their torment."

My head sags in relief. I run a hand through my hair, overwhelmed in sudden gratitude. In a sudden flood of unworthiness that someone I barely know would sacrifice so much. Something told me she would be willing, but this was more than I could have hoped for. More than I could have asked for.

"Now, I must confess," Oleria says, a delicate line appearing between her brows, "I do not know how I would accomplish such a feat. And I assume you need it done today."

"Faradir loves to be served by humans—"

"I have no human servants."

That shouldn't surprise me, but it does. I nod. "We can still work with that. There are many human refugees in the Small Cities that can be hired for a single day. You do not have to explain everything to them, so long as they are aware of the danger. I'd offer one of my servants, but the High King keeps a catalog of who works for me. Once you have a servant, you'll need to sneak them into the kitchens as a server. And—"

"How would I do that?"

"Simple. Poison one human who usually brings out the High King's food. Poison several, if you have to. It doesn't hurt to have more than one servant on our side. I have a few different ideas for how to get our servant—or servants—to replace the poisoned ones. And I don't mean to poison to kill. Just something to make them fall asleep. *Nampir* perhaps. Then, when dinner is being served, you stay hidden—I'll tell you where—and *after* the taster has tested for poison, have your servant bring you the platter of food. I'd recommend using *yirop* for its weakness-inducing properties."

"Do I poison the High King to kill? And what do I poison of his food?"

"No, no, personally, I would recommend you poison to incapacitate. It'll be harder for him to detect if it's smaller amounts. Poison everything on his plate, in his goblet too. But just enough to weaken him, not enough to raise his alarm."

"Then I sneak up and stab him?" Her face turns a little green.

I'm asking so much of her. *So much.* "How are you with a bow?"

"Better than with a knife," she replies with a nervous laugh.

I'll need to engage another layer or two of contingency plans. I've had several options brewing in my brain in case this doesn't work. I prefer this option, though, because it involves less human blood.

Less full-out *war*.

"We'll work through the plan," I assure her. "We'll make it foolproof."

"You are betting a lot on my talents of subterfuge."

She means it as a joke, but uncertainty flashes across her face. Perhaps I wouldn't have noticed it if not for being married to Stella.

"I trust you, Oleria. I know you can do this. The fact that you are a half human living in a fae's world, and that you've gotten away with it your entire life, tells me everything I need to know."

Her wings give a little shiver. She looks away, and for a moment, I am afraid she will tell me she cannot do this.

Instead, she squeezes her eyes shut and growls in a voice of such determination, "I will play my part to end Faradir's reign."

My mouth draws into a thin line. "Then let us bargain, Princess Oleria."

The tattoo that imprints on the back of my hand is a pair of beautiful wings.

CHAPTER 59

THE PRINCESS

I STAND IN the hallway, hidden behind the wall, listening to Ash and Oleria's conversation. I'm still wearing my nightgown, my hair in disarray from sleep. I close my eyes, a knot of dread sliding downward from my throat, until it lands like a rock in my gut.

Today is Lulythinar. And if we do not kill Faradir today, then Ash is bound by magic to destroy the human lands tomorrow.

I know we can do it.

That doesn't mean we will.

I say a silent prayer. Then it's time to get ready for the day.

"It's the tailor at the door, my lord," Edvear tells Ash while we eat our noon meal together. Ash looks up at me, his eyes rounded in remembrance—echoing my surprise.

"We were supposed to see him yesterday," I say.

"He couldn't come yesterday due to the uproar." Edvear's face is pulled back under its usual mask, as if yesterday he hadn't nearly had a breakdown.

I pity him, the burdens he bears, the losses he's endured.

Ash gives me one look. It's quick, but weighty. He turns to Edvear. "Tell him to come in."

Edvear bows and hurries to obey. Ash rises from the table, leaving his unfinished breakfast. He comes to me, cups the back of my head with a hand that bears a new tattoo, presses a kiss to my forehead. "I'll be in my study."

I'm dabbing my mouth with my napkin and rising when the tailor enters, bearing a garment bag over one shoulder and a measuring tape hanging from around his neck. He adjusts his spectacles, runs a finger along his curled mustache, and gives me a professional once-over as though ensuring the gown he made looks as well as he expected it to.

Then he's all clipped business. "I've brought your white dress. If you would come inspect it and ensure it is to your liking, I would be grateful."

When I come to his side, he is unbuttoning the garment bag, revealing layers of creamy white fabric that fall to the floor without a single crease or seam.

I barely see it, though, because the moment Edvear goes back to the kitchen, the tailor presses a tiny sheet of paper into my hand. "As you will see, I did scallops on the ends of the sleeves for a more romantic look," the tailor says, as if he has done nothing suspicious in the slightest.

"It's beautiful," I say, distracted. "It is exactly to my liking."

"Perfect. I shall be off then. Many need last-minute alterations for their Lulythinar gowns. Be sure to summon me if you need the same, though from a glance, I do not think that will be necessary."

"Thank you," I manage to say before he leaves. Then, looking down at the paper in my hand, my heart gives a painful thump in my chest. I unfold it and read.

After dark. West garden gate. Rap three times on the nearest tree. The Ivy Mask will meet you there.

I put the note away when Edvear comes into the living room, offers me an obligatory compliment on the dress, and whisks it away to my wardrobe.

The directions on the paper . . . they mean the servant's exit of the garden outside my window, don't they? This Ivy Mask must be associated with the Small City we visited.

"What did he say?"

I nearly jump straight out of my skin. "Stop being so quiet, Ash! You scared me!"

He takes my shoulders, rubs them gently, and peers at the paper I show him. "This Ivy Mask is new to me. I wonder if he is fae or human."

"Human," I reply promptly.

"Why do you say that?"

"Because—and I mean this with no offense to you—fae don't risk their lives for humans."

"If this Ivy Mask gets humans back to their own world, I have a difficult time believing he would be a human. Caphryl Wood is a dangerous place for humans."

I shrug. "I think he's human."

And then we stop speaking, as though we both realize that this discussion is only a distraction from the impending dread we feel.

"Dottie probably wants to start preparations for tonight," I mumble. Even though the dining room is only a few steps away, and our unfinished meal sends steaming wafts of a spiced, creamy sauce filled with pheasant and root vegetables, I am not inclined to go finish it.

"I have a few things to attend to. Final preparations for tonight." Ash tightens his grip on my shoulders, and despite his words, he makes no move to let me go.

I crane my neck, looking back at him.

One way or another, everything will change tonight. He takes me by the waist, pulls me to himself, and angles a fierce kiss against my mouth.

Then we part ways. Just for a few hours.

Still, it feels as though it will be forever.

CHAPTER 60
THE PRINCE

EDVEAR BRINGS A mirror into my study for me to get ready with, since apparently a whole army of maids are working on Stella in the washroom. I do not know how women do it—the hours and hours of primping and polishing. I would go utterly mad.

As I comb through my hair, button my tunic, and place my circlet crown atop my head, a heavy weight settles on my shoulders.

I am both afraid and hopeful, nervous and confident. I worked with Oleria to iron out the rest of the details for our plan, and every time I run over each aspect in my head, I grow surer that it will work. Last I heard, three servants had been successfully planted in the kitchens. I tried to anticipate everything that could go wrong. We have fail safes in place for if the servants get discovered, if Oleria cannot get to her hiding spot in time, if the High King detects the poison, and even if she shoots him and misses.

All outcomes end with Faradir dead by the end of the Lulythinar celebration. Even if I must kill him myself.

Because, as much as I do not want to give up my throne, as aware I am of the utter destruction that will ensue if Faerieland is left with no heir, my first priority is protecting my wife. My second is nullifying the bargain I made with Faradir to decimate the human lands, which will only happen if he is dead. Only after those things do I care about my throne.

I am not fearful because I think my plan will fail; I know it can succeed. I think it *will* succeed. What I fear, rather, is what Faradir intends to do tonight.

He will strike. He *will* retaliate.

Blood will flow this Lulythinar.

I buckle my sword to my belt. I wrap my fingers around its familiar hilt, slowly draw the blade out of its sheath. Light catches its lethal tip and flashes in the mirror.

One way or another, this ends tonight.

⚜ ⚜ ⚜

The sun is setting when Edvear comes to my study and tells me that Stella is ready and waiting in the living room.

My heart gives a nervous pitter-patter as I walk through the hallway toward where I hear a swish of skirts. Am I more nervous to see my wife in this moment than I am about this entire night? I might be.

I come around the corner, and there she is.

She smiles shyly at me, her hands clasped behind her—as though she is as nervous as I am. For a moment, I just stand there. Stunned speechless once more by the beauty of this girl I had married without a clue to what she looked like. The difference tonight is that she looks utterly and completely *herself.*

Her gown is iridescent purple. It turns blue when the light catches it, making it breathtaking. Her shoulders are bare, but purple and silver filigree bands on her arms are fastened to glossy, transparent sleeves that shine with the same iridescent beauty. The skirt flares out at a flattering angle, emphasizing her waist and the loveliness of her figure. Her hair is left down, to my delight Its only ornament is the silver and diamond tiara.

I swallow hard. "Hello."

She pulls out what she was hiding behind her back—and sets it on her face. "It even comes with a mask!"

The mask is the same iridescent purple as the dress, sweeping away from her eyes in elaborate, bejeweled butterfly wings.

My mind immediately goes to the wings tattooed on the back of my hand. Cold rushes down my spine. I try to shove the sensation away as I step forward, catch her hand, and bring it to my lips.

"You truly are an angel," I murmur.

"Don't be dramatic! A simple *you look lovely* will suffice!"

My mouth turns up. "You look lovely."

"You look lovely too."

That wrings a chuckle out of me. I roll my eyes and step close enough to tilt her chin up for a kiss.

She stops me with a glare. "I'm not mocking you. You are a very beautiful man. Especially so when you wear your crown."

"I thought I was supposed to be handsome."

"Well, you're that too. But human men are sometimes handsome, and yet I have never seen one that I also thought was beautiful. Unlike you. There is this elegance to the features of your face that just makes you . . . well, beautiful!"

"Quit talking about my face and let me kiss yours," I reply, then claim her soft, sweet lips before she has another chance to protest.

This will not be our last kiss, I vow to myself as I pull away, take her hand, and lead her with a pounding heart out of the safety of our chambers, toward the beginning of the Lulythinar Masquerade.

CHAPTER 61
THE PRINCESS

THE PALACE GREENS are a dazzling array of costumed fae dancing in concentric circles under every tree. Lumiral globes float through the air like tiny glowing stars, illuminating the vast spreads of food on overflowing tables and a dizzying number of sloshing goblets.

I turn away from the pockets of debauchery happening in the shadows as the full moon rises overhead. Instead, I watch the beautiful dances, the elaborate costumes on display, the children running around with friends and pets. One magnificent fae is dressed like a weeping willow, with long trailing leaves and a twisted trunk. Another is sea green, like a foamy ocean, with white hair and bright green-blue eyes. Ash isn't wearing a costume or a mask, and even my pouting mouth couldn't make him change his mind on that front.

Not that I truly mind. I like being able to see his beautiful face.

Faradir sits on his temporary throne—the one that is brought for him for these outdoor celebrations. His long golden hair falls over one shoulder, and he wears a new crown of shooting sunbeams. His mask is made of living fire that licks around his eyes, always burning but never devouring.

He finds us right away. I hold his gaze for a moment, then find something more interesting to give my attention to. I don't have to be concerned about him just yet, and I intend to enjoy the celebration with my husband.

Two long fangs from a dripping wolf's maw snap in my face. I flinch, stumbling backward into Ash, even though only a second later, I realize it's just a mask. Still, my heart thumps wildly in my chest, and I become aware of the weight of curiosity pressing into me from all sides.

Do I have a reputation now, having come back from two supposed poisonings? Perhaps they think Ash is crazy to bring me here at all. Perhaps they believe me truly resurrected, not believing the High King's proclamation last night that I wasn't dead.

Children stare at me openly, and I could swear some of them do so in adoration. They don't flinch when a tall fae with a mask of writhing tentacles passes between them. One of those tentacles reaches for my face, only to withdraw at Ash's fierce glare.

"It seems you have become a legend," my husband murmurs to me.

"I think I liked it better when they ignored me," I reply with an uneasy chuckle.

But then the singing starts, and every thought in my brain is suddenly gone.

At first, I'm not sure where it is coming from—it is so all-encompassing, so all-surrounding. Until I realize it is coming from *everyone*, even Ash beside me.

The dancing has stopped. The eating, the drinking, the conversation, even the debauchery—all of it stops. Instead, every person sings, tilting their head back toward the sky, toward the moon climbing its way to its zenith.

The song washes over me, sung in words I do not know or understand, but the meaning of which transcends language. It is light, ethereal, but even as I listen, it deepens, broadens, until I can do nothing but look up at the sky with tears streaming down my cheeks.

I look at Ash, find him looking at me with warm, glittering eyes. He bends down to my ear, whispers softly enough not to disturb the soaring music around us.

"It is a song of thanksgiving. Gratitude that no matter what these last hundred years held, that we are still here, that the moon still rises, that the tides still come and go."

"I know," I whisper back, closing my eyes and letting the glory of the song sweep through me. I've always known fae music held magic, magic that could turn me foolish and make me forget my own name. Magic that could make me a slave.

And yet, part of me wonders if that is not as terrible as I once believed it to be. Perhaps it is a thing that can be used for evil. But in this moment? I see how it can be so soul-transcendingly good.

"Come with me," Ash says.

He takes my hand, leads me quietly through the throngs of singing people. I follow gladly, even when the farther he leads me, the quieter the music gets. We reach a cliff that overlooks the sea, and Ash bids me look over the climbing vines to the reflective water below.

"It's beautiful," I whisper.

He only smirks. "Wait a moment."

I wait, listening to the strains of music as stars glitter across the waves breaking against the rocky shore. Then, suddenly, at the swell of the music, the waves surge upright. I gasp, my hands landing on the railing and holding tight as the water splits in half. It starts at the shore, peeling back like two curtains, water reaching as high as the face of the cliff.

"What is happening?" I cry.

"It happens every Lulythinar, so the people of the sea can sing too."

The words are hardly out of his mouth before a multitude of voices from below join the chorus. I nearly stagger at the beauty, the expanding strength of the music.

But then Ash is tugging me again, and I barely let myself be dragged away. The song stays with us as we go. Not even when we enter the garden I love so much—the one outside my window—is the music entirely gone.

Ash pulls me through the tunnel of roses, and I laugh as I try to keep up with him without tripping on my gown. My mask goes askew, but I don't have a free hand to fix it.

We reach the overgrown gazebo. Singing wafts on the breeze that tangles in my hair and ruffles the sleeves on my gown. Ash stops before the steps and sweeps me a grand bow. I grin even as I blush.

"Would you dance with me, fair maiden?" he asks, pinning me with that heated gaze like a thousand fallen stars. When he looks at me like that, he holds more sway over me than fae music ever could.

He extends his hand to me.

I place my fingers in his. He presses a kiss to them.

Then he draws me up the steps, into the dark gazebo. With a sweep of his arm, a host of tiny glows like fireflies light up the small space, the cracked stone and twisted vines.

"Dance with me," Ash whispers.

"Always, my prince," I reply.

He reaches out, straightens my mask. Then he takes my waist and draws me to his chest so not even an inch separates us. His mouth hovers above my hairline, warm and intimate as we begin dancing. Slowly, to the strains of the song on the air.

"I want to know everything about you, Ash," I say to him, tilting my face up. "I want to know everything about your people. Next Lulythinar, I want to sing their song with them."

With my glamours, I can find a place to belong where I am not always at risk.

Ash's grip on my waist tightens. His chest expands against mine, and then his exhale stirs my hair. "Then you shall."

At some point, we stop dancing. We simply stand together, our gazes melding into one. Ash slowly lifts a hand, places a knuckle beneath my chin, and tilts my face up to his. I close my eyes as the warmth of his mouth descends to mine.

Ash stops.

A shock goes through him.

I freeze. My eyes fly open. On instinct, I look behind us, searching for a threat. When I find nothing, I turn back to Ash.

He gazes at the back of his hand, his face turned white as the rising moon.

"A-Ash?"

"The wings are gone," he rasps. He lifts desperate eyes to mine.

"Isn't that a good thing?" I say, unable to keep the shaking out of my voice. "Don't the tattoos only vanish when the bargain is fulfilled?"

"Or when it becomes impossible to fulfill them."

"Oleria," I gasp.

We're running down the rose archway before everything finally processes in my brain. Ash's face is hard, grim.

"But . . . she wasn't supposed to do anything yet!"

His grip on mine is like iron. He doesn't run too quickly for me to keep up, but it doesn't take long before I'm panting, anyway.

I hate to ask, but I must know. "Does this mean . . . Does this mean she's dead?"

The flintlike set of his jaw answers my question. Sickness swells in my stomach and nearly gets the better of me.

The music stopped.

The realization hits me like a blow. When did it stop? The moment Ash's tattoo disappeared?

Ash comes to an abrupt halt when we reach the palace greens. I almost go racing past him, but he pulls me back and tucks me behind him.

At once, I realize how loud my breathing is.

How quiet everything else is.

How still the hundreds of fae around us have gone.

Sweat slides down my back as I peer around Ash to see what has made him stop short. My blood runs cold as ice.

Because there, in front of Faradir's throne, is a young woman I've never seen before, shuddering on her knees, her wrists bound behind her. Perhaps I wouldn't have recognized her at all, if it hadn't been for the pair of ripped, bloodied wings lying in the dust at Faradir's feet.

The High King looks up, directly at me.

He smiles.

CHAPTER 62
THE PRINCE

A T FIRST, I can hardly think past the horror, can hardly suppress the urge to vomit right here in front of all Valehaven.

The second thought?

Someone betrayed me.

This wasn't a failure of my plan. Someone betrayed us. One of the human servants we hired?

It doesn't matter who.

All I know is that everything—*everything*—is compromised.

If everything is compromised, then there is only one thing I'm sure of.

Stella needs to get out of here.

Before I can move a muscle, the words Faradir is speaking register in my mind. "She is yours as a token of our continued goodwill. She will make a beautiful bride, even without her wings."

That is when I notice who stands opposite, where Oleria shivers in pain. Or rather *what*.

He stands as tall as the Neverseen King, with rows of black braids down his head and back, garments made of fur and an iron breastplate against his mostly bare chest. In one hand, he holds a massive staff with a glowing blue orb that casts a strange color onto Faradir's golden smile.

He is a fae . . . from beyond the Veil.

Oleria—I need to save her! I need to intervene somehow—

But even before I've drawn my sword, the barbarian fae takes Oleria's arm, pulls her to her feet, scoops her over his shoulder . . . and vanishes with a flash of the orb on the end of his staff.

She's gone.

Oleria is gone. Beyond the Veil. I have no way of reaching her unless I beg the Neverseen King to let me use one of his portals—a request that would certainly be denied. Even then, I might end up at the right place, but the wrong time by thousands of years.

She is well and truly . . . *gone.*

"Oleria," Stella chokes, tears streaming down beneath her mask.

I don't bother looking at Faradir, don't bother glamouring my expression. There is room now for only one thought.

I grab Stella's waist and pull her up the marble steps into the palace.

Faradir's voice rings out behind me. "Will you truly show your cowardice by fleeing?"

I couldn't care less what he thinks of me—what all Valehaven thinks of me. Stella's life is all that matters.

"Ash!" she demands, grabbing my sleeve and yanking it hard. "Don't you *dare* do what I think you're doing!"

I keep moving, ignoring her. Violence simmers beneath my silence, ready to explode. Resolve hardens to stone in my gut. If the High King wants a war, then I will give him a war. I will destroy this place we call home. I will make all of Faerieland run with blood.

I will show Faradir why he was right to fear me.

Finally, I will become the monster he intended me to be.

But first I'm getting Stella out of here.

A sharp pain on my scalp makes me whirl, expecting an attack. There is no one—no one has followed us. It is Stella who grips a fistful of my hair. Her mask is gone, and beneath it lies the stormiest expression I've ever seen on her.

"I am not leaving," she snarls.

"Yes, you are," I growl back. "Our plans are compromised. Everything is lost. One of those human servants must have betrayed Oleria—and who knows what information Faradir got from her before he ripped off her wings?"

"Oleria wouldn't have betrayed you. Even to save her wings."

"We must operate under the assumption that the High King knows everything." Everything, like that the Neverseen King has returned to the Bridge and is not going to attack. That his coup was only another trick of mine and not a real threat.

Stella hasn't relinquished her grip on my hair, as if that alone is keeping me from killing everyone in the near vicinity. I hold her wrist in one hand, and our furious faces are hardly an inch apart.

"Now is not the time to be stubborn," I growl. "Let me spare you. Let me get you out of this."

She doesn't flinch from me for even a second. "You made a bargain. What about the human lands? What about my people, Ash? We cannot give up!"

"I'm not giving up. I'm going to kill Faradir, and that will end both of my bargains. But you must leave before I do." I let go of her wrist, grab her face in both of my hands. At some point, a tear slipped down my cheek, and it drips from my jaw onto the polished floor of this abandoned corridor. "Stella, I'm begging you. Don't make me watch you die. Get out of here. Go to the Ivy Mask. Let him take you to safety. Please, Stella."

Her brow turns to flint, her usually sweet eyes flashing like a furnace. "No. I'm going to stay."

"My lord! My lord!" a gasping voice cries from the other end of the hallway. "What has happened? Princess Oleria—"

Edvear's voice nearly makes me startle and break the war I wage against Stella and her will of iron.

Then I close my eyes. Instead of Oleria kneeling at the High King's feet, I see Stella. Bleeding out on the grass. Her lovely gown torn. Her hair stained red.

I will not let that happen.

I open my eyes, take one last look at the sweet curves of Stella's face, the stubborn line of her full lips. Her beautiful brown eyes that have captivated me from the start. I'm so glad I can know for certain that she carries no child, no matter how many times my resolve almost broke.

"I love you," I tell her.

Her eyes widen in horror. "Ash, no—"

Then I pry her fingers out of my hair and shove her toward Edvear, who is quick to grab her before she can run back to me. "Get her out of here. Get her to the Ivy Mask."

"Ash!" Stella screams, fighting Edvear's hold on her even as he drags her away. "Don't be an idiot! What about the things we talked about? *Ash!*"

Just an hour ago, her cries would have pierced my heart. Now, they fall like dust over a boulder. She deserves to live. She can use her life to hate me, to curse my name—I don't care. I just want her to be alive.

I wrap my fingers around the hilt of my sword. My first stop: the Nothril Court to get my only friend.

After that?

It's time to bring an end to the High King's reign once and for all.

And bring down all of Faerieland with him.

CHAPTER 63
THE PRINCESS

I REFUSE TO cry. Somehow, I knew this was going to happen. I knew last night was our last together. My heart is a conflicted storm of hating myself for being so petty as to not want Ash to see my gray skin after being poisoned, for not asking for the one thing we had deprived ourselves of since our marriage—while simultaneously being glad to be spared the multiplied pain that I would have suffered if I had.

At some point, I give up screaming for my husband. I give up fighting Edvear. He's not as strong as Ash, but he is still far stronger than I am.

This is how things are unfolding, apparently.

Edvear is taking me to the servant's exit in the gardens where Ash and I were dancing. I'm leaving Faerieland. The Ivy Mask will take me . . . somewhere. Probably not to the human worlds, if they're to be razed tomorrow. But if I am taken back there, perhaps I can warn my family in time to help them escape?

I need to be careful about this and not lose my senses, even if I want to just curl into a ball and cry over my broken heart. So I keep my eyes and ears alert, my glamours ready at a second's notice.

I want to take Ash by the shoulders and give him a ferocious shake.

Instead, Edvear is quickly navigating through the palace corridors, taking the quietest routes with the least traffic. Still, when we come upon a split in the hallway and Edvear takes me toward the right, I frown.

"Shouldn't we be going left?" I ask, glad I always paid attention when Ash walked me places.

"There's a short-cut in this direction," Edvear replies.

And my nose instantly fills with iron-stink.

It's so shocking, I almost give into a coughing fit. Instead, my face contorts—and would have given me away, if Edvear was looking at me.

Shock hits me, followed by such a profound certainty, it almost makes me freeze in my tracks.

Edvear lied to me.

He isn't taking me to the Ivy Mask.

He was the one who betrayed us. Not one of the human servants Ash and Oleria hired from the Small Cities.

I know why he did it, too. His words from yesterday ring through my head.

"Surely there has been enough death! How much more can we endure? How much more until we admit the High King is the High King and we cannot gainsay him?"

Dread fills me to the brim. I need to get out of here. I need to somehow escape Edvear, and I cannot let him realize that I know he betrayed me. First, however, I need to find out the truth of his betrayal.

"Did you see what happened to Oleria?" I manage to ask. "Through your mirror?"

"Yes. Milton was there, spying for us again. I couldn't believe my eyes. It was horrifying."

All of it rings true.

Which means . . . it's possible he didn't betray us of his own will. Perhaps he was forced to.

But perhaps the knowledge about Oleria wasn't the first thing he leaked.

His grip doesn't slacken on my arm as we reach a spiral staircase that opens into the night air. Surely, he doesn't think me so stupid as to believe climbing a staircase will get us closer to the garden on the ground level?

I need to get out of this.

Think, Stella. Think, think!

"We're almost there," Edvear says as the full moon comes into view over the stair's railing. It is almost to its zenith. I swallow against my parchment-dry throat.

Then I give a sudden sharp yank on Edvear's arm and throw my weight into him, pressing him against the stair railing. It catches him off-guard just long enough for me to pin him.

And to press a glamoured knife to his throat.

The knife isn't real, but I send the illusion of its sharpness, its lethal tip into Edvear's mind.

He freezes, his cat's eyes dilating in sudden fear.

"What have you told the High King?" I demand.

His hands come up in surrender, his face crumpling. "He was going to kill every one of my staff! I didn't know what to do! I never wanted—"

Heavy, armed footsteps sound at the top of the staircase, beginning their descent. I react immediately, covering Edvear's mouth with my hand and losing the glamoured knife.

They cannot find us like this, with me in control of Edvear, knowing the truth behind his betrayal. Not unless I want the High King to carry out his threats.

I could run, but the outcome would assuredly be the same.

"Grab me," I mouth, and when he doesn't move, I take his hand and clamp it down on my upper arm. Then I begin pulling against him, struggling just enough to get his eyes to widen, his mind to catch up with what I'm doing. Horror makes his mouth fall open, but he

481

still does what I say, and begins dragging me the rest of the way up the steps.

Two winged guards meet us only a second later, and I don't even have a chance to look back at Edvear before they grab me, lift me straight off my feet, and carry me between them the rest of the stairs. I struggle against their relentless grips even as my heart races and I wonder if I have just sacrificed myself for nothing.

Then, just before a strange, black door at the top of the staircase is opened, I tilt back my head and scream at the top of my lungs: "Ash!"

I'm flung to my hands and knees on a woven red rug, and the door behind me slams shut.

My hair falls around my face as I breathe in and out. I become aware of a second, softer breathing rhythm in the room with me.

Slowly, I lift my head.

The golden, sandaled feet before me are framed by heavy, white-gold robes that land in elegant folds on the ground. Two bejeweled hands hold the carved lion heads of the chair's armrests.

Above that is a broad chest, an elegant neck, and a mask of living fire. The sunbeam crown glows upon his glorious head. Faradir takes one graceful hand, removes his mask, and lets his bright blue eyes fall to me. "Hello, Princess Stella."

Coldness like winter rushes through my blood.

"I have bided my time until now," the High King says. He reaches out one long-fingered hand, takes my jaw, and tilts it up so my neck is taut, exposed. I cannot even swallow, not as he strokes one line down my throat with his sharp nail.

I keep my eyes locked on his, on the way his gaze runs over me. My skin crawls, but I refuse to look away. Refuse to cower. Refuse to let my guard down for even a second.

He wants something from me. That is why he has brought me here, why he hasn't slaughtered me already. If he wants something from me, then that means I have power. And if I have even a shred of power, then I am not completely helpless.

Not even before the High King of the Fae.

So I let him hold my jaw. I let him touch my neck—as if he wants to sink his fangs into my vulnerable flesh. I say nothing.

And I wait.

"You seem to have woven some spell over my son," Faradir says, clicking his tongue. "Which shouldn't surprise me, considering that you"—he gives the tip of my nose a little tap—"have glamour magic, don't you?"

I keep my mouth firmly shut.

The room isn't particularly large. Neither is it especially beautiful, or perhaps it only seems that way when a fae of such magnificent beauty such as the High King sits between its four dark mahogany walls. Vases with liquid of various levels line a row of shelves toward the back of the room. A table sits to one side with discarded cloths, discarded trays, little measuring beakers, funnels, and worn notebooks. Behind the rickety chair Faradir sits in now is another shelf that bears what almost seem like bits of flesh, floating in liquid.

Is this some sort of poison study room?

If it belongs to the High King, I can see where Ash got his fascination with poisons.

Faradir's lips pull to one corner. "It seems he's wrought a spell over you, too. Humans can be so disgustingly loyal. But I suppose you must be. Sad little race of sad little men, with your sad little wars and your sad little kingdoms."

I suppose he would see us that way, grand and glorious being that he is. He's probably glad I'm keeping my sad little mouth closed.

"Let us discuss business then, shall we, little mortal?" He doesn't loosen his grip on my jaw, but tilts my face to the side, as if to get a better view of my rounded ear. "Would you bargain with me for the fate of your world?"

"I do not think it would be very smart of me to bargain with you," I reply honestly.

That earns me a wide, sharp-toothed grin. "Prudent girl. Though, we must agree, the human lands are in a bit of a predicament, are they not? Your Prince Trenian will leave at dawn to destroy your world and your family. Unless, of course, he kills me to negate the bargain. Which, I think you can understand, will be challenging for him to accomplish when I have his favorite little morsel of mortal flesh. Do not presume that I have missed how he looks at you—like he'd devour you whole if he could."

"Demeaning me will not make me more likely to bargain with you."

"The shrinking flower has a sharp tongue." Faradir gives another grin, releases my face, and leans back in his chair. It gives a long, protesting *creeeeak*.

I take that opportunity to get my feet beneath me and rise. No use kneeling when I don't have to.

Faradir regards me as I smooth the front of my gown and tuck my hair out of my face. Vaguely, I process that my tiara is gone. It must have fallen out of my hair at some point.

"The current state of your human lands is dire," Faradir continues. "Your options are to let tomorrow happen, or to bargain with me to prevent it. So, human princess, which will it be?"

Is he actually tempting me? If I somehow bargained with him to spare my people, could I buy Ash time to win his throne? So he wouldn't be forced to try to kill Faradir tonight and give up the crown?

And Faradir does have a point: he has me.

If I refuse to give an inch here, then he may decide I'm of no use to him and kill me, or brutalize me to shock Ash. But, to bargain with the High King is to give up what little power I have now, and probably not even get what I want in return.

Stalling seems like my best option. Maybe Edvear will find Ash and tell him where I am.

I dare not glamour myself, so I try to hide my shaking hands in the folds of my skirts. "What do you want from me in exchange for you vowing to relinquish all claim to the human lands, from now until the end of your reign?"

"There are a great many things you could offer. Your youth and vitality. Your will. Your marriage to my son. You could vow your loyalty to me instead of Trenian. You could offer even yourself. Do not underestimate what you have to give, my dear."

I take that to mean: *Do not underestimate what I can take from you.* "I see. And I should assume that you will try to trick me with this bargain?"

Faradir waves one hand. "Of course. That's how bargains work. But I do not have endless time to waste with you. Do you want your human lands spared?"

I see no point in denying that.

"Then what if I propose this: I will nullify my bargain with Trenian—with his agreement, of course—in exchange for you to belong to me for just one day, from sun up to sun down."

"Which bargain?" I ask, immediately suspicious.

He smiles. "We can clarify that it is the bargain to raze the human lands. Surely a few hours of your freedom are worth a continent of lives."

If I were the only one affected, I would be tempted. Faradir would use that time to publicly torment, humiliate, and torture me. But I wouldn't be the true target: it would be Ash.

After seeing how he reacted to Oleria's fate, I know the aftermath of this would be so much worse. Ash is already planning to kill Faradir and end their line tonight, which he told me will result in war across Faerieland and the dissolving of any barriers between the fae and the humans.

The human lands are no safer if I take this bargain.

Still, I pretend I'm turning the offer over in my mind, willing the sweat sliding down my back to not give away how terrified I am. "Will this bargain keep my people safe from the swords and magic of the fae?" I ask, testing his reply.

"While I cannot predict everything that might—"

"I assumed the answer was no," I say, cutting him off. "Thank you for confirming."

For once, he looks taken aback.

I remember the story Ash told me last night. How Faradir only wanted to play games he could win.

I've made a mistake.

Faradir's eyes dilate suddenly. Instinctively, I throw myself backward against the door. Frantically, I reach for the knob, as if I could open it and run for freedom.

Faradir's forearm slams into the wall above my head. His bare hand closes around my throat, cutting off my scream. I scrabble with my useless fingers at his hold. It tightens, tightens, until I cannot breathe.

"I have tried to be patient with Trenian, and with you, but enough is enough," he snarls into my face, his incisors lengthening to fangs. "I am your High King, and you belong to me. Your life is in my hands. Submit to your King."

He loosens his grip on my throat just enough for me to drag in one desperate gulp of air before blackness completely swallows my vision.

"You may be their High King," I rasp, choking. "But you were never mine."

His furious blue eyes are the last thing I see.

CHAPTER 64

THE PRINCE

M Y THOUGHTS ROAR so loudly in my head I am nearly blind as I take one of the portal gates to the Nothril Court. When the black-as-night spires pierce the stars and Lulythinar moon above me, I barely hear the words the guards hurl at me, demanding to know on what business I have come.

I don't answer them. I just open my mouth and bellow: "Prince Rahk!"

My voice echoes off carved stone walls that shine like obsidian. More guards come spilling out of the palace, ready to defend their lord and lady at a moment's notice.

But they do not lay a hand on me, and I lay none on them.

"Prince Rahk!" I bellow once more.

The main palace entrance swings open. Long, tied back silver hair comes into view first, then polished black armor, the hilts of two swords sticking up over his shoulders, and finally, Rahk's clenched fists. He shoves aside the guards in his way and hurries down the palace steps to where I wait.

"Ash," Rahk says, his eyes sharp even as his face remains stoic. Even if we weren't blood sworn, it's nice for once to not be wasting time constantly referring to each other by titles.

"Come with me," I say, and it's more of a plea than a demand.

Rahk casts one glance back at the palace—which tells me he will likely face retribution from Lord and Lady Nothril if he returns to Valehaven now. That is his only hesitation before he marches with me back toward the gate that will transport us through time and space to Valehaven.

"Where is Stella?" he asks under his breath once the guards are far enough behind us, and there is genuine fear in his tone.

"The Ivy Mask is taking her to the Small Cities."

"What is the Ivy Mask?"

"No time. I am going to kill Faradir tonight and I need—"

Rahk grabs my shoulder, yanks me back just before I reach the stone archway at the edge of this side of Caphryl Wood.

"You cannot do that," he growls, the mask falling enough to reveal the sudden fury twisting his face. "You will plunge us all into war if you give up your throne."

"I don't care about that," I snarl back, baring my teeth at him. "There will be war one way or another, but there will be no peace until Faradir is dead. He *ripped off* Oleria's wings and sold her in slavery to one of the barbarian kings beyond the Veil. I sent my wife away—forever. You know I'm bound by blood to destroy the human lands at dawn if Faradir is not dead. What choices do I have, Rahk? Tell me!"

Rahk's brow thickens with his horror, but he doesn't back down. "Get him to break one of his own laws—like you'd originally planned!"

"What do you think I've been trying to do these last dozens of years?" I shoot back. "Why do you think I married Stella? My father is not stupid, and the difference between us is that I have things I care about, and he doesn't. He's practically invincible."

488

"Stop right there," Rahk demands, putting out his arm to keep me from stepping through the portal to Valehaven. "You're losing your rationale. The High King—"

"He ripped her wings off! *He ripped her wings off.* He killed Calver for no reason except that I refused to marry Listhra. He blinded Hylath because I married Stella. Now look what he has done to Oleria. Can you imagine what he would do if he got his hands on Stella? He would shred her to pieces!" My voice collapses in on itself, and I cover my mouth with my shaking hand.

"Keep your love for Stella," Rahk says, his voice softening just slightly. "That is what makes you strong. But you need to stop believing the High King has all the power in this situation. There *is* something he wants, the one thing he has protected at all costs. The one thing he would sacrifice his only heir to keep."

I shake my head, my hair flying into my eyes. "The only thing he cares about is his throne, and it's not as though I can drag his throne before him and threaten to chop off its legs if he doesn't behave!"

"Stop being purposefully dense. Thinking like that will land us in war. A war that will lead to all of us being killed and Faradir still reigning. Stop reacting and start thinking."

Then Rahk marches through the portal, which bursts into bright white light as it swallows him whole. I stand there a second longer, huffing, both furious and yet . . . maybe a twinge hopeful. A touch less desperate.

Why did Rahk have to get all the steadiness, all the unmovable strength of the two of us? All I got was rebellion and a flare for drama.

Then again . . . perhaps that is what I need to bring down Faradir. A knee that refuses to bend and something to mislead him—a way to lie to him while always telling the truth.

Stella's face flashes before my eyes, beautiful and distraught and fiercely stubborn.

I shouldn't have sent her away.

Mountains of Ildrid, I was a fool to let my fear overcome me like that. She has that brilliant glamour magic of hers, and it might be just what I need. Though not in the way I'd originally planned.

The gate swallows me in dazzling white as I step through it.

My feet land on smooth rocks beside a moonlit sea of glass. Rahk waits a pace away, his face like the jagged edges of a cliff. His presence always gives me the courage I need to press onward.

Our eyes meet, and then I'm in motion, heading toward the bridge that leads to the cliffside palace of Valehaven. "We need to find Stella and bring her back. Immediately."

Rahk gives one nod and falls into stride beside me, matching my quick pace. Hardly a minute later, before we've cut through shrubbery instead of minding the winding path, Rahk's head whips up, his nose twitching.

"Edvear is coming," he announces.

"Toward us?"

"Toward us. He's running."

Dread like nothing I've ever faced clamps hold of me, chains my ankles and wrists. "Oh, Great Kings have mercy."

I break into a run, Rahk on my heels, my desperation quickening my pace. Then I come to a sudden halt, nearly getting rammed in the chest by Edvear's horns. I grab him by the shoulders, restraining him as he gasps for air.

"Lady Stella!" he chokes out, his cat eyes ringed in white. "The High King!"

"He has her?" I demand, then shake Edvear when he doesn't answer fast enough. "Does he have her?"

"Yes!" And then he crumples into sobs as his limbs give out.

"How?" I tighten my grip on him, so he hangs like a ragdoll with his buckled knees. "How did he get her?"

"It doesn't matter—" Rahk tries to tell me.

Because we both know. We both knew the moment Edvear started crying.

"What. Did. You. Do?" I seethe, bringing my face to Edvear's. "You betrayed her."

"The High King made me!" Edvear cries out in pain as my nails dig into his skin. "I never—"

"Let him go, Ash!"

"When?" I demand. "Tell me when!"

"While you were gone at the Small City!" Edvear babbles. "He caught me, and I thought he was going to kill me but—"

Rahk grabs my wrist in one hand, my elbow in the other, and pulls at a threatening angle. "Let. Him. Go."

I ignore him and his threat to break my arm. "You betrayed Oleria! You're the reason she—she—"

"He was going to kill Hylath and Milton and all the rest!" Edvear screams back. "I had no choice!"

"No wonder the High King accepted my second bargain!" I'm shaking with rage, with hurt, with hatred. "Because he knew he had you all along to tell him *everything* I was planning! I *trusted* you!"

"Enough!" Rahk bellows and throws his body into mine with such force I am nearly knocked off my feet. I scramble to catch my balance, lose my grip on Edvear, and the moment I'm almost recovered Rahk barrels into me once more and smashes me against a marble column.

"Go save your wife," he hisses at me when our faces are only a few inches apart. "What's done is done. Edvear is a victim, not the enemy. So get ahold of yourself!"

I only stand there frozen for half a second, but it feels like time slows, expands, enough for the panic in my mind to crystallize. How can I not keep my balance? Every time I think I'm in control, think I have what it takes to overcome the High King, something shoves me to the ground. It's as though I walk a tightrope of courage with gaping chasms of fear on either side, ready to swallow me the moment I take a misstep. And I'm *always* taking missteps.

It's not hatred of the High King that is stronger than my love for you, Stella. It's my fear of losing you.

491

Ash, I believe with all my soul that there is always, always happiness on the other side of heartbreak.

I love you, Stella.

Stella has been my guiding line, my north star since the moment she came into my life. She is the calm to my chaos, the reason to my madness, the hope to my despair.

I am better for having known you, Ash. Can you believe that? For me?

The memory of Stella's face replaces that of Rahk's before me. That first moment I saw her, when I removed her veil. The sheer terror always reflected in her gaze—it's gone, and has been for some time.

Maybe I need to accept that, despite the heartbreak of our circumstances, we have made each other better. And that even if this ends worse than I could have imagined, I wouldn't give up the time we had with each other. I *won't* give it up.

My father may take her away from me, but he can never take away my love for her, or the happiness I found with her. Those are mine, forever.

"I won't touch Edvear," I growl. "We need to get to Stella."

With that, Rahk releases me, and we break into the fastest run of our lives to the place where my gut just *knows* Faradir has taken her.

We reach the massive double doors with the carved oak and its spreading foliage and vast root network. The usual guards are gone, probably out celebrating on the blood-soaked lawn.

"I'm with you," Rahk murmurs under his breath.

Those soft words bring a sudden lump to my throat.

"He's fighting to keep his tyranny," Stella once told me. *"But you are fighting for freedom. For hope, for peace. For me, Ash. That's why you won't break."*

When I drag my composure under control and shove open those double doors, it is for those that I love. Stella, Rahk, my mother, Oleria, Hylath, Calver, my whole staff—even Edvear—the Small Cities, and for those who sang the Call of Lulythinar earlier. It is for what is good in the human lands and in all of Faerieland that I step into that room and face the man sitting on his throne. The man I call Father.

He smiles at me, long and slow and triumphant.

In his palm is a small, clear globe of glass, and trapped inside, shrunken to the size of one of my fingers . . . is Stella.

CHAPTER 65
THE PRINCE

STELLA'S TINY HANDS bang against the glass, her mouth moves—like she's trying to shout to me, to tell me something. Not a sound escapes the globe.

The High King gives it a little toss. My stomach bottoms out. He catches it. Tosses it again. I swallow hard.

"Prince Rahk, it pains me that my son has brought you into this," Faradir says. "I would hate for you to become collateral damage."

Rahk, ever collected, executes a bow and replies, "I have no intention of becoming thus, High King. Thank you for your concern."

Faradir gives Stella another toss. When he catches her again, her face is the color of murky seawater. She presses a brazing hand against the wall of the globe, the other to her stomach. "Suit yourself. Prince Trenian, I believe we have something to discuss."

I itch to draw my blade, to send it hurtling through the air into Faradir's heart. But he would be expecting that, and at any moment, he can smash the globe to the floor and kill Stella.

I must be careful about this.

"I believe we do, my High King," I say. I do not give the false nonchalance and carelessness I usually wear. Instead, I'm serious, forthright. "What is it that you want from me?"

"Compliance."

I draw in a breath through my teeth. "Compliance for what?"

"Everything."

I nod slowly. "You wish to tell me you want something, and for me to say 'Yes, Father,' and go do it?"

Teeth glitter back at me. "Exactly."

I narrow my eyes at him, take one step closer to him. To Stella in the palm of his hand. "Do you want that more or less than you want me dead?"

He thinks for a moment. Gives Stella another toss that sends my gut swooping once more. "Less."

Rahk is doing his best to not react, but he blinks just a smidge too quickly at that. Many people know Faradir is corrupt. Few people know how deep his hatred truly runs.

Or perhaps . . . how deep his *fear* runs.

"Easier to start over with a new heir than risk this one turning on you," I say, and take another step closer.

Pointed incisors gleam at me. "You understand."

"I understand some," I correct, tilting my head to one side. "But there are other things I do not understand. For example, why you have trapped my wife in a spell instead of killing her."

Faradir shifts in his throne, sitting up instead of lounging. "Ah! That. Yes." He takes the globe, brings it closer to his eye, and gives it a quick shake.

My blood pounds as Stella hits the roof of the globe, then falls back to the bottom . . . and stays there. My hands clench into fists. For a second, my vision goes black with fury. I breathe, count to five, and wait until the darkness clears and there is Faradir once more, sitting on his throne with his crown on his brow.

"I need her for a few things," says Faradir. "One of which is your wedding. She's going to be one of the witnesses."

I blink. "My wedding?" My mind goes back to that bargain we made the day I went to the human lands for a bride. I wrack my brain for the exact wording of the bargain, for what I might have missed—

Great Kings.

Every drop of blood leaches from my face.

Faradir taps his chin. "I do believe the wording of the bargain specified you must be in *possession* of a wife by Lulythinar. And . . ." He looks up at the ceiling, where there is only the tiniest sliver of sky visible before the moon fills the skylight. "We're only moments away from the deadline. Last I checked, you might be married, but *I* have possession of your wife."

I barely keep my feet rooted to the floor instead of staggering backward.

Rahk's head whips to the side even before a small door to our right opens, and two women enter.

One I barely even know. It's Lady Iluna, the one friend of Listhra that Faradir didn't kill. One of the women who tormented Stella. I may not know her, but I certainly hate her.

The other is the High King's wife, and she shuffles with bunched shoulders to his throne. She bows and says the only words I've ever heard her say: "I brought her as you requested, my lord High King."

"Let us begin the bonding at once!" declares the High King, standing and gesturing with a broad grin to me and Iluna. "Kneel before the throne, both of you. Trenian, I recommend you comply if you don't want anything bad to happen to your human." Then he frowns, glancing at his wife, still bowing at his feet. "Get up, woman. You're dismissed. Three witnesses are more than enough."

She does as he says, keeping her head down as she makes toward the door she just came through. Iluna kneels as bidden before the High King. I start to follow, my hands shaking, my mind spinning, my heart reeling. How did I miss that when we made the bargain?

In possession of a wife.

497

He put that clause in the bargain for this very reason: so he could trump and null my choice if he didn't approve. And I didn't—

Wait.

In possession of a wife.

I look up at the ceiling, the hairline crack of the moon counting down the seconds till midnight. Then I look at Stella, who is struggling to pull herself to her feet in Faradir's globe. She meets my gaze, and her face softens, her brown eyes big and emotive as always. She mouths: "I love you."

And in that moment, I want for nothing but to be deserving of that love.

I never should have shoved her away. Never should have let my fear get the better of me.

I should have fought harder for her.

I will fight now.

Shoving to my feet, I break into a sprint, rip a small knife out of my belt, and throw it as hard as I can.

It thumps into the open door the High King's queen is about to slip through, slamming it shut. I'm there a second later, grabbing the poor woman and yanking her back to the throne. She lets out a tiny, pitiful scream.

"Release her!" the High King bellows, marching down the steps to his throne. "Release her!"

"I don't believe you specified *whose* wife I had to be in possession of at Lulythinar," I growl.

Faradir reacts faster than I expect. One second, I hold his wife, the next I realize he intends to kill her in the last seconds before Lulythinar. But I lose my footing to either block the knife or throw Faradir's wife out of the way.

A cry of pain splits the room.

It's not a woman's.

My shock stuns me for a second, and I do not realize what has just happened. Not until Iluna cries, "Prince Rahk!"

That's when I see the knife in his shoulder, the tightness of his body at the pain.

I almost lose my grip on the queen, whose name no one knows. *Rahk.* He threw himself in front of the blade. Not to spare the queen. To spare me.

"Rahk!" I croak, shoving the queen behind me while I keep my grip tight on her arm. "Rahk!"

"I'm fine," he grunts, breathing hard, lifting his iron cold eyes to where Faradir stands at the foot of his throne, an open-mouthed Stella clutched tightly in his hand.

At once, a slight burning bands around my wrist. I look down just in time to watch the broken crown tattoo disappear.

Then I look up to the ceiling, where the moon perfectly fills the skylight.

The bargain is fulfilled.

Faradir clenches his jaw, closes his eyes. Then he opens them in a flash, and for a second, deep voids of black stare back at me before they return to jewel blue. He swivels his attention from me to Rahk as he gets to his feet, his bloody hand pressed beneath the knife still in his shoulder.

"You shouldn't have done that," Faradir says softly. "Your parents will never forgive me for what I'm now forced to do."

I shove the queen away from me. "Get out of here," I hiss at her. "And summon the Court."

I don't wait to make sure she's going to be safe as she turns and runs as fast as she can, silent tears streaming down her cheeks. My hand finds my hilt.

Rahk lets go of his wound, backing up as he draws one of his swords. The High King shakes his head, as though mournfully. He yanks an axe from the array of golden weapons behind his throne.

"You have been so loyal, young prince," Faradir says, his tone smooth as honey before it turns hard as rock. "Kneel before me and take your punishment like a man. Do not prove yourself a coward."

"I do not see how protecting your queen constitutes disloyalty," Rahk replies, blood dripping down his arm. His voice has only the barest quaver from the pain. He continues backing up slowly, making the High King follow him. "If you tell me what crime I have committed, I will submit to my punishment."

"You have defied me," Faradir hisses. "You and—"

He lets out a bloodcurdling scream as bright blue blood gushes onto the polished marble floor. My arcing sword clatters to the ground. I am already in motion, diving to catch the falling globe—and the hand still holding it—before it shatters.

I barely catch it, wrench Faradir's severed hand from around it, and clutch the blood-smeared globe to my heart. "I've got you," I whisper, hardly able to draw a full breath. This close, I can see the purple bruises ringing her tiny neck, marring her beautiful face. But her eyes are open, blinking at me—in horror, shock, perhaps—and her hands are braced against the edges of the globe. She mouths my name. "Forgive me for leaving you," I tell her quickly, then shove her in my pocket and throw myself behind one of the pillars ringing the throne to dodge Faradir's flying axe.

"You are the bane of my existence!" he roars at me. A blast of pure magic hurtles from his one remaining hand and hits the pillar, toppling it in seconds. I scramble out of the way of the blackened rubble, struggling to get my footing as blast after blast rains down on me. "I have never hated anyone so much as I hate you!"

The words ring true, and I am too busy trying not to die to be hurt by them. Rahk's words echo in my ear, telling me not to kill the High King. No matter how much I want to hurl my own magic right back at him.

Instead, I get up and run.

"Don't run from me!" Faradir bellows as I sprint for the double doors. Rahk is already there, holding one open for me. "Don't you dare run like a coward! Face me, you spineless spawn of mine!"

I keep running.

The doors to the throne room blast open from a shockwave that leaves them broken and smoking. Rahk ducks into another hallway while I take refuge behind the massive winged statue overgrown with greenery.

I cup the glass globe through the fabric of my tunic, say a prayer, and then the head is blasted off my statue.

My feet are in motion again. I long for my sword, but it's gone, back in the throne room. Over the sound of my own loud breathing and the crashes from Faradir's magic bolts, I can hear him chasing me.

I run into the first door I find, fling it open, and throw myself inside. Everything is blue.

"You've got to be kidding me," I growl under my breath as my eyes land on a perfectly formed glass peacock lying in two pieces on the floor. The waterfall of blue silk partitions what was once Listhra's reception room from her personal rooms. I haven't even made it to that partition before the door is blasted down behind me.

I turn, panting.

The High King storms into the room through the smoking, blackened rubble of the door. His single fist glows with brilliant light, ready to decimate me.

"Do you enjoy being trapped like a dog?" Faradir snarls at me, striding toward me.

I stand on the threshold of that blue silk, my foot ready to take one step back. Waiting. Deliberating with myself.

I will *not* say his title. No matter how much my instinct screams for me to obey the law.

Not yet.

"You do not know how many times I have cursed that geas on the throne that prevents me from siring more than one heir," says Faradir, slowly approaching with his one hand raised. Ready to end me.

"I know you hate me," I pant. "I hate you too."

Keep him distracted. Don't say his title. Let him forget the law.

"It appears we're even on that front, at least." His stump of a wrist drips blue blood on the floor. It matches the aesthetic of the room.

I dare not let my hand stray to where Stella rests in my pocket. My heel inches back. My mind wars with itself, remembering Rahk's demand. I clench my jaw. If I must make this sacrifice, then I swear by the Great Kings, it will be my last. "I didn't always. I remember a time when I loved you."

His teeth flash in an ugly, maniacal grin. "Well, I hated you the first moment I smelled you. You reeked of your mother."

The lie fills my nostrils with iron. Neither of us flinches, but the truth remains exposed.

Faradir takes three aggressive steps toward me. Blood smears his face, stains his teeth as he speaks. "Tell me you are sorry for the misery you've wrought in my life, and perhaps I won't kill you now."

I hold my ground. *Keep him distracted.* "I know why you hate me."

"Because you are a rebel to your core! You tried me as a child, and you have never once stopped to aid me, to strengthen our throne. All you have done is seek to weaken me. To weaken *us*."

We're only a foot apart now. I'm unarmed, and my magic alone cannot stand against his great well of power as High King. He holds a ball of magic in his one good fist, ready to annihilate me. His only heir. His key to a strong and peaceful reign.

"That's just it, Father." Now that we're so close, so near destroying each other, I don't have to shout. I can speak quietly, and he will hear. But the words don't matter so much as the distraction. "You hate me because I am too much like you. Because I am a threat to you. You never wanted a son you could be proud of. You wanted a son you could control."

Faradir's good hand darts out. Searing heat grips my throat. I choke. Panic surges in my blood. I nearly retaliate, nearly react and summon my own magic.

"I know what you're trying to do, *Prince Trenian*," Faradir hisses just as I lift my heel. "But if you're so determined to steal my throne, then I will steal yours."

It's only a split second. A split second for him to shove me aside as he throws himself across the threshold into the next room. A split

502

second to realize that he intends to trick me exactly like I am trying to trick him. My mouth opens in sudden hope, the title bursting from my mouth in one last effort to save my throne: "High King!"

But it's already too late. He's already in the next room.

Immediately, I feel a well brimming with latent magic inside me—a well I wasn't even aware of—wash away completely. I didn't realize I would sense the loss so keenly. But I do. There's no question now.

I broke a law of Valehaven. I did not address the High King upon entering a new room.

And now, I can never be High King.

I stagger to regain my balance, pulling myself into a defensive position. Except, it's not necessary.

Faradir is on his hand and knees, eyes wide and bulging in their sockets, his usually silken hair wild and tangled. *He feels it too.* Blood stains his golden robes. Slowly, he lifts his face to me. "What did you do?"

A smile twists my lips as I reach into my pocket, and pull out the forgotten globe where Stella is busy kicking the glass to break it. "It seems you forgot I wasn't the only person in the room. You failed to address my wife by her title. Congratulations: you are no longer the High King of Faerieland."

I counter the blast of magic that comes flying my way, and the bolts collide in a blinding explosion between us. Stella doubles over, covering her eyes from the brightness. I hold her close to my breast, about to slip her back into my pocket so I can end Faradir once and for all.

But when the light clears, Faradir is gone. I'll bet on my mother's grave he's going back to the throne room. That scene is going to be hideous.

First, I pull the globe out. I cover it with my other hand, growl a string of spells, and the glass vanishes like a puff of smoke. Light bursts from the tiny Stella, almost as blinding as the magic blast from a few seconds ago. Then she's on her knees before me, back to her normal size, coughing and heaving.

CHAPTER 66
THE PRINCESS

MY HEAD SPINS, my body trembles, but all I can think is:
He gave up his throne.

"Are you alright?" Ash's hands are running over me, pulling my tangled hair back from my face as I dry heave. Tears fill his voice. "I thought I'd lost you!"

I want a sappy reunion as much as he does, but there's no time. The moment my brain stops whirling, I push to my feet, saying, "I'm fine! My head just hurts from that spell." And my neck hurts from being choked, my whole body from being tossed around and banged into the walls of my temporary prison over and over again. That's not important, though. Ash grabs my arm to steady me, and distantly, I realize I lost a sleeve at some point. That's also not important. "Ash! You gave up your throne! Faradir might be going to reclaim it even now!"

I dive toward the burned remains of Listhra's door, only for Ash to catch me around the waist and haul me back. He sticks his own head out first, checks to ensure there's no one ready to blast us to pieces, and then the moment he turns his face to tell me it's safe, I'm bolting out again.

"Faradir cannot reclaim the throne, not unless he intends to conquer all twelve of the Courts, which he will certainly try to—"

I stop, turn toward him, grab him by the front of his shirt, and yank him into a fierce kiss. He seems stunned at first, not reacting except in shock. Then his arms are around me, his hand in my hair, his mouth moving even fiercer against mine.

"I will sacrifice a hundred thrones, a thousand crowns for you," Ash whispers. "And I won't regret it for an instant."

I let myself enjoy those pretty words, let myself bask in the warmth of his affection, the sincerity of his earlier apology for sending me away, the brilliance of his maneuvers.

Then it's time to face the throne room, and the chaos that has broken loose with a rogue former king and a throne that no one can sit in. Ash and I quickly pick our way through the rubble of the smashed door and enter a room full of turmoil.

Faradir hunkers like a cornered animal at the steps of his throne, a spear in one hand and a crossbow in the other. His fangs break through his glamours as he snarls at the last people I hoped to see right now: Lord and Lady Nothril. They stand at the front of the shrieking crowd, shouting at Faradir.

"Give me the crown," Lady Nothril demands.

"It will not accept you!" Faradir snarls back. The crown lies on the second stair beside the throne. When Lord Nothril reaches for it while Lady Nothril argues with Faradir, it sparks an angry red, and he pulls back with a sharp hiss.

"I told you!" shrieks Faradir.

Ash shoves to the front of the crowd. He picks up his fallen sword on the way, and my heart triples its rhythm as it becomes clear what he intends to do.

Rahk stops him, grabbing him by the arm and hissing under his breath: "You lost your throne? This was the one thing you were *not supposed to do*! What have you done to our people?"

Suddenly, something tugs at me. I turn, looking for who touched me. But no one is close enough. Despite the madness, the fae give Ash and I a wide berth. They view the two of us with a mingle of horror, curiosity, shock, and awe.

The tugging returns.

I frown. It's not physical. It's . . .

I lift my eyes. Tilt my head to one side. The cacophony around me fades to a soft hum. My awareness focuses on the source of that tugging: the throne itself.

I step away from Ash. All I see is that large, golden throne, the discarded crown beside it. It strikes me suddenly how lonely that throne and crown are by themselves. My heart twists in my chest, wringing at their pain of separation. Of loss.

Certainty builds inside me as I keep approaching. I can ease that pain, that loss. Because I *am* what is lost. Something in my blood calls to the throne before me.

I bend to pick up the crown. Immediately, it glows, as though with joy. *Wear me,* the crown seems to beg. *Please, you must bring harmony to me again.*

So I do. I place the crown on my head. It adjusts to fit me, and a rush of rightness, of joy and . . . and *power* courses through me in a torrent.

The throne pulses, aching, calling me to sit in it. Begging me to make all of us one again.

So I sit.

And it is like a symphony plays in perfect harmony. Like birdsong and forest rain, the beauty of flowers and the flap of hummingbird wings. A golden well opens up from the throne, rushing into me, but not overwhelming. I am merely a channel, one that can reach into this tremendous well and spread it to all the worlds.

Now I understand why they say humans are nothing but a cycle of death. There is such abundant life in this flood of magic.

What a heartbreaking shame it is that Faradir wasted all this life and goodness.

Faradir.

Ash!

I open my eyes. Before me is a throne room full of slack-jawed, utterly silent fae. There is Lord and Lady Nothril, looking like they were just blasted in the face by Faradir's magic. Beside them, still with that knife in his shoulder, is Prince Rahk, who for once cannot disguise the shock on his face. Faradir has gone white as a sheet, restrained by Ash—as though he tried to attack me.

Wait.

Am I . . . sitting on the throne of all Faerieland? Did I just put this crown on my head? What am I even doing?

Oh Mountains of Ildrid.

I nearly slide off the throne in my shock. But all around me, there is a brilliant, almost blinding light.

Vines shoot from the floor, from the throne, from the ceiling, from the walls, twisting around the marble pillars, bursting into an overwhelming rainbow of iridescent blossoms. They wrap around the throne until I am sitting amid a chorus of sweet fragrance and hundreds of flowers.

The water in the stream around the throne shoots high into the air as flocks of doves appear out of nowhere and fill the throne room, their wings a beating choir of magnificence. At once, the water spirals in a mesmerizing dance, rising to the ceiling and then splashing back down in a shower of thousands of tiny droplets.

A loud, ear-splitting holler of triumph slices through the silence left behind after the water and the doves.

It's Ash. He wears the most dazzling grin I have ever seen and throws punches into empty air—as if he is made of pure elation and cannot contain it. He hollers again, his crystalline voice carrying through the vast chamber. "All hail High Queen Stella!"

High what?

"But how?" Faradir, still pale, clutches the stump of his arm. "She is not of the bloodline!"

A slow, triumphant grin spreads across Ash's face as he climbs over the absolute *forest* of vines surrounding the throne, until he reaches the steps. "She carries my blood. My blood from *before* I broke a law and disqualified myself from the throne."

I think I'm going to pass out. And yet, the moment I have that thought, strength surges into me from the throne, fortifying my limbs, clearing my vision, until I am not sure where the magic ends and where I begin.

"All hail High Queen Stella!" Ash cries once more. He drops where he is on the steps, kneeling before me.

"Ash!" I breathe. *I cannot be High Queen! There's no version of reality where they will accept me as their ruler!*

He is too busy kneeling before me to respond.

Shuffling rips my attention from my husband to the masses before me. My breath catches in my lungs.

Prince Rahk sinks into a deep bow. Then one by one, shock shifts to awe, and each fae in the room bends their knee and bows. *To me.*

I don't know what to do or say. More fae come streaming in through the open doors, and each one joins in suit. Even Lord and Lady Nothril bow, until the entire room is on their knees before me.

Except one.

Faradir remains standing where he is, murderous hatred twisting his beautiful face, contorting his broad shoulders. Ash gets to his feet and draws his sword. Faradir's gaze shifts from me to him, and that hatred only redoubles. Ash points the blade at his father's neck. His voice rings out: "Bow to your High Queen."

"I bow to no one," Faradir snarls.

Ash takes another step, presses the blade against Faradir's throat. "Bow. To. Your. Queen. *Now.*"

A shift takes place in my mind. The shock melts away. The denial follows. There is still a strong feeling of inadequacy—though I'm not sure anyone can sit on the throne of all Faerieland and feel adequate—but there's also necessity.

I'm not the High Queen of Faerieland because I want to be, or because I'm the best candidate.

I'm the High Queen of Faerieland because Ash paid in blood for his people to be free.

It is not for me that I open my mouth and issue my first order. It is for Ash, for his mother, for Hylath, for Dottie, for Edvear, for Oleria, for Rahk, for Mama Bagogs, for the Small Cities, and everyone who has suffered under the wrath of Faradir.

I lift my chin, narrow my eyes, and let the well of power rush through me as I say to Faradir: "Bow."

Surprise rolls through his shoulders. He looks at me—*really* looks at me. He doesn't see the human girl his son brought home to spite him. He doesn't see the pawn, or the girl who learned to stand before him without shrinking.

He sees the woman I truly am. The woman Ash fell in love with. The woman who sits on his throne.

And he bows.

Uproar bursts around the room, a mingling of alarm to see the former monarch brought so low, and something that sounds almost like *cheering*. Lord and Lady Nothril protest with vocal shouts, while others dance in celebration.

It distracts me.

I miss the way Faradir gathers magic in a glowing ball in his one good hand, hidden by the bloody stump of the other.

"No!" Ash arcs his sword downward, but not fast enough.

Faradir shoots the blast of magic. It hurtles toward me like a shooting star.

I don't even know what I do. I'm not sure it's *me* that does anything. I throw up a hand toward the ball of light, and energy rushes through me. Pulsing energy from the throne, from the deep wellspring of magic now available to me.

One minute, a ball of light is about to incinerate me.

The next, a blackened spot of cracked marble is all that is left of Faradir.

Ash turns saucer-wide eyes at me. "Did you just do that?"

My eyes are equally wide. "I think I might have?"

Ash lets out another holler, so out of place with the increasing number of kneeling fae, protesting fae, and the fact that I just killed his father. He runs up the steps to me, catches my face in both hands.

"You, my darling, are magnificent beyond anything I could have ever imagined," he tells me, and then kisses me in front of all my new subjects.

CHAPTER 67
THE PRINCESS

I T IS DAWN before I leave the throne room. The rest of the night is a fog. Ash and I barely claim control of the palace guard before the Nothril Court does, and the entire celebration turns wild with rejoicing and bloody rebellion. At first, it seemed like the night would devolve into full-blown war between the Courts glad to be rid of Faradir and the Courts horrified at the prospect of being ruled by a human.

But after striking dead three more rebels, and Rahk leading the palace guard to regain control of Valehaven, I solidify my right to rule—regardless of the dissenters.

On Ash's request, I've already abolished the law dictating that all fae must greet those with titles upon entering a room. The result? The loyal fae flooded the throne room to celebrate the end of Faradir's reign with an absolutely disastrous amount of spirits.

I do remember the singing, though. My favorite was the song everyone sang as the sun began to rise, marking a new century. A new age.

Ash ensures Rahk's wound is taken care of. At my request, Edvear is brought forward, and I issue him an official pardon. Ash quickly forgives him and asks for forgiveness in return for whatever happened between them earlier. He is not reinstated as Ash's steward, but Rahk accepts his service, and he will start once the Nothril prince brings Mama Bagogs back to her Small City home.

To my surprise, Edvear brings Hylath—guided by her daughter—to pay her respects to me. Maybe it is just the exhaustion, but I cry at the sight of her and tell her so many times how much I cannot wait to have her back as my maid, if she would like that. Her drooping, severed eye tendons perk upright, and she sticks out her long tongue and wiggles it. Which I take to mean that she would love it.

After her, a new figure approaches. For the first time, the widow of the former High King meets my gaze, her eyes soft and hopeful. Her voice is quiet, but her grip is strong when I extend my hand to her. "Thank you, Majesty."

I swallow hard. "What is your name?"

"Elara," she breathes.

"Anything that you desire, it is yours, Elara," I tell her, and squeeze her hand when the corners of her mouth lift in a smile.

I am more than ready to sink into the hot bath Dottie drew for me. It's a good thing she's there to help me, because I nearly fall asleep several times. Just as she helps me from the tub, a knock sounds on the door. I hold the towel close to my chest and, because I know who it is, say, "Come in!"

Ash enters, his hair wet from his own bathing. He wears a long silken bathrobe and an exhausted, beaming smile.

"I can take it from here," he tells Dottie, who obediently scurries away to give us some privacy.

Once the door shuts, Ash crosses the distance between us, pulls me to his chest, tilts my face up, and claims my mouth in a long, lingering kiss.

"I don't know about this whole ruling business," I whisper as his kisses move to my ear. "I will do it; I know I must. But I'm still learning so much about your people and—"

His finger lands on my lips. "You will never be alone. I will be with you whenever you need me. Any aid, any help, any guidance that you require—or that you simply desire—you shall have it. Though now I'm not even sure you'll need my protection! Now, now, don't blush and look away." He catches my chin again, drags my gaze back up to his. His eyes soften. "You are my brilliant, beautiful queen, and I am your loyal subject."

"Loyal subject?" I whack him across the chest. "You and your drama! You will be at *least* a prince consort. Is a king consort a real title?"

"I don't particularly care if it is," Ash replies, catching my wrist. "You can give me whatever funny courtly names you so desire, but my favorite shall always and ever be husband."

With that, he scoops me up into his arms, towel and all. Suddenly, I'm not so tired anymore. Not as he carries me to our bedroom and shuts the door. He kisses me until I've forgotten everything but his lips and mine, and the complete and utter freedom that is now ours.

"As your advisor," Ash murmurs against my neck, "I should like to suggest that your first act as High Queen of Faerie be to prioritize the creation of an heir."

I cannot help my bright, beaming grin. "As the High Queen of Faerie, I think I will take your advice, my dear, handsome advisor."

And so, for the first time, there is no fear between us. Only hope, promise, and sweet belonging.

EPILOGUE
THE PRINCE

I T WAS ON Stella's order that a chair of equal elegance be placed beside her throne. I insisted I didn't need one, that I was more than delighted to take my usual place—leaning against one of the grand pillars and either cleaning my fingernails or causing mischief. She only arched an eyebrow at me and said, "Maybe I just don't like being parted from my advisor. He gives *very* good advice."

What can I say? I had no will to protest after that declaration.

So I sit in my throne beside hers while she holds court. There has been much, much to do following the end of Faradir's reign and the beginning of a new era in Faerieland. Rah< and I have already put down a rebellion in one of the lesser courts. That has been the only blatant rebellion since Lulythinar, but there have been plenty of subtler messages, especially from the Nothril Court. There is much reform Stella and I long to make, but we cannot rush anything if we are to peacefully maintain the loyalty of the Courts.

Slowly and surely, we are rebuilding a stronger world for many centuries to come. In time, we will make Faerie a safer place for the humans who end up trapped here. For humans, for us, for our people.

But not just for them.

My gaze drifts from the fae coming forward one by one, bringing their problems to their High Queen. It shifts to where Stella sits on her throne in a gown of pure white, with scalloped sleeves, and her crown of gold rests on her lovely head. This is the best part about having my own place to sit beside her: I get to look at her all day, enjoy every little play of emotion across her face. Her focus, her concern, her surprise, her joy, even her anger. There are fewer observers in the gallery than when my father reigned, for the simple fact that Stella's reign is far more boring. It is full of problem solving and practical, level-headed considerations. There are no heads rolling for random reasons, no explosions of temper, no trickster bargains.

I never thought I'd come to love the throne room. But as of late, my view has improved.

Stella listens to the latest fae babble about how the rebellion ruined his millow farmland, and now he's lost his livelihood. Her hand gently glides absent-mindedly over her rounded middle while her brow puckers, and she nods in understanding to the case brought before her.

Then she turns to me.

Her face pinks when she catches my expression, catches that I've been watching her instead of listening. Her voice is dry when she asks, "Do you have any advice to impart? My inclination is to use the crown's funds to hire laborers from the Small Cities to repair the damage, if there be no fae available for the task."

She understands well by now that fae do not like to be hired. They think of it as temporary—or permanent—slavery, regardless of the fact that they are compensated. Humans, on the other hand, have no such compunction.

My lips pull into a smirk. "Brilliant, Your Worship."

She rolls her eyes and goes back to work, but there's a glint in her eye, a warmth to her mouth, and she gives our unborn heir an extra caress.

One of Stella's favorite things to do now is to ask for advice on things she neither needs advice on, nor is at all relevant to my areas of expertise. This morning, she asked if I advised her to wear white or blue. Yesterday, she asked for which variety of human food we should serve at our next banquet. The day before, she asked if I advised a limit on kissing. At least on that front, I have a strong opinion.

As the hours of work drag by, I find myself struggling with restlessness more than usual. Part of me thinks I am not at all suited to the sheer boredom and tedium of ruling, whether in part or full, but the rest of me knows it's because I'm excited for tonight. I have a surprise for Stella that I think she is going to *love*.

Hopefully.

The doors to the throne room open, and where I expect more fae to enter with cases to bring to their High Queen, it's a human who enters. He speaks to the crier, who announces: "King Roland of Aursailles, here on behalf of peace!"

Stella's attention sharpens, but she makes no move as the doors open wide, and a human entourage enters. I push upright from my slouching position, old anger turning my gut hot as one of the two kings I despised most in the world comes down the center aisle. His gaze is transfixed on Stella, his jaw slackening as he beholds her in her full glory.

He realizes now who among his daughters is *actually* the fairest.

Coward that he is, he's brought one of said daughters. She wears no veil—of course, because that was not a real tradition. I have no idea which sister this one is, only that she carries a toddler at her hip and looks much older than Stella does. Another man stands at her side, and the crier announces him as King Ilbert of Enslington.

"Amelia!" Stella gasps.

Amelia's mouth drops open. "Isabelle Louise!"

I drum my fingers impatiently on my armrest. "Will the humans not bow before the High Queen of Faerie?"

Stella's eyes are wet as she watches King Ilbert and her sister kneel before her. King Roland, his hairline having receded in the time since our last parting, hesitates. As though he cannot believe the woman he stands before now is his daughter.

I'm about to bark a threat that he bows before my wife *or else*, when Stella shifts her attention from her beloved sister to her father, and the tears vanish. They're replaced with boredom.

"If you have come on terms of peace, I see not why you should care to insult me," she replies, lifting her chin. "I know why you're here, and if you want me to oblige your request, you will offer the due respect."

My mouth twists into an utterly delighted, wicked grin. Stella was born to be Queen of Faerie.

King Roland bows.

Stella gestures with one hand. "You may speak."

Roland looks like he would rather eat his own shoe than obey his daughter, but he rises. "We ask for the land stolen by the Long Lost Wood to be returned. It was part of our bargain."

"The bargain you made with *me* that when I was High King, I would return your land?" I interject. "That is not a transferable bargain, I'm afraid."

Roland grinds his jaw, likely to keep from shouting that I am a liar and a trickster.

Stella tilts her head to one side. "You think that, because I am your blood, you can request anything of me that you wish? Would you have asked High King Faradir thus?"

We both know it's not a fair question. He wouldn't have even come to Faerieland—wouldn't have been able to find the door to Valehaven without a fae escort—if Faradir was still on the throne. But fairness isn't the issue.

"No, Your Majesty," Roland says. His face contorts, as if the words are physically painful.

Amelia steps forward, holding her child close. King Ilbert tries to pull her back, but she pushes his hand away. "Please, Your Majesty, we mean no insult by coming. It is only that with the expansion of the fae borders, we have lost many."

Stella knows this better than any of them, and if it was me on that throne, I likely would be offended. But she doesn't take insult. She softens, her gaze warming at the little boy her sister carries.

"I will do what I can to restore the borders," Stella announces, only the slightest hint of emotion in her voice. "It will take time, and it may not be safe for humans immediately. But you have my word that I will begin the process."

And so, Stella dismissed the party, only to request a temporary recess a few moments later.

"You're High Queen," I remind her. "You don't make requests. You order."

She looks at me, eyes bright, her hands shaking just a smidge. "I forget sometimes."

I continue hearing her subjects' requests while she sneaks off to have a few private moments with her beloved sister before they are parted once more.

THE HIGH QUEEN

IT TAKES TREMENDOUS effort to keep my composure together as I hurry down the hallway after my sister. My hands won't stop shaking. I didn't believe I would ever see her again!

Once, I traversed these hallways in fear. Now, the only thing I fear is that things will be different between my sister and I after all that has happened.

I round a corner, and there she is. Walking with King Ilbert, her son on her hip, as my guards escort them out of the palace. My throat tightens with emotion, but my voice doesn't waver as I call out: "Stop!"

Amelia's head whips back to me.

She's not the same girl I left behind in Aursailles. She's older. Matured. Her form isn't twiglike like when she was eighteen. Her face has changed too, the glow of youth replaced by a sense of gravity and responsibility.

But her eyes still shine with warmth and heart, and if anything, she is only more beautiful.

Those eyes light with hope when I break into a run. She shoves her son to King Ilbert's chest, and he takes the child as she bursts into a full sprint.

We collide in an explosion of tears and laughter. The lilac scent of her curls takes me back to the home we once shared. Our embrace is almost violent in its force, and I wouldn't have it any other way as I press kisses to her forehead, her cheek, her hair.

"I never thought I'd see you again!" I cry.

"I cried for *months* after you left!" she replies, wiping more tears off her cheek. "And then word came that the High King had been dethroned—you wouldn't believe the conflicting reports we heard! Some said the High King had stepped off his throne. Others claimed Prince Trenian murdered him in cold blood to take it. But then the reports began converging on something I didn't believe could be *possible*. I didn't truly believe it until I walked into that throne room!"

"I still have my days of disbelief myself," I say, laughing.

Her cheeks, pink as a raspberry, stretch wide as her enthusiasm takes over. "You haven't aged a day since I last saw you! I wish we brought Ilbert's portrait painter with us so he could have captured Father's face when he walked into that throne room!" Her words dissolve into giggles. Amelia hasn't changed nearly as much as I believed. "He was *shocked* to see how magnificent you are!"

The praise makes me uncomfortable, but I would be lying if I said I hadn't enjoyed making Father squirm just a little bit. "Well, look at you! Look at your *son*!"

"Isn't he the most perfect thing you've ever seen?" she says, grinning back at the boy sleeping contentedly in Ilbert's arms. "And you're expecting your own!"

My hand falls to my stomach as warmth fills me. I open my mouth to reply, but Amelia keeps prattling on.

"I doubt you've heard, so I'll tell you now! But Yvonne—you know how she was betrothed to that horrible man? Well, it was literally during their wedding banquet, after he'd made himself unbelievably drunk, that he just dropped dead! They say it was heart problems, but I think she might have killed him. Which, if she did, I applaud!"

"What?" I say, the only thing I can say in my shock.

"And then she ran off with one of his knights! A dashingly handsome one, apparently. No one has seen them since! It was the scandal of the century! As for Vivienne, she married that old king and had his heir before he died. She's somehow made herself regent until her son comes of age. Can you believe it? As far as I know, Jacquelle is doing just fine in her marriage. Who knew she and I would be the only normal ones of the five of us?"

"Who knew, indeed," I say. "I wish them all the best. Please tell them hello for me, if you see them."

King Ilbert quietly approaches until he stands behind Amelia. There is fondness in the gaze he turns upon her, and respect in the one he turns upon me. "We will tell them."

Amelia and I look at each other once more. She blinks hard to keep the tears at bay, and I chew my lip to do the same.

"This won't be the last time we see each other," I say. "I cannot allow you to come and go from Faerieland whenever you please, but I will find a way for us to see each other. *Regularly.*"

"I need to meet that baby of yours," she says, smiling as her tears slip free. "And I need to meet him *before* he's an adult, for heaven's sakes!"

"You will," I promise.

When we have said our last goodbyes, I return to the throne room with wet cheeks but firm shoulders. Ash's penetrating gaze runs over my face. I smile at him. "I missed her."

And then I get back to work.

THE PRINCE

AT LAST, THE SUN declines, casting the throne room in alternations of shadow and brilliant golden beams. The doors are shut, the crowds dispersed, and I am finally left alone with my wife. Her head swivels to me, her long hair falling over her shoulder.

"You were somewhere else most of today," she accuses, pointing one dainty finger at me.

"Guilty as charged," I reply, unable to stop my grin. I'm much too excited for her surprise.

She stands, comes to a stop in front of my wide-spread knees. "You must tell me why. I'm agog to know. You've had a glint in your eye all day."

"Probably because I've been looking at you," I say, and pull her into my lap.

She squirms, attempting to get back on her feet, but I wrap both arms around her and keep her firmly where I want her. "Ash!" she hisses, her face turning my favorite shade of pink. "A guard could walk in here at any moment. This is hardly a dignified position for a queen!"

I pull her closer, savoring the feel of her swollen belly under my hand, and whisper: "I have a surprise for you."

She stops struggling immediately and whips her attention to me. "A surprise? What is it?"

"It's back in our quarters. It was delivered today."

"Delivered? Did you get me new clothes because these are starting to be too tight? I cannot *believe* how big I'm getting already!"

"It has nothing to do with that."

"Then what? I cannot imagine what you would have had delivered! New poisons? No, no, you would never trust someone to handle those for you. Hmm, let me—"

I place a finger over her lips, stopping her as my grin widens. "Why don't we go see what it is? Instead of sitting here and risking a guard entering and seeing how undignified their queen is?"

"Don't throw my words back at me," she huffs, but she can't hide her smile either.

When we get back to our quarters, I tell Stella to sit down and wait. She insists she's been sitting all day and would rather stand. I shrug and agree, only on the condition that she closes her eyes. She frowns but obliges.

"Milton!" I call.

He comes at once from the back rooms of our quarters, a carrier under one arm that jolts at random intervals. "At last! It's been whining all day."

"Whining?" Stella asks, puckering her brow. "It sounds like five sets of nails scraping against something!" Then both eyebrows rise nearly to her hairline. "Did . . . did you get me a *pet*?"

"Keep your eyes closed!" I say by way of answer, taking the carrier from Milton and setting it on the ground. "Are you ready? Three, two, one—"

I turn the latch for the carrier, and the idalpuff shoots out like a squirrel with a jay on its tail. Stella opens her eyes. She only catches a glimpse of the creature's shiny, luminescent shell before it launches itself straight into her arms.

Her mouth and eyes go wide at the same time, and she lets out a terrified shriek. "It's a giant bug!"

The idalpuff has grabbed its front legs onto her dress, and begins crawling up in its overenthusiastic attempt to make friends. Stella lets out another scream, falls back on the couch, and tries to shove the creature off her.

I scoop it up quickly, holding the cool, wriggling body in my arms. "I'm so sorry! I thought—these are common pets of ours! I thought you'd like the color of its shell—"

Stella's shrieks have turned to tears of laughter. "It looks like a massive rainbow cockroach!"

"Cockroach?" I cock my head to one side, glance down at the adorable little face, the way it has wrapped all sixteen of its legs around my arm and elbow. "What is that?"

She covers her face with one hand, presses the other to her heart. She's still laughing, wiping away tears, and squints at the creature in my arms. "A cockroach is a pest humans hate. What—forgive my reaction, I was only startled—what *is* that?"

Relief warms my belly. So maybe I didn't *completely* misjudge her. I kneel in front of her, holding up my arm where the idalpuff hangs, its antenna twitching, reaching for Stella. "It's an idalpuff. It has a smooth shell, and each one is different. This one reminded me of you, with all the many colors."

"People like crunchy pets?"

"Crunchy?" I ask in horror. "You don't *eat* them!"

The idalpuff rears back, affronted.

"No, no! I don't mean that. I mean that our pets are usually, you know—they're soft! Fluffy! Like a fluffy little lap dog or a cat." Still, she reaches out a tentative hand toward the creature, and when it doesn't react, slides one finger across its beautiful shell. It responds by letting go of my arm with one of its legs and curling it around her finger.

She's going to be in love with it in no time.

"Well, it *does* have some fluff," I tell her, lifting it a little more. "See? It's got a fluffy tail."

At that, Stella bursts out in full-blown laughter. "It *is* a fluffy tail! What a strange creature! You're like a giant, beautiful cockroach with a lamb's tail!"

"It's friendly," I say, holding it out.

She gives me a dubious expression, then carefully takes the creature by the sides of its shell and sets it in her lap. Immediately, it rolls over on its back, exposing its belly and wriggling all its legs at her. It lets out a happy chirp. "Do I scratch it?"

"It'll love you forever if you do."

She starts to reach toward it, pulls back slightly, gives me a grimace. "It's so *buggy.*" Then she overcomes her qualms and gently starts scratching. The creature lets out a chorus of chirps, rolling and wriggling under Stella's attentions. She's laughing before she can help it, and I am once again restored to confidence that I know my wife well.

"There's one more surprise," I tell her with a wink.

She looks up from her new pet. "There's more?"

"Oh yes. Come along, now!"

She skips after me out the door—the idalpuff at her heels—unable to hold back her grin as I take her outside. Through the garden she loves so much, out the gate. Glowing paths curve through the forest, but she carefully steps past each until I find the right one. The one I created just for this.

"I still can't believe these were right in front of my face," she grumbles, making me smirk.

It's only a few minutes' walk on the path. To a blank plot of land, freshly tilled, beside a quiet pool. Her mouth drops even before I say the words. "Is this what I think it is?"

"If you are thinking it's your own secret garden, ready to be planted with whatever you'd like—including those herbs you've got overgrowing their pots—then yes."

She lets out a scream of excitement, barrels into me with a forceful kiss, and immediately begins surveying the spot I picked out for her. "I'm going to grow vegetables!" she declares. "Because while I love all the beautiful flowers, there is a criminal lack of potatoes in this place!"

I spend the evening in deep amusement as Stella and the creature she decided to name Richard—I assume because it's sort of close to *roach*—get to know each other while she plans her garden. There are many stern warnings to behave from Stella, and even more ecstatic burbling replies from Richard that are often paired with aggressive attempts at affection.

There's still plenty we're figuring out about each other. Vastly more to figure out about ruling all of Faerieland together. In a few months, there will be yet more to learn when our first child enters our world.

I'm not afraid of any of it. Where once the future was perpetually dark clouds and a crimson sky, now it is bright stars and a shining sun. Every day is a new beginning, one full of fresh hope and ever-growing love.

For the first time in ages, I am truly excited about tomorrow.

Want to find out who Prince Rahk's love interest is and read the moment they meet for the first time?

Download the scene here:

AnastasisBlythe.com/Rahk

COMING SOON:

BRIDE OF THE MIDNIGHT PRINCE

An undercover vigilante.
The fae prince sworn to hunt her down.
Can they survive the most impossible of arrangements: marriage?

MORE FROM ANASTASIS BLYTHE

THE ZHENINGHAI CHRONICLES

Maiden of Candlelight and Lotuses
Guardian of Talons and Snares
Warrior of Blade and Dusk
Princess of Shadows and Starlight
Captive of Twilight and Treachery
Daughter of Darkness and Dreams

THE KING AND THE ASSASSIN

The Assassin Bride
The Neverseen King (Coming Soon)

BRIDES OF THE FAE

Bride of the Fae Prince
Bride of the Midnight Prince (coming soon)

ABOUT THE AUTHOR

Anastasis Blythe makes her home in central Texas with her husband. When she's not writing, she gardens, accompanies local bands and choirs on piano, rescues feral cats, and tries to keep up with the laundry. She loves exploring the world through reading, walks in nature, and thoughtful conversations.

To stay connected with her, be sure to sign up for her newsletter at AnastasisBlythe.com/Rahk.

Connect with Anastasis online at:
Website - AnastasisBlythe.com
Instagram - @AnastasisBlythe
Facebook - Anastasis Blythe
Goodreads - Anastasis Blythe

Milton Keynes UK
Ingram Content Group UK Ltd.
UKHW011318280624
444891UK00012B/32/J

LETTER
TO A YOUNG
MARRIED MAN

ABDUSSABUR
KIRKE

Lifeboat Press

Contact: abdussabur.kirke@gmail.com

LETTER
TO A YOUNG
MARRIED MAN

It would be wrong to address the subject of marriage without mentioning the social framework within which people today find themselves. The two are inextricably linked, and in the scheme of things, society at large is more important because you can help marriages by helping society at large, but you cannot heal the ills of society solely by doctoring on your own marriage. Allah's Laws have been broken and trampled on in the broader social realm, especially in matters of financial transaction, while the laws of marriage are still at least partly intact. Priorities must be laid down.

Notwithstanding, the institution of marriage in its exalted, evolved and complete form is basically dead. It remains for modern-day Muslims to re-discover and revive it, for this will certainly not happen from outside the realm of Islam. For those who take up this significant task there will be the satisfaction of helping to re-weave the shredded fabric of society itself, since in that reciprocal pattern, family is the warp of the cloth. Yet as I have already said and as I will say again later, the point of action, the main handle by which changes can be made lies not within the family zone but out beyond.

I shall not attempt to address the science of how matches between men and women are made. It is too important and too much its own subject, one which demands serious consideration and which is, perhaps even more than the damaged practice of marriage, encrusted and compromised by layers of prejudice and fiction. It has its own experts and masters who should be consulted.

Furthermore it is necessary to state that a written discussion can merely act as a help, a support, for the real learning and real transmission, which takes place in the company of, and witnessing the actions and states of, those who have made a success out of marriage. There is no substitute for that. Knowledge is not from books.

Nor can adequate space be given to the immense subject of the benefits of marriage and the gifts that men and women are to each other. Shaykh Abdalqadir as-Sufi has written at length on this matter. In a 2011 article entitled 'Ikhwanism' he says:

> "Half the Deen from Aisha is not a compliment, it is her station as a high expert of Fiqh, treated as such by 'Umar ibn al-Khattab when Amir al-Mumineen. Aisha, who, at the crisis of the Fitnat al-Kubra, made the correct judgment which saved the Deen, and who was the one beside the Messenger at his end, Allah's mercy and peace on him.
>
> It is the Ikhwan's failure to appreciate the unique power and otherness of women – which transcends even motherly affection and tenderness – we mean, that faculty to see truth in the event and be the voice of truth, that leaves them in the wrong and in defeat."

With all the above provisos, there are nevertheless certain things which can be said to men about marriage that will hopefully be of some benefit.

◊

Marriage, for man, is above all else about the man himself. It is not about a "Couple" as some ultimate entity in itself, it is about two or more different people who share a particular proximity by virtue of an agreement they have. Within that is a whole dimension of human life.

Therefore each man must look to himself. As he is well in his life-configuration, so his marriage will benefit, and as he is unhappy with the world at large, so his marriage will reflect that.

There are certain prerequisites which I would like to survey briefly and which involve the man – although they may also concern women. Islam itself is the first. Without it, there is little chance, and Allah knows best. Marriage between a Muslim woman and a non-Muslim man is not valid; it is not considered. And although it is permitted for a Muslim man to marry a non-Muslim woman, that is only if she is one of the People of the Book. Whether that category even includes today's Christians is a matter of contention which is beyond the scope of this discussion. In the setting of our current society, most such marriages are fraught with difficulties. There is much to be said for a man to make entry into Islam a precondition before accepting to marry a woman who is not yet Muslim.

Secondly, each man must have good company. This is absolutely essential: regular male company of the highest calibre. Rasul, sallallahu alayhi wa sallam, said, "Look carefully at those whom you take as companions."

A man's keeping the company of other men is what makes him and keeps him manly, and only if he guards his masculinity can he expect to enjoy that sought-after symbiosis, nowadays so elusive, which men seek in women.

The natural extension of this is that man must treat his wife as a woman and not like another of his man-friends. They are not the same. One of the most damaging doctrines of this age is that of "gender equality", a contradiction in language, logic and life. Men have been made to feel guilty about this.

Women require particular courtesies from men, as do one's male peers, one's children, one's superiors and one's parents, and a man does well to acquaint himself with them. A man does not conduct himself with women in the same way as he does with other men – neither physically nor socially. In one sense a man's wife is just a woman among women, and he is well advised to extend his manners to her accordingly, albeit with certain differences, which involve honour and intimacy, not laxness or carelessness or over-familiarity. Good manners are even more important in the home than outside it. Even more courtesy, greater fineness, more care, more impeccability, greater concern, more overlooking and more honouring are required, along with a good eye, a friendly smile and a good tongue, with good words dispensed liberally. The same applies to dealings with children.

This may at times seem at odds with the natural familiarity that comes of living together, some-times in a confined space, and sharing intimate relations. But there are few arts more worth mastering.

To enjoy togetherness, you must be apart. Men and women must have separation; this is pure chemistry. For anyone who can afford it and arrange it, separate bedrooms are worth considering. To expect to attain to a happy marriage by renouncing one's personal privacy and invading someone else's is a delusion further compounded by economic pressures.

It is now commonplace, in magazines, talk shows and online, for anxious couples to ask: why have we lost the sparkle we once had? It is because our society has lost the science of what marriage is for and what should be expected of it. Many or most of the kuffar have even abandoned marriage altogether because they see it as meaningless. To a large degree the meaning of marriage is children, more of which later. But people are no longer taught these and other meanings, so they become disappointed because soon they are looking only backwards, daring not to confront the great unknown space ahead once the initial chemistry of lust has run its course. Therein lies one of the advantages enjoyed by arranged marriages – provided, that is, the arrangers are concerned about the good of others and not merely clinging to their own tribe or race.

Notwithstanding, Allah in His Wisdom has endowed men and women with a propensity for passionate and lustful attraction, that exquisite near-madness which can blot out all reason, but which can in turn contain its own wisdom. It is called "falling in love", a questionable description. It is in any case an experience that must be contained by good counsel. For a man facing marriage, therefore, he should look to his counsellors and choose them well.

Marriage can have difficult periods but it is dangerous to get caught up in arguments. The Muslim must have a good opinion with Allah of all that befalls him. The despairing question "Why is life like this for me?" – this is not a good question. Rather: "Alhamdulillahi wa Shukrulillah. Allah has not made things like this for nothing. What is required of me?" Then the jewels, the treasures begin to surface, and of all the terrains which we cross, marriage can be among the richest.

◊

Marriage, however, is not the greatest transaction there is. It is not the be-all-and-end-all, and woe to the one who makes his family his raison d'être. This is the next foundation: that man must have a higher project than just his family or his own enrichment, however alluring both these projects

10

may appear. If he turns entirely towards his family he is turning inwards, and they will turn back against him – especially the children. It has become in vogue among Muslims everywhere to dedicate one's life to one's own children and their upbringing, thinking one is contributing to the furtherance of society, but then parents are dismayed when the children later cast aside what was offered to them.

It is better for children to see a father engaged, even if only modestly, in service to the wider community. This is the riddle which the family poses to the man. The wife and the children say: "Come to us! Come to us!" But until he refuses, saying, "I must go forth!" – until then he finds them as sirens on whose rocks the vessel of his attentions is dashed. When he does finally set off to sea he finds them as lighthouses and safe harbours of sweet homeward return. Unless, that is, his voyaging is merely to avoid his marriage and that part of himself which it brings to light.

Similar must also be said about the pursuit of wealth as a life-aim. It is a betrayal of the wife, and ultimately she will not thank him for it, even if to some men it will seem that wealth-acquisition is all that is required of them. This is not to say you should not be wealthy or work to earn money – on the contrary: marriage is aided by wealth. But wealth, like the family itself, must submit to the man's pursuit, his glorious pursuit of Allah's pleasure in serving others at his utmost and highest potential. This is worth fighting for; it is a panacea for the sicknesses we suffer.

When it comes to wealth and work, it is as well for men in long-term salaried employment to reflect on the nature of that contract, one which in many ways may affect his marriage, and not only that. In terms of the Shari'ah of Islam, the contract of the salaried employee in some important respects resembles the position of the owned slave. He is not his own man. This is another topic that deserves deeper reflection and research than this essay can afford; nevertheless it is important for a man to have a realistic idea of the dynamics of duty within which he finds himself and to have engaged in some honest reflection on the subject. If employment is akin to slavery, then modern humans exhibit a historically peculiar tendency to abandon their freedom en masse, avidly signing away ownership of their time and their labour while at the same time entering into debt transactions that seem to bind them tighter and tighter to the yoke. Fear of provision looms large. Belief in the Divine Provider is not an intellectual matter. It is something as natural to man as eating and breathing – yet it has become something rare. Add pervasive usury as the underlying basis of both industry and private purchases, factor in a need for controlled mass labour, and the resultant socio-economic pattern is the employment contract as normality. Slavery has been abolished – but pseudo-enslavement has never been so widespread. The size of the salary does not change the nature of the contract; indeed the well paid slave is in danger of complete acquiescence. The natural labour transactions belonging to the free man beckon to us, calling us back to our primal ground, as does the practice of real trade. If this seems out of reach then we must strive for it – in groups.

Another mark of our times is that men and women mingle much more liberally than has previously been the case, including in the world of work, a development which has been touted as "progress" and whose result has been the freeing-up of the female half of humankind for the aforementioned employment – and the relegation of a woman's great gift of motherhood to non-work status; hence expressions such as "working mum," which is a grave insult to mothers who are not employees. But what is germaine to our topic is that men, thrust into a "liberated" world of work, are in danger on many fronts. What is perhaps more frightening than the risk of slipping up with a colleague in a moment of passion and boredom is the prospect of becoming desensitized to woman as something that is *other*, that being the basis of the erotic experience. Liberal mingling, trumpeted as an expression of sexual freedom, is in fact in complete opposition to good sexual chemistry. Man is becoming numb by overexposure. Furthermore if he spends much of his day with other women – especially in potentially intimate, private situations – then even in the unlikely (and frankly strange) event that the thought never crosses his mind, it will certainly cross the mind of his wife, and how could it not? These are some of the meanings of the strong discouragement of such situations by Muslims. It is not puritanical. It is wisdom. The fierce, free man will disdain such a situation in which he loses his manhood by degrees and punishes his wife to boot.

The other, overriding matrix within which we are bound up is the matrix of usury, by which we mean riba in its full definition, not just excess interest. In his poem 'Usura' Ezra Pound says about it:

It stayeth the young man's courting
It hath brought palsey to bed, lyeth
between the young bride
    and her bridegroom.

So much has been written on the subject, and war on usury is so central to the Deen of Islam, that it suffices here to say that every Muslim, man and woman, must be engaged against it in some way. This is not about "Islamic banking" (as if such a thing could exist). Social action against riba, the activating factor of which is allegiance to an Amir, is pro-marriage because it connects men and women to a cause more important than the outcome of their marriage. This gives perspective and stops infighting. The married individuals do then indeed become a Couple, but only once they turn away from one another and face the common foe.

◊

The next support upon which marriage is built is the Prayer, done together – man, women,

15

children who are of age. It is the husband's responsibility to organise this without conflict, and to make it easy and not difficult. Children whom Allah does not oblige because they are not of age, should not be treated harshly by us. It is better for them to see their parents praying and want to be like them. When they reach puberty, counsel them with wisdom like you would any other young person, telling them about Allah's Vastness and Mercy, the wonders of Islam and the greatness of the Deen, the superiority of the Muslim life-pattern and the high station they as children have with Allah just for having been chosen to be Muslims. Your children are, after all, just young people like any others, over whom you have been given the honour of guardianship.

Familiarity breeds contempt. Your sons and daughters are not your friends, mates or buddies. Instead be concerned for them, first as small and defenceless beings whom you have been given charge of, then later as emerging adults with their own personalities and intellects. Their job when young is to play and be in their natural form, so leave them alone and let their mothers and women in general look after them. It is demeaning for a man to be overly engrossed in his young children and to be stuck at home. Leave the house! Your main job as man is to guard the periphery: hand over authority for everyday childrearing affairs to your women. Speak to your children kindly and with respect, and listen with attentive interest to what they have to say. Praise their efforts. Express pride and satisfaction with their good qualities, which it is your job to recognise and foster. What they are deficient in, overlook – or provide them with real help. Discuss with your wife how to help them, but don't discuss that with them when they are small.

The place which plays host to the family transaction above all others is the meal table. One does not wander alone to the fridge and walk around the kitchen, the house or the streets eating one one's feet like lost people do. Gather your wife and children to the meal table, even if it is just a cloth on the floor. At the meal table, speak of good things and not unpleasant things, as our Messenger guided us, Allah's blessings and peace be upon him. Ask your children and wife about their day, tell them what was interesting about yours, and maintain a high opinion of them. Nothing will serve you better in that than remembering Allah's Words in Surat ar-Rahman:

"Every day He is engaged in some new affair."

It is better not to criticise your wife's food, but rather to have the courage to ask her with enthusiasm for the things you like, and be subtle in putting across the things you don't like.

◊

The marriage contract itself is simple, and around it lie the words of Sayyiduna Ali, may Allah be pleased with him: "As for your duties towards others: scrupulousness. As for their duties towards you: overlook."

A man is responsible for: feeding, clothing and putting a roof over his wife's head, and the children's too. If he does this, or at least struggles to do it, there is hope. If he neglects it or does not consider it his responsibility, or thinks it a duty automatically shared with his wife, then he cannot reasonably expect to be treated with respect or listened to by his wife or children. This is one of the foundations, which is why it is rarely wise for a young man to marry if he has no means. In those countries that have such things, unemployment benefits and state income support are a terrible danger for families, especially the man, since they short-circuit the natural order. The State becomes the provider and robs men of that Fitri role.

Where the woman or women have wealth and the man has less, he is fortunate and at the same time at risk, because of the danger of slumping into an unnatural reliance on the woman's income. He ought not to forget how things lie in the balance. If what she gives is a Sadaqa then take it with civility or politely decline.

Shaykh Muhammad ibn al-Habib said that the key to a happy marriage is to invite guests into your home. Do not allow yourself to be stopped in this by poverty, lack of status or possessions, or sour grapes about not being invited yourself. Or even worse, the idea that you "don't have enough time". Invite people and feed them. It is medicine for you and them.

◊

The permissibility of having more than one wife is one of the glories of Islam and a generosity from Allah, for men and also for women – although the man who tries to persuade his women of these benefits is barking up the wrong tree. He should either take the step or keep quiet.

No-one may deny its permissibility like the modernists do as they crawl on their faces to please the kuffar, while the kafir man, crushed beneath the post-Christian boot of monogamous guilt, envies Muslim sanity.

Still, given the configuration of prevailing social mores, economic constraints and even architectural factors that have locked man down in a Guantanaman cage of impossibilities, he must take stock of his predicament and tread with wisdom and patience.

One of the traps a man can fall into is to have theories about multiple-wife marriage. The married man obsessed with the *idea* of another wife or wives is just like the single man obsessed with the *idea* of what marriage is about in the first place. Firstly, the best women may instinctively steer clear of him. Secondly, if he acts on the idea alone, he risks a failed adventure. Thirdly, he is disappointed because the reality is quite unlike the fantasy.

Few men surrounded by the present-day capitalist/humanist miasma make a real success of marrying more than one woman, although some do, and these are important examples, given how crucial the Muslim marriage model obviously is to society's survival. When it boils down to it, the man-woman dynamic remains fundamentally the same, multiple wives or not. I once asked a great man who had two wives the classic, innocent question: What is it like? He said: "Much the same as any marriage."

That is why it is folly for a man to try to resurrect a shipwrecked marriage by adding some new woman he lusts after to the equation. Some even do that – or worse, threaten it by tactless verbal pronouncements to the existing wife – to deal a death-blow to a marriage they have not found the courage either to repair or to end by honourable means.

In this as in all things, counsel is key. The passions are dangerous, and man is ill-equipped on his own.

It should be borne in mind that each extra wife and group of children adds to the aforementioned financial obligations, and this must be considered with honesty to oneself and concern for the good of others.

Shaykh Abdalqadir, may Allah have mercy on him, often told us that if a man takes more than one wife, then he should beware making it the same as having a mistress. He quotes Jane Arden, talking about extra-marital affairs: "It's no fun unless you're cheating on Mum!" That is the cop-out, the low possibility. The high possibility is that, living together in one extended household or compound or group of connected or neighbouring dwellings, a family can emerge which is greater than the sum of its parts, and greater than the isolated existence to which many modern families see themselves condemned. It is also a changed economy.

Those who do it must do it well, with courage, tact, wisdom, and as much correctness as possible, acting in the knowledge that they are rebuilding on natural foundations something that has been shattered.

◊

To a man it may at times seem, both prior to marriage and within it, as if the most important function of the union is to legitimise sexual relations. To modern man this strong and natural drive to intercourse is complicated and perverted by media and corporation marketing which presents a picture of women exposed, combined with a message of "freedom" and an invitation to consume and buy. All this tantalises man. The irony is that eroticism is not about what you can see, but what you cannot. Today's man is unbalanced by a lot of exposure to what he cannot have. This leads to what D. H. Lawrence described as "Sex in the head".

In marriage, meanwhile, sex is more like a fine ingredient in fine cuisine: it is only one of several things, and you cannot be nourished by it alone. The mainstay in fact is that staple which is everyday life, and it is as well to look to that and its quality, enjoy it, and make an art out of it.

The art of the everyday consists in courtesy, good manners, and appropriate dealings with one another. The Rasul, sallallahu alayhi wa sallam, said about marriage: "You play with her and she plays with you, and you laugh with her and she laughs with you, and you joke with her and she jokes with you." This is one of the modes of marriage; one of its sweetnesses. Good manners include smiling when you meet her, complimenting her, and setting aside times for her. The Prophet, may Allah bless him and grant him peace, said that smiling is a Sadaqa. Fortunate is the wife whose husband has a kind smile.

◊

Authority in marriage is like authority anywhere else: it is a matter of legitimacy. If what you require of someone is out of concern for the best for yourself and others, then it carries weight. To demand and not to be listened to is to squander your credit; it is better to stay quiet and wait until circumstances change. The art of leadership includes knowing when and what you can ask of people, and when to leave it. That includes women and children. Authority also refers back to the earlier subject: that the man's project must be beyond his family. That is his legitimacy. It lets everything else fall into place.

◊

We ask Allah, the Great, the Generous, to make us people who enjoy the sweetness of marriage and the treasures of marriage and the greatness of marriage. Amin.

Milton Keynes UK
Ingram Content Group UK Ltd.
UKHW041318181024
2256UKWH00011B/26

CW0149631

Original title:

Field of Fables

Copyright © 2025 Creative Arts Management OÜ

All rights reserved.

Author: Harris Montgomery

ISBN HARDBACK: 978-1-80566-666-0

ISBN PAPERBACK: 978-1-80566-951-7

ISBN HARDBACK: 978-1-80566-666-0
ISBN PAPERBACK: 978-1-80566-951-7

Secrets in the Meadow Mist

In the morning's light, things dance,
Squirrels serenade, a nutsy romance.
Grasshoppers play hopscotch with glee,
While a wise old owl sips mint tea.

A rabbit in a tie joins the race,
With carrot cake as his winning ace.
Bumblebees wear tiny capes,
As laughter unfolds in funny shapes.

Shadows of Mythic Creatures

Under moonlight, fairies prance,
With twinkling toes, they seize their chance.
Trolls trade tales while sipping stew,
Claiming they once caught a ghost, it's true.

A dragon, sleek, with a belly ache,
Refused his lunch of flamed-up cake.
A gnome in a hat pulls funny tricks,
As laughter echoes through tree-top bricks.

Tales Sprouting from the Earth

From the soil sprout stories bright,
A talking turnip says, 'What a sight!'
Carrots giggle, bursting with pride,
As radishes roll, trying to hide.

Potatoes wear glasses, reading the news,
While onions cut jokes—they share their views.
Tomatoes toss puns, ripe for a bite,
As the garden bursts forth into joyous delight.

Beneath the Glade's Embrace

In the shade, where the shadows play,
A fox strums tunes on a leafy tray.
Mice wear shoes made of acorn caps,
While frogs in tuxedos take silly naps.

Underneath the bark, stories twist,
A raccoon insists he can't be kissed.
With chuckles and giggles, the woodland sings,
A comedy show, with feathers and wings.

The Radiance of Forgotten Voices.

In a land where shadows tease,
Loud laughter dances with the breeze.
A squirrel claims the crown with pride,
While wise old owls can barely glide.

Beneath the bramble, secrets play,
As frogs hold court at the end of day.
Whiskers twitch, a parade of dreams,
In whimsical chatters, nothing's as it seems.

Whispers of the Wandering Woods

The trees have tales of giddy sprites,
Who juggle acorns in wild sights.
A rabbit with a bowtie neat,
Dances 'round on tiny feet.

A bear in jammies snores with glee,
While busy ants sip honey tea.
The whispers swirl like picnic plans,
Disguised in leaves, they clap their hands.

Chronicles Beneath the Canopy

A raccoon writes with inky paws,
Unraveling all the nature's laws.
He giggles as he pens his tale,
Of cheeky winds that skip the gale.

A chipmunk's punk rock, loud and clear,
While lightning bugs throw lights in cheer.
The canopy hums a rhythmic song,
As laughter mingles all day long.

The Tales That the Blooms Tell

Petals gossip in hues so bright,
About the gardener's latest plight.
"Don't let the bees steal your sweet snacks!"
Chortles a bloom with polka-dot slacks.

A sunflower winks at a dandelion,
"Tell me, friend, what's your next plan?"
Bubbles of laughter drift through the air,
As they weave a tapestry, vibrant and rare.

Melodies of the Mystic Meadow

In a meadow where giggles bloom,
Bouncing bunnies clear the gloom.
Silly squirrels wear tiny hats,
Juggling nuts with acrobatic chats.

Butterflies dance in polka dots,
Whispering secrets in quirky lots.
A singing snail plays jazz on a leaf,
While wiggly worms share comic relief.

The daisies chuckle, it's quite absurd,
As ants compete in a game of word.
A frog in a tux does a little tap,
Making the crickets giggle and clap.

With every note and every cheer,
This meadow sings, it's crystal clear.
A whimsical world where laughter flows,
In a place where anything goes.

The Odyssey of the Overgrown Trail

In a forest thick with tangled vines,
A raccoon rolls in glorious designs.
He wears a crown made of pinecones,
Leading the squirrels in humorous tones.

A chatty parrot yells out a joke,
As foxes gather and start to poke.
With wobbly legs, they dance like fools,
In a parade of nature's quirky rules.

A turtle slows down, like he's sipping tea,
While rabbits race by, all wild and free.
Their laughter echoes through the tall grass,
As they teeter and totter, letting time pass.

So follow the trail that twists and bends,
Where every corner brings new friends.
With a hop, skip, and a silly tail,
Join the adventure on this overgrown trail.

The Ballad of the Blooming Hills

On blooming hills where laughter sprouts,
Giggling daisies whisper doubts.
A clumsy goat gets tangled in blooms,
While bees make music with their sweet tunes.

A plump pig dons a flowered bow,
Dancing around with a splishy-splash show.
The chickens clap in a feathery beat,
As butterflies join in, oh what a feat!

In the bright sun, a cat tries to prance,
Falling head over paws in a silly dance.
With every tumble, the hilltops cheer,
Echoing joy for all to hear.

So frolic and play on these vibrant hills,
With laughter and whimsy that always thrills.
In this land of fun, let time stand still,
For joy is the treasure in the blooming hills.

A Tidal Wave of Folklore

In a land where tales do leap,
The fish tell secrets as they sleep.
Waves of laughter end the day,
With crabs who dance and whales who play.

A pirate parrot squawks with glee,
As mermaids sip their minty tea.
Octopus writes a novel bright,
With squid as stars, oh what a sight!

The sea cows sing in harmony,
With jokes about big feet and tea.
Dolphins leap in splendid arcs,
Sharing tales of their past sparks.

With tides that twist and tales that spin,
Each wave brings laughter from within.
So gather 'round, it's quite the show,
This ocean's jest, where dreams will flow.

The Lost Art of Woodland Storytelling

In the woods where squirrels jest,
A toad's the bard; we love him best.
He croaks of kings and golden crowns,
While owls wear ties and act like clowns.

The bunnies weave their stories tight,
Each fable shared beneath moonlight.
A raccoon's laugh rings loud and clear,
He cooks up tales with roots and beer.

Woodpeckers drum a quirky beat,
As hedgehogs dance on tiny feet.
The pines, they lean to hear the game,
Of woodland tales that never wane.

So grab a seat, this tale's for you,
A forest fest with stories new.
With every laugh, the trees will sway,
In humor's grip, we lost our way.

Celestial Narratives Among the Trees

Stars winking in the night's embrace,
A fox with dreams begins to race.
He recounts how the moon got its glow,
With tales of a midnight disco show.

The owls wear glasses, wise and neat,
And dance on branches to the beat.
They tell of comets made of cheese,
And sunflowers with secret keys.

Shooting stars drop hints of fun,
About a prince who couldn't run.
He tripped on clouds, a sight to see,
While shadows giggled silently.

So sit back now, enjoy the lore,
These cosmic tales, you'll want much more.
For laughter's echo fills the trees,
In the celestial breeze, we're all at ease.

Treasures Beneath the Sapphire Sky

Underneath the endless blue,
A turtle spins her yarns for you.
She hides her treasures in the sand,
With stories told, oh so well planned.

Sea stars gossip of the past,
About a crab who raced too fast.
A jellyfish with glowing pride,
Tells how she danced with the ocean tide.

The gulls play tricks, they swoop and dive,
Their tales of fish keep humor alive.
With every splash, the stories swirl,
As dolphins twirl in life's grand whirl.

So let the waves be your guide,
With laughter rolling like a tide.
Beneath the sky, let fables fly,
For treasures bloom where seagulls cry.

Whimsy in the Wilderness

In a wood where rabbits dance,
A squirrel tells a funny chance.
With acorns dropped in silly glee,
And raccoons snickering by a tree.

A fox wearing socks, oh what a sight,
Chasing shadows in the fading light.
The trees giggle with a whispering breeze,
While frogs perform their comedic tease.

The Parables of Pebbled Paths

Along the pebbled, winding way,
A turtle joins in on the play.
With a rabbit racing in a hat,
They laugh together, imagine that!

Each pebble holds a tale so bright,
Of squirrels juggling through the night.
A wise old owl gives a cheeky wink,
As mice debate what cheese to stink.

Tales Told by Twinkling Twilight

In twilight's glow, the stories rise,
Of fireflies wearing tiny ties.
The stars above chuckle and blink,
As crickets play and start to wink.

A hedgehog plays the marimba well,
While a bumblebee tries to tell.
With puns and jokes that swirl and spin,
The night is alive with laughter within.

Sagas of the Singing Shrubs

The shrubs break into joyous song,
With melodies both silly and strong.
A garden party under the moon,
Where flowers spin and start to swoon.

With petals slapping in tune with glee,
And a daisy dressed up for the spree.
They share their quirks in a floral chat,
As whimsical stories tumble and spat.

A Patchwork of Fabled Dreams

In a garden of whispers, frogs wear crowns,
Telling tales of the moon with silly frowns.
Turtles race in suits, what a sight to see,
While squirrels hold court, sipping nutty tea.

A parrot recites, the jokes from the sky,
As sunbeams shimmer, and giggles float by.
Each creature a jester, in feathery glee,
Painting laughter in shades of wild esprit.

Traces of Imagination

Balloons held by hedgehogs, floating so high,
Chasing clouds like marshmallows, oh, what a fly!
Dancing daisies wear shoes of luscious bright green,
Their petals all twirling, a whimsical scene.

A fox in a tux, at a grand dinner dance,
Serves soup made of stars; won't you give it a chance?
With owls as the band, they play tunes that delight,
As the sunset chuckles, painting day into night.

The Alchemy of Nature's Stories

A wise old owl brews tea from the sun,
While rabbits mix laughs, just for a pun.
The trees share tall tales, they bend and they sway,
In the chorus of blossoms, they giggle and play.

Caterpillars juggle, with grace and with flair,
While bees gossip sweetly, their buzz fills the air.
Each story a potion, a silly affair,
Concocting delight with a sprinkle of care.

Harmony in the Rustling Leaves

Leaves laugh like children, tickled by breeze,
As crickets conduct a symphony with ease.
A parade of ants don hats made of grass,
In marching formation, they strut and they pass.

The sun winks and whispers, "Join in the fun!"
While shadows join hands, they dance and they run.
Together they twirl in a comedy show,
A whimsical world where giggles just flow.

The Dreamscape of Dandelions

Dandelions dance in the air,
A swirl of wishes, without a care.
Bees wear tiny caps, quite absurd,
While ants hold a meeting, just being heard.

A rabbit hops in a fancy vest,
Claiming he's cooler than all the rest.
Rainbows giggle with popcorn skies,
As ladybugs throw candy surprise.

Harmony Beneath the Ferns

Under ferns where the crickets sing,
A frog plays piano, it's quite the fling.
Squirrels wear shades, sipping on tea,
While owls debate who's the wisest of glee.

The wind hums a tune, quite offbeat,
As lizards dance to its funky beat.
A snail in a hat tries to keep pace,
With mushrooms that twirl in a jolly race.

Fables of the Fleeting Breeze

The breeze tells tales of socks on trees,
Of flying fish and mountain bees.
A squirrel's secret, a nutty decree,
While clouds play chess with a shade of glee.

Picnics float by on giddy floats,
As ants wear hats and parades their coats.
Kites in the air boast of their flair,
With frogs giving speeches on life's great care.

Narratives in Nature's Embrace

In cozy nooks where stories cling,
A raccoon croons about his bling.
The trees make puns on the silly plight,
While rabbits argue who hops just right.

The moon chuckles at owls' wise quests,
As flowers giggle in their fancy vests.
Every whisper of the wind recalls,
The funny fables that nature sprawls.

The Allure of Yesteryears

In a land where rubber chickens roam,
And every tree hums a silly poem.
The mice wear glasses, read all day,
While cats debate the best cheese to sway.

A clock ticks backward, time's a jest,
Each tick a giggle, a playful quest.
Where shoes lace themselves for a dance,
And socks hold a grand parade at a glance.

Whimsical Whispers of Wonder

A frog with a top hat leaps on a page,
Telling tales of a royal wage.
Squirrels with sass run their own cafe,
Serving acorns in a most charming way.

Clouds shaped like cats drift through the sky,
Winking at birds as they flutter by.
The sun wears shades, a diva's delight,
While the moon throws a party in the night.

Tides of Ancient Narratives

Once upon a wave, a fish wore a crown,
He ruled over pancakes, a breakfast town.
Turtles tell tales in slow, grand style,
While seagulls squawk jokes that make you smile.

A crab in a tuxedo conducts the sea,
In a ballet of bubbles, a sight to see.
Mermaids giggle, sharing secret schemes,
Dreaming of some wild, watery dreams.

The Elixir of Enchantments

A wizard's brew of giggles and glee,
Pours laughter as sweet as honeybee.
Fairies in gardens play hide and seek,
With flowers that bloom when the sky is bleak.

A dragon dons glasses to read bedtime lore,
While trolls throw stones, then ask for more.
The potion of joy fills the air with cheer,
Creating a world where all jokes appear.

Fables of the Ancient Grove

In an ancient grove, where laughter sings,
A turtle wears boots and runs on springs.
A rabbit with glasses reads tales of delight,
While squirrels play chess deep into the night.

The owl is the judge of a comical jest,
While hedgehogs throw pies; they do it the best.
They giggle and dance, a whimsical show,
In the heart of the grove, where wild fables grow.

The Path of Storytellers

On the path where the storytellers tread,
A cat counts the stars from her cozy bed.
With socks on his paws, a fox tells a tale,
While monkeys swing high with a wink, never pale.

A frog croaks in rhythm, a curious croon,
As bees join the chorus beneath a full moon.
Each step that they take is a twist and a spin,
In this trail of whispers, where laughter begins.

Sylvan Murmurs and Magical Echoes

In the heart of the woods, where secrets reside,
A frog claims the throne with a flip and a glide.
The fox paints the sky with stories absurd,
As whispers of giggles flutter like a bird.

Squirrels in hats hold a riddle contest,
While a deer in a tutu shows off her best.
They dance through the thickets, under branches so wide,
In this echo of joy where imagination can't hide.

Colors of Elysian Dreams

In a meadow of dreams where the colors collide,
A parrot invents tales with whimsical pride.
With purple and green, a giraffe paints a tune,
While fireflies glow like the stars in June.

A rabbit in polka dots jigs in the sun,
With a sprinkle of laughter, life's just begun.
Together they weave tales of joy and delight,
In the colors of dreams that sparkle so bright.

Fantasies Under the Starlit Canopy

In a land where socks dance free,
And the moon hums a silver spree.
Bunnies wear hats, quite absurd,
While owls recite a funny word.

Stars wink down with playful glee,
As wishes ride on the breeze.
Turtles tell tales of old,
While crickets rub their legs, so bold.

Bubbles float, holding secrets tight,
And ghosts giggle with delight.
Each shadow tells a chuckling joke,
As laughter sparks and dreams provoke.

In this wonder beneath the night,
Every corner holds a frightful fright.
Yet silly friends make it all okay,
In this whimsical, starlit play.

The Lorekeepers' Sanctuary

In a realm where stories sprout,
And even the stars twist about.
A dragon plays hide and seek,
While a mouse dons a pilgrim's peak.

Tales are told over sweetmead cups,
As silly hats adorn wise pups.
Legends of shoes that can fly,
And bicycles that cheerfully sigh.

Cats recite in a dramatic tone,
While a parrot mimics a groan.
Wizards mix potions of fun,
In this sanctuary—never undone.

With each turn, a new giggle rings,
As crickets join in with their strings.
Every creature joins the song,
In this place where we all belong.

Birch Songs and Pine Whispers

Where birch trees giggle in the breeze,
And pine needles sway with careless ease.
A squirrel wears spectacles of glass,
While ants form a band in the grass.

Each leaf rustles tales of delight,
As shadows dance in the fading light.
Chipmunks jive with a bouncy flair,
While hedgehogs giggle beyond compare.

A brook sings songs of frothy cheer,
And water lilies sway, oh so near.
In this carnival of woodsy fun,
Chasing butterflies, everyone runs.

At twilight's blush, the laughter grows,
Under the watch of stars' soft glow.
Nature spins dreams with a wink,
As the creatures gather 'round for a drink.

Chronicles Under the Oak

Under the oak, stories unfold,
Of brave little mice who are bold.
With capes made from leaves, a sight to see,
And ants marching on in harmony.

The history is told by the breeze,
As beetles debate with utmost ease.
Bugs in tuxedos gather to feast,\nOn crumbs left behind
by the mischievous beast.

Each acorn holds secrets of fun,
As the sun dances, it has begun.
Fireflies twinkle like gems in the gloom,
While frogs croak out a cheerful tune.

Thus the chronicles weave their jest,
Of creatures big and small, all blessed.
In this realm, laughter reigns supreme,
As we wander through this whimsical dream.

Ballads in the Blossom

In a garden where the tulips dance,
Bees wear hats, they take a chance.
A ladybug sings a funny tune,
While grasshoppers leap to the moon.

Petunias gossip about the breeze,
They share secrets with the honeybees.
A squirrel juggles acorns with flair,
As butterflies twirl without a care.

The daisies tell of the sun's bright laugh,
While worms sign autographs on a gaff.
Tulip cups hold tea for all,
In bloom and jest, they have a ball.

At dusk, the roses play charades,
Tickling each other, a game that invades.
With petals and laughter, they find their place,
In this blossom land, a whimsical space.

Footprints in the Mystic Soil

A rabbit hops with a top hat on,
He claims he's the wizard of the lawn.
With carrot spells and a cheeky grin,
He makes the flowers giggle and spin.

The shadows stretch, and the mushrooms sway,
As toadstools tell tales of yesterday.
A fox in socks makes quite the fuss,
In muddy puddles, he rides the bus.

The crickets chirp in a grand parade,
While slugs wear glasses that they exchanged.
Each track a story, each step a jest,
In this enchanted realm, we feel so blessed.

A whisper of magic floats through the air,
In the soil, secrets, and mischief, we share.
With giggles and mischief, each creature enjoys,
Playing among all their whimsical toys.

The Weaving of Verdant Tales

In the meadow, threads of green unite,
Weaving tales under starlit night.
The daisies knit scarves of pure delight,
While crickets crack jokes 'til morning light.

A spider spins stories from high above,
Of lost socks and a playful dove.
With each tale, a new stitch appears,
And the flowers chuckle, oh, what cheers!

A wise old owl perched on a tree,
Writes the antics of a busy bee.
In this tapestry, laughter is sewn,
Each petal a verse in a song well-known.

Together they dance on the winds of lore,
In a world of wonder, they always explore.
Every weave holds a chuckle or two,
A humorous quilt, forever fresh and new.

Silhouettes of Timeless Whispers

The shadows of rabbits play tag at dusk,
With whispers of mischief, they dance in husk.
Owl in a hat flashes twilight eyes,
While fireflies glow, sharing chuckling sighs.

In the crook of night, the frogs recite,
Poems of warts and a fairy's slight.
Laughter erupts underneath the stars,
As ants parade home with tiny guitars.

A breeze chuckles as it tickles the reeds,
Shaking secrets from mischievous seeds.
Old trees tell tales of long-ago fun,
With echoes that jump like a springing gun.

And when dawn edges in with a shy glow,
All the creatures wear smiles, putting on a show.
In silhouettes cast by the sun's first light,
They blend joy and tales in the softest flight.

Whispers of Enchanted Tales

In a grove where squirrels sing,
Tales of cheese and dancing bling.
A rabbit hops with shoes so bright,
Chasing shadows in the night.

Mice are warming up the stage,
Telling tales of cheese and age.
With every joke, the trees respond,
Dancing roots, they carry on.

A wise old owl starts to hoot,
Jokes so funny, they bear fruit.
All the creatures start to laugh,
Roasting marshmallows by the path.

Then a fox with a silly hat,
Tries his hardest to chase a cat.
But the laughter steals the show,
In this place where dreams can grow.

Echoes Beneath the Canopy

Beneath the branches, whispers rise,
Silly tales in moonlit skies.
A frog appears with a crown so grand,
Claiming rule of this merry land.

The raccoons bring a feast so bright,
Juggling cookies in delight.
Each one trips and spills a treat,
Then dances away on small, quick feet.

A kazoo plays a silly tune,
As shadows dance beneath the moon.
Chipmunks gather in a line,
Singing loudly, oh so fine.

Tickled grass starts to giggle,
As the creatures start to wiggle.
In the laughter, all is fun,
In the night, the joy has won.

Chronicles of Forgotten Dreams

In a meadow where laughter flows,
A snail races, oh how it slows.
While a beetle tries to cheer,
With a trumpet, loud and clear.

A grasshopper that loves to joke,
Danced with glee, fell in a smoke.
The flowers giggled with delight,
As they argued who was bright.

A turtle wearing shades so cool,
Claimed he was the fastest fool.
But they all knew deep inside,
In this place, there's no need to hide.

They share dreams beneath the sun,
As they laugh and wildly run.
In a swirl of giggles, they weave,
Crafting stories you wouldn't believe.

The Garden of Woven Legends

In a garden full of buzzing bees,
Blooms that sneeze and giggle with ease.
A spider spins a yarn so grand,
Of tangled tales across the land.

The daisies chat in sunny hues,
Playing dress-up in bright blues.
A dandelion blows a puff,
Saying, "Funny is always enough!"

A bear with honey on his paws,
Starts to dance, without a cause.
While a hedgehog joins the show,
With acrobatics, just for flow.

Underneath the laughing Moon,
They share stories, none too soon.
In this laughter, life feels right,
With legends woven through the night.

Ciphers of the Forest Floor

In the grass, a squirrel sits,
Reading maps that fit like mitts.
He waves to ants with tiny hats,
Discussing snacks and acrobats.

A raccoon plays a game of charades,
With mushrooms whispering in cool glades.
The trees still laugh, their secrets shared,
While giggles dance through sunlit air.

A rabbit hops like he's on air,
Decoding signs without a care.
With carrots acting out each plot,
He wonders why they taste so hot!

The forest speaks in playful tones,
With toe-tapping roots and dancing stones.
Each leaf a page of wild delight,
Beneath the sun, they frolic bright.

Riddles Beneath the Bramble

Under bramble, riddles play,
Where shadows twist and twirl away.
A badger tells a pun so sly,
While beetles scoff and wink their eye.

An owl hoots out a riddle rare,
With laughter echoing through the air.
Each branch a perch for witty puns,
That sprinkle mirth like falling suns.

A snail declares, 'I'm rather quick!'
While hummingbirds dispute the trick.
With every turn, a giggle blooms,
In this world of playful rooms.

The bramble weaves its stories deep,
Where silly tales can make us leap.
In laughter spun from nature's thread,
The forest chuckles as we tread.

Lullabies of the Wandering Spirits

Whispers drift on a gentle breeze,
As spirits tell tales beneath the trees.
They sing of joy in playful rhymes,
Tickling the leaves with timeless chimes.

A ghost with a grin brings forth delight,
Dancing shadows in the night.
While a fox joins in with a prancing jig,
Their laughter echoes, soft yet big.

Each haunting note suggests a dance,
To frolic in the moonlight's glance.
With every bob and whimsical swing,
The forest hums, a happy thing.

These lullabies of carefree souls,
Wrap us in warmth as the darkness rolls.
Together, they make the darkness bright,
With tales of fun in the still of night.

Secrets Carved in the Bark

Trees stand tall with secrets near,
Carved into bark that can make you cheer.
A raccoon reads, with eyes so wide,
"Don't tickle the toad, he's full of pride!"

A wise old owl checks his notes,
Among the squirrels with acorn coats.
"Beware the squirrels, they're quite the trick,
Their pranks can leave you feeling sick!"

The forest giggles with tales of old,
Turned into bark that's worn and bold.
Each carving holds a giggling fight,
Between the laughter and the light.

So if you wander, take a glance,
At whispering trees that love to prance.
For secrets shared will always hark,
In the joyous dance of the carved bark.

The Epic of the Evening Dew

In the garden where gnomes play,
Dewdrops giggle, new friends to sway.
A snail slides by with a jaunty hat,
Chasing a butterfly, oh, imagine that!

At dusk, the crickets put on a show,
With tap shoes shining, they steal the glow.
The daisies clap, their petals in cheer,
While frogs croak jokes that all can hear!

The moon winks down; it's quite the sight,
Worms doing the worm in the silver light.
A moth tells tales of wild, wild dreams,
As stars snicker with twinkling beams.

So gather 'round, both beast and bug,
With laughter shared, give your heart a shrug.
In this realm where whimsy abounds,
Funny magic in the night resounds.

Murmurs of the Mossy Hollow

In the hollow where shadows dance,
Mossy blankets invite a prance.
Toadstools bob with a silly grin,
As hedgehogs spin, let the fun begin!

Wise old owls debate the stars,
While fireflies blink like little cars.
A rabbit juggles acorns galore,
As mice shout, "More!" from the forest floor.

The brook whispers secrets, so sweet and clear,
Of tales untold that only it hears.
Squirrels wear spectacles, looking so bright,
Pondering snacks for their long, cozy night.

But laughter echoes through every nook,
In this mossy land, take a good look.
Who knew such joy could be found right here,
With giggles and grins, we hold so dear!

Fables Cast in Twilight Shadows

As twilight drapes a shimmering hue,
The shadows gather, a curious crew.
A raccoon dons a cape made of leaf,
While singing loudly, beyond belief!

A fox with charm, in a dapper suit,
Tells tales of a dance with a giant root.
The owls hoot laughter and flip a coin,
While rabbits sway to an old-time join.

Glowworms shimmer, the floor lights up,
As critters gather for a tea party cup.
Lemons on daisies, sugar on grass,
Where every fable feels like a blast!

A marmoset hops, spreading pure cheer,
In shadows and fables, we hold so dear.
With a wink and a giggle, we write tonight,
In this mirthful realm of pure delight!

The Rhyme of Rustling Reeds

In the reeds where the whispers play,
Tickled by breezes, they sway and sway.
Each leaf a laughter, a story to sing,
With spiders weaving in their tiny bling!

A duck wearing glasses, with a book in tow,
Reads poetry loud to the pond below.
Fish chuckle softly, they wiggle their tails,
In this world where everything prevails.

A turtle's slow motion is truly a feat,
As frogs leap high to a jazzy beat.
With crickets tapping their tiny feet,
The evening's laughter feels complete.

Join in the rhyme, let your spirit soar,
With rustling reeds, who could ask for more?
For in this realm of glee and jest,
Every little creature knows it's blessed!

Murals of the Gentle Breeze

A chicken danced with glee in a hat,
While squirrels plotted schemes with a cat.
They painted mustaches on frogs quite grand,
And laughed as the flowers took a stand.

The sun wore sunglasses, quite an affair,
As bees played music with hornets in the air.
The grass joined in, swaying left and right,
While ants marched on, ready for a fight.

A rabbit painted rainbows on a tree,
The humor in nature, wild and free.
Each leaf told tales, some silly, some wise,
As the breeze whispered secrets, beneath sunny skies.

With laughter echoing, the party grew,
The daisies danced, wearing skirts of dew.
In this quirky world, nothing seems wrong,
Where fun and joy are the heart of the song.

Starlight and Silver Shadows

The moon wore pajamas, spots and stripes,
While owls threw a dance exclusive to types.
A hedgehog juggled acorns with flair,
And fireflies sparkled, twinkling in the air.

In a treehouse, the raccoons made pie,
With berries stolen—oh my, oh my!
The frogs played poker, their eyes all a-glow,
While melodies drifted where soft breezes blow.

Silly shadows stretched under the starry beams,
As laughter echoed like sweet, playful dreams.
The world turned bright in a whimsical kiss,
Where the night wore joy like a laugh-filled bliss.

With starlit giggles and dreams taking flight,
Creatures of wonder danced through the night.
Who would have thought, in this echoing glow,
That such silly tales would steal the show?

The Echoing Song of Lost Souls

A ghost with a lollipop wandered the street,
Telling tall tales of his last trick-or-treat.
His friends were all giggles, a slippery bunch,
While a pumpkin head joined them for lunch.

They tripped on their tales, what a sight!
As ghoulish laughter filled up the night.
They held a séance on a surfing board,
Spooking each other with a tuba chord.

Banshees sang 'I'm too cool!' in a choir,
While mummies broke dance moves, never tires.
The lost souls proclaimed their newfound fame,
In a realm where fright turned into a game.

With echoes of chuckles and mishaps galore,
The phantoms made friends with a bear at the door.
A party of shadows, both silly and bright,
Crafted a symphony of joy in the night.

Songs From the Old Stones

The stones started singing, what a surprise!
With whispers of laughter and twinkling eyes.
Granite made jokes about sand's silly prance,
While quartz did the tango, leading the dance.

Every pebble giggled, rolling along,
As minerals gathered to join in the song.
With boulders as drums, and flint playing wood,
They formed quite a band, understood?

The hills echoed back with a rascally cheer,
As laughter rolled down, it was clear.
The chorus of rocks, both merry and spry,
Was a sight that made even the clouds sigh.

From glens to the crags, their melody swirled,\nIn this
playful realm, the stones rocked the world.
Who knew that such wisdom was wrapped in a jest?
In the songs of the stones, humor gleamed best.

The Song of the Unseen Creatures

In the dusk, a whisper sings,
Of critters wearing fanciful wings.
A beetle dances, quite absurd,
While mockingbirds join, in laughter blurred.

A squirrel with a top hat prances by,
Cracking jokes that make the trees sigh.
The shadows giggle at their own game,
As shadows and silliness are never the same.

A rabbit juggles acorns with ease,
While frogs croon ballads, aiming to please.
Chirps and cackles fill the air,
In this world of giggles, with joys to share.

And if one strains to hear the cheer,
It's the unseen critters, drawing near.
With laughter ringing, loud and clear,
They remind us to dance, to shout, to cheer!

Enigmas of the Emerald Grove

In the grove, mysteries abound,
Where giggles of trees can be found.
A gnome misplaces his favorite shoe,
While fairies debate a dance or two.

A wise old owl cracks a dad joke,
While mischievous mice giggle and poke.
The vines entwine in a playful way,
Creating mazes where critters play.

A turtle wears glasses, just for looks,
Reading about secret, hidden nooks.
And every time a leaf drops down,
Squirrels burst out, wearing a crown.

With whispers of laughter, echoing bright,
The grove unfolds with pure delight.
Each riddle and jest is part of the dance,
Where the silly and secret take every chance!

Chronicles of the Canopied Sky

Under the canopy, laughter swirls,
As clouds play tag with the breezy curls.
A crow recites poems, full of wit,
While a parrot mimics, just a bit.

The sun and moon, in friendly brawl,
Share silly stories, illuminating all.
Stars snicker softly, twinkling bright,
As the twilight plays with shadows' height.

A blushing raccoon takes a bow,
As jests float down from the trees somehow.
While bats swoop low with friendly grins,
In this high-flying party where laughter begins.

With every flap and every cheer,
The chronicles flow, drawing near.
For in this sky, endless and free,
The humor finds roots on laughter's tree!

Lullabies of the Lush Landscape

In the meadow, where daisies twirl,
A snail sings softly, causing a whirl.
With a wink and a nudge, he takes the lead,
While crickets join in, planting the seed.

The grasshoppers hop, making a scene,
As ladybugs twirl in shades of green.
A raccoon drum circles, fast and loud,
His rhythm inviting every critter crowd.

The willows sway, singing along,
Their whispers turning chaos to song.
And as the sun paints the world in blush,
Every blossom dances in a gentle hush.

So under the stars, in giggles so sweet,
These lullabies make the night complete.
With each sleepy note, gently they weave,
A tapestry of laughter, dreams to believe!

Myths in the Morning Dew

In the morning glow, the snail wears shoes,
A hat of lettuce, sipping on the blues.
Squirrels tap dance with a cheeky grin,
While sleepy clouds are dreaming about the din.

The frogs play cards in a lily-pad suite,
Celebrating victories with a chorus so sweet.
A T-Rex joins for a game of charades,
While the sun peeks in, and laughter cascades.

Daisies gossip about the bees' new tricks,
As chipmunks plot to steal the best picks.
The grass is all ears for tales just spun,
While the dew drops waltz, oh, what fun!

On a breeze of giggles, the stories fly,
Through fields of whimsy, where shadows sigh.
These comical creatures, in mischief and jest,
Fill the morning dew with their flamboyant fest.

Voices of the Whispering Wind

The wind tells tales of a fish in a hat,
Who swore he could dance like a charming diplomat.
He twirled on the banks, causing quite a stir,
While crickets chirped, "What a ludicrous blur!"

A wise old owl hoots riddles at night,
While a band of mice sings with all their might.
They jam on cheese with a rhythm divine,
As bushes rustle with laughter and wine.

The breeze carries whispers from trees that giggle,
Of unicorns playing a magical wiggle.
They prance through the glades with sparkles and glee,
While clouds shake their fluff, "Come join us, you'll see!"

So when the wind blows, just lend it your ear,
For the stories it carries are always sincere.
Amidst all the chuckles, the meadows will sway,
And you'll find humor in nature's ballet.

Enigma of the Hidden Glens

Deep in the glens where shadows do play,
A raccoon in glasses reads newspapers all day.
He snickers at headlines and sips on a spritz,
While the mushrooms gossip about the latest bits.

The squirrels have formed a detective squad,
To solve the mystery of a missing façade.
With magnifying glasses and nut-shaped clues,
They crack silly jokes as they share their views.

A turtle so slow once placed on a bet,
Won a race against a hare, oh what a set!
They laughed till they cried, shared a feast of wild greens,
In a world full of giggles, where laughter convenes.

In these winding paths, where nonsense is king,
The secrets they hold are the funniest thing.
So tiptoe through laughter, it's where you belong,
In the enigma of glens, where silliness throngs.

Lost Legends of the Wildwood

They say a bear once knitted a sweater of fluff,
For a kitten who boasted of being quite tough.
With buttons of acorns and stitches of grass,
They strutted through woods, the oddest of class.

A fox with a top hat sings opera at dawn,
While lizards in tuxedos dance 'til they're drawn.
They twirl on the leaves, and the critters all cheer,
For the lost legends echoing, far and near.

In a hollowed-out tree, a genie resides,
Who grants wishes in riddles, with no place to hide.
"I'll trade you a wish for a tale that you weave,"
And giggles erupt at the tricks that they leave.

So wander the wildwood, with humor to chase,
For lost legends linger in every warm place.
With whimsical wonders and stories untold,
The laughter you find is a treasure to hold.

Tales from the Roots of Time

In a land where squirrels hold court,
A wise old owl plays the sport.
Rabbits race with shoes on their toes,
While hedgehogs boast of stealthy woes.

A frog recites with flair and glee,
Of battles won by a bumblebee.
The ants march on, a frantic line,
Believing they're dancing, pure divine.

At dusk, the crickets serenade,
While fireflies paint the twilight shade.
A turtle gossip's through the breeze,
With tales that flutter like the leaves.

So gather 'round, a merry crowd,
For laughter echoes, bright and loud.
In this realm where laughter blooms,
Even shadows wear silly costumes.

The Dreams Woven Into Twilight

In the twilight where giggles roam,
A hedgehog dreams of a cozy home.
With pillows made of soft, sweet grass,
He whirls and twirls, a noble sass.

A cat in boots, so stylish and neat,
Takes a leap while tapping his feet.
He juggles fish in a laughing stream,
While frogs croon sweetly, or so it seems.

The clouds play tricks with silver seams,
As butterflies craft their finest dreams.
A raccoon tells of treasures rare,
With marshmallow fluff tangled in hair.

As night descends, the stars will scream,
In a spectacle that reignites the theme.
For in this realm where whimsy grows,
Laughter dances, and joy overflows.

Legends Bathed in Moonlight

Under the moon, a goat sings loud,
Proud of his echo, he charms the crowd.
A wolf in a cloak of dazzling green,
Recites his tales with a wink, unseen.

The mice in capes join a wild chase,
Playing tag in this enchanting space.
While owls scrabble for shiny things,
And gather treasures the night sky brings.

Baboons jump high, with raucous mirth,
As giraffes twirl in the moonbeams' hearth.
The night is bright with each silly plan,
Where laughter leaps, and no one can.

So tell your tales beneath the stars,
Where every note plays sweet guitars.
In moonlit laughter, joy does swell,
In stories shared, all's well, all's well.

A Tapestry of Untold Stories

In a tapestry spun from threads of cheer,
A rabbit juggles carrots, oh dear!
With every flop and every fall,
He giggles hard, the star of the ball.

A parrot squawks about grand schemes,
Of chocolate rivers and whipped cream dreams.
While wise old tortoises slow and steady,
Plan the pranks, their minds always ready.

The sunset paints the sky a hue,
Where mice argue if cheese is blue.
As shadows gather, tales take flight,
And laughter echoes into the night.

So heed these stories, woven bright,
In whimsical antics of pure delight.
As every creature joins the fun,
In this crafted world, we have begun.

Bark-Scribbled Myths

In a tree where squirrels chat,
A raccoon wears a fancy hat.
Tales spun from an acorn's tale,
The wise owl's facts never fail.

Frogs debate on lily pads,
Who's the fairest among the lads?
A gossip vine trails all around,
Spreading laughter where it's found.

The chipmunk sings a silly tune,
Underneath the smiling moon.
While fireflies dance in giggly rays,
Nature's party lights amaze!

So gather 'round, dear woodland friends,
For each tale, a giggle sends.
In the shade of the whispering trees,
Myths grow wild, like giggling leaves.

The Language of the Longing Leaves

Leaves converse in whispers sweet,
Ticklish words from roots to feet.
Dancing in the autumn breeze,
They trade secrets with such ease.

A squirrel's jokes make branches sway,
As they prance and jump and play.
'Why did the nut cross the street?'
The punchline's got you on your feet!

Mice scurry with tales of cheese,
Reciting rhymes that aim to please.
In the rustling, laughter spreads,
Telling tales where fun embeds.

So if you stroll where crickets sing,
Listen close; let laughter ring.
The world speaks in a playful way,
Nature's humor at play today.

Parchments of the Prairie

Amid the grass, a story blooms,
With ants composing tiny tunes.
The bison boast of ancient quests,
While daylight fades with silly jest.

A prairie dog hops in delight,
Dressed in shades of day and night.
'Why was the corn so shy?' it said,
'It had ears but never fled!'

The clouds peek in for a quick look,
Closing the chapter of shy books.
Gusts of giggles swirl and lift,
As laughter's breeze becomes a gift.

So wander through this wondrous scene,
Where every leaf has something keen.
On parchment skies, stories play,
In a jolly jumble, bright and gay.

Myths Among the Misty Thickets

In the thicket, shadows frolic,
Where tales become a bit symbolic.
A fox in glasses reads all night,
While porcupines throw a quill fight.

Turtles tell of racing dreams,
As will-o'-the-wisps sing their themes.
Each whisper feasts on funny lore,
Unlocking giggles like never before.

The mist curls round with secret glee,
Hiding laughter beneath the tree.
There's magic in the thick of veils,
Where stories flow like playful trails.

So join the critters, hear their song,
In thickets where we all belong.
For in the mist, with jesters' tricks,
The world's a stage of funny picks.

Sagas Written on the Wind

In the breeze, tall tales flit,
A talking cow plays a trumpet bit.
Rabbits wear caps, dancing with glee,
While owls share gossip beneath the old tree.

A squirrel can juggle, what a sight!
He boasts he can twirl through day and night.
Balloons float by, with mice in tow,
Each twist of fate makes the laughter grow.

Goblins in masks play peek-a-boo,
While fairies chase shadows, just like you do.
Every gust brings a chuckle to share,
Turning drab days into a comical affair.

And as the sun sets, the crow does a dance,
Inviting all creatures to join in the prance.
With each fluttering leaf, the stories expand,
In this whimsical world, wasn't it grand?

The Enchanted Pathway

Upon the path of chuckling stones,
A parrot tells secrets in quirky tones.
Frogs with top hats hop with flair,
While mushrooms giggle at the woodland air.

A hedgehog in boots, with style and grace,
Sprints for a race, oh what a pace!
Snails in a carriage glide slowly near,
To witness the fun with a smile and cheer.

A raccoon spins tales of lost pirate gold,
While mischief unfolds, never growing old.
Wind chimes laugh, tinkle and sway,
Casting spells of joy along the way.

As dusk paints hues on the playful scene,
The path is aglow, almost like a dream.
Creatures of laughter gather to play,
In this joyful lane, where silliness stays.

Whimsy Among the Wildflowers

In the meadow where daisies tell jokes,
Grasshoppers gossip, sharing the hoax.
Butterflies flutter in vibrant attire,
Winking at petals that seem to conspire.

Ladybugs giggle, rolling in clover,
As bumblebees buzz, their daydreams hover.
Dandelion puffs share giddy plans,
While ants march in line, choreographed dance.

A sunbeam slips in, dressed up with flair,
Tickling each flower, spreading joy in the air.
Even the pond holds a shimmering grin,
Reflecting the laughter, where stories begin.

Evening drapes colors, a whimsical sight,
With laughter and giggles fading into night.
The wildflowers whisper as stars start to gleam,
In this lovely patch, where dreams dance and beam.

Reflections of the Forgotten

In a pond once still, now laughter's embrace,
Frogs leap and croak, stealing the space.
A turtle with spectacles, wise but aloof,
Tells tales of the woods as if they were proof.

Old benches are filled with chairs made of fluff,
As crickets recite their melodious stuff.
Lost in the whispers of creatures so sly,
Each ripple of water whispers a sigh.

The tree roots intertwine, conspire and smile,
While shadows play tricks, if just for a while.
A jester bat swoops, bringing laughter around,
In the quietest corners, yonder joy is found.

With twilight descending, a riddle takes flight,
Reflecting the stories, from day into night.
In this world of mirth, all echoes grow bold,
Breathing life into whispers, forgotten yet told.

Dreamweavers of the Vale

In a valley where dreams roam free,
A frog wears glasses, sipping tea.
A snail races with some flair,
While butterflies dance in the air.

A rabbit scribbles on a tree,
'Carrots are the key!' says he.
But every time he takes a bite,
His whiskers tickle with delight.

A wise old owl hoots with glee,
'Plant some giggles, let them be!'
The moonlight chuckles, stars align,
In this vale of dreams, all is fine.

So come and join the merry chase,
Where nonsense twirls with utmost grace.
In this vale, with laughter holds,
A tapestry of tales retold.

Beneath the Canopy of Colors

Beneath a tree, a parrot sings,
In polka dots and fuzzy things.
A chameleon, quite absurd,
Changes shades when it hears a word.

A squirrel hosts a tea party,
With acorn snacks, oh so hearty!
The guests are moss and dandelions,
In a world full of twisted lines.

A balloon floats, giggling high,
While ladybugs sip lemonade dry.
The sun winks down, a jester's grin,
As laughter rolls on the butterfly wind.

So come and bask in hues so bright,
Where squirrels play games till the night.
In this joyful, colorful spree,
The trees will dance just for thee.

Chronicles of Serene Solitude

In a quiet nook by the brook,
A cat reads tales in a cook's book.
The fish laugh at their own jokes,
While the wind whispers, tickling folks.

A hedgehog knits with tangled glee,
Stitching stories for you and me.
The mushrooms gather in a line,
Sipping tea with sweetened brine.

The shadows play in a giggling race,
As crickets hop, keeping pace.
In this calm, where silliness sways,
Even the stars gleefully gaze.

So if you find peace with a grin,
Join the fun where laughter's been.
In solitude, the silly sings,
A world of nonsense, joy it brings.

The Legend of the Hidden Glade

In a glade where giggles grow,
A gopher tells tales in a show.
His friends all cheer and join in song,
As the sun hums right along.

A turtle wears a pirate hat,
Exclaims, 'I'll sail!' then sits on that!
The butterflies plot a silly heist,
Stealing smiles, oh what a feast!

A deer prances, twirls with flair,
While rabbits juggle with great care.
The clouds above dance, twirling free,
In this glade, pure jubilee!

So venture forth and come to play,
Where laughter blooms in bright array.
In the hidden glade, joy's parade,
Holds legends sweet that never fade.

www.ingramcontent.com/pod-product-compliance
Ingram Content Group UK Ltd.
Pitfield, Milton Keynes, MK11 3LW, UK
UKHW010435170125
4146UKWH00047B/68